Advanced Numerical Modeling and Data Assimilation Techniques for Tropical Cyclone Prediction

Advanced Numerical Modeling and Data Assimilation Techniques for Tropical Cyclone Prediction

Edited by

U.C. Mohanty
School of Earth, Ocean and Climate Sciences
Indian Institute of Technology Bhubaneswar
Bhubaneswar, India

and

Sundararaman G. Gopalakrishnan
Hurricane Research Division
Atlantic Oceanographic Meteorological Laboratory, NOAA
Miami, Florida

Springer

CP

A C.I.P. Catalogue record for this book is available from the Library of Congress.

ISBN 978-94-024-0894-2 (HB)
ISBN 978-94-024-0896-6 (e-book)

Copublished by Springer,
P.O. Box 17, 3300 AA Dordrecht, The Netherlands
with Capital Publishing Company, New Delhi, India.

Sold and distributed in North, Central and South America by Springer,
233 Spring Street, New York 10013, USA.

In all other countries, except SAARC countries—Afghanistan, Bangladesh, Bhutan, India, Maldives, Nepal, Pakistan and Sri Lanka—sold and distributed by Springer, Haberstrasse 7, D-69126 Heidelberg, Germany.

In SAARC countries—Afghanistan, Bangladesh, Bhutan, India, Maldives, Nepal, Pakistan and Sri Lanka—sold and distributed by Capital Publishing Company, 7/28, Mahaveer Street, Ansari Road, Daryaganj, New Delhi, 110 002, India.

www.springer.com

Printed on acid-free paper

Printed in India.

Preface

Tropical cyclones (TCs), also known as hurricanes in the Atlantic and typhoons in the west Pacific, are one of the deadliest and costliest natural disasters known to humankind. On an average, about 80-90 TCs form globally every year. The Northwest Pacific Ocean is the most active basin, with an average frequency of 20-30 TCs annually. The genesis of TCs over the North Indian Ocean (including the Bay of Bengal and Arabian Sea, together known as the Indian seas) is bimodal, with the primary maximum of cyclonic activity occurring in the post-monsoon season (October–December) and the secondary maximum occurring in the pre-monsoon season (April–May). The typical duration of a TC over the North Indian Ocean is 5–6 days, with severe cyclonic storm intensity persisting for 2–3 days in comparison to six days for the global average. Yet, this subcontinent accounts for the highest number of cyclone-related fatalities. Of the 23 recorded most deadly TCs over the last 300 years (more than 10,000 fatalities), 20 were formed over the Bay of Bengal. Recent studies show that the majority of TC-related deaths are attributed to inland freshwater flooding and storm surge. Although average annual deaths from TCs affecting the USA have decreased over the 20th century, property damages and loss due to TCs have always been on an increase. Five TCs, Hurricanes Sandy (2012), Katrina (2005), Ike (2008), Wilma (2005) and Andrew (1992) have alone resulted in net damages exceeding 300 billion dollars.

Research and operational agencies around the globe have been continually striving towards improving TC forecasts to save life and property. Although TC track forecasts have improved significantly in recent years, intensity and rainfall forecasts remain a foremost challenge. The high casualties due to TCs may be attributed to various factors including geographic conditions, the limitation of observations in data-sparse ocean regions to initialize and verify models, prediction systems for improved guidance, i.e., intensity and structure forecasting, inaccurate initial conditions from global forecasting systems, and understanding of physical processes within and around the TCs.

The National Oceanic and Atmospheric Administration (NOAA) has made significant advances in TC intensity forecasting through development of the high-resolution Hurricane Weather Research and Forecasting (HWRF) model and its associated data assimilation techniques pioneered at the National Centers for Environmental Prediction (NCEP) and the Atlantic Oceanographic and Meteorological Laboratory (AOML) under the Hurricane Forecast Improvement Program (HFIP). For instance, intensity forecasts from HWRF have improved by about 25% since 2012. The HWRF model has also evolved to become a global TC forecasting system. Nevertheless, much work remains to be done: from TC genesis to rapid intensification to weakening over the ocean and decay over land—our knowledge and understanding is limited. Questions still remain on how to improve numerical models and data assimilation techniques to further improve predictions of these killer storms, as well as questions about the key observations available to improve guidance.

To share state-of-the-art understanding and forecasting techniques acquired in recent years by NOAA and its partnering institutions in the US and India, a six-day advanced workshop and training session titled the "Advanced Indo-US Workshop and Colloquium on Modeling and Data Assimilation for Tropical Cyclone Predictions" was organised jointly by the Indian Institute of Technology (IIT) Delhi and IIT Bhubaneswar, at Bhubaneswar, Odisha, India during 9–14 July 2012. Sponsors included the Indo-US Science and Technology Forum (IUSSTF), Council of Scientific and Industrial Research (CSIR), Ministry of Earth Science (MoES), IIT Delhi, IIT Bhubaneswar, NOAA, and the National Science Foundation (NSF). Their contributions are duly acknowledged.

The week-long workshop brought together ten eminent scientists from NOAA and US universities, one from Australia, and five from India. It also brought together four NSF-sponsored students from the US and 27 young scientists and students from various organisations across India including the India Meteorological Department (IMD), National Centre for Medium Range Weather Forecasting (NCMRWF), Indian National Centre for Ocean Information Services (INCOIS), Indian Institute of Tropical Meteorology (IITM), Indian Air Force, Indian Navy, Indian Space Research Organisation (ISRO), Indira Gandhi Centre for Atomic Research (IGCAR), IITs, and various Indian universities. The workshop consisted of 36 hours of lectures and 10 hours of colloquium, with extended hours for discussion and interaction amongst the resource experts and participants. The workshop accomplished its primary goal of capacity building and infrastructure sharing under the NOAA (US)-MoES (India) agreement on weather and climate sciences. In addition, the exchange of state-of-the-art knowledge and prediction techniques immensely benefited not only the participants but also the resource experts from all countries involved.

In consideration of the significant findings and recommendations presented at the workshop, it was decided to publish a special textbook covering several important aspects of TCs such as observations, analysis and prediction. The textbook was envisioned to serve as a teaching and reference material resource at universities/academic institutions for post-graduate students and researchers around the world. It has been designed to provide a broad outlook on recent advances in observations, assimilation and modeling of TCs with detailed and advanced information on genesis, intensification, movement and storm surge prediction. Specifically, it focusses on (i) state-of-the-art observations for advancing TC research, (ii) advances in numerical weather prediction for TCs, (iii) advanced assimilation and vortex initialization techniques, (iv) ocean coupling, (v) current capabilities to predict TCs, and (vi) advanced research in physical and dynamical processes in TCs. It is also expected to stimulate critical thinking for cyclone forecasters and researchers, managers, policy makers, and graduate and post-graduate students to carry out future research in the field of TCs.

As editors of this special textbook, we are highly thankful to all the authors for their efforts and cooperation in bringing out this publication. We are sincerely thankful to Dr. Jian-Wen Bao, Prof. Noel Davidson, Dr. Alexandre O. Fierro, Prof. T.N. Krishnamurti, Dr. Frank D. Marks, Jr., Prof. Michael T. Montgomery, Prof. Dev Niyogi, Dr. Vijay Tallapragada, Dr. Da-Lin Zhang, and Prof. Fuquing Zhang for making their valuable time available for the lecture series in India and for dedicating considerable time and effort to compiling their lecture notes into textbook materials. In addition, Dr. Sujata Pattanayak, Dr. Ashish Routray and Dr. Richard M. Yablonsky are acknowledged for their contributions and considerable effort in preparation of the book chapters. Thanks are due to Indian experts, viz. Prof. S.K. Dube, Dr. C.M. Kishtawal, Dr. M. Mohapatra, and Dr. S.K. Roy Bhowmik, who spared their valuable time to provide lectures at the workshop. We are also sincerely thankful to all reviewers, viz. Prof. D.V. Bhaskar Rao, Dr. Someswar Das, Dr. John P. George, Dr. S.C. Kar, Dr. Frank Marks, Dr. D.R. Sikka and Dr. (AVM) Ajit Tyagi for their continued efforts in assessing and adding value to this material. We along with Prof. D. Niyogi and Prof. T.N. Krishnamurti gratefully acknowledge the U.S. National Science Foundation grant AGS-1239642 (Program Director: Dr. A. Bamzai) which supported the participation and visit of graduate students and postdoctoral scientists from the US in the workshop. We are also sincerely thankful to Prof. S. Tripathy, Head, School of Earth, Ocean and Climate Sciences, and his colleague, Dr. Sandeep Pattanaik, for their countless efforts in hosting the workshop in a grand manner at IIT Bhubaneswar. Dr. Krishna K. Osuri has contributed significantly in editorial process during preparation of the textbook materials and his dedicated efforts are duly acknowledged. Our special thanks to Capital Publishing Company for their untiring editorial efforts to bring the monograph to its present state.

U.C. Mohanty
Sundararaman G. Gopalakrishnan

Contents

Part 5: Advances in Tropical Cyclone Research

Part 1

State-of-the-art Observations for Advancing TC Research

1

Advancing the Understanding and Prediction of Tropical Cyclones Using Aircraft Observations

Frank D. Marks, Jr.
NOAA/AOML Hurricane Research Division
Miami, FL 33149
frank.marks@noaa.gov

1. Introduction

Tropical cyclones (TCs) were observed by both operational and research aircraft since the first flight piloted by U.S. Army Air Force Lt. Col. Joseph P. Duckworth on 17 July, 1943 (Sumner, 1943). The U.S. military conducted the first dedicated research flight into a TC (Wexler, 1945; Wood, 1945), and a subsequent flight examined the upper troposphere of a 1947 Atlantic TC (Simpson, 1954). In the early 1950s, somewhat regular TC research was conducted during operational military reconnaissance flights originating in Bermuda and Guam (Simpson, 1952; 2003).

Strong hurricanes Carol, Edna and Hazel in 1954 killed nearly 200 people and devastated large areas of the Middle Atlantic States, New England and Southern Ontario (Davis, 1954). These disasters, and hurricanes Carol and Diane in 1955, prompted Congress to allocate money to the Weather Bureau the following year to form the National Hurricane Research Project (NHRP). In August 1956, meteorologists at NHRP began what would become an annual TC research flight programme (Simpson, 1981; Gentry, 1981). The capabilities of airborne meteorological instrumentation, data recorders, and data processors were limited, so the information gathered during the first flight was not available for analysis for several months. During the first 20 years of the research programme, NHRP and two subsequent TC research laboratories used aircraft whose scientific instrumentation capabilities gradually improved (Gentry, 1981).

In the mid-1970s, the National Oceanic and Atmospheric Administration (NOAA) purchased two customized WP-3D (P-3) aircraft. Projected purposes of the P-3s were observation of TC structure and dynamics, participation in TC modification experiments, and monitoring of TC formation, all with the goal of improving TC forecasts and mitigating damage and loss of life; the P-3s were also to engage in research of other vital meteorological and oceanographic programmes. The NOAA Research Flight Center (now the Aircraft Operations Center) received the first P-3 in 1976 and the other in 1977. Since then, NOAA maintained the P-3s to be among the elite meteorological research aircraft in the world.

2. P-3 Aircraft

When the P-3s were first procured, basic understanding of TC structure and behaviour had been deduced from previous research and reconnaissance flights, as well as other conventional data such as rawinsondes and land-based radar reflectivity. The TC was then considered to be a nearly circular and nearly axisymmetric vortex with nearly constant wind speeds extending to at least 400 hPa and with deviations from axisymmetry considered to be gusts. The eyewall was thought to be vertically aligned up to 5-10 km above the surface, and funnel-shaped above. The eye was observed to have warm, moist air below an inversion, with clear, dry air above and ice clouds streaming inward near the top. Both the eyewall and rainbands were understood to have convergent airflow in low levels and outflow above. Rainbands were seen to move inward toward the eye and to cause intensification of eyewall convection. Numerical modelling of TCs was limited to idealized two-dimensional studies due to a lack of adequate computer power, adequate observations of the dynamic and thermodynamic fields in three dimensions, and techniques to assimilate the data into model initial conditions (Simpson and Riehl, 1981).

Since this time, the P-3s proved to be outstanding observing platforms and, partially as a result, the description and understanding of TC behaviour and structure was revolutionized. The National Hurricane Research Laboratory and its successor, the Hurricane Research Division (HRD), conducted missions in 168 TCs in the Atlantic and Eastern Pacific Oceans and near Australia (Table 1). Data were obtained on the micro- to the synoptic-scale, and data analyses leading to many new insights about TC structure, dynamics, thermodynamics, and environmental interactions. The real-time use of the information by the National Hurricane and Environmental Modeling Centers (NHC and EMC, respectively) of the National Centers for Environmental Prediction (NCEP) and resultant research helped to increase the accuracy of wind, flood and storm surge forecasts and severe weather warnings and resulted in significant improvements to operational numerical model guidance for TC track forecasts.

Advances in the study of TC using aircraft observations came about through improvements (scientific and technological) in the ability to

Table 1. List of 167 TCs investigated by, and number of missions of, each NOAA P-3 (42 and 43) during the first 35 years of dual-aircraft operations

Year	*Storm*	*42*	*43*	*Year*	*Storm*	*42*	*43*
1976	Bonny	1		**1987**	Arlene		1
1976	Frances	1		**1987**	Emily	2	3
1976	Gloria	3		**1987**	Floyd	2	3
1977	Anita	4	1	**1987**	TD14	2	
1977	Heather	1		**1988**	Chris	3	2
1978	Debra		1	**1988**	Florence	2	2
1978	Ella	1	3	**1988**	Gilbert	8	5
1978	Greta		2	**1988**	Joan		1
1979	**Kerry**	4		**1988**	Keith		1
1979	**Rosa**	2		**1989**	Dean	1	1
1979	David	4	5	**1989**	Gabrielle	2	2
1979	Frederic	4	3	**1989**	Hugo	1	6
1980	Allen	5	4	**1989**	Jerry		3
1980	Georges		1	1990	Gustav	4	5
1981	Dennis	2	2	1990	TD11	1	2
1981	Floyd	1	1	1990	Marco	1	1
1981	Gert	3	3	1991	Bob	2	
1981	Irene	2	2	1991	Claudette	1	2
1981	Katrina	1	2	1991	Fabian	1	1
1982	Alberto		1	*1991*	*Fefa*	*2*	
1982	Debby	3	2	*1991*	*Guillermo*	*3*	
1982	*Olivia*	*2*	*2*	*1991*	*Jimena*	*1*	*2*
1983	Alicia	3	2	1992	Andrew	1	1
1983	Barry	1	1	*1992*	*Tina*	*5*	*3*
1983	Dean	1	1	1993	Emily	4	5
1983	*Raymond*	*1*	*1*	1994	Chris	1	3
1983	*Tico*	*1*	*1*	1994	Debby	3	2
1984	Arthur	3		*1994*	*Olivia*	*3*	*3*
1984	Diana	5	6	1994	Gordon		2
1984	*Norbert*	*3*	*3*	1995	Erin		1
1984	Isidore	4	4	1995	Felix		2
1984	Josephine	3	4	1995	Jerry	1	
1985	Bob	1		1995	Iris	3	3
1985	Danny	1	2	1995	Luis	2	2
1985	Elena	4	7	1995	Marilyn	1	
1985	Isabel	3	3	1995	Opal		1
1985	Gloria	4	5	1996	Bertha		1
1985	Juan	1	3	1996	Dolly	3	4
1986	Bonnie	1		1996	Edouard	4	5
1986	Charley	2	1	1996	Fran	1	1
1986	Earl	2	2	1996	Hortense	1	1
1986	*Newton*	*1*	*1*	1996	Josephine	1	
1986	*Paine*	*2*	*2*	1996	Lili	4	

(*Contd.*)

Table 1. (*Contd.*)

Year	*Storm*	*42*	*43*	*Year*	*Storm*	*42*	*43*
1996	Marco		1	2005	Dennis	3	4
1997	Danny		1	2005	Emily		1
1997	*Guillermo*	*4*	*4*	2005	Gert	1	3
1997	Erika		4	2005	Katrina		5
1998	Bonnie	6	5	2005	Ophelia	8	5
1998	Danielle	2	2	2005	Rita	2	6
1998	Earl		1	2005	Wilma	3	
1998	Georges	7	6	2006	Ernesto	2	
1998	Mitch		2	2006	Helene	4	
1999	Bret	1	2	2007	Erin	1	
1999	Dennis	4	3	2007	Felix	2	2
1999	Floyd	4	4	2007	Ingrid	3	3
1999	TD11		2	2007	Karen	1	1
1999	Irene		2	2007	TD10	1	1
1999	Lenny		2	2007	Noel		2
2000	Debby	1		2008	Dolly	4	2
2000	Gordon	1		2008	Fay	4	3
2000	Helene		3	2008	Gustav	3	6
2000	Keith	1	1	2008	Ike	6	5
2000	Leslie	1	2	2008	Kyle	4	4
2000	Michael	1		2008	Paloma		3
2001	Barry	1		2009	Ana		1
2001	Chantal	1		2009	Bill		5
2001	Erin	1		2009	Danny		5
2001	Gabrielle	2		2010	Alex	1	
2001	Humberto	3	2	2010	TD02	3	
2001	Iris	1		2010	Bonnie	1	
2001	Michelle	3		2010	TD05	1	
2002	Edouard	1	1	2010	Earl	5	6
2002	Gustav	2		2010	Karl	2	2
2002	Hanna	1	1	2010	Richard		1
2002	Isidore	5	6	2010	Tomas	3	2
2002	Lili	2	5	2011	Irene	4	3
2003	Fabian	4	4	2011	Ophelia		1
2003	Isabel	6	4	2011	Rina	4	
2003	Odette		3	*2011*	*Dora*		*3*
2004	Frances		6	*2011*	*Hilary*		*2*
2004	Ivan	4	6	2012	Isaac	10	
2004	Jeanne	4	3	2012	Leslie	3	
2005	Arlene		1	2012	Rafael	2	
2005	Cindy	1		2012	Sandy	7	

Australian TCs are in bold characters; East Pacific TCs are in italic characters.

observe different aspects of the storms. Early studies provided the basis for understanding TC structure and evolution (for an excellent overview see Dorst, 2007). Technological advances in aircraft in-situ and remote sensing observing capabilities, particularly the development of airborne Doppler radars, revolutionized our depiction of TC structure and dynamics (for a review see Marks, 2003). Further technological advances in aircraft and instrumentation design over the last 30 years provided the tools necessary to greatly expand our knowledge of TCs (for an excellent overview of these accomplishments see Aberson, 2006).

Sophisticated instrumentation installed on and developed for the P-3s is unique among meteorological airborne platforms. At delivery, each P-3 was outfitted with a wide variety of tools to observe the TC and its environment (Jorgensen, 1984a). The most advanced meteorological equipment on the P-3s consisted of three digital radars that are still used. Two record the reflectivity signal: a 5.5-cm-wave-length [lower fuselage (LF)] radar extends below the fuselage and measures the horizontal distribution at all azimuth angles; a 3-cm-wavelength radar is in the tail (TA) and determines the reflectivity distribution along rays oriented either perpendicular to the aircraft track or at angles fore and aft within 25° of the aircraft heading. A third radar with a 5.4-cm wavelength is in the nose and scans horizontally back and forth; the pilots use it to avoid turbulent weather, and the data were not recorded since 1987. Important additions to the radar systems first occurred in 1980 when a prototype Doppler signal processing system was added to the TA radar on one P-3 (Jorgensen et al., 1983) and again in 1988–89 when refined radar data systems were installed on both aircraft.

Both P-3s measured flight-level temperature, pressure and moisture and had a then-state-of-the-art inertial navigation system allowing for flight-level wind speed calculations with 0.1–0.3 m s^{-1} accuracy. Additionally, both P-3s had 24 chutes in the fuselage for external ejection of airborne expendable bathythermographs (AXBTs) to measure ocean temperature to a depth of ~300 m. The P-3s also were equipped with Particle Measuring System probes mounted on the wingtips to distinguish between water and ice particles and estimate particle size (Black and Hallett, 1986).

The aircraft had three main data systems: one for navigational and flight-level meteorological data, one for radar data, and a third for cloud microphysics data. Observations were recorded on magnetic tapes, a capability that greatly aided timely data processing and analysis.

As electronic equipment and expendable probes became more accurate, smaller, faster, and lighter, new instrumentation enhanced the potential to gather comprehensive TC research datasets. For example, P-3 instruments now include the airborne infrared radiation thermometer (AIRT) for remotely determining flight-level temperature in clear air (Barnes et al., 1991), a stepped frequency microwave radiometer (SFMR) for estimating surface wind speed (Uhlhorn and Black, 2003), and Global Positioning System (GPS) navigation. Omega dropwindsondes (ODWs; Govind, 1975) were used first in 1982, primarily to study the TC environment. The GPS dropwindsonde (Hock and Franklin, 1999), first used in 1996, provided reliable wind and thermodynamic profiles in the TC eyewall for the first time. Each aircraft also carries a workstation for airborne data processing and can now transmit crucial data via satellite to the Global Telecommunications System (Griffin et al., 1992). All of this makes the two P-3s the most advanced and comprehensive of all meteorological research aircraft currently in use.

Five key technological advances were instrumental in improving observations of TCs in the last 30 years: (1) the development of new and improved in-situ flight-level instrumentation (Willoughby et al., 1982; Jorgensen, 1984a); (2) the development of new and improved airborne expendables such as the GPS dropwindsondes to sample the atmosphere (Hock and Franklin, 1999) and upper ocean (Shay et al., 1989); (3) the development of Stepped-Frequency Microwave Radiometer (SFMR, Uhlhorn and Black, 2003); (4) the development of new and improved airborne Doppler radar systems (Marks and Houze, 1984); and (5) development of new and improved telecommunications systems to enable real-time transmission of all of these data directly to the ground for use by operational forecasters and to initialize numerical models (Griffin et al., 1992). Each of these technologies started as tools used by researchers to better understand and describe processes critical to estimation of a TC's location, intensity and structure. Many were subsequently transitioned into operational use to improve situation awareness for the forecasters, and more recently to initialize and evaluate numerical models.

3. Flight-level Observations

Observations along P-3 flight tracks put into the context of airborne radar observations substantially improved basic understanding of TC structure and its relationship to intensity change. Successive aircraft penetrations of individual TCs enabled investigation of their evolution beyond what is possible during short, infrequent encounters with conventional land-based observing platforms. To this end, Willoughby and Chelmow (1982) developed an objective algorithm to determine TC centre positions necessary for the accurate placement of data into a storm-relative coordinate framework, enabling much of the following TC structure research.

At the beginning of the P-3 era, the TC was considered to be an axisymmetric vortex with convergence at low levels and divergence high in the eyewall and rainbands, upward motion within these features, and compensating subsidence in the eye and outside the vortex. Flight-level observations obtained from the P-3 allowed for refinement of descriptions of the axisymmetric structure of TCs, and also for the first understanding of higher-wavenumber features.

Using flight-level and airborne radar reflectivity observations, Jorgensen (1984a, b) provided a more accurate description of the symmetric eyewall and rainband structure than was previously available. The largest updrafts were located radially inward from the wind and reflectivity maxima in the eyewall that slope radially outward with height (Jorgensen, 1984a). Adjacent rainbands were more cellular than the eyewall and lacked consistent updrafts and distinct wind speed maxima at flight level (Jorgensen, 1984a). Jorgensen (1984b) documented the TC eyewall structural evolution during coordinated eyewall penetrations by the P-3s and a NOAA C-130. An eyewall radius–height schematic (Fig. 1) from these studies depicted the eyewall and its associated wind speed maximum, the relative positions of clouds and radar features, the regions of in- and outflow, and the location and strength of up- and downdrafts. Although the schematic did not describe asymmetries, and Marks and Houze (1987) subsequently showed that the maximum tangential wind was not inside the reflectivity maximum at all levels, the diagram still is representative of mean mature axisymmetric TC structure.

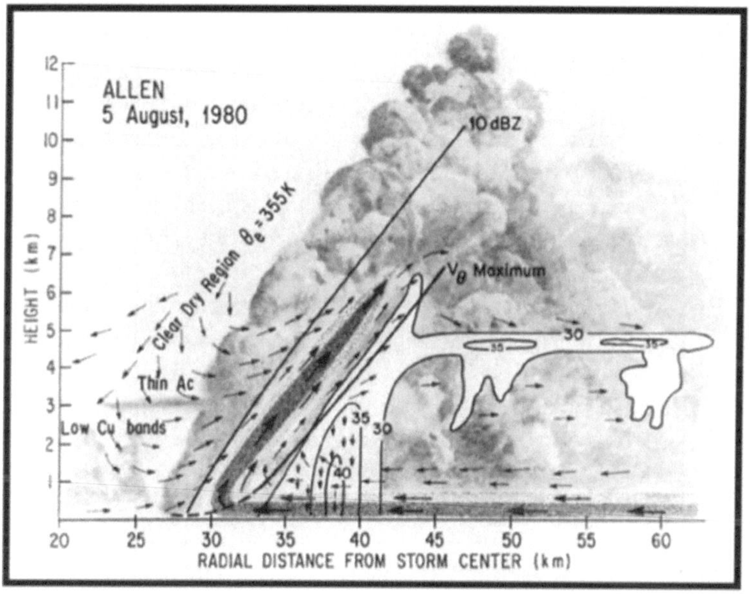

Fig. 1. Schematic cross section depicting the locations of the clouds and precipitation, radius of maximum wind, and radial-vertical airflow through the eyewall of Hurricane Allen on 5 August 1980, from Jorgensen (1984b). The downward motion in the eye was based on estimates of radial divergence, and the downdrafts within the high rainfall rate region were inferred from vertical velocity time series from the individual passes. Reflectivity (dBZ) contours were based on a composite using 15 passes, although there was considerable pass-to-pass variability. The slope of the clouds on the inside edge of the eyewall was based on radar minimum detectable signal analysis, aircraft altimeter readings, photography and observer notes. Dark shading denotes the location of the largest radial and vertical velocities.

Willoughby (1990a) and Willoughby and Rahn (2004) constructed a database of over 900 radial flight-level profiles of basic meteorological parameters in TCs. The axisymmetric tangential wind above the boundary layer in TCs was in approximate gradient balance (Willoughby, 1990b), explaining the outward slope of the eyewall with height, descent of air within the eye, and lower-tropospheric inflow outside.

Willoughby et al. (1982) used P-3 flight-level and reflectivity data to describe the evolution of concentric eyewalls that contract due to latent heat release (Shapiro and Willoughby, 1982; Willoughby, 1990a). Figure 2 shows the classical evolution of concentric eyewall features and their impact on TC intensity in Hurricane Gilbert (1988). Outer convective rings formed around the preexisting one, contracted and strangled the original eyewall causing weakening. TCs with a single, vigorous and symmetric convective ring strengthened rapidly, whereas TCs with weak or asymmetric convective rings underwent episodic intensification.

Willoughby et al. (1984) described the flow associated with rainbands that rotate about and were connected to the inner convective ring and that intensify and initiate the eyewall cycle by developing into a new, outer eyewall. Barnes et al. (1983) described the mesoscale thermodynamic and kinematic structure of one such large rainband. The frequency of cellular, convective precipitation decreased, and stratiform precipitation increased, along the rainband toward where the rainband spiraled toward the eyewall. Air with relatively equivalent potential temperature in the subcloud layer, the result of outflow from strong convective activity within the rainband, was observed outside the rainband, especially between the rainband and the eyewall. This air could stabilize the eyewall lapse rate and decrease TC intensity. Samsury and Zipser (1995) described the relationship between the rainbands and wind speed maxima associated with outer eyewalls. Almost all such maxima were associated with mesoscale rainband features (an outer eyewall), but >70% of rainbands did not have a pronounced wind speed maximum at flight level.

Data from P-3 aircraft was also able to show the extent of mixing of air between the eye and eyewall and its implications. Using eye soundings, Willoughby (1998) found that the air aloft in the eye was clear, warm and dry, and overlays an inversion below which moist, cloudy air resides. The air above the inversion remains there for long periods, subsides only a few km, and mixes weakly with moist air from the eyewall. The air below the inversion comes from the low-level inflow and mixes easily with the eyewall. Kossin and Eastin (2001), using flight-level and dropwindsonde data, suggested that barotropic instability could cause more mixing between the eyewall and eye than Willoughby (1998) found.

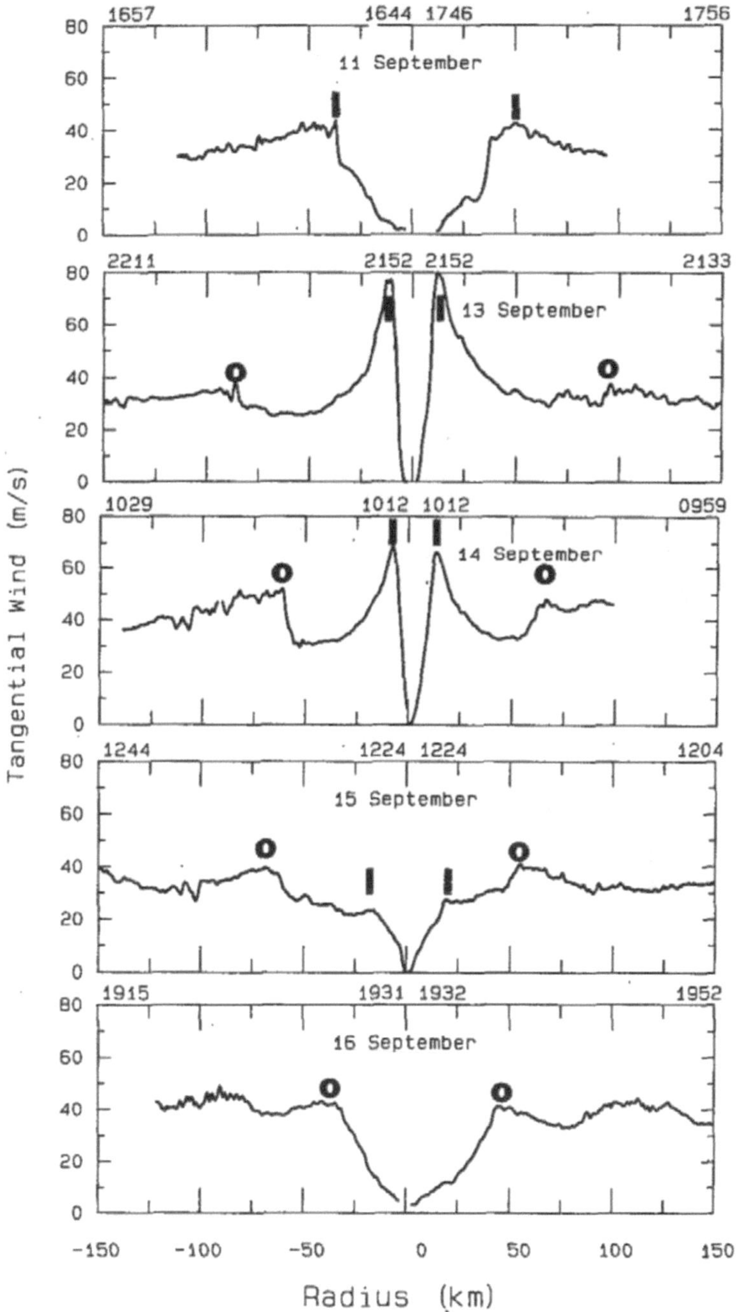

Fig. 2. Flight-level tangential wind speed from the south–north traverses through the centre of Hurricane Gilbert (1988). Bold I and O denote the location of the inner- and outer-eyewall wind maxima, respectively. Times (UTC) at the beginning of each radial pass are at the top of the panels. From Black and Willoughby (1992).

4. Airborne Expendables

a. GPS Dropwindsondes

As a TC approaches land, successful evacuations and damage mitigation measures, based on accurate forecasts, can reduce property damage and loss of lives. In 1982, TC numerical modelling was in its infancy, and the best track guidance available to forecasters was statistical. However, to help provide improved synoptic analyses and numerical forecasts to optimize warning areas and increase warning lead times, HRD began to test the hypothesis that additional vertical wind and thermodynamic profiles in the TC environment (the "synoptic flow") would provide improved operational guidance to human forecasters (e.g., Fig. 3).

The data obtained during the 19 P-3 research missions conducted for this purpose through 1996 helped to reduce errors in TC track forecasts significantly (as much as 30%) in global and regional dynamical models (Burpee et al., 1996). In addition, Shapiro and Franklin (1995) used an analysis of Hurricane Gloria (1985) synoptic flow data to diagnose the potential vorticity structure of a TC and to describe its dynamics, wind balance and interaction with the

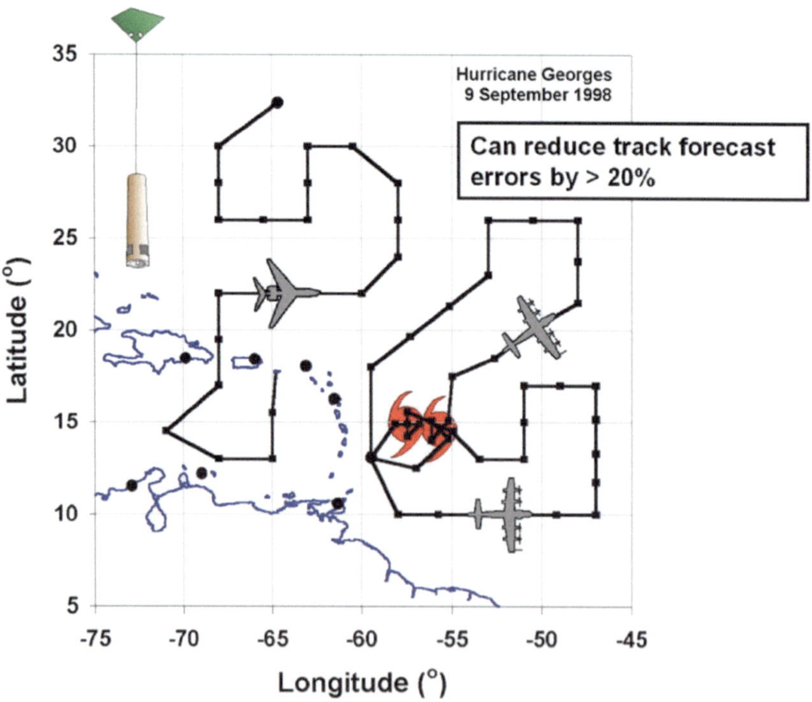

Fig. 3. Synoptic surveillance flight tracks for the NOAA G-IV and two P-3 aircraft in Hurricane Georges 1998. The hurricane symbols denote the storm locations during the time of the two P-3 inner core-sampling missions. Blue line denotes the coastline of South America and the Windward and Leeward Islands that bound the Caribbean Sea.

environmental flow. Franklin et al. (1996) used analyses of all the synoptic flow dropwindsonde data to document the existence of beta gyres and investigate their impact on TC motion.

The success of synoptic flow flights led to the development of a new generation of dropwindsonde (Hock and Franklin, 1999) and the acquisition of a high-altitude jet aircraft (the Gulfstream-IVSP hereafter G-IV) as a dedicated operational platform to conduct synoptic surveillance missions. These operational flights, along with concurrent research into optimal targeting and sampling strategies and data assimilation, continue to improve NCEP operational global model track forecasts significantly (Aberson, 2003). A complete G-IV flight and the required dropwindsondes cost about $40,000, far less than the estimated average of $1 million needed to evacuate just one mile of coastline for landfall (Whitehead, 2003).

The improved dropsondes, relying on GPS navigation and improved humidity sensors, could now be deployed in clouds and TC eyewalls, which HRD did in Hurricane Guillermo (1997). Dropwindsondes released from the P-3s in 1997 provided the first high-resolution wind and thermodynamic profiles in a TC eyewall. Research and operational aircraft since obtained more than 1000 such profiles (e.g. Fig. 4). Franklin et al. (2003) showed that the

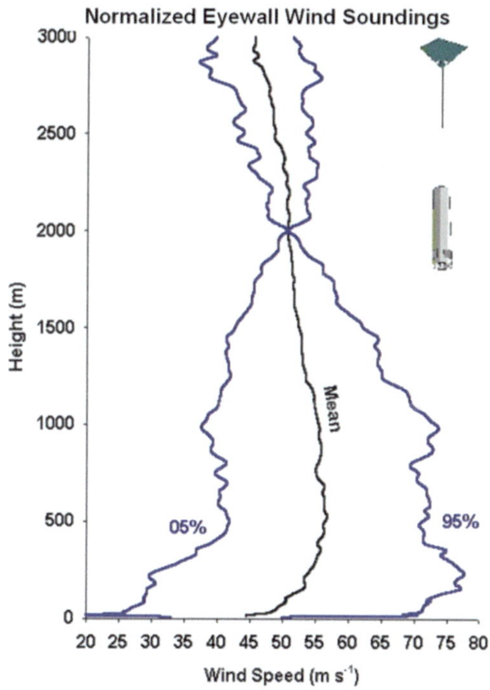

Fig. 4. Frequency distribution of the horizontal wind speed (m s^{-1}) with altitude from 1200 eyewall dropwindsondes. The black line denotes the mean of the sample and the two blue lines the 95 and 5 percentile of the distribution.

mean surface to 700 hPa (the standard flight level for reconnaissance aircraft in TCs) wind speed ratio was about 90% in the eyewall in several hundred over-ocean profiles. They determined an adjustment factor for TC flight-level wind interpretation: without additional information, NHC operationally assesses surface maximum 1-min wind speed estimates to be about 90% of the peak 10-s wind speed observed at 700 hPa. This new understanding of TC boundary layer wind and better methodology for adjusting flight-level wind speed measurements to the surface also is applied to historical TC reanalysis.

These data also became the standard for surface wind estimates in TCs. To take advantage of these developments and to insure that high-quality measurements could be received in real-time by NHC, HRD scientists and programmers developed on-board quality control software for the dropsondes and studied the most efficient targeting of them (Aberson, 2003).

b. Expendable Ocean Probes

Expendable oceanographic instruments used in TC research can be divided into the following three categories: profilers, drifters and floats. Snapshots of the three-dimensional upper-ocean thermal, salinity and current structure under a TC are measured with airborne expendable bathythermographs (AXBT), current profilers (AXCP), and conductivity–temperature–depth sensors (AXCTD) deployed on one of the NOAA aircraft. Each instrument deploys a sensor that descends from a surface float at a rate of 4-5 m s^{-1} and measures the water temperature, conductivity and current induced electric fields between two horizontally separated sensors spaced about 5 cm apart (Sanford et al., 1982). The electromagnetic current measuring technique acquires observations of the baroclinic currents at intervals of 0.3 m relative to an unknown reference velocity. Flight patterns are designed (Jaimes and Shay, 2009) to sample mesoscale ocean features such as warm and cold core eddies that a TC may move over (e.g., Fig. 5).

Figure 6 shows drawings of some drifters and floats, and a schematic of their operation. Drifters aim to follow the ocean current at 15-m depth while measuring both near-surface atmospheric and upper-ocean properties. A small surface float supports a much larger drogue centered at 15-m depth. The large drogue causes the drifter to nearly follow the horizontal water motion at approximately 15 m. A transmitter in the surface drifter sends data to the Argos satellite system. The same signals are used to track the drifter. The standard drifter measurements are position and near-surface temperature. Minimet drifters are also designed to estimate wind speed using the sound level at 8 kHz (Nystuen and Selsor, 1997) and wind direction using a vane on the surface float. Evaluation of the accuracy of this approach at TC wind speeds is still under way. ADOS drifters additionally measure the temperature profile to 100-m depth.

The three varieties of floats are shown in Fig. 6. All floats operate by mechanically changing their volume, and thus their density, in order to control

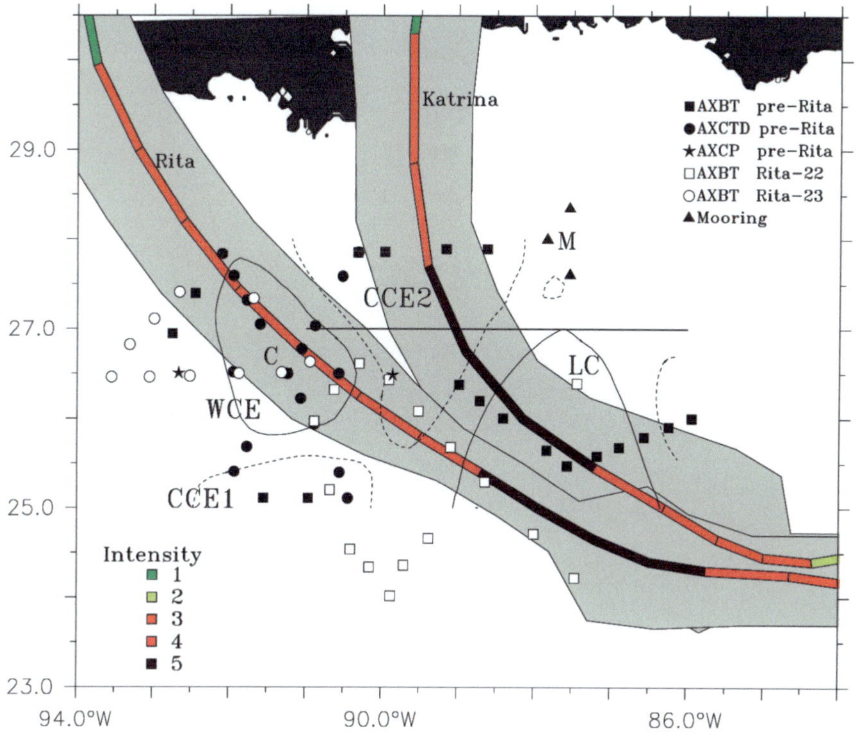

Fig. 5. Airborne profilers deployed in September 2005 relative to the track and intensity of Hurricanes Katrina and Rita (coloured lines, with colour indicating intensity as per the legend) over the LC system. The light gray shading on the sides of the storm tracks represents two times the radius of maximum wind. The contours are envelops of anticyclonic (solid: WCE and LC) and cyclonic (dashed: CCE1 and CCE2) circulations. A set of AXBTs was deployed after Hurricane Rita (26 September), following a sampling pattern similar to pre-Rita (or post-Katrina on 15 September).

their depth. By making themselves light, they can profile to the surface, thereby extending an antenna out of the water, enabling them to obtain a GPS fix and relay data to and receive instructions from their shore-based operators. The EM-APEX floats (Fig. 6, green lines) operated as profilers, continuously cycling while measuring temperature, salinity and velocity. Profiles extended from the surface to 200 m with profiles to 500 m every half inertial period. During the storm, the top of the profiles terminated at 50 m. The Lagrangian floats (D'Asaro, 2003) profiled only before and after the storm (Fig. 6, black line). During the storm, they remained neutrally buoyant, following the three-dimensional motion of water parcels in the highly turbulent upper boundary layer. They measured temperature, salinity and gas concentration. The SOLO floats profiled temperature, salinity and oxygen from the surface to approximately 200 m (Fig. 6, blue line) while hovering at about 40 m for a period of time during each dive interval to remotely measure surface waves

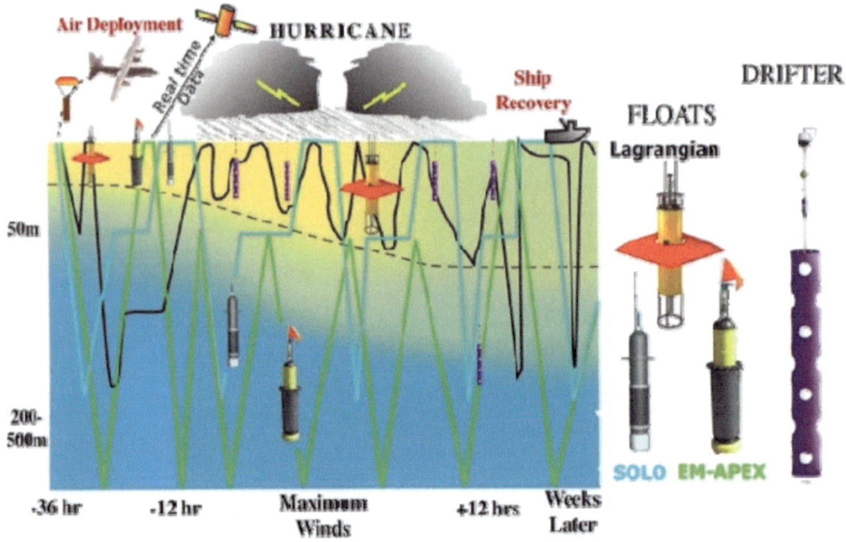

Fig. 6. Drawings of the three varieties of floats and a surface drifter deployed.

and the depth of the bubble layer created by surface wave breaking, using a compact sonar and 0–50–kHz ambient sound with a passive hydrophone. The floats were programmed to repeat its dive interval every four hours.

Black (1983) showed that, of 19 TCs, those over an oceanic mixed layer depth (MLD) <50 m produced crescent-shaped decreases of as much as a few kelvin in the sea surface temperature. The cooling was concentrated in the right rear quadrant (relative to motion in the Northern Hemisphere), with the maximum occurring a short distance outside the eyewall. This implies that surface air flowing inward over relatively cool water can lessen the enthalpy flux from the ocean into the atmosphere, thereby becoming cooled and reducing buoyancy in the eyewall. The left front quadrant had only a very small SST decrease. Black et al. (1988) and Shay et al. (1992) further documented the significant impact of TCs on the upper ocean. Cione and Uhlhorn (2003) showed that SST reduction near the TC centre is modest (~0.2–1.2 K), and that much of the observed 4–6-K cooling occurs long after TC passage. The magnitude of the SST change was statistically linked to subsequent intensity changes, suggesting that SST measurements in regions over which the TC core will pass, and during passage, will aid in intensity forecasts. These issues were investigated in flights supporting the Coupled Boundary Layer Air–Sea Transfer (C-BLAST) experiment during the 2000–04 Atlantic hurricane seasons (Black et al., 2007) and the Impact of Typhoons on the Ocean in the Pacific (ITOP) experiment in 2010 (D'Asaro et al., 2011; Pun et al., 2011).

Careful analysis of data from dropwindsondes and ocean expendable probes released from the P-3s is elucidating the ways in which TCs interact with the underlying ocean to cause intensity changes. Cione et al (2000) and

Cione and Uhlhorn (2003) found that low-level inflow into the TC centre was not isothermal, that surface moisture was not constant, and that the air near the surface was not in thermodynamic equilibrium with the sea. The surface air temperature drops as the air flows toward the TC centre; this drop occurs far from the eyewall where the largest pressure drops occurred, so the change is not adiabatic. This allows surface heat and moisture fluxes to be much higher near the eyewall than further away; the fluxes are thus greater than previously believed, and more energy is provided to the TC from the underlying ocean. Since the eyewall is nearly saturated, the surface air temperature near the TC centre was relatively constant with distance from the centre despite the potential for spray-induced evaporational cooling and adiabatic expansion. Black and Holland (1995) and Wroe and Barnes (2003) confirmed these observations and worked out the energetics within rainbands and near the eyewall. They also confirmed the result of Gamache et al. (1993) that unsaturated convective downdrafts near and under the eyewall may dry and evaporatively cool the air near the surface.

5. Stepped Frequency Microwave Radiometer (SFMR)

Accurate specification of 1-min maximum surface wind speed and its location and the distribution of surface gale (18 m s^{-1}) and hurricane (33 m s^{-1}) force winds are essential for operational NHC intensity and structure forecasts, applications such as storm surge modelling (Jelesnianski et al., 1992) and insured loss estimation models (Malmquist and Michaels, 2000). The maximum 1-minute surface wind speed is the official NHC intensity estimate and determines to a large degree the extent of emergency preparations and evacuations needed for each TC landfall.

A new suite of airborne remote-sensing instruments is available on the NOAA P-3 aircraft for the purpose of measuring surface winds in and around TCs. The major new instrument is the SFMR first tested in Hurricane Allen in 1980, with the first real-time retrieval of winds on board the aircraft in Hurricane Earl in 1985, and the first operational transmission of winds to NHC in Hurricane Dennis in 1999 (Uhlhorn and Black, 2003). SFMR measures brightness temperatures (T_B) at six frequencies that are used to estimate the emissivity of the ocean surface, derive estimates of the wind speed, wind stress, rain rate, and to calculate two-way attenuation due to precipitation (e.g., Fig. 7). The SFMR infers sea surface wind speed from the increased T_B due to foam coverage, which is a function of the local energy input from the wind, and perhaps the local wave environment.

Surface wind speeds in TCs derived from SFMR measurement of microwave emissions from the wind-driven sea surface are well correlated with in situ surface wind measurements by GPS dropwindsonde. SFMR wind speeds are compared with independent measurements from Global Positioning

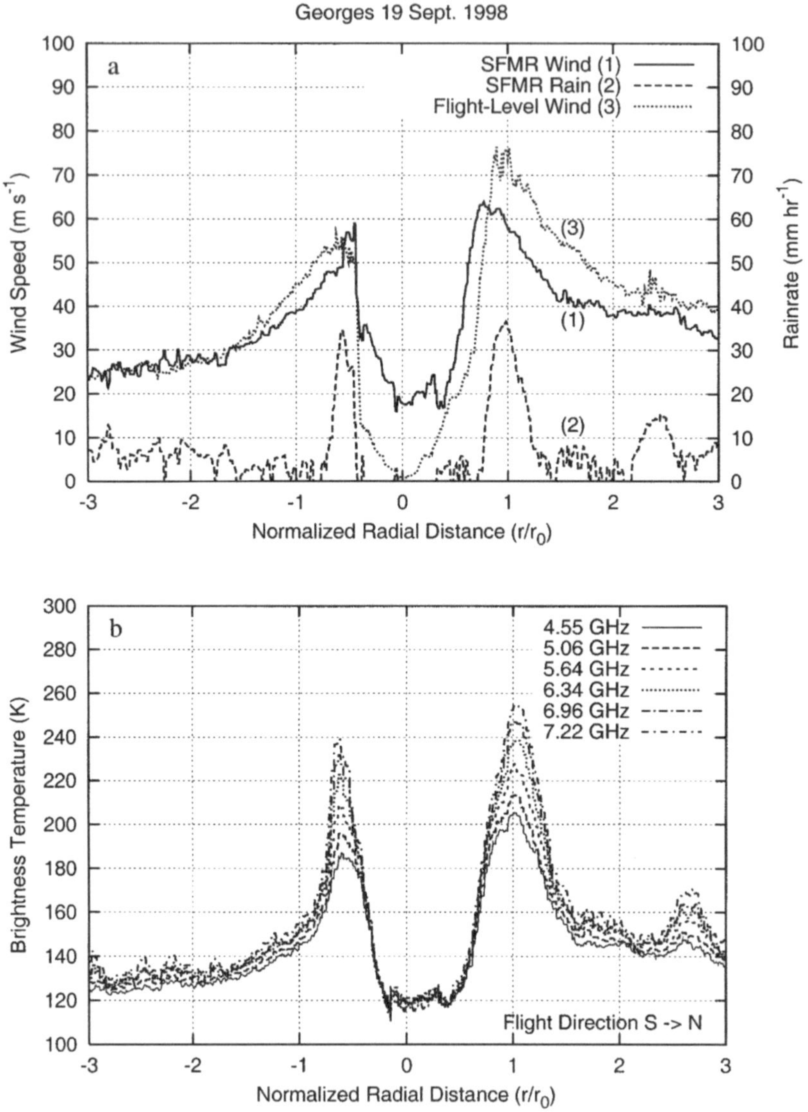

Fig. 7. (a) SFMR and flight-level wind speeds (m s^{-1}) and SFMR rain rate (mm h^{-1}) and (b) T_B as a function of normalized radial distance (r/r_0) from the centre of Hurricane Georges, where $r_0 = 27$ km is defined by the flight-level wind maximum in the north eyewall. Flight direction is south (left) to north (right). In (b), T_B generally increases with higher frequency.

System (GPS) dropwindsondes from the 1998, 1999, and 2001 hurricane seasons. SFMR yields wind speed measurements that are overall within ~4 m s^{-1} rms of the dropwindsonde-estimated surface wind, and within ~5 m s^{-1} of the direct 10-m wind speed measurement. Hence, the SFMR provides independent estimates of surface winds at a horizontal resolution of ~10 s

(1.5 km) along the flight track, suitable for correct location and estimation of maximum sustained wind speeds, as well as high-resolution mapping of the wind field. Currently the SFMR is the only reliable passive microwave instrument for measuring surface winds >20 m s^{-1}.

The SFMR surface wind estimates are operationally available to the hurricane specialists at NHC for use in forecast guidance, and in determining surface wind estimates at different locations surrounding the storm. These surface wind estimates made a major impact on our understanding of high-wind, air-sea fluxes in conjunction with sea spray, surface wave and oceanic mixing processes.

6. Airborne Doppler Radar

Radar played an important role in studies of TCs since it was developed in the 1940s. The spiral echo pattern that is characteristic of tropical cyclones was one of the first meteorological features described by radar. These early studies established radar as an ideal remote sensing tool for studies of the structure and evolution of tropical cyclones. Since then our understanding of tropical cyclones and their characteristics improved dramatically, in part due to the use of radars, both non-coherent and coherent. Advances in radar technology and data processing go hand in hand with advances in our understanding. These advances rapidly accelerated over the last twenty years with the advent of operational ground-based Doppler radar networks in the United States, India, Japan, China, Taiwan, the French islands of La Réunion and Martinique.

In the last 25 years, technological improvements such as the P-3 tail airborne Doppler radar on the P-3s, the operational Weather Service Radar 1988-Doppler (WSR-88D) radar network, and portable Doppler radars produced a new generation of TC data that gave scientists an unprecedented opportunity to document the dynamics of TCs and hence led to improved understanding of these devastating storms. Airborne Doppler radar data sets collected in the 1980s represented our best information on the kinematic structure of TCs (e.g., Marks and Houze, 1984, 1987; Marks et al., 1992; Houze et al., 1992; Gamache et al., 1993), while airborne and ground-based Doppler radars were instrumental in determining the kinematic structure of the outer rainbands (e.g., Ishihara et al., 1986; Powell, 1990a, b; Barnes et al., 1991; Tabata et al., 1992; Ryan et al., 1992; Barnes and Powell, 1995). These data sets revolutionized our perception of the kinematic structure of a mature TC from an axisymmetric vortex to one that recognizes the significance of asymmetric motions to track and intensity changes.

The NOAA P-3 airborne Doppler datasets were used for improved understanding of symmetric vortex and major asymmetries. The addition of a second airborne Doppler radar on the other P-3 enabled true dual-Doppler analyses and the ability to study the temporal evolution of the kinematic structure over 3–6 h. The advent of the WSR-88D Doppler radar network, and construction of portable Doppler radars that can be moved to a location near TC landfall have also generated new and unique datasets enabling improved understanding of (1) severe weather events associated with landfalling TCs, (2) boundary layer wind structure as the storm moves from over the sea to over land, and (3) spatial and temporal changes in the storm rain distribution. The P-3 airborne Doppler and WSR-88D data were also instrumental in developing a suite of operational single Doppler radar algorithms to objectively analyze a TC's wind field by determining the storm location and defining the primary, secondary and major asymmetric circulations (e.g., Lee et al., 1994, 1999, 2000). These algorithms are used operationally on the P-3 aircraft and on the ground at NHC.

Using the first Doppler radar measurements obtained in a TC, Marks and Houze (1984) described for the first time important asymmetric TC structure characteristics. Data from orthogonal flight legs from one aircraft were combined into a three-dimensional pseudo-dual-Doppler horizontal wind analysis assuming that features were stationary during the observation time (e.g., Fig. 8). This allowed for the first description of mesocyclones associated with potentially destructive local wind speed maxima superimposed on the basic flow. Marks and Houze (1987) first documented the three-dimensional kinematic asymmetric eyewall structure with another pseudo-dual-Doppler analysis. They extended the composite structure shown in Fig. 1 and documented the outward sloping tangential wind maximum comprising the primary circulation and deep low-level inflow, rising motion in the eyewall sloping outward with height and outflow above 10 km that characterizes the secondary circulation (e.g., Fig. 9). Convective-scale up and downdrafts within the eyewall were inferred using Doppler measurements at vertical incidence after the removal of aircraft vertical motion and estimated particle fall speeds. With the analysis, particle trajectories could be followed throughout the circulation to describe the TC water budget. They concluded that this organized structure makes the TC more predictable than other tropical convective systems and allows for accurate numerical modelling of TCs. Marks et al. (1992) refined and extended this analysis, deriving vertical motion from mass continuity, and providing the first complete documentation of the three-dimensional TC core wind field at 10-25-km radial distance and 1-km scales in the horizontal and vertical, respectively. Asymmetries in the radial wind that varied with height were described for the first time, and these asymmetries were shown to contribute to the manner in which different air masses entered and exited the vortex and thus aided in intensity change.

Reasor et al. (2000) first documented TC mesoscale eyewall temporal evolution with seven detailed true dual-Doppler analyses during a 3-h period, focussing on the temporal evolution of asymmetries and the response of the vortex to increasing vertical shear of the horizontal wind. They provided the first evidence of vortex Rossby waves outside the eyewall, an important confirmation of analytical and numerical studies of TC intensity change (Montgomery and Kallenbach, 1997; Moller and Montgomery, 1999, 2000).

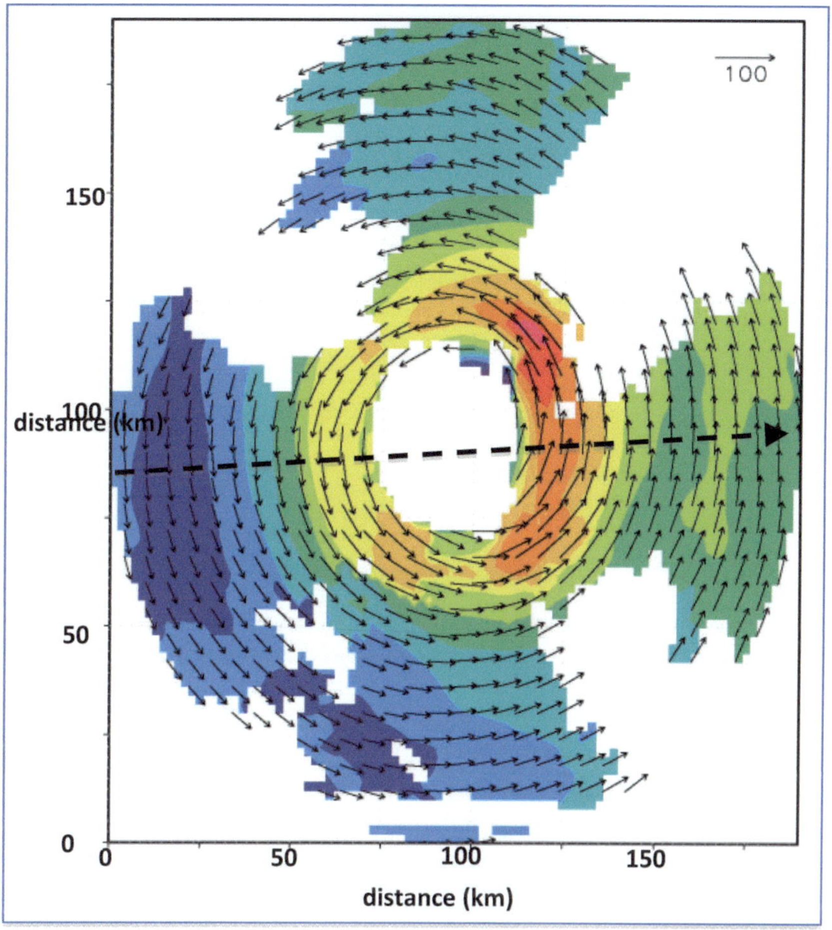

Fig. 8. Airborne Doppler-analyzed horizontal wind field at 2-km altitude in Hurricane Katrina, 28 September 2005. Colour denotes wind speed (m s^{-1}) and the arrows denote the wind direction. The dashed black line denotes flight track of the aircraft.

Rogers et al. (2012) presented the axisymmetric structure of mature tropical cyclones using composites of tail Doppler radar measurements from 40 radial penetrations in a large number of TCs. The Doppler radar composite document multiscale structure including the primary and secondary circulations, eyewall slope, decay of the tangential wind with height, low-level inflow layer and region of enhanced outflow, radial variation of convective and stratiform reflectivity, eyewall vorticity, and rainband signatures in the radial wind and vorticity fields. Rogers et al. (2012) also documented statistics of convective-scale fields (e.g., vertical velocity, reflectivity, vorticity) with contoured frequency by altitude diagrams (CFAD) of both vertical velocity and radar reflectivity. The radar composites can be used to further evaluate the

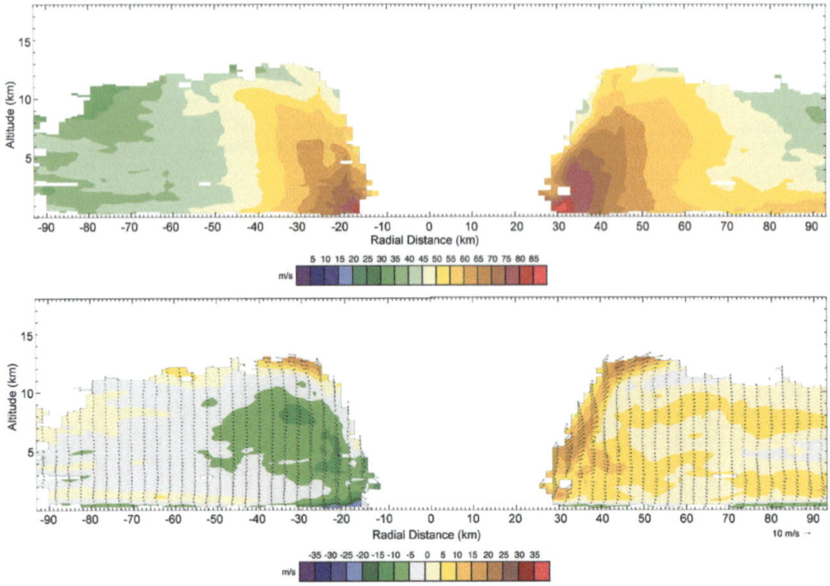

Fig. 9. Airborne Doppler-analyzed vertical wind field across the centre of Hurricane Katrina, 28 September 2005. Colour in the top panel denotes the tangential wind speed (m s^{-1}) and in the bottom panel the radial wind (m s^{-1}). Arrows in the bottom panel denote the wind direction in the vertical plane.

structures above the boundary layer in order to identify model biases in the convection, cumulus and microphysics schemes.

The next step is the assimilation of Doppler radar data into mesoscale numerical models to improve our situation awareness and analysis. As model resolution shrinks with advances in computing, the scale of information necessary to initialize the model approaches that of a Doppler radar. Based on the success in using the airborne Doppler radar to describe the axisymmetric TC structure and major asymmetries it seems that the TC problem is the most tractable to attempt such an effort. In a TC, the dynamic constraints are strongly controlled by the high vortex angular velocity. In a TC the time-scale of the divergent component of the wind, which typically dominates the variance in the Doppler winds, is one order of magnitude smaller than that of the mean vortex. Data assimilation techniques need to accommodate such differences and be designed such that the aliasing of convective motions on the resultant initial condition is limited. The problem seems tractable and should be a high priority over the next 5-10 years.

The tail Doppler radar was also used to derive the turbulence kinetic energy (TKE) in mature TCs (Lorsolo et al., 2010). The TKE is developed using the variance of the radial velocity of ~150 m spatial resolution about the mean Doppler radial velocity of a 10 km (across flight track) × 1.5 km (along flight track) × 0.15 km (vertical) grid cell centered on the flight track. These TKE data can be used to further evaluate the TKE-type planetary boundary layer scheme in HWRF.

7. Improved Use of Observations: Hurricane Forecast and Improvement Programme (HFIP)

In order to increase the pace of forecast improvement NOAA developed the Hurricane Forecast Improvement Project (HFIP) whose goal is to decrease the error in forecast guidance for track and intensity by 20% in 5 years and 50% in 10 years (HFIP Plan). In order to achieve such an ambitious goal HFIP is pursuing a strategy that is focussed on making use of the aircraft observations to provide better initial analyses of the storm's structure for the models to start from. Recent results from the programme show that assimilation of aircraft observations, particularly airborne Doppler winds, shows promise of reaching HFIP intensity goals (Gall et al., 2013). As part of HFIP the NOAA Intensity Forecast Experiment (IFEX) aircraft observing efforts into TCs are geared towards (1) collecting observations that span the TC life cycle in a variety of environments for model initialization and evaluation; (2) developing and refining measurement strategies and technologies that provide improved real-time monitoring of TC intensity, structure and environment; and (3) improving the understanding of physical processes important in intensity change for a TC at all stages of its life cycle. This data collection is part of the development of modelling systems, and involves research into data assimilation, understanding and parameterization of physical processes, and model evaluation and validation with comprehensive datasets collected by airborne platforms.

a. Improving Initialization

To demonstrate the importance of the aircraft observations to model initialization and forecast guidance two research groups—one at the Pennsylvania State University using the ARW model and working with HRD (Zhang et al., 2011a), and the other at HRD using the HWRF model (Aksoy et al., 2012)—showed that assimilating data collected by the NOAA P-3 tail-Doppler radar in TCs during the annual Intensity Forecast Experiment (IFEX, Rogers et al., 2006, 2013) run by HRD, as well as other data collected by the P-3 and Air Force C-130 hurricane hunter aircraft (including flight level and dropsonde data) can significantly improve forecasts of intensity. The results compared to the 5-Day Statistical Hurricane Intensity Forecast Model (SHIFOR5) are shown in Fig.10 for most cases for which radar was taken in storms during 2008-2010.

Note that for all forecast lead times from 36 to 96 hours inclusion of aircraft data improves the forecasts by 30% and as much as 50% over SHIFOR5 and for those lead times are close to or exceed the 20% HFIP goal and approach the 50% goal for these cases. This is the first time that dynamical models showed this level of skill for intensity forecasts. It is likely that this result is because the high-resolution data near the core better defines the environment flow and storm structure within the TC and hence the impact, even out to five days. These analyses avoid the use of filtering and bogusing to create initial conditions that are used in most regional TC models. Note also that there is a problem in the early part of the forecast where the model is still adjusting to the initial conditions. Fixing this initial problem is a major current focus of HFIP.

Note that even the HWRF model without adding aircraft data (red line in Fig. 10) gives a substantial improvement relative to the baseline. This improvement suggests that the newer regional models (HWRF and ARW) could provide an improvement as compared to the HFIP baseline (likely

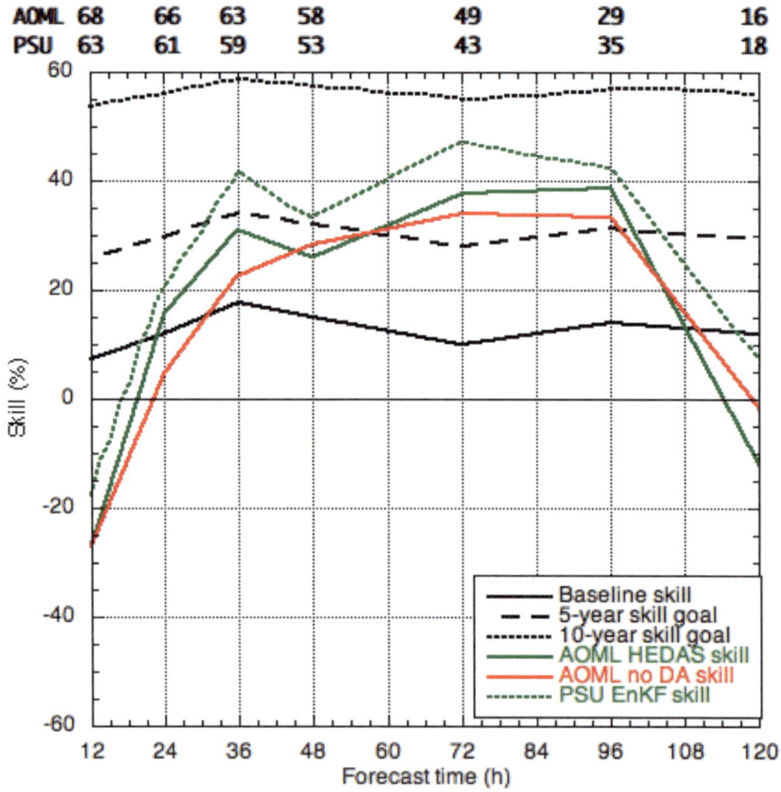

Fig. 10. Impact of aircraft data (including the tail Doppler radar) on intensity forecasts made by the HRD (AOML) 9/3 HWRF using HEDAS (solid green) and the PSU ARW using a EnKF data assimilation system (dashed green) compared to SHIFOR5 as a percent improvement over SHIFOR5. The black lines (solid, dashed and dotted) show the baseline skill and the 20% and 50% improvement goals respectively. The red line shows a homogeneous comparison to the solid green line but with no aircraft data included in the initialization. The PSU results only included the tail Doppler radar data while the HRD results include all aircraft data (flight level data dropsondes, SMFR surface wind data and tail Doppler radar data). Most cases with radar data from 2008-2010 are included. Number of cases for the HRD and PSU results at each forecast lead-time is shown across the top.

because of higher resolution) and that adding aircraft data, when available, could add another 10%-20%. This would meet (at some forecast periods) the HFIP interim (five-year) goal of 20% improvement, for those initializations for which tail Doppler radar and other aircraft data are collected. For other times, HFIP is pursuing the use of high-resolution satellite data near the TC.

b. Improved Model Evaluation

At the same time that HFIP is demonstrating the importance of the aircraft observations to model initialization and forecast guidance, Rogers et al. (2012) and Zhang et al. (2011b) are using the aircraft observations collected during IFEX missions since 2005 and earlier TC flights with the P-3 aircraft to improve our understanding of the TC structure and physical processes responsible for that structure and its evolution. A model developmental framework for improving the physical parameterizations using quality-controlled and post-processed aircraft observations was developed, with steps that include model diagnostics, physics development, physics implementation and further evaluation. Model deficiencies are first identified through model diagnostics by comparing the simulated axisymmetric multi-scale structures to observational composites. New physical parameterizations are developed in parallel based on in-situ observational data from specially designed TC field programmes. The new physics package is then implemented in the model, which is followed by further evaluation. The developmental framework was found to be successful in improving the surface layer and boundary layer parameterization schemes in the operational Hurricane Weather Research and Forecast (HWRF) model.

Previous numerical studies demonstrated that simulated TC intensity and structure are sensitive to the choice of planetary boundary layer schemes that parameterize turbulent fluxes and vertical mixing process in the boundary layer (e.g., Braun and Tao, 2000; Nolan et al., 2009a, b; Kepert, 2010a, b; Smith and Thomsen, 2010). In numerical models, turbulent transport in the atmospheric boundary layer is generally parameterized through eddy diffusivity. However, observational estimates of eddy diffusivity are rare, especially for high wind conditions (>15 m s^{-1}). Using the flight-level data during the low-level eyewall penetrations of TCs Allen (1980), Hugo (1989) and Zhang et al. (2011b) made the first attempt to estimate momentum fluxes, turbulent kinetic energy and vertical eddy diffusivity in the intense eyewall at around 500 m altitude. They found that the vertical eddy diffusivity in the intense eyewall region increases with wind speed with a maximum value of ~100 m^2 s^{-1}. Comparing the observed values of K_m (Fig. 11a) with those used in the HWRF boundary layer parameterization (Fig. 11b) shows that there is pronounced high bias of K_m in the model. This significant difference resulted in a bias in the boundary layer parameterization scheme of the

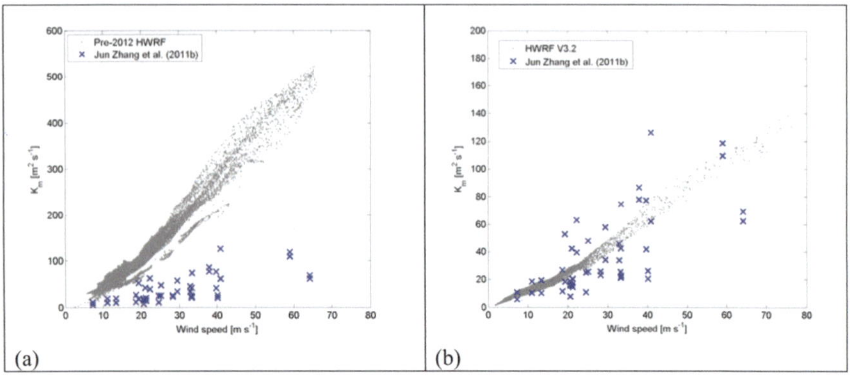

Fig. 11. Plot of vertical eddy diffusivity for momentum flux as a function of wind speed from observations and an idealized simulation with the HWRF model used operationally (a) prior to 2012 seasons, and (b) in the 2012-version operational HWRF model upgraded based on observations as reported by Zhang et al. (2011b).

model towards boundary layers that are too deep. Analysis of these data sets was instrumental in improvements to the vertical diffusion and air-sea exchange of momentum and enthalpy in the operational HWRF model that was transitioned into operations (Gopalakrishnan et al., 2012).

8. Future Observing Technologies

In the future NOAA will be working with NASA on the development and testing of unmanned aerial systems (UAS) such as the Global Hawk (http://airbornescience.nasa.gov/aircraft/Global_Hawk). In 2010 the Global Hawk operated in a number of TCs during NASA's Genesis and Rapid Intensity Project (GRIP, Braun et al., 2013) which was conducted coincident with NOAA's IFEX and the NSF-sponsored Pre-Depression Investigation of Cloud-Systems in the Tropics (PREDICT, Montgomery et al., 2012). During these missions the Global Hawk was able to fly over two TCs (Earl and Karl) for more than 15 hours providing an unprecedented look at the temporal evolution of the storm. Starting in 2012 NASA began a five-year effort to further test the Global Hawk in TC environments under the auspices of the Hurricane and Severe Storm Sentinel (HS3, http://www.espo.nasa.gov/hs3/) Mission. The NASA Global Hawks are ideal platforms for investigations of TCs, capable of flight altitudes greater than 55,000 ft. and flight durations of up to 30 h. HS3 will utilize two Global Hawks (Fig. 12), one with an instrument suite geared towards measurement of the environment and the other with instruments suited to inner-core structure and processes. Field measurements will take place for one month each during the hurricane seasons of 2012-2014.

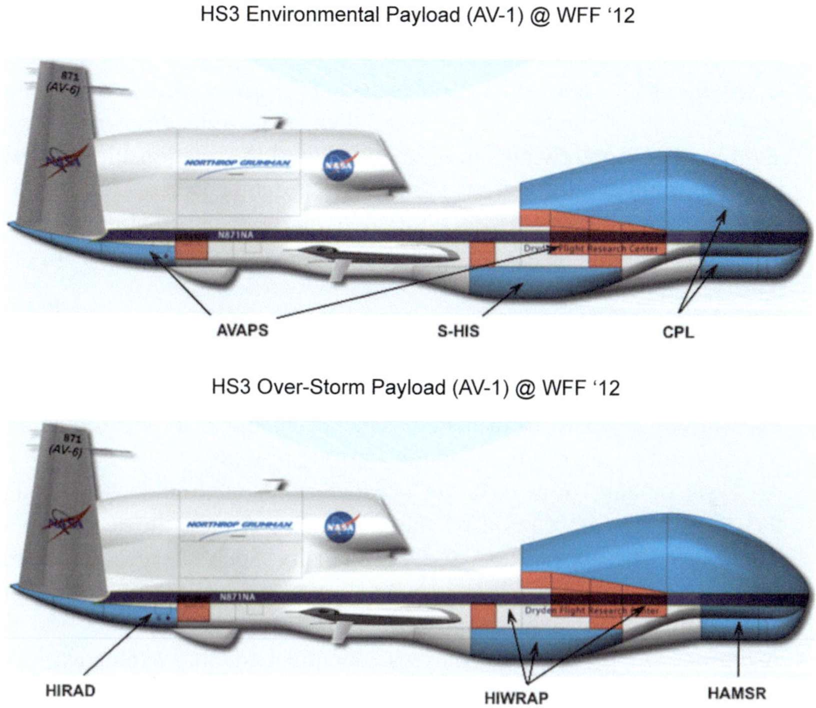

Fig. 12. NASA Global Hawk configurations for the HS3 Mission in 2013.

9. Conclusions

The historical objective of the NOAA Hurricane Field Program was the collection of data required to support analytical and theoretical TC studies designed to improve the understanding of storm structure and behaviour. Whereas the P-3s initially provided storm location and intensity information, the gradual increase in the amount of aircraft data transmitted to the ground led to wider support of operations (Burpee et al., 1994). Improvements to telecommunications will allow for increases in the type and density of transmitted data, including upper-ocean temperature profiles and Doppler radar wind observations for assimilation in operational numerical models.

Research in TCs utilizing observations collected by the NOAA P-3 aircraft provided much of the current understanding of TC structure, motion and intensity change. More than 200 refereed articles and numerous conference and technical reports were published utilizing data obtained by the P-3 aircraft in TCs. Some of the most important achievements and contributions to operational forecasts include:

- GPS dropwindsonde wind, pressure, mass, temperature and humidity profiles in the TC environment led to significant improvements to TC track forecasts. This finding led to the acquisition of the NOAA G-IV for operational surveillance missions.

- Dropwindsondes released in 1997 provided the first look at vertical wind profiles in the boundary layer. Since then, data from dropwindsondes and the SFMR greatly enhanced knowledge of the relationship between winds at flight level and at the surface where damage occurs.
- Analysis of near-surface thermodynamic observations from dropwindsondes and AXBT data defined the structure of the air-sea interface in the TC.
- Instruments developed for the P-3s, such as SFMR and the GPS dropwindsonde, provided important documentation of TC structure for research. In addition, such instrumentation became integral to the TC warning process. Both instruments are installed on operational TC reconnaissance aircraft to aid in the warning process.
- Analysis of data from airborne oceanic expendables released by the P-3s aided the description of the feedback between the temperature structure of the upper-ocean and TC intensity. Such research is leading to improved implementation of physical parameterizations in operational and research coupled air-sea numerical TC models.
- Airborne Doppler radar analyses documented the symmetric TC structure, including concentric eyewall cycles that can modulate intensity of mature TCs. Study of TC asymmetric structure was only in its infancy when the P-3 era began, and analysis of data collected during the P-3 flights allowed numerous studies of this structure and its importance in TC evolution. This led to increased understanding of internal dynamical processes causing intensity changes. The recent upgrade of the G-IV to include a Doppler radar system underscores the importance of these observations. Doppler wind analyses are currently made in real time onboard the P-3s and transmitted to NCEP in preparation for assimilation into next-generation TC forecasting models. This will enable, for the first time, accurate initialization of the TC vortex in operational numerical models and holds great promise for improvements in TC intensity forecasts.

Future aircraft observing efforts into TCs will be geared toward improving track, intensity, and rainfall forecasts by providing these data in order to improve current operational numerical modelling systems and develop new ones, and to understand intensity change and rainfall processes during the entire storm life cycle. This data collection is part of the development of modelling systems, and involves research into data assimilation, understanding and parameterization of physical processes, and model evaluation and validation with comprehensive datasets collected by airborne platforms.

Acknowledgements

The author wishes to acknowledge Dr. Balachandran Sethurathinam, Director, RMC Chennai for providing initial inputs for this work. The author acknowledges NOAA's Hurricane Forecast Improvement Project that supported this work. Thanks are due to Dr. Hua Chen for providing the internal review of this work.

REFERENCES

Aberson, S.D., 2003: Targeted observations to improve operational tropical cyclone track forecast guidance. *Mon. Wea. Rev.*, **131**, 1613–1628.

Aberson, S.D., M.L. Black, R.A. Black, R.W. Burpee, J.J. Cione, C.W. Landsea and F.D. Marks, 2006: Twenty-five years of tropical cyclone research with the NOAA P-3 aircraft. *Bull. Amer. Meteor. Soc.*, **87**, 1039–1055, doi: http://dx.doi.org/10.1175/BAMS-87-8-1039.

Aksoy, A., S. Lorsolo, T. Vukicevic, K.J. Sellwood, S.D. Aberson and F. Zhang, 2012: The HWRF Hurricane Ensemble Data Assimilation System (HEDAS) for high-resolution data: The impact of airborne Doppler radar observations in an OSSE. *Mon. Wea. Rev.*, **140,** doi: http://dx.doi.org/10.1175/MWR-D-11-00212.1

Barnes, G.M., E.J. Zipser, D.P. Jorgensen and F.D. Marks, 1983: Mesoscale and convective structure of a hurricane rainband. *J. Atmos. Sci.*, **40**, 2125–2137.

Barnes, G.M., J.F. Gamache, M.A. LeMone and G.J. Stossmeister, 1991: A convective cell in a hurricane rainband. *Mon. Wea. Rev.*, **119**, 776–794.

Barnes, G.M., E.J. Zipser, D. Jorgensen and F.D. Marks, 1983: Mesoscale and convective structure of a hurricane rainband. *J. Atmos. Sci.*, **40**, 2125–2137.

Barnes, G.M. and M.D. Powell, 1995: Evolution of the inflow boundary layer of Hurricane Gilbert (1988). *Mon. Wea. Rev.*, **123**, 2348–2368.

Black, M.L. and H.E. Willoughby, 1992: The concentric eyewall cycle of Hurricane Gilbert. *Mon. Wea. Rev.*, **120**, 947–957.

Black, P.G., 1983: Ocean temperature changes induced by tropical cyclones. Ph.D. thesis, The Pennsylvania State University.

Black, P.G., R.L. Elsberry, L.K. Shay, R.P. Partridge and J.F. Hawkins, 1988: Atmospheric and oceanic mixed layer observations in Hurricane Josephine obtained from air-deployed drifting buoys and research aircraft. *J. Atmos. Oceanic Technol.*, **5**, 683–698.

Black, P.G. and G.J. Holland, 1995: The Boundary Layer of Tropical Cyclone Kerry (1979). *Mon. Wea. Rev.*, **123**, 2007–2028.

Black, P.G. et al., 2007: Air-sea exchange in hurricanes: Synthesis of observations from the coupled boundary layer air-sea transfer experiment. *Bull. Amer. Meteor. Soc.*, **88**, 357–374.

Black, R.A. and J. Hallett, 1986: Observations of the distribution of ice in hurricanes. *J. Atmos. Sci.*, **43**, 802–822.

Braun, S.A. et al., 2013: NASA's Genesis and Rapid Intensification Processes (GRIP) Field Experiment. *Bull. Amer. Meteor. Soc.*, **94**, 345–363. doi: http://dx.doi.org/10.1175/BAMS-D-11-00232.1

Braun, S.A. and W.-K. Tao, 2000: Sensitivity of high-resolution simulations of Hurricane Bob (1991) to planetary boundary layer parameterizations. *Mon. Wea. Rev.*, **128**, 3941–3961.

Burpee, Robert W. et al., 1994: Real-Time Guidance Provided by NOAA's Hurricane Research Division to Forecasters during Emily of 1993. *Bull. Amer. Meteor. Soc.*, **75**, 1765–1783.

Burpee, R.W., J.L. Franklin, S.J. Lord, R.E. Tuleya and S.D. Aberson, 1996: The impact of Omega dropwindsondes on operational hurricane track forecast models. *Bull. Amer. Meteor. Soc.*, **77**, 925–933.

Cione, J.J. and E.W. Uhlhorn, 2003: Sea surface temperature variability in hurricanes: Implications with respect to intensity change. *Mon. Wea. Rev.*, **131**, 1783–1796.

Cione, J.J., P.G. Black and S.H. Houston, 2000: Surface observations in the hurricane environment. *Mon. Wea. Rev.*, **128**, 1550–1561.

D'Asaro, E., P. Black, L. Centurioni, P. Harr, S. Jayne, I.-I. Lin, C. Lee, J. Morzel, R. Mrvaljevic, P.P. Niiler, L. Rainville, T. Sanford and T.Y. Tang, 2011. Typhoon-ocean interaction in the western North Pacific: Part 1. *Oceanography*, **24,** 24–31, http://dx.doi.org/10.5670/oceanog.2011.91.

D'Asaro, E.A., 2003: Performance of autonomous Lagrangian floats. *J. Atmos. Oceanic Technol.*, **20**, 896–911.

Davis, W.R., 1954: Hurricanes of 1954. *Mon. Wea. Rev.*, **82**, 370–373.

Dorst, N.M., 2007: The National Hurricane Research Project: 50 years of research, rough rides, and name changes. *Bull. Amer. Meteor. Soc.*, **88**, 1566–1588, doi: http://dx.doi.org/10.1175/BAMS-88-10-1566.

Franklin, J.L., S.E. Feuer, J. Kaplan and S.D. Aberson, 1996: Tropical cyclone motion and surrounding flow relationships: Searching for beta gyres in omega dropwindsonde data sets. *Mon. Wea. Rev.*, **124**, 64–84.

Franklin, J.L., M.L. Black and K. Valde, 2003: GPS dropwindsonde wind profiles in hurricanes and their operational implications. *Wea. Forecasting*, **18**, 32–44.

Gall, R., J. Franklin, F.D. Marks, E.N. Rappaport and F. Toepfer, 2012: The Hurricane Forecast Improvement Project. *Bull. Amer. Meteor. Soc.*, **94**, 329–343. doi: http://dx.doi.org/10.1175/BAMS-D-12-00071.1.

Gamache, J.F., R.A. Houze Jr. and F.D. Marks Jr., 1993: Dual-aircraft investigation of the inner core of Hurricane Norbert. Part III: Water budget. *J. Atmos. Sci.*, **50**, 3221–3243.

Gentry, R.C., 1981: History of hurricane research in the United States with special emphasis on the National Hurricane Research Laboratory and associated groups. Preprints, 13th Technical Conf. on Hurricanes and Tropical Meteorology, Miami Beach, FL, Amer. Meteor. Soc., 6–16.

Gopalakrishnan, S., F.D. Marks, J. Zhang, X. Zhang, J.-W. Bao and V. Tallapragada, 2013: A study of the impact of vertical diffusion on the structure and intensity of tropical cyclones using the high-resolution HWRF system. *J. Atmos. Sci.*, **70**, 524–541, doi: http://dx.doi.org/10.1175/JAS-D-11-0340.1

Govind, P.K., 1975: Dropwindsonde instrumentation for weather reconnaissance aircraft. *J. Appl. Meteor.*, **14**, 1512–1520.

Griffin, J.S., R.W. Burpee, F.D. Marks and J.L. Franklin, 1992: Real-Time Airborne Analysis of Aircraft Data Supporting Operational Hurricane Forecasting. *Wea. Forecasting*, **7**, 480–490.

Hock, T.F. and J.L. Franklin, 1999: The NCAR GPS Dropwindsonde. *Bull. Amer. Meteor. Soc.*, **80**, 407–420.

Houze, R.A., F.D. Marks Jr. and R.A. Black, 1992: Dual-aircraft investigation of the inner core of Hurricane Norbert. Part II: Mesoscale distribution of ice particles. *J. Atmos. Sci.*, **49**, 943–963.

Ishihara, M., Z. Yanagisawa, H. Sakakibara, K. Matsuura and J. Aoyagi, 1986: Structure of typhoon rainband observed by two Doppler radars. *J. Met. Soc. Japan*, **64**, 923–939.

Jaimes, B. and L.K. Shay, 2009: Mixed layer cooling in mesoscale oceanic eddies during Hurricanes Katrina and Rita. *Mon. Wea. Rev.*, **137**, 4188–4207.

Jelesnianski, C.P., J. Chen and W.A. Shaffer, 1992: SLOSH: Sea, lake and overland surges from hurricanes. NOAA Tech. Rep. NWS 48, Silver Spring, MD.

Jorgensen, D.F., 1984a: Mesoscale and Convective-Scale Characteristics of Mature Hurricanes. Part I: General Observations by Research Aircraft. *J. Atmos. Sci.*, **41**, 1268–1286.

Jorgensen, D.F., 1984b: Mesoscale and convective-scale characteristics of mature hurricanes. Part II: Inner core structure of Hurricane Allen (1980). *J. Atmos. Sci.*, **41**, 1287–1311.

Jorgensen, D.F., P.H. Hildebrand and C.L. Frush, 1983: Feasibility test of an airborne pulse-Doppler meteorological radar. *J. Climate Appl. Meteor.*, **22**, 744–757.

Kepert, J.D., 2010a: Slab- and height-resolving models of the tropical cyclone boundary layer. Part I: Comparing the simulations. *Quart. J. Roy. Meteorol. Soc.*, doi: 10. 1002/qj.667.

Kepert, J.D., 2010b: Slab- and height-resolving models of the tropical cyclone boundary layer. Part II: Why the simulations differ. *Quart. J. Roy. Meteorol. Soc.*, doi: 10. 1002/qj.685.

Kossin, J.P. and M.D. Eastin, 2001: Two distinct regimes in the kinematic and thermodynamic structure of the hurricane eye and eyewall. *J. Atmos. Sci.*, **58**, 1079–1090.

Lorsolo, S., J.A. Zhang, F.D. Marks and J. Gamache, 2010: Estimation and Mapping of Hurricane Turbulent Energy Using Airborne Doppler Measurements. *Mon. Wea. Rev.*, **138**, 3656–3670.

Malmquist, D.L. and A.F. Michaels, 2000: Severe storms and the insurance industry. Storms, Vol. I, R.A. Pielke Jr. and R.A. Pielke Sr. (Eds), Routledge Press, 54–69.

Marks, F.D., 2003: State of the Science: Radar View of Tropical Cyclones. *In:* Radar and Atmospheric Science: A Collection of Essays in Honor of David Atlas. R.M Wakimoto and R.C. Srivastava (Eds), *Meteorological Monographs*, **52**, AMS, Boston, MA, 33–74.

Marks, F.D. and R.A. Houze, 1984: Airborne Doppler Radar Observations in Hurricane Debby. *Bull. Amer. Meteor. Soc.*, **65**, 569–582.

Marks, F.D. and R.A. Houze, 1987: Inner core structure of Hurricane Alicia from airborne Doppler radar observations. *J. Atmos. Sci.*, **44**, 1296–1317.

Marks, F.D., R.A. Houze Jr. and J.F. Gamache, 1992: Dual-aircraft investigation of the inner core of Hurricane Norbert. Part I: Kinematic structure. *J. Atmos. Sci.*, **49**, 919–942.

Moller, J.D. and M.T. Montgomery, 1999: Vortex Rossby waves and hurricane intensification in a barotropic model. *J. Atmos. Sci.*, **56**, 1674–1687.

Moller, J.D. and M.T. Montgomery, 2000: Tropical cyclone evolution via potential vorticity anomalies in a three-dimensional balance model. *J. Atmos. Sci.*, **57**, 3366–3387.

Montgomery, M.T. and R. Kallenbach, 1997: A theory for vortex Rossby-waves and its application to spiral bands and intensity changes in hurricanes. *Quart. J. Roy. Meteor. Soc.*, **123**, 435–465.

Montgomery, M.T. et al., 2012: The Pre-Depression Investigation of Cloud-Systems in the Tropics (PREDICT) Experiment: Scientific Basis, New Analysis Tools, and Some First Results. *Bull. Amer. Meteor. Soc.*, **93**, 153–172. doi: http://dx.doi.org/10.1175/BAMS-D-11-00046.1

Nolan, D.S., J.A. Zhang and D.P. Stern, 2009a: Evaluation of planetary boundary layer parameterizations in tropical cyclones by comparison of in-situ data and high-resolution simulations of Hurricane Isabel (2003). Part I: Initialization, maximum winds, and outer core boundary layer structure. *Mon. Wea. Rev.*, **137,** 3651–3674.

Nolan, D.S., D.P. Stern and J.A. Zhang, 2009b: Evaluation of planetary boundary layer parameterizations in tropical cyclones by comparison of in-situ data and high-resolution simulations of Hurricane Isabel (2003). Part II: Inner core boundary layer and eyewall structure. *Mon. Wea. Rev.*, **137,** 3675–3698.

Nystuen, J.A. and H.D. Selsor, 1997: Weather classification using passive acoustic drifters. *J. Atmos. Oceanic Technol.*, **14**, 656–666.

Powell, M.D., 1990a: Boundary layer structure and dynamics in outer hurricane rainbands. Part I: Mesoscale rainfall and kinematic structure. *Mon. Wea. Rev.*, **118**, 891–917.

Powell, M.D., 1990b: Boundary layer structure and dynamics in outer hurricane rainbands. Part II: Downdraft modification and mixed layer recovery. *Mon. Wea. Rev.*, **118**, 918–938.

Pun, I.F., Y.-T. Chang, I.-I. Lin, T.Y. Tang and R.-C. Lien, 2011: Typhoon-ocean interaction in the western North Pacific: Part 2. *Oceanography*, **24,** 32–41, http://dx.doi.org/10.5670/oceanog.2011.92.

Reasor, P.D., M.T. Montgomery, F.D. Marks and J.F. Gamache, 2000: Low-wavenumber structure and evolution of the hurricane inner core observed by airborne dual-Doppler radar. *Mon. Wea. Rev.*, **128**, 1653–1680.

Rogers, R. et al., 2013: NOAA's Hurricane Intensity Forecasting Experiment (IFEX): A Progress Report. *Bull. Amer. Meteor. Soc.*, **94**, 859–882. doi: http://dx.doi.org/10.1175/BAMS-D-12-00089.1.

Rogers, R., S. Lorsolo, P. Reasor, J. Gamache and F.D. Marks, 2012: Multiscale Analysis of Tropical Cyclone Kinematic Structure from Airborne Doppler Radar Composites. *Mon. Wea. Rev.*, **140**, 77–99. doi: http://dx.doi.org/10.1175/MWR-D-10-05075.1.

Rogers, R. et al., 2006: The Intensity Forecasting Experiment: A NOAA Multiyear Field Program for Improving Tropical Cyclone Intensity Forecasts. *Bull. Amer. Meteor. Soc.*, **87**, 1523–1537. doi: http://dx.doi.org/10.1175/BAMS-87-11-1523

Ryan, B.E., G.M. Barnes and E.J. Zipser, 1992: A wide rainband in a developing tropical cyclone. *Mon. Wea. Rev.*, **120**, 431–437.

Samsury, C.E. and Edward J. Zipser, 1995: Secondary Wind Maxima in Hurricanes: Airflow and Relationship to Rainbands. *Mon. Wea. Rev.*, **123**, 3502–3517.

Sanford, T.B., R.G. Drever, J.H. Dunlap and E.A. D'Asaro, 1982: Design, operation and performance of an expendable temperature and velocity profiler (XTVP). Rep. APL-UW 8110, Applied Physics Laboratory, University of Washington. (Available from Applied Physics: Laboratory, University of Washington, 1013 NE 40th Street, Seattle, WA 98105.)

Shapiro, L.J. and J.L. Franklin, 1995: Potential vorticity in Hurricane Gloria. *Mon. Wea. Rev.*, **123**, 1465–1475.

Shapiro, L.J. and H.E. Willoughby, 1982: The response of balanced hurricanes to local sources of heat and momentum. *J. Atmos. Sci.*, **39**, 378–394.

Shay, L.K., R.L. Elsberry and P.G. Black, 1989: Vertical Structure of the Ocean Current Response to a Hurricane. *J. Phys. Oceanogr.*, **19**, 649–669.

Shay, L.K., P.G. Black, A.J. Mariano, J.D. Hawkins and R.L. Elsberry, 1992: Upper-ocean response to Hurricane Gilbert. *J. Geophys. Res.*, **97**, 20,227–20,248.

Simpson, R.H., 1952: Exploring the eye of Typhoon "Marge," 1951. *Bull. Amer. Meteor. Soc.*, **33**, 286–298.

Simpson, R.H., 1954: Hurricanes. *Scientific American*, **109**, 22–37.

Simpson, R.H., 1981: Implementation phase of the National Hurricane Research Project 1955–1956. Preprints, 13th Technical Conf. on Hurricanes and Tropical Meteorology, Miami Beach, FL, Amer. Meteor. Soc., 1–5.

Simpson, R.H. (Ed.), 2003: Hurricanes! Coping with Disaster. Amer. Geophys. Union.

Simpson, R.H. and H. Riehl, 1981: The Hurricane and Its Impact. Louisiana State University Press.

Smith, R.K. and G.L. Thomsen, 2010: Dependence of tropical-cyclone intensification on the boundary layer representation in a numerical model. *Quart. J. Roy Met. Soc.*, **136**, 1671–1685.

Sumner, H.C., 1943: North Atlantic hurricanes and tropical disturbances of 1943. *Mon. Wea. Rev.*, **71**, 179–183.

Tabata, A., H. Sakakibara, M. Ishihara, K. Matsuura and Z. Yanagisawa, 1992: A general view of the structure of Typhoon 8514 observed by dual-Doppler radar: From outer rainbands to eyewall clouds. *J. Met. Soc. Japan*, **70**, 897–917.

Uhlhorn, E.W. and P.G. Black, 2003: Verification of Remotely Sensed Sea Surface Winds in Hurricanes. *J. Atmos. Oceanic Technol.*, **20**, 99–116.

Wexler, H., 1945: The structure of the September, 1944, hurricane when off Cape Henry, Virginia. *Bull. Amer. Meteor. Soc.*, **26**, 156–159.

Whitehead, J.C., 2003: One million dollars per mile? The opportunity costs of hurricane evacuation. *Ocean Coastal Manag.*, **46**, 1069–1083.

Willoughby, H.E., 1990a: Temporal changes in the primary circulation in tropical cyclones. *J. Atmos. Sci.*, **47**, 242–264.

Willoughby, H.E., 1990b: Gradient balance in tropical cyclones. *J. Atmos. Sci.*, **47**, 265–274.

Willoughby, H.E., 1998: Tropical cyclone eye thermodynamics. *Mon. Wea. Rev.*, **126**, 3053–3067.

Willoughby, H.E. and M.B. Chelmow, 1982: Objective determination of hurricane tracks from aircraft observations. *Mon. Wea. Rev.*, **110**, 1298–1305.

Willoughby, H.E., J.A. Clos and M.G. Shoreibah, 1982: Concentric Eye Walls, Secondary Wind Maxima, and The Evolution of the Hurricane vortex. *J. Atmos. Sci.*, **39**, 395–411.

Willoughby, H.E., F.D. Marks and R.J. Feinberg, 1984: Stationary and moving convective bands in hurricanes. *J. Atmos. Sci.*, **41**, 3189–3211.

Willoughby, H.E. and M.E. Rahn, 2004: Parametric representation of the primary hurricane vortex. Part I: Observations and evaluation of the Holland (1980) model. *Mon. Wea. Rev.*, **132**, 3033–3048.

Wood, F.B., 1945: A flight into the September, 1944, hurricane off Cape Henry, Virginia. *Bull. Amer. Meteor. Soc.*, **26**, 153–156.

Wroe, D.R. and G.M. Barnes, 2003: Inflow layer energetics of Hurricane Bonnie (1998) near landfall. *Mon. Wea. Rev.*, **131**, 1600–1612.

Zhang, F., Y. Weng, J. Gamache and F.D. Marks, 2011a: Performance of convection-permitting hurricane initialization and prediction during 2008-2010 with ensemble data assimilation of inner-core airborne Doppler radar observations. *Geophys. Res. Lett.*, **38**, doi: http://dx.doi.org/10.1029/2011GL048469.

Zhang, J.A., F.D. Marks, M.T. Montgomery and S. Lorsolo, 2011b: An Estimation of Turbulent Characteristics in the Low-Level Region of Intense Hurricanes Allen (1980) and Hugo (1989). *Mon. Wea. Rev.*, **139**, 1447–1462. doi: http://dx.doi.org/10.1175/2010MWR3435.1.

2

Use of Satellite Observations in Tropical Cyclone Studies

C.M. Kishtawal
Space Applications Centre
Indian Space Research Organization
Ahmedabad, India – 380015
cmk307@yahoo.com

1. Overview

The launch of first meteorological satellite, TIROS-1 on 1st April, 1960 started a new era of space-based observations, and the capability of meteorological satellites has come a long way since then. Monitoring of tropical cyclones is a major application of weather satellites and almost all the operational centres worldwide depend on satellite observations for monitoring and prediction of tropical cyclone movements. Factors which determine the usefulness of a satellite observation for cyclone monitoring are the sensor characteristics, spatial resolution of observations, and the orbital parameters, e.g. orbit type (geostationary, polar, etc.), orbital altitude and inclination, and swath. In general the satellite observations can be classified according to sensor characteristics. Meteorological satellites make use of visible (0.4 to 0.7 μm), infrared (1 to 100 μm) and microwave (0.3 to 30 cm) region of electromagnetic spectrum for obtaining observations of different atmospheric and terrestrial parameters. The following subsections briefly describe the usefulness of these satellite sensors for the study of tropical cyclones.

2. Visible/Infrared Sensors

Visible/Infrared Imagers

The earliest satellite observations were made by using visible/infrared wavelengths and even today these observations are regarded equally important for operational meteorological applications. Almost all geostationary satellites occupying different slots along the equatorial belt of the globe use visible/infrared sensors to obtain regular observations of earth-atmosphere system at every 30 minutes. Operational meteorological polar satellites like NOAA also make global observations in visible/IR channels providing better spatial and spectral resolution yet poorer temporal resolution than geostationary satellites. The most primary application of these observations is the delineation of clouds in the atmosphere, and can be used for detecting the regions of cyclogenesis (based on the development of organized cloud clusters), the location, movement and areal extent of cyclones. Dvorak (1984) developed an important and now widely used technique for determining the cyclonic intensity based on the observations of certain features associated with the cloud fields of tropical cyclones. This technique used originally only the visible (VIS) imagery. Later it incorporated infrared (IR) and other sensors data. It describes tropical cyclone development in terms of day-to-day changes in the cloud pattern of the storm and its environment.

The VIS and IR techniques differ mainly in the cloud features that are used in the analysis and the way in which they are measured. The enhanced and digital infrared techniques when applied to storms of very severe strength rely almost entirely on measurements of the 'EYE' temperature and the temperature of the clouds around the eye for the intensity determination. Higher end applications of visible/IR observations include determination of cloud top temperature, cloud height, cloud thickness, classification of clouds into convective/stratiform types, and the determination of cloud level winds, known as Cloud Motion Vector (CMV) winds, based on the motion of clouds. The above parameters can be determined at regular time intervals from geostationary satellites, and are extremely useful in operational monitoring of tropical cyclones. Arkin and other researchers developed techniques for determining the rainfall using infrared observations, though such estimates are not highly accurate for short time intervals (e.g. less than six hours) due to the fact that the rainfall determination techniques based on visible/IR observations are "physically indirect". This means that in visible/IR frequencies, one can only observe the cloud top, and not the underlying region where the rain actually occurs.

More accurate quantitative estimates of rainfall can be made using microwave frequencies, and this aspect is described in the next subsection. Another useful parameter determined by visible/IR measurements is the sea surface temperature (SST) which may indicate the regions where intense convection may develop, or a moving convective system may intensify or decay. Determination of SST normally requires more than one channel in infrared region (e.g. a split window channel: 10.5-11.5 and 11.5-12.5 μm) for making the corrections for atmospheric absorbers of radiation, like water vapour. Observations of global SST are made by NOAA satellites operationally. However, the accuracy and availability of visible/IR based SST is seriously affected by the presence of clouds, and hence it is difficult to obtain these measurements in the close vicinity of tropical cyclones. SST can be determined in the presence of clouds using microwave techniques.

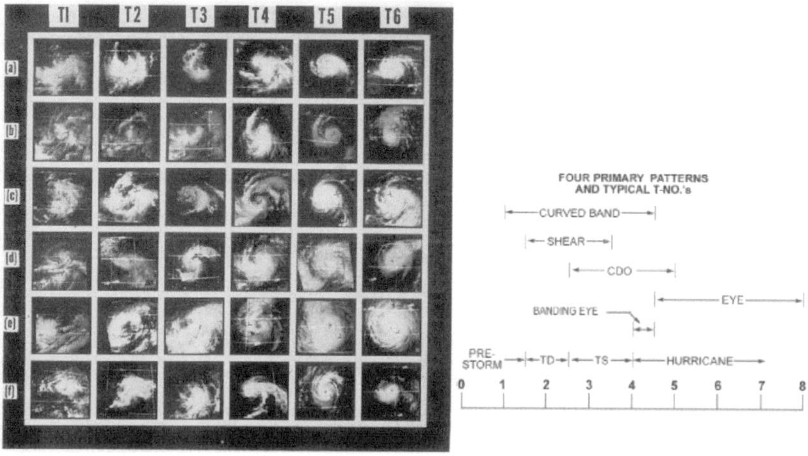

Fig. 1. Tropical cyclone cloud patterns as seen in visible/infrared images from geostationary or polar satellites. Intensity estimation based on cloud pattern is the basis of Dvorak's technique that is still used operationally for cyclone intensity estimation. (Figure courtesy: Chris Velden and co-authors, *Bull. Amer. Met. Soc.*, Sep. 2006.)

"Sounders" are the instruments used for measuring vertical profiles of some atmospheric parameters, like temperature or humidity, and when such instruments are placed on a satellite, they can provide three dimensional distribution of these parameters. Sounders work on the principle of absorption/emission of electromagnetic radiation by various gases in the atmosphere. Depending upon the variation of physical temperature and density of absorbing gases along the vertical direction, the emitted radiance shows a Gaussian type variation with height. The height at which the peak of this Gaussian curve (known as weighting function) occurs, is different for different wavelengths of infrared, because, the absorption efficiency of a gas is a function of wavelength. Thus, it is possible to determine the physical temperature at some height by mathematically inverting the above physical process, and temperature at different heights can be known by taking observations at different wavelengths. Retrieval of humidity profiles is also similar in principle, but employs a slightly different procedure. High Resolution Infrared Sounder (HIRS) onboard NOAA satellites is such an operational sounder which uses 13 spectral channels (associated with CO_2 absorption bands, seven channels placed around 15 μm, and six around 4.3 μm) for temperature sounding, and three channels (associated with water vapour absorption at 8.2, 7.3 and 6.7 μm) for determining water vapour concentration in three broad vertical layers. Unfortunately the quality of infrared sounding is seriously affected by the presence of clouds, and the retrievals are not possible if the satellite field of view is totally covered by the clouds. Such situations are quite common in the domain of tropical cyclones. However, outside the cyclone, the retrievals are possible, and are very useful in determining the environmental influences which may affect the cyclone track and intensity.

3. Microwave Sensors

Microwave frequencies offer a significant advantage over visible/Infrared frequencies that the clouds are transparent in microwave and the oceans act as a cold background. This means that the determination of atmospheric parameters can be made both in clear as well as in cloudy conditions using microwave frequencies. Also, precipitation size rain drops interact strongly with microwave radiation either by means of absorption/emission (at lower frequencies like 10 GHz, 18 Ghz, etc.) or by scattering processes (at higher frequencies e.g. 85 GHz). Thus the estimation of rainfall rates based on microwave observations becomes "physically direct" in nature, unlike visible/IR techniques. These features of microwave sensors make them a very useful observational tool for the study of tropical cyclones. Microwave sensors can be further subdivided into two types, passive and active sensors. While passive sensors make use of microwave radiation emitted from earth and atmosphere, the active sensors deploy their own source of microwave radiation, and analyze the patterns of reflected or scattered radiation for retrieval of geophysical parameters like wind velocity, and sea level height.

3.1 Passive Microwave Sensors

Best examples of passive microwave sensors are Special Sensor Microwave/ Imager (SSM/I), TRMM Microwave Imager (TMI), Multifrequency Scanning Microwave Radiometer (MSMR, onboard Indian OceanSAT-1), and MADRAS onboard Indian low earth satellite Megha-Tropiques. The operating frequencies of these sensors range from 6 GHz to 150 GHz. Lower frequencies like 6 and 10 GHz are used mainly for detection of SST, though they also contribute significantly to the estimation of surface wind speed and rainfall rates. Highly accurate estimates of SST (r.m.s. accuracy better than 0.5 K) are now being made using 6.6 GHz channel of TMI (Chelton et al., 2000). Availability of SST in cloudy situations can be extremely helpful in cyclone studies, as the SST variations along the cyclone path may determine the course of its evolution. Surface level wind speed is another useful parameter that can be derived using passive microwave methods. The accuracy of wind speed retrievals by SSM/I is estimated to be better than 2 ms^{-1}. However, if rainfall rate exceeds about 10 mm/h, the retrieval of surface wind speed becomes extremely difficult, and the retrievals are usually not available in such situations.

In case of cyclones, such retrievals are generally not available within 200 km distance from the centre. Also, as the microwave emission saturates rapidly at higher surface wind speeds (speed >20 ms^{-1}), the retrieved surface winds may show large disagreement with the actual values for some strong winds associated with tropical cyclones. Vertically integrated water vapour (termed as "precipitable water", or PW here onwards) can be retrieved using passive microwave techniques owing to the fact that water vapour is a strong absorber of microwave radiation at about 21 GHz frequency. The accuracy of PW retrievals is estimated to be better than 10%, or about 0.4 gm/cm^2. Similar to wind speed, retrieval of integrated water vapour also becomes difficult under heavy precipitation; however the threshold value of precipitation is much higher in case of PW. In general, the estimation of PW can be made over a large part of a typical tropical cyclone, except within a distance of about 100 km from the centre. The combined patterns of surface wind and PW are good indicators of the nature of moisture flux transport, which is very crucial for maintenance and intensification of a tropical cyclone.

Calculating rainfall rates from microwave sensors requires some fairly complicated calculations. The basis of these calculations is in Planck's radiation law, which describes how much energy a body radiates given its temperature. Water surfaces such as oceans and lakes have an additional property which is very important. The surfaces emit only about one half the microwave energy specified by Planck's law and therefore appear to have only about half the real temperature of the surface. Water surfaces therefore look very "cold" to a passive microwave radiometer. Raindrops, on the other hand, appear to have a temperature that equal their real temperature. They appear warm to a passive microwave radiometer and therefore offer a contrast against "cold" water surfaces. The more raindrops, the warmer the whole scene appears, and research over the last three decades now make it possible to obtain fairly accurate rainfall rates based on the temperature of the microwave scene. Land is very different from oceans in terms of the emitted microwave radiation, appearing to have about 90 percent of its real temperature. In this case, there is little contrast to observe the "warm" raindrops. Certain properties of rainfall, however, still can be inferred. The high frequency microwaves (85.5 GHz) measured by TMI are strongly scattered by ice present in many raining clouds. This reduces the microwave signal at the satellite and offers a contrast against the warm land background.

3.2 Active Microwave Sensors

An active microwave sensor transmits pulsed microwave energy to illuminate the desired area. The return signal is received by the same antenna and is processed. The measurement of radar sea return began during World War II, but the object of early experiments was establishing the size of the radar to determine the speed of the winds at sea using a satellite scatterometer which was proposed in 1996. Since that time, many experiments have been conducted to ascertain the way in which the radar signal responds to variations in the surface of the sea caused by variations in local winds and the theory of radar return from the sea has been advanced significantly. Most common examples of satellite active microwave sensors are scatterometer and altimeter.

Wind vector is an indispensable parameter, along with a few others, for the prediction of storm surges caused by cyclones which hit the coasts. It is to be noted that a significant portion of damage associated with tropical cyclones is caused by storm surges. Repetitive measurements of surface wind field over the large oceanic regions are thus necessary for the above studies and applications. Wind scatterometer is the only device that can provide ocean surface wind vectors. The operating frequencies of these scatterometers range from 5 to 14 GHz. The physical basis for the measurement technique is the Bragg scattering of microwave energy from centimetre length capillary waves of the oceans created by the action of the surface winds.

The strength of the radar backscatter is proportional to the capillary wave amplitude, which is assumed to be in equilibrium with the wind friction speed u^*. The wind backscatter can be derived from the scatterometer back scatter. At present ASCAT onboard European satellite METOP and ISRO's Oceansat-2 Scatterometer (OSCAT) are operational scatterometers that measure ocean surface winds with a speed accuracy of 2 ms^{-1} and direction accuracy of 20°. However, many more satellites with scatterometer, etc. will be launched within next five years.

Another important active microwave sensor is "radar altimeter", which is basically a nadir pointing radar system which measures time return echoes from ocean surface and is designed to measure significant wave height, surface wind speed, sea surface elevation, which relates to ocean currents, surface geoids and tides. It can be a useful tool for measuring the sea surface undulations, and intense winds caused by tropical cyclone movements. Altimeters can also provide information of the vertical structure of the ocean if complemented by historical hydrographic data. Given the relatively slow-time scale of mesoscale ocean features of a few km/day, altimetry measurements provide the surface data for the detection and location of warm mesoscale oceanic features, usually identified as positive SHA values. The current altimeter missions are TOPEX, JASON, ENVISAT and ERS-2. The accuracy of sea level height measurements by TOPEX satellite is about 5–6 cm.

One of the unique instruments in the category of active microwave sensors is the TRMM (Tropical Rainfall Measuring Mission) Precipitation Radar, the first space-borne instrument designed to provide three dimensional maps of storm structure. These measurements yield invaluable information on the intensity and distribution of the rain, on the rain type, on the storm depth and on the height at which the snow melts into rain. The estimates of the heat released into the atmosphere at different heights based on these measurements can be used to improve models of the global atmospheric circulation. The Precipitation Radar has a horizontal resolution at the ground of about five kilometres and a swath width of ~250 kilometres. One of its most important features is its ability to provide vertical profiles of the rain and snow from the surface up to a height of about 20 kilometres. The Precipitation Radar is able to detect fairly light rain rates down to less than one millimetres per hour. At intense rain rates, where the attenuation effects can be strong, new methods of data processing have been developed that help correct for this effect. The

Precipitation Radar is able to separate out rain echoes for vertical sample sizes of about 250 metres at nadir. Observations from TRMM Rain Radar are highly valuable for calibration and validation of other rain-measuring sensors, and analysis of three dimensional rain structures within the cyclone. These observations can be crucial for understanding the physical processes responsible for intensification of tropical cyclones.

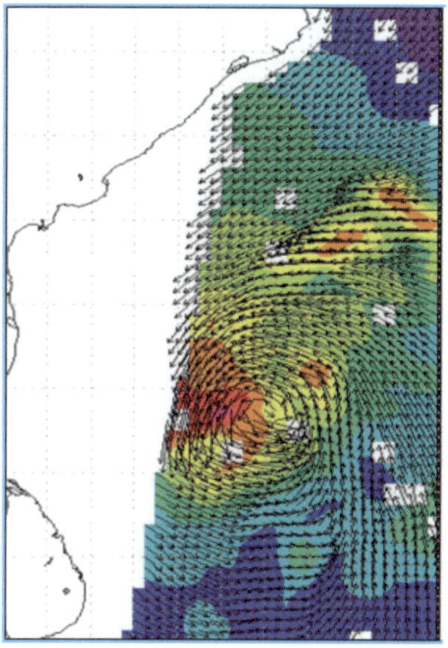

Fig. 2. QuikScat Scatterometer winds for a cyclone over the Bay of Bengal.

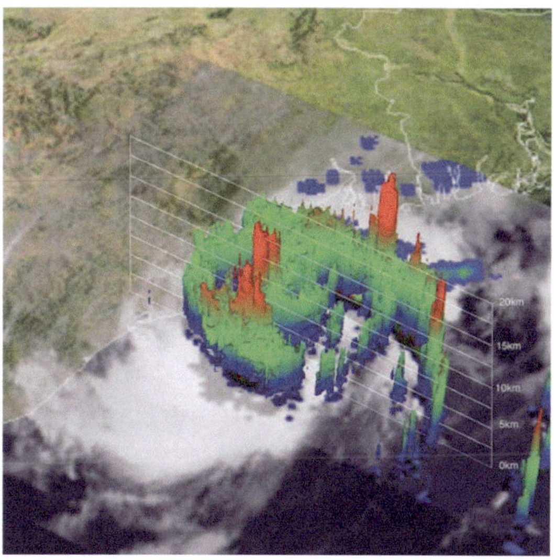

Fig. 3. Three dimensional structure of rain-bands within tropical cyclone Bijlee (15-17 April, 2009). (Figure courtesy: pmm.nasa.gov)

3.3 Microwave Sounders

The atmospheric sounders in the microwave region operate in oxygen and water vapour resonance bands. The temperature profile retrieval is achieved by utilizing the radiometric measurements in 60 GHz complex or at 118 GHz single resonance line of oxygen absorption while humidity profile is derived from measurements at frequencies 183.3 GHz water vapour resonance region. The example of temperature sounders is Microwave Sounding Unit (MSU), and Advance Microwave Sounding Unit (AMSU-A) onboard NOAA satellites, while AMSU-B is used for humidity sounding. Velden analyzed MSU data over tropical cyclones and detected a "warm core" region at about 300 mb the intensity of which was a strong indicator of the intensity of the cyclone. Intensity estimates then can lead to the determination of maximum value of tangential wind.

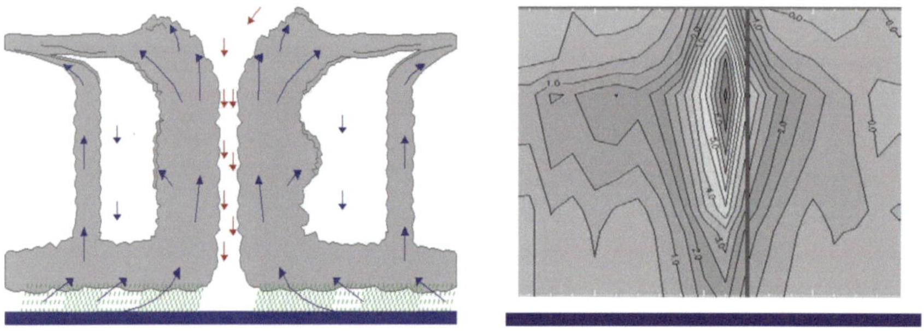

Fig. 4. (a) Cross section of a mature cyclone indicating general direction of motion of air mass within a cyclone. (b) Observations from microwave sounder AMSU-A show temperature anomalies (°K) for the cyclone cross section. The magnitude of the warm anomaly near the eye region has high correlation with cyclone intensity.

4. Some Interesting Applications of Satellite Data in Tropical Cyclone Research

Satellite observations have vital role in cyclone research and operational monitoring and prediction. Today, NWP models (global as well as mesoscale models) worldwide assimilate all the satellite observations mentioned in the previous sections. A brief description of a few interesting applications of satellite observations in context of tropical cyclones is given below.

4.1a Prediction of Tropical Cyclogenesis Based on Infrared Images

"Tropical cyclogenesis" defines an event when a tropical low pressure system develops into a tropical cyclone. Tropical cyclogenesis is a rare phenomena where only about 15% of lows finally develop into cyclones. Prediction of cyclogenesis has significant operational value because a successful prediction of cyclogenesis provides sufficient time to operational forecasters and emergency managers to plan and prepare for impending hazard. Over the period of past few years, researchers developed innovative ideas to detect cyclogenesis using satellite observations. One of such techniques (Venkatesh and Mathew, 2004) uses a sequence (available at an interval of 30 minutes from geostationary satellites) of satellite infrared observations. This technique is based on the theory that mid-level mesoscale vortices interact in a tropical low pressure system. If this interaction results in merger of vortices, the second stage is reached. This larger vortex increases in strength due to the air-sea interaction mechanism, and may lead to cyclogenesis. The above phenomenon can be detected in satellite IR images, because the mesoscale vortices are generally accompanied by deep convection. The above authors define an index called the "merger index" denoted as

$$\mu = r^*(n)/(R_{CG}/L) \quad (1)$$

where $r^*(n)$ is the critical radius for merger, R_{CG} is the average distance of mesoscale vortices from their common centroid and L is the mean radius of the vortices system. The merger index is monitored using real-time satellite images from geostationary satellites and the merger index is continuously monitored. Tropical cyclogenesis is indicated by a high value of merger index (e.g. $\mu > 0.75$). This technique is illustrated in Fig. 5

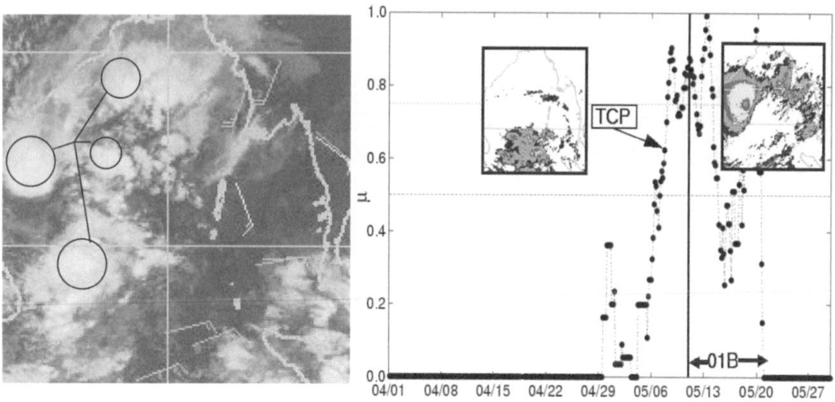

Fig. 5. (a) Identification of mesoscale vortices (circles) in satellite IR images. (b) Example of time-evolution of merger index for a real-time situation. X-axis in this figure is time.

Another interesting way to capture cyclogenesis is to analyze surface level wind fields derived from Microwave Scatterometers. One of the earliest indications of cyclogenesis is the low level cyclonic vorticity that appears about 3-4 days ahead of the time when a low level system develops into a cyclone. Microwave frequencies can penetrate clouds even in light precipitation situations, and hence microwave observations of surface winds become valuable input for the analysis of cyclogenesis. This technique is mentioned below.

4.1b Prediction of Tropical Cyclogenesis Using Scatterometer Observations

Observations from ISRO's recently launched satellites like OceanSat-2 and Megha-Tropique provide valuable observations to predict and monitor tropical cyclones in real time. OceanSat-2 derived wind products are found to be very useful for predicting the tropical cyclogenesis in all the cyclone active basins of the globe. For tropical cyclogenesis prediction the wind vectors derived by the OceanSat-2 onboard scatterometer (OSCAT) are used to develop a technique based on wind pattern matching. The technique is based on the premise that there is some degree of similarity in low level wind circulation among developing systems, which can be utilized to distinguish them from non-developing systems. This similarity of wind patterns has been measured quantitatively by computing the "matching index" between the given wind pattern and the wind signatures of developing systems available from the past observations (QuikScat derived wind vectors). A database of the past observations for wind pattern matching was formed using the early stages of wind patterns of tropical cyclones developed during 2000-2008 using QuikScat data. Different database have been prepared for each basin for the matching of wind patterns. The mean prediction lead time of the technique was 63 hours. Probability of Detection (POD) of the technique was 100%, while the False Alarm Ratio (FAR) was 2%. The OSCAT winds showing the signatures of tropical cyclogenesis of Tropical cyclone "Thane" have been shown in Fig. 1. The winds associated with genesis of cyclone are marked as a block in the figure. The earliest prediction of TC Thane by OSCAT was given on 23rd December, 06:18 Z whereas the operational agencies reported the system as tropical storm on December 25, 18Z. Thus the cyclone was detected 60 hours before its formation.

4.2 Initialization of Cyclonic Vortex in NWP Models

One of the most important applications of satellite observations is to specify the position intensity and size of tropical cyclones correctly in the initial conditions of NWP models. Mesoscale models generate high resolution initial conditions using coarse resolution analyses provided by global models. Due to coarse resolution, these analyses fields can not represent the location and intensity of cyclones with the accuracy desired by high resolution models. Vortex initialization is a process to represent the dynamic and thermodynamic structure of cyclonic vortex as realistically as possible in a high resolution initial condition to be used by the mesoscale models. Satellite observations provide reasonably accurate estimates of cyclone location and intensity that can be utilized for vortex initialization. Figure 6 shows an example where

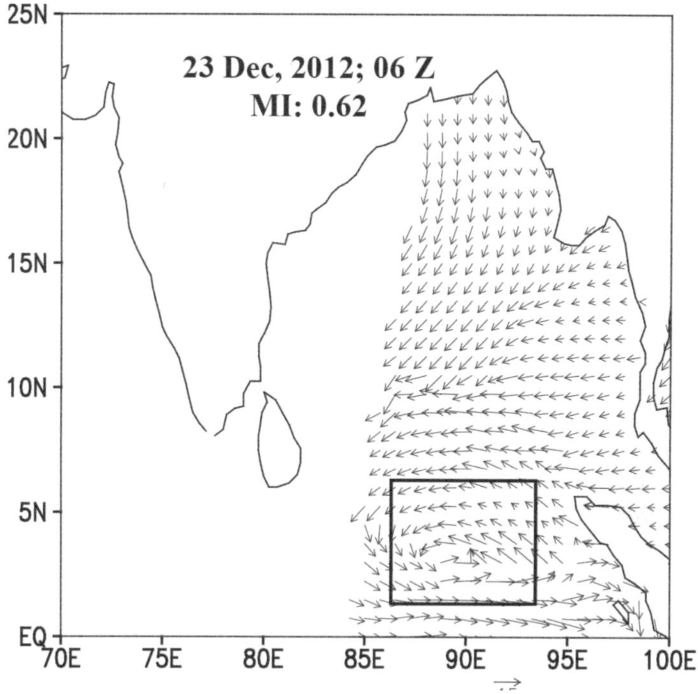

Fig. 6. Tropical cyclogenesis prediction by OSCAT derived wind fields.

satellite observations of surface winds from QuikScat scatterometer were used to modify the cyclonic circulation in model initial fields. Results from a number of studies confirm the fact that proper representation of cyclonic vortex in models leads to significantly improved prediction of cyclone track and intensity (Fig. 7).

4.3 Assimilation of Satellite Observations of Atmospheric Ozone in NWP Models

There are several benefits of the assimilation of ozone satellite data for numerical weather forecast. The retrieval of temperature profiles from infrared sounding instruments is influenced by ozone. Cold, stratospheric ozone absorbs and emits infrared radiation, and thereby reduces the radiance or brightness temperature as observed by the satellite instrument. Ozone has a strong influence on both short and long-wave radiation and the temperature in the middle atmosphere. An accurate knowledge of the ozone distribution is expected to lead to improvements of these aspects. Further, the time evolution of ozone contains information on the wind field that transports ozone. High quality, detailed information of the ozone distribution is available on an almost continuous basis from 1979. The successful Total Ozone Mapping Spectrometer (TOMS) (on four different satellites) measures the total column of ozone with a good resolution of about 50 km, and with a nearly global

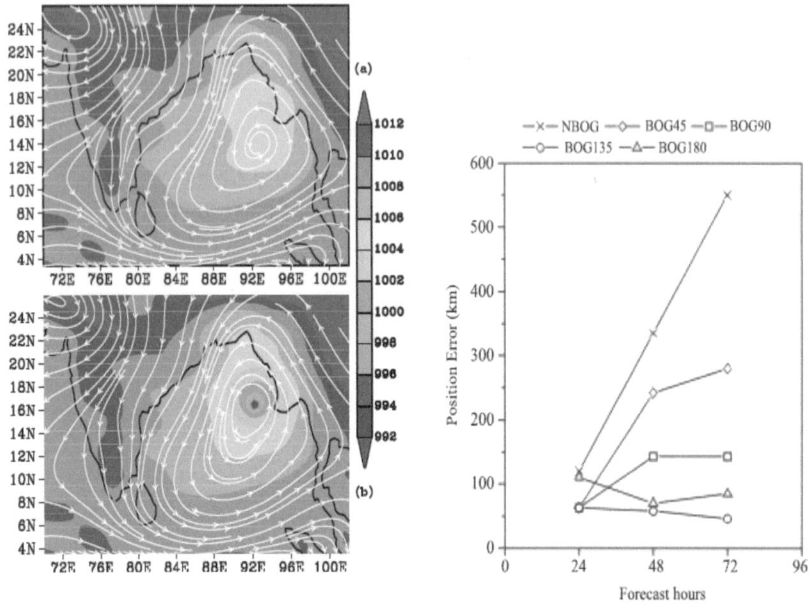

Fig. 7. (a) 850 hpa wind streamlines and surface pressure before (upper panel) and after (lower panel) vortex initialization. (b) Reduction of track prediction errors using vortex assimilation (experiments BOG45, BOG90, BOG135, and BOG180) with respect to control (experiment NBOG). (Figure courtesy: R. Singh et al. J. Earth Syst. Sci., 114, No. 4, August 2005.)

coverage each day (e.g. McPeters et al., 1998). The Solar Backscatter Ultra Violet (SBUV, SBUV/2) instruments have been flown on the NOAA satellites for basically the same time period, and provide additional vertical profile information (e.g. Bhartia et al., 1996). Since 1995 the European spectrometer GOME (Global Ozone Monitoring Instrument) on the ESA ERS satellite has collected eight years of ozone data (Burrows et al., 1999). SCIAMACHY (Scanning Imaging Absorption Spectrometer for Atmospheric Cartography) is part of the Envisat payload, launched in 2002 (Bovensmann et al., 1999). Zou and Wu (2005) showed that variations of total ozone amounts are closely related to the formation, intensification and movement of cyclones. High correlation between TOMS ozone and vertically integrated potential vorticity fields was found in a hurricane environment.

REFERENCES

Bhartia, P.K., R.D. McPeters, C.L. Mateer, L.E. Flynn and C. Wellemeyer, 1996: Algorithm for the estimation of vertical ozone profiles from the backscattered ultraviolet technique. *Journal of Geophysical Research*, **101**, 18793–18806.

Bovensmann, H., J.P. Burrows, M. Buchwitz, J. Frerick, S. Nol, V.V. Rozanov, K.V. Chance and A.P.H. Goede, 1999: SCIAMACHY: Mission objectives and measurement modes. *J. Atmos. Sci.*, **56**, 127–150.

Burrows, J.P., M. Weber, M. Buchwitz, V. Rozanov, A. Ladstatter-Weibenmayer, A. Richter, R. De-beek, R. Hoogen, K. Bramstedt, K.-U. Eichmann, M. Eisinger and D. Perner (1999). The Global Ozone Monitoring Experiment (GOME): Mission concept and first results. *J. Atmos. Sciences*, **56(2)**, 151–175.

Dvorak, V.F., 1984: Tropical cyclone intensity analysis using satellite data. NOAA Tech. Rep. 11 (Available from NOAA/NESDIS, 5200 Auth Rd., Washington, DC 20333).

Dvorak, V.F., 1995: Tropical clouds and cloud systems observed in satellite imagery: Tropical cyclones. Workbook Vol. 2. NOAA/NESDIS (Available from NOAA/ NESDIS).

McPeters, R.D., P.K. Bhartia, A.J. Krueger, J.R. Herman, C.G. Wellemeyer, C.J. Seftor, G. Jaross, O. Torres, L. Moy, G. Labow, W. Byerly, S.L. Taylor, T. Swissler and R.P. Cebula, 1998: Earth Probe Total Ozone Mapping Spectrometer (TOMS) Data Products User's Guide. NASA Technical Publication 1998-206895, NASA Goddard Space Flight Center, Greenbelt, Maryland 20771.

Venkatesh, T.N. and J. Mathew, 2004: Prediction of tropical cyclone genesis using a vortex merger index. *Geophysical Research Letters*, **31**, L04105.

Zou, X. and Y. Wu, 2005: On the relationship between Total Ozone Mapping Spectrometer (TOMS) ozone and hurricanes. *J. Geophys. Res.*, **110**, D06109, doi: 10.1029/2004JD005019.

PART 2

Advances in Numerical Weather Predictions for Tropical Cyclones

3

Overview of the NOAA/NCEP Operational Hurricane Weather Research and Forecast (HWRF) Modelling System

Vijay Tallapragada
Environmental Modeling Center (EMC),
NCEP/NOAA/NWS, College Park, MD, USA
vijay.tallapragada@noaa.gov

1. Introduction

Prediction of track and intensity of a tropical cyclone (TC) is one of the many challenging problems in meteorology, but very important for issuing timely warning for many agencies engaged in disaster preparedness and mitigation. Hurricane[1] forecasting is one of the most important areas of modelling at Environmental Modeling Center (EMC) at the National Centers for Environmental Prediction (NCEP) of National Oceanic and Atmospheric Administration (NOAA). The Hurricane Weather Research and Forecast (HWRF) modelling system for hurricane prediction, developed by scientists at EMC, in collaboration with various academic and government agencies, is a sophisticated, high-resolution, atmosphere-ocean coupled model designed to resolve inner-core features of hurricanes and to improve our understanding of multi-scale, spatio-temporal interactions of storm circulations within the large-scale environment. The HWRF modelling system became operational at NCEP starting with the 2007 hurricane season. It provides high-resolution track, intensity and structure forecast guidance to the operational forecasters at the National Hurricane Center (NHC) and Central Pacific Hurricane Center

[1]The term 'hurricane' denotes a particular category of tropical cyclones. Likewise, the term 'typhoon' specifically denotes severe tropical cyclones that develop in the western North Pacific. In this text, the terms 'hurricane', 'tropical cyclone', and the abbreviation 'TC' are used interchangeably.

(CPHC) for all tropical cyclones in their area of responsibility (Atlantic and Eastern Pacific, and Central Pacific, respectively). Recent efforts supported by NOAA's Hurricane Forecast Improvement Project (HFIP) allowed the HWRF team to expand the applications of the HWRF modelling system (henceforth to be referred to simply as 'HWRF') for tropical cyclone forecasts for all Northern Hemispheric tropical ocean basins, including North Indian Ocean and North Western Pacific Ocean.

Development of HWRF began in 2002 at EMC in collaboration with the NOAA's Geophysical Fluid Dynamics Laboratory (GFDL) and the University of Rhode Island (URI). The initial strategy for HWRF development was to take advantage of the advancements made in improving track prediction through a focused collaboration between EMC, GFDL and URI and to transition those modelling advancements to HWRF. This strategy ensured comparable track skill to the GFDL forecasts for both the Eastern North Pacific and Atlantic (including Caribbean and Gulf of Mexico) basins. Additionally, features of the GFDL hurricane model that led to demonstrated skill for intensity forecasts, such as ocean coupling, upgraded air-sea physics, and improvements to microphysics, were also captured in the newly developed HWRF system.

While the HWRF was developed at NCEP as a highly advanced ocean coupled, land coupled, atmospheric forecasting system, it was not until 2012 that a series of foundational improvements especially in terms of resolution and physics were incorporated to the operational system. The original HWRF model operated at a resolution of 27 km for the static domain and 9 km for the single movable nest. Meanwhile, scientists at the Atlantic Oceanographic and Meteorological Laboratory's Hurricane Research Division (HRD) developed an experimental research version of HWRF called HWRFx that was run in both, idealized (Gopalakrishnan et al., 2011; Bao et al., 2012; Gopalakrishnan et al., 2013) as well as in real-time mode (Zhang et al., 2011; Yeh et al., 2012; Gopalakrishnan et al., 2012a) to target the intensity change problem at higher resolution ($\leq$3 km). Central to the development of the high resolution HWRF is the advancement of the surface and boundary layer parameterization schemes. For the first time, inner core data collected by NOAA WP-3D research aircraft were used as the basis to redesign the boundary layer parametrization for high-resolution hurricane applications (Gopalakrishnan et al., 2013). Significant improvements to modelled boundary layer structure as well as size predictions were demonstrated with those advancements. This modelling system was supported by NOAA's Hurricane Forecast Improvement Project (HFIP).

Based on "R2O type", HIFP demonstration experiments that illustrated significant impacts of high resolution for TC predictions, scientists at EMC worked with NOAA research, in particular, HRD, and other academic partners and implemented major changes to original operational version of HWRF resulting in a new operational HWRF for the 2012 hurricane season. The central improvement was the triple-nest capability (27:9:3) that included a cloud-resolving innermost grid operating at 3 km horizontal resolution along with several improvements to the physics schemes based on observational findings and advanced vortex initialization data assimilation techniques for better representation of the inner core structure of the storms. Apart from obtaining significant improvements in the track forecast skill compared to previous versions, 2012 version of the operational HWRF has conclusively demonstrated the positive impact of resolution on storm size and structure forecasts (Tallapragada et al., 2013).

The 2013 version of the operational HWRF made further significant improvement in track, intensity and structure prediction of tropical cyclones by taking advantage of the high-resolution capability built in the 2012 HWRF. One of the highlights of the 2013 HWRF configuration retrospective T&E, performed on a vast sample of three hurricane seasons (2010-2012), was the remarkable intensity forecast skill (Tallapragada et al., 2013c). Results indicated that HWRF outperformed the statistical models and NHC's official forecasts for tropical cyclone intensity (subjectively made by the hurricane specialists) by about 15%, with track forecast skills competitive to the best performing NCEP GFS global model skills. Historically, statistical models have been more skillful than dynamical models for hurricane intensity prediction. These HWRF results demonstrate, for the first time, the potential of an operational dynamical model as a viable hurricane intensity prediction tool.

Starting with the 2012 typhoon season, experimental real-time HWRF forecasts for all tropical cyclones in the Western Pacific basin are provided to the US Navy's Joint Typhoon Warning Center (JTWC) in support of their operations (Tallapragada et al., 2013a). HWRF is also implemented at the India Meteorological Department (IMD) in New Delhi for operational tropical cyclone forecasts over the North Indian Ocean basin. This was accomplished through an Implementation Agreement (IA) between NOAA and India's Ministry of Earth Sciences (MoES). Several international tropical cyclone forecast agencies, including Vietnam's Institute for Meteorology, Hydrology and Environment (IMHEN), China's Shanghai Typhoon Institute (STI), Taiwan's Central Weather Bureau (CWB), and Oman's Directorate General of Meteorological Affairs and Air Navigation (DGMAAN) have also adopted HWRF for their operational needs. The concept of an operational hurricane model available as a research tool and supported through a dedicated community modelling framework at NOAA's Developmental Testbed Center (DTC) allowed further expansion and outreach of the HWRF modelling system to many academic and research organizations across the world, facilitating accelerated model development, research and transition to operations (Gopalakrishnan et al., 2013).

This article provides an overview of the evolution of the current operational HWRF and detailed description of its components, and model performance for various tropical cyclone basins of the world. Section 2 provides a description of atmospheric model components of the operational HWRF followed by special emphasis on the sophisticated vortex relocation and initialization procedure in Section 3 along with data assimilation techniques in Section 4. State-of-the-art hurricane specific physics employed in the operational HWRF is provided in Section 5. Performance of the operational HWRF in various tropical cyclone basins is given in Section 6. Finally, Section 7 gives a summary and outlines future plans for advancing the operational HWRF in addressing comprehensive tropical cyclone forecast problems for the next generation.

2. Atmospheric Model Components of HWRF

HWRF is composed of the WRF model software infrastructure, the Non-Hydrostatic Mesoscale Model (NMM) dynamic core, the Princeton Ocean Model for Tropical Cyclones (POM-TC) for the Atlantic and Eastern Pacific basins, and the NCEP coupler that facilitates communication between the atmospheric and ocean model components using Multiple Program Multiple Data (MPMD) architecture for computational efficiency. HWRF employs a suite of advanced physics developed for tropical cyclone applications. These include modified GFDL surface physics to account for air-sea interaction over warm water and under high wind conditions, GFDL land surface model and radiation, modified Ferrier Microphysics (Ferrier, 1994), modified NCEP GFS planetary boundary layer, and GFS SAS deep and shallow convection. Figure 1 illustrates all components of HWRF, which also include the WRF Pre-Processor System (WPS), and prep_hybrid (a routine used to process spectral coefficients of GDAS and GFS in their native vertical coordinates). HWRF also contains the NCEP Unified Post Processor (UPP), the GFDL vortex tracker, and the regional HWRF Data Assimilation System (HDAS). HDAS is a sophisticated initialization of both the ocean- and the storm-scale circulation, and is based on Gridpoint Statistical Interpolation (GSI) technique. It is a hybrid combination of an Ensemble Kalman Filter (EnKF) and a three-dimensional variational (3DVAR) filter. Additionally, unlike other NCEP forecast systems which run continuously throughout the year, HWRF forecasts are launched for operational use only when NHC/CPHC determines that a disturbed area of weather has the potential to evolve into a depression anywhere over their area of responsibility. After an initial HWRF run is triggered, new runs are launched in cycled mode at 6-h intervals, until either the storm dissipates after making landfall, becomes extra-tropical, or degenerates into a remnant low (typically identified when convection becomes disorganized around the centre of circulation). Currently, the HWRF runs in NCEP operations four times daily producing 126-h forecasts of track, intensity, structure and rainfall to meet NHC/CPHC operational forecast and warning process objectives.

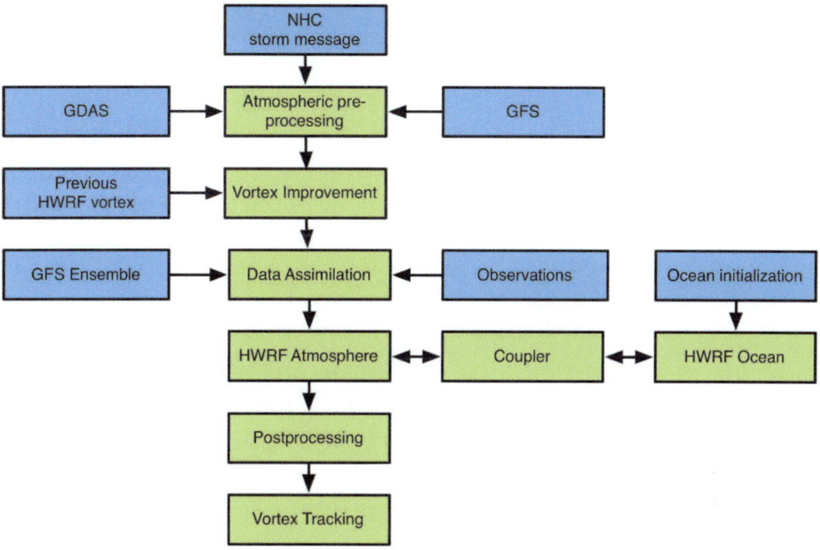

Fig. 1. A simplified overview of the HWRF. Components include the atmospheric initialization, the vortex improvement, the GSI data assimilation, the HWRF atmospheric model, the atmosphere-ocean coupler, the ocean initialization, the POM-TC ocean model, the post processor, and the vortex tracker.

2.1 Dynamic Core and Model Equations

As described in the NMM scientific documentation (Janjic et al., 2010), the WRF-NMM is a non-hydrostatic model formulated on a rotated latitude-longitude, Arakawa Egrid, with a vertical pressure-sigma hybrid vertical coordinate system. The rotated latitude-longitude coordinate is transformed in such a way that the coordinate origin is located in the centre of the parent domain, and the x-axis and y-axis are aligned with the new coordinate equator and the prime meridian through the domain centre, respectively. In order to deal with multi-scale forecasting, a horizontal mesh refinement capability was developed for this system. All interpolations from the parent to the nested domain are achieved on a rotated latitude-longitude Egrid. The nested domain can be freely moved anywhere within the parent domain, yet the rotated latitude-longitude lines of the nested domain will always coincide with the rotated latitude-longitude lines of the parent domain at a fixed parent-to-child gridsize ratio of 3:1.

Let s denote a generalized mass based terrain following vertical coordinate that varies from 0 at the model top to 1 at the surface (e.g., Eckermann, 2009). Let π be the hydrostatic pressure, and let π_{sfc} and π_T be the hydrostatic pressures at the surface and at the top of the model atmosphere, respectively. Then, the difference in hydrostatic pressure between the base and the top of the model column is $\mu = \pi_{sfc} - \pi_T$. Here, π_T is a non-negative constant, whereas

π_{sfc} is a function of time and horizontal position. In the hybrid coordinate the hydrostatic pressure is computed from the formula

$$\pi(x, y, s, t) = \pi_T + \sigma_1(s)\Pi + \sigma_2(s)\mu(s, y, t), \tag{1}$$

where Π is the constant depth of the hydrostatic pressure coordinate layer at the top of the model atmosphere, σ_1 is zero at the top and the bottom of the model atmosphere, and σ_2 increases from 0 to 1 from top to bottom.

The hypsometric equation

$$\frac{\partial \Phi}{\partial \pi} = -\alpha \tag{2}$$

relates the geopotential Φ to the hydrostatic pressure. Assuming that the atmosphere is dry, the specific volume is related to the temperature T and pressure p by the ideal gas law $\alpha = RT/p$, R being the gas constant. Note that the ideal gas law does not involve the hydrostatic pressure but rather the actual pressure, p, hereafter referred to as non-hydrostatic pressure. Using the ideal gas law, from (2)

$$\frac{\partial \Phi}{\partial s} = -\frac{RT}{p}\frac{\partial \pi}{\partial s}. \tag{3}$$

Upon integration of (3) from the surface, where the geopotential is denoted by Φ_{sfc}, to an arbitrary level s,

$$\Phi = \Phi_{sfc} + \int_s^1 \frac{RT}{p}\frac{\partial \pi}{\partial s'} ds' \tag{4}$$

Using (2), the third equation of motion may be written as

$$\frac{dw}{dt} = g\left(\frac{\partial p}{\partial \pi} - 1\right). \tag{5}$$

Defining the ratio of the vertical acceleration and gravity g,

$$\varepsilon \equiv \frac{1}{g}\frac{dw}{dt}, \tag{6}$$

(5) may be rewritten as

$$\frac{\partial p}{\partial \pi} = 1 + \varepsilon, \tag{7}$$

which defines the relationship between the hydrostatic and the non-hydrostatic pressures. Integrating (7) with respect to π, one obtains the non-hydrostatic pressure at an arbitrary hydrostatic pressure, or on a coordinate surface s,

$$p = \int_{\pi_T}^{\pi} \frac{\partial p}{\partial \pi'} d\pi' = \int_0^s (1+\varepsilon)\frac{\partial \pi'}{\partial s'} ds'. \tag{8}$$

As can be seen from (7) and (8), should ε vanish, the pressure and the hydrostatic pressure become equivalent.

In the hydrostatic s coordinate system, the time derivative of a fluid property q following the motion of an air parcel may be written as

$$\frac{dq}{dt} = \left(\frac{\partial q}{\partial t}\right)_s + \mathbf{v} \cdot \nabla_s q + \left(\dot{s}\frac{\partial \pi}{\partial s}\right)\frac{\partial q}{\partial \pi}. \tag{9}$$

Here, $\dot{s}$ is the vertical velocity and the subscripts indicate the variable that is kept constant while the differentiation is performed.

The non-hydrostatic continuity equation takes the form

$$w = \frac{1}{g}\left[\left(\frac{\partial \Phi}{\partial t}\right)_s + \mathbf{v} \cdot \nabla_s \Phi + \left(\dot{s}\frac{\partial \pi}{\partial s}\right)\frac{\partial \Phi}{\partial \pi}\right] + W(\lambda, \varphi, t), \tag{10}$$

i.e., reduces to the definition of the vertical velocity w, with W denoting an integration constant that may depend on horizontal coordinates and time. For simplicity of exposition, in subsequent derivations we assume that W is zero. The familiar hydrostatic mass continuity equation

$$\left[\frac{\partial}{\partial t}\left(\frac{\partial \pi}{\partial s}\right)\right]_s + \nabla_s \cdot \left(\mathbf{v}\frac{\partial \pi}{\partial s}\right) + \frac{\partial}{\partial s}\left(\dot{s}\frac{\partial \pi}{\partial s}\right) = 0 \tag{11}$$

also follows from the nonhydrostatic continuity equation.

Using the material surface boundary conditions $\dot{s} \equiv ds/dt$ at $s = 0$ and $s = 1$, one may obtain two equations from (2.10). The first one gives the tendency of the hydrostatic surface pressure

$$\frac{\partial \mu}{\partial t} = -\int_0^1 \nabla_{s'} \cdot \left(\mathbf{v}\frac{\partial \pi}{\partial s'}\right) ds', \tag{12}$$

and the second one is used to calculate the vertical velocity term

$$\left(\dot{s}\frac{\partial \pi}{\partial s}\right)_s = -\left(\frac{\partial \pi}{\partial t}\right)_s - \int_0^s \nabla_{s'} \cdot \left(\mathbf{v}\frac{\partial \pi}{\partial s'}\right) ds'. \tag{13}$$

Using the relations (2) and (7), in the case of a non-hydrostatic atmosphere one obtains

$$-\frac{1}{\rho}\nabla_z p \equiv -(1+\varepsilon)\nabla_s \Phi - \alpha \nabla_s p. \tag{14}$$

Using (14), the inviscid non-hydrostatic equation for the horizontal part of the wind takes the form

$$\frac{d\mathbf{v}}{dt} = -(1+\varepsilon)\nabla_s \Phi - \alpha \nabla_s p + f\mathbf{k} \times \mathbf{v}. \tag{15}$$

Again, for vanishing ε, (15) reduces to the form used in hydrostatic models.

The first law of thermodynamics for adiabatic processes has the form

$$c_p \frac{dT}{dt} = \alpha \frac{dp}{dt} \tag{16}$$

in which c_p is the specific heat at constant pressure. In hydrostatic models, the derivative dp/dt is replaced by the derivative of hydrostatic pressure $d\pi/dt$, often denoted by the Greek letter omega (ω). For this reason, the right hand side of the equation is frequently referred to as the "omega–alpha" term. The derivative of pressure can be separated into a component ω_1 which reduces to the hydrostatic expression when ε vanishes, and a component ω_2 which vanishes with vanishing ε. Note that, generally, $p = p(x, y, \pi, t)$. Then

$$\frac{\partial p}{\partial t} = \left(\frac{\partial p}{\partial \pi}\right)_t \frac{\partial \pi}{\partial t} + \left(\frac{\partial p}{\partial t}\right)_\pi = (1+\varepsilon)\frac{\partial \pi}{\partial t} + \left(\frac{\partial p}{\partial t}\right)_\pi, \tag{17}$$

where the subscripts indicate the variable that is kept constant while the differentiation is performed. In addition, as can be seen from (7),

$$\left(\dot{s}\frac{\partial \pi}{\partial s}\right)\frac{\partial p}{\partial \pi} = (1+\varepsilon)\left(\dot{s}\frac{\partial \pi}{\partial s}\right). \tag{18}$$

Thus, dp/dt is written in the form

$$\frac{dp}{dt} = \omega_1 + \omega_2 \tag{19}$$

where

$$\omega_1 = (1+\varepsilon)\frac{\partial \pi}{\partial t} + \mathbf{v}\cdot\nabla_s p + (1+\varepsilon)\left(\dot{s}\frac{\partial \pi}{\partial s}\right), \tag{20}$$

or taking into account (13),

$$\omega_1 = \mathbf{v}\cdot\nabla_s p - (1+\varepsilon)\int_0^s \nabla_{s'}\cdot\left(\mathbf{v}\frac{\partial \pi}{\partial s'}\right)ds'. \tag{21}$$

Note that the contribution of the second term of the pressure gradient force in (14) to the kinetic energy generation is compensated by the contribution of the horizontal advection of pressure in (20). The second part of ω is defined by

$$\omega_2 \equiv \frac{\partial p}{\partial t} - (1+\varepsilon)\frac{\partial \pi}{\partial t}. \tag{22}$$

Note that the term (22) vanishes for vanishing ε.

In view of the separation of omega into two parts, the thermodynamic equation is separated into two parts as well,

$$\left(\frac{\partial T}{\partial t}\right)_1 = -\mathbf{v}\cdot\nabla_s T - \left(\dot{s}\frac{\partial \pi}{\partial s}\right)\frac{\partial T}{\partial \pi} + \frac{1}{c_p}(\alpha\omega_1) \tag{23}$$

and

$$\left(\frac{\partial T}{\partial t}\right)_2 = \frac{1}{c_p}(\alpha\omega_2). \tag{24}$$

With the aid of (21), (23) may be rewritten as

$$\left(\frac{\partial T}{\partial t}\right)_1 = -\mathbf{v}\cdot\nabla_s T - \left(\dot{s}\frac{\partial \pi}{\partial s}\right)\frac{\partial T}{\partial \pi} + \frac{\alpha}{c_p}\left[\mathbf{v}\cdot\nabla_s p - (1+\varepsilon)\int_0^s \nabla_{s'}\cdot\left(\mathbf{v}\frac{\partial \pi}{\partial s'}\right)ds'\right]. \quad (25)$$

Again, when ε vanishes, (23) and (25) take the form used in hydrostatic models, and the equation for the second part (24) takes the trivial form $(\partial T/\partial t)_2 = 0$.

The non-hydrostatic system of equations is closed by applying the operator (9) to the continuity equation (10) in order to obtain the vertical acceleration *dw*/*dt*. Then, from (6),

$$\varepsilon = \frac{1}{g}\frac{dw}{dt} = \frac{1}{g}\left[\left(\frac{\partial w}{\partial t}\right)_s + \mathbf{v}\ \nabla_s w + \left(\dot{s}\frac{\partial \pi}{\partial s}\right)\frac{\partial w}{\partial \pi}\right]. \quad (26)$$

The parameter ε is the central point of the extended, non-hydrostatic dynamics. Assume for a moment that ε is zero. Then, Eqs. (3), (11), (15) and (25), together with the gas law, represent the set of equations describing the hydrostatic, inviscid, adiabatic atmosphere. However, the presence of non-zero ε in (7), (15) and (25) demonstrates in a transparent way where, how, and to what extent relaxing the hydrostatic approximation affects the familiar hydrostatic equations. Note that the system of equations developed above bears a close relation to the system discussed by Laprise (1992).

On the synoptic scales, ε is small and approaches the computer round–off error. However, in case of vigorous convective storms, or strong vertical accelerations in flows over steep obstacles, the vertical velocity can reach the order of 10 ms^{-1} over a period on the order of 1000 s. This yields an estimate of the vertical acceleration on the order of 10^{-2} ms^{-2}, and consequently, ε on the order of 10^{-3}. As can be seen from (7), for this value of ε the non-hydrostatic deviation of pressure can reach 100 Pa. Bearing in mind that the typical synoptic scale horizontal pressure gradient is of the order of 100 Pa over 100 km, this suggests that significant local non-hydrostatic pressure gradients and associated circulations may develop on small scales. Nevertheless, ε remains much smaller than unity in atmospheric flows, and therefore, the non-hydrostatic effects in (7), (15) and (25) are of a higher order magnitude. An important consequence of this situation for the discretization is that high accuracy of computation of ε does not appear to be of paramount importance, since the computational errors are of even higher order than ε.

As can be seen from (3), the geometric height *z* is uniquely defined by the hydrostatic and nonhydrostatic pressures π and *p*, and temperature *T*. Thus, if these three variables are known, the vertical velocity *w* can be computed using the definition (9), or the non-hydrostatic continuity equation (10). Hence, *w* (and consequently ε) cannot be considered as an independent prognostic variable. Nevertheless, for internal consistency, the vertical velocity *w* still must satisfy the prognostic vertical equation of motion (5).

In meso and large scale atmospheric flows $\varepsilon \ll 1$, and therefore the hydrostatic approximation can be safely used. Experiments have shown that impact of non-hydrostatic dynamics becomes detectable at resolutions <10 km, and important at the resolution of about 1 km.

So far the equations for the dry atmosphere have been discussed. In real data applications, the presence of moisture and condensate affects the density and heat capacity of the air which also should be taken into account. In order to do that, the temperature T in (3) and (4) is replaced by virtual temperature

$$T_v = T(1 + 0.608q - c), \tag{27}$$

where q is specific humidity and c is condensate.

Thus the complete list of variables dealt with by the model dynamics includes

Mass variables:

- μ, π hydrostatic pressure (column mass weight) [Pa]
- p non-hydrostatic pressure [Pa]
- T temperature [K]
- q specific humidity (advection only) [kg/kg]
- c total water condensate (advection only) [kg/kg]
- various tracers (advection only)

Wind variables:

- u, v wind components [m/s].

It should be noted that, although HWRF uses the same dynamic core as the NMM in the Arakawa Egrid (NMM-E) developed at NCEP, HWRF was customized for hurricane/tropical cyclone forecast applications, and is very different from other operational models that employ NMM-E, such as the High-Resolution Windows (HRW) and the Short-Range Ensemble Forecast (SREF) System. HWRF also differs substantially from the North American Mesoscale (NAM) model, which now employs the NMM dynamic core in the ArakawaB-grid (NMM-B). HWRF is an atmosphere-ocean model configured with a parent grid and two telescopic, high-resolution, movable two-way nested grids that follow the storm, using a unique physics suite and diffusion treatment.

2.2 Moving Nest Algorithm

HWRF, within the NMM dynamic core, and under the WRF model software framework, supports moving, one-way or two-way interactive nests. The model can handle multiple domains at the same nest level, and/or multiple nest levels (telescoping) with twoway interaction. In the current implementation of the nesting algorithm, only horizontal refinement is available. The nested grids adopt a ratio of 3:1 to refine the resolution of the parent grid based on ArakawaE-grid staggering structure. Correspondingly, the time step ratio between the parent and nested grids is 3:1 as well. The mass points of the nested grids are aligned with those of the parent grid within which they are nested. The coincidence of grid points between the nested and parent domains simplifies remapping and feedback procedures. The design of nested grids also conforms to the parallel strategy within the WRF advanced software framework (Michalakes et al., 2004) and enhances code portability of the model for various applications. HWRF also inherits the time controlling capabilities of WRFNMM, allowing HWRF to initialize and terminate the integration of nested grids at any time during the model run.

The use of enhanced resolution only in the region of a TC is a practical solution commonly adopted in the tropical cyclone NWP community to reduce computational costs. In order for the TC to be always contained within the highest-resolution domain, that domain must follow the storm. The nest motion for tropical cyclones and tropical depressions is currently based on the GFDL Vortex Tracker. Nine fields are calculated and smoothed, and their *fix locations* (the extremes of the fields) are calculated: minimum wind speed at 10 m, 850 hPa and 700 hPa; maximum 10-m vorticity at 10 m, 850 hPa and 700 hPa; minimum geopotential height at 700 hPa and 850 hPa, and minimum mean sea-level pressure (MSLP).

Once all fix locations have been calculated, the standard deviation of the fix locations with respect to the domain centre is calculated. Fix locations that are far from the domain centre are discarded. The maximum allowed distance differs by parameter and varies with time, based on prior fix-to-centre standard deviations. Once the final set of parameters is chosen, the mean fix location is used as the storm centre. The standard deviation of the fix parameters is stored for discarding fix parameters at later time steps. Lastly, the nest is moved if the storm location is more than two parent grid points in the y-direction (rotated north-south) or more than one parent grid point in the x-direction (rotated east-west). One should note that, while numerous fields are passed between domains at each time step, both before and after the grid motion, the interpolation and hydrostatic mass balancing are also applied in the region of the leading edge of the moving nest as described in Section 2.4. This latest nest movement algorithm is a significantly enhanced technique compared to original dynamic pressure-based nest motion (Gopalakrishnan et al., 2011) or the centroid based method developed for high-resolution grids (Zhang et al., 2012), and provides robust procedure for storm tracking applications within HWRF.

2.3 Terrain Treatment and Topography

Terrestrial properties are important external forces on the dynamics and thermodynamics of any numerical model. The impact of terrain effects on TC track, intensity and structure has been recognized in many previous studies (e.g. Lin, 2007). Therefore, careful treatment of static terrestrial conditions, such as terrain and land-sea contrast, is necessary to contain any possible contamination or computational noise in the solution resulting from improper adjustment in the resolution of terrestrial information from coarse to fine.

The terrain treatment in HWRF is tailored to the TC problem by using high-resolution topography to take into account the detailed topographic effects of the complex islands and landmasses. To ensure that the movable nests always have access to highresolution topography, WPS is used to interpolate topography information from prescribed highresolution terrain datasets to the applicable grid resolution over the entire parent domain before the start of the forecast. For example, in a typical operational forecast of 27 km grid resolution, with two centre-aligned, movable nests of 9 km and 3 km resolution, respectively, terrestrial data are generated at all three resolutions for the entire static parent domain. This way, the nested grids always have access to high-resolution topographic information as they move during the forecast.

Topography is the only static dataset generated in high resolution. For pragmatic considerations, all other static terrestrial information for the nests is scaled from the parent domain. The terrain within the nest is smoothed before being used. The grids are smoothed through a four-point weighted average with a special treatment at the four corners. Points in the inner part of the domain are smoothed using

$$\overline{t_{hr}}(i,j) = 0.25 \times [t_{hr}(i+1,j) + t_{hr}(i,j+1) + t_{hr}(i-1,j) + t_{hr}(i,j-1)] \quad (28)$$

where t_{hr} is the high-resolution terrain in a nest. Points along the boundary use a modified equation. For example, for the western boundary, the equation is

$$t_{hr}(1,j) = 0.125 \times [t_{hr}(1,j,-1) + t_{hr}(1,j+1)] + 0.75 \times t_{hr}(1,f) \quad (29)$$

The smoothed topography in the corner points also follows a modified formula. For example, the average terrain at the southwest point is given by

$$\overline{t_{hr}}(1,1) = 0.75 \times t_{hr}(1,1) + 0.125 \times t_{hr}(2,2) + 0.0625 \times [t_{hr}(2,1) + t_{hr}(1,2)] \quad (30)$$

The high-resolution terrain conforms to land-sea mask binary categories, that is, only land points have terrain assigned.

2.4 Fine Grid Initialization and Mass Balancing

For the high-resolution nests, all variables except topography are initialized from the corresponding variables of the parent grid. Potential problems related to singularities (due to high-resolution terrestrial information in the nested domain) can arise. To alleviate these, the land variables, such as land-sea mask, soil temperature and vegetation type, are exclusively initialized via a nearest-neighbour approach. This occurs at all points for the initial time step and at the leading edge of a moving nest during integration.

A three-step, hydrostatic mass balance is applied to obtain the temperature, geopotential and moisture fields for the nest initialization. The first step is to horizontally interpolate coarse-resolution data from the parent grid to the fine-resolution grid of the nest. The second step is to apply the high-resolution terrain and the geopotential to determine the surface pressure on the nest. The pressure values in the hybrid surfaces of the nest are then calculated. The final step is to compute the geopotential, temperature and moisture fields on the hybrid surfaces of the nest using linear interpolation in a logarithm of pressure vertical coordinate. This procedure is illustrated schematically in Fig. 2.

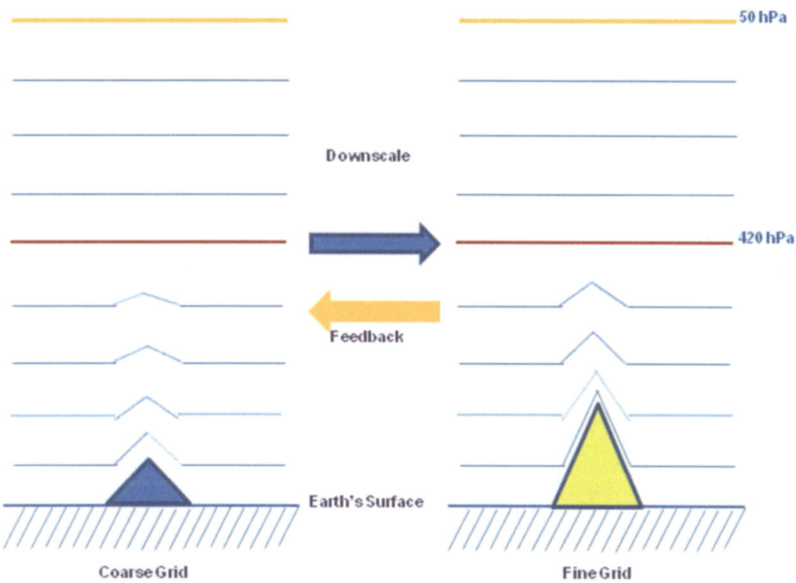

Fig. 2. An illustration of the vertical interpolation process and mass balance. Hydrostatic balance is assumed during the interpolation process.

The zonal and meridional components of the wind are obtained by first performing a horizontal interpolation from the parent grid points to the nest grid points using a bi-linear algorithm over the diamond-shaped area indicated in grey in Fig. 3a. The wind components are then linearly interpolated in the vertical from the parent hybrid surfaces onto the nest hybrid surfaces. Note that, while the hybrid levels of the nest and parent in sigma space coincide, the nest and the parent do not have the same levels in pressure or height space. This is due to the differing topography resolution, and consequently, different surface pressure between the parent and the nest.

2.5 Lateral Boundary Conditions

Figure 4 illustrates a sample E-grid structure, in which the outermost rows and columns of the nest are termed the prescribed interface, and the third rows and columns are termed the dynamic interface. The interface is prescribed to be identical to the parent domain, but interpolated to the nest grid points.

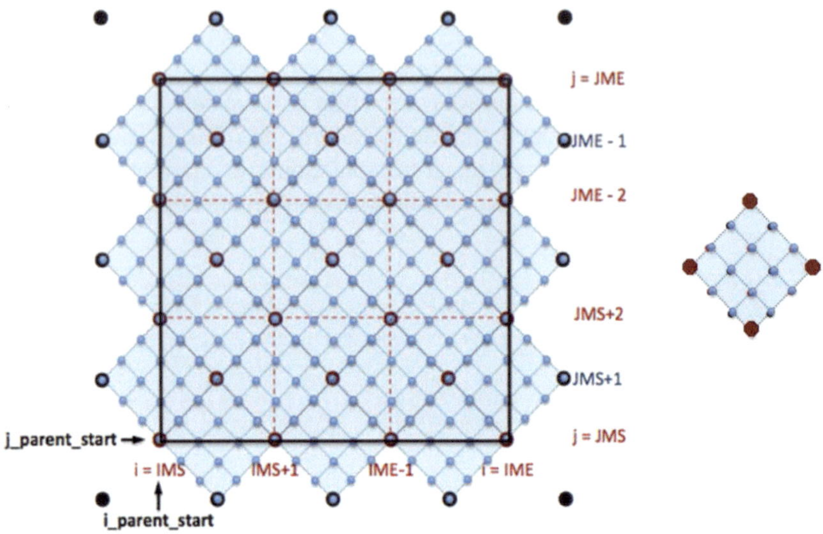

(a) (b)

Fig. 3. The schematic E-grid refinement—dot points the represent mass grid. Big and small dots represent coarse and fine resolution grid points, respectively. The black square in (a) represents the nest domain. The diamond square in (b) is composed of four big dot points representing the bilinear interpolation control points.

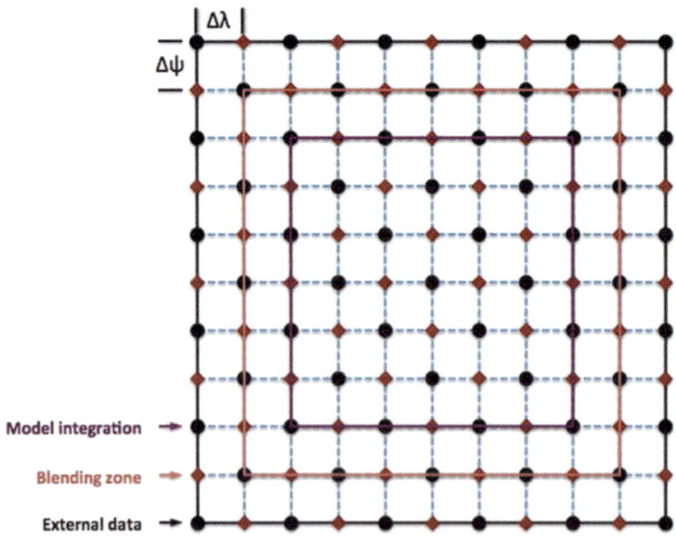

Fig. 4. Lateral boundary condition buffer zone—the outermost column and row are prescribed by external data from either a global model or regional model. The blending zone is an average of data prescribed by global or regional models and those predicted in the HWRF domain. Model integration is the solution predicted by HWRF. $\Delta\psi$ and $\Delta\lambda$ are the grid increment in the rotated latitude-longitude coordinate.

The dynamic interface is obtained from internal computations within the nest. The second rows and columns are a blend of the first and third rows/columns. Because the prescribed interface is well separated from the dynamic interface in the Egrid structure, nested boundaries can be updated at every time step of the parent domain exactly the same way as the parent domain boundary is updated from the external data source. This is done using bi-linear interpolation and extrapolation using the mass balance procedure described in Section 2.4. This approach is simple, yet produces an effective way of updating the interface without excessive distortion or noise.

2.6 Nest-to-Parent Feedback

Feedback from the nest domain to the parent domain is an important process for a hurricane forecast model. It reflects the multiple-scale physical interactions that occur within the hurricane environment. This is done using the mass balance procedure of Section 2.4, except that the parent pressure is retained, and horizontal averaging, rather than horizontal interpolation, is used. The nine grid points of the nest surrounding each grid point of the parent are averaged before applying the mass balance process. Feedback is applied with a relaxation factor of 0.5, replacing the parent data with the average of the parent and nest data.

$$P\_new = 0.5(P\_old + N\_new) \tag{31}$$

3. HWRF Vortex Initialization and Relocation

With the advancement in observational technology (e.g., weather satellites, buoys and Doppler radar), there is considerable improvement in the quantum of observational data around a TC. But because TC genesis invariably occurs far out to sea, over warm tropical oceans, a major difficulty arises in collecting sufficient data to accurately define it in the initial analysis fields. Many forecast centres rely on the prediction generated by high-resolution numerical models for cyclone track forecast. Even then, adequately representing a TC in the initial analysis for use in numerical weather prediction (NWP) models is a major problem.

At most NWP centres a 'bogusing' scheme is thus employed to force a TC vortex into the numerical analysis. This is typically done by using a vortex with suitable horizontal and vertical structure to derive a set of bogus observations for inclusion in analysis/assimilation cycle. Bogusing methods vary between the centres but most involve a symmetric vortex with some added asymmetry to take into account current movement of cyclone and environmental flow. There are primarily three bogusing methods that are widely used in operational models, as summarized by Peng et al. (1993). The first is to bogus observational data before the objective analysis is carried out. Examples of this type of bogusing are those used in the US NCEP Global Forecast System

(GFS) model (Lord, 1991), in the US Navy Operational Global Atmospheric Prediction System (NOGAPS), and in the United Kingdom Meteorological Office (UKMO) global model (Heming et al., 1995). The second approach is to add a more complex vortex circulation defined by an analytical expression after the objective analysis but before the model initialization. Examples of this type of bogusing are those used in the Quasi-Lagrangian Model (QLM) (Mathur, 1991; Prasad and Rama Rao, 2003) and Typhoon Model of the Japan Meteorological Agency (JMA) (Ueno, 1995). The third approach is to bogus a 'spinup' vortex generated by the same forecast model, instead of using an analytical one. Examples of this are the multiple nested tropical cyclone model of the GFDL (Kurihara, 1995) and the Typhoon-Track Forecast System (TFS) of the Central Weather Bureau (CWB) in Taiwan (Peng et al., 1993). In addition to the different methods, both the horizontal and vertical structures of the axisymmetric vortex vary considerably between the centres even for the same method.

The HWRF vortex initialization consists of several major steps: definition of the HWRF domain based on the observed storm centre position; interpolation of the atmospheric fields generated by HWRF Data Assimilation System (HDAS, see Section 4) onto the HWRF parent domain; removal of the global model vortex and insertion of a modified mesoscale vortex obtained from the previous cycle's HWRF 6-h forecast (if available), from HDAS, or from a synthetic vortex (cold start). The modification of the mesoscale hurricane vortex in the first guess field is a critical aspect of the initialization problem. Modifications include corrections to the storm size and to the three-dimensional structure based on observed parameters, including radius of maximum wind (RMW), radius of 34-kt winds (R34) and/or radius of outermost closed isobar (ROCI), maximum sustained 10-m winds (intensity), and minimum MSLP. Each of these corrections requires careful rebalancing between the model winds, temperature, pressure and moisture fields. An advancement of HWRF over the GFDL model bogus vortex initialization is the capability of HWRF to run in-cycle to improve the three dimensional structure of the TC vortex. This capability provides a significant opportunity to add more realistic structure to the forecast storm and is a critical step towards advancing hurricane intensity/structure prediction.

The 2013 operational initialization of hurricanes in HWRF as described by Liu et al. (2013) involves several steps to prepare the analysis at various scales. The environmental fields are derived from the 6-h forecast from Global Data Assimilation System (GDAS) or GFS analysis, enhanced through the HDAS. The vortex-scale fields are generated by inserting a vortex, corrected using TCVitals data, onto the large-scale fields. The vortex may originate from HDAS, from the previous HWRF forecast, or from a bogus calculation, depending on the storm intensity and on the availability of a previous HWRF forecast. Additionally, if inner core observations are available and inner core data assimilation is turned on, storm-scale data assimilation is performed.

Finally, the analyses are interpolated onto the HWRF outer domain and two inner domains to initialize the forecast.

The original design for the HWRF initialization (Liu et al., 2006a) was to continually cycle the HWRF large-scale fields, applying the vortex relocation technique (Liu et al., 2000, 2006b) at every model initialization time. However, the results were problematic. Large-scale flows can drift, and the errors increased as cycles passed. To address this issue, the environmental fields from the HDAS analysis are now used at every initialization time.

3.1 HWRF Cycling System

The location of the HWRF parent and nest domains is based on the observed hurricane's current and projected centre position. If a storm is moving, the nest domain will not be in the same location for subsequent cycles. Once the domains have been defined, the HDAS analysis and a vortex replacement strategy are used to create the initial fields. If a previous 6h HWRF forecast is available and the observed intensity of the storm is greater than or equal to 16 ms^{-1}, the vortex is extracted from that forecast and corrected to be included in the current initialization. If the previous HWRF forecast is not available, or if the observed storm has a maximum wind speed of less than 16 ms^{-1}, the HDAS vortex is corrected and added to the current initialization.

The vortex correction process involves the following steps, partially represented in Fig. 5.

1. Interpolate the 6-h GDAS/GFS forecast fields onto the HWRF parent grids.
2. Perform a one-way hybrid ensemble-3DVAR GSI analysis, using largescale observation data and the GFS 80-member ensemble background error correlation, to create HDAS analysis fields for the HWRF parent domain (see Section 4).
3. Remove the vortex from the HDAS analysis. The remaining largescale flow is termed the *environmental field.*
4. Determine which vortex will be added to the environmental fields. Check the availability of the HWRF 6-h forecast from the previous run (initialized 6 h before the current run) and the observed storm intensity.
5. If the forecast is not available (cold start):
   (i) if the observed storm maximum wind speed is greater than, or equal to, 30 ms^{1}, use a bogus vortex; or
   (ii) if the observed maximum wind speed is less than 30 ms^{-1}, use a corrected HDAS vortex.
6. If the forecast is available (warm start):
   (i) if the observed maximum wind speed is equal to or more than 16 ms^{-1}, extract the vortex from the forecast fields and correct it based on the TCVitals; or
   (ii) if the observed maximum wind speed is less than 16 ms^{-1}, use a corrected HDAS vortex.

7. Add the vortex obtained in step (4) to the environmental fields obtained in step (3).
8. Interpolate the data obtained from step (5) onto the outer and ghost domains. If performing inner core data assimilation (optional in HWRF as detailed in Section 4), assimilate the inner core data on the ghost domain. This domain is created for inner core data assimilation only, has the same resolution as the inner-most nest (0.02°), and is about three times larger than the inner nest). Finally, merge the data from the ghost domain onto the outer and inner nest domains.
9. Run the HWRF forecast model.

As noted above, a bogus vortex is only used in the initialization of strong storms (intensity greater than 30 ms^{-1}). This is in contrast with previous HWRF implementations. For instance, in the 2012 operational HWRF initialization, a bogus vortex was used for all cold start runs. Generally speaking, a bogus vortex does not produce the best intensity forecast. Also, cycling very weak storms (less than 16 ms^{-1}) without inner-core data assimilation often leads to large errors in intensity forecasts. To reduce the intensity forecast errors for cold starts and weak storms, the corrected HDAS vortex is used in the 2013 operational HWRF. These changes improve the intensity forecast for the first several cycles, as well as for weak storms (less than 16 ms^{-1}).

The vortex correction adjusts its location, size and structure based on the TCVitals: storm location (data used: storm centre position); storm size (data

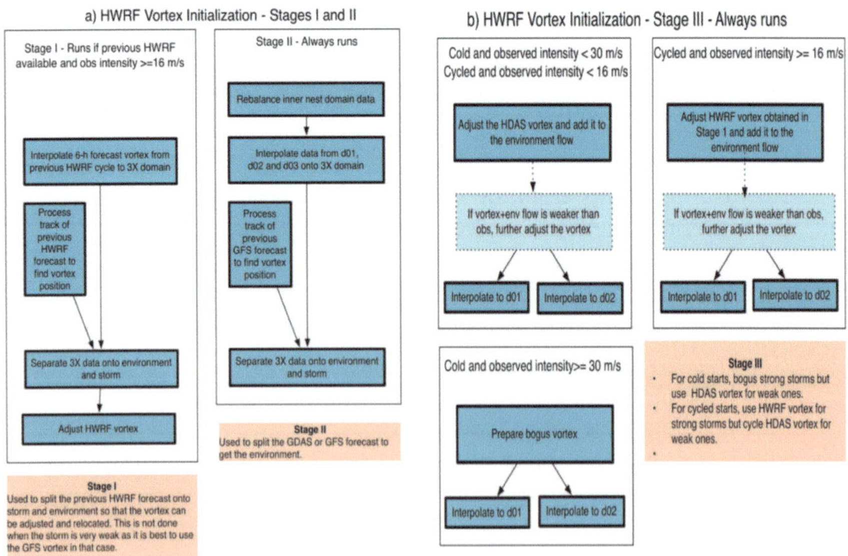

Fig. 5. Simplified flow diagram for HWRF vortex initialization describing the split of the HWRF forecast between vortex and environment, the split of the background fields between vortex and analysis, and the insertion of the corrected vortex in the environmental field after data assimilation.

used: radius of maximum surface wind speed. 34-kt wind radii, and radius of the outermost closed isobar); and storm intensity (data used: maximum surface wind speed and, secondarily, the minimum sea level pressure).

3.2 Bogus Vortex Used to Correct Weak Storms

The bogus vortex discussed here is primarily used to cold-start strong storms (i.e., observed intensity greater than or equal to 30 ms^{-1}) and to increase the storm intensity when the storm in the previous HWRF 6-h forecast is weaker than suggested by the observation data. It is also employed when: (a) the observed RMW is over 16 times greater the one present in the HWRF 6-h forecast, or (b) if the storm intensity is greater than 64 and the intensity correction factor β (defined in Section 3.3) is greater than 0.7. This is in contrast with previous HWRF implementations (Liu et al., 2012), in which a bogus vortex was used in all cold starts. This change has been shown to significantly improve the intensity forecasts in the first 1-3 cycles of a storm (Tallapragada et al., 2014).

The bogus vortex is created from a two-dimensional axisymmetric synthetic vortex generated from a past model forecast. The two-dimensional vortex only needs to be recreated when the model physics has undergone changes that strongly affect the storm structure. We currently have two composite storms, one created in 2007 for strong deep storms, another one created in 2012 for shallow and medium depth storms. For the creation of the two-dimensional vortex, a forecast storm (over the ocean) with small size and near axisymmetric structure is selected. The three-dimensional storm is separated from its environment fields, and the two-dimensional axisymmetric part of the storm is calculated. The two-dimensional vortex includes the hurricane perturbations of horizontal wind component, temperature, specific humidity, and sea-level pressure. This two-dimensional axisymmetric storm is used to create the bogus storm. To create the bogus storm, the wind profile of the two-dimensional vortex is smoothed until its RMW or maximum wind speed matches the observed values. Next, the storm size and intensity are corrected following a procedure similar to the cycled system. The vortex in medium-depth and deep storms receives identical treatment, while the vortex in shallow storms undergoes two final corrections: the vortex top is set to 400 hPa and the warm core structures are removed.

3.3 Improvement of Vortex Representation: Storm Size Correction

Before describing the storm size correction, some frequently used terms will be defined. *Composite vortex* refers to the two-dimensional, axisymmetric storm which, once created, is used for all forecasts. The *bogus vortex* is created from the composite vortex by smoothing and performing corrections to size and/or intensity. The *background field*, or *guess field*, is the output of the vortex initialization procedure, to which inner core observations can be added through data assimilation. The *environment field* is defined as the HDAS analysis field after removing the vortex component. For hurricane data assimilation, a high-quality background field is needed. Storms in the background field (this background field can be the GFS analysis or, as in the operational HWRF, the previous 6-h forecast of GDAS) may be too large or too small, so the storm size needs to be corrected based on observations. Two parameters are used to correct the storm size: the radius of maximum winds (RMW) and the radius of the outermost closed isobar (ROCI).

The storm size correction can be achieved by stretching/compressing the model grid. Consider a storm of incorrect size in cylindrical coordinates. Assume the grid size is linearly stretched along the radial direction

$$\alpha_i = \frac{\Delta r_i^*}{\Delta r_i} = a + br_i, \tag{32}$$

where a and b are constants. The distances from the storm centre before and after the grid is stretched are denoted by r and r^*, respectively. Index i represents the ith grid point.

Let r_m denote the RMW for the storm in the background field, and let R_m denote the ROCI of the same field (note: the minimum sea-level pressure is always scaled to the observed value before calculating this radius). Let r_m^* and R_m^* be the observed RMW and ROCI (which can be redefined if α in Equation (32) is set to be a constant). If the high-resolution model is able to resolve the hurricane eyewall structure, r_m^*/r_m will be close to 1. Therefore, we can set $b = 0$ in Equation (32) and $\alpha = r_m^*/r_m$ is a constant. However, if the model is unable to resolve the eyewall structure well (r_m^*/r_m will be smaller than R_m^*/R_m) within the background fields. Equation (32) must be used to stretch or compress the grid.

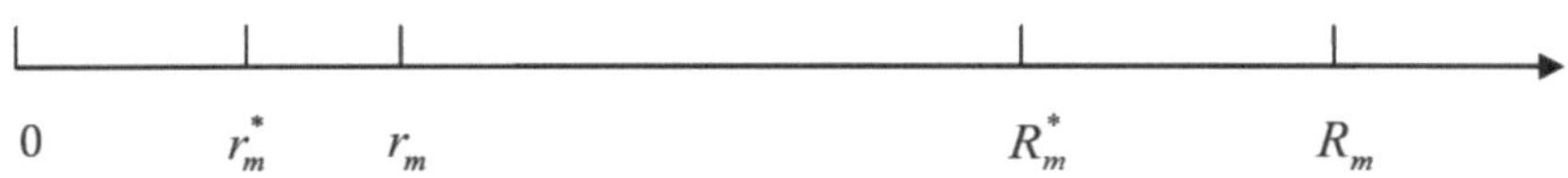

Integrating Equation (32) gives

$$r^* = f(r) = \int_0^r \alpha(r)dr = \int_0^r (a + br)dr = ar + \frac{1}{2}br^2 \tag{33}$$

We stretch or compress the grid such that

At $$r = r_m, r^* = f(r_m) = r_m^* \tag{34}$$

At $$r = R_m, f(R_m) = R_m^* \tag{35}$$

This results in stretching or compression components as

$$ar_m + \frac{1}{2}br_m^2 = r_m^* \tag{36}$$

$$aR_m + \frac{1}{2}bR_m^2 = R_m^*. \tag{37}$$

Solving for a and b, we have

$$a = \frac{r_m^* R_m^2 - r_m^2 R_m^*}{R_m r_m (R_m - r_m)}, \quad b = 2\frac{R_m^* r_m - R_m r_m^*}{R_m r_m (R_m - r_m)} \tag{38}$$

Therefore,

$$r^* = f(r) = \frac{r_m^* R_m^2 - r_m^2 R_m^*}{R_m r_m (R_m - r_m)} r + \frac{R_m^* r_m - R_m r_m^*}{R_m r_m (R_m - r_m)} r^2 \tag{39}$$

A special case is α being constant, so that

$$\alpha = \alpha_m = \frac{r_m^*}{r_m} = \frac{R_m^*}{R_m} \tag{40}$$

where $b = 0$, and the storm size correction is based on only one parameter, RMW, in the initial implementation of the operational HWRF model in 2007 and in the subsequent implementations through 2010. ROCI was calculated, but never used. Since the 2011 upgrade, a second parameter, ROCI (or R34 for hurricanes), was added by Kevin Yeh (HRD, personal communication). Specifically, in the 2010 HWRF initialization, Equation (40) was used for storm size correction, and b was set to zero in Equation (28). In the revised configuration, both a and b are calculated using Equation (38).

Storm size correction can be problematic because the eyewall size produced by the model can be larger than that observed, and the model does not support small observed eyewalls. For example, the observed RMW for 2005's Hurricane Wilma was 9 km. The modeled RMW was over 20 km. If the modelled RMW is compressed to 9 km, the eyewall would collapse and significant spin-down would occur. Thus, the minimum value for a storm eyewall is currently set to 19 km. The eyewall size in the model is related to model resolution, model dynamics, and model physics. In the storm size correction procedure, the observed RMW is not matched. Rather, it is replaced with the average model value and the observation. The correction is limited to a percentage of the model value. In the 2013 version, the limit (retained from the 2012 version) is set as follows: 10% if r_m^* is less than 20 km; 10-15% if r_m^* is between 20 km and 40 km; and 15% if r_m^* is greater than 40 km. For ROCI (R34 if storm intensity is greater than 64 knots), the correction is limited to 15% of the model value.

Even with the current settings, major spin-down may occur if the eyewall size is small and lasts for many cycles, due to the consecutive reduction of the storm eyewall size in the initialization. To address this, size reduction is stopped if the model storm size (measured by the average radius of the filter domain) is smaller than ROCI.

3.3.1 Surface Pressure Adjustment

The surface pressure must be re-adjusted after the storm size correction to maintain the gradient wind balance. In the HWRF vortex initialization, only the surface pressure of the axisymmetric part of the storm is corrected. The governing equation for the axisymmetric components along the radial direction is

$$\frac{\partial u}{\partial t} + u\frac{\partial u}{\partial r} + w\frac{\partial u}{\partial z} - v(\frac{v}{r} + f_0) + \frac{1}{\rho}\frac{\partial p}{\partial r} = F_r \tag{41}$$

where u, v and w are the radial, tangential and vertical velocity components, respectively. F_r is friction and $F_r \approx -C_d \frac{u}{H_B} v$, where H_B is the top of the boundary layer. F_r can be estimated as $F_r \approx -10^{-6}v$ away from the storm centre, and as $F_r \approx -10^{-5}v$ near the storm centre. Neglecting the small terms, Equation (41) is close to the gradient wind balance.

Since the hurricane component is separated from its environment, the contribution from the environment flow to the average tangential wind speed can be neglected. Henceforth, the tangential velocity will refer only to the vortex component.

We define the gradient wind stream function ψ as

$$\frac{\partial \psi}{\partial r} = \frac{v^2}{rf_0} + v \tag{42}$$

and

$$\psi = \int_{\infty}^{r} (\frac{v^2}{rf_0} + v)dr. \tag{43}$$

In rotated coordinates, Equation (42) can be rewritten as the following,

$$\frac{\partial \psi}{\partial r} = \frac{\partial \psi}{\delta r^*}\frac{\delta r^*}{\delta r} = \alpha \frac{\delta \psi}{\delta r^*}$$

$$\frac{v^2}{rf_0} + v = \frac{v^2}{r^*}\frac{r^*}{rf_0} + v = \frac{v^2}{r^*}\frac{f(r)}{rf_0} + v \; (r = r/r^*).$$

Therefore, the gradient wind stream function becomes

$$\psi = \int_{\infty}^{r^*} \frac{1}{\alpha(r^*)}\left[\frac{v^2}{r^*}\frac{f(r^*)}{r(r^*)f_0} + v(r^*)\right]dr^*. \tag{44}$$

A new gradient wind stream function can also be defined for the new vortex as

$$\frac{\partial \psi^*}{\partial r^*} = \frac{v^2}{r^* f_0} + v, \tag{45}$$

where v is a function of r^*. Therefore,

$$\psi^* = \int_{\infty}^{r^*} \left(\frac{v^2}{r^* f_0} + v\right) dr^* . \tag{46}$$

assuming the hurricane sea-level pressure component is proportional to the gradient wind stream function at model level 1 (~40 m), i.e.,

$$\Delta p(r^*) = c(r^*)\, \psi\, (r^*) \tag{47}$$

and

$$\Delta p^*(r^*) = c(r^*)\, \psi^*\, (r^*), \tag{48}$$

where $c(r^*)$ is a function of r^* and represents the impact of friction on the gradient wind balance. If friction is neglected, $c\,(r^*) = 1.0$, and there is gradient wind balance.

From equations (47) and (48),

$$\Delta p^* = \Delta p \frac{\psi^*}{\psi}, \tag{49}$$

where $\Delta p = p_s - p_e$ and $\Delta p^* = p_s^* - p_e$ are the hurricane sea-level pressure perturbations before and after the adjustment, and p_e is the environment sea-level pressure. Note that the pressure adjustment is small due to the grid stretching. For example, if α is a constant in Equation (41), it can be shown that Equation (44) becomes

$$\psi = \int_{\infty}^{r^*} \left(\frac{v^2}{r^* f_0} + \frac{1}{\alpha} v\right) dr^* . \tag{50}$$

This value is very close to that of Equation (46) since the first term dominates.

3.3.2 Temperature Adjustment

After the surface pressure is corrected, the temperature field must be corrected to maintain thermodynamic balance. Consider the vertical equation of motion. Neglecting the Coriolis, water load and viscous terms,

$$\frac{dw}{dt} = -\frac{1}{\rho}\frac{\partial p}{\partial z} - g \tag{51}$$

The first term on the right-hand side is the pressure gradient force, g is gravity, and dw/dt is the total derivative (or Lagrangian air parcel acceleration) which, in the largescale environment, is small compared to the other terms. Therefore,

$$-\frac{1}{\rho}\frac{\partial p}{\partial z} - g = 0$$

or

$$\frac{\partial p}{\partial z} = -\frac{p}{RT_v} g \tag{52}$$

assuming an ideal gas. Applying equation (52) to the environmental field and integrating from surface to model top gives

$$\ln\frac{p_s}{p_T} = \frac{g}{R}\int_0^H \frac{dz}{\overline{T}_v} \tag{53}$$

where H and p_T are the height and pressure at the model top, respectively. $\overline{T}_v$ is the virtual temperature of the environment.

The hydrostatic equation for the total field (environment field + vortex) is

$$\ln\frac{p_s+\Delta p}{p_T} = \frac{g}{R}\int_0^H \frac{dz}{(\overline{T}_v+\Delta T_v)}, \tag{54}$$

where Δp and ΔT_v are the sea-level pressure and virtual temperature perturbations for the hurricane vortex. Since $\Delta p << p_s$ and $\Delta T_v << \overline{T}_v$, Equation (54) can be linearized as

$$\ln\frac{p_s}{p_T}(1+\frac{\Delta p}{p_s}) = \frac{g}{R}\int_0^H \frac{dz}{(\overline{T}_v+\Delta T_v)} \approx \frac{g}{R}\int_0^H \frac{dz}{\overline{T}_v}(1-\frac{\Delta T_v}{\overline{T}_v}). \tag{55}$$

Subtracting Equation (53) from Equation (55) gives

$$\ln(1+\frac{\Delta p}{p_s}) \approx -\frac{g}{R}\int_0^H \frac{\Delta T_v}{\overline{T}_v^{\,2}}dz,$$

or

$$\frac{\Delta p}{p_s} = -\frac{g}{R}\int_0^H \frac{\Delta T_v}{\overline{T}_v^{\,2}}dz. \tag{56}$$

Multiplying Equation (56) by $\Gamma(r^*) = \psi^*/\psi$ (Γ is a function of x and y only) gives

$$\frac{\Gamma\Delta p}{p_s} = -\frac{g}{R}\int_0^H \frac{\Gamma\Delta T_v}{\overline{T}_v^{\,2}}dz. \tag{57}$$

A simple solution to equation (57) is chosen, i.e., the virtual temperature correction is proportional to the magnitude of the virtual temperature perturbation. So the new virtual temperature is

$$T_v^* = \overline{T}_v + \Gamma\Delta T_v = T_v + (\Gamma-1)\Delta T_v. \tag{58}$$

In terms of the temperature field

$$T^* = \overline{T} + \Gamma\Delta T = T + (\Gamma-1)\Delta T \tag{59}$$

where T is the three-dimensional temperature before the surface pressure correction, and ΔT is perturbation temperature for vortex #1.

3.3.3 Water Vapour Adjustment

Moisture adjustment is one of the most challenging aspects of vortex initialization. This is mainly due to lack of appropriate background moisture analysis that dominates inner core regions. A very simple water vapour adjustment is used in the HWRF vortex initialization.

Assume the relative humidity is unchanged before and after the temperature correction

$$\mathrm{RH} = \frac{e}{e_s(T)} \approx \frac{e^*}{e_s^*(T^*)} \tag{60}$$

where e and e_s (T) are the vapour pressure and the saturation vapour pressure in the model guess fields, respectively, and e^* and e_s^* (T) are the vapour pressure and the saturation vapour pressure respectively, after the temperature adjustment. Using the definition of the mixing ratio,

$$q = 0.622\frac{e}{p-e} \tag{61}$$

at the same pressure level and from Equation (60)

$$\frac{q^*}{q} \approx \frac{e^*}{e} \approx \frac{e_s^*(T^*)}{e_s(T)}. \tag{62}$$

Therefore, the new mixing ratio becomes

$$q^* = \frac{e^*}{e}q \approx \frac{e_s^*}{e_s}q \approx q + (\frac{e_s^*}{e_s} - 1)q. \tag{63}$$

From the saturation water pressure

$$e_s(T) = 6.112\exp\left[17.67\frac{(T-273.16)}{(T-29.66)}\right] \tag{64}$$

it can be shown that

$$\frac{e_s^*}{e_s} = \exp\left[\frac{17.67 \times 243.5(T^* - T)}{(T^* - 29.66)(T - 29.66)}\right]. \tag{65}$$

Substituting Equation (65) into (63) gives the new mixing ratio after the temperature field is adjusted.

3.4 Improvement of Vortex Representation: Storm Intensity Correction

In general, the storm in the background field has a different maximum wind speed compared to the observations. The storm intensity must be corrected based on observations, which is discussed in detail in this section.

3.4.1 Computation of Intensity Correction Factor, ß

Consider the general formulation in Cartesian coordinates (x, y, z); where u_1^* and v_1^* and are the background horizontal velocity, and u_2 and v_2 are the vortex horizontal velocity to be added to the background fields. Let

$$F_1 = \sqrt{(u_1^* + u_2)^2 + (v_1^* + v_2)^2} \tag{66}$$

and

$$F_2 = \sqrt{(u_1^* + \beta u_2)^2 + (v_2^* + \beta v_2)^2}\,. \tag{67}$$

Function F_1 is the wind speed if a vortex is simply added to the environment (or background fields). Function F_2 is the new wind speed after intensity correction. There are two possibilities:

Case I: F_1 is larger than the observational maximum wind speed. Set u_1^* and v_1^* to be the environment wind component; i.e., $u_1^* = U$ and $v_1^* = V$ (the vortex is removed and the field is relatively smooth); and $u_2 = u_1$ and $v_2 = v_1$ are the vortex horizontal wind components from the previous cycle's 6-h forecast (vortex #1, which contains both the axisymmetric and asymmetric parts of the vortex).

Case II: F_1 is smaller than the observational maximum wind speed. In this case, the vortex is added back into the environment fields after the grid stretching, i.e., $u_1^* = U + u_1$ and $v_1^* = V + v_1$. Here, u_2 and v_2 are chosen to be an axisymmetric composite vortex (vortex #2) having the same RMW as that of the first vortex.

In both cases, it can be assumed that the maximum wind speed for F_1 and F_2 are at the same grid point. To find β, the grid point where F_1 is at its maximum must first be located. Denote the wind components at this model grid point as u_1^m, v_1^m, u_2^m, and v_2^m (for convenience, the superscript m is dropped), so that

$$(u_1^* + \beta u_2)^2 + (v_1^* + \beta v_2)^2 = v_{obs}^2 \tag{68}$$

where v_{obs} is the 10 m observed wind converted to the first model level.

Solving for β, gives

$$\beta = \frac{-u_1^* u_2 - v_1^* v_2 + \sqrt{v_{obs}^2 (u_2^{\,2} + v_2^{\,2}) - (u_1^* v_2 - v_1^* u_2)^2}}{(u_2^{\,2} + v_2^{\,2})}\,. \tag{69}$$

For wind speed correction, first, the maximum wind speed from Equation (66) is calculated by adding the vortex into the environment fields. If the maximum of F_1 is larger than the observed wind speed it is classified as Case I and the value of β is calculated. If the maximum of F_1 is smaller than the observed wind speed, it is classified as Case II. The reason it is classified as Case II is so the asymmetric part of the storm is not amplified. Amplifying it may negatively affect the track forecasts. In Case II, the original vortex is added to the environment fields after the storm size correction, then a small portion of an axisymmetric composite storm is added. The composite storm portion is calculated from Equation (69). Finally, the new vortex three-dimensional wind field becomes

$$u(x, y, z) = u_1^*(x, y, z) + \beta u_2(x, y, z)$$

$$v(x, y, z) = v_1^*(x, y, z) + \beta v_2(x, y, z)$$

3.4.2 Surface Pressure, Temperature and Moisture Adjustments

Similar to the adjustments required for surface pressure, three-dimensional temperature and moisture fields after the storm size correction, a second round of adjustments for these fields are performed after the storm intensity correction. If the background fields are produced by high resolution models (such as in HWRF), the intensity corrections are small and the correction of the storm structure is not necessary. The guess fields should be close to the observations, therefore,

In Case I β is close to 1;
In Case II β is close to 0.

In Case I, β is close to 1. Following the discussion in Section 3.3.1, the gradient wind stream function ψ is defined as

$$\frac{\partial \psi}{\partial r} = \frac{v_2}{rf_0} + v_2 \tag{70}$$

and

$$\psi_2 = \int_{\infty}^{r} \left(\frac{v_2^2}{rf_0} + v_2\right) dr. \tag{71}$$

The new gradient wind stream function is

$$\psi^{new} = \int_{\infty}^{r} \left(\frac{\beta v_2^2}{rf_0} + \beta v_2\right) dr. \tag{72}$$

The new sea-level pressure perturbation is

$$\Delta p^{new} = \Delta p \frac{\psi^{new}}{\psi_2} \tag{73}$$

where $\Delta p = p_s - p_e$ and $\Delta p^{new} = p_s^{new} - p_e$ are the hurricane sea-level pressure perturbations before and after the adjustment and p_e is the environment sea-level pressure.

In Case II, β is close to 0. Define

$$\psi_1 = \int_{\infty}^{r} \left(\frac{v_1^2}{rf_0} + v_1\right) dr, \tag{74}$$

and the new gradient wind stream function is

$$\psi^{new} = \int_{\infty}^{r} \left[\frac{(v_1 + \beta v_2)^2}{rf_0} + (v_1 + \beta v_2)\right] dr. \tag{75}$$

and the new sea-level pressure perturbation is calculated as,

$$\Delta p^{new} = \Delta p \frac{\psi^{new}}{\psi_1}. \tag{76}$$

Equations (73) and (76) are expected to be close to the observed surface pressure. However, if the model has an incorrect surface pressure-wind relationship, these equations may indicate a large surface pressure difference from the observation. In 2013 HWRF, the pressure-wind relationship is further improved, and the limit can be set to 10% of the observation Δp_{obs} without producing large spin up/spin down problems.

The correction of the temperature field is as follows,

In Case I,

$$\Gamma = \frac{\psi^{new}}{\psi_2}. \tag{77}$$

Then the following equation is used to correct the temperature fields.

$$T^* = T_e + \Gamma \Delta T_1 = T + (\Gamma - 1)\,\Delta T_1 \tag{78}$$

In Case II,

$$\Gamma = \frac{\psi^{new}}{\psi_1} \tag{79}$$

and

$$T^* = T_e + \Delta T_1 + (\Gamma - 1)\Delta T_2 = T + (\Gamma - 1)\,\Delta T_2, \tag{80}$$

where T is the three-dimensional background temperature field (environment +vortex #1), and ΔT_2 is the temperature perturbation of the axisymmetric composite vortex. The corrections of water vapour in both cases are the same as those discussed in Section 3.3.3.

It is worth mentioning that the storm intensity correction is, in fact, similar to data analysis. The observation data used is the surface maximum wind speed (single point data), and the background error correlations are flow dependent and based on the storm structure. The storm structure used for the background error correlation is vortex #1 in Case I, and vortex #2 in Case II (except for water vapour which still uses the vortex #1 structure). Vortex #2 is an axisymmetric vortex. If the storm structure in vortex #1 could be trusted, one could choose vortex #2 as the axisymmetric part of vortex #1. In HWRF, the structure of vortex #1 is not completely trusted when the background storm is weak, and therefore an axisymmetric composite vortex from old model forecasts is employed as vortex #2.

4. HWRF Data Assimilation System (HDAS)

After the vortex initialization procedure, data assimilation is performed in HWRF using the Gridpoint Statistical Interpolation (GSI) analysis system. GSI is a unified global/regional three-dimensional variational data assimilation (3DVAR) system. The analysis in GSI is performed in model grid space, allowing more flexibility in constructing background error covariances, and simplifying the unification of global and regional applications (Wu et al., 2002; Kleist et al., 2009).

Major changes were made to the data assimilation procedure in the 2013 version of HWRF. The procedure was upgraded from a 3DVAR-only scheme to a hybrid EnKF-3DVAR scheme. In the 2012 version, the background error covariance was obtained through the National Meteorological Center (now NCEP) method, which was isotropic and static. By introducing the hybrid method, flow-dependent background error covariances are estimated from the short-term ensemble forecast and incorporated into the 3DVAR framework of GSI through the extended control variable method (Lorenc, 2003; Buehner, 2005). The hybrid method has been shown to be better than EnKF alone, especially when the ensemble size is small or when large model error is present (Wang et al., 2007b).

4.1 Hybrid EnKF-3DVAR Algorithm

The following description of the algorithm follows Wang (2010).

The analysis increment, denoted as $\mathbf{x}'$, is a sum of two terms:

$$\mathbf{x}' = \mathbf{x}_1' + \sum_{k=1}^{K} (\alpha_k \circ \mathbf{x}_k^e), \tag{81}$$

where $\mathbf{x}_1'$ is the increment associated with the GSI static background covariance and the second term is the increment associated with the flow-dependent ensemble covariance. In the second term $\mathbf{x}_k^e$ is the kth ensemble perturbation normalized by $(K-1)^{\frac{1}{2}}$, where K is the ensemble size. The vector α_k, $k = 1, \ldots, K$, contains the extended control variables for each ensemble member. The second term represents a local linear combination of ensemble perturbations, and α_k is the weight applied to the kth ensemble perturbation.

The cost function minimized to obtain $\mathbf{x}'$ is

$$J(\mathbf{x}_1', \alpha) = \beta_1 (\mathbf{x}_1')^T \mathrm{B}_1^{-1} (\mathbf{x}_1') + \beta^2 (\alpha)^{\mathrm{T}} \mathrm{A}^{-1} (\alpha)$$
$$+ (\mathbf{y}^{0\prime} - \mathbf{H}\mathbf{x}')^T \mathrm{R}^{-1} (\mathbf{y}^{0\prime} - \mathbf{H}\mathbf{x}') + Jc, \tag{82}$$

where $\mathbf{B}_1$ is the static background error covariance matrix, β_1 is the weight applied to the static background error covariance, $\mathbf{A}$ defines the spatial correlation of α, β_2 is the weight applied to the ensemble covariance, $\mathbf{Y}$ is the innovation vector, $\mathbf{R}$ is the observational and representativeness error covariance matrix, $\mathbf{H}$ is the observation operator, and J_c is the constraint term.

Wang (2007a) proved the equivalence of using (81) and (82) to find the solution to that, with the ensemble covariance explicitly included as part of the background covariance. It is also shown in this paper that matrix $\mathbf{A}$ determines the covariance localization on the ensemble covariance. The same conjugate gradient minimization algorithm used for the 3DVAR scheme is used to find the optimal solution for the analysis problem (81) and (82), except the control variable and the background covariance are extended as

$$\mathbf{x} = \begin{pmatrix} x_1' \\ \hline \alpha \end{pmatrix}, \tag{83}$$

and

$$\mathbf{B} = \begin{pmatrix} \frac{1}{\beta_1}\mathbf{B}_1 & 0 \\ 0 & \frac{1}{\beta_2}\mathbf{A} \end{pmatrix}. \tag{84}$$

More information about the hybrid algorithm can be found in Wang (2010). In the iteration algorithm, two outer loops with 50 iterations each are used for HWRF. The outer loop consists of more complete (nonlinear) observation operators and quality control. Usually, simpler observation operators are used in the inner loop. Variational quality control, which is part of the inner loop, is not used for HWRF.

The analysis variables for HDAS are (i) the streamfunction; (ii) the unbalanced part of velocity potential; (iii) the unbalanced part of temperature; (iv) the unbalanced part of surface pressure; (v) the pseudo-relative humidity or normalized relative humidity; (vi) the ozone mixing ratio; (vii) the cloud condensation mixing ratio; and (vii) the satellite bias correction coefficients. Ozone and cloud variables are not analyzed in HWRF. The definition of the normalized relative humidity allows for a multivariate coupling of the moisture, temperature, and pressure increments, as well as flow dependence (Kleist et al., 2009). Therefore, this option is used for HWRF.

The 2013 HDAS is a one-way hybrid system. It uses the GDAS at T254L64 resolution to provide ensemble perturbations (Fig. 6). It is "one-way" in that it does not include an HWRF ensemble forecast and the associated EnKF analysis components. The HWRF hybrid analysis does not feed back to the updated ensemble perturbations through EnKF, as is the case with GDAS. The first guess of the HWRF hybrid analysis at each analysis time is the global GDAS 6-h forecast at T574L64 resolution after vortex relocation is performed in the global model, the same first guess used in global hybrid analysis.

In the 2013 HWRF, the method to combine data assimilation and vortex initialization has been modified. Most of the time, observations within the vortex area are quite sparse. Without good data coverage, applying the GSI analysis after the vortex initialization can negatively affect the vortex structure and wind-mass balance. To address this issue in the 2012 HWRF, conventional data within 1200 km of the storm centre were not assimilated. However, inner core observations, such as NOAA P3 tail Doppler radar (TDR) radial winds and aircraft (US Air Force WC-13J and NOAA P3) reconnaissance data (flight level observations, dropsonde, and surface wind speed data), can be used to retrieve key structures of the TC vortices. The capabilities of assimilating TDR and aircraft reconnaissance data have been added into GSI and are available in the HWRF v3.5a. Based on results obtained from retrospective testing for the 2010-2012 hurricane seasons, TDR data was assimilated into the operational 2013 HWRF. Other aircraft reconnaissance data (e.g., flight level observations, dropsondes, and surface wind speeds obtained with the SFMR[2]), are not assimilated in operations due to a negative impact on intensity forecasts.

TDR and reconnaissance data are not always available. If inner core observations are not available, the GSI hybrid analysis is performed only on the HWRF outer domain (75° × 75°, 0.18° horizontal resolution). This outer domain analysis then constitutes the TC environment. In the vortex initialization procedure, the original vortex from the outer domain analysis is replaced by the relocated, size- and intensity-corrected HWRF or HDAS vortex, or by a bogus vortex (see Section 3.2). The fields on the vortex initialization domain (30° × 30°, 0.02° horizontal resolution) are then interpolated into the outer domain and two inner nests to initialize the forecast. This procedure is represented by the blue arrows in Fig. 6.

If inner core observations are available, the hybrid analysis is performed on the intermediate ghost domain (20° × 20°, 0.02° horizontal resolution), and the GSI analysis on the ghost domain is done after vortex initialization. This is represented by the red arrows in Fig. 6. Note that, when inner core data are assimilated, the hybrid analysis on the outer domain is not used as input to the

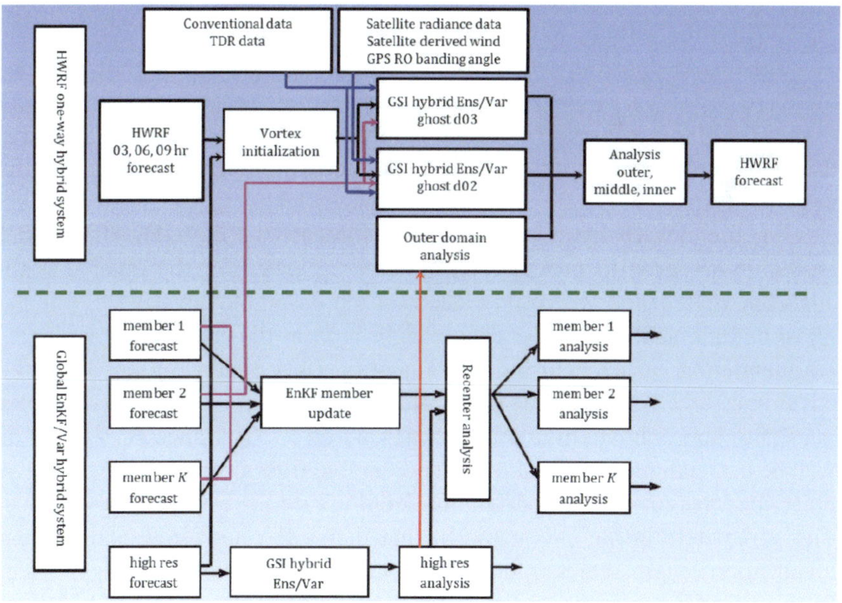

Fig. 6. Flow diagram of HWRF and GFS hybrid data assimilation systems. Processes described by the black and magenta lines always run, while those described by the red and blue lines run when inner core data assimilation is performed or is not performed, respectively.

[2]Stepped Frequency Microwave Radiometer

vortex initialization. Instead, the vortex initialization uses the GDAS forecast as input. This is because including observations is done in the GSI hybrid analysis performed in the ghost domain after the vortex initialization. After data assimilation, the ghost domain analyses are interpolated onto the HWRF outer domain and two inner domains to initialize the forecast. For the HWRF outer domain, a blending zone is added around the ghost domain boundary area, so that the model fields gradually change from the values of the ghost domain analysis to the values of the HWRF outer domain analysis.

To collect inner core observations, the aircraft has to penetrate the targeted TC a number of times, following a prescribed pattern. The pattern depends on storm intensity and each reconnaissance mission takes several hours. To account for the distribution of the inner core observations in the data assimilation time window, a technique called First Guess at Appropriate Time (FGAT) is used. In traditional 3DVAR data assimilation schemes, observations are assumed to be valid at the analysis time. With FGAT, observations are compared with the first guess at the observation time to obtain the innovation. The first guess at observation time is obtained by interpolating the two closest time levels of background fields within GSI. The GDAS forecasts, with vortex relocation applied, can only be computed at three-hour intervals. Therefore, the preprocessing steps and the vortex initialization are generated at the time of HWRF initialization, and also at plus and minus three hours.

The GDAS forecast with vortex relocation applied is preferred, because there are occasions in which two storms are close to each other and present in the HWRF outer domain at the same time. It would be undesirable to have the initial position error of the non-targeted storm to impact the forecast of the targeted storm.

For the HWRF hybrid analysis, 6-h forecasts of the GFS 80-member ensemble are used to provide ensemble covariance. For the outer domain analysis, 75% of the weight is placed on the ensemble covariance. For the ghost domain analysis, 80% of the weight is from the ensemble covariance. As mentioned earlier, matrix **A** in Equation (82) effectively conducts the covariance localization on the ensemble covariance. In GSI, a recursive filter is used to approximate the static background error covariance $\mathbf{B}_1$, as well as **A**. The correlation length scale of the recursive filter used to approximate **A** prescribes the covariance localization length scale for ensemble covariance. The parameters in the GSI namelist that define the horizontal and vertical localization length scales refer to the recursive filter efolding length scale. In most EnKF applications, the correlation function given by Eq. (4.10) of Gaspari and Cohn (1999) is used for covariance localization. The Gaspari-Cohn type of localization length scale refers to the distance at which the covariance is forced to be zero, which is roughly equivalent to the e-folding length scale divided by 0.388.

For the HWRF outer domain analysis, the horizontal localization length scale is set to 600 km from the 1st through the 26th model level (approximately

300 hPa). Above the 26th model level, the horizontal length scale gradually increases, reaching 900 km at the model top. For the ghost domain analysis, the horizontal localization length scale does not change vertically and is set to 150 km. This smaller value is used because the higher horizontal resolution can resolve small-scale structures. The vertical localization length scale for the outer domain is 0.5 in units of ln p, where p is the pressure in units of cb. Note that if the vertical localization length scale is measured in units of ln p, s_ens_v is expressed as a negative value and the length scale is the absolute value of s_ens_v. For the ghost domain analysis, the vertical localization length scale is 10 in vertical grid units for storms weaker than category 1 and is 20 grid units for storms equal to or stronger than category 1. When using GSI with HWRF, 'wrf_nmm_regional' and 'uv_hyb_ens' in the GSI namelist should be set to 'true'. The use of 'uv_hyb_ens=.true.' means that ensemble perturbations contain the zonal and meridional components of the wind instead of stream function and velocity potential.

Conventional observations assimilated in the HWRF outer and ghost domains include radiosondes; dropsondes; aircraft reports (AIREP/PIREP, RECCO, MDCRS-ACARS, TAMDAR, AMDAR); surface ship and buoy observations; surface observations over land; pibal winds; wind profilers; radar-derived Velocity Azimuth Display (VAD) wind; WindSat scatterometer winds; and integrated precipitable water derived from the Global Positioning System. Dropsonde wind data and surface pressure data are not assimilated near the storm centre because they negatively impact the forecast. Dropsonde wind data are excluded within a radius of 111 km or three times the RMW, whichever is larger. Surface pressure data are excluded within a radius of 200 km or the ROCI, whichever is larger. MSLP data from the TCVitals are only assimilated in the HWRF outer domain. MSLP data may not impact the targeted storm significantly, because the vortex in the outer domain analysis is later replaced by the modified vortex through the vortex initialization procedure. However, the TCVitals MSLP may help describe non-targeted storms in the outer domain. MSLP is not assimilated in the ghost domain because it degrades the wind forecast, which is of primary importance for the NHC. This is a result of an inconsistency between the forecast and observed wind-pressure relationship.

The NOAA P3 TDR, using the Fore-Aft Scanning Technique (FAST) probes the three-dimensional wind field in the inner cores of the hurricanes (Gamache et al., 1995). The antenna is programmed to scan as much as 25° fore or aft of the plane perpendicular to the fuselage. Major quality control of the radial velocity data, including: (i) removing the projection of the aircraft motion on the observed Doppler velocity; (ii) removing the reflection of the main lob and side lobs off the sea surface; (iii) removing noise; and (iv) unfolding, are conducted aboard the P3 aircraft before the data are sent to the ground. The TDR data in Binary Universal Form for the Representation of meteorological data (BUFR) format contain quality-controlled radial velocities averaged over eight gates along the radial direction. Further data thinning to the model resolution and quality control of the TDR radial velocities are performed before the data are assimilated. Figure 7 is an example of the TDR radial velocity data assimilated for Hurricane Earl at 12 Z on August 29, 2010. The observation error, including the representative error, of the radial velocity data is set to be 5 ms^{-1}. When the difference between the observation and the background field is more than 10 ms^{-1}, the observation error gradually increases to 10 ms^{-1}. Observations differing from the background more than 20 ms^{-1} are rejected. Positive impact of assimilating TDR data was found in a HWRF experiment performed during the 2012 hurricane season. Most of the time, the data within the vortex area are quite sparse. Without enough data coverage and not good enough background error covariance, using the data within the vortex area can easily hurt the vortex structure and wind-mass balance.

Satellite radiance observations and satellite wind estimates are also not assimilated. The main issue with radiance data is that the HWRF model top is too low (50 hPa). The channels that can be used for HWRF need to be carefully selected. There are also issues associated with bias correction. HWRF model was known to have a wrong surface pressure-wind relationship. Given a strong wind speed (e.g., 120 knots), the temperature anomaly at the

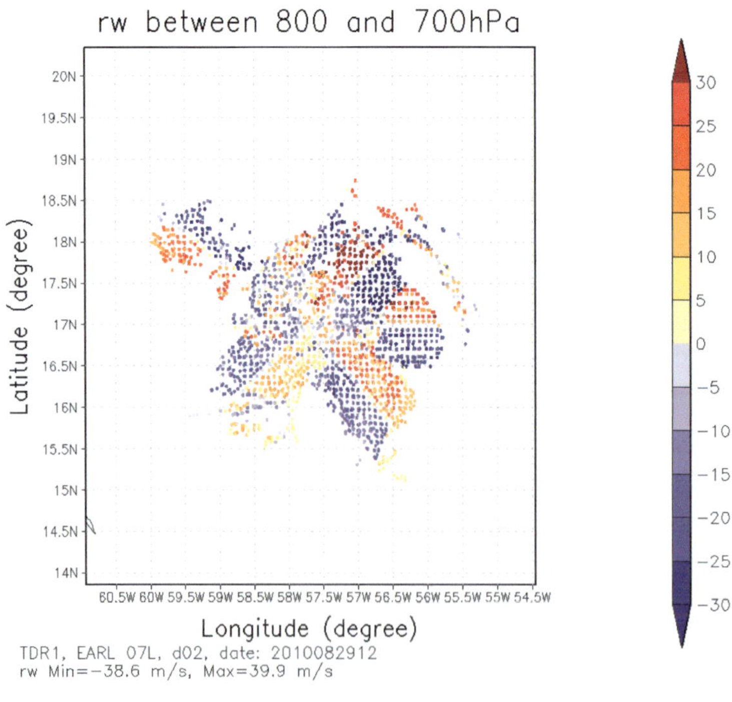

Fig. 7. NOAA TDR radial velocities between 800 hPa and 700 hPa assimilated at 12 Z on August 29, 2010.

hurricane core region in the model is much warmer than observation. If the observation data is used (without modification), the model-consistency of the vortex will be destroyed (or the maximum wind speed will be significantly reduced, if model-consistency is maintained). Further research is needed to be able to better use radiance data in HWRF. Negative impact on both track and intensity was found when assimilating satellite wind estimates. Quality control is one of the issues with satellite wind data. More test and analysis need to be done to understand how to improve the use of the satellite wind data.

5. Physics in the HWRF Model

Representation of the physical processes governing tropical cyclones is a very important part of the hurricane modelling system. The need to accurately capture multi-scale interactions, coupled with a lack of comprehensive observations to validate various physical formulations, and the various approximations and empirical formulations used in the existing physical parameterization schemes make it difficult to configure most compelling physics packages for operational hurricane forecast models. The HWRF system was initially designed to utilize the strengths of the WRF software system, the well tested NMM dynamic core, and the physics packages of the GFDL and GFS forecast systems. The rationale for using the GFDL model physics suite in HWRF was to take advantage of proven physical parameterization schemes from many years of hurricane applications in the GFDL hurricane model (Bender et al., 2007). While GFDL's physics served well for their initial purpose, modifications and/or upgrades to those schemes were needed due to the differences in between GFDL and HWRF regarding model dynamics and framework. Furthermore, upgrades were needed to make best use of the recent findings based on observational studies. This section outlines the physical parameterizations used in the operational HWRF model.

In the WRF framework, the physics section is insulated from the rest of the dynamics solver by the use of physics drivers. These drivers are split between the following solver-dependent steps: pre-physics pre-parations and post-physics modifications of the tendencies. The physics preparation involves filling arrays with physics-required variables, such as temperature, pressure, heights, layer thicknesses, and other state variables in MKS units (i.e., metres, kilograms, seconds) at half-level and full levels. The velocities are destaggered so that the physics code is independent of the dynamical solver's velocity staggering. Since HWRF uses the Egrid on the rotated latitude-longitude projection of the WRFNMM dynamic core, this velocity de-staggering involves interpolating the momentum variables from the velocity to the mass grid points. Physics packages compute tendencies for the unstaggered velocity components, potential temperature, and moisture fields. The solver-dependent post-physics step re-staggers the tendencies as necessary. It then couples the tendencies to the coordinate metrics, and converts to variables or units appropriate to the dynamics solver. As in other regional models, the physics tendencies are generally calculated less frequently than dynamic tendencies for computational expediency. Horizontal diffusion, which may also be considered part of the physics, is not described in this section (see NMM scientific documentation by Janic et al., 2004). The WRF system has been expanded to include all HWRF physics and, for each category, the operational HWRF employs a specific choice within the WRF spectrum of physics options. As mentioned, the HWRF physics initially followed the physics suite used by the benchmark operational GFDL hurricane model, but in the last few years several modifications have been introduced.

5.1 Surface Layer Physics Parameterization

The surface layer is the portion of the flow field most affected by the ocean or land surface in terms of momentum and enthalpy fluxes, and usually the depth of a surface layer is regarded as lower 10% of the boundary layer. This forms an important mechanism for transferring airsea interactions and parameterization of surface layer can have large sensitivity on the hurricane intensity forecasts. The HWRF surface physics is based on Monnin-Obkuhov similarity theory and the surface fluxes are defined by bulk method. In bulk method, surface fluxes are proportional to difference of quantity between surface and atmosphere right above, and exchange coefficients for heat and momentum. Because the main energy sources and sinks of tropical cyclones are sensible/latent heat fluxes over warm oceans, and momentum flux (dissipation) over land, the determination of surface fluxes plays critical role in predicting accurate hurricane intensity.

The surface fluxes of sensible heat, latent heat and momentum in HWRF model are defined as

$$H_o = \rho C_p C_h |V| \, (\theta_S - \theta_a) \tag{85}$$

$$L_o = \rho L C_h |V| \, (Q_S - Q_a) \tag{86}$$

$$\vec{\tau_o} = \rho C_d |V| \vec{V} \tag{87}$$

where H_o, L_o and τ_o are sensible, latent and momentum fluxes between surface and atmosphere, respectively. C_h and C_d denote enthalpy exchange coefficient and drag coefficient. Subscription '*s*' denotes the surface while subscription '*a*' represents the air above it. Others are as in conventional use. As can be seen in Equations 85-87, the fluxes are proportional to exchange coefficients (C_d and C_h), wind speeds and the difference of values between surface and atmosphere.

In this formulation, the neutral drag coefficient C_d is defined as:

$$C_d = \kappa^2 \left(\ln \frac{z_m}{z_0} \right)^{-2}, \tag{88}$$

where κ is the von Karman constant (= 0.4), z_0 is the roughness length for momentum, and z_m is the lowest model level height. The neutral heat and humidity coefficients (assumed equal, C_k) are expressed as

$$C_k = \kappa^2 \left(\ln \frac{z_m}{z_0} \right)^{-1} \left(\ln \frac{z_T}{z_0} \right)^{-1} \tag{89}$$

where z_T is the roughness length for heat and humidity. The modelled *Cd* and *Ch* from HWRF are compared with available observations in Fig. 8.

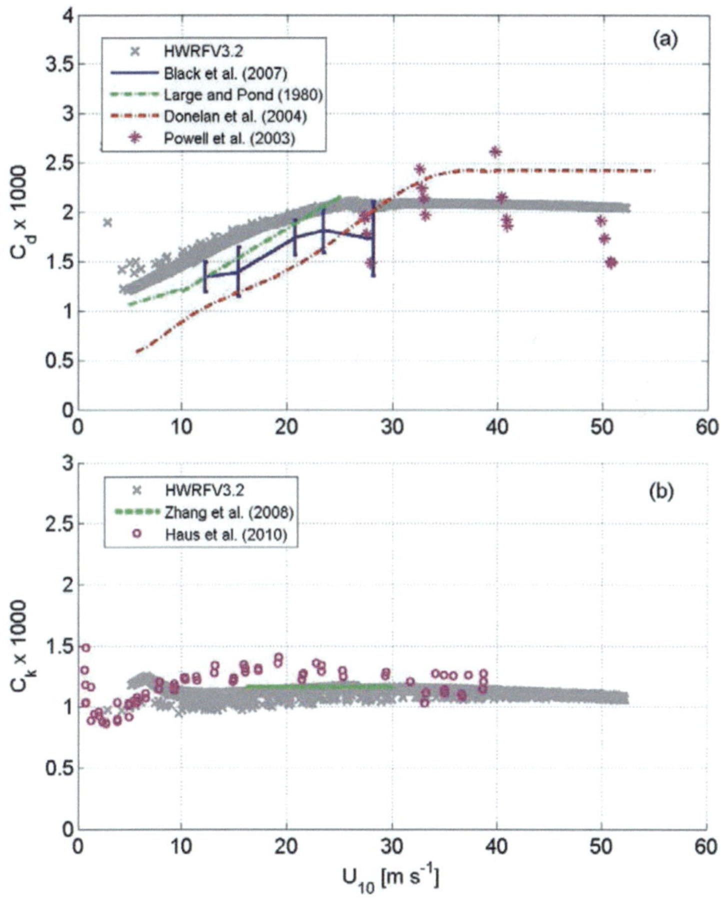

Fig. 8. Over-water drag coefficients for (a) momentum (C_d) and (b) enthalpy (C_h) coefficients at a height of 10 m above ground level (Gopalakrishnan et al., 2013). Values for HWRF (gray crosses) are compared against observational estimates by Zhang et al., 2008 (green dashes) and Haus et al., 2010 (pink circles).

The parameterization of heat fluxes involves consideration of micro-scale physical processes near the sea surface, including spray production and advection, the characteristics of the interfacial sub layers, and the character of the surface sea state, including wave breaking and currents. Our understanding of these processes is still severely limited. Therefore, the effective values of C_k at hurricane wind speeds could still be problematic.

5.2 Planetary Boundary Layer (PBL) Parameterization

The PBL parameterization is responsible for vertical sub-grid scale fluxes due to eddy transports in the whole atmospheric column, not just the boundary layer. Thus, when a PBL scheme is activated, no explicit vertical diffusion (e.g. Smagorinsky-type diffusion) is activated with the assumption that the PBL scheme will handle this process. Horizontal and vertical mixing are therefore treated independently. The surface fluxes are provided by the surface layer and land-surface schemes. The PBL schemes determine the flux profiles within the well-mixed boundary layer and the stable layer, and thus provide atmospheric tendencies of temperature, moisture (including clouds), and horizontal momentum in the entire atmospheric column. Most PBL schemes consider dry mixing, but can also include saturation effects in the vertical stability that determines the mixing. Conceptually, it is important to keep in mind that PBL parameterization may both complement and conflict with cumulus parameterization. PBL schemes are one-dimensional, and assume that there is a clear scale separation between sub-grid eddies and resolved eddies. This assumption will become less clear at grid sizes below a few hundred metres, where boundary layer eddies may start to be resolved, and in these situations the scheme should be replaced by a fully three-dimensional local sub-grid turbulence scheme, such as the Turbulent Kinetic Energy (TKE) diffusion scheme.

The HWRF code uses the same non-local scheme as the GFDL operational hurricane model (Hong and Pan, 1998) which is based on Troen and Mahrt (1986), was implemented in GFS in 1995, and is similar to the current GFS scheme. Historically the GFS PBL scheme has been found to give reasonable tropical cyclone tracks for the global GFS and GFDL hurricane models when packaged with the Simplified Arakawa-Schubert cumulus scheme (Han and Pan, 2011). Modifications to PBL scheme used in HWRF model are discussed extensively in Chapter 23 of this book.

The GFS PBL scheme is a first-order vertical diffusion parameterization that uses the surface bulk-Richardson approach to iteratively estimate the PBL height starting from the ground upward.The PBL height (h) depends on the virtual temperature profile between the surface and the PBL top, on the wind speed at the PBL top, and on the critical Richardson number (Ri_c), and is given by

$$h = \mathrm{Ric} \times \frac{\theta_{vg} U^2(h)}{g(\theta_v(h) - \theta_s)}, \tag{90}$$

where θ_{vg} and $\theta_v(h)$ are the respective virtual potential temperatures at surface and at the top of the PBL, $U(h)$ is wind speed at the top of the PBL, and θ_s is the surface potential temperature. Once the PBL height is determined, a preliminary profile of the eddy diffusivity is specified as a cubic function of the PBL height. This value is then refined by matching it with the surface layer fluxes. The process above determines the local component of the eddy diffusivity, which can be expressed as

$$K_c(z) = \kappa\left(\frac{u_*}{\Phi_m}\right)\left[\alpha\left(1-\frac{z}{h}\right)^2\right] \tag{91}$$

where z is the height above ground, Φ_m is a wind profile function evaluated at the top of the surface layer, and α is a parameter that controls the eddy diffusivity magnitude (Zhang et al., 2012).

Additionally, a counter-gradient flux parameterization based on the surface fluxes and on the convective velocity scale (Hong and Pan, 1998) is also included. This non-local effect incorporates the contribution of large-scale eddies driven by surface layer conditions. The overall diffusive tendency of a variable C can be expressed as

$$\frac{\partial C}{\partial t} = \frac{\partial}{\partial z}\left[K_c\left(\frac{\partial C}{\partial z}\right) - Y_c\right] \tag{92}$$

where $\partial C/\partial z$ and γ_c are the local and non-local parts, respectively.

In addition, the GFS boundary layer formulation also considers dissipative heating (the heat produced by molecular friction of air at high wind speeds) (Bister and Emanuel, 1998). Previous studies have shown that the class of PBL schemes used in HWRF (GFS/MRF/YSU) often produces too deep a PBL when compared to observations in the hurricane environment (Braun and Tao, 2000, Zhang et al., 2012). Because the magnitude of the eddy diffusivity coefficient in the HWRF PBL scheme is directly proportional to the PBL height, a deep PBL causes stronger vertical, which in turn leads to a more diffuse and larger storm. In order to reduce the feedback mechanism of the HWRF PBL, a few modifications were made in the 2013 HWRF upgrades. One was the use of a variable Ri_c instead of the constant value of 0.25 employed in previous implementations. This modification was based on Vickers and Mahrt (2004), who performed several tests by using different values of Richardson numbers and Ri_c in order to investigate the sensitivity of the PBL height. They concluded that a Ri_c which varies with the surface Rossby number (Ro) produced the best solution compared to all methods of modification of the Richardson number. The surface Rossby number and the dependent Ri_c fit to it using observational data are given by

$$Ro = (U_{10}/fz_0),\ \text{and}\ Ri_c = 0.16\,(10^{-7}Ro)^{-0.18} \tag{93}$$

where U_{10} is the wind speed at 10 m above ground level (AGL) and f is the Coriolis parameter, introduced for dimensional purposes. Vickers and Marhrt

(2004) stated that there is no evidence that *Ro* depends on *f*, so the HWRF model uses a typical value of *f*, 10^4 s^1. Because the Vickers and Mahrt (2004) study did not use observational data for hurricanes, the archived datasets of HRD/AOMLwere used to confirm that this Ri_c definition is applicable to hurricane conditions. By using the variable Ri_c method, the PBL height was matched seamlessly around the hurricane and its environment. Although the variable Ri_c method reduced the diffusivity in hurricane regions, analysis showed that further reduction of diffusivity was still needed to match the observational data.

5.3 Land Surface Model: The GFDL Slab Model

LSMs use atmospheric information from the surface layer scheme, radiative forcing from the radiation scheme, and precipitation forcing from the microphysics and convective schemes, together with internal information on the land's state variables and landsurface properties, to provide heat and moisture fluxes over land points and sea-ice points. These fluxes provide a lower boundary condition for the vertical transport done in the PBL schemes (or the vertical diffusion scheme in the case where a PBL scheme is not run, such as in large-eddy mode). Landsurface models have various degrees of sophistication in dealing with thermal and moisture fluxes in multiple layers of the soil and also may handle vegetation, root, and canopy effects and surface snow-cover prediction. In WRF, the LSM provides no tendencies, but updates the land's state variables which include the ground (skin) temperature, soil temperature profile, soil moisture profile, snow cover, and possibly canopy properties. There is no horizontal interaction between neighbouring points in the LSM, so it can be regarded as a one-dimensional column model for each WRF land gridpoint, and many LSMs can be run in a stand-alone mode when forced by observations or atmospheric model input. One of the simplest land models involves only one soil layer (slab) and predicts surface temperature only. In this formulation, all surface fluxes (both enthalpy and momentum) are predicted by the surface layer routines. HWRF uses such a simple land model: the GFDL slab option.

The GFDL slab model was developed by Tuleya (1994) based on Deardorff (1978). This simple one-level slab model, together with the GFDL radiation package, completed the requirement for realistic tropical cyclone behaviour over land during the development of the GFDL hurricane model. The surface temperature, T^*, is the only predicted parameter in this system.

$$\partial T^*/\partial t = (-\sigma T^{*4} - Shfx - Levp + (S + F))/(\rho_s c_s d), \tag{94}$$

where *Shfx* is the sensible heat flux, *Levp* is the evaporative flux, (*S*+*F*) is the net downward radiative flux, ρ_s, c_s and *d* are the density, specific heat and damping depth of the soil, respectively.

The surface wetness is assumed to be constant during the model forecast, with initial values based on the host model GFS analysis. Note that this simple model is able to realistically simulate the development of the 'cool pool' land temperature under landfalling tropical storms, thereby drastically reducing the surface evaporation over land leading to rapid decay over land.

5.4 Radiation Parameterization: GFDL Radiation Scheme

Radiation schemes provide atmospheric heating due to radiative flux divergence and surface downward longwave and shortwave radiation for the ground heat budget. Longwave radiation includes infrared or thermal radiation absorbed and emitted by gases and surfaces. Upward longwave radiative flux from the ground is determined by the surface emissivity that in turn depends upon land-use type, as well as the ground (skin) temperature. Shortwave radiation includes visible and surrounding wavelengths that make up the solar spectrum. Hence, the only source is the Sun, but processes include absorption, reflection and scattering in the atmosphere and at surfaces. For shortwave radiation, the upward flux is the reflection due to surface albedo. Within the atmosphere, radiation responds to model-predicted cloud and water vapour distributions, as well as specified carbon dioxide, ozone and (optionally) trace gas concentrations and particulates. All the radiation schemes in WRF currently are column (one-dimensional) schemes, so each column is treated independently, and the fluxes correspond to those in infinite horizontally uniform planes, which is a good approximation if the vertical thickness of the model layers is much less than the horizontal grid length. This assumption would become less accurate at high horizontal resolution, especially where there is sloping topography. Atmospheric radiation codes are quite complex and computationally intensive and are therefore often invoked at less frequent intervals than the rest of the model physics. The HWRF radiation parameterization used in operations is that from GFDL (see below) and is similar to the one used in the NAM. Compared to extra-tropical phenomena, hurricanes are less dependent on radiative fluxes except when migrating out of the tropics and/or progressing over land. Radiation-cloud interactions may be more important than direct radiative impacts, except during extra-tropical transition.

The longwave radiation scheme (Eta GFDL longwave scheme) follows the simplified exchange method of Fels and Schwarzkopf (1975) and Schwarzkopf and Fels (1991), with calculation over spectral bands associated with carbon dioxide, water vapour, and ozone. Included are Schwarzkopf and Fels (1985) transmission coefficients for carbon dioxide, a Roberts et al. (1976) water vapour continuum, and the effects of water vapour-carbon dioxide overlap and of a Voigt line-shape correction. The Rodgers (1968) formulation is adopted for ozone absorption. Clouds are randomly overlapped. More recent versions of the GFDL longwave radiation scheme, such as the one used in the NAM model but not adopted in HWRF, contain parameters for urban effects, as well as surface emissivities that can be different than 1.0.

The shortwave radiation in HWRF is a GFDL version of the Lacis and Hansen (1974) parameterization. Effects of atmospheric water vapour, ozone (both from Lacis and Hansen 1974), and carbon dioxide (Sasamori et al., 1972) are employed. Clouds are randomly overlapped. Shortwave calculations are made using a daylight-mean cosine solar zenith angle for the specific time and grid location averaged over the time interval (given by the radiation call frequency). The newest version of the GFDL shortwave radiation scheme, used for example in the NAM model but not adopted in HWRF, contains parameters for urban effects.

5.5 Cumulus Parameterization: The Simplified Arakawa-Schubert (SAS) Scheme

Cumulus parameterization schemes, or convective parameterization schemes, are responsible for the sub-gridscale effects of deep and/or shallow convective clouds. These schemes are intended to represent vertical fluxes unresolved by gridscale microphysics schemes such as updrafts, downdrafts, and compensating motion outside the clouds. In its early development, convective parameterization was believed to be necessary to avoid possible numerical instability due to simulating convection at coarse resolutions. The schemes operate only on individual vertical columns where the scheme is triggered and provide vertical heating and moistening profiles. Some schemes additionally provide cloud and precipitation field tendencies in the column, and additionally some schemes, such as the one used in HWRF, provide momentum tendencies due to convective transport of momentum. The schemes all provide the convective component of surface rainfall.

Cumulus parameterizations are theoretically only valid for coarser grid sizes (e.g., greater than 10 km), where they are necessary to properly release latent heat on a realistic time scale in the convective columns. While the assumptions about the convective eddies being entirely sub-gridscale break down for finer grid sizes, sometimes these schemes have been found to be helpful in triggering convection in 510 km grid applications and accurately predicting rainfall patterns. Normally, they should not be used when the model itself can resolve the convective eddies (grid spacing less than approximately 5 km). In the 2013 operational implementation of HWRF, the cumulus parameterization is activated only in the parent domain and in the coarser nest (27 km and 9 km horizontal grid spacing, respectively). No convective parameterization is used in the 3 km horizontal grid spacing inner nest.

HWRF uses the Simplified Arakawa-Schubert (SAS) cumulus parameterization also employed, with some modifications, in the GFS (Pan and Wu, 1995; Hong and Pan, 1998; Han and Pan, 2011) and GFDL models. This scheme, which is based on Arakawa and Schubert (1974), was simplified by Grell (1993) to consider only one cloud top at a specified time and location and not the spectrum of cloud sizes, as in the computationally expensive original scheme. Since 2011, the GFS and HWRF models use the newly upgraded SAS scheme, which no longer considers a random distribution of cloud tops but one cloud top value in a grid box from various entrainment ensemble averaged parameters. The scheme was revised to make cumulus convection stronger and deeper by increasing the maximum allowable cloud base mass flux and having convective overshooting from a single cloud top.

In addition to the deep convection scheme, the shallow convection parameterization was incorporated in the operational GFS and HWRF models in 2011 and 2012, respectively. The parameter used to differentiate shallow from deep convection is the depth of the convective cloud. When the convective thickness is greater than 150 hPa, convection is defined as deep; otherwise it is treated as shallow. In the HWRF model, precipitation from shallow convection is prohibited when the convection top is located below the PBL top and the thickness of the shallow convection cloud is less than 50 hPa. These customizations were made in order to remove widespread light precipitation in the model domain over the open ocean areas. Note that because the shallow convection scheme requires knowledge of the PBL height, it needs to be run in conjunction with a PBL parameterization that provides that information. In the current code, only the GFS PBL scheme has been tested to properly communicate the PBL height to the HWRF SAS parameterization.

In SAS, convection depends on the cloud work function, a quantity derived from the temperature and moisture in each air column of the model, which is similar to the Convective Available Potential Energy (CAPE). When the cloud work function exceeds a certain critical threshold, which takes into account the cloud base vertical motion, the parameterizations are triggered and the mass flux of the cloud is determined using a quasi-equilibrium assumption. As the large-scale rising motion becomes strong, the cloud work function is allowed to approach zero (therefore approaching neutral stability).

Revisions specific to SAS scheme employed in HWRF include the cloud top level determined by the parcel method to be the level where the parcel becomes stable with respect to the environment. Detrained water is separated into condensate and vapour, with the condensate used as a source of prognostic cloud condensate above the level of the minimum moist static energy. In contrast to HWRF, the GFDL hurricane model version of SAS does not export condensate to the rest of the model.

In the current implementation of SAS, the mass fluxes induced in the updrafts and the downdrafts are allowed to transport momentum. The momentum exchange is calculated through the mass flux formulation in a manner similar to that of heat and moisture. The effect of the convection-induced pressure gradient force on cumulus momentum transport is parameterized in terms of mass flux and vertical wind shear (Han and Pan, 2006). As a result, the cumulus momentum exchange is reduced by about 55% compared to the full exchange in previous versions.

5.6 Microphysics Parameterization: The Ferrier Microphysics Scheme

Microphysics parameterizations explicitly handle the behaviours of hydrometeor species by solving prognostic equations for their mixing ratio and/or number concentration, so they are sometimes called explicit cloud schemes (or gridscale cloud schemes) in contrast to cumulus schemes, which parameterize subgridscale convection. The adjustment of water vapour exceeding saturation values is also included inside the microphysics. The treatment of water species such as rain, cloud, ice and graupel was first utilized in the development of cloud models, which simulated individual clouds and their interactions. Gradually, as it became more computationally feasible to run at high grid resolutions, microphysics schemes were incorporated into regional atmospheric models. At high enough resolution (~1 km or less), convective parameterization of cloud processes may not be needed because convection can be resolved explicitly by a microphysics scheme. In the simpler microphysics schemes (single moment schemes), such as the one used in HWRF, only the mixing ratios of the water species are carried as predicted variables, while the number concentration of the variables is assumed to follow standard distributions. If number concentrations are also predicted, the schemes are coined "double moment". A further sophistication in microphysics schemes is introduced if the water species are predicted as a function of size. This added level of complexity is coined a "bin" scheme. The present HWRF model, like the NAM and GFDL models, uses the Ferrier scheme, which is simplified so that the cloud microphysical variables are considered in the physical column, but only the combined sum of the microphysical variables, the total cloud condensate, is advected horizontally and vertically. A possible upgrade of HWRF microphysics would be to extend the Ferrier scheme to handle advection of cloud species. Note that in the last year modifications were made to the HWRF model such that it can now be run in research mode with a variety of microphysics packages, including the Thompson and WRF single-moment 6-class (WSM6) parameterizations.

The present HWRF Ferrier microphysics scheme is based on the Eta Grid-scale Cloud and Precipitation scheme developed in 2001 and known as the EGCP01 scheme (Ferrier, 2002). The WRF model has three versions of the Ferrier microphysics: one for general applications (used in the old WRF-based NAM model), one for higher resolution (used operationally in the 4 km grid spacing High-Resolution Windows application), and one tailored for the tropics (used in HWRF). The latter duplicates some features used in the GFDL model implementation. For example, the number concentration of droplets is set to 60 cm^{-3} and 100 cm^{-3} in the HWRF and old WRF-based NAM versions, respectively. In addition, the onset of condensation above the planetary boundary layer in the parent grid of the tropical Ferrier is set to 97.5%, while the standard value of 100% relative humidity is used throughout the domain in the old WRF-based NAM version. These changes for the tropics were implemented primarily to obtain a more realistic intensity distribution in HWRF and GFDL forecasts. The scheme predicts changes in water vapour and condensate in the forms of cloud water, rain, cloud ice and precipitation ice (snow/graupel/sleet). The individual hydrometeor fields are combined into total condensate, and it is the water vapour and total condensate that are advected in the model. This is done for computational expediency. Local storage arrays retain first-guess information of the contributions of cloud water, rain, cloud ice and precipitation ice of variable density in the form of snow, graupel or sleet.

The density of precipitation ice is estimated from a local array that stores information on the total growth of ice by vapour deposition and accretion of liquid water. Sedimentation is treated by partitioning the time-averaged flux of precipitation into a grid box between local storage in the box and fall out through the bottom of the box. This approach, together with modifications in the treatment of rapid microphysical processes, permits large time steps to be used with stable results. The mean size of precipitation ice is assumed to be a function of temperature following the observational results of Ryan et al. (1996). Mixed-phase processes are now considered at temperatures warmer than –40°C (previously –10°C), whereas ice saturation is assumed for cloudy conditions at colder temperatures.

Changes were made in three parameters of the Ferrier microphysics scheme for the 2012 HWRF upgrades. Although these changes did not significantly alter the HWRF forecasted track and intensity of a storm, they contributed to produce better mixing ratio forecasts, allowing the creation of more realistic large-scale forecast products, in particular simulated infra-red and water vapour satellite images. The three parameters changed were the maximum allowable ice concentration (NLI_{max}), the number concentration of cloud droplet (NCW) and snow fall speed. NLI_{max} was increased from 2 to 20 L^{-1} and NCW was increased from 60 to 250 cm^{-3}. The reason for the NCW increase was the reduction of the unrealistic widespread spotty drizzle/light rain in the open ocean of the parent domain. Finally, the fall speed of snow particles in the regions of warmer-than-freezing temperature (0°C) was also increased to mimic higher order moment schemes.

While the model physics parameterizations are categorized in a modular way, it should be noted that there are many interactions between them via the model state variables (potential temperature, moisture, wind, etc.) and their tendencies, and via the surface fluxes. The surface physics, while not explicitly producing tendencies of atmospheric state variables, is responsible for updating the land-state variables as well as updating fluxes for ocean coupling. Note also that the microphysics does not output tendencies, but updates the atmospheric state at the end of the model time step. The radiation, cumulus parameterization and PBL schemes all output tendencies, but the tendencies are not added until later in the solver, so the order of call is not important. Moreover, the physics schemes do not have to be called at the same frequency as each other or at the basic model dynamic time step. When lower frequencies are used, their tendencies are kept constant between calls or time interpolated between the calling intervals. In contrast to HWRF, note that the GFDL hurricane modelling system calls all physics packages once per time step except for radiation. The land-surface and ocean models, excluding simple ones, also require rainfall from the microphysics and cumulus schemes. The boundary-layer scheme is necessarily invoked after the land-surface scheme because it requires the heat and moisture fluxes.

6. Performance of the Operational HWRF

Although HWRF was originally developed and implemented to provide real-time operational forecast guidance to the NHC for the North Atlantic and North Eastern Pacific basins, recent developments expanded the application of HWRF model for other oceanic basins including Western Pacific and North Indian Ocean. US Navy's JTWC has officially requested forecast guidance from operational HWRF for the North Western Pacific basin. International collaborations with various operational forecasting agencies in India, Vietnam, China, Taiwan and Oman led to further evaluation of operational HWRF performance in the respective oceanic basins of those counties. Due to annual upgrades and continuous improvements obtained for the model performance, it is difficult to establish the forecast performance skills of operational HWRF based on a single season or single oceanic basin. However, progress of HWRF performance is continuously monitored by the forecasting agencies in order to assess the forecast skill improvements and for providing valuable feedback on model deficiencies and areas for improvement. A brief history of HWRF performance is presented in this section for different oceanic basins.

6.1 HWRF Model Performance for the North Atlantic and Eastern Pacific Basins

Since the original implementation in 2007, HWRF has steadily improved its performance in terms of hurricane track, intensity and structure forecast guidance. A significantly enhanced version of HWRF was implemented on July 25, 2013. Implementation of the 2013 version of the operational HWRF model includes improvements in the areas of data assimilation, high-resolution nesting algorithms and model physics. HWRF Data Assimilation System (HDAS) is upgraded to include regional one-way hybrid global Ensemble Kalman Filter (EnKF) and regional 3-Dimensional Variational (3DVAR) data assimilation scheme, which incorporates flow-dependent global ensemble forecast error statistics in the variation data assimilation scheme. Compared with the static error covariances used in traditional 3DVAR scheme, the error covariances estimated from the short range ensemble forecasts better represent the 'error of the day'. Apart from conventional data, for the first time, HWRF will be assimilating in real-time the high-density inner-core

Tail Doppler Radar (TDR) data collected from NOAA P3 reconnaissance aircraft. The nesting techniques employed in the HWRF model have been upgraded to include a sophisticated storm tracking algorithm based on nine different atmospheric variables providing more accurate storm centre location compared to the mass centroid method used in the past. The retrospective tests had confirmed that the upgraded storm tracking algorithm rarely lost a storm, giving the full benefit of using high resolution around a tropical storm. The nest-parent interpolations have also been revised for more accurate computation of up-scale and down-scale feedback, apart from providing better vertical interpolation and mass adjustments. Finally, the planetary boundary layer (PBL) scheme is upgraded to improve PBL structure both in the hurricane region and its environment.

As a result, the 2013 HWRF model was found exceeding the track forecast skill in the North Atlantic basin by more than 15% and, most importantly, the intensity forecast skills by about 20% (Fig. 9) compared to the 2012 HWRF model. Evaluation of retrospective forecasts for three hurricane seasons (2010-2012) indicated that the 2013 HWRF model exhibits better intensity forecast skill than NHC official forecasts for the Atlantic basin tropical cyclones (NHC, personal communication). The HWRF model intensity errors

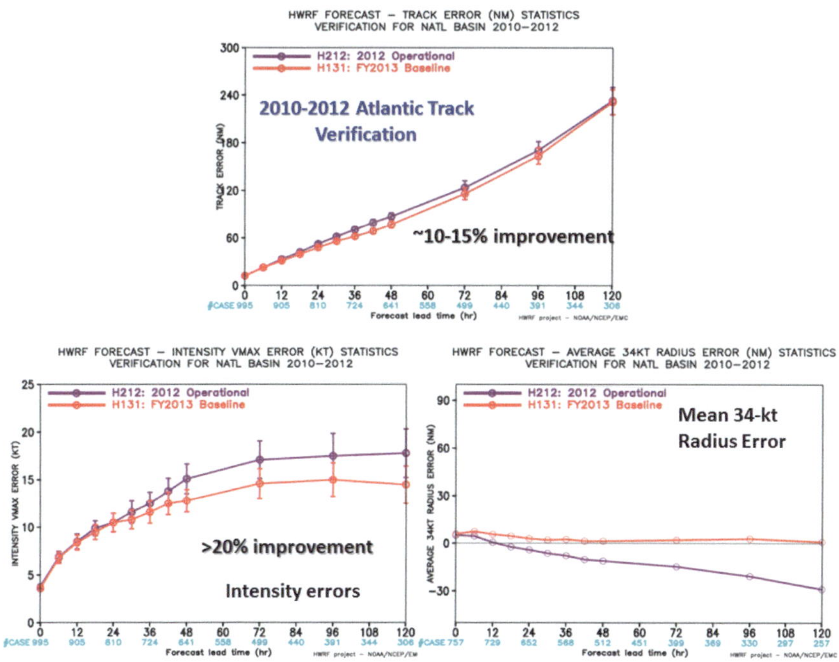

Fig. 9. Performance of 2013 HWRF configuration for the North Atlantic basin. Top left: Track errors; Bottom left: Intensity errors and Bottom right: Structure errors. Sample size includes all tropical cyclones formed in the North Atlantic basin for the years 2010, 2011 and 2012. Number of cases verified at each forecast hour is shown for each plot along the x-axis.

are much smaller than the statistical model errors at all forecast periods. This remarkable intensity prediction skill from a deterministic dynamical model compared to official or statistical models is unprecedented. In addition to the improvements in intensity forecasts, the track forecasts for the North Atlantic basin also showed significant improvements from 2013 HWRF compared to 2012 HWRF. The hurricane structure forecasts as measured by mean radius of 34-kt winds (tropical storm force winds) showed that the 2013 HWRF configuration has near-neutral bias for the entire forecast period and a remarkable improvement compared to 2012 operational HWRF configuration.

Similar improvements were noted for the North Eastern Pacific basin from 2013 operational HWRF configuration compared to 2012 version of HWRF (Fig. 10). These improvements are consistent with the real-time performance goals set by the HWRF team at EMC, and are a result of enormous testing and evaluation activities supported by NOAA's HFIP and other collaborators. Operational implementation criteria for hurricane models include testing of the proposed upgrades for at least 2 to 3 past seasons, and are evaluated annually by operational forecasters at NHC before recommending for implementation. It is important to note that the operational hurricane models are evaluated by NHC based on post-season best track data sets and the metrics for improvement are part of the US Government Performance and Results Act (GPRA) goals.

Figure 11 shows steady improvement obtained from operational HWRF since its initial implementation in 2007. The track errors in the Atlantic

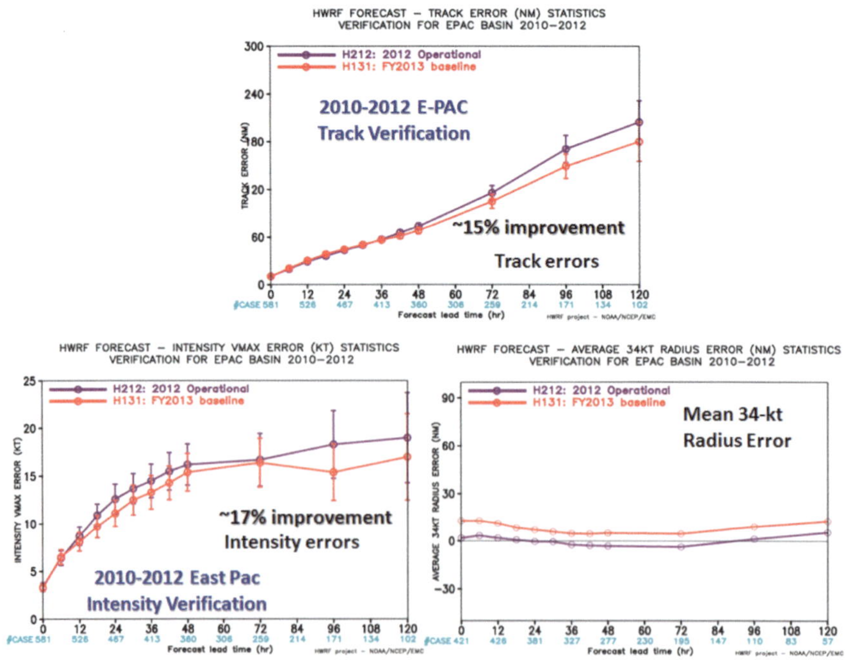

Fig. 10. Same as Fig. 9 except for the Eastern Pacific basin.

were reduced by about 30% in 2012 compared to 2007 while more than 50% reduction is noted for intensity errors. Similar or greater improvements were noted in the North Eastern Pacific basin. HWRF Model performance in operations depend on various factors including computational limitations, timely completion of pre-implementation testing and evaluation and choice of model upgrades based on retrospective testing. No operational model changes are allowed during the season, which limits the ability of the model developers to fix any scientific problems noted during the season. Despite these factors, HWRF has shown the much intended progress during the past few years, and far exceeded the baseline performance (defined by NHC as the mean error from top performing models for a period of five years) that is the benchmark for HFIP goals (Gall et al., 2012).

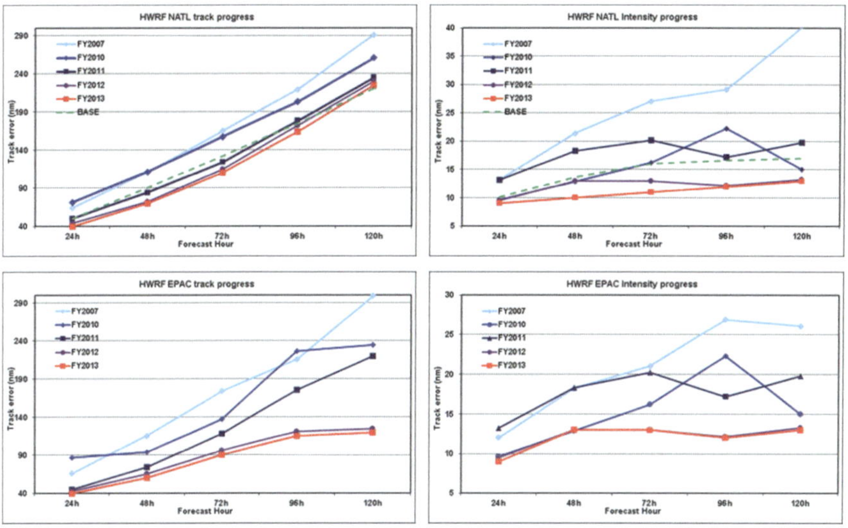

Fig. 11. Track and intensity errors from operational HWRF at 24-hr forecast intervals for all years since the initial implementation in 2007 through 2012. Top panels are for North Atlantic basin and bottom panels are for North Eastern Pacific basin."Base", as shown for the North Atlantic in dashed line, is the mean error of track and intensity as reported by NHC based on 5-year sample from best performing operational models.

6.2 HWRF Model Performance for the North Western Pacific Basin

With support from NOAA's Hurricane Forecast Improvement Project (HFIP), starting with the 2012 typhoon season, the hurricane team at NCEP/EMC has been providing experimental real-time guidance to the forecasters at US Navy's Joint Typhoon Warning Center (JTWC) for all tropical cyclones of the North Western Pacific (WPAC) basin. JTWC has decided to include the 2013 version HWRF model to their official consensus model, which is unprecedented for non-operational models in that basin (Tallapragada et al., 2013).

To demonstrate the overall performance of the HWRF model in the WPAC basin, Fig. 12 shows the homogeneous verification of the HWRF forecasts in terms of track and intensity forecast errors during the 2012 typhoon season and storms in the 2013 typhoon season through July. These verifications are relative to two other operational regional models run at US Navy's Fleet Numerical Meteorology and Oceanography Center (FNMOC) including Navy's version of the GFDL model (GFDN), and the experimental Coupled Ocean Atmosphere Mesoscale Prediction System for Tropical Cyclones (COAMPS-TC) model, which became the primary US Navy tropical cyclone forecast model starting with 2013 typhoon season (Jim Doyle, personal communication). The NCEP GFS global forecasts (AVNO) are also included as a reference. Of the three regional models, HWRF appears to have consistently better track forecast skill at all lead times with the forecast errors from 1day to 3-day comparable to the AVNO forecasts at 95% confidence interval. Although HWRF shows better performance for intensity at all lead times, it performed best for the first 24 hours owing to better vortex initialization procedure described in Section 3.

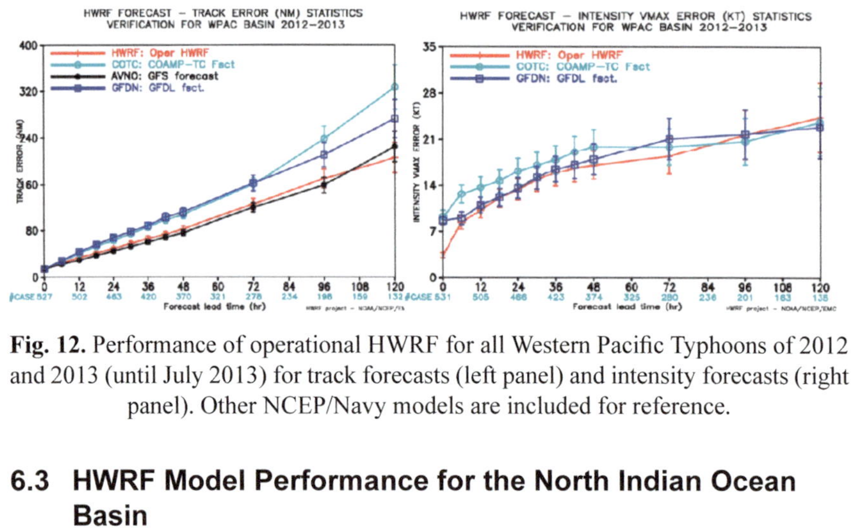

Fig. 12. Performance of operational HWRF for all Western Pacific Typhoons of 2012 and 2013 (until July 2013) for track forecasts (left panel) and intensity forecasts (right panel). Other NCEP/Navy models are included for reference.

6.3 HWRF Model Performance for the North Indian Ocean Basin

A bilateral agreement was signed between NOAA and India's Ministry of Earth Sciences (MOES) to improve tropical cyclone forecasting over the Indian seas. In order to share NOAA's advanced understanding and forecasting techniques acquired in the last few years, scientists from NCEP/EMC helped India Meteorological Department (IMD) and its partnering research institution, Indian Institute of Technology, New Delhi (IIT-D), to implement the HWRF modelling system along with the advanced vortex initialization and data assimilation schemes. Independent model evaluation efforts from IMD (Rama Rao et al., 2013) and EMC (Zhang et al., 2013) have demonstrated that the HWRF model for the North Indian Ocean has superior forecast skill compared to operational models at IMD and NCEP.

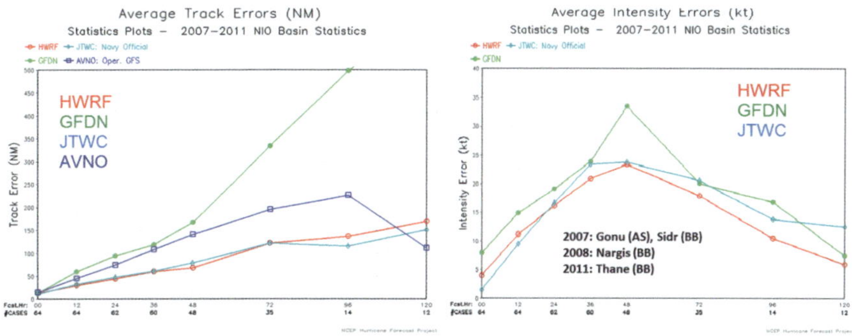

Fig. 13. Performance of operational HWRF implemented for North Indian Ocean basin.Track errors to the left and intensity errors to the right for a sample of 64 cases of Bay of Bengal and Arabian Sea storms formed between 2007 and 2011 are shown here in comparison to NCEP GFS and Navy GFDN. JTWC provides official forecasts for this basin for Navy's purpose.

Track and intensity errors from a sample of 64 cases of Bay of Bengal and Arabian Sea storms for the period between 2007 and 2011 are shown in Fig. 13. HWRF has outperformed the NCEP GFS and Navy GFDN models for both track and intensity, and is found to be very close to the official forecasts from JTWC. These results prompted for operational implementation of HWRF by IMD starting with the 2011 tropical cyclone season. Based on 5 tropical storms over the Indian seas, Das et al. (2015) demonstrated that the operational HWRF implemented at IMD has the average track errors of 135 km in 24 hours, 173 km in 48 hours and 323 km in 72 hours forecast. The HWRF track forecast errors illustrated an improvement of 9% in 24 hours, 34% in 48 hours, 16% in 72 hours over IMD operational forecast. HWRF intensity forecasts have also shown marginal improvement of 5 to 8% over IMD operational forecasts in 48 hours to 72 hours.

Following the successful implementation of HWRF for the Indian Seas, an advanced workshop on HWRF was jointly conducted by NOAA and Indian Institute of Technology, Delhi (IIT-D) in July 2012 at the Indian Institute of Technology, Bhubaneshwar, with support from India's MoES and Indo-US Science and Technology Forum (IUSSTF). This book chapter is part of the proceedings from that workshop. Under the auspices of NOAA-MoES Implementation Arrangement (IA) and Memorandum of Understanding (MoU), IMD and NOAA are in continuous collaboration in sharing the modelling and data assimilation technologies for advancing tropical cyclone forecast skills, and for enhanced research in tropical cyclone related studies.

7. Summary and Future Work

An overview of the operational Hurricane Weather Research and Forecast (HWRF) modelling system is presented in this chapter. Operational HWRF is developed at NCEP/EMC of NOAA and has evolved into the first and foremost community tropical cyclone modelling system for research and operations. HWRF is currently available to the research community through NOAA's DTC which also provides user support and facilitates research to operations (R2O).

HWRF became operational in 2007 and has been providing operational tropical cyclone forecast guidance for all storms in the Northern Atlantic and North Eastern Pacific basins for NHC's operational forecasters. The annual development and upgrade cycle resulted in continuous advancements to the operational HWRF system. Starting in 2012, a very high resolution HWRF model operating at 3 km resolution was implemented for operations. *A major accomplishment was the development of a third nest, at 3 km resolution, to cover the inner core and most of surrounding circulation of the storm. This would make the HWRF model the highest resolution hurricane model ever implemented for operations in the National Weather Service.* Further enhancements obtained in 2013 implementation makes HWRF model the first regional dynamical model that outperformed statistical models and NHC official forecasts for hurricane intensity. In addition to the great improvement of forecast skills noted in 2012 HWRF (10-15% for intensity and 15-20% for tracks), the 2013 HWRF model was found to exceed the track forecast skill by more than 15% and, most importantly, the intensity forecast skills by about 10-15% compared to the 2012 HWRF model. Evaluation of retrospective forecasts for three hurricane seasons (2010-2012) indicated that the 2013 HWRF model exhibits better intensity forecast skill than NHC official forecasts for the Atlantic basin tropical cyclones. This remarkable intensity prediction skill from a deterministic dynamical model compared to official or statistical models is unprecedented.

The success of HWRF model performance comes from carefully planned development, testing and evaluation (T&E) activities supported by NOAA's HFIP programme. Recent advancements and accomplishments were possible due to multi-agency collaborations that led to accelerated progress in the last two years. Apart from extensive T&E and systematic model diagnostics, use of available observations in calibrating model physics, improved data assimilation techniques and feedback from national and international agencies all have shown the desired impact in improving the HWRF modelling system.

In addition to NHC, the HWRF model also produces tropical cyclone numerical prediction for the North Western Pacific and Indian Ocean basins in real-time for use by forecasters at JTWC, and India Meteorological Department (IMD) has officially adopted this model for operational forecast guidance in the Bay of Bengal and Arabian Sea regions. Several other international agencies including Vietnam, Oman, China and Taiwan have also been working with NOAA in implementing HWRF for their respective operational needs.

Future developments of the HWRF model are planned on implementing more sophisticated physics packages applicable for high-resolution, assimilation of satellite data in all sky conditions, and better utilization of flight level data from reconnaissance aircraft. Efforts will be also continued to develop and upgrade HWRF ensemble prediction system. Expanding the ocean model to other basins and coupling HWRF to land, wave, surge, and hydrology models are other developmental paths EMC has taken to develop the next generation comprehensive hurricane forecast modelling system. Figure 14 provides a schematic of next generation HWRF modelling system in operations, which includes downstream applications for landfall related rainfall, surge and inundation forecast products.

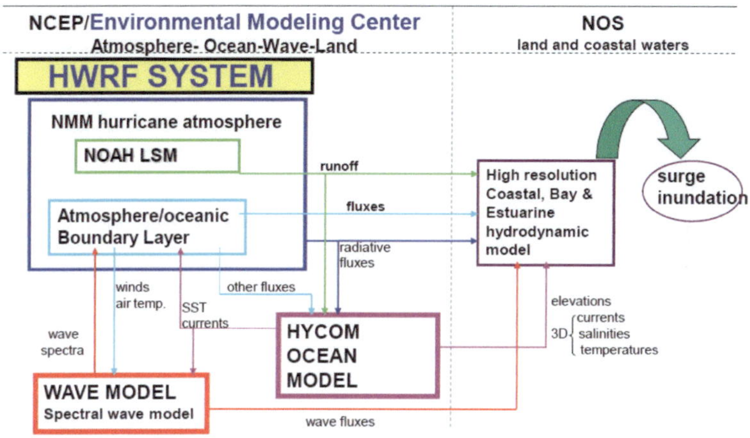

Fig. 14. Schematic showing a comprehensive hurricane modelling system for addressing the next generation forecast problems. Several components include coupling to land surface, waves, ocean, hydrology, storm surge and inundation models. Arrows indicate flow of data among different components.

Moving away from storm-centric approach, EMC/NCEP and HRD/AOML are collaborating on a basin-scale modelling system with multiple moveable nests to address more complex multi-scale interactions associated with tropical cyclones. The basin-scale approach will help alleviating the current drawbacks, especially related to tropical cyclogenesis, and extended range predictions from operational HWRF. This will also help developing a selfcycled regional data assimilation system along with improved inner core data assimilation. A real-time example of basin-scale HWRF system with multiple moveable nests is shown in Fig. 15.

A single deterministic hurricane model can only provide one of the many possibilities of tropical cyclone evolution. It is imperative that we need adopt the ensemble approach for representing the uncertainty and improve the predictability of the tropical cyclone path, intensity, structure and rainfall forecasts. HWRF team at NCEP started experimenting with a 20-member high-resolution ensemble forecasts starting with the 2013 hurricane season. As expected, the ensemble mean outperformed the single deterministic forecast from operational HWRF. Future plans include further development of HWRF ensembles for operational purposes.

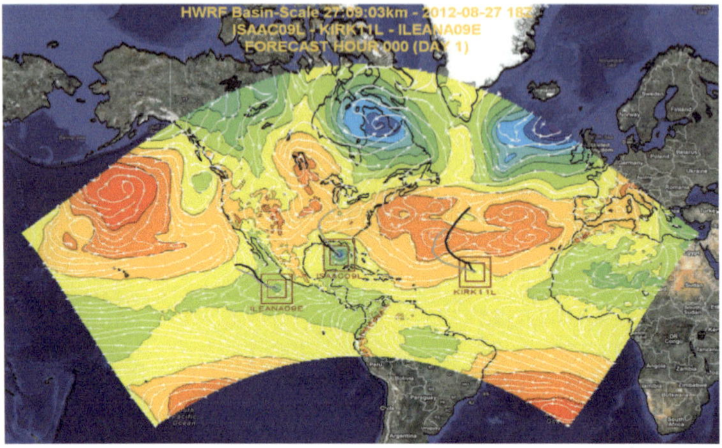

Fig. 15. A real-time example of Basin-Scale HWRF with multiple moveable nests placed around various co-existing tropical cyclones in the North Atlantic and Eastern Pacific basins. Individual telescopic nests were placed around hurricanes Kirk, Isaac and Ileana and are integrated simultaneously in a single model forecast cycle.

Improvements in physics, data assimilation, numerical and computational methods, ensemble techniques, product generation and diagnostics remain the focus areas of hurricane model development in NOAA. Specifically, improvements in hurricane vortex initialization to reduce the rapid growth of intensity error in early forecast period (0-24 hr), either through static or dynamic/physical initialization; development and testing of high resolution hurricane physics suites for multi-scale applications; development of advanced diagnostics tools for hurricane models in addition to conventional diagnostics methods; study of tropical cyclone rapid intensification (RI) and predictability of RI in HWRF; advanced quality control and assimilation techniques for aircraft reconnaissance data; vortex-shear interactions with focus on resilience of model vortex to environmental shear; downstream applications of HWRF model, with special emphasis on storm related rainfall, flooding, storm surge, hydrology and inundation, through improved land surface representation; impact of aerosols and/or sea spray on hurricane intensity forecasts; and enhanced techniques for coupling of atmosphere-ocean-wave-land-hydrology. Community involvement in the model development activities should accelerate the transition of research into operations and that is where collaboration among all tropical cyclone researchers will yield significant improvements to the operational tropical cyclone forecast capabilities, resulting in saving lives and property in all the regions impacted by tropical cyclones in the world.

Acknowledgements

The author acknowledges NOAA's Hurricane Forecast Improvement Project that supported this work. Thanks are due to Dr. Zavisa Janjic, NCEP for providing the complete set of equations and Dr. Rama Rao, IMD, New Delhi for providing inputs. The author also acknowledges the support provided by Indo-US Science and Technology Forum, NOAA's Hurricane Forecast Improvement Project and various national and international collaborators who have helped improving the operational HWRF model.

REFERENCES

Arakawa, A. and Wayne Howard Schubert, 1974: Interaction of a Cumulus Cloud Ensemble with the Large-Scale Environment, Part I. *J. Atmos. Sci.*, **31**, 674–701.

Bao, J.-W., S.G. Gopalakrishnan, S.A. Michelson, F.D. Marks and M.T. Montgomery, 2012: Impact of Physics Representations in the HWRFX on Simulated Hurricane Structure and Pressure–Wind Relationships. *Mon. Wea. Rev.*, 140, 3278–3299.

Das, A.K., Y.V. Rama Rao, Vijay Tallapragada, Zhan Zhang, S.K. Roy Bhowmik and Arun Sharma, 2015: Evaluation of the Hurricane Weather Research and Forecasting (HWRF) model for tropical cyclone forecasts over the North Indian Ocean (NIO). *Natural Hazards*, **75**, 1205–1221.

Fels, S.B. and M.D. Schwarzkopf, 1975: The simplified exchange approximation: A new method for radiative transfer calculations. *J. Atmos. Sci.*, **32**, 1475–1488.

Fels, S., J.T. Kiehl, A.A. Lacis and M.D. Schwarzkopf, 1991: Infrared cooling rate calculations in operational general circulation models: Comparisons with benchmark computations. *Journal of Geophysical Research*, **96(D5)**, 9105–9120.

Ferrier, B.S., Y. Lin, T. Black, E. Rogers and G. DiMego, 2002: Implementation of a new grid-scale cloud and precipitation scheme in the NCEP Eta model. Preprints. 15th Conference on Numerical Weather Prediction. American Meteorological Society, San Antonio, TX.

Gopalakrishnan, S., F. Marks Jr., X. Zhang, Jian-Wen Bao, Kao-San Yeh and R. Atlas, 2011: The Experimental HWRF System: A Study on the Influence of Horizontal Resolution on the Structure and Intensity Changes in Tropical Cyclones Using an Idealized Framework. *Monthly Weather Review*, **139(6)**, 1762–1784.

Gopalakrishnan, S., S. Goldenberg, T. Quirino, X. Zhang, F.D. Marks Jr., K.-S. Yeh, R. Atlas and V. Tallapragada, 2012a: Toward Improving High-Resolution Numerical Hurricane Forecasting: Influence of Model Horizontal Grid Resolution, Initialization, and Physics. *Weather and Forecasting*, **27(3)**, 647–666.

Gopalakrishnan, S., Q. Liu, T. Marchok, S. Sheinin, Vijay Tallapragada, Mingjing Tong, R. Tuleya, R. Yablonsky and X. Zhang, 2012b: Hurricane Weather Research and Forecasting (HWRF) Model Scientific Documentation. Edited by L. Bernardet.

Gopalakrishnan, S., F. Marks, J. Zhang, X. Zhang, J.-W. Bao and V. Tallapragada, 2013: A study of the impact of vertical diffusion on the structure and intensity of tropical cyclones using the high-resolution HWRF system. *J. Atmos. Sci.*, **70(2)**, 524–541.

Grell, G., 1993: Prognostic evaluation of assumptions used by cumulus parameterizations. *Mon. Wea. Rev.*, **121**, 764–787.

Han, J. and H.-L. Pan, 2011: Revision of convection and vertical diffusion schemes in the NCEP global forecast system. *Weather Forecasting*, **26**, 520–533.

Heming, J.T., J.C. Chan and A.M. Radford, 1995: A new scheme for the initialisation of tropical cyclones in the UK Meteorological Office global model. *Meteor. Appl.*, **2**, 171–184.

Hong, S.-Y. and H.-L. Pan, 1998: Convective Trigger Function for a Mass-Flux Cumulus Parameterization Scheme. *Mon. Wea. Rev.*, **126**, 2599–2620.
Kurihara, Y., M.A. Bender, R.E. Tuleya and R. Ross, 1995: Improvements in the GFDL Hurricane Prediction System. *Mon. Wea. Rev.*, **123(9)**, 2791–2801.
Liu, Q., S. Lord, N. Surgi, Y. Zhu, R. Wobus, Z. Toth and T. Marchok, 2006b: Hurricane relocation in global ensemble forecast system. *Preprints,* 27th Conf. on Hurricanes and Tropical Meteorology, Monterey, CA, Amer. Meteor. Soc., P5.13.
Liu, Q., T. Marchok, H.-L. Pan, M. Bender and S. Lord, 2000: Improvements in Hurricane Initialization and Forecasting at NCEP with Global and Regional (GFDL) models. *NCEP Office Note* 472.
Liu, Q., N. Surgi, S. Lord, W.-S. Wu, S. Parrish, S. Gopalakrishnan, J. Waldrop and J. Gamache, 2006a: Hurricane Initialization in HWRF Model. *Preprints*, 27th Conference on Hurricanes and Tropical Meteorology, Monterey, CA.
Lord, S.J., 1991: A bogusing system for vortex circulations in the NMC global forecast model. *Amer. Met. Soc.*, 329–330.
Mathur, M.B., 1991, The National Meteorological Center's Quasi-Lagrangian model for Hurricane Prediction. *Mon. Wea. Rev.*, **119(6)**, 1419–1447.
Pan, H.-L. and W.-S. Wu, 1995: Implementing a mass flux convective parameterization package for the NMC medium range forecast model. NMC Office Note 409.
Pan, H.-L., 2003: The GFS atmospheric model. NCEP Office Note No. 442.
Peng, M.S., B.F. Jeng and C.P. Chang, 1993: Forecast of Typhoon motion in the vicinity of Taiwan during 1989-90 using a dynamical model. *Wea. Forecasting*, **8**, 309–325.
Prasad, K. and Y.V. Rama Rao, 2003: Cyclone track prediction by quasi-Lagrangian limited area model. *Meteol. Atmos. Phys.*, **83**, 173–185.
Roberts, R.E., J.E.A. Selby and L.M. Biberman, 1976: Infrared continuum absorption by atmospheric water vapor in the 8–12-μm window. *Appl. Opt.*, **15**, 2085–2090.
Rodgers, C.D., 1968: Some extensions and applications of the new random band model for molecular band transmissions. *Q.J.R. Meteorol. Soc.*, **94**, 99–102.
Ryan, B.F., 1996: On the global variation of precipitating layer clouds. *Bull. Amer. Meteor. Soc.*, **77**, 53–70.
Sasamori, T., J. London and D.V. Hoyt, 1972: Radiation budget of the Southern Hemisphere. *Meteorol. Monogr.*, **35**, 9–22.
Schwarzkopf, M.D. and S. Fels, 1985: Improvements to the algorithm for computing CO_2 transmissivities and cooling rates. *Journal of Geophysical Research*, **90(C10)**, 10,541–10,550.
Schwarzkopf, M.D. and S. Fels, 1991: The simplified exchange method revisited: An accurate, rapid method for computation of infrared cooling rates and fluxes. *J Geophys Res*, **96(D5)**, 9075–9096.
Ueno, M., 1995: A study on the impact of asymmetric components around tropical cyclone center on the accuracy of bogus data and the track forecast. *Meteorol. Atmos. Phys.*, **56**, 125–134.
Yeh, K.-S., X. Zhang, S. Gopalakrishnan, S. Aberson, R. Rogers, F. Marks and R. Atlas, 2012: Performance of the experimental HWRF in the 2008 Hurricane Season. *Natural Hazards*, **63(3)**, 1439–1449.
Zhang, X., T.S. Quirino, K.-S. Yeh, S. Gopalakrishnan, F.D. Marks Jr., S.B. Goldenberg and S. Aberson, 2011: HWRFx: Improving Hurricane Forecast with High-Resolution Modeling. *Computing in Science and Engineering*, **13(1)**, 13–21.

4

Physical Processes in Tropical Cyclone Models

Jian-Wen Bao
NOAA/Earth System Research Laboratory, Boulder, Colorado
jian-wen.bao@noaa.gov

This chapter is the first of the three-lecture series on the physical processes and their descriptions in numerical models of tropical cyclones. There are two sections in this chapter. Section 1 is intended to provide an overview of the physical processes required to be included in Tropical Cyclone (TC) models, and Section 2 presents results from a recent study on the sensitivity of simulated TC intensification to parameterized physical processes. Materials presented in Section 1 are derived from Pielke (2001), Jakob and Miller (2003), and Tao and Moncrieff (2003), while those presented in Section 2 are based on Bao et al. (2012a).

1. An Overview of Physical Processes Required to be Included in TC Models

1.1 Importance of Cloud Parameterization in TC Models

Various influences that clouds have on the evolution of TCs need to be simulated accurately by the atmospheric models that are used for the prediction and simulation of tropical cyclones. TC prediction and simulation models seek numerical solutions to the partial differential equations that govern atmospheric motions. Various numerical techniques can be applied to achieve this goal, but all of them ultimately involve splitting the area over which the model is applied into 'grid boxes' of finite size in both the horizontal and the vertical. While the continuous differential equations describe atmospheric motions on all spatial and temporal scales, their discrete form can only describe processes on spatial scales of the order of twice the grid length. Physical processes that occur in TCs cover a wide range of spatial scales, from micrometres in the condensation and evaporation of individual cloud droplets, through a few hundred metres in the case of fair weather cumulus clouds, up to several hundred kilometres for the cloud systems associated with TC eyewall and spiral rain bands. Hence, describing the detailed dynamics of individual clouds requires model grid-box sizes of the order of metres or less. Current computing power, as well as difficulties in finding the necessary initial conditions at those spatial scales, prohibit the use of such grid-box sizes in TC models. In reality, typical horizontal grid lengths in contemporary TC models range from around 1 km in regional scale models to about 30 km in global scale models.

To describe the main effects clouds have on the atmosphere as outlined above, the following cloud related quantities need to be known:

- horizontal coverage of cloud, normally referred to as cloud fraction;
- vertical extent of the clouds;
- sources and sinks of cloud condensate including condensation;
- evaporation/sublimation, and conversion into precipitation and fallout;
- phase of the condensate;
- particle size and shape;
- in-cloud distribution of condensate; and
- amounts of heat, water vapour, and momentum that are transported in convective clouds.

There are a number of problems that need to be overcome in the parameterization of clouds. First, there exists a variety of cloud types, such as stratocumulus clouds at the top of convective boundary layers, vast cloud systems associated with extratropical disturbances, deep convective systems that may or may not be organized, and upper-tropospheric cirrus clouds. These different cloud types are formed, maintained and dissipated by different physical processes, such as convection, small-scale turbulence, large-scale ascent or descent, and cloud microphysical processes that lead to the generation of precipitation. Many of these processes are poorly understood and act on scales smaller than those resolved in a large-scale model, which makes them the subject of physical parameterization themselves. Furthermore, the radiative effects of clouds depend on a large number of different cloud parameters that all need to be described accurately to ensure their correct treatment in the radiation parameterization. It is worthwhile pointing out that, because of their distinctive properties, cumulus and cumulonimbus clouds have been recognized as being of particular importance. This has led to (an artificial) separation of the description of the vertical transport and condensation effects from the radiative effects of convective clouds in what is now known as cumulus convection parameterizations. As will be briefly discussed below, current efforts in improving cloud parameterizations involve attempts to overcome this artificial process splitting. Furthermore, the details of the radiative transfer in clouds are normally dealt with in radiation parameterizations. Thus, typical cloud parameterizations need to (1) describe the generation and dissipation of clouds and the precipitation formed in them and (2) provide the radiation parameterization with the necessary information to evaluate the cloud effects on the radiative fluxes, most prominently the area coverage and cloud condensate content.

1.2 Cloud Microphysics Parameterization

Most recently, the attention in cloud parameterization has shifted significantly towards the treatment of cloud microphysics. This has been facilitated by increased computing power and the availability of sophisticated microphysics parameterizations from cloud-resolving and meso-scale numerical models. Although increased sophistication in describing precipitation processes in GCMs is certainly justified, the transplantation of a microphysics scheme from a cloud-resolving model to a GCM is not without problems. This is mainly due to the scales at which the input variables of the microphysical scheme are available and to the difference in time steps used by the different models. Microphysical processes are highly nonlinear and their parameterization has to rely on the knowledge of the local amount of condensate. In GCMs, only the grid-mean value (or cloud-mean value if cloud fraction is a model variable) for condensate is known. This has led to the need for significant modifications to microphysical constants in the parameterizations in order to achieve reasonable cloud condensate and precipitation amounts. The detailed treatment of microphysical process would also require the use of very short model time steps.

With increasing horizontal and vertical resolution, the concept of cloud fraction becomes less important and grid-point values of cloud condensate are more representative for local conditions. Hence, in numerical models with horizontal resolutions of less than a few kilometres, it is common to apply more complex and physically more realistic parameterizations of cloud microphysics. This is usually achieved by introducing additional condensate species (e.g., graupel, hail) and a more realistic description of the microphysical processes themselves (e.g., a separate description of nucleation and deposition; a description of riming).

1.3 General Concepts in Sub-Grid Cloud Parameterization

Processes that act on scales smaller than these grid sizes are normally referred to as subgrid-scale processes and are, per se, not represented in the solution of the finite difference equations. Many of these processes do, however, affect the dynamic and thermodynamic state of the atmosphere on the grid resolved scales. Obvious examples are the large amounts of water vapour, heat and momentum that are transported by sub-grid turbulent and convective motions within the TC eyewall and spiral rain bands. Even away from the convective eyewall and spiral rain bands, one finds from observations that often the area is only partially covered with cloud. Since this has important consequences, especially for the radiative cloud effects, almost all cloud parameterizations describe the fractional coverage of a model grid box with cloud as one of their key parameters. Since an explicit description of the subgrid-scale processes is prohibited, only their statistical effects on the grid box-mean state can be taken into account. Since the numerical solution of the model equations allows the atmospheric state to be known only on scales of the order of the grid box size, the description of these statistical effects has to be expressed in terms of those scales. The technique involved is generally referred to as sub-grid cloud parameterization (i.e., sub-grid convective parameterization scheme).

1.4 Other Physical Processes in TC Models

1.4.1 Sub-Grid Turbulent Mixing

While large eddies are resolved in cloud models, eddies much smaller than the grid-scale must be parameterized. An implicit assumption is that the small scales approximate to an inertial subrange where the energy spectrum is in statistical equilibrium, with an energy cascade from the resolved scales to the dissipation scales. The most sophisticated turbulence parameterization presently used is a third-moment closure. Typical cloud models used simple K-type (first-order) turbulence closure or determine the coefficient K, diagnostically or prognostically, from the turbulence kinetic energy (TKE) equation (one-and-a-half order). In the prognostic TKE method, thermodynamic stability, deformation, shear stability, diffusion, dissipation and transport of subgrid energy are included. In the diagnostic method, deformation and stability are used for computing the k coefficient.

1.4.2 Radiation and Cloud Interaction

Emission and absorption by water vapour and cloud droplets are represented by two-stream long-wave radiative transfer schemes. Broadband methods for long-wave radiation combine the effects of reflection, emission, and transmission by cloud droplets and air molecules. The treatment of short-wave radiation is also based on broadband approximations. One key issue is how to parameterize cloud optical properties (optical thickness), especially in the presence of the ice phase, in view of the important impact of radiative heating and cooling profiles within clouds. Note that only limited observations are available upon which to base parameterizations for ice clouds. The use of a fully explicit microphysics scheme (liquid and ice) and a fine horizontal resolution can provide relatively realistic cloud optical properties, which are crucial for determining the radiation budgets. With high spatial resolution, each atmospheric layer is considered either completely cloudy (overcast) or clear. No partial cloudiness is assumed.

1.4.3 Ocean Surface Fluxes

Two types of surface flux schemes are typically used. The first is a simple bulk aerodynamic formula where the transfer coefficients for momentum, sensible heat and latent heat fluxes are functions of wind speed only. The second type is more complex but, nevertheless, primarily a bulk approach. The transfer coefficients for momentum, sensible heat and latent heat fluxes are based on the Monin-Obukhov similarity theory of the atmospheric surface layer. The parameters, such as the roughness lengths, are closely related to the sea surface characteristics and the turbulence characteristics. At very low wind speeds the similarity profile is singular, a problem addressed by adding a convective velocity to yield nonzero fluxes under windless conditions. The exchange coefficients in the simple bulk aerodynamic formula method and in the second bulk flux algorithm differ in two ways. First, in the lower wind speed regime (less than 4 ms^{-1}), the exchange coefficients in the complex bulk scheme increase with decreasing wind speed in order to account for the convective exchange at low wind speeds. Second, the coefficients in the simple bulk aerodynamic formula linearly increase with respect to the wind speed, but decrease if the wind speed is greater than 5 ms^{-1} in the more complex bulk schemes. These differences in the exchange coefficients can affect rainfall amounts and boundary layer structure.

1.4.4 Land Surface Processes

Detailed interactive land surface process models of the heterogeneous land surface (soil and vegetation) and adjacent near-surface atmosphere have recently been applied in TC models to account for the effect of the land surface on the evolution of TCs once they make landfall. A land surface model usually has three elements: (1) a soil module that includes at least several water reservoirs (i.e., plant internal storage, dew/intercepted precipitation, surface material, a top soil root layer, a subsoil root layer, and two deeper layers that regulate longer-term variability of the soil hydrology); (2) a surface slab of vegetation, litter, and other loose material that shades the soil and acts as the source for sensible heat flux, and intercepts precipitation and dew; and (3) the surface layer of the atmosphere (up to the lowest grid level of the model to which it is coupled), within which the fluxes of sensible heat and water vapour are calculated.

2. Sensitivity of Simulated TC Intensification to Parameterized Physical Processes

This part of the chapter presents results from a study, in which a series of idealized experiments with the NOAA Experimental Hurricane Research and Forecasting Model (HWRFX) are performed for the purpose of evaluating the sensitivity of HWRFX to commonly used BL and cloud parameterization of microphysical processes. The HWRFX model is a version of the National Centers for Environmental Prediction's (NCEP) Hurricane Weather Research and Forecasting (HWRF) system specifically modified at the Hurricane Research Division (HRD) of the Atlantic Oceanographic and Meteorological Laboratory (AOML) and the Earth System Research Laboratory (ESRL) to study the intensity change problem at the finest model grid resolution operationally feasible at this time for forecasting. The model is initialized with a weak axisymmetric vortex disturbance in an idealized tropical environment that is favourable for the vortex amplification. The initial mass and wind fields associated with the weak vortex disturbance are obtained by solving the nonlinear balance equation for the given wind distributions of the initial vortex, and the prescribed background thermal profile. We employ the foregoing metrics, Vmax and Pmin as well as the azimuthally-averaged structure of the simulated tropical cyclone, as a way of quantifying the sensitivity of the intensification process to variations in physical parameterization schemes.

The remaining paper is organized as the following order. The setup of the HWRFX sensitivity experiments is described first. Then, our current understanding of the basic dynamics of tropical cyclone intensification is reviewed to provide a meaningful context for the sensitivity experiments. Next, the results from the sensitivity experiments are presented and compared. Finally, a summary and discussion of the results are provided, along with their implication for tropical cyclone model evaluation.

2.1 Experimental Design

The sensitivity experiments presented here are run with a parent domain (about 55 × 55 degrees) at a horizontal resolution of 9 km with a single moving nest (about 8 × 8 degrees) at 3 km horizontal resolution. There are 43 stretched pressure-sigma hybrid levels in the vertical direction with the top level set to 50 hPa.

To initialize the idealized vortex in all the experiments, the non-linear balance equation in the pressure-based sigma coordinate system described in Wang (1995) is solved within the rotated latitude-longitude E-grid framework on an f plane located at 12.5°N. The mass field is derived from the wind field corresponding to an axisymmetric cyclonic vortex of maximum surface tangential wind set to 15 ms^{-1} at 90 km from the vortex centre that is embedded in a quiescent flow. The temperature and humidity profiles of the far-field are based on Jordan's Caribbean sounding (Gray et al., 1975). In all of the experiments, the sea surface temperature is set to 302 K (approximately 29 degrees Celsius).

The physical configurations used in all the sensitivity experiments are summarized in Table 1. One of these configurations (Experiment 1) is very close to the operational HWRF system, in which the HWRF version of the Global Forecast System (GFS) surface and BL formulations are used to parameterize the sea-to-air flux transport and the subsequent mixing in the atmosphere. The Ferrier scheme is used to provide latent heating due to the microphysical processes of clouds in the atmosphere, and the Simplified Arakawa and Schubert scheme (SAS, see Pan and Wu, 1995) is used to parameterize sub-grid-scale cumulus-cloud activity. The National Center for Atmospheric Research (NCAR) long- and short-wave radiation schemes are used. To maintain consistency between our results at 9 and 3 km and those in Gopalakrishnan et al. (2010), the SAS convection scheme is used at 3 km resolution along with the Ferrier scheme for grid-resolved cloud microphysical processes. While it remains a research subject whether a SCP scheme should

Table 1. The suite of sensitivity experiments

Experiment number and Name (colour symbol designation in the Vmax and Pmin time series figures shown in section 2.3)	*Description of physics options*
1. GFS/SAS/FER/ NCAR (Red)	GFS BL and surface scheme, SAS convective scheme on both grids, Ferrier microphysics scheme, NCAR Rapid Radiative Transfer Model longwave radiation scheme, Dudhia shortwave radiation scheme
2. MYJ/SAS/FER/ NCAR (Black)	MYJ BL and surface scheme, SAS convective scheme on both grids, Ferrier microphysics scheme, NCAR Rapid Radiative Transfer Model longwave radiation scheme, Dudhia shortwave radiation scheme
3. MYJ/BMJ/FER/ NCAR (Gray)	MYJ BL and surface scheme, Betts-Miller-Janjic convective scheme on both grids, Ferrier microphysics scheme, NCAR Rapid Radiative Transfer Model longwave radiation scheme, Dudhia shortwave radiation scheme
4. GFS/BMJ/FER/ GFD (Orange)	GFS BL and surface scheme, Betts-Miller-Janjic convective scheme on both grids, Ferrier microphysics scheme, GFDL radiation scheme
5. GFS/BMJ/FER/ NCAR (Pink)	GFS BL and surface scheme, Betts-Miller-Janjic convective scheme on both grids, Ferrier microphysics scheme, NCAR Rapid Radiative Transfer Model longwave radiation scheme, Dudhia shortwave radiation scheme
6. GFS/SAS/FER/ NCAR MOD-DRAG (Brown)	GFS BL and surface scheme, SAS convection scheme on both grids, Ferrier microphysics scheme, NCAR Rapid Radiative Transfer Model longwave radiation scheme, Dudhia shortwave radiation scheme, realistic drag coefficient consistent with recent observations
7. GFS/SAS/WS5/ GFD (Light blue)	GFS BL and surface scheme, SAS convection scheme on both grids, WRF Single-Moment 5-class microphysics scheme, GFDL radiation scheme
8. GFS/SAS/WS6/ GFD (Magenta)	GFS BL and surface scheme, SAS convection scheme on both grids, WRF Single-Moment 6-class microphysics scheme, GFDL radiation scheme
9. GFS/SAS/Thom/ GFD (Yellow)	GFS BL and surface scheme, SAS convection scheme on both grids, WRF Thomson microphysics scheme, GFDL radiation scheme
10. GFS/SAS/FER/ GFD (Green)	GFS BL and surface scheme, SAS convection scheme on both grids, Ferrier microphysics scheme, GFDL radiation scheme
11. GFS/noSAS/ FER/GFD (Purple)	GFS BL and surface scheme, SAS convection scheme on 9-km grid, no convective scheme on 3-km grid, Ferrier microphysics scheme, GFDL radiation scheme

Explanations of the individual physics options are available at http://www.dtcenter.org/HurrWRF/users/docs/scientific_documents/HWRF_final_2-2_cm.pdf and http://www.dtcenter.org/HurrWRF/users/online_tutorial/tutorial02222010.php).

be turned on in the 3-km grid and, if it should, whether the SAS scheme is an appropriate one to use remains a subject of research. As shown by one of the sensitivity experiments in this study, turning on the SAS scheme in the 3-km grid basically modifies the latent heating distribution that is realized by the explicit microphysical process only and, thus, influences the intensity and structure and simulated TC. The effect of radiation is investigated by changing the long- and short-wave radiation scheme from the NCAR scheme to the Geophysical Fluid Dynamics Laboratory (GFDL) scheme.

The results from all the experiments listed in Table 1 are divided into three groups to allow for the examination of the sensitivity of the simulated tropical cyclone development (i) to BL-mixing, subgrid convection and radiation (Experiments 1-5), (ii) to the surface drag (Experiments 6 and 7), and (iii) to the bulk microphysics and subgrid convection schemes (Experiments 8-11). The BL mixing schemes determine the vertical subgrid turbulence mixing and vertical diffusion within the atmospheric boundary layer and the free atmosphere above. The two schemes used in this study, i.e., the MYJ and the GFS BL schemes, are both one-dimensional, and representative of two types of BL mixing parameterizations widely used in numerical weather prediction models: the turbulent kinetic energy (TKE) scheme and the flux scaling scheme. The drag and enthalpy exchange coefficients used in the MYJ BL scheme are calculated using the surface layer scheme developed by Janjic (1996, 2002), and is based on the Monin-Obukhov similarity theory and includes parameterizations of a viscous sub-layer. In the GFS BL scheme, on the other hand, the drag and enthalpy exchange coefficients are calculated in a nearly identical way as in the operational GFDL hurricane model, which is a bulk parameterization based on the Monin-Obukhov similarity theory. In both of the BL schemes, the drag and enthalpy exchange coefficients increase with 10-m wind speed. Two subgrid convection schemes are permutated in the experiments: the SAS scheme and the Betts-Miller-Janjic (BMJ) scheme (Janjic, 1994, 2000). In addition to the NCAR long- and short-wave radiation schemes, the GFDL schemes for long- and short-wave radiation are used in the sensitivity experiments. The surface drag coefficient calculation for the sensitivity experiments is described in the results section.

A total of four different microphysics schemes are used in the sensitivity experiments: the Ferrier scheme, the WRF single-moment 5-class (WSM5) scheme, the WRF single-moment 6-class (WSM6) scheme, and the Thompson double-moment 6-class scheme (Skamarock et al., 2008). The Ferrier scheme predicts changes in water vapour and condensate in the forms of cloud water, rain, cloud ice, and precipitation ice (snow/graupel/sleet) (see Skamarock et al., 2008). The individual hydrometeor fields are combined into total condensate for advection calculation. The WSM5 scheme has five prognostic equations of microphysical processes for vapour, rain, snow, cloud ice and cloud water, and allows super-cooled water to exist, and a gradual melting of snow as it falls below the melting layer. The WMS6 scheme extends the WSM5 scheme to include graupel and its associated processes. The Thompson double-moment scheme includes six prognostic equations of moisture species plus number concentration for ice as prognostic variables. Since only water vapour and total condensate are advected in Ferrier scheme, the horizontal and vertical advection of hydrometeor species in all the non-Ferrier schemes is consistently treated in accordance with the assumptions made in the Ferrier scheme (i.e., keeping the partition of hydrometeors unchanged during the advection of the total condensate) such that the differences in all the microphysical processes are kept in the parameterization of cloud microphysical processes.

2.2 Metrics for Comparing the Sensitivity Experiments

In order to understand the salient characteristics of the model solution and improve the model performance, it is important to choose metrics that carry dynamical information for the examination of sensitivity experiments. The results from all the sensitivity experiments summarized in Table 1 will be first compared in terms of the intensity of the simulated tropical cyclone using the Vmax, Pmin and PWR. Then, the structures of the simulated tropical cyclone in the sensitivity experiments will be compared in terms of azimuthally-averaged winds and acceleration. The latter metrics for comparison are necessary to complement the intensity comparison. This is because observations suggest that the intensity in terms of the Vmax and the Pmin of tropical cyclones are not strongly correlated with their size (e.g., Weatherford and Gray, 1988), and simply using the intensity metric to compare the sensitivity experiments is insufficient to provide useful information for understanding and improving the model performance. Thus, in this section, a brief review is provided to summarize the advantage and weakness of these metrics.

The intensity of tropical cyclones is operationally described in terms of minimum sea-level pressure of the cyclone centre (i.e., Pmin) and the local maximum surface wind speed (i.e., Vmax). The advantage of using the Pmin in operational forecasts and climatological records is that it can be obtained reliably from dropsonde measurements or direct observations at aircraft reconnaissance flight level. On the other hand, the Vmax is a difficult quantity to measure by its definition (i.e., 10-m level, 1-min sustained), despite the fact that it is often related to the destructive energy[1] and the societal impact of tropical cyclones due to its close link to the structure of tropical cyclones. For this reason, pressure-wind relationships (i.e., PWRs) were developed to facilitate the need for describing tropical cyclone intensity in terms of the Vmax in accordance with the Pmin (for useful historical perspectives see Knaff and Zehr, 2007; Holland, 2008). Although these PWRs were attempts to describe the mean relationship between the Pmin and the Vmax, the actual relationship between them is a function of multiple factors related to the tropical cyclone

[1] Strictly speaking, the destructive energy of tropical cyclones is related to storm integrated kinetic energy, which is closely linked to storm structure.

environment and structure that varies from case to case. Consequently, there is considerable scatter about any given PWR derived from observations (see, e.g., Fig. 7 and Table 1 in Holland (2008) for a statistical summary). This hinders the effectiveness of using the observationally-derived PWRs to gain insight into the scattering of the PWRs from the sensitivity experiments, particularly when the sensitivity spread lies within the uncertainties associated with the observed PWR.

There is also a fundamental weakness in using the observationally-derived PWRs to compare with those from the sensitivity experiments. The development of the PWRs based on observations is, historically, motivated by the assumption that the gradient-wind balance is dominant in the overall dynamics of tropical cyclone intensification. Although the gradient-wind balance embedded in the observationally derived PWRs appears to contain all of the essential dynamics of the vortex tangential wind above the BL, it is fundamentally invalid in the BL. The gradient-wind balance is formally invalid in the BL because, firstly, it assumes that the radial inflow makes a negligible impact to the dynamics there. This is not the case in the BLs of tropical cyclones (see, e.g., Montgomery et al., 2006 and also the scale analysis by Smith and Montgomery, 2008 and Smith and Vogl, 2008). Secondly, the Vmax in the PWR is the locally, instantaneous maximum wind near the surface, while the wind required by the gradient-wind balance is axisymmetric. Thus, it seems inherently problematic to seek a physically meaningful understanding of the sensitivity experiments if only the model-simulated PWRs are compared.

On the other hand, since the gradient-wind balance is a good approximation to the azimuthally averaged dynamics above the BL (e.g., Bui et al., 2009 and refs. therein), it seems physically appropriate to use the azimuthally-averaged structure of the simulated tropical cyclone to compare the sensitivity experiments. Examples of useful parameters for illustrating the cyclone structure are the azimuthally-averaged radius of the 15 ms^{-1} (slightly less than the lower threshold of gale force winds, 17 ms^{-1}) and radius of the 35 ms^{-1} tangential velocity (slightly greater than the 33 ms^{-1} definition of hurricane strength). In fact, in some of previous studies (e.g., Xu and Wang, 2010a and 2010b), the radius of the damaging force (50 knots) wind is used often as a size parameter of the simulated vortex.

Using the azimuthally-averaged structure to compare the sensitivity experiments takes advantage of the current understanding of the dynamics of the primary and secondary circulations associated with an axisymmetric vortex. The essential premise of this understanding is that the primary circulation of a tropical cyclone vortex is so strong that the mean axisymmetric dynamics broadly control the dynamics of tropical cyclone intensification and structural evolution (see Willoughby, 1995, Bui et al., 2009 and section 2 of Montgomery and Smith, 2011). These two spin-up mechanisms have been identified to coexist during the intensification of the primary circulation (see section 7 in Montgomery and Smith, 2011).

In the first spin-up mechanism, the intensification of the primary circulation can be explained by the convergence driven by the aggregate diabatic heating in the eyewall region associated with the rotating deep convection and the material conservation of the absolute angular momentum. The secondary circulation of tropical cyclones can be understood dynamically as the response of a balanced axismmetric vortex above to lateral and vertical forcing distributions associated with diabatic processes and their interaction with the environment and the lower boundary. Such a response can be described by diagnostic solutions of Eliassen's balanced vortex equations (see, e.g., Shapiro and Willoughby, 1982; Bui et al., 2009 and section 2 of Montgomery and Smith, 2011). The surface forcing associated with the surface drag and the turbulent sensible and latent heat fluxes can further distort the primary and secondary circulations that intensify with this spin-up mechanism, as seen in the work of Willoughby (1979), Schubert and Hack (1982), Shapiro and Willoughby (1982) and Bui et al. (2009).

The second spin-up mechanism is associated with the intensification of the primary and secondary circulations within the BL. It requires that the radial pressure gradient increases with time, which, in turn, requires spin up of the tangential wind at the top of the BL by the first mechanism. This mechanism becomes progressively important in the eyewall region as the vortex intensifies. During the intensification, although absolute angular momentum is not conserved in the BL, the largest wind speeds anywhere in the vortex can be achieved in the boundary layer. This occurs when the radial inflow is sufficiently large enough to move air parcels close to the vortex centre without a large loss of absolute angular momentum. This mechanism is coupled to the first one through BL dynamics because the radial pressure gradient of the BL is determined by the overall vortex flow above the BL as discussed in section 2.6 of Montgomery and Smith (2011). This spin-up mechanism explains why the maximum azimuthally-averaged tangential wind speeds in the model simulations (e.g., Smith et al., 2009) are located near the top of the boundary layer.

The major caveat of using other metrics such as those proposed in the foregoing discussion for comparing the sensitivity experiments is that it requires more observational information about the azimuthally averaged structure of real tropical cyclones (e.g., see Rogers et al., 2011, personal communication, also see http://www.ofcm.gov/ihc10/Presentations/Session08/s08-03HWRFx_update_Marks_v3.ppt) to help discriminate which of the sensitivity experiments is more realistic. Although general structural information about tropical cyclones have been reasonably well documented now (e.g., Houze, 2010), information about the structure of individual tropical cyclones are not as readily available as estimates of Vmax and Pmin. However, information on historical tropical cyclones events derived from aircraft observations is available at AOML/HRD to provide the needed structural evaluation of the HWRFX model simulations in future studies.

2.3 Results

In this section, we focus on the results from the sensitivity experiments for the idealized tropical cyclone intensification scenario. The results of the sensitivity experiments are compared in terms of time series of the minimum mean sea-level pressure (Pmin) at the centre of the cyclone, the spatially local maximum 10-m wind speeds (Vmax) around the centre and the PWR from the 3-km grid. The structures of the simulated tropical cyclone from the sensitivity experiments are compared also in terms of the axisymmetric mean primary and secondary circulations created from hourly model output for the entire period of the simulation.

2.3.1 Sensitivity to Boundary Layer Mixing, Subgrid Convection and Radiation (Experiments 1-5)

During intensification, the enthalpy from the sea surface driven by thermal disequilibrium across the air-sea interface is transported upward and consumed by the tropical cyclone. Such transport is accomplished primarily by turbulence- and convection-induced vertical mixing processes in the BL. In view of these, the first set of five sensitivity experiments (Experiments 1-5 in Table 1) are used to reveal the dependency of the HWRFX-simulated tropical cyclone intensification on the choice of BL mixing, subgrid convection and radiation parameterization schemes. Figure 1 depicts the time series of Pmin and Vmax from these experiments. The time series indicates the relative sensitivity of the simulated intensification to variation in the BL scheme (Experiments 1 and 2), the subgrid convective scheme (Experiments 1 and 5, 2 and 3) and the radiation scheme (Experiments 4 and 5).

Figure 1 shows that Experiments 1 and 2, which differ only in BL schemes, have similar development trends in the spin-up stage of the first 36 h. However, after 36 h, the simulation with the GFS scheme persistently produces a greater Vmax than the simulation using the MYJ scheme. Although the differences between the two schemes are not only in the formulations of the vertical diffusion but also in the details of how the surface momentum and enthalpy flux are calculated, a separate study (Bao et al., 2012b) has shown that the differences in the intensity are more associated with the differences in the surface flux calculation, while the differences in the structure are more associated with the differences in the vertical diffusion. Figure 1 shows also that the differences in the subgrid convection scheme produce more differences in the simulated tropical cyclone intensification than the differences in the BL schemes (see the differences between Experiments 1 and 5 versus Experiments 1 and 2, or Experiments 2 and 3 versus Experiments 1 and 2). Also, when the BMJ convective parameterization scheme is used, the tropical cyclone develops much more slowly than when the SAS scheme is used, although both simulations reach similar intensities towards the end of the simulations. The differences made by the two different radiation schemes (Experiments 4 and 5) are the smallest among the five experiments. It is interesting to note

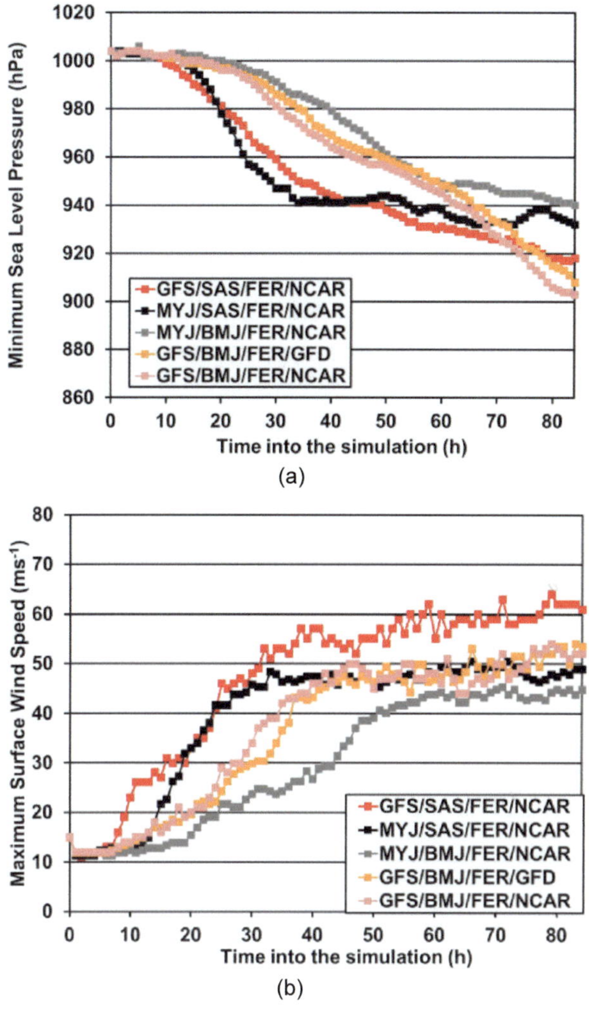

Fig. 1. Top panel (a) is the minimum sea level pressure (hPa). The bottom (b) is the maximum surface wind speed (ms^{-1}). The red lines are Experiment 1 (GFS/SAS/FER/NCAR), the black lines are Experiment 2 (MYJ/SAS/FER/NCAR), the gray lines are Experiment 3 (MYJ/BMJ/FER/NCAR), the orange lines are Experiment 4 GFS/BMJ/FER/GFD and the pink lines are Experiment 5 (GFS/BMJ/FER/NCAR).

in Fig. 1 that due to the differences in the inner-core size of the simulated tropical cyclone (defined as RMW), the intensity looks more similar in terms of Pmin than Vmax.

It is important to note that the two different BL schemes produce a significant difference in the asymptotic behaviour of Pmin. Figure 1 shows that Vmax from all five experiments levels off and becomes quasi-steady after 60 h into the simulation. While Pmin from all the experiments continues to decrease after 60 h, the decrease slows down in the two experiments using the

MYJ BL scheme, but remains almost the same in all the experiments using the GFS BL scheme. It is expected that for a steady environment such as the one prescribed in all the experiments herein, the simulated cyclone should eventually reach a quasi-steady state in which the inner-core size and intensity of the simulated cyclone undulate with small amplitudes[2] and the overall intensification of the simulated cyclone ceases.

Since the major difference in the asymptotic behaviour of Pmin shown in Fig. 1 occurs with the different BL schemes, this would suggest that the BL scheme plays an important role in determining how the simulated cyclone reaches the quasi-steady state. It is obvious that the experiments using the GFS BL scheme takes longer to approach a quasi-steady state than those using the MYJ BL scheme. The implication of this particular result to the evaluation of hurricane prediction models is believed to be non-trivial since it is unknown in theory how long it takes for a weak initial vortex, such as the one used in this study, to reach the quasi-steady state corresponding to the prescribed environment. It is unknown also what primary structural characteristics (e.g., spatial distributions of wind and thermal properties) the quasi-steady cyclone should have. These considerations suggest that other metrics are necessary to evaluate the behaviour of the HWRF model with changes in physics, particularly when Pmin continues to decrease with time while Vmax becomes quasi-steady.

Figure 2 compares the five experiments in terms of the PWR. Until Vmax >55 ms^{-1} and Pmin <940 hPa, Experiment 1 produces a correspondence in the trend of Vmax and Pmin that is close to the statistical mean obtained by Knaff and Zehr (2007) and within the spread of uncertainty (see more discussion below). The two experiments with a different subgrid convection scheme (Experiment 5) and a different radiation scheme (Experiment 4) than the one used in Experiment 1 produce similar correspondences in the trend of Vmax and Pmin until Vmax >50 ms^{-1} and Pmin <950 hPa. The use of the MYJ BL scheme (Experiments 2 and 3) leads to a significantly different trend of PWR than Experiment 1. Overall, the MYJ scheme tends to produce a smaller Vmax for a given Pmin than the GFS scheme. More data points from the experiments using the MYJ scheme for Vmax greater than 40 ms^{-1} and Pmin lower than 940 hPa are clustered together than those from the experiments using the GFS BL scheme. This clustering is a manifestation of the fact that the simulated cyclone in the experiments using the MYJ scheme approaches the quasi-steady state faster than that from the experiments using the GFS BL scheme.

---

[2] Such undulation is associated with replenishment cycles of the primary eyewall in which the inner-core structure of the simulated vortex undergoes rapid changes due to the development of outer rainbands and their ensuing coalescence into the existing primary eyewall, and/or the formation of a secondary outer eyewall that sequentially undergoes contraction and replaces the primary eyewall.

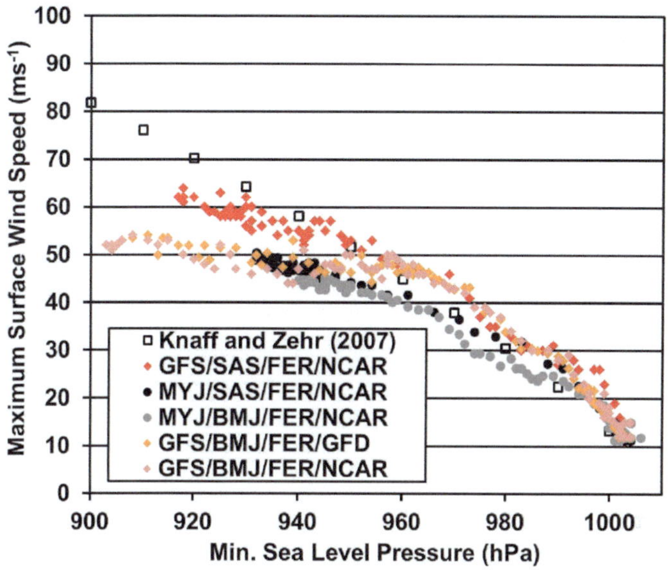

Fig. 2. Minimum sea level pressure (hPa) versus maximum surface wind speed (ms^{-1}). The red filled diamonds are Experiment 1 (GFS/SAS/FER/NCAR), the black filled circles are Experiment 2 (MYJ/SAS/FER/NCAR), the gray filled circles are Experiment 3 (MYJ/BMJ/FER/NCAR), the orange filled diamonds are Experiment 4 (GFS/BMJ/FER/GFD), the pink filled diamonds are Experiment 5 (GFS/BMJ/FER/NCAR) and the open squares are from Knaff and Zehr (2007). The open red triangles (Experiment 1, GFS/SAS/FER/NCAR) and the open black circles (Experiment 2, MYJ/SAS/FER/NCAR) are the minimum sea level pressure (hPa) versus the surface maximum azimuthally averaged tangential wind speed.

It should be pointed out that caution needs to be applied when comparing the model-simulated PWRs with the statistical mean, such as the one obtained by Knaff and Zehr (2007). Firstly, the datasets used in the derivation of the PWRs have uncertainty. Knaff and Zehr (2007) and Holland (2008) pointed out that the datasets used to produce the various statistical mean PWRs have a large scatter. For example, careful examination of Fig. A1 in Knaff and Zehr (2007) and Fig. 7 in Holland (2008) reveals that, for Vmax >40 ms^{-1}, the scatter about the mean is as large as 10 ms^{-1}. It is seen from our Fig. 2 that the largest differences in the Vmax are caused by the differences in the BL parameterization schemes, which are within the magnitude of about 10 ms^{-1} and comparable with the uncertainty in the datasets used in the derivation of the statistical PWRs. Secondly, there is a discrepancy between the model output sampling and observational data sampling to define the Vmax. The Vmax from the model is defined as instantaneous maximum 10-m wind at the beginning of each hour of the simulation. While by definition the observed Vmax is the maximum 10-m, 1-min sustained wind, there is a quite large disparity in data sampling with different instrumentation techniques. Thirdly, the gradient-wind balance is applied as a guidance in the derivation of the statistical mean PWRs (see more detailed discussion of this in Knaff and Zehr, 2007); however, as discussed earlier, such a balance is not valid in the BL near the surface where the mean radial inflow is no longer negligible in mature tropical cyclones as is required in the gradient balance approximation (Smith and Montgomery, 2008). In fact, when the surface maximum azimuthally-averaged tangential winds are compared with the statistical mean PWR in Fig. 2, the differences between the experiments with the GFS (open red triangles in Fig. 2) and MYJ (the open black circles in Fig. 2) BL schemes are much smaller than those shown by the Vmax. This strongly suggests that any PWR-based metric for comparing results from physical processes sensitivity experiments is inadequate to reveal physically meaningful information in a conclusive manner[3]. It suggests also that the Pmin and Vmax are less dynamically coupled for Vmax >50 ms^{-1} than for smaller Vmax. Therefore, it is necessary to use other metrics to compare the sensitivity experiments. In fact, we will demonstrate below that structural metrics such as Hovmöller diagrams of the maximum azimuthally averaged tangential wind speed are better than the Vmax and Pmin to reveal the differences in the sensitivity experiments.

Figure 3 shows the Hovmöller diagrams of the azimuthally-averaged instantaneous tangential wind speed at 1 km above the model sea surface and at hourly outputs from 0 to 84 h into the simulations. The structure at 1 km was chosen to minimize the effects of friction. The most prominent sensitivity feature shown in Fig. 3 is the differences in the azimuthally-averaged maximum tangential wind speed and the radius of the 35 ms^{-1} contour. These results indicate also that the structure of the HWRFX-simulated tropical cyclone is more sensitive to changes in either the BL scheme or the subgrid convection scheme than the radiation scheme. Particularly, the MYJ BL scheme produces a more compact simulated cyclone with stronger 1-km tangential winds than the GFS BL scheme (note that this is opposite to what is seen with the surface winds), and the BMJ subgrid convection scheme produces slower intensification than the SAS convection scheme. The overall size of the simulated cyclone characterized by 35 ms^{-1} radii in all the experiments increases with time, but the rate of increase is more sensitive to the choice

---

[3] It is worth mentioning that although the gradient wind balance was used as a guidance to derive the statistical PWR, the actual statistical PWR is not completely constrained by the gradient wind balance due to the statistical regression. Consequently, the statistical PWR may represent some of the overall dynamical relation between the mass and wind field in TC vortices above the BL that is different from the gradient balance approximation. Such dynamical relation may include other information beyond Vmax and Pmin, such as the vortex size and the intensity tendency as a function of environment and geographical location. While it is still unclear what this additional information entails physically and mathematically, to certain degree, the statistical PWR may be regarded as the best fit to the "observed" actual relation between the hurricane mass and wind field. Therefore, it is still useful to use it for model evaluation.

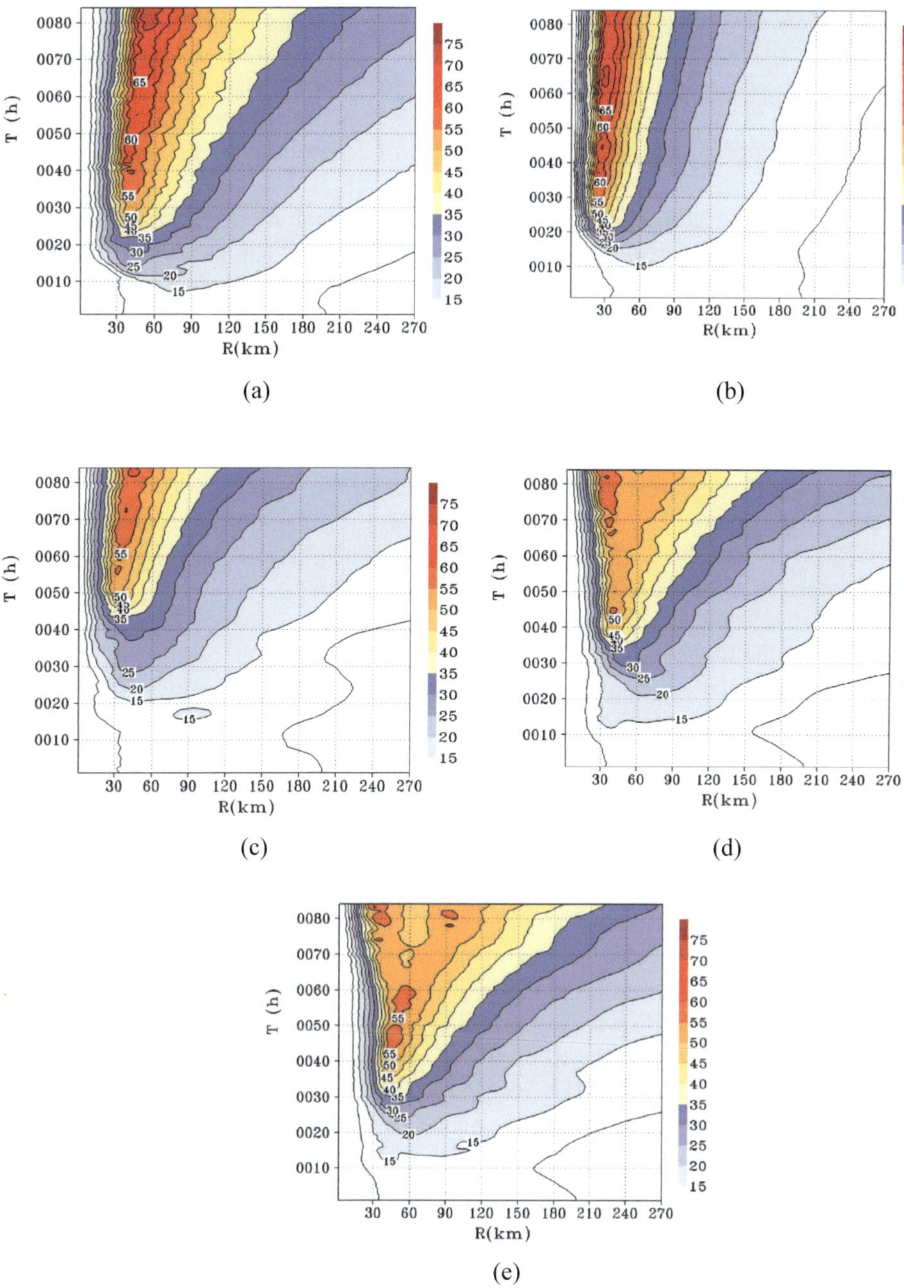

Fig. 3. Hovmöller diagrams of the azimuthally-averaged tangential wind speed (ms^{-1}) for Experiment 1 (GFS/SAS/FER/NCAR) (a), Experiment 2 (MYJ/SAS/FER/NCAR) (b), Experiment 3 (MYJ/BMJ/FER/NCAR) (c), Experiment 4 (GFS/BMJ/FER/GFD) (d) and Experiment 5 (GFS/BMJ/FER/NCAR) (e) at 1 km above the surface. The shaded and contour intervals are 5 ms^{-1}.

of the BL scheme than either the subgrid convection or the radiation scheme. The MYJ BL scheme leads to a smaller inner-core size-increase with time than the GFS BL scheme. The reason for this difference is due to the fact that the MYJ BL scheme produces smaller vertical eddy diffusivity than the GFS BL scheme does, and the detailed causality will be explored in a subsequent paper.

Figure 3 is a good example to show the advantage of structure metrics such as the Hovmöller diagram in shedding light on the comparisons of the intensity in terms of Vmax and Pmin. In fact, Fig. 3 suggests a close connection between the inner-core size increase and the departure of the simulated Vmax-Pmin relation from the statistical one for Pmin <940 hPa. This example also illustrates one of the problems in tropical cyclone model evaluation using only Vmax to represent the overall intensity of the simulated tropical cyclone. Despite these Hovmöller diagrams showing that the cyclone intensity in terms of the maximum wind speed at 1 km above the model sea level in the experiment using the MYJ BL scheme is greater than that in the run using the GFS BL scheme, the corresponding Vmax shown in Fig. 1 (defined as the local peak 10-m wind around the centre of the cyclone) is smaller.

Figure 4 depicts the azimuthally and 12-h averaged radius-height cross sections of the tangential wind contours superimposed on the vectors of the secondary circulation, contours of the radial wind speed, and the material acceleration in the tangential direction, with the friction effect included for all five experiments. The time averaging is done over 60-72 h, representing the quasi-steady stage of the simulated Vmax, and is required in order to better represent the quasi-steady state by reducing the small temporal undulations in the simulated structure. The intensity of the tangential circulation following the area enclosed by the 30 ms^{-1} and 50 ms^{-1} red contours, as well as the tangential material acceleration, varies with the choices of the BL. The MYJ BL scheme results in a greater maximum tangential wind speed and material acceleration in the eyewall region than the GFS BL scheme (compare Fig. 4a with Fig. 4b and Fig. 4c with Fig. 4e), but smaller areas enclosed by the 30 ms^{-1} and 50 ms^{-1}. That is, the MYJ BL scheme tends to produce a stronger, but smaller vortex than the GFS BL scheme. Additionally, the radius of the maximum tangential wind speed above the low-level inflow is smaller in the experiments with the MYJ BL scheme than in the experiments with the GFS BL scheme. However, the near-surface tangential winds in the experiments with the MYJ BL scheme are weaker than in the experiments with the GFS BL scheme, which is consistent with what is shown in Fig. 1 where the 10-m winds are weaker with the MYJ BL than with the GFS BL scheme. This result clearly indicates that the MYJ BL scheme produces a greater vertical shear of tangential winds than the GFS BL scheme. Furthermore, the combination of the MYJ BL scheme and the BMJ subgrid convection scheme produces the weakest near-surface tangential winds because this combination produces an overall weaker storm, which is also consistent with Fig. 1. On the basis of

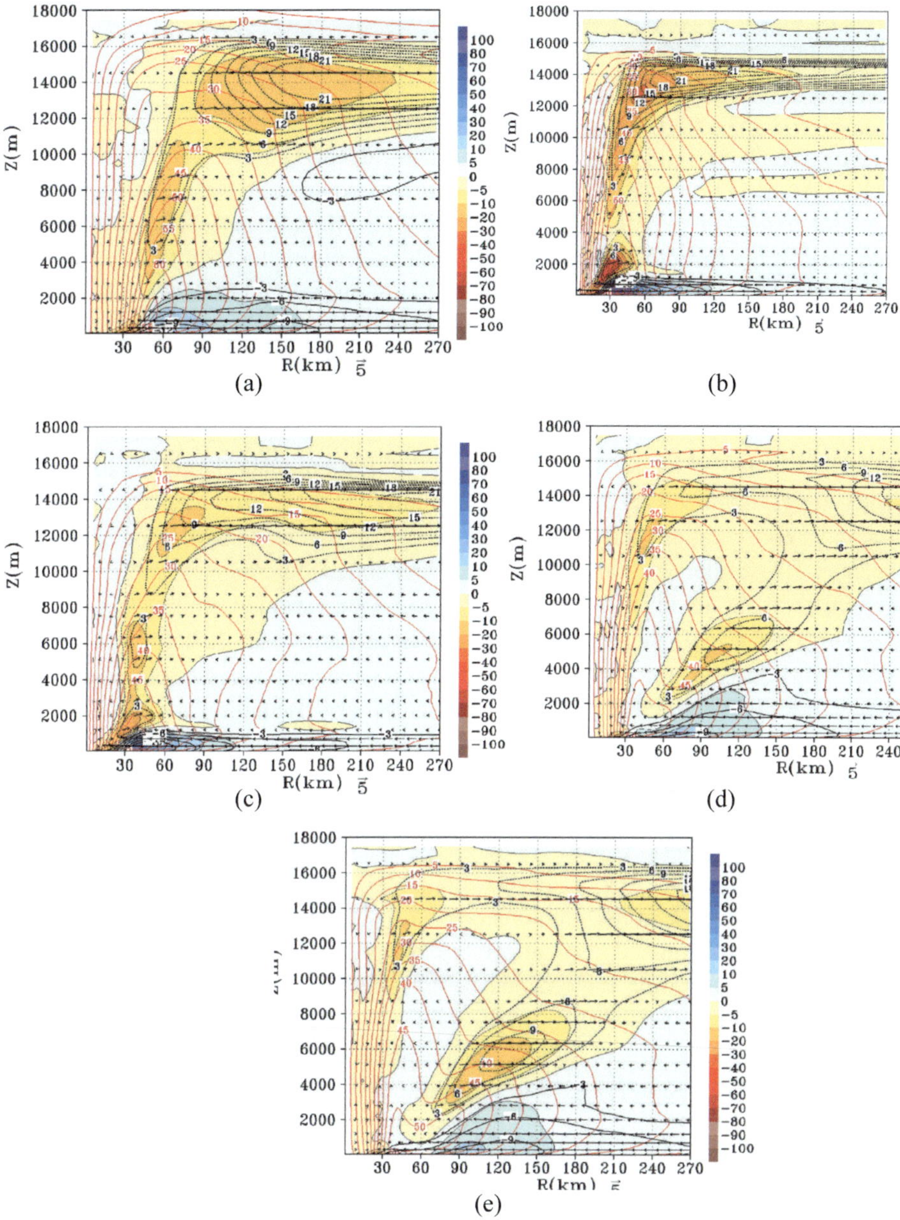

Fig. 4. Azimuthally and 60-72 h averaged radius-height cross sections of the tangential wind contours (in red colour) superposed on top of the secondary circulation vectors, radial wind speed (black contours, ms^{-1}), and the net tangential forcing (frictional effect included) in units of ms^{-1} h^{-1} (colour shaded) related to the primary circulation term in equation 2 for Experiment 1 (GFS/SAS/FER/NCAR) (a), Experiment 2 (MYJ/SAS/FER/NCAR) (b), Experiment 3 (MYJ/BMJ/FER/NCAR) (c), Experiment 4 (GFS/BMJ/FER/GFD) (d), and Experiment 5 (GFS/BMJ/FER/NCAR) (e). The positive contribution towards the spin-up process is indicated by the blue end of the spectrum.

these results, it is evident that the largest differences in the tangential winds in all five experiments are, to a great degree, restricted to the inner core.

As for the sensitivity of the simulated secondary circulation, Fig. 4 indicates that when the SAS convection scheme is used, the GFS BL scheme produces (Experiments 1, 4 and 5) a deeper inflow layer than the MYJ scheme (Experiments 2 and 3). The return flow above the inflow in the eyewall region in the simulation with the GFS BL scheme is also weaker than that in the simulation with MYJ BL scheme. While the GFS BL scheme produces a stronger and deeper upper-level outflow than the MYJ BL scheme (Experiments 1 and 2), the MYJ BL scheme produces a narrower eyewall indicating stronger vertical motion above the low-level inflow than the GFS BL scheme. Comparing Experiment 1 with Experiment 5 indicates that the SAS subgrid convection scheme leads to a more vertical eyewall than the BMJ scheme, while the comparisons of Experiments 1 and 5, along with Experiments 2 and 3, show that the BMJ scheme produces a weaker secondary circulation corresponding to a slower intensification process as seen in Hovmöller diagrams. It should be pointed out that the sensitivity shown thus far in Figs 3-4 reflects the fact that various combinations of the subgrid and BL mixing schemes lead to different 3-D diabatic heating distributions associated with the formation and replacement cycles of primary circulations, and the interaction between the primary circulation and the outer rainbands. The physical mechanism connecting the variation in 3-D diabatic heating distribution and variation in the simulated TC intensity and structure has been revealed and documented in a series studies by Wang (2009), Hill and Lackmann (2009), Xu and Wang (2010a and b), and Fudeyasu and Wang (2011).

Despite the foregoing demonstrated sensitivity of the intensity and structure of the simulated tropical cyclone to various BL schemes, it remains a question to determine the reasons behind the differences. There are two possible factors contributing to the differences. Firstly, it should be kept in mind that the differences in the BL schemes are not only in the formulations of the vertical diffusion but also in the parameterizations used in the surface momentum and enthalpy flux calculations. Secondly, the overall dynamical response of the numerical model varies with different realizations of sub-grid turbulent mixing in the model, particularly in the BL inflow layer above the surface (see. e.g., Kepert, 2012; Gopalakrishnan et al., 2012). In fact, it is demonstrated in a separate study (Bao et al., 2012b) that the differences in the intensity as shown above are more associated with the differences in the surface flux calculation, while the differences in the structure are more associated with the differences in the vertical diffusion. It is also shown in this separate study that the use of horizontal diffusion and the divergence damping term in the dynamical solver of the model, which possesses a degree of numerical artifact, contributes significantly to the sensitivity of the model solution to various BL schemes, particularly in terms of the inner-core size and the simulated PWR when Vmax becomes quasi-steady.

The sensitivity of the model solution to various physical representations can be seen also in the simulated inflow in the mid-to-upper troposphere between 4 km and 8 km in Fig. 4. The sensitivity to radiation scheme may be due to the cloud-radiation feedback to the total diabatic heating distribution as is discussed in Fovell et al. (2010). It appears from Fig. 4 that the mid-to-upper tropospheric weak inflow is more sensitive to the BL schemes (Experiments 1 and 2) and the subgrid convection schemes (Experiments 1 and 5) than the radiation schemes (Experiments 4 and 5). When the MYJ BL scheme is used, the sensitivity to the subgrid convection schemes is not as great as when the GFS BL scheme is used. Several axisymmetric model studies (e.g., Ooyama, 1982; Willoughby, 1979; Yamasaki, 1977) have illustrated the importance of this mid-level inflow layer to the slow evolution of a tropical cyclone vortex. Ooyama (1982) pointed out that the deep-layer inflow is, in essence, all that is needed for intensification of cyclonic rotation because at that level surface friction is not affecting the evolution process. Willoughby (1979) pointed out that the development of this weak mid-level inflow may be related to inner core warming. Recently, Fudeyasu and Wang (2011) pointed out that the mid-level inflow often develops in response to diabatic heating in outer rainbands and affects the size of the simulated TC. In the perspective of the two-mechanism spin-up processes (Montgomery and Smith, 2011), this inflow may enhance the convergence of angular momentum above the BL. However, despite the sensitivity of the mid-to-upper tropospheric weak inflow to various BL schemes, the time series of Vmax and Pmin are apparently most sensitive to the differences in the subgrid convection schemes.

It is shown (Figs 4b and 4c) that the strong BL inflow associated with the MYJ BL scheme results in strong tangential acceleration near the surface, despite the effects of friction that acts to oppose the acceleration in the tangential direction. It is also worth pointing out that the MYJ BL scheme produces the strongest tangential winds in the inner core of the cyclone right above the maximum inflow, regardless of what subgrid convection scheme is used. Moreover, although the MYJ BL scheme generates a smaller radius of the maximum tangential winds than the GFS BL scheme, the intensity of the cyclone measured by the Vmax shown in Fig. 1, defined as local maximum 10-m wind, indicates that the GFS BL scheme tends to produce a stronger mature cyclone. These results, on the one hand, are consistent with previous intensity-forecast studies using numerical weather prediction models (see, for example, Braun and Tao, 2000), which show a strong sensitivity of the model forecast to BL parameterization schemes. On the other hand, these results strongly suggest that in addition to the Vmax and Pmin that are conventionally used for model evaluation, observations of tropical cyclone structure (in terms of parameters such as the radii of the gale-force wind speed and the hurricane-force wind), together with the knowledge of the radius of the maximum tangential winds, should be used in model evaluation and validation.

2.3.2 Sensitivity to the Surface Drag (Experiment 6)

This section focuses on the sensitivity of the idealized tropical cyclone intensification to changes in the surface drag coefficient. Some of previous theoretical and numerical studies of the sensitivity to the surface exchange coefficients in axisymmetric models found that the intensity decreases markedly with increasing drag coefficient (e.g., Emanuel, 1995; Craig and Gray, 1996). In contrast, whereas a vortex intensifies in axisymmetric model when there is no surface drag (see, e.g., Craig and Gray, 1996), Montgomery et al. (2010) showed that no vortex intensification occurs in a three-dimensional model with zero surface drag (despite persistent sea-to-air fluxes of moisture to maintain deep convective activity). They showed also that the intensification rate and maximum intensity of the three-dimensional vortex increases with increasing surface drag coefficient until a certain threshold value is attained and then the intensification rate decreases.

In both the GFS and the MYJ BL scheme, the drag coefficient exhibits a steady increase with wind speed following the use of the Charnock formulation in the determination of the surface roughness from the momentum flux. Unfortunately, this wind-speed dependency has not been corroborated by any observations for the marine boundary layer in the extreme wind conditions of tropical cyclones. Recent observations of Powell et al. (2003), Donelan et al. (2004), French et al. (2007), and Black et al. (2007) show that the drag coefficient exhibits a steady increase to a wind speed of approximately 30 ms^{-1} and then levels off for higher winds. To see how this behaviour of the drag coefficient affects the cyclone intensification, Experiment 6 is carried out in which a realistic wind-dependency of the drag coefficient is specified in the GFS BL scheme. In this experiment, since the exchange coefficient required for the determination of the surface sensible and latent heat fluxes is dependent on the drag coefficient, the impact the modified drag coefficient on the surface sensible and latent heat fluxes is accounted for automatically in the surface flux calculation in the GFS BL scheme and consequently the heat exchange coefficient levels off at about 30 ms^{-1}.

Time series of Pmin and Vmax from the experiment are shown in Fig. 5 in comparison with those from Experiment 1. In general, the modified drag coefficient produces a very similar intensification rate of the simulated tropical cyclone in terms of Vmax to the original one in the GFS BL scheme when Vmax exceeds 30 ms^{-1}. Although the decreasing rate of Pmin with the modified drag coefficient is smaller in the experiment than that in Experiment 1, the PWR from the experiment depicted in Fig. 6 exhibits a significantly different trend than Experiment 1. It shows that the realistic wind-dependency of the modified drag coefficient improves the PWR, indicating strongly that the surface drag coefficient is a key parameter in controlling the PWR in the

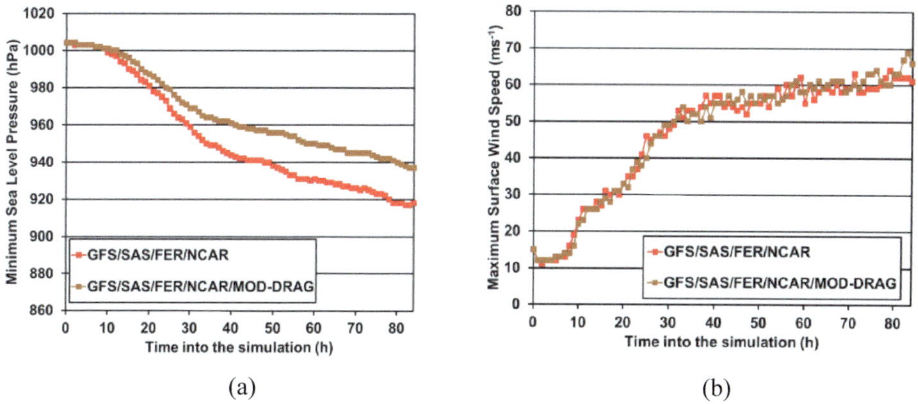

Fig. 5. Left panel (a) is the minimum sea level pressure (hPa). The right (b) is the maximum surface wind speed (ms^{-1}). The red lines are Experiment 1 (GFS/SAS/FER/NCAR) and the brown lines are Experiment 6 (GFS/SAS/FER/NCAR/MOD-DRAG).

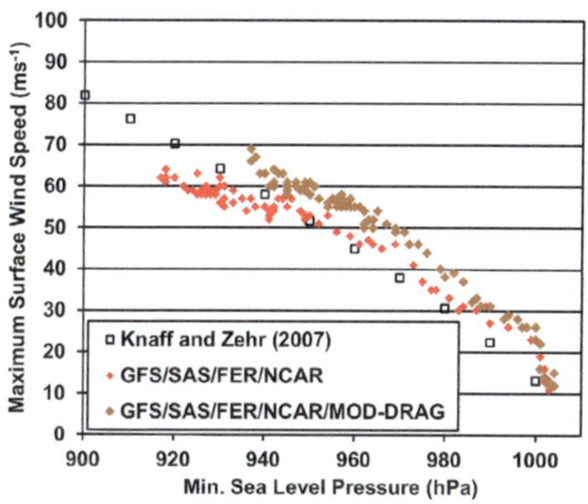

Fig. 6. Minimum sea level pressure versus maximum surface wind speed. The red filled diamonds are Experiment 1 (GFS/SAS/FER/NCAR), the brown filled diamonds are Experiment 6 (GFS/SAS/FER/NCAR/MOD-DRAG), and the open squares are from Knaff and Zehr (2007).

HWRFX model when the GFS BL scheme is used. Overall, the differences made by the change in the drag coefficient are within the uncertainty of about 10 ms^{-1} in the Knaff and Zehr (2007) PWR, but the slope of the PWR is greatly improved for Pmin <970 mb. This result is consistent with the role that the surface friction plays in disrupting gradient-wind balance and the conservation of angular momentum in the BL and, thus, is strongly related to the fact that smaller surface friction (associated with a more realistic surface drag coefficient in our case) lessens the disruption of the gradient-wind balance in the BL.

The use of the realistic wind-dependency of the surface exchange coefficients also makes noticeable differences in the structure of the cyclone that are shown in the Hovmöller diagrams of the azimuthally-averaged tangential wind speed at 1 km above the surface for the experiment (Fig. 7). Comparing Fig. 7 with Fig. 3a, the prominent differences made by the realistic wind-dependency of the surface exchange coefficients are in the structure of the eyewall that is characterized by the RMW and the 35 ms^{-1} contour. Overall, when the surface drag coefficient is modified to follow the observed wind-dependency, the tangential winds within the eyewall become weaker and the reduced winds are accompanied by a decrease in the agradient force associated with surface friction during spin-up. Figure 8 shows the azimuthally-averaged radius-height cross section of the tangential wind contours superimposed on the vectors of the secondary circulation from Experiment 6 (average between 60 and 72 h); shown also are the radial wind speed and the net forcing in the tangential direction. Comparing this cross section with those for Experiment 1 shown in Fig. 4a, it is seen that the modified surface drag coefficient results in a decrease in the low-level acceleration of the tangential wind in the eyewall region above the boundary layer inflow, as well as a decrease of the intensity and depth of boundary layer inflow in the eyewall region. It can be seen also that as the surface drag increases the RMW and the slope of the eyewall near the sea surface increase.

As discussed earlier, the spin up of the primary circulation is caused by the radial convergence of absolute angular momentum both above and in the (frictional) boundary layer. In the former case, the absolute angular momentum is approximately conserved materially, while in the latter it is reduced as air parcels spiral inwards in the boundary layer on account of surface drag. Although it is recognized that the aggregate heating produced by the convection in the inner core is the primary forcing responsible for the radial convergence of absolute angular momentum above the boundary layer inflow, it does not seem possible for us to provide a simple link between the surface heat fluxes and the local cloud buoyancy necessary to support

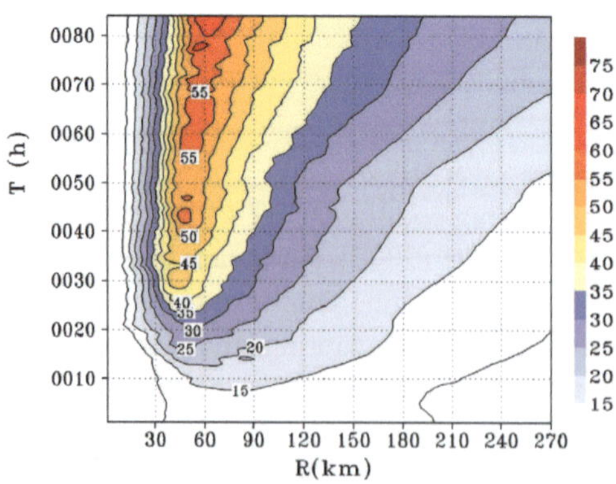

Fig. 7. Hovmöller diagrams of the azimuthally-averaged tangential wind speed at 1 km above the surface for Experiment 6 (GFS/SAS/FER/NCAR/MOD-DRAG). The shaded and contour intervals are 5 ms^{-1}.

a deep overturning circulation. Nevertheless, as a step towards a more complete understanding, the significant change in the maximum tangential wind speed shown in Fig. 8 (~10 ms^{-1}) between the surface-drag and the control experiments is found to be associated with a reduction in the radial convergence of angular momentum (both in and above the boundary layer) due to the decreased convection in terms of grid-resolved vertical flux of water vapour in the eyewall region (not shown). The decreased convection is related in part to the decrease in the surface enthalpy flux associated with the decreased surface drag in the eyewall region.

2.3.3 Sensitivity to the microphysics and sub-grid convection schemes (Experiments 7-11)

It has long been accepted that the primary energy source for the development and maintenance of a tropical cyclone is the latent heat that is transferred from the sea surface through the turbulent flux of water vapour and later released by condensation in convective clouds (Montgomery and Smith, 2011 and refs. therein). The structural evolution of the simulated tropical cyclone is also found to be sensitive to the details in the treatment of microphysical processes (see, e.g., Lord et al., 1984 for the effects of cloud ice and Smith et al., 2009 for the effects of warm rain process). As there has been a trend for tropical cyclone modellers to run numerical weather prediction models with higher spatial resolution using cloud microphysical processes for describing convective clouds (either combining or dispensing with subgrid convection schemes), an immediate question to address is how sensitive the model-simulated intensification and structural evolution are to the differences in the details of cloud microphysical processes and subgrid convection. Therefore, five more experiments (Experiments 7-11) are carried out and analyzed to explore this question.

Figure 9 shows the time series of Pmin and Vmax from the five experiments (in comparison with those from Experiment 1). Inspecting Fig. 9 reveals that the intensity evolution of the simulated tropical cyclone is sensitive to the choice of the microphysical process. In particular, the use of various microphysical processes results in noticeable differences of both Pmin and Vmax, but smaller than that for BL or drag in the intensification rate of the simulated cyclone during the first 60 h of model integration. That is, the sensitivity shown in Fig. 9 is not as widespread as the other sensitivities shown previously. It is worth noting that exclusion of the subgrid convection scheme on the inner nest (Experiment 11) reduces the intensification rate. This result is consistent with the finding reported by Zhu and Smith (2002) that a gestation period before the intensification is required to bring the boundary layer airflow near saturation, and the use of the parameterization scheme for sub-grid convection helps shorten the gestation period before the intensification. As seen below, the exclusion of the SAS subgrid scheme on the inner nest also affects the structure of the simulated TC.

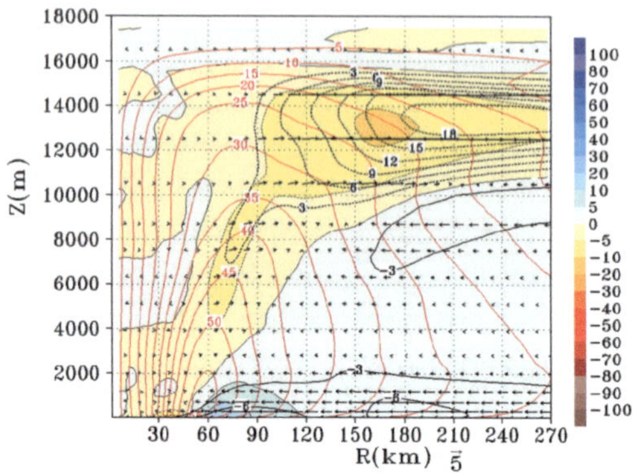

Fig. 8. Azimuthally and 60-72 h averaged radius-height cross section of the tangential wind contours (in red colour) superposed on top of the vectors of the secondary circulation, radial wind speed (black contours, ms^{-1}), and the net tangential forcing (frictional effect included) in units of ms^{-1} h^{-1} (colour shaded) related to the primary circulation term in equation 2 for Experiment 6 (GFS/SAS/FER/NCAR/MOD-DRAG). The positive contribution towards the spin-up process is indicated by the blue end of the spectrum.

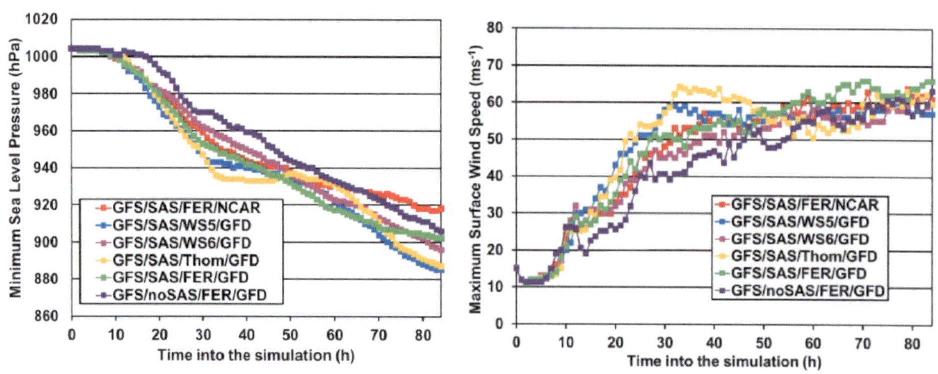

(a) (b)

Fig. 9. Left panel (a) is the Pmin (hPa). The right (b) is the Vmax (ms^{-1}). The red lines are Experiment 1 (GFS/SAS/FER/NCAR), the blue lines are Experiment 7 (GFS/SAS/WS5/GFD), the magenta lines are Experiment 8 (GFS/SAS/WS6/GFD), the dark-yellow lines are for Experiment 9 (GFS/SAS/Thom/GFD), the green lines are for Experiment 10 (GFS/SAS/FER/GFD) and the purple lines are for Experiment 11 (GFS/noSAS/FER/GFD).

The PWR from Experiments 7-10 appears (Fig. 10) to be more variable than those shown for the sensitivities to the BL mixing and the surface drag coefficient, particularly when the surface winds exceed 40 ms^{-1}. However, the scatter is comparable with the uncertainty in the mean PWR. These results indicate also that the variability of the simulated cyclone intensification to variations in the parameterized microphysical processes is comparable to the variability of the simulated intensification to the variations in either the BL

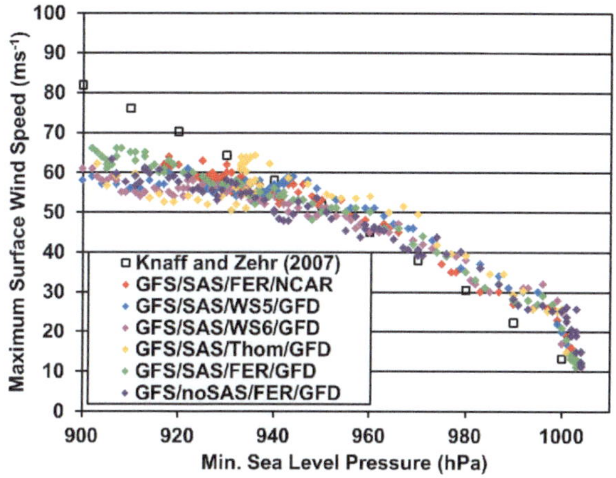

Fig. 10. Minimum sea level pressure versus maximum surface wind speed. The red filled diamonds are Experiment 1 (GFS/SAS/FER/NCAR), the blue filled diamonds are Experiment 7 (GFS/SAS/WS5/GFD), the magenta filled diamonds are Experiment 8 (GFS/SAS/WS6/GFD), the dark-yellow filled diamonds are for Experiment 9 (GFS/SAS/Thom/GFD), the green filled diamonds are for Experiment 10 (GFS/SAS/FER/GFD), the purple filled diamonds are for Experiment 11 (GFS/noSAS/FER/GFD), and the open squares are from Knaff and Zehr (2007).

scheme or the surface drag coefficient. There are two quantitative aspects in the sensitivity results. First, different treatments of microphysical processes in the conversion of various hydrometeor species lead to different intensification during the first 50 h of the model integration: the differences can be as big as, e.g. 15 ms^{-1} between the WSM5 and WSM6 scheme at 30 h into the simulation. Second, the differences in the fluctuation of Vmax after the simulated vortex reaches the quasi-steady state (from 50 h on into the simulation) can be as big as 10 ms^{-1}. However, like the other experiments, the Vmax from these experiments are all smaller than that shown in Knaff and Zehr (2007) for the Pmin <940 hPa, but greater for the Pmin >950 hPa. However, the scatter shown in the model simulated PWRs have the similar magnitude to that of the uncertainty in the datasets used to obtain the Knaff and Zehr (2007) PWR (about 10 ms^{-1}). Therefore, there is again a need to use other metrics in order to more precisely discriminate the sensitivities.

Figure 11 shows the Hovmöller diagrams of the azimuthally-averaged tangential wind speed at 1 km above the surface hourly from 0 to 84 h for the five experiments. It appears that in addition to the overall intensity, the variation in the microphysical process significantly affects the structural characteristics of the simulated cyclone. The differences in the characteristics of azimuthally-averaged tangential wind are noticeable and, particularly when comparing with Fig. 3a, the structural and intensity evolution of the tangential winds as characterized by the RMW and the 35 ms^{-1} contour shows noticeable sensitivity to the choice of the microphysical process

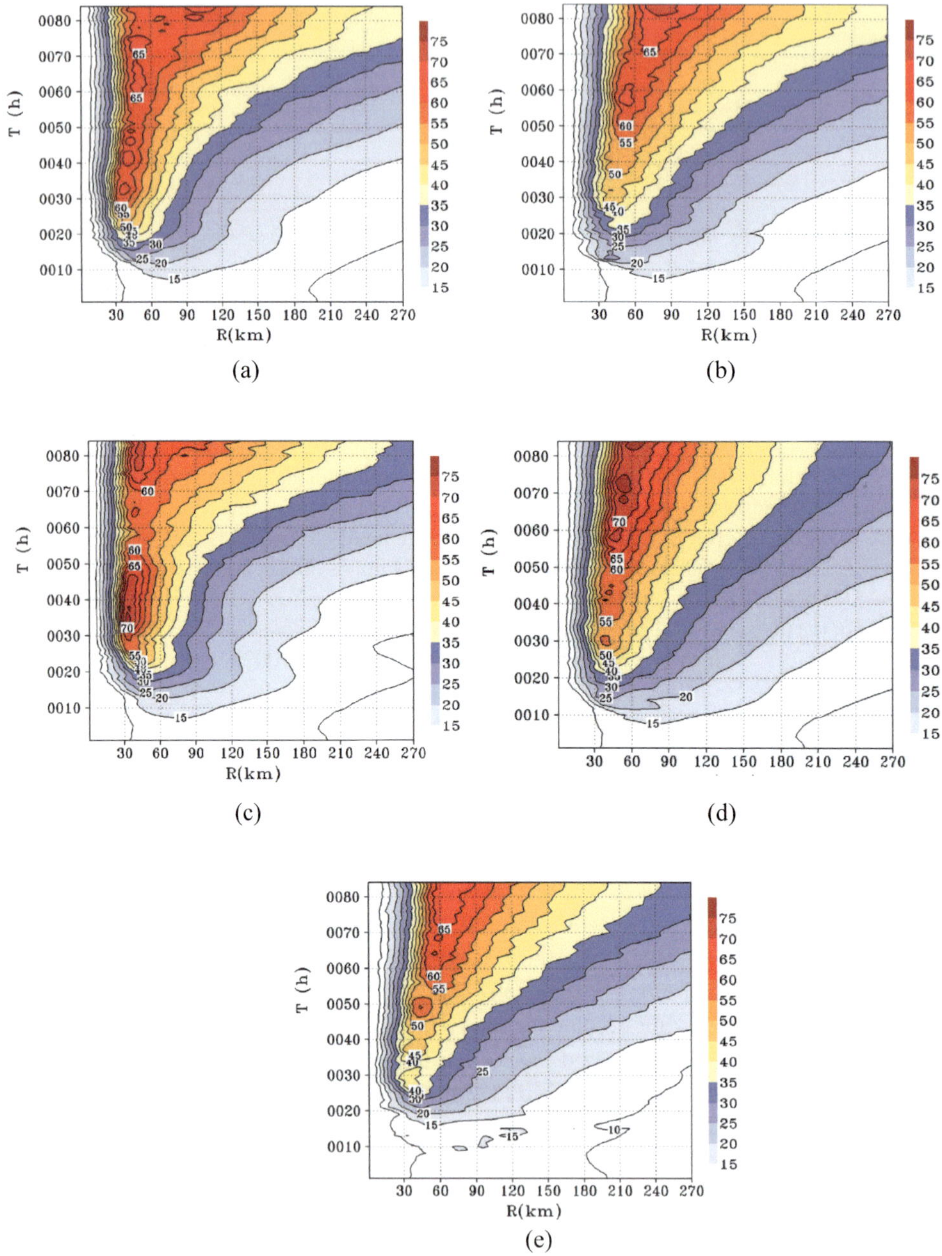

Fig. 11. Hovmöller diagrams of the azimuthally-averaged tangential wind speed at 1 km above the surface for Experiment 7 (GFS/SAS/WS5/GFD) (a), Experiment 8 (GFS/SAS/WS6/GFD) (b), Experiment 9 (GFS/SAS/Thom/GFD) (c), Experiment 10 (GFS/SAS/FER/GFD) (d), and Experiment 11 (GFS/noSAS/FER/GFD) (e). The shaded and contour intervals are 5 ms^{-1}.

used. Since the differences in the structural evolution shown among these experiments are comparable to those discussed between Experiment 4 and Experiment 5 (Fig. 3), one can conclude that the influence of the variation in the radiation scheme on the simulated cyclone development depends on the choice of the microphysical process. This can be explained by the fact that different microphysical processes produce different vertical distributions of hydrometeors in clouds, leading to variation in the simulated interaction of clouds and radiation, as elaborated on by Fovell et al. (2010).

Figure 12 shows the 60-72 h averaged radius-height cross sections of the tangential wind superimposed on the vectors of the secondary circulation, the radial wind speed, and the net forcing in the tangential direction. Overall, different microphysical processes produce noticeable differences in the structures of tangential wind speed in terms of the shape of the area enclosed by the 30 m s^{-1} contour in spite of rather small differences in Vmax. Also, the use of different microphysical processes appears to affect the radius of the maximum winds near the surface, possibly due to the cloud-radiation feedback effect on the vortex structure as revealed by Fovell et al. (2010). Comparing all five experiments, it is clearly seen that, relative to the Ferrier microphysical process, the other microphysical processes produce smaller tangential acceleration in the low-level inflow and more deceleration above the inflow in the eyewall region, resulting in different characteristics of the low-level inflow and the outflow in the eyewall region. It appears that the variations in the microphysical process have a stronger impact on the secondary circulation than the variations in either the BL scheme or the surface drag coefficient. The impact of the microphysics appears to be less on Vmax than on the structure.

The sensitivity of the simulated tropical cyclone development to variation in the microphysical process can also be interpreted in terms of the role of diabatic heating in the development and maintenance of the secondary circulation. It is seen in Fig. 13 that even though all the experiments produce a thermal structure qualitatively similar to the well-recognized characteristics of a warm core vortex (see, e.g., Houze, 2010), there are great differences in the details of the distribution of θ_e, relative humidity and moisture flux when different microphysical processes are used. Particularly, the distributions of latent heating as indicated by the 98% relative humidity and 7 g m^{-2} s^{-1} contours of moisture fluxes that serve as the proxy of the maximum diabatic heating are drastically different among the five experiments. Careful examination of the spin-up of the simulated vortex (not shown) indicates that the aforementioned differences in the diabatic heating distribution are due to the differences in the evolution of spiral rainbands outside the eyewall among the five experiments. Wang (2009) and Xu and Wang (2010a and b) revealed that the diabatic heating outside the eyewall strongly influences the overall intensity and structure of the simulated TC.

It is quite interesting to note that there are more differences in Experiments 7-11 in terms of the structure than in terms of the Vmax, Pmin and PWR.

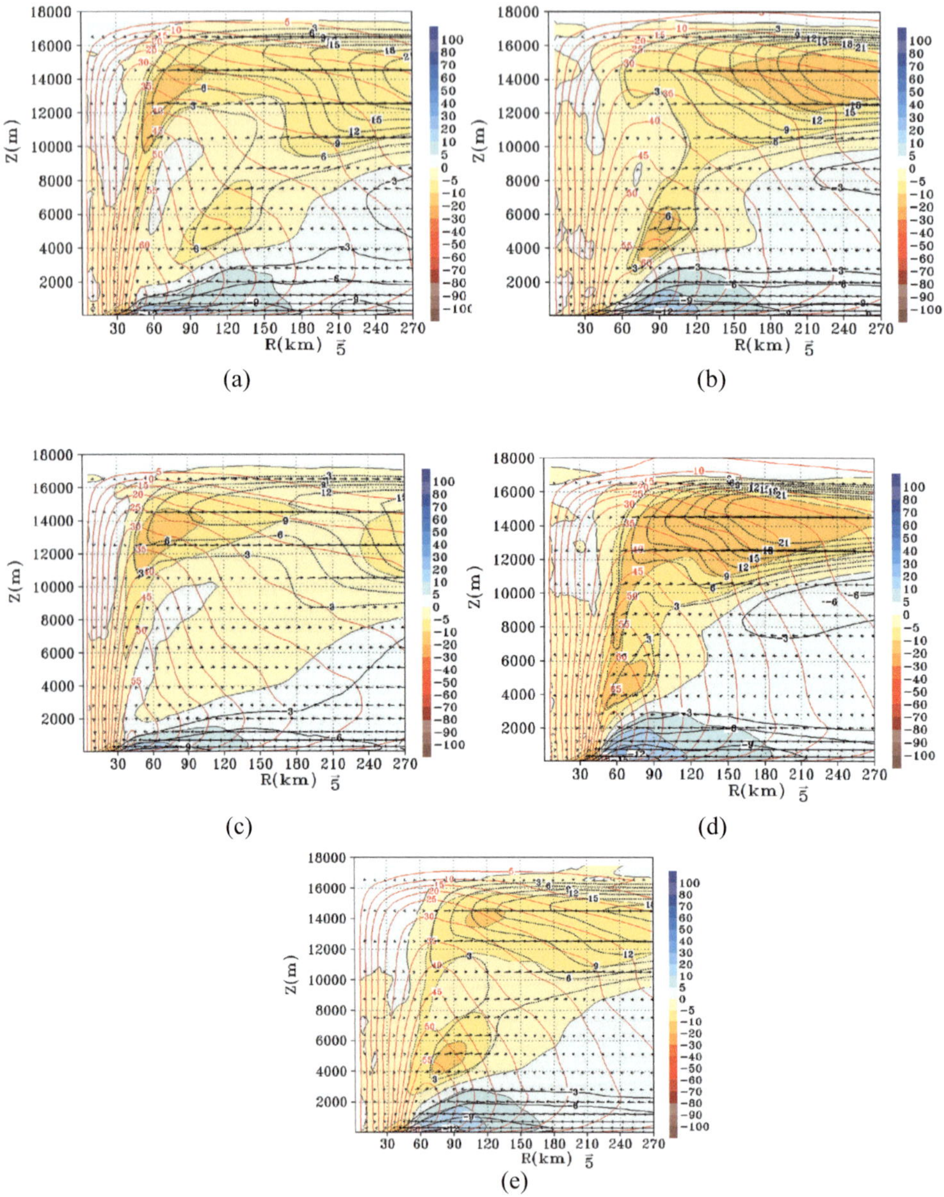

Fig. 12. Azimuthally and 60-72 h averaged radius-height cross sections of the tangential wind contours (in red colour) superposed on top of the vectors of the secondary circulation, radial wind speed (black contours, ms^{-1}), and the net tangential forcing (frictional effect included) in units of ms^{-1} h^{-1} (colour shaded) related to the primary circulation term in equation 2 for Experiment 7 (GFS/SAS/WS5/GFD) (a), Experiment 8 (GFS/SAS/WS6/GFD) (b), Experiment 9 (GFS/SAS/Thom/GFD) (c), Experiment 10 (GFS/SAS/FER/GFD) (d), and Experiment 11 (GFS/noSAS/FER/GFD) (e). The positive contribution towards the spin-up process is indicated by the blue end of the spectrum.

This can be possibly interpreted by factors affecting the amplitudes of low-wavenumber eyewall asymmetries and the areal coverage of the surface sensible and latent heat fluxes. As shown by Fierro et al. (2009) for example, smaller amplitude of low-wavenumber eyewall asymmetries and larger thermal radial gradients are favourable for tropical cyclone intensification in terms of the Vmax and Pmin, while smaller surface fluxes limit the intensification. The smaller differences among Experiments 7-11 in terms of Vmax, Pmin and the PWR suggest strongly that the differences in the microphysical processes must have produced compensating differences in the structural properties and the surface fluxes, causing the change in the Vmax, Pmin and the PWR to be relatively smaller for different microphysical processes than the noticeable structural differences.

The drastic differences shown in Fig. 13 can also be understood dynamically as the differences in the response of a balanced axismmetric vortex to different horizontal and vertical forcing distributions. Diabatic heating distribution is the result of the interaction between cloud physical representation (including the subgrid component) and the dynamical model's response to diabatic heating realized by the cloud physical representation. Such a response can be described by diagnostic solutions of Eliassen's balanced vortex equations (see, e.g., Shapiro and Willoughby, 1982; Bui et al., 2009 and section 2 of Montgomery and Smith, 2011). The dynamic essence of the response is summarized as the following. A forcing above the surface, such as diabatic heating and turbulent mixing of momentum, induces a transverse circulation around the source. Two circulations will form around the forcing in which ascent occurs in the centre with return subsidence immediately outside; an inflow layer and an outflow layer will be established. Parcels leaving a forcing region are able to move horizontally or vertically depending on the inertial and static stabilities and baroclinicity. Strong inertial and static stabilities will constrain the circulation to near the vicinity of the forcing. Baroclinicity will tilt the transverse circulation as parcels tend to follow the sloping isentropic surfaces (as a response to a momentum forcing) or the angular momentum surfaces (as a response to heating). The feedback of cloud-radiation interaction and the surface forcing associated with the surface drag and buoyancy flux can further distort the circulations (see, e.g., Fovell et al., 2010; Smith et al., 2011). Thus, the sensitivity of the HWRFX-simulated tropical cyclone intensification simply reflects the dependence of the primary and secondary circulations to variations in the diabatic heating and its feedback with vertical moisture flux associated with different microphysical processes. Such dependence is dynamically intrinsic and has been revealed in TC simulations using models other than the HWRFX (see, e.g., Fovell and Su, 2007; Fovell et al., 2009), particularly in a series of studies of how the outer rainbands affect the simulated TC intensity and structure by Wang (2009), Hill and Lackmann (2009), Xu and Wang (2010a and b), and Fudeyasu and Wang (2011).

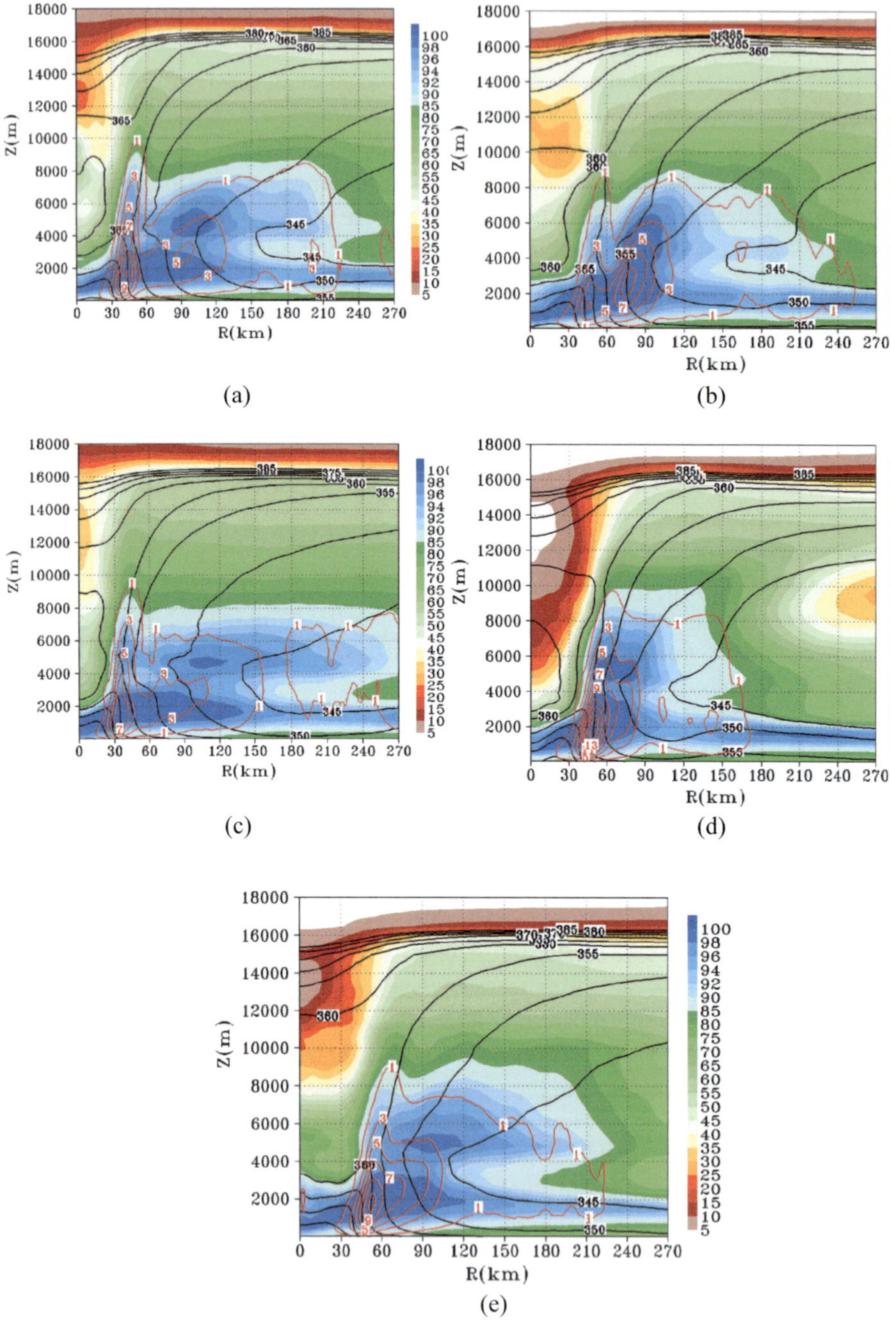

Fig. 13. The azimuthally and 60-72 h averaged radius-height cross section of, relative humidity (shaded colours), θ_e (black contours) and moisture fluxes (red contours) for Experiment 7 (GFS/SAS/WS5/GFD) (a), Experiment 8 (GFS/SAS/WS6/GFD) (b), Experiment 9 (GFS/SAS/Thom/GFD) (c), Experiment 10 (GFS/SAS/FER/GFD) (d), and Experiment 11 (GFS/noSAS/FER/GFD) (e).

3. Summary and Discussion

Three major results are obtained from the comparisons of the sensitivity experiments. First, different boundary layer parameterization of physical processes for vertical sub-grid turbulence mixing lead to differences not only in the intensity evolution in terms of the Vmax and Pmin, but also in the structural characteristics of the simulated tropical cyclone. Second, the surface drag coefficient is a key parameter that controls the PWR and the agradient force near the surface. Third, different microphysical and subgrid convection parameterization processes, due to different realization of diabatic heating distribution, lead to significant variations in the vortex structure. All of these results indicate that the current uncertainties in the BL mixing, surface drag, parameterization of microphysical processes have comparable impacts on the intensity and structure of the HWRFX model-simulated tropical cyclones.

These findings suggest that the differences of the sensitivity experiments measured in terms of the Vmax and Pmin, along with the corresponding PWR, are not as revealing as the structural metrics in terms of azimuthally-averaged tangential winds and the secondary circulation. The structural metric of the azimuthally-averaged radii of maximum tangential winds appears to be much more effective in highlighting the sensitivity of the model simulated tropical cyclone intensity to variation in parameterization of physical processes. In fact, structural metrics have long been used in the research community for research model verifications using reanalysis products along with measurements obtained from various observational instruments such as radars, satellites and dropsondes, while operational models have long been evaluated only in terms of Vmax and Pmin. These findings illustrate the drawback of using the operational metrics of Vmax and Pmin to evaluate the performance of the HWRFX model because they are not representative of the structure of the model-simulated cyclones. Practically speaking, the details of the model-simulated structure are important to forecasting coastal and inland flooding. While the Vmax and the Pmin are used widely as the essential parameters for verification of operational tropical cyclone forecasts, this study clearly demonstrates that other metrics are essential to evaluate dynamical impacts associated with the change in model physics. We suggest, therefore, that in order to effectively evaluate and provide useful recommendation for the HWRF model improvement, structural metrics should be used in addition to the operational metrics. These structural metrics should not be restricted to the ones in terms of azimuthal means we have used in this study, and they should enable the use of all available observations of TC structure.

The sensitivity results reported herein can be understood largely from the axisymmetric perspective of the two coexisting and mutually-dependent mechanisms of the idealized tropical cyclone spin-up process: the radial convergence of absolute angular momentum above the BL driven by the convection in the eyewall region; and the radial convergence of absolute angular momentum within the BL induced by the vortex intensification above the BL. That is, diabatic heating in the eyewall region induces convergence above the BL and intensifies the primary circulation due to the conservation of absolute angular momentum. As the axisymmetric vortex above the BL intensifies, the presence of surface friction induces radial inflow in the BL, which generally strengthens as parcels are accelerated down the radial pressure gradient towards the developing eyewall. Depending on the relative strength of the frictional torque and the generalized Coriolis force, the increasing radial inflow may generate a tangential wind that is stronger than that found above the boundary layer despite the loss of absolute angular momentum en route to the eyewall. The second spin up mechanism becomes progressively more important as the vortex develops and cannot be captured by axisymmetric balanced dynamics. From this perspective, the sensitivity of the HWRFX-simulated tropical cyclone intensification shown here simply reflects the dynamical dependence of the primary and secondary circulations to variations in the boundary layer and diabatic forcing associated with different parameterization of physical processes. Such a perspective points to the usefulness of using metrics beyond Vmax and Pmin for the evaluation of this and other operational models.

The availability of high performance computers at relative low cost makes it now possible to numerically forecast tropical cyclones in near real-time with complex physical parameterizations at horizontal grid resolution of the order of one kilometre. It is also tempting for models suitable for operational tropical cyclone prediction to be run at horizontal grid resolutions of the order of a few hundred metres for process studies. However, it remains a question as to how the quantitative aspects of the results from model sensitivity studies such as this one are affected as the model resolution increases. More importantly, whether or not the model solution with the current physics configuration will eventually converge as the model resolution continues increasing is still a subject of research. Nevertheless, the results from this study indicate that there is a need to develop a physically sound strategy to answer the question: given the availability of various choices of physics parameterizations, what is the "optimal combination" of the physics parameterizations in operational numerical models that should be used to forecast tropical cyclone intensity?

The current inability to evaluate the simulated structure of tropical cyclones using observations makes it difficult to determine the "optimal choice" of physics parameterizations. This difficulty has implications for the predictability of tropical cyclone intensification using current models. To tackle the difficulty, it would require the research and operational communities to expand the current metrics of model evaluation to include structural parameters that can be reliably derived from observations. Some early efforts could be focused on using available observations from the past operational reconnaissance measurement and field programmes such as the coupled boundary-layer air-sea transfer experiment CBLAST. These observations might enable one to identify what is needed to modify the current operational HWRF model physics package to yield physically sound structure as well as the intensity without degrading the track prediction. When it is possible to tailor future observations to better meet specifics for operational model physics package improvement, more rigorous assessments would be required to identify and overcome fundamental shortcomings of the operational HWRF model physics though combined observational and modelling studies.

Acknowledgements

The author acknowledges NOAA's Hurricane Forecast Improvement Project that supported this work. The author acknowledges the support provided by Indo-US Science and Technology Forum, and various national and international collaborators who have helped improving the operational HWRF model.

REFERENCES

Bao, J.-W., S.G. Gopalakrishnan, S.A. Michelson et al., 2012a: Impact of physics representations in the hwrfx model on simulated hurricane structure and wind-pressure relationships. *Mon. Wea. Rev.*, in press.

Bao, J.-W., S.G. Gopalakrishna, S.A. Michelson, F.D. Marks and M.T. Montgomery, 2012b: Impact of horizontal and vertical diffusion on idealized intensification in the HWRF Model. *Mon. Wea. Rev.*, in preparation.

Bui, H.H., R.K. Smith, M.T. Montgomery and J. Peng, 2009: Balanced and unbalanced aspects of tropical cyclone intensification. *Q. J. R. Meteor. Soc.*, **135**, 1715-1731.

Black, P.G. et al., 2007: Air–sea exchange in hurricanes: Synthesis of observations from the Coupled Boundary Layer Air–Sea Transfer Experiment. *Bull. Amer. Meteor. Soc.*, **88**, 357-374.

Braun. S.A. and W.-K. Tao, 2000: Sensitivity of high-resolution simulations of Hurricane Bob (1991) to planetary boundary layer parameterizations. *Mon. Wea. Rev.*, **128**, 3941-3961.

Brown, D.P., J.L. Beven, J.L. Franklin, E.S. Blake, 2010: Atlantic Hurricane Season of 2008. *Mon. Wea. Rev.*, **138**, 1975-2001.

Bryan, G.H. and R. Rotunno, 2009: The maximum intensity of tropical cyclones in axisymmetric numerical model simulations. *Mon. Wea. Rev.*, **137**, 1770-1789.

Charney, J.G. and A. Eliassen, 1964: On the growth of the hurricane depression. *J. Atmos. Sci.*, **21**, 68-75.

Craig, G.C. and S.L. Gray, 1996: CISK or WISHE as a mechanism for tropical cyclone intensification. *J. Atmos. Sci.*, **53**, 3528-3540.

Donelan, M.A., B.K. Haus, N. Reul, W.J. Plant, M. Stiassnie, H.C. Graber, O.B. Brown and E.S. Saltzman, 2004: On the limiting aerodynamic roughness of the ocean in very strong winds. *Geophys. Res. Lett.*, **31**, L18306, doi:10.1029/2004GL019460.

Emanuel, K.A., 1995: Sensitivity of tropical cyclones to surface exchange coefficients and a revised steady-state model incorporating eye dynamics. *J. Atmos. Sci.*, **52**, 3969-3976.

Fierro, A.O., R.F. Rogers, F.D. Marks and D.S. Nolan, 2009: The Impact of horizontal grid spacing on the microphysical and kinematic structures of strong tropical cyclones simulated with the WRF-ARW model. *Mon. Wea. Rev.*, **137**, 3717-3743.

Fovell, R.G. and H. Su, 2007: Impact of cloud microphysics on hurricane track forecasts. *Geophys. Res. Lett.*, **34**, L24810, doi:10.1029/2007GL031723.

Fovell, Robert G., Kristen L. Corbosiero and Hung-Chi Kuo, 2009: Cloud microphysics impact on hurricane track as revealed in idealized experiments. *J. Atmos. Sci.*, **66**, 1764-1778.

Fovell, R.G., K.L. Corbosiero, A. Seifert and K. N. Liou, 2010: Impact of cloud-radiative processes on hurricane track. *Geophys. Res. Lett.*, **37**, L07808, doi:10.1029/2010GL042691.

French, Jeffrey R., William M. Drennan, Jun A. Zhang and Peter G. Black, 2007: Turbulent Fluxes in the Hurricane Boundary Layer. Part I: Momentum Flux. *J. Atmos. Sci.*, **64**, 1089-1102.

Fudeyasu, H. and Y. Wang, 2011: Balanced contribution to the intensification of a tropical cyclone simulated in TCM4: Outer core spin-up process. *J. Atmos. Sci*, **68**, 430-449.

Gopalakrishnan, S.G. et al., 2010: The experimental HWRF system: A study on the influence of horizontal resolution on the structure and intensity changes in tropical cyclones using an idealized framework. *Mon. Wea Rev.*, **139**, 1762-1784.

Gopalakrishnan, S.G., F. Marks, J. Zhang, X. Zhang, J.-W. Bao and V. Tallapragada, 2011: A study of the impacts of vertical diffusion on the structure and intensity of the tropical cyclones using the high resolution HWRF system. *Mon. Wea. Rev.*, **139(6)**, 1762-1784.

Gray, W.M., E. Ruprecht and R. Phelps, 1975: Relative humidity in tropical weather systems. *Mon. Wea. Rev.*, **103**, 685-690.

Harper, B.A., 2002: Tropical cyclone parameter estimation in the Australian region: Wind–pressure relationships and related issues for engineering planning and design—A discussion paper. Systems Engineering Australia Party Ltd. (SEA) for Woodside Energy Ltd., SEA Rep. J0106-PR003E.

Hausman, S.A., 2001: Formulation and sensitivity analysis of a nonhydrostatic, axisymmetric tropical cyclone model. Atmospheric Science Paper 701, Dept. of Atmospheric Science, Colorado State University.

Hill, K.A. and G.M. Lackmann, 2009: Influence of environmental humidity on tropical cyclone size. *Mon. Wea. Rev.*, **137**, 3294-3315.

Holland, G., 2008: A revised hurricane pressure–wind model. *Mon. Wea. Rev.*, **136**, 3432-3445.

Houze, R.A., Jr., 2010: Clouds in tropical cyclones. *Mon. Wea. Rev.*, **138**, 293-344.

Janjic, Z.I., 1996: The surface layer in the NCEP Eta Model. The Eleventh Conference on Numerical Weather Prediction, Norfolk, VA, 19–23 August; Amer. Meteor. Soc., Boston, MA, 354-355.

Janjic, Z.I., 2000: Comments on "Development and Evaluation of a Convection Scheme for Use in Climate Models". *J. Atmos. Sci.*, **57**, 3686.

Janjic, Z.I., 2002: Nonsingular Implementation of the Mellor–Yamada Level 2.5 Scheme in the NCEP Meso model. NCEP Office Note, No. 437.

Jakob, C. and M. Miller, 2003: Parameterization of Physical Processes. Encyclopedia of Atmospheric Sciences. (Eds) J.R. Holton, J.A. Curry and J.A. Pyle, Vol. 1, 1692-1717, Academic Press, New York.

Kepert, J.D., 2011: Choosing a boundary-layer parameterization for tropical cyclone modeling. *Mon. Wea. Rev.* (early view online).

Knaff, J.A. and R.M. Zehr, 2007: Reexamination of tropical cyclone wind–pressure relationships. *Wea. Forecasting*, **22**, 71-88.

Koba, H., T. Hagiwara, S. Asano and S. Akashi, 1990: Relationships between CI number from Dvorak's technique and minimum sea level pressure or maximum wind speed of tropical cyclone. *J. Meteor. Res.*, **42**, 59-67.

Kossin, J.P. and C.S. Velden, 2004: A pronounced bias in tropical cyclone minimum sea-level pressure estimation based on the Dvorak technique. *Mon. Wea. Rev.*, **132**, 165-173.

Liu, Y., D.-L. Zhang and M.K. Yau, 1999: A multiscale numerical study of Hurricane Andrew (1992). Part II: Kinematics and inner-core structures. *Mon. Wea. Rev.*, **127**, 2597-2616.

Lord, S.J., H.E. Willoughby and J.M. Piotrowicz, 1984: Role of a parametrized ice-phase microphysics in an axisymmetric, non–hydrostatic tropical-cyclone model. *J. Atmos. Sci.*, **41**, 2836-2848.

Marks, F.D., R.A. Houze and J.F. Gamache, 1992: Dual-aircraft investigation of the inner core of the Hurricane Norbert. Part I: Kinematic structure. *J. Atmos. Sci.*, **49**, 919-942.

Montgomery, M.T., M.E. Nicholls, T.A. Cram and A.B. Saunders, 2006: A vertical hot tower route to tropical cyclogenesis. *J. Atmos. Sci.*, **63**, 355-386.

Montgomery, M.T., R.K. Smith and S.V. Nguyen, 2010: Sensitivity of tropical cyclone models to the surface exchange coefficients. *Quart. J. Roy Met. Soc.*, **136**, 1945-1953.

Montgomery, M.T. and R.K. Smith, 2014: Paradigms for tropical-cyclone intensification. *Aust. Met. Ocean. Soc. Journal*, **64**, 37-66.

Nguyen, S.V., R.K. Smith and M.T. Montgomery, 2008: Tropical cyclone intensification and predictability in three dimensions. *Q. J. R. Meteorol. Soc.*, **134**, 563-582.

Ooyama, K.V., 1982: Conceptual evolution of the theory and modeling of the tropical cyclone. *J. Meteor. Soc. Japan*, **60**, 369-380.

Pan, H.-L. and W.-S. Wu, 1995: Implementing a mass flux convective parameterization package for the NMC medium range forecast model, NMC office note, 409, available at: http: //www.emc.ncep.noaa.gov/officenotes/FullTOC.html#1990, 1995.

Pielke, R.A., 2001: Mesoscale Meteorological Modeling, Int. Geophys. Ser., vol. 78, 2nd ed., Elsevier.

Powell, M.D., P.J. Vickery and T.A. Reinhold, 2003: Reduced drag coefficient for high wind speeds in tropical cyclones. *Nature*, **422**, 279-283.

Schubert, H.V. and J.J. Hack, 1982: Inertial stability and tropical cyclone development. *J. Atmos. Sci.*, **39**, 1687-1697.

Shapiro, L.J. and H.E. Willoughby, 1982: The response of balanced hurricanes to local sources of heat and momentum. *J. Atmos. Sci.*, **39**, 378-394.

Skamarock, W.C., J.B. Klemp, J. Dudhia, D.O. Gill, D.M. Barker, W. Wang and J.G. Powers, 2008: A description of the Advanced Research WRF Version 3. NCAR Tech Notes-475+STR.

Smith, R.K. and S. Vogl, 2008: A simple model of the hurricane boundary layer revisited. *Q. J. R. Meteorol. Soc.*, **134**, 337-351.

Smith, R.K. and M.T. Montgomery, 2008: Balanced boundary layers in hurricane models. *Q. J. R. Meteorol. Soc.*, **134**, 1385-1395.

Smith, R.K., M.T. Montgomery and S.V. Nguyen, 2009: Tropical cyclone spin-up revisited. *Q. J. R. Meteorol. Soc.*, **135**, 1321-1335.

Smith, R.K. and G.L. Thomsen, 2010: Dependence of tropical cyclone intensification on the boundary layer representation in a numerical model. *Q. J. R. Meteor. Soc.*, **136**, 1671-1685.

Smith, R.K., M.T. Montgomery and G.L. Thomsen, 2011: Sensitivity of tropical cyclone models to the surface drag coefficient in different boundary-layer schemes. *Q. J. R. Meteorol. Soc.*, **140(680)**, 729-1128.

Tao, W.-K. and M.W. Moncrieff, 2003: Parameterization of Physical Processes. *In:* Encyclopedia of Atmospheric Sciences. (Eds) J.R. Holton, J.A. Curry and J.A. Pyle, Vol. 1, 539-548, Academic Press, New York.

Wang, Y., 1995: On an inverse balance equation in sigma-coordinates for model initialization. *Mon. Wea. Rev.*, **123**, 482-488.

Wang, Y., 2002: An explicit simulation of tropical cyclones with a triply nested movable mesh primitive equation model: TCM3. Part II: Model refinements and sensitivity to cloud microphysics parameterization. *Mon. Wea. Rev.*, **130**, 3022-3036.

Wang, Y., 2009: How do outer spiral rainbands affect tropical cyclone structure and intensity? *J. Atmos. Sci.*, **66**, 1250-1273.

Weatherford, C.L. and W.M. Gray, 1988: Typhoon structure as revealed by aircraft reconnaissance. Part II: Structural variability. *Mon. Wea. Rev.*, **116**, 1044-1056.

Willoughby, H.E., 1995: Mature structure and evolution. *In:* Global Perspectives on Tropical Cyclones, pp. 21-62. WMO/TDNo 693 (ed.) R.L. Elsberry, World Meteorological Organization, Geneva.

Xu, J. and Y. Wang, 2010a: Sensitivity of tropical cyclone inner-core size and intensity to the radial distribution of surface entropy flux. *J. Atmos. Sci.*, **67**, 1831-1852.

Xu, J. and Y. Wang, 2010b: Sensitivity of the simulated tropical cyclone inner-core size to the initial vortex size. *Mon. Wea. Rev.*, **138**, 4135-4157.

Willoughby, H.E., 1979: Forced secondary circulations in hurricanes. *J. Geophys. Res.*, **84**, 3173-3183.

Yamasaki, M., 1977: A preliminary experiment of the tropical cyclone without parameterizing the effects of cumulus convection. *J. Meteor. Soc. Japan*, **55**, 11-30.

Yanai, M., 1964: Formation of tropical cyclones. *Rev. Geophys.*, **2**, 367-414.

Zhang, D.-L., Y. Liu and M.K. Yau, 2001: A multiscale numerical study of Hurricane Andrew (1992). Part IV: Unbalanced flows. *Mon. Wea. Rev.*, **129**, 92-107.

Zhu, H. and R.K. Smith, 2002: The importance of three physical processes in a minimal three-dimensional tropical cyclone model. *J. Atmos. Sci.*, **59**, 1825-1840.

5

Sub-Grid Vertical Turbulent Mixing in the Atmospheric Boundary Layer

Jian-Wen Bao
NOAA/Earth System Research Laboratory, Boulder, Colorado
jian-wen.bao@noaa.gov

This chapter is the second of the three-lecture series on the physical processes and their descriptions in numerical models of tropical cyclones. There are two sections in this lecture. Section 1 is intended to provide an overview of the parameterization approaches used to account for sub-grid vertical turbulent mixing in the model-simulated atmospheric boundary layer, and Section 2 demonstrates an example of how one should carefully realize specific details of the parameterization in the implementation of the approaches. Materials presented in Section 1 are derived from Holtslag (2003), Pielke (2001), Garratt (1992) and Stull (1988), while those presented in Section 2 are based on Bao et al. (2008).

1. Parameterization of Sub-Grid Vertical Turbulent Mixing

1.1 Description of Sub-Grid Motions

Numerical atmospheric models for the forecasting and studying of tropical cyclones are based on integration of the basic equations governing atmospheric motions, which are the gas law, the equation of continuity, the first law of thermodynamics, the conservation equations for momentum, and equations expressing the conservation of moisture. Since these equations must be solved numerically on a finite grid mesh, there is a need to separate the sub-grid turbulence from resolved-scale motions. Let E denotes an atmospheric variable, such as specific humidity or a horizontal wind component. If $\overline{E}$ represents a grid-resolved value of E, an instantaneous value of E would differ from $\overline{E}$. Thus, we have

$$E = \overline{E} + e \tag{1}$$

In principle, the sub-grid fluctuations around the grid-resolved motion reflect all the turbulent motions on scales smaller than the smallest resolved one. To make the mathematical handling of e tractable, it must satisfy the so-called "Reynolds postulates". If the Reynolds postulates are fulfilled, then the averaging for the product of two variables provides

$$EF = \overline{EF} + \overline{ef} \tag{2}$$

The second term on the right-hand side of Eqn (2) is known as the sub-grid turbulent covariance. Similarly, the turbulence variance of a quantity is given by $\overline{E^2} - \overline{E}^2$ (which is the square of the standard deviation).

If in Eqn (2) the variable E represents one of the velocity components (U, V, W in the x, y, z direction, respectively), then $\overline{EF}$ is the grid resolved flux of F and the second term on the right-hand side of Eqn (2) represents a sub-grid turbulent flux of F. For example, $\overline{uc}$ and $\overline{wc}$ are the horizontal and vertical turbulent fluxes of the variable C, respectively. Here u and w are the sub-grid turbulent fluctuations of the horizontal and vertical velocities. Near the surface, the grid-resolved vertical wind is usually small, and thus the total vertical fluxes are normally dominated by the sub-grid turbulent contributions.

1.2 Grid-resolved Variables and Sub-Grid Turbulent Fluxes

Turbulence is a prominent feature in the atmosphere, particularly in the atmospheric boundary layer (ABL). Turbulence length scales vary typically from a few millimetres to a few hundreds of metres, which is smaller than the grid spacing in models of the atmosphere. However, these small-scale motions have a strong impact on the evolution of the mean flow and therefore a parameterization scheme is needed in models that simulate the flow in the atmosphere. A mathematical framework to describe the effect of turbulence on the mean flow can be obtained by separating variables into a mean and a fluctuating part: the so-called Reynolds decomposition. The resulting equations for the mean flow contain variances and covariances of fluctuating quantities that are unknown. This is known as the closure problem. Expressing the unknown terms into mean flow variables is called turbulence parameterization.

The challenge of accurately simulating the sub-grid turbulent flux in the ABL is the prediction of the temporal variation of the vertical and horizontal structures of the ABL, by the integration of the atmospheric governing equations, in response to the influence of the dynamical processes acting above the ABL and at the Earth's surface. As mentioned above, the variables in the governing equations are split into grid-resolved motions and sub-grid fluctuations as in Eqn (1). Inserting this into the basic governing equations and averaging over the grid-spacing provides a set of equations for the behaviour of the grid-resolved variables. It is these grid-resolved variables that are used explicitly as the prognostic variables in atmospheric models. This can be demonstrated as follows. The general character of any of the budget equations dealing with atmospheric motions is

$$\frac{DE}{Dt} = S_i. \tag{3a}$$

Here S_i represents the subsequent sources and sinks for the variable E (such as radiation or chemistry effects). The notation $\frac{D}{Dt}$ represents the total rate of change for the variable E by local changes $\left(\frac{\partial}{\partial t}\right)$, and changes transported with the fluid motion in the three directions. As such, we have

$$\frac{\partial E}{\partial t} + U\frac{\partial E}{\partial x} + V\frac{\partial E}{\partial y} + W\frac{\partial E}{\partial z} = S_i. \tag{3b}$$

Here U, V and W are the wind components in the three directions x, y and z, respectively. If each variable is split into the grid-resolved component and the sub-grid fluctuation, then Eqn (3b) provides—after Reynolds-averaging, some algebraic manipulations, and simplifying assumptions—a budget equation for the mean variable E. This reads as

$$\frac{\partial \bar{E}}{\partial t} + \bar{U}\frac{\partial \bar{E}}{\partial x} + \bar{V}\frac{\partial \bar{E}}{\partial y} + \bar{W}\frac{\partial \bar{E}}{\partial z} = \bar{S}_i - \frac{\partial \overline{ue}}{\partial x} - \frac{\partial \overline{ve}}{\partial y} - \frac{\partial \overline{we}}{\partial z}. \tag{4}$$

We may note that in the derivation of Eqn (4) single terms representing sub-grid fluctuations have disappeared (as above in Eqn (2)). However, terms involving the product of two sub-grid fluctuations do remain. Thus, because the basic governing equations are nonlinear, the budget equations for the grid-resolved variables contain terms involving sub-grid motions. Closure formulations or parameterizations are needed to introduce mixing by the subgrid motions into the equations for the larger-scale motions (as resolved by the model).

A special and simple form of Eqn (4) arises for horizontally homogeneous conditions. In such cases the terms including horizontal derivatives are negligible. If in addition the mean vertical wind is small, and if there are no other sources and sinks, then Eqn (4) becomes

$$\frac{\partial \bar{E}}{\partial t} = \frac{\partial \overline{we}}{\partial z}. \tag{5}$$

This equation is known as the one-dimensional, vertical diffusion equation. It shows that the local time rate of change for a grid-resolved variable (such as temperature or wind) at a certain height is given by the divergence of the turbulent (corresponding heat or momentum) flux in the vertical direction. Thus, information on the turbulent fluxes may produce a local forecast of the variation of a mean variable (but only under the aforementioned simplifications). However, normally the other terms in Eqn (4) are also relevant, in particular the terms with mean wind speed (the so-called 'advection terms'). This means that in general the budget equations for momentum, heat, and the various scalars are closely coupled in full-blown atmospheric model.

1.3 A Diagnostic Local Mixing Parameterization

To solve the budget Eqn (4) for all the grid-resolved variables involved in atmospheric models, the terms involving sub-grid turbulent fluxes need to be parameterized. This means that the sub-grid turbulent fluxes need to be expressed in terms of available grid-resolved model variables. Until this is achieved, the atmospheric model equations cannot be numerically integrated. For environmental and atmospheric models the parameterization most frequently used is known as first-order closure, or *K*-theory. In this theory it is assumed that the flux $\overline{we}$ of a variable *E* in the vertical direction Z is down the vertical gradient of the mean concentration of *E* per unit mass. Thus

$$\overline{we} = -K_e \frac{\partial E}{\partial z} \tag{6}$$

where K_e is the so-called "eddy diffusivity" or mixing coefficient for the variable *E*. Similarly, the horizontal fluxes can be represented in terms of horizontal gradients. Note that the corresponding eddy diffusivities typically are not constant but depend generally on properties of the flow and the variable of interest. This also means that normally no analytic solutions are possible, not even for the simple case in which Eqns (5) and (6) are combined.

We note that the dimension of an eddy diffusivity is a length scale *l* times a velocity scale. These are proportional to the products of effective eddy sizes and eddy velocities in the corresponding directions. Often a diagnostic expression is used for the eddy diffusivity, on the basis of what is called "mixing length theory" (in analogy with molecular diffusion). The result reads as

$$K_e = l^2 S\alpha(R_i), \tag{7}$$

where *S* is vertical wind shear (that is the variation of mean horizontal wind with height). *l* is mixing length. Various functions have been proposed in the literature for $\alpha(R_i)$ where R_i is the gradient Richardson number defined by

$$R_i = \frac{g}{\overline{\Theta_v}} \frac{\partial \overline{\Theta_v} / \partial z}{(\partial \overline{U} / \partial z)^2 + (\partial \overline{UV} / \partial z)^2} \tag{8}$$

where g is the acceleration due to gravity and $\overline{\Theta_v}$ the grid-resolved virtual potential temperature. The specification of the length scale l is not at all straightforward, except near the surface, where so-called "surface-layer similarity theory" (see Stull, 1988 and Garratt, 1992) provides that $l \propto z$. A frequently used form for l is:

$$\frac{1}{l} = \frac{1}{\kappa z} + \frac{1}{\lambda} \tag{9}$$

where λ is a turbulent length scale, which should be valid for the turbulence far above the surface, and κ is the so-called von Karman constant. We note that λ has a rather empirical nature and consequently there is no agreement on its specification in the literature. Equations (7) and (9) are diagnostic equations, which indicate that the eddy diffusivity varies with height, wind speed, stability, etc. In combination with the flux parameterization of Eqn (8), it follows that the flux at a certain height depends on the local gradient of the mean variable involved. Consequently the approach is referred to as a "diagnostic local mixing approach". Such an approach is mostly suitable for relatively homogeneous conditions with neutral and stable stratification, and is not so suitable for cases with convection (see nonlocal mixing parameterizations below).

1.4 Prognostic Mixing Parameterizations

A physically realistic alternative to the diagnostic approach is to relate the eddy diffusivity of Eqn (7) to the actual turbulent kinetic energy of the flow, by using the prognostic turbulent kinetic energy equation and an appropriate choice for the turbulent length scale. It is important to realize that the kinetic energy of atmospheric motion per unit of mass E is given by the half of the sum of the velocities squared in the three directions (as in classic mechanics), i.e.,

$$E = (U^2 + V^2 + W^2)/2.$$

As with respect to Eqn (2), we can separate between the mean kinetic energy E of the resolved atmospheric motions and the sub-grid turbulent kinetic energy (*TKE* or e) of the sub-grid fluctuating motions. Thus e is given by

$$e = (\overline{u^2} + \overline{v^2} + \overline{w^2})/2.$$

The prognostic equation for e reads in its basic form as:

$$\frac{De}{Dt} = -\overline{uw}\frac{\partial \bar{U}}{\partial z} - \overline{vw}\frac{\partial \bar{V}}{\partial z} + \frac{g}{\overline{\Theta_v}}\overline{v\theta_v} - \frac{\partial \overline{we}}{\partial z} + \delta - \varepsilon \tag{10}$$

where $\frac{De}{Dt}$ is the total variation of e with time (the sum of local variations and those transported with the mean air motion). The two terms on the immediate right-hand side of Eqn (9) represent the shear production of turbulence. These depend primarily on vertical variations of wind or, near the ground, on wind speed and surface roughness. The terms are almost always positive. The third term in Eqn (10) represents the rate of production or breakdown of turbulence by buoyancy effects (such as heat convection). It depends directly on density effects, which can be written in terms of the virtual potential temperature $\overline{\Theta_v}$, and its turbulent flux $\overline{w\theta_v}$. The term δ in Eqn (9) represents divergence and pressure redistribution terms. These have a tendency to cancel near the surface. Finally, the term ε reflects the molecular dissipation of turbulence into heat, and this term is always positive. In fact, ε is typically proportional to e/τ, where τ is the characteristic time scale for the turbulent mixing process. Using Eqn (10), turbulent kinetic energy can be calculated for given grid-resolved profiles when the corresponding sub-grid turbulent fluxes are calculated using Eqn (6) for all fluxes involved. In this approach the diffusivities are typically calculated with equations of the form

$$K_e = al\sqrt{e} \tag{11}$$

where a is a constant depending on the variable of interest. The length scale l is typically calculated with a similar type of diagnostic equation such as Eqn (9).

This approach is known as the "TKE-length scale approach" and it is an example of the so-called 1.5-order closure. Sometimes a prognostic equation is used for the length scale as well, but such an approach is more popular in engineering applications than in the atmospheric sciences. It can be shown that Eqn (7) is a solution of Eqns (10) and (11) in stationary conditions and when other simplifications are made such as the neglect of the influences by advection and turbulence divergence in the TKE equation. It should be cautioned that the 1.5-order closure scheme has a pathological behaviour associated with the singularity of the TKE equation and the specification of the length scale. The so-called nonsingular realization constraints need to be imposed in practical application (see, e.g., Helfand and Labraga, 1988 and the example shown in Part 2 below).

A more advanced prognostic mixing parameterization scheme is known as "second-order closure" approach. In such an approach, prognostic equations are developed for both the sub-grid fluxes and variances themselves. These equations, although possessing a structure very similar to that of Eqn (9) for TKE, have new unknown terms and must be related to the other variables in the model equations. Describing these terms always involves assumptions. As such, the second-order closure approach involves many more equations than the 1.5-order closure and is therefore computationally more time-consuming.

A note of caution is worth mentioning. One might expect that a model with the 1.5- or the second-order approach would produce more realistic results than a model with a first-order closure. However, in practice this is often not the case, because of complex model interactions and the difficulty of representing all the relevant details in sub-grid turbulent motions with sufficient accuracy. That is why diagnostic approaches still remain popular. Nevertheless, second-order equations are useful for gaining insight in the governing physics, and after simplification useful extensions of the basic parameterizations may be achieved.

1.5 Non-Local Mixing Parameterizations

We have so far discussed a popular approach for parameterizing sub-grid vertical mixing that is popularly applied in numerical weather prediction model. Other approaches can be found in the literature and have been widely used in models for weather, climate and air quality. As an example, we now discuss an approach that has been applied in global and regional models for parameterizing vertical sub-grid turbulent mixing in the ABL with strong atmospheric convection. Observations have indicated that in the convective ABL the turbulent flux of a conserved quantity is typically not proportional to the local gradient alone as predicted by Eqn (6). In fact, in a large part of the ABL, the grid-resolved gradients are small in conditions with dry convection, in particular for potential temperature. Therefore, the sub-grid vertical turbulent fluxes should depend mostly on the mixing characteristics of the large eddies across the ABL.

Existing theories can be found to deal with the parameterization of sub-grid mixing in the convective ABL. According to these theories, modified *K*-theory should be used to allow for the influence of convection, for example, by including additional terms at the right hand side of Eqn (6). The resulting formulation reads as

$$\overline{we} = K_e \frac{\partial E}{\partial z} + \overline{we_{nl}} \tag{12}$$

where $\overline{we_{nl}}$ is the non-local flux representing the influence of the sub-grid large-eddy mixing in the boundary layer with convection. The formulation of the latter is not so straightforward in the general case. In one possible realization of the approach, the eddy diffusivity K_e in Eqn (11) can be suitably specified with the 1.5-order approach using the full kinetic energy equation, and a length scale formulation, which may depend on the actual height and the depth h of the boundary layer. In the case of a clear, stationary boundary layer dominated by dry convection, it appears that the eddy diffusivity is also well described by a profile function (Holtslag and Moeng, 1991)

$$K_e = w_* h \left(\frac{z}{h}\right)^{4/3} \left(1 - \frac{z}{h}\right)^2 \tag{13}$$

where w_* is known as the convective velocity scale for the clear boundary layer and is defined as

$$w_* = \left(\frac{g}{\Theta_v} \overline{w\theta_{v0}} h\right)^{1/3} \tag{14}$$

where $\overline{w\theta_{v0}}$ is the surface sensible heat flux heating the boundary layer from below. For dry convection, the non-local flux correction term of Eqn (12) is typically proportional to the surface turbulent flux of the variable in question. That is,

$$\overline{we_{nl}} = \alpha(z / h)\, \overline{w\theta_{v0}} \tag{15}$$

where $\alpha(z/h)$ is a dimensionless function of z/h. Note that in practice the exponents of (z/h) and $(1 - z/h)$ are empirically determined to fit the vertical profile of K_e to observations. Information about the vertical profile of K_e from large-eddy simulations models is also often used to refine the values of the exponents (see, e.g., Holtslag and Moeng, 1991).

Other non-local flux parameterizations are also discussed in the literature, including mass flux approaches and 'transilient' turbulence closures. In a mass flux approach the turbulent layer is divided into updrafts and downdrafts, each with a typical concentration for the variable of interest. The updrafts and downdrafts can be defined on the basis of the sign of vertical velocity or more restrictive definitions (in the case of clouds). Using such an approach, the updraft and downdraft variables are connected to a number of equations, which can be closed when the horizontal exchange ('lateral entrainment') of mass and other variables between the up- and downdrafts is treated well. It is noted that the mass flux approach has been quite successful for representing cumulus and stratocumulus in atmospheric models (Emanuel, 1994).

1.6 Summary and Final Remarks

This lecture is intended to give a brief summary of the basic approaches for the modelling and parameterization of sub-grid turbulent mixing atmospheric models for tropical cyclone prediction. Emphasis has been given to the treatment of atmospheric turbulent mixing in models with a horizontal spatial grid distance of order 10 to100 km, a vertical grid distance of 10 to 100 m, and resolving time scales of order 10 min or larger. These approaches are popularly used in regional and global-scale models for the forecasting and study of weather, climate, and air quality. Some of the approaches can be adapted for turbulence above the boundary layer, such as turbulence in clouds or in elevated shear layers. However, important exceptions have been documented in the literature when so-called "mesoscale processes" have their influence – for instance, when extensive fields of stratocumulus are present, or in the case of deep cumulus clouds and thunderstorms. The atmospheric model equations can also be applied on much smaller spatial and temporal scales than discussed here, for instance by using vertical and horizontal grid spacing on orders of 10 to 100 m, and timesteps on order of a second or less. This type of modelling is nowadays known as "very high resolution simulation" and has become a powerful and popular tool in the last decade to help understand and parameterize sub-grid turbulent mixing in operational weather and climate models with vertical and horizontal grid spacing on orders of 1 to a few kilometres, and timesteps on order of a few seconds to one minute. It is important to realize that in the case of very high resolution simulations the simplifying assumptions leading to Eqn (2) are normally not valid. Much work still needs to be done before we have a full understanding of the complexity of atmospheric turbulence and before we have a more unified treatment of sub-grid turbulent mixing with respect to different horizontal grid spacings. It is hoped that a better understanding of atmospheric turbulence will also contribute to our capability of refining and unifying the parameterizations of sub-grid turbulent mixing in numerical weather prediction models, particularly for modelling of the ABL in response to the different types of surfaces and meteorological forcing above the ABL that are found in reality.

2. A Non-sigular Realization of the 1.5-Order TKE Scheme with Prognostic Mixing Length

One of the popular methods to simulate vertical turbulent mixing in the atmospheric boundary layer (ABL) is to use the Mellor-Yamada level 2.5 second-moment closure (MY closure) turbulence model (Mellor and Yamada, 1974, 1982), which is centered on the prognostic equation for the turbulent kinetic energy (TKE, denoted as $q^2/2$). An important closure parameter in the turbulence model is the so-called length scale (denoted as l). Although the length scale is essential to the level 2.5 second-moment closure, there is no unique approach to its specification. The simplest approach to specifying the length scale is to assume that l is proportional to the distance from the surface within the ABL (see, e.g., Blackadar, 1962; Mellor and Yamada, 1982; Janjić, 1994). While this method works well in many applications, it is diagnostic and its treatment of the length scale far above the surface has a rather empirical nature (Holtslag, 2003). Therefore, it is more desirable to use a prognostic equation with minimum empiricism to predict the length scale. A vertical turbulent mixing scheme of the MY closure uses a prognostic equation for specifying l is often called a two-equation scheme, in contrast with a one-equation scheme in which l is diagnosed.

Various approaches to deriving the prognostic equation for l have been proposed. Unlike the TKE equation that is derived from the Reynolds-averaged Navier-Stokes equations, the equation for l cannot be derived systematically from the governing laws of fluid motion. Essentially, it is cast as a transport equation for a quantity involving l that is analogous to the TKE equation with a different set of closure constants. In general, the equation for l can be written as a transport equation for a quantity involving q and l: $c^p q^m l^n$, where c is a constant, and p, m and n are integers (see, e.g., Burchard, 2002 and Kantha, 2004). It has been suggested that for a homogeneous shear layer all the approaches to deriving the equation for l are equivalent (Burchard, 2002; Umlauf and Burchard, 2003; Kantha, 2004), and thus no one approach is fundamentally superior to another. It should be noted, however, that the ABL is not a homogeneous shear layer, and the non-uniform thermal stratification in the ABL plays an important role in controlling vertical turbulent mixing. Therefore, various length-scale equations may perform differently for the ABL.

In this chapter, we present the implementation of a two-equation vertical turbulent mixing scheme in the National Center for Atmospheric Research and the Pennsylvania State University mesoscale model (MM5-V3). The prognostic equation for l in this scheme is formulated in terms of the turbulence transport of twice the TKE. This two-equation scheme was first proposed by Mellor and Yamada (1974, 1982) for general applications in geophysical flows, and later revised by Kantha and Clayson (1994) and Burchard (2001) to simulate the vertical turbulent mixing in the ocean. Although this scheme has been applied extensively in simulating oceanic mixed layer, this is the first effort to implement it in an atmospheric mesoscale model.

Three major issues are encountered during the implementation. The first issue is related to the closure constants in the length-scale equation. The values of the closure constants in the length-scale equation are determined *a priori* according to turbulence measurements or direct turbulence simulations. In oceanic applications, measurements of simple turbulent flows generated in laboratories are commonly used to determine these closure constants (see, e.g., Mellor and Yamada, 1982; Umlauf and Burchard, 2003). It has been recently recognized that there exist constraining relationships among the closure constants that can be utilized to determine the closure constants according to comprehensive turbulence measurements (see, e.g., Burchard, 2002). Although in many oceanic applications the closure constants determined from the laboratory measurements have appeared to work well, uncertainties remain in the constants' values and continuing efforts are still being made to refine them. It is also not clear whether the closure constants that have worked well in oceanic applications are well suitable for simulating vertical turbulent mixing in the ABL. Therefore, how to properly determine the closure constants that are suitable for applications in the ABL needs to be investigated.

The second issue is related how to ensure physically sensible performance of the scheme. Vertical turbulent mixing schemes based on the MY closure have been known to require the use of the so-called realizability constraints on the shear and stability parameters as well as on the length scale to perform reasonably and robustly (see, e.g., Mellor and Yamada, 1982; Galperin et al., 1988; Helfand and Labraga, 1988; Burchard, 2002). The equation for the length scale itself does not provide a guarantee that the predicted l is always in agreement with the realizability constraints such that the modelled turbulent mixing is physically meaningful for a given flow situation. Therefore, properly established constraints are required to limit the solution of the length-scale equation. Although various approaches to imposing the constraints on the length scale have been proposed for oceanic applications, they are not as theoretically appealing as that proposed by Janjić (2002). Janjić's approach is innovatively different than the others in that the constraints on the length scale are derived based on requiring that the analytical solution of the TKE equation be nonsingular, and thus there is no need to constrain the shear and stability parameters like the other approaches. However, this approach has not been applied to a two-equation scheme.

The third issue is concerned with numerical integration of the scheme. In oceanic applications, the scheme is discretized implicitly in time with centered in space differencing for numerical integration. It is found that the numerical solution of the equations for both the TKE and the length scale are apparently contaminated with computational modes. To make the results from the two-equation scheme physically meaningful, the computational modes need to be eliminated or significantly reduced. The numerical scheme proposed by Janjić (2002) is found to be effective to that end.

The rest of the chapter is organized as follows. The two-equation vertical turbulent mixing scheme is described in section 2.1. Section 2.2 presents results from the investigation on how to determine the closure constants in the length-scale equation. How to apply Janjić's approach to establishing the constraints on the length scale and to numerical integrating the two-equation scheme is presented in section 2.3.

2.1 The Vertical Turbulent Mixing Scheme

The vertical turbulent mixing scheme consists of two equations: the equation for the TKE and the equation for l. The TKE equation is written as

$$\frac{\partial\left(\frac{\eta^2}{2}\right)}{\partial t}-\left(\frac{\partial}{\partial z}\right)\left[l\,q\,S_q\left(\frac{\partial}{\partial z}\right)\left(\frac{\eta^2}{2}\right)\right]=P_s+P_b-\varepsilon, \tag{16}$$

where

$$\eta^2=2e=(\overline{u^2}+\overline{v^2}+\overline{w^2}),\ P_s=-\overline{wu}\left(\frac{\partial U}{\partial z}\right)-\overline{wv}\left(\frac{\partial V}{\partial z}\right)\ \text{and}$$
$$P_b=\beta g\,\overline{w\theta_v} \tag{17}$$

$$\varepsilon=\frac{\eta^3}{B_1 l} \tag{18}$$

$$-\overline{wu}=K_m\frac{\partial U}{\partial z},-\overline{wv}=K_m\frac{\partial V}{\partial z},-\overline{w\theta_v}=K_H\frac{\partial\Theta_v}{\partial z} \tag{19}$$

$$K_m=lqS_M,\ K_H=lqS_H, \tag{20}$$

$$S_M(6A_1A_2G_M)+S_H(1-3A_2B_2G_H-12A_1A_2G_H)=A_2 \tag{21a}$$

$$S_M(1+6A_1^2G_M-9A_1A_2G_H)-S_H(12A_1^2G_H+9A_1A_2G_H)$$
$$=A_2(1-3C_1) \tag{21b}$$

$$G_M = \left(\frac{l^2}{q^2}\right)\left[\left(\frac{\partial U}{\partial z}\right)^2 + \left(\frac{\partial V}{\partial z}\right)^2\right], G_H = -\left(\frac{l^2}{q^2}\right)\beta g \frac{\partial \Theta}{\partial z}, \tag{22}$$

where U, V and Θ_v are the velocity components and virtual potential temperature of the mean flow, and the rest of symbols are of the same meaning as in Mellor and Yamada (1982).

The success of (16)-(22) in the simulation of vertical turbulent mixing is critically dependent on well-calibrated closure constants (A_1, A_2, B_1, B_2, C_1, S_q) and an appropriate specification of the length scale. The values of the constants in (16)-(21) are chosen as those by Janjić (2002), i.e.,

$$(A_1, A_2, B_1, B_2, C_1, S_q \text{ and } \beta) = (0.65988838, 0.65742096, 11.877992, 7.226971, 0.00083092297, 0.2 \text{ and } 1.270)$$

Note that they are different than those proposed in Mellor and Yamada (1982) that are also used in Kantha and Clayson (1994). They are chosen here because they are used in the original vertical turbulent mixing scheme (hereafter referred to as the ETA scheme) in the mesoscale model in which this two-equation scheme is implemented. This original scheme, developed by Janjić (1990, 2002) for the National Centers for Environmental Prediction's (NCEP) operational limited area model, is the same as this two-equation scheme except that the length scale is diagnosed.

Following Mellor and Yamada (1982) and Kantha and Clayson (1994), the equation for l is expressed as a transport equation of $\eta^2 l$, i.e.,

$$\frac{\partial}{\partial t}(\eta^2 l) - \frac{\partial}{\partial z}\left[\eta l S_l \frac{\partial}{\partial z}(\eta^2 l)\right] = P_{sl} + P_{bl} - \varepsilon_l, \tag{23}$$

where $P_{sl} = lE_1P_s$, $P_{bl} = lE_2P_b$ and $\varepsilon_l = F\dfrac{\eta^3}{B_1 l}$.

In the above equation, E_1, E_2 and F are the closure constants. These constants are defined in the following paragraphs. Note that in both Mellor and Yamada (1982) and Kantha and Clayson (1994), F is not a constant, but a wall function that approaches a constant far from the surface. It seems that the inclusion of the wall function is not absolutely necessary if E_1 is chosen properly. As discussed later in section 2.2, if E_1 and E_2 are chosen to be constants, F should be a constant according to the validity of (23) in a neutral surface layer. Rotta (1951) was first to propose an equation of transport form like (23) to describe the evolution of $\eta^2 l$ according to the two-point correlation equations derived from the Navier-Stokes equations (see also Kantha, 2004).

The determination of the closure constants in (23), like that for the TKE equation, is empirical. We recommend that the following set of values be used for the results that are in the best agreement with those from the ETA scheme:

$$(E_1, E_2, F, S_l) = (2.75, 3, 3.0, 0.2).$$

As discussed later, they are determined based on the close analogy between the TKE equation and the length-scale equation and the properties of the turbulent flow in the neutral surface layer. This analogy is regarded as physically appropriate if the length scale is defined as the characteristic length scale of the largest energy-containing eddies and is related to the distance that these eddies travel in the vertical direction before losing their initial TKE due to turbulent mixing and buoyancy effects.

It should be pointed out that the values of the constants in both Mellor and Yamada (1982) and Kantha and Clayson (1994) were determined based on reliable measurements of decaying grid-generated homogeneous turbulence or turbulence near a wall without thermal stratification. The values that they recommended are

$$(E_1, E_2, S_l) = (1.8, 1.8, 0.2) \text{ and } F = E_3\left[1 + E_4\left(\frac{l}{\kappa L}\right)^2\right],$$

where $(E_3, E_4) = (1.0, 1.33)$, L is the distance from the boundary and κ is the von Karman constant. It is assumed that the applicability of the constants obtained in such a way can be extrapolated to other flow situations of more complexity with thermal stratification such as in the ABL where reliable and comprehensive measurements are relatively difficult to make. However, the validity of this assumption has not been theoretically and experimentally well established, leading to the uncertainties in the determination of the closure constants. Recently, Umlauf and Burchard (2003) proposed an analytical approach in which the properties of the two equations for the TKE and the length scale are used as extra constraints to calibrate the constants. To include the buoyancy effects for stably-stratified flows, Freedman and Jacobson (2003) proposed how to modify the constants in the length scale equation that are calibrated with measurements of flows without thermal stratification. Despite the extensive effort to optimize the values of these closure constants for flow situations of different complexity, consensus has not been reached on the possibility of seeking a set of universal values suitable for the simulation of vertical turbulent mixing in geophysical flows, reflecting the fact that it is still an ongoing research how to properly specify the length scale as an essential part of parameterized turbulence modelling.

Boundary conditions are required to solve (16) and (23). In our implementation of the scheme in MM5-V3, the lower boundary condition for the TKE equation is provided using the surface flux algorithm used in the ETA scheme (Janjić, 1990, 1994). This surface flux algorithm is based on the well accepted Monin-Obukhov similarity theory with the correction for the free convective limit proposed by Beljaars (1994). The lower boundary condition for $\eta^2 l$ is 0.

2.2 The Closure Constants of the $\eta^2 l$ Equation

The physical meaning of the length scale is of fundamental importance to the simulation of vertical turbulent mixing in the ABL. In the literature on the simulation of vertical turbulent mixing in the ABL, the length scale l is most often interpreted as the characteristic length scale of the largest energy-containing eddies (see, e.g., Moeng and Wyngaard, 1989; Wyngaard, 1992). It is directly related to the distance that these eddies travel in the vertical direction before they loss their initial TKE due to turbulent mixing and buoyancy effects. In some references, the length scale is also linked to the so-called integral length scale (see, e.g., Kantha and Clayson, 2000, p. 16). Strictly speaking, the integral length scale is associated with unbounded isotropic turbulence and is three-dimensional. The length scale l discussed here is associated turbulence in the atmospheric boundary layer, which is much more complicated than isotropic turbulence due to the existence of thermal stratification in the vertical direction. The nature of the closure as described in (16)-(20) requires that the direction of l be along the vertical wind shear and thermal gradient.

The closure constants in the length-scale equation must be determined empirically. Baumert and Peters (2000) indicate that the original values proposed by Mellor and Yamada (1982), and later adapted by Kantha and Clayson (1994), are not able to simulate the most basic properties of weakly stratified shear flows. They showed that for the well-understood near-wall flow situations without thermal stratification, there exist constraints on the determination of the values of the closure constants. Here we use a different approach to address the issues related to how to determine these constants.

The equation for the length scale, (23), can be rewritten as

$$l\frac{\partial \eta^2}{\partial t}+\eta^2\frac{\partial l}{\partial t}=E_1 l\,K_m\,S^2-E_2\,l\,K_H\,N^2-F\frac{\eta^3}{B_1}+\frac{\partial}{\partial z}\left[\eta\,l\,S_l\frac{\partial}{\partial z}(\eta^2 l)\right] \quad (24)$$

where

$$S^2=\left(\frac{\partial U}{\partial z}\right)^2+\left(\frac{\partial V}{\partial z}\right)^2,\ N^2=\beta\,g\frac{\partial \Theta_v}{\partial z} \quad (25)$$

Substituting (16) into (25) and using the so-called Kolmogorov-Prandtl relationships (i.e., $\eta l S_q = K_m/\sigma_m$ and $\eta l S_i = K_m/\sigma_l$ where both σ_m and σ_l are empirically determined constants) yield

$$\eta^2\frac{\partial l}{\partial t}=(E_1-2)l\,K_m\,S^2-(E_2-2)l\,K_H\,N^2-(F-2)\frac{\eta^3}{B_1}$$

$$+\frac{\partial}{\partial z}\left[K_m/\sigma_l\frac{\partial}{\partial z}(\eta^1 l)\right]-l\frac{\partial}{\partial z}\left[K_m/\sigma_m\frac{\partial}{\partial z}(\eta^2)\right]. \quad (26)$$

Now let us apply (26) into a steady neutral surface layer in which $\partial l/\partial t = 0$, $N^2 = 0$, $S = u_*/\kappa z$, $\eta^2 = 2c_\mu^{-1/2} u_*^2$ and $l = \kappa z$ (where u_* is the surface friction velocity, κ is the von Karman constant and c_μ is an empirical constant). It is straightforward to verify that if $\sigma_l = 1$, the following relationship holds for the neutral surface layer:

$$\kappa^2 = \frac{1}{2} c_\mu^{-1/2} (F - E_1) . \tag{27}$$

This equation imposes an important constraint on the relationship between E_1 and F. It also implies that only the difference between E_1 and F controls the balance among the production, dissipation and diffusion of the length scale, l, and therefore $\eta^2 l$ in the neutral surface layer. This implication is important for the determination of both E_1 and F as shown later because no matter how F is determined, E_1 must be determined through (27) for it to be dynamically consistent with F. A consequence of applying (27) to the determination of E_1 and F is that if F involves a wall function such that it depends on z, E_1 should also be dependent on z such that the difference between them is independent of z. On the other hand, F must be a constant according to (27) if E_1 is chosen as a constant.

Baumert and Peters (2000) proposed $E_1 = 2$ according to the hypothesis introduced by Tennekes (1989) that the *integral* length scale does not, based on dimensional grounds, depend on the homogenous shear. It should be noted that setting $E_1 = 2$ implies that the evolution of the length scale is controlled by the buoyancy and dissipation terms. Given the fact that turbulence grows in a stratified unstable flow as well as in a weakly stratified stable flow, one has to use a positive E_2 for the unstable flow and a negative E_2 for the stable flow (see, e.g., Baumert and Peters, 2000). Although this may not pose great difficulty in the simulation of vertical turbulent mixing in the oceanic mixed layer where the stratification is often weakly stable, it is definitely not desirable for the ABL because with a diurnal cycle, it is common for the stratification within the ABL to vary between stable stratification and unstable stratification. Moreover, although Tennekes' hypothesis may be acceptable for idealized unbounded turbulence with homogenous shear, it is perhaps too simple to be applicable to turbulence in the ABL where the shear is hardly homogenous and the buoyancy effect on turbulence is much stronger than most of the engineering flows in laboratories (more discussion on this point is provided in Kantha et al., 2005).

We wish to argue that in the simulation of vertical turbulent mixing in the ABL, if the length scale is defined as the characteristic length scale of the largest energy-containing eddies and is regarded as being related to the distance that these eddies travel in the vertical direction before they loss their initial kinetic energy due to turbulent mixing and buoyancy effects, its evolution as described by (26) should be closely analogous to that of the TKE in that the deviation from the balance between the turbulence production (by both shear and buoyancy) and the turbulence dissipation controls the change of the length scale. During the initial development of turbulence, the characteristic length scale of the largest energetic turbulent eddies increases with the intensification of turbulence. The greater the characteristic length scale is, the farther the largest energy-containing eddies are able to move in the vertical direction before they completely mix with the surrounding environment. The intensity of turbulence and therefore l cannot increase indefinitely because turbulent mixing tends to smooth out the vertical velocity shear and unstable stratification that are favourable for turbulence production. Therefore, the evolution of the characteristic length scale of the largest energy-containing eddies is governed by the same processes that control the evolution of the TKE. In order for the length-scale equation (26) to be valid for both unstable and weakly stable stratifications with the same set of closure constants, it is required that all the parameters before the turbulence production terms (i.e., the shear production and the buoyancy production) and the dissipation term be positive. With this requirement, (26) indicates that the lower bounds of the closure constants should be 2.

Requiring $F > 2$ is not in agreement with the conventional value of F (which is close to 1.0 corresponding to the decay exponent of ~1.0; see, e.g., Durbin and Pettersson-Reif, 2001; Umlauf and Burchard, 2003) that is derived from the experimental data of decay exponent for homogenous, isotropic turbulence. If the behaviour of the vertical turbulent mixing in the ABL is the same as in the turbulence in thermally unstratified flows, the values of all the closure constants can be determined uniquely from simple measurements in laboratories that isolate each term. However, there is no observational evidence to indicate whether or not this is indeed the case.

The simulation of vertical turbulent mixing in weather prediction models perhaps should be different than the simulation of engineering turbulence because the former emphasizes the effect of the largest energy-containing (i.e., flux carrying) eddies on vertical turbulent mixing in a thermally stratified environment. Because the two-equation turbulence model is not exact and the thermal stratification strongly affects vertical turbulent mixing in geophysical flows, it may not be practically possible to make the closure constants universally applicable to geophysical and engineering flows. Take the night-time atmospheric boundary layer for example. The largest eddies travel in the vertical direction against the restoring force caused by the stable stratification. According to the scaling argument made by Wyngaard (1992), this implies that in an equilibrium, stably stratified ABL, the distance (i.e., l) that the largest energy-containing eddies travel is proportional to the square root of the eddies' initial TKE. Thus, if the thermal stratification is held constant, l increases (or decreases) as q increases (or decreases). When $F > 2$ is chosen, the behavior of our length-scale equation appears to be in agreement with this argument, but apparently contradictory to the prediction based on the data of homogeneous, isotropic turbulence. However, if F is chosen to be smaller than

2 as done conventionally for engineering-turbulence modelling, the prediction of (26) is that the decrease of q due to the turbulence dissipation will lead to the increase of l, which is contrary to the above argument.

Currently, there is no practical way to observationally/experimentally determine the decay rate of the length scale as a function of the turbulence dissipation rate in the ABL. This study suggests that it is not unreasonable to set $F = 3.0$. Because typically $c_\mu = 0.09$, $\kappa = 0.4$ (see, e.g., Durbin and Pettersson-Reif, 2001) and E_1 should be slightly smaller than F, this leads to $E_1 \approx 2.9$. In accordance with the consistent analogy between the length-scale equation and the TKE equation, E_2 should be close to E_1, assuming that the Prandtl number is very close to unity. Results from our numerical experiments indicate that $E_2 = 3$ is acceptable when E_1 varies between 2.5 and 3.0. Hence, the set of values recommended in section 2.1, i.e., $(E_1, E_2, F, S_l) = (2.75, 3, 3.0, 0.2)$, are obtained for the results that are in the best agreement with those from the ETA scheme.

Summarizing the above discussions, the close analogy between the evolutions of the length-scale and the TKE should be used to determine the constraints on the values of the closure constants in the length-scale equation. This is based on the interpretation that l is the characteristic length scale of the largest energy-containing eddies, which is directly related to the turbulent mixing of these eddies in the vertical direction. It is required that the values of E_1, E_2 and F be greater than 2. E_1 should be slightly smaller than F, and E_1 and E_2 should be close to each other. The actual values of E_1 and E_2 are determined according to F. The ultimate validation of the value of F, and those of the closure constants, must rely on the comparison of the predictions of the scheme and the observations of the turbulence in the ABL.

2.3 Constraints on the Length Scale

In general, the two-equation vertical turbulent mixing scheme does not ensure that the quantities of the TKE and l are both always within the physically meaningful range for a given flow situation. It does not even guarantee that both quantities are positive definite. Thus, constraints on the two quantities should be established to prevent nonphysical results. While it is straightforward to set the lower bounds for the two quantities (they should not be negative), setting the upper bounds is not straightforward.

It has been recognized for a long time that because the closure hypothesis and the specification of the length scale involve empirically determined closure constants, the MY closure has performance problems (see, e.g., Mellor and Yamada, 1982; Galperin et al., 1988; Burchard, 2002). To alleviate the problems requires that the shear and stability parameters as well as the length scale be constrained properly. A proper specification of the limiting constraints on the shear and stability parameters as well as the length scale is very critical for the successful application of the MY closure. Various approaches to imposing the constraints on the length scale have been proposed for oceanic applications to constrain the shear and stability parameters along with the length scale. However, these approaches are not as theoretically appealing and straight forward as that proposed by Janjić (2002). Janjić's approach is innovatively different than the others in that it only applies constraints on the length-scale by requiring that the solution of the TKE equation be nonsingular. The effectiveness of this approach was illustrated in the implementation of the ETA scheme. The essence of the approach can be summarized as the following.

When the thermal stratification is unstable corresponding to a negative gradient Richardson number (R_i), the TKE production and dissipation, i.e., the right-hand side of (16) should be nonnegative and bounded. Following Janjić (2002) and using the definitions in (19) and (22), the contribution of the TKE production and dissipation can be written as

$$\left(\frac{\partial \eta^2/2}{\partial t}\right)_{prod/dissp} = [S_M G_M + S_H G_H - 1/B_1]\,(\eta^3/l). \tag{28}$$

Substituting the definition of G_H and G_M along with S_H and S_M obtained by solving (21a) and (21b), (28) can be written in the following form

$$l\frac{\partial}{\partial t}\left(\frac{1}{\eta}\right) = -\left\{\frac{\left[\alpha\left(\frac{l}{\eta}\right)^4 + \beta\left(\frac{l}{\eta}\right)^2\right]}{\left[\gamma\left(\frac{l}{\eta}\right)^4 + \delta\left(\frac{l}{\eta}\right)^2 + 1\right]} - \frac{1}{B_1}\right\}, \tag{29}$$

where

$$\alpha = -[9A_1A_2^2(\beta g)^2 g_H^2 + 3A_1A_2(3A_2 + 3B_2C_1 + 18A_1C_1 - B_2)g_M(\beta g)g_H]$$

$$\beta = [A_1(1-3C_1)g_M - A_2(\beta g)g_H]$$

$$\gamma = [9A_1A_2^2(12A_1 + 3B_2)(\beta g)^2 g_H^2 + 2A_1^2A_2(B_2 - 3A_2)g_M(\beta g)g_H]$$

$$\delta = 3[2A_1^2 g_M + A_2(7A_1 + B_2)(\beta g)g_H]$$

$$g_h = \frac{\partial \theta_v}{\partial z}$$

$$g_M = \left(\frac{\partial U}{\partial z}\right)^2 + \left(\frac{\partial V}{\partial z}\right)^2$$

In order for the right-hand side of (28) or (29) to be bounded and non-negative when the stratification is unstable or neutral, it requires that

$$l < \eta\,(P_1)^{1/2}\ (\text{for } R_i < 0), \tag{30}$$

where P_1 is the non-negative root of the following equation:

$$\gamma\left(\frac{l}{\eta}\right)^4+\delta\left(\frac{l}{\eta}\right)^2+1=0.$$

It is known that an equilibrium solution may exist when the thermal stratification is neutral or stable (i.e., $G_H \leq 0$), but the gradient Richardson number (R_i) is smaller than a critical value (R_{ic}) (see, e.g., Gerrity et al., 1994). This critical value of the gradient Richardson number can be obtained by letting the right-hand side of (28) or (29) be zero. It is equivalent to the following:

$$\frac{\left[\alpha\left(\frac{l}{\eta}\right)^4+\beta\left(\frac{l}{\eta}\right)^2\right]}{\left[\gamma\left(\frac{l}{\eta}\right)^4+\delta\left(\frac{l}{\eta}\right)^2+1\right]}-\frac{1}{B_1}=0, \tag{31}$$

which can be rewritten as

$$\left(\frac{\eta}{l}\right)^4+\chi\left(\frac{\eta}{l}\right)^2+\kappa=0 \tag{32}$$

with

$$\chi=[3A_2(7A_1+B_2)+A_2+B_1](\beta g)\,g_H\,[6A_1^2-A_1+B_1(1-3C_1)]\,g_M$$

$$\kappa=[9A_1A_2^2B_1+9A_1A_2^2(12A_1+3B_2)](\beta g)^2g_H^2$$

$$[3A_1A_2B_1(3A_2+3B_2C_1+18A_1C_1-B_2)+18A_1^2A_2(B_2-3A_2)]\,g_M(\beta g)g_H$$

By definition, when $R_i > R_{ic}$, η/l vanishes, and thus $\kappa = 0$. This implies

$$R_{ic}=\frac{3[A_1A_2^2B_1+A_1A_2^2(12A_1+3B_2)]}{A_1A_2(3A_2+3B_2C_1+18A_1C_1-B_2)+6A_1^2A_2(B_2-3A_2)} \tag{33}$$

For the empirical constants listed in section 2.1, it is obtained that $R_{ic} = 0.505$.

Since η and l may not vanish when $G_H \leq 0$ and the gradient Richardson number $R_i < R_{ic}$, a constraint on l needs to be established accordingly. Again, following Janjić (2002), one can derive such a constraint by analyzing the diagnostic equation for $\overline{w^2}/\eta^2$ in the MY closure. Appealing to the expression for $\overline{w^2}/\eta^2$ in the Mellor and Yamada level 2.5 turbulence model (Mellor and Yamada, 1982), Janjić showed that q/l depends on $\overline{w^2}/\eta^2$ through the following equation:

$$\xi\left(\frac{\eta}{l}\right)^4 + \sigma\left(\frac{\eta}{l}\right)^2 + \omega = 0, \tag{34}$$

where

$$\xi = (1 - 3R_s)$$

$$\sigma = \{[18A_1^2 C_1 g_m + (9A_1 A_2 + 3A_2 B_2)(\beta g) g_H] - 3R_s \delta\}$$

$$\omega = \{[27 A_1 A_2^2 B_2 (\beta g)^2 g^2_{\mathrm{H}} + 54A_1^2 A_2 B_2 C_1 g_M (\beta g) g_{\mathrm{H}}] - 3R_s \gamma\}$$

$$R_s = \overline{w^2} / \eta^2$$

It can be shown then that the smallest possible value of $\overline{w^2} / \eta^2$ (corresponding to $\eta/l = 0$) for the above given closure constants is 0.144. It can also be verified that the greater $\overline{w^2} / \eta^2$ is, the greater η/l. Therefore, the constraint on l is

$$l < \eta\, (P_2)^{1/2} \ \ (\text{for } R_i \geq 0), \tag{35}$$

where P_2 is the nonnegative solution of (34) with $R_s = \overline{w^2} / \eta^2 = 0.144$.

It is important to emphasize that the constraints derived above are based on the TKE equation and thus irrelevant to how the length scale is specified. In other words, no matter what prognostic or diagnostic approach to obtaining the length scale, it should be limited by the same constraints as expressed by (30) and (35), so long as (16)-(22) are used for the TKE prediction.

Acknowledgements

The author acknowledges NOAA's Hurricane Forecast Improvement Project that supported this work. The author acknowledges the support provided by Indo-US Science and Technology Forum and various national and international collaborators who have helped improving the operational HWRF model.

REFERENCES

Bao, J.-W., S.A. Michelson, L. Kantha and J.M. Brown, 2008: Implementation of a two-equation vertical turbulent mixing scheme in a mesoscale atmospheric model. NOAA Technical Memorandum OAR PSD-313.

Baumert, H. and H. Peters, 2000: Second-moment closures and length scales for weakly stratified turbulent shear flows. *J. Geophys. Res.*,**105**, 6453-6468.

Beljaars, A.C.M., 1994: The parameterization of surface fluxes in large-scale models under free convection. *Quart. J. Roy. Meteor. Soc.*, **121**, 255-270.

Blackadar, A.K., 1962: The vertical distribution of wind and turbulent exchange in neutral atmosphere. *J. Geophys. Res.*, **67**, 3095-3102.

Burchard, H., 2001: On the q2l equation by Mellor and Yamada (1982). *J. Phys. Oceonogr.*, **31**, 1377-1387.

Burchard, H., 2002: Applied Turbulence modeling in marine waters, Springer-Verlag, Berlin.

Cuxarc J., P. Bougeault, P. Lacarre, J. Noilhan and M.R. Soler, 1994: A comparison between transilient turbulence theory and the exchange coefficient model approach. *Bound.-Layer Meteor.*, **67**, 251-267.

Durbin, P.A. and B.A. Pettersson-Reif, 2001: Statistical Theory and Modeling for Turbulent Flow. John Wiley and Sons, Chichester, U.K.

Emanuel, K.A., 1994: Atmospheric Convection. Oxford University Press.

Fletcher, C.A.J., 1991: Computational Techniques for Fluid Dynamics, Vol.1: Fundamental and General Techniques, Springer-Verlag, Berlin/New York.

Freedman, F.R. and M.Z. Jacobson, 2003: Modification of the standard ε-equation for the stable ABL through enforced consistency with Monin-Obukhov similarity theory. *Boundary-Layer Meteorol.*, **106**, 383-410.

Galperin, B., L.H. Kantha, S. Hassid and A. Rosati, 1988: A quasi-equilibrium turbulent energy model for geophysical flows. *J. Atmos. Sci.*, **45**, 55-62.

Garratt, J.R., 1992: The Atmospheric Boundary Layer. Cambridge University Press.

Gerrity, J.P., T.L. Black and R.E. Treadon, 1994: The Numerical Solution of the Mellor-Yamada Level 2.5 turbulent kinetic energy equation in the Eta model. *Mon. Wea. Rev.*, **122**, 1640-1646.

Helfand, H.M. and J.C. Labraga, 1988: Design of a nonsingular level 2.5 second-order closure model for the prediction of atmospheric turbulence. *J. Atmos. Sci.*, **45**, 113-132.

Holtslag, A.A.M., 2003: Boundary layers/Modeling and Parameterization. *In:* Encyclopedia of Atmospheric Sciences. J.R. Holton, J.A. Curry and J.A. Pyle (eds). Vol. 1, 253-261, Academic Press, New York.

Holtslag, A.A.M. and C.H. Moeng, 1991: Eddy diffusive and counter-gradient transport in the convective boundary layer. *J. of. Atmos. Sci.*, **48**, 1690-1698.

Janjić, Z.I., 1990: The step-mountain coordinate: Physical package. *Mon. Wea. Rev.*, **118**, 1429-1443.

Janjić, Z.I., 1994: The step-mountain eta coordinate model: Further developments of the convection, viscous sublayer and turbulence closure schemes. *Mon. Wea. Rev.*, **122**, 927-945.

Janjić, Z.I., 2002: Nonsingular Implementation of the Mellor–Yamada Level 2.5 Scheme in the NCEP Meso model. *NCEP Office Note*, No. 437, 61 pp.

Kantha, L.H., 2004: The length scale equation in turbulence models. *Nonlin. Processes Geophys.*, **11**, 83-97.

Kantha, L.H. and C.A. Clayson, 1994: An improved mixed layer model for geophysical applications. *J. Geophys. Res.*, **99**, 25,235-25,266.

Kantha, L.H. and C.A. Clayson, 2000: Small Scale Process in Geophysical Fluid Flows. Academic Press.

Kantha, L.H., J.-W. Bao and S. Carniel, 2005: A note on Tennekes hypothesis and its impact on second moment closure models. *Ocean Modelling*, **9**, 23-29.

Mellor, G.L. and T. Yamada, 1974: A hierarchy of turbulence closure models for planetary. *J. Phys., Oceanogr.*, **5**, 718-728.

Mellor, G.L. and T. Yamada, 1982: Development of a turbulent closure model for geophysical fluid problems. *Rev. Geophys. Space Phys.*, **20**, 851-875.

Moeng, C.-H. and J.C. Wyngaard, 1989: Evaluation of turbulent transport and dissipation closures in second-order modeling. *Jour. Atmos. Sci.*, **46**, 2311-2330.

Pielke, R.A., 2001: Mesoscale Meteorological Modeling, Int. Geophys. Ser., vol. 78, 2nd ed. Elsevier.

Rotta, J.C., 1951: Statistische Theorie nichthomenger Turbulenze, 1. *Z. Phys.*, **129**, 547-572.

Stull, R.B., 1988: An Introduction to Boundary Layer Meteorology. Kluwer Academic Publishers.

Tennekes, H., 1989: The decay of turbulence in plane homogeneous shear flow. *In:* Lecture Notes on Turbulence. J.R. Herring and J.C. McWilliams (eds). World Scientific, 32-35.

Tennekes, H. and J.L. Lumley, 1982: A First Course Turbulence (2nd ed). MIT Press.

Umlauf, L. and H. Burchard, 2003: A generic length-scale equation for geophysical turbulence. *J. Mar. Res.*, **61**, 235-265.

Wyngaard, J.C., 1992: Atmospheric turbulence. *Annu. Rev. Fluid Mech.*, **24**, 205-233.

6

Air-Sea Turbulent Flux Parameterizations in Tropical Cyclone Models

Jian-Wen Bao
NOAA/Earth System Research Laboratory, Boulder, Colorado
jian-wen.bao@noaa.gov

This chapter is the last of the three-lecture series on the physical processes that are accounted for in numerical models of tropical cyclones. There are two sections in this chapter. Section 1 is intended to provide an overview of the physical processes required to be included in TC models near the air-sea interface under tropical cyclones, and Section 2 presents results from a recent study on the sensitivity of simulated TC intensification to parameterized physical processes. Materials presented in Section 1 are derived from an overview book chapter by Bao et al. (2002), while those presented in Section 2 are based on Bao et al. (2011).

1. An Overview of Physical Processes near the Air-Sea Interface

1.1 Parameterizations of Air-Sea Fluxes under High Wind Conditions

While air-sea interaction involves processes on a wide range of spatial and temporal scales (e.g. Gill, 1982), it is the turbulent processes transferring momentum, heat, moisture, gaseous chemicals, and aerosols across the air-sea interface that physically couple the atmosphere and ocean together. When using a coupled atmosphere-wave-ocean model to simulate air-sea interaction in tropical cyclone prediction models, it is necessary to parameterize these processes using the model-resolvable variables that drive and influence them. Despite the fact that, for a given problem in air-sea interaction, some of the parameterized processes may be more important than others, it is a minimal requirement for the coupled model to have the parameterizations of surface momentum, and sensible and latent heat fluxes across the air-sea interface (e.g., Bao et al., 2000). Such parameterizations are formulated based on the so-called Monin-Obukhov (M-O) similarity theory (e.g., Fairall et al., 1996) in which many of the physical processes involved in air-sea turbulent transfer of momentum, sensible heat or latent heat flux are embedded within a transfer coefficient for the particular flux. Like most parameterizations in atmospheric models, the parameterizations of turbulent transfer processes at the air-sea interface are deduced from observations, and they are more or less approximations to the processes occurring in nature.

The sea surface is undulated by waves that are a mixture of swell and locally generated wind waves. The complexity of the sea-surface undulation strongly depends on the interaction between wind and waves that is accomplished by wind-wave momentum exchange. Therefore, air-sea momentum flux is dependent on sea state. For the M-O theory, this means that the roughness length used in the determination of momentum flux is sea-state dependent. When the wind speed exceeds 15 m/s, sea-spray droplets from wave breaking can modulate the way in which the thermal fluxes across the air-sea interface are partitioned, and ultimately affect the quantity of the thermal fluxes. Thus, the roughness lengths that are used in the M-O theory to calculate the thermal fluxes are also sea-state dependent. The sea-state dependent air-sea fluxes of momentum and thermal energy are the major forces driving the dynamics of both the atmospheric boundary layer (ABL) and the oceanic mixed layer (OML). As a consequence, the air-sea temperature difference becomes sea-state dependent under high winds, which in turn affects the thermal fluxes across the air-sea interface. Therefore, coupled atmosphere-wave-ocean models, when used under high wind conditions, should include these sea-state dependent processes.

Parameterizations of air-sea momentum, sensible and latent heat fluxes have been developed for use in coupled atmosphere-wave-ocean models to take into account the effects of waves in the specification of roughness length for momentum, and the contribution of sea spray to sensible heat and latent heat fluxes. The development and validation of these parameterizations have been based on observations taken in both the tropics and mid-latitudes at wind speeds below 25 m/s. Therefore, in numerical simulations of air-sea interaction using coupled atmosphere-wave-ocean models under high wind conditions (>30 m/s), extrapolation of the parameterizations of air-sea sensible and latent heat fluxes is inevitable. Thus, an important step in the development of a coupled atmosphere-wave-ocean model is to test the performance of the parameterizations under high wind conditions.

Over the past 20 years, TC modellers have begun to take advantage of the progress made in the parameterizations of air-sea interaction processes. Efforts have been made to test and evaluate state-of-the-art parameterization schemes of air-sea interaction in both idealized and realistic scenarios. As an example of such efforts, we will describe the work carried out at ESRL/PSD in which a regional coupled atmosphere-ocean modelling system is developed to serve as a research platform to test the parameterizations of air-sea fluxes that have been developed based on the physical understanding.

1.2 Sea Surface Roughness Parameterization

Assume that the total stress close to the surface can be expressed as the sum of contributions from surface wave and turbulence: $\tau = \tau_w + \tau_t$, where τ, τ_w and τ_t are the total stress, wave-induced stress, and atmospheric turbulence-induced stress, respectively. The mean wind profile above the wave surface is assumed to be logarithmic. Following Janssen (1991), we assume that the effective roughness length of the developed (old) waves can be modelled by means of the Charnock relation

$$z_0 = \alpha \frac{\tau}{\rho_a g}, \tag{1}$$

where α is a constant, g is the acceleration of gravity, and ρ_a is the air density. The effective wave-induced roughness length of the developing (young) waves is modelled by a parametric height, z_l, so that the mean wind profile under neutral stratification is given by

$$U(z) = \frac{\sqrt{\tau / \rho_a}}{\kappa} \ln \frac{z + z_l}{z_0 + z_l}, \tag{2}$$

where κ is the von Kármán constant. Consider the steady-state stress balance of airflow over sea waves. Following Janssen (1991), the atmospheric turbulence-induced stress at $z = z_0$ is given by

$$\tau_t(z_0) = \left(\frac{z_0}{z_0 + z_1} \right)^2 \tau \tag{3}$$

in which it is assumed that τ_t is modelled by

$$\tau_t(z) = l \left| \frac{\partial U(z)}{\partial z} \right| \frac{\partial U(z)}{\partial z}, \tag{4}$$

where $U(z)$ is the wind profile given by (2), and l is the mixing length (given by κz for neutral stratification). Substituting (3) into the expression of the total stress and using (1) then gives

$$z_l = \alpha \frac{\tau}{g \rho_a} \left(\frac{1}{\sqrt{1 - \tau_w / \tau}} - 1 \right). \tag{5}$$

Although this parameterization is derived for neutral stratification, it is easy to show that the equation for z_l holds under non-neutral stratification if the wind profile follows the Monin-Obukhov similarity theory and can still be expressed as

$$\frac{\kappa(z+z_l)}{\sqrt{\tau/\rho_a}}\frac{\partial U(z)}{\partial z}=\frac{\kappa z}{l}=\varphi\,(z/L), \tag{6}$$

where φ is the well-known Monin-Obukhov similarity function, and L is the Monin-Obukhov length.

1.3 The Parameterization of Roughness Lengths for Sensible and Latent Heat Fluxes

A few parameterization schemes of the roughness lengths for surface sensible and latent heat fluxes have been developed based on measurements taken in both laboratories (e.g., Kader and Yaglom, 1972) and the natural environment (e.g., Brutsaert, 1979, 1982; Brutsaert and Sugita, 1996; Zilitinkevich et al., 2001). According to the approaches used in the parameterizations, these schemes can be categorized into four groups: (i) direct data fit with respect to a non-dimensional parameter such as the roughness Reynolds number (R_{re}) (e.g., Fairall et al., 2001); (ii) assuming that the flow in the molecular sublayer is instantaneously smooth (e.g., Makin and Mastenbroek, 1996); (iii) use of the surface renewal model where the ratio of the temperature or the moisture roughness length to the momentum roughness length is functionally dependent on the roughness Reynolds number (R_{re}) and the Prantdl number (P_r) (e.g., Donelan, 1990; Zilitinkevich et al., 2001), i.e., $z_{0T,q}/z_0 = F\,(R_{re}, P_r)$; and (iv) assuming that the heat or moisture transfer coefficient at a reference height (z_r) above the surface is constant based on observations, such as Large and Pond (1982), Smith (1988) and DeCosmo et al. (1996). Therefore, great disparity exists in the parameterizations of z_{0T} and z_{0q}.

It should be noted that among all the formulas mentioned above, only those of Fairall et al. (2001), Large and Pond (1982) and Zilitinkevich et al. (2001) were derived using observations taken over the sea. The rest of the formulas were established based on observations taken over the land with different surface characteristics, or in controlled laboratory experiments. The use of these formulas over the sea is based on the assumption that the heat and mass transfer for smooth flow and fully rough flow over the sea are similar, respectively, to the heat and mass transfer for the smooth surface and for the bluff-rough surface over the land (Garratt, 1992). It is worth noting that this assumption has never been validated by observations.

1.4 The Spray Contribution to Sensible and Latent Heat Fluxes from the Ocean

When the wind speed is in excess of approximately 15 m/s, a substantial amount of sea spray is produced by breaking waves, bursting bubbles, and wind gusts (e.g., Kraus and Businger, 1994; Andreas et al., 1995). This results in the development of a near-surface layer that can be characterized as a droplet evaporation zone. Both theory and observations suggest that at high wind speeds evaporation from sea spray is significant. Above the droplet evaporation zone, the droplet contribution to the flux takes the form of an enhanced turbulent flux of water vapour and a corresponding change in the turbulent flux of sensible heat that is dependent on the enhanced water vapour flux. Simple scaling models to parameterize the spray effect on sensible and latent heat fluxes have been developed based on recent observations (see, e.g., Andreas, 1992; 1998, Fairall et al., 1994). The thermodynamic effect of sea spray in these models is characterized as a competition between their evaporation rate and their lifetime in the air. Two factors determine the lifetime of sea spray in the air: the fall velocity of the droplets and their vertical transport by turbulence. The evaporation rate of sea spray is controlled by the amount of heat available for evaporation and interactions between droplets of different sizes. The heat for the spray evaporation comes from three sources: (1) upward transfer of heat from the ocean by turbulence and the spray droplets, (2) downward heat transfer from the atmosphere by ABL turbulence and, when cumulus convection exists, convective downdrafts, and (3) dissipation of TKE. These heat sources adjust internally until some equilibrium is reached, with the spray evaporation acting as a moisture source and as a heat source or sink to the ABL. As a result, the ABL structure is adjusted accordingly and this adjustment feeds back on the dynamic processes above the ABL. Obviously, a coupled air-sea model with a state-of-the-art ABL parameterization is ideal for investigation of the net effect of sea spray on air-sea interaction.

2. Parameterizations of Sea-Spray Impact on the Air-Sea Momentum and Heat Fluxes

The omnipresence of sea spray in the marine surface boundary layer (MSBL) has long been observed under hurricane-strength winds. In contrast, the effect of sea spray on the air-sea momentum and heat fluxes at hurricane-strength winds and how the effect is parameterized in numerical weather prediction (NWP) models are relatively new subjects of very active investigation. High surface winds (>30 ms^{-1}) generate large re-entrant sea-spray droplets (Fairall et al., 2009) which tend to increase the sea-air enthalpy transfer and, thus, have a positive feedback to the intensification of hurricanes (Andreas and Emanuel, 2001). On the other hand, sea spray is generated at the expense of the momentum of turbulence at the air-sea interface, which inevitably causes the air-to-sea turbulent momentum flux to change. The combined feedback effect of sea spray on the overall air-sea momentum and enthalpy fluxes has not been well-understood since such a combination, in general, adds to the complexity of air-sea interaction. It is interesting to note that due to the difficulty in making observations in the hurricane surface layer, the evaluation of the importance of sea-spray effect has been performed using NWP model simulations of tropical cyclones and the results from these simulations are, to a great degree, dependent on their representation in numerical models (e.g., Bao et al., 2000; Wang et al., 2001).

Powell et al. (2003) presents an analysis of the air-sea momentum transfer as a function of 10-m wind speed for high winds in tropical cyclones using the wind profiles measured by Global Positioning System (GPS) sondes. They find that the surface drag coefficient levels off as the wind speeds increase above 35 ms^{-1}. This finding is contrary to surface flux parameterizations that are currently used in NWP models. They speculate that sea spray may play an important role in this behaviour. Analysis of the measurements from the Slope to Shelf Energetics and Exchange Dynamics (SEED) project (Jarosz et al., 2007) confirms the trend of the dependence of the drag coefficient on 10-m winds that is revealed by Powell et al. (2003). Makin (2005) and Andreas (2004) further confirm that the consideration of the sea spray effect on the air-sea momentum transfer does lead to the trend of the dependence of the drag coefficient on 10-m winds that is similar to that presented in Powell et al. (2003). Despite the technical challenges in making direct observations of the modification of sea spray to air-sea heat and moisture fluxes at high winds, recent measurements of bulk air-sea fluxes of momentum and heat from the Coupled Boundary Layer Air-Sea Transfer (CBLAST) (Black et al., 2007) strongly suggest a different trend of the dependence of the heat exchange coefficient on 10-m winds that is widely used in NWP models (Zhang et al., 2008). All these studies point out the need to account for the effect of sea spray in the calculations of air-sea momentum and enthalpy fluxes in NWP models.

We have proposed that the effect of sea spray be considered as an additional modification to the stratification of the profiles of wind, temperature and moisture in the MSBL, and thus they can be parameterized in NWP models using a common framework of parameterizations of air-sea fluxes: the Monin-Obukhov similarity theory. The proposed parameterization scheme assumes that wave breaking and sea spray generation takes place on the subgrid scale, and the air drag that accelerates sea spray is part of the air-sea interfacial stress. Sea spray generation due to wave breaking is associated with turbulent gusty flows that are related, but not equal, to the grid resolved mean flows through the so-called subgrid Reynolds stress in the governing equations for the grid-resolved flows.

2.1 Dynamical and Thermal Effect of Sea Spray

The dynamical aspects of sea spray can be addressed through the budget equation of TKE in the MSBL, which in the absence of sea spray, is governed approximately by

$$-\langle uw\rangle\frac{\partial U}{\partial z}+\frac{g\langle\theta_v w\rangle}{\Theta_v}-\frac{\partial}{\partial z}\left(\frac{\langle Ew\rangle}{2}+\frac{\langle pw\rangle}{\rho}\right)-\varepsilon=\frac{DE}{Dt}, \tag{7}$$

where z is the vertical coordinate, $\langle\ \rangle$ denotes mean (following the Reynolds convention), E is twice the TKE, U is the mean horizontal velocity of air motion, u and w are the eddy components of U and the vertical air motion, Θ_v and θ_v are the mean and eddy components of virtual potential temperature, p and ρ are the pressure and density of air (including moisture), ε is the dissipation rate of TKE, and g is the acceleration of gravity. The virtual potential temperature used here is defined as dry potential temperature multiplied by $(1 + 0.608q - q_l)$, where q is the mass fraction of water vapour in the air (i.e., the specific humidity) and q_l is the mass fraction of liquid water. The right hand side of Eq. (7) is the total derivative of E. Here details of the complex wavy geometry of the sea surface are ignored by treating the quantities of interest in an average sense (Mueller and Veron, 2009).

When sea spray is present in the MSBL, the budget equation of TKE is rendered into the following form

$$-\langle uw\rangle\frac{\partial U}{\partial z}+\left(\frac{g\langle\theta_v w\rangle}{\Theta_v}+\frac{g\langle\theta_{v,sp} w\rangle}{\Theta_v}\right)-\frac{\partial}{\partial z}\left(\frac{\langle Ew\rangle}{2}+\frac{\langle pw\rangle}{\rho}\right)-(\varepsilon+g\,\sigma V_f S)=\frac{DE}{Dt} \tag{8}$$

where $\theta_{v,sp}$ is the perturbation in virtual potential temperature caused by heat transfer between air (of virtual potential temperature Θ_v) and sea spray (of mean virtual potential temperature $\Theta_{v,sp}$ at the droplet surface), $\sigma = (\rho_w - \rho)/\rho$ is the relative excess of the density of sea water ρ_w over the density of air, V_f is the fall speed of sea spay droplets, and S is the volumetric concentration of sea spray mass. The last term including V_f represents the rate at which turbulent energy is consumed for spray droplets to be suspended in the flow against gravity (see, e.g., Adams and Weatherly, 1981). This equation shows that the impact of sea spray on the budget equation of TKE is taken into account by adding an additional term to the buoyancy production and an additional term in the dissipation due to the suspension of sea spray by turbulence.

After a simple rearrangement of the above equation using the friction velocity u_* defined as $\sqrt{-\langle uw\rangle}$, one can obtain the following dimensionless budget equation of TKE

$$\frac{\kappa z}{u_*}\frac{\partial U}{\partial z}+\frac{\kappa z}{u_*^3}\left(\frac{g\langle\theta_v w\rangle}{\Theta_v}+\frac{g\langle\theta_{v,sp} w\rangle}{\Theta_v}-g\sigma V_f S\right)-\frac{\kappa z}{u_*^3}\frac{\partial}{\partial z}\left(\frac{\langle Ew\rangle}{2}+\frac{\langle pw\rangle}{\rho}\right)-\frac{\kappa z}{u_*^3}\varepsilon=\frac{\kappa z}{u_*^3}\frac{DE}{Dt} \tag{9}$$

where κ is the von Karman constant. The second term of the left-hand side can be expressed as $-z/L$, where L is defined as

$$L = -\frac{u_*^3}{\kappa\left(\frac{g\langle\theta_v w\rangle}{\Theta_v} + \frac{g\langle\theta_{v,sp} w\rangle}{\Theta_v} - g\sigma V_f S\right)} \tag{10}$$

and has a dimension of length.

An important aspect of the dimensionless TKE budget equation is that each of the dimensionless terms in the equation should depend only on a dimensionless parameter z/L. In fact, in the absence of sea spray (i.e., $\frac{g\langle\theta_{v,sp} w\rangle}{\Theta_v} = 0$ and $g\sigma V_f S = 0$), this equation connects the scaling argument of the Monin-Obukhov similarity theory and the dynamical aspect of the budget equation of TKE in the surface boundary layer. This equation suggests that the parameterizations of the feedback effect of sea spray on the momentum and heat fluxes can be based on the following assumption: the turbulent eddies responsible for turbulence mixing are mostly generated by shear and the variations of static stability due to sea spray loading modify the log profiles of wind, temperature and moisture corresponding to the so-called neutral stratification. As noted by Stull (1997), Monin-Obukhov similarity theory is based on the same assumption that takes into account the feedback effect (i.e., modifications) of variations in static stability caused by turbulent heat fluxes to the log profiles of wind and temperature. Given the prominent role that the Monin-Obukhov similarity theory plays in the calculation of the momentum and heat fluxes in the MSBL, it is natural to consider the feedback effect of sea spray on the fluxes by assuming that the Monin-Obukhov similarity theory is still valid in the sea-spray laden MSBL, provided the length scale parameter is chosen properly. Therefore, if we define $L_{MO} = -\frac{u_*^3}{\kappa\frac{g\langle\theta_v w\rangle}{\Theta_v}}$, and Θ_v and θ_v are the mean and eddy components of virtual potential temperature, Θ and θ are the mean and eddy components of potential temperature and Q and q are the mean and eddy components of specific humidity, the conventional Monin-Obukhov similarity relations [with $\varphi_m(z/L_{MO})$, $\varphi_h(z/L_{MO})$ and $\varphi_c(z/L_{MO})$ being functions of z/L_{MO}],

$$\frac{\kappa z}{u_*}\frac{\partial U}{\partial z} = \varphi_m(z/L_{MO}), \frac{\kappa z u_*}{-\langle\theta w\rangle_*} = \varphi_h(z/L_{MO}), \text{ and}$$

$$\frac{\kappa z u_*}{-\langle q w\rangle_*}\frac{\partial Q}{\partial z} = \varphi_c(z/L_{MO}) \tag{11}$$

can be extended to include the sea-spray effect by replacing the conventional Monin-Obukhov length L_{MO} with L as defined previously (Eq. 10). The parameterizations of the feedback effect of sea spray on the momentum and heat fluxes come down to the parameterizations of the sea-spray modification to the fluxes and mean profiles calculated using the conventional Monin-Obukhov similarity relations. That is, in the presence of sea spray and ignoring the wavy geometric effect of water surface, both the formulations of the MSBL momentum, sensible heat and moisture fluxes have the following functional form:

$$u_* = \frac{\kappa(U - U_0)}{\ln(z/z_0) - \Psi_m(z/L)}, \quad \frac{-\left(\langle \theta w \rangle + \langle \theta_{sp} w \rangle\right)}{u_*} = \frac{\kappa(\Theta - \Theta_0)}{\ln(z/z_{0T}) - \Psi_h(z/L)}, \text{ and}$$

$$\frac{-\left(\langle q w \rangle + \langle q_{sp} w \rangle\right)}{u_*} = \frac{\kappa(Q - Q_0)}{\ln(z/z_{0q}) - \Psi_h(z/L)} \tag{12}$$

where z_0, z_{0T} and z_{0q} are the reference heights for the zero mean wind, the sea surface temperature and moisture which are slightly above the wave height, θ_{sp} is the perturbation in potential temperature caused by the sensible heat release from sea spray, q_{sp} is the perturbation in specific humidity due to the evaporation from sea spray, and the subscript 0 of a variable indicates the value of the variable assumed at the corresponding "surface" reference height (z_0, z_{0T}, or z_{0q}). It should be pointed out that the premise of the argument behind these functional forms is that the Monin-Obukhov similarity relations hold in the possible range of surface winds above the wave height and, in particular, the existence of spray over breaking waves does not invalidate them. This premise is convenient since Monin-Obukhov similarity formulations for surface momentum and heat fluxes are commonly used in numerical weather prediction models.

It is further assumed that the wind in the hurricane surface boundary layer is strong enough to render $z \ll |L|$, i.e., the vertical profiles of wind, temperature and moisture are slightly departed from those associated with the neutral stability and the departure is linearly dependent on z/L (see p. 50 Garratt, 1992). Then, the stability functions in the above flux formulations can be approximately partitioned into two parts: one is related to L_{MO} and the other is related to L_{sp} (i.e., $L^{-1} = L^{-1}_{MO} + L^{-1}_{sp}$,

where $L_{MO} = -\dfrac{u_*^3}{\kappa \dfrac{g\langle \theta_v w \rangle}{\Theta_v}}$ and $L_{sp} = -\dfrac{u_*^3}{\kappa\left(\dfrac{g\langle \theta_{v,sp} w \rangle}{\Theta_v} - g\sigma V_f S\right)}$). That is,

the assumption leads to the following approximations:

$$\psi_m(z/L) \approx \psi^{(1)}_m(z/L_{MO}) + \psi^{(2)}_m(z/L_{sp}) \quad (13a)$$

and

$$\psi_h(z/L) \approx \psi^{(1)}_h(z/L_{MO}) + \psi^{(2)}_h(z/L_{sp}), \quad (13b)$$

where superscripts (1) and (2) denote the two partitioned functions that are dependent on z/L_{MO} and z/L_{sp} respectively. In sub-sections 2.2 and 2.3, parameterizations of the sea-spray modification to the heat fluxes and surface momentum are developed separately based on these approximations.

In summary, the challenge of parameterizing how sea spray modifies the surface momentum and heat (and moisture) fluxes lies in the determination of the stability functions associated with L_{sp}. Nonzero stability functions indicate non-neutral stratification and result in departures of the wind speed, temperature and moisture profiles from the so-called neutral ones. Since it is shown above that the effect of sea spray can be accounted for in the stability functions for the surface momentum and heat (and moisture) fluxes, the parameterizations of the effect can be derived separately from the determination of the departures from the neutral profiles using the principle of the enthalpy conservation and the steady budget equation of TKE. The parameterized effect is then implemented together in NWP models consistently according to Eqs. (12) and (13) to account for the combined effect of sea spray on the surface momentum and heat (and moisture) fluxes.

2.2 Parameterization of the Sea-spray Effect on the Heat Fluxes

Consider a vertically uniform air column of depth D above the sea surface that has a unit horizontal area. In NWP models, D refers to the lowest model level where surface fluxes are defined and calculated. When the Monin-Obukhov similarity theory is used to calculate the momentum and heat (and moisture) fluxes between the sea surface and elevation $z = D$, Eqs. (12) and (13) in section 2.1 indicate that the overall thermal feedback effect of sea spray on the heat fluxes is to induce changes to the profiles of wind speed, temperature and moisture of the air below the level $z = D$. Using δT_a and δQ_a to denote the average changes of the air temperature and moisture in the column, the goal of the parameterization of the thermal feedback effect of sea spray is to relate δT_a and δQ_a to the mean wind speed (U), temperature (T_a) and moisture (Q_a) in the column. Following Andreas and Emanuel (2001), the principle of enthalpy conservation is used to determine δT_a and δQ_a.

Let T_s denote the sea surface temperature and assume that the column consists of dry air mass m_d and water vapour mass m_v. If the wind speed at the top of the column is high enough to convert a small amount of sea-water mass δm_i into spray droplets, the conservation of enthalpy in the air-sea coupled system requires that the following equation holds:

$$[c_{pd}\, m_d + c_{pv}\,(m_v + f_e\, \delta m_i)]\; \delta T_a + L_v f_e\, m_i$$
$$= c_w \delta m_i\, [(T_s - T_a) + (1 - f_e)(T_a - T_w)], \tag{14}$$

where f_e is the fraction of δm_i that evaporates, c_{pd}, c_{pv} and c_w are the heat capacity of dry air, water vapour and sea water, T_w is the mean wet-bulb temperature in the column, and L_v is the latent heat of vaporization.

If we let ρ_d and ρ_v be the mean densities of dry air and water vapour in the column, Eq. (14) can be written as:

$$[c_{pd}\, \rho_d + c_{pv}\,(\rho_v + f_e\, \delta m_i/D)]\; \delta T_a + L_v f_e\, \delta m_i/D$$
$$= c_w \delta m_i/D[(T_s - T_a) + (1 - f_e)(T_a - T_w)]. \tag{15}$$

Eq. (15) immediately leads to the expression of the air temperature perturbation due to the spray injection and evaporation as follows:

$$\delta T_a = \frac{c_w \delta m_i \,/\, D\,[T_s - T_a) + (1 - f_e)(T_a - T_w)] - L_v\, f_e\, \delta m_i \,/\, D}{c_{pd}\, \rho_d + c_{pv}(\rho_v + f_e\, \delta m_i \,/\, D)} \tag{16}$$

Corresponding to this change in temperature is the change in water vapour mixing ratio, $\delta Q_a = f_e \delta m_i/D$ (or the change in dew temperature, δT_d, which can be obtained using δQ_a from the Clausius-Clapeyron equation). The mass variable is computed from the specified droplet mass flux, F_m, as $\delta m_i = F_m\, \delta t$ where δt is an adjustment time scale for the droplet evaporation layer.

After both δT_a and δQ_a are computed according to the formulations described above, the direct turbulent and sea-spray mediated sensible and latent heat fluxes (denoted with suffixes *dir* and *sp* respectively) at $z = D$ can be written as

$$H_{s,\,dir} = \rho_a\, c_{pa}\, C_H\, U\, [T_s - (T_a + \delta T_a)], \tag{17a}$$

$$H_{L,\,dir} = \rho_a\, L_v\, C_E\, U\, [Q_s\,(T_s) - Q_s\,(T_d + \delta T_d)], \tag{17b}$$

$$H_{s,\,sp} = \rho_w\, c_w\, F_M\, [T_s - (T_w + \delta T_w)], \tag{17c}$$

$$H_{L,\,sp} = \rho_w\, L_v\, F_E\, [Q_s\,(T_a + \delta T_a) - Q_s\,(T_d + \delta T_d)], \tag{17d}$$

where ρ_a is the density of moist air, ρ_w is the density of sea water, c_{pa} and c_w are the heat capacity of air (including moisture) and sea water, $Q_s(T)$ is the grid-box mean saturation specific humidity at temperature T, $H_{s,\chi}$ and $H_{L,\chi}$ denote the sensible and latent heat fluxes respectively where χ denotes the association with direct turbulence (*dir*) or sea spray (*sp*), C_H and C_E are the exchange coefficients for the direct turbulent transfer of sensible and latent heat at $z = D$, and F_M and F_E are the spray droplet mass and water-vapour fluxes where $f_e = F_E/F_M$. Both factors F_M and F_E are specified from the droplet spectrum (see Fairall et al., 1994). Subscript *dir* denotes the direct turbulent fluxes without the feedback effect of sea spray which, as specified by Eqs. (17a) and (17b), are connected with the intensity of turbulence in the surface layer through the following relations among the heat exchange coefficient C_H (and C_E), the drag coefficient C_D, and the friction velocity u_*:

$$C_H = C_D^{1/2} \frac{\kappa}{\ln(z / z_{0T}) - \Psi_h(z / L_{MO})} \quad \text{and} \quad u_* = C_D^{1/2} U.$$

The friction velocity u_*, as a general measure of the intensity of turbulence in the surface layer and usually obtained according to the Monin-Obukhov similarity theory, is influenced by the suspension of sea spray. The parameterization of the sea-spray influence on u_* is discussed in the next section.

2.3 Parameterization of the Sea-spray Effect on the Momentum Flux

The effect of sea spray on the surface momentum flux is realized through its enhancement of airflow stratification—airflow lamination under gravity—due to mass loading (see, e.g., Barnes, 2008), and thus it can be dealt with following Barenblatt (1996) and Lykossov (2001). Ignoring the thermal effect of sea spray, the solution of the simultaneous steady equations of droplet mass conservation and TKE budget leads to the following expression of friction velocity at the level of D, the depth of the column referred to in section 3 (or the lowest model level) above the mean sea level (see details in Appendix A of Bao et al., 2011):

$$u_* = \frac{\kappa U}{\ln\left(\frac{D}{z_0}\right) - \Psi_s}, \quad -\Psi_s = \begin{cases} \omega^{-1} \ln\left(1 + \frac{\alpha\omega^2}{1-\omega}\left[\left(\frac{D}{z_0}\right)^{\gamma} - 1\right]\right) & \text{for } \omega \neq 1 \\ \ln\left[1 + \alpha \ln\left(\frac{D}{z_0}\right)\right] & \text{for } \omega = 1 \end{cases}, \qquad (18)$$

where $\gamma = 1 - \omega$, $\omega = \frac{V_f}{\kappa u_*}$, $\alpha = \frac{\beta g \kappa^2 z_0 \sigma S_0}{u_*^2}$, $S_0 = S(z_0)$, V_f is the mean fall speed of droplets in the MSBL, σ is defined the same as in Eq. (8), and β is a constant associated with a linear approximation of the non-dimensional vertical gradient of mean wind. It is assumed for simplicity that z_0 is the level of wave breaking (i.e., spray generation). It should be pointed out that the Monin-Obukhov similarity theory is only valid at heights well above z_0 where the wave breaking takes place as revealed by the results from an explicit simulation of the impact of sea spray on the TKE budget in the surface boundary layer (Bianco et al., 2011).

Numerical implementation of the parameterizations of the sea-spray effect in NWP models requires a proper interface between the direct air-sea turbulent fluxes, which are computed using the Monin-Obukhov similarity theory, and the sea-spray effect parameterizations. In theory, calculating surface momentum and buoyancy fluxes in NWP models using the Monin-Obukhov similarity formulations requires solving the equations iteratively due to the dependence of the fluxes on the stability functions, which in turn are dependent on the length scale (L_{MO}) that is defined by the surface momentum and buoyancy fluxes. Since the generation rate of sea spray is dependent on the surface momentum flux, the calculation of the sea spray effect should be part of the iterative procedure for the surface flux calculation. One iterative procedure commonly used in NWP models is to use the surface fluxes obtained from the previous time step as the first guess for the solution at the current time. To follow this procedure, it is recommended that the above mentioned parameterizations of the sea-spray effect be called in the code of NWP models right after the direct air-sea turbulent fluxes are calculated, so that the current spray-modified friction velocity, sensible heat and moisture fluxes are used at the next time step in the determination of the Monin-Obukhov similarity stability functions that are required in the calculations of the direct air-sea fluxes. This will allow the interaction between the direct fluxes and the spray-mediated fluxes. Special care should also be taken in the interface when the NWP models are coupled with wave prediction models.

2.4 The Physics of the Parameterized Sea-spray Effect

Closing the previously discussed parameterization scheme of the effect of sea spray on air-sea exchange processes requires information on the production rate of droplets at various radii as a function of the wind speed or wave states. In this study, this information is provided through the droplet source function based on the physical model of droplet generation developed by Fairall et al. (2009). This physical model incorporates the most recent laboratory measurements and the current understanding of the turbulence energetics in wave breaking. It predicts the size spectrum of sea-spray droplets produced by wave breaking as a function of the surface wind stress and the wave state. When implemented in an NWP model that is not coupled with a surface wave model, the wave state is diagnostically determined by the surface wind stress (Fairall et al., 2009).

Figure 1 illustrates the parameterized effect of sea spray on the 10-m drag coefficient (C_D) and enthalpy exchange coefficient C_k, along with their ratio (C_k/C_D), as functions of 10-m wind speed. All the exchange coefficients that include the effect of sea spray are calculated according to the formulations summarized in Bao et al. (2011) by prescribing 10-m wind speeds and using the bulk surface flux formulations described in Fairall et al. (1996) with modifications to include the parameterized effect of sea spray according to Eqs. (17) and (18) for 10-m wind speeds greater than 10 ms^{-1}. The wave state required for calculating the sea spray generation is specified diagnostically by the surface wind stress. The counterparts of these coefficients and ratio

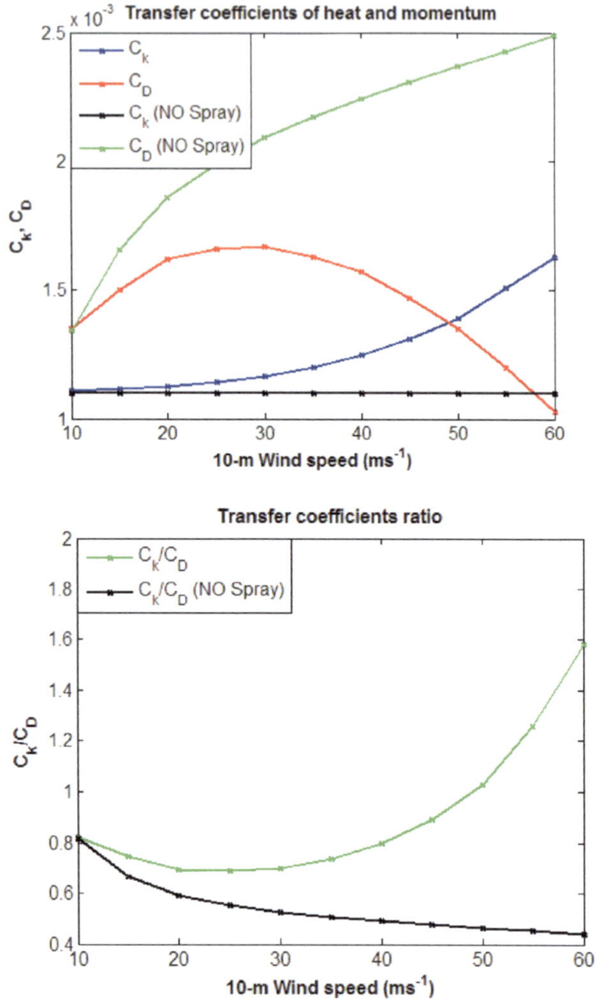

Fig. 1. Right panel shows the 10-m neutral drag coefficient C_D and enthalpy exchange coefficient C_k as functions of 10-m wind speed (Bao et al., 2011), while the left panel depicts the corresponding ratio of C_k/C_D.

without the effect of sea spray are also shown in Fig. 1 for comparison. The formulation for C_D without the sea spray effect combines the Fairall et al. (1996) formulation for low winds and the Moon et al. (2007) formulation for high winds based on the Powell et al. (2003) observational analysis. The formulation for C_k (assuming $C_k = C_H = C_E$) without the sea spray effect shown in Fig. 1 is based on the Fairall et al. (1996) formulation for low winds and the extrapolation of the fitting function of the CBLAST observations for high winds.

The prominent feature shown in Fig. 1 is that the increase of sea-spray mass generation with wind leads to a reduction (or an increase) in the drag coefficient (or the heat exchange coefficient) at hurricane strength winds. The behaviour of C_D with sea-spray effect as a function of the 10-m wind is consistent with recent observations (Powell et al., 2003; Black et al., 2007; Drennan et al., 2007; French et al., 2007; Zhang et al., 2008) and the newly developed surface resistance law (Makin, 2005; Andreas, 2004) over rough seas associated with extreme winds, both of which show that the drag coefficient is reduced at hurricane strength winds. The impact of sea spray on both C_D and C_k on the 10-m wind speed is insignificant until the 10-m wind speed is well above 30 ms^{-1}. The dependence of the ratio of C_k/C_D on the 10-m wind speed is consistent with C_H/C_D shown in Zhang et al. (2008) for 10-m winds below 30 ms^{-1} and in agreement with the theoretical requirement for a tropical cyclone to intensify (Emanuel, 1995) for 10-m winds above 30 ms^{-1}.

The physical behaviour of the scheme is consistent with the results from the explicit modelling of the TKE budget of the spray-laden surface boundary layer flow (Fairall et al., 1994; Kepert et al., 1999; Bianco et al., 2011) and the parameterized size distribution of sea-spray droplets (Fariall et al., 2009). When the 10-m wind speed is below 30 ms^{-1}, the droplets tend to be small in size and tend to evaporate substantially, and thus tend to cool the spray-filled layer. When the 10-m wind speed is well above 30 ms^{-1}, the size of droplets tend to be so big that they do not have enough time to evaporate that much before falling back into the sea. When the large droplets are still in the air, they have sufficient time to release sensible heat to the ambient air (Andreas, 1990, 1995), increasing the buoyancy of the surface layer and enhancing the turbulent mixing. On the other hand, the suspension of sea-spray droplets reduces the buoyancy and makes the surface layer more stable, lowering the friction velocity and the downward turbulent mixing of momentum is reduced. The overall effect of these two processes is to increase the enthalpy exchange coefficient and reduce the drag coefficient at high winds (>30 ms^{-1}). The fact that sea spray affects the intensity of turbulence at hurricane strength winds has an implication in the quantitative evaluation of the feedback effect of sea spray. That is, although the reduction (or increase) in the drag coefficient (or the heat exchange coefficient) at winds greater than 30 ms^{-1} depends only on the total mass and size distribution of sea spray mass generation and is independent of which bulk scheme is used, the total momentum and enthalpy fluxes are dependent on how the direct turbulent fluxes are calculated.

It should be pointed out that the parameterizations of the sea spray effect are different from the previous ones (such as Fairall et al., 1994; Andreas, 2004; Makin, 2005) in that the influence of sea spray on the air-sea momentum, sensible heat and moisture fluxes is simultaneous. It is also worth noting that, although the qualitative characteristics of the above parameterizations are physically and dynamically reasonable in comparison with the observational evidence from Black et al. (2007) and Zhang et al. (2008), the quantitative aspects of the parameterizations are subject to uncertainties due to the lack of observations to validate the spray generation function and the effect of droplets on the air-sea momentum and enthalpy fluxes. Further, by using the Monin-Obukhov similarity framework, the derivation of the parameterizations uses the assumptions about the steadiness and horizontal homogeneity of turbulent flow in the surface boundary layer of tropical cyclones. Admittedly, these assumptions may not be valid in the hurricane surface boundary layer. Despite the fact that the Monin-Obukhov similarity theory is widely accepted in the meteorological community for surface turbulent flux calculations in NWP models, the steadiness and horizontal homogeneity assumptions may impose a fundamental limitation on the validity of using the Monin-Obkhov similarity theory in NWP models for tropical cyclone prediction. Thus, the validity of our parameterization scheme is inevitably affected by such a limitation.

There are also intrinsic uncertainties in the physical model on which the formulation of the spray generation function is based. In the parameterization of the sea-spray impact on the surface momentum flux (Bao et al., 2011), the formulation for the so-called fall speed is used to estimate the fall speed at a given elevation for a given size of droplet. It may not be appropriate to assume that all the droplets will reach the steady fall speed before they fall back to the sea. Detailed observations are required to reduce these uncertainties. It is expected that the quantitative aspects of the parameterizations will continue evolving as future observations, along with the increasing theoretical knowledge of the interaction between turbulence and sea-spray droplets, become available. Nevertheless, the conceptual framework presented here will continue to be useful for properly taking into account the effect of sea spray in the air-sea momentum and heat transfer that is formulated in accordance with the Monin-Obukhov similarity theory.

2.5 Sensitivity of an NWP Model to the Parameterizations

In order to demonstrate the impact of the aforementioned sea-spray parameterizations on an intensifying tropical cyclone, sensitivity simulations are conducted using the 3.0.1.1 version of the Weather Research and Forecast Model (WRF) initialized with an idealized vortex. The WRF model is a nonhydrostatic, fully compressible mesoscale model (Skamarock et al., 2008). A cyclone-centered moving grid of 3 km grid spacing and 241×241 grid points, which is nested in a steady outer 9-km grid, is used for all the simulations. The outer 9-km grid has 481×481 grid points. There are a total of 43 vertical levels with the model top at 50 mb. To initialize the idealized vortex, the non-linear balance equation in the pressure-based sigma coordinate system described in Wang (1995) is solved within the WRF-grid framework on an *f* plane located at 12.5ºN. The mass field is derived from the wind field corresponding to an axisymmetric cyclonic vortex of maximum surface tangential wind set to 15 ms^{-1} at 280 km from the vortex centre that is embedded in a quiescent flow. The far field temperature and humidity are based on Jordan's Caribbean sounding (Gray et al., 1975). In all of the experiments, the sea surface temperature was set to 302.16 K. All the experiments were run for seven days. The physics options used in the simulations on both grids include the Yonsei University atmospheric boundary layer scheme, the WRF 6-class single-moment microphysics scheme which has prognostic equations for water vapour, cloud water, rain, cloud ice, graupel and snow, the Dudhia simple shortwave scheme, and the Rapid Radiative Transfer Model (RRTM) longwave scheme. No convective parameterization scheme is used. It is worth mentioning that in the surface flux scheme coupled with the Yonsei University atmospheric boundary layer scheme, the roughness length, z_0, for momentum flux is specified using the Chanock relationship, while the roughness length, z_{0h}, for the sensible and latent heat fluxes is defined through a friction-velocity Reynolds number (defined as $u_* z_{0h}/v$, where u_* is the friction velocity and n is the molecular viscosity of air). Four stability regimes are considered in the formulations of the stability functions. Details of all the physics schemes including corresponding references are available in Skamarock et al. (2008).

The impact of the parameterized sea-spray effect on the simulated minimum sea-level pressure and maximum surface wind speed (defined at the lowest model level which is about 30 m above the sea surface) is demonstrated by comparing three sensitivity runs (Fig. 2). The first run, in which the sea-spray parameterization is not included, is named as the control. In the second run (named the no impact on heat run), only the feedback impact of sea spray on the surface momentum flux (i.e., the friction velocity u_*) is taken into account. The full impact of sea spray on both the surface momentum and

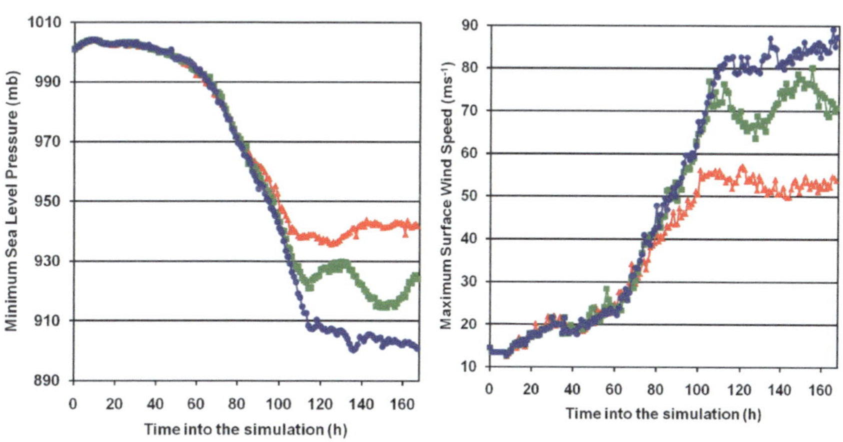

(a) (b)

Fig. 2. (a) Minimum sea-level pressure (mb) and (b) maximum surface wind speed (ms^{-1}). The control run (without the sea spray parameterization) is in red; the no impact on heat run (with the feedback of sea spray to the momentum flux only) is in green; and the full impact run (with the feedback of sea spray to both momentum and heat fluxes) is in blue.

heat fluxes is taken into account in the third run (named the full impact run). There are significant differences in the intensity of the simulated hurricane. In all three runs, the simulated cyclone intensification rate starts to slow down and level off by 100 hours after the initial time. During this period of reduced intensification rate (100-160 hours), the minimum sea level pressure is about 40 mb lower and the maximum surface wind speed is about 30 ms^{-1} faster in the full impact run than in the control run. In the no impact on heat run, the maximum surface wind speed is about 20 ms^{-1} faster than the control run, while the minimum sea level pressure is 20 mb lower.

The differences made by the inclusion of the sea-spray effect in the surface sensible heat, moisture and momentum fluxes are shown in Figs 3-5 valid at

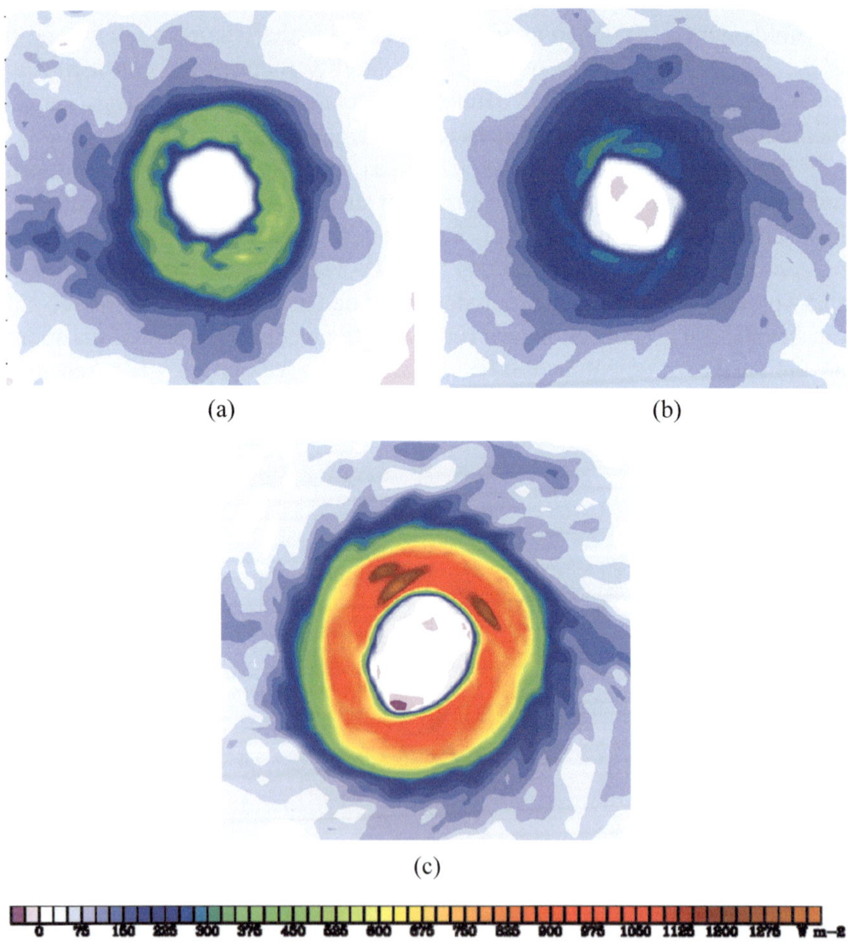

Fig. 3. The surface sensible heat flux (W m^{-2}) valid at 120 hours into the simulation from (a) the control run, (b) the no impact on heat run, and (c) the full impact run. The size of the plotting area is 300 km × 300 km with short tick marks separated by 3 km and long tick marks by 30 km.

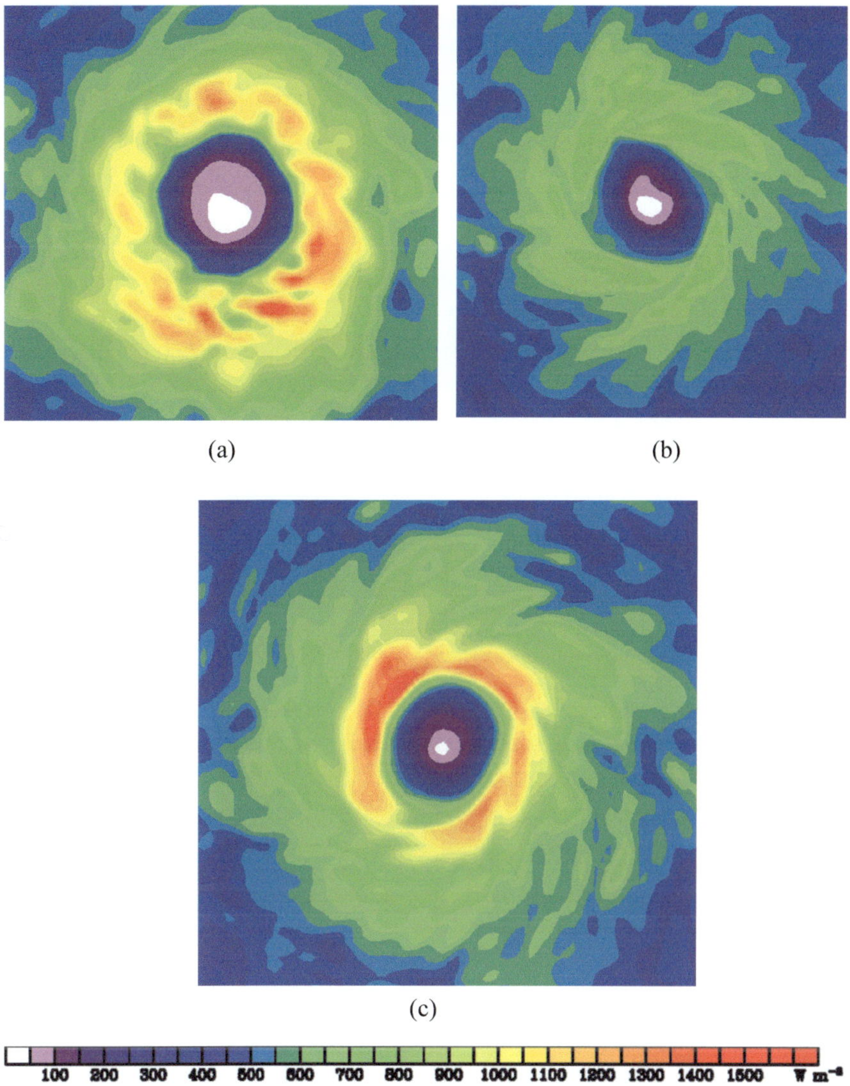

Fig. 4. The surface latent flux (W m^{-2}) valid at 120 hours into the simulation from (a) the control run, (b) the no impact on heat run, and (c) the full impact run. The size of the plotting area is 300 km × 300 km with short tick marks separated by 3 km and long tick marks by 30 km.

120 hours after the initial time. The momentum flux is shown in terms of the friction velocity. At this hour, the surface fluxes are significantly different due to the feedback of sea spray to the cyclone intensification. In the no impact on heat run, the surface heat fluxes are reduced corresponding to the reduction in the turbulence intensity associated with the reduced surface momentum flux (see Figs 3-5b compared to Figs 3-5a). In the full impact run, the effect of

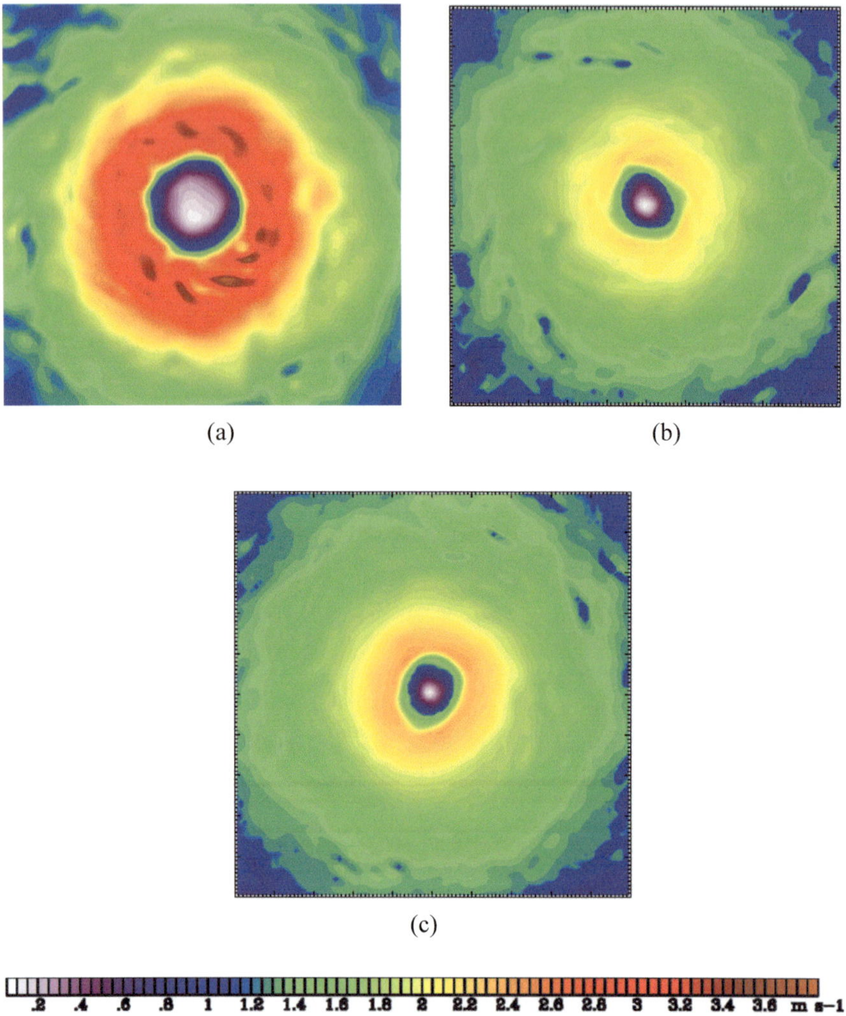

Fig. 5. The surface momentum flux in terms of the frictional velocity (m s^{-1}) valid at 120 hours into the simulation from (a) the control run, (b) the no impact on heat run, and (c) the full impact run. The size of the plotting area is 300 km × 300 km with short tick marks separated by 3 km and long tick marks by 30 km.

sea spray increases the maximum sensible and latent heat fluxes significantly near the centre of the cyclone (see Figs 3c and 4c). The maximum sensible heat flux in the full impact run is about two times greater than the control and the maximum latent heat flux is about 10% greater than the control. The differences in the sensible and latent heat fluxes also indicate that there is an increase of effective areal air-sea interface caused by spray droplets. The surface momentum flux in terms of the friction velocity from the full impact run is reduced by about 35% compared to the control run. The increase in the heat fluxes and decrease in the momentum fluxes are consistent with the trend in the dependence of the drag and heat exchange coefficients on wind speed that is shown in Fig. 1. These results support the notion that sea spray plays an important role in turbulent mixing near the air-sea interface. It should be noted that the differences in the surface fluxes are also the consequence of the difference in the near-surface wind distribution that is caused by the suspension and evaporation of spray droplets.

Figure 6 compares the hourly time series of the sensible heat flux, latent heat flux, and frictional velocity at the surface for the three runs that are

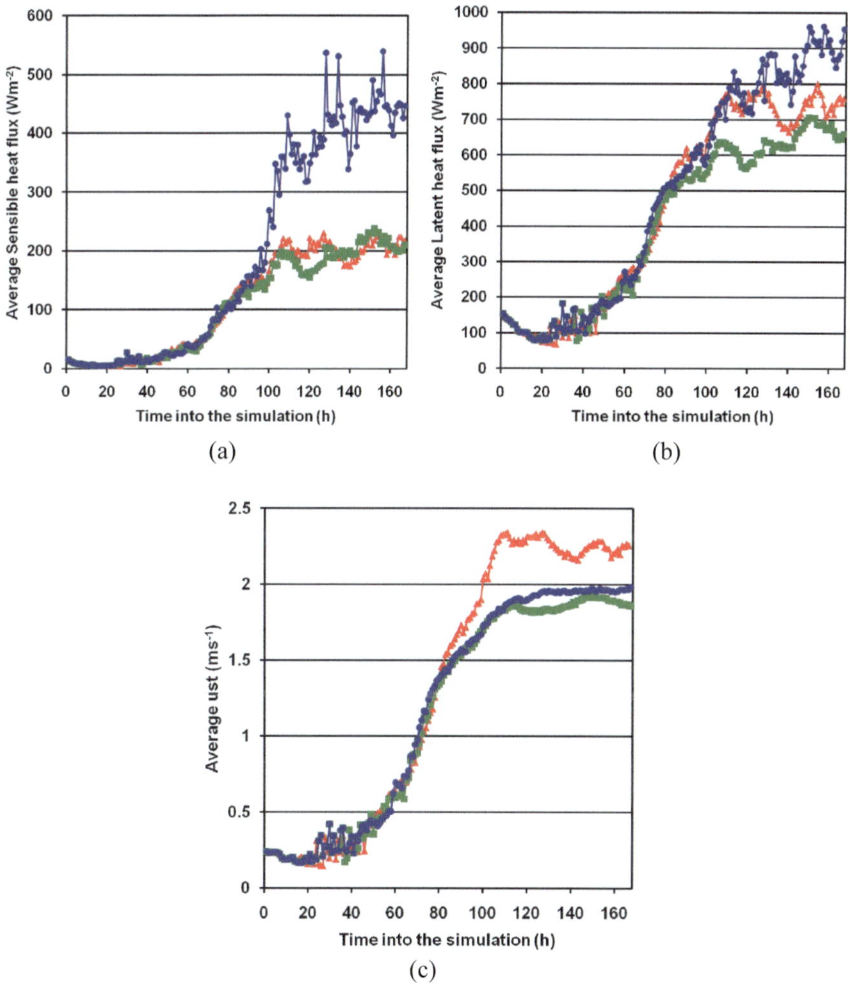

Fig. 6. Time series of (a) sensible heat flux (Wm^{-2}), (b) latent heat flux (Wm^{-2}), and (c) friction velocity (ms^{-1}) at surface averaged over a 150 km × 150 km box that is centered on the cyclone. The red line is the control run, the blue line is the full impact run, and the green line is the no impact on heat run.

averaged over a 150 km × 150 km box centered on the cyclone centre. It is seen that the full impact run has the greatest mean sensible and latent fluxes after about 100 hours, while the no impact on heat run has the least mean latent heat flux, but similar mean sensible heat flux as compared to the control run. Both runs with sea spray included have lower frictional velocities than the control, with the no impact on heat run having the lowest frictional velocity. These time series confirm the overall effect of sea spray on the surface fluxes when the surface wind speed exceeds 40 ms^{-1} and the sea spray production is significant enough to reduce the momentum flux but enhance the sensible and latent heat fluxes. The time series of the friction velocity clearly indicates that the mass loading of sea spray in the MSBL reduces the intensity of turbulent mixing, leading to the friction velocity in both runs with sea-spray effect being smaller than the control run.

Figure 7 illustrates the impact of sea spray on the azimuthally averaged structure of the ABL near the simulated cyclone centre averaged from 100 to 160 hours into the simulation for the three runs. This is the time period in all three simulations where the cyclone intensity begins to level off. Here, following Smith et al. (2009), the ABL near the cyclone center is defined as the layer of the near surface inflow. In all three runs, the maximum tangential wind speed is within the upper part of the ABL. However, there are differences in the depth in the ABL near the maximum tangential wind speed. The ABL depth within the eyewall is greater in the control and the full impact runs than in the no impact on heat run (about 1.2 km in the control and the full impact runs versus 1 km in the no impact on heat run). This suggests that the ABL depth in the eyewall tends to be decreased by the increase in the flow stability due to the sea-spray mass loading, but be increased by the spray-enhanced total enthalpy flux from sea to air. That is, the effect of spray mass loading

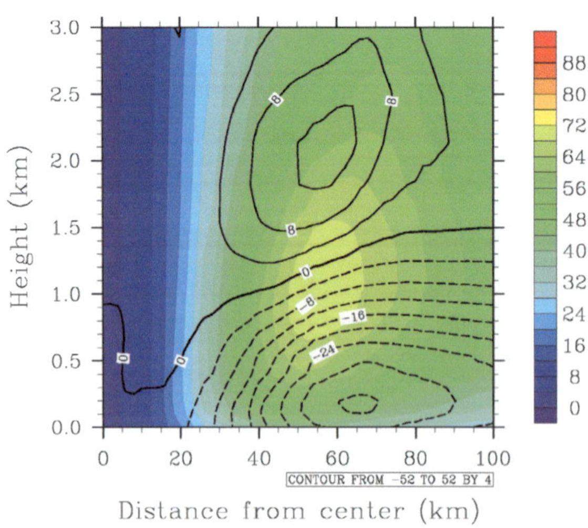

(a)

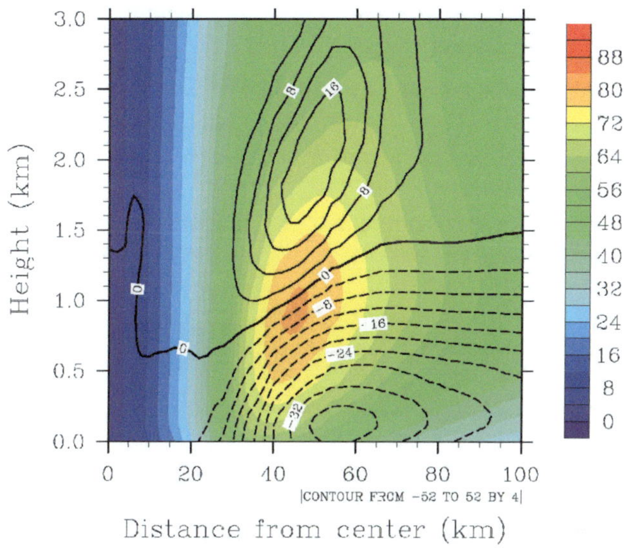

(b)

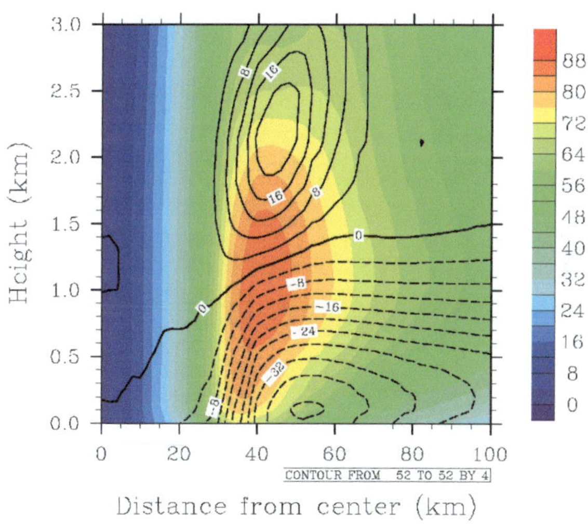

(c)

Fig. 7. Radius-height cross sections of azimuthally-averaged tangential and radial wind components in the lowest 3 km averaged over 100-160 h after the initial time for the (a) control run, (b) no impact on heat run, and (c) full impact run. The colour shaded contours are the tangential wind component (4 ms^{-1} interval), while the contour lines are the radial wind component (4 ms^{-1} interval and the negative indicates the direction toward the cyclone centre).

and enthalpy flux enhancement on the ABL depth in the eyewall is opposite. It is also obvious that the sea spray decreases the radius of maximum near-surface tangential wind speed, but increases the intensity of the radial inflow.

This leads to a physical interpretation for the positive impact of sea spray on the simulated cyclone intensification: the mass-loading of sea spray decreases the surface drag, which reduces the dissipation of the cyclone rotation, while the sensible and latent heat fluxes from the sea to the air are enhanced by sea spray and increase the buoyancy for deep convection in the cyclone, which intensifies the cyclone.

Figure 8 depicts the azimuthally and temporarily averaged radius-height cross sections of equivalent potential temperature (θ_e) within the lowest 3

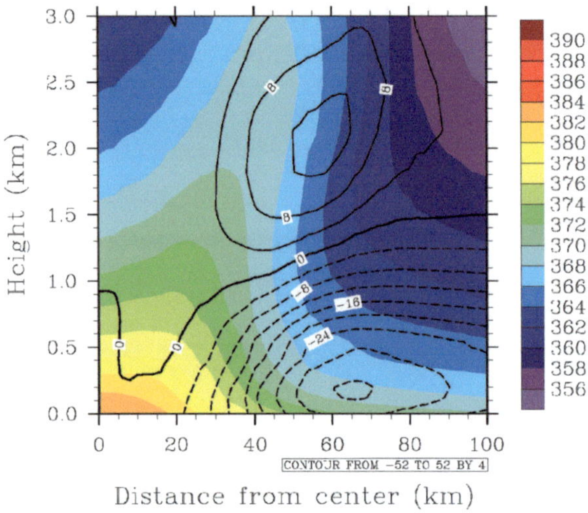

(a)

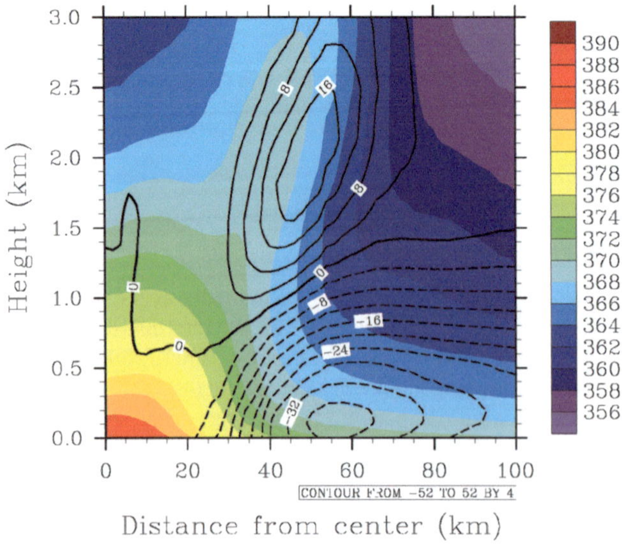

(b)

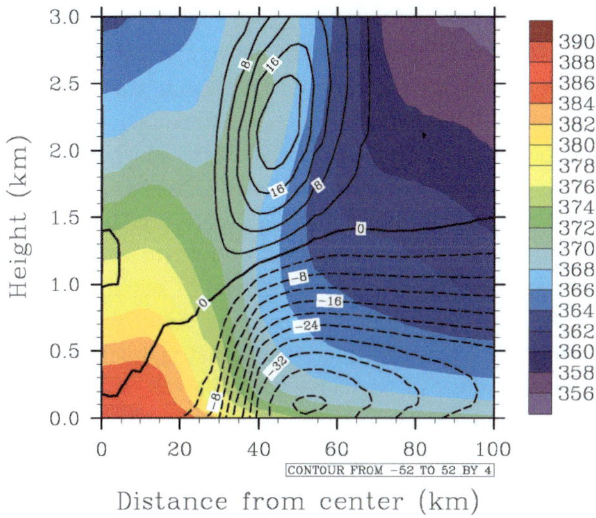

(c)

Fig. 8. Radius-height cross sections of azimuthally-averaged equivalent potential temperature and radial wind components in the lowest 3 km averaged over 100-160 h after the initial time for the (a) control run, (b) no impact on heat run, and (c) full impact run. The color shaded contours are equivalent potential temperature (2 K interval), while the contour lines are the radial wind component (4 ms^{-1} interval and the negative indicates the direction toward the cyclone centre).

km around the eyewall region for the three runs. These cross sections exhibit characteristics of a mature hurricane that are consistent with the low-level thermal structure of a mature hurricane eyewall shown by, e.g., Rotunno and Emanuel (1987), Persing and Montgomery (2003) and Houze (2010). The change in the slope of θ_e surfaces between the inflow and the outflow is the result of convection in the eyewall transporting higher θ_e air from the surface into the eyewall (characterized by the maximum of the outflow above the inflow in the MSBL). The radial gradient of θ_e near the surface corresponds to the radial pressure gradient associated with the cyclone intensity. When the radial gradient of θ_e increases, the surface pressure at the cyclone centre decreases and the winds at the surface strengthen. Thus, the differences in the radial gradient of θ_e near the surface clearly show that not only the intensity of the cyclone, but also its near-surface thermal structure, is strongly influenced by the sea-spray modification to the air-sea momentum and enthalpy fluxes.

3. Summary and Conclusions

This part of the chapter presents a parameterization scheme suitable for numerical weather prediction models to account for the effect of sea spray on the surface momentum and heat fluxes. This scheme is based on the conventional Monin-Obukhov similarity theory, which is a common framework for computing air-sea fluxes in numerical weather prediction models. According to the budget equation of TKE in the spray-filled surface layer, the effect of sea spray is considered in the scheme as a modification to the stratification of the profiles of wind, temperature and moisture in the MSBL. The scheme assumes that the Monin-Obukhov similarity relations hold in the MSBL above the elevation where wave breaking takes places and the resulting sea spray droplets are ejected into the air.

The scheme predicts that the overall impact of sea-spray droplets on the mean winds depends on the wind speed at the level of sea-spray generation. As the wind speed increases, the average droplet size and the total mass of droplets increases, leading to an increase in the overall wind speed in the surface layer above the level of sea-spray generation. Thermodynamically, the size of droplets at hurricane-intensity winds is so large that they do not have enough time to evaporate that much before falling back into the sea. When the large droplets are still in the air, they have sufficient time to release sensible heat to the ambient air and increase the buoyancy of the surface layer and enhancing the turbulent mixing. Mechanically, the suspension of sea-spray droplets reduces the buoyancy and makes the surface layer more stable. As a consequence, the friction velocity is lowered and the downward turbulent mixing of momentum is reduced. As the cyclone continues intensifying, the reduced friction velocity, however, leads to an increased shear between the surface and the flow above the surface layer, which in turn increases the shear-induced vertical mixing in the surface layer. Also, enhanced air-sea enthalpy flux provides necessary energy to further intensify the cyclone, which in turn induces more vertical mixing.

The testing of the scheme in the simulation of an intensifying idealized tropical cyclone using the WRF model shows that the inclusion of the scheme produces significant differences in the surface sensible heat, moisture and momentum fluxes at winds greater than 30 ms^{-1}. Overall, the scheme increases the cyclone intensity in terms of maximum surface wind and minimum sea-level surface pressure. A physical interpretation for the positive impact of sea spray on the simulated cyclone intensification is that the mass-loading of sea spray decreases the surface drag, which reduces the dissipation of the cyclone rotation, while the sensible and latent heat fluxes from the sea to the air are enhanced by sea spray to increase the buoyancy for deep convection in the cyclone, thus intensifying the cyclone.

We wish to caution that the quantitative aspects of the sea-spray parameterization scheme have considerable room for improvement, particularly when observational technology advances enough to provide reliable and accurate measurements of sea-spray generation and its interaction with the near-surface turbulence above surface waves. Thus, although the WRF model's response to the sea-spray parameterization scheme in the idealized simulation is physically consistent in terms of the scheme's prediction of the influence of sea spray on the air-sea momentum and enthalpy fluxes, the quantitative aspects of the sea-spray impact on the development of tropical cyclones are inevitably inconclusive. It is still unknown how in reality the physics of sea spray affect the ABL structure and turbulence characteristics associated with the inflow boundary layer and the outer rainbands of hurricanes that have been documented observationally (Powell, 1990a, 1990b; Barnes and Powell, 1995; Zhang et al., 2009). A key shortcoming of the sensitivity experiments presented in this chapter is the lack of fully coupled air-sea interaction physics in the sensitivity simulations. Further WRF simulations with the combined sea-spray physics and the oceanic vertical mixing physics are currently being investigated, and the results revealing the sea-spray effect on the air-sea interfacial characteristics such as those described by observations in the hurricane environment (Barnes and Powell, 1995; Cione et al., 2000) will be topics for future research.

Hurricane development is dynamically controlled by several processes, one of which is air-sea interaction. The parameterized sea-spray effect is just one important aspect of the air-sea interaction processes. Wave breaking, which is a critical component of surface wave dynamics and a root cause for sea-spray generation, is parameterized in the current scheme to be dependent on surface wind stress only. The dynamic link between sea-spray generation and wave state as output from a coupled atmosphere-wave-ocean modelling system is not addressed in this study. While it is conceptually straightforward, how to numerically couple the current parameterization scheme of sea-spray effect with the wave model in a fully coupled modelling system, in which wave breaking, sea-spray generation and air-sea energy transfer are interacting dynamically, remains a challenging research subject.

Acknowledgements

The author acknowledges NOAA's Hurricane Forecast Improvement Project that supported this work. The author acknowledges the support provided by Indo-US Science and Technology Forum, and various national and international collaborators who have helped improving the operational HWRF model.

REFERENCES

Adams, C.E. and G.L. Wetherly, 1981: Some Effect of Suspended Sediment Stratification on an Oceanic Bottom Boundary Layer. *J. Geophys. Res.*, **86**, 4161-4172.

Andreas, E.L., 2004: Spray stress revisited. *J. Phys. Oceanogr.*, **34**, 1429-1440.

Andreas, E.L., 1995: The temperature of evaporating sea spray droplets. *J. Atmos. Sci.*, **52**, 852-862.

Andreas, E.L., 1990: Time constants for the evolution of sea spray droplets. *Tellus*, **42B**, 481-497.
Andreas, E.L. and K.A. Emanuel, 2001: Effect of sea spray on tropical cyclones intensity. *J. Atmos. Sci.*, **58**, 3741-3751.
Andreas, E.L., J.B. Edson, E.C. Monahan, M.P. Rouault and S.D. Smith, 1995: The spray contribution to net evaporation from the sea: A review of recent progress. *Boundary-Layer Meteorology.*, **72**, 3-52.
Bao, J.-W., C.W. Fairall, S.A. Michelson and L. Bianco, 2011: Parameterizations of Sea-Spray Impact on the Air-Sea Momentum and Heat Fluxes. *Mon. Wea. Rev.*, **139**, 3781-3797.
Bao, J.-W., J.M. Wilczak, J.K. Choi and L.H. Kantha, 2000: Numerical simulations of air-sea interaction under high wind conditions using a coupled model: A study of hurricane development. *Mon. Wea. Rev.*, **128**, 2190-2210.
Bao, J.-W., S.A. Michelson, J.M. Wilczak and C.W. Fairall, 2002: Test of parameterizations of air-sea energy fluxes in a regional coupled atmosphere-ocean modeling system under high wind conditions. *In:* Atmosphere-Ocean Interactions, Advances in Fluid Mechanics. W. Perrie (ed.). WIT Press, Boston, pp. 115-153.
Barenblatt, G.I., 1996: Scaling, Self-Similarity, and Intermediate Asymptotics. Cambridge Univ. Press, Cambridge, U.K.
Barnes, G.M., 2008: Atypical thermodynamic profiles in hurricanes. *Mon. Wea. Rev.*, **136**, 631-643.
Barnes, G.M. and M.D. Powell, 1995: Evolution of the inflow boundary layer of Hurricane Gilbert (1988). *Mon. Wea. Rev.*, **123,** 2348-2368.
Bianco, L., J.-W. Bao, C.W Fairall and S.A. Michelson, 2011: Impact of sea spray on the surface boundary. *Boundar.-Layer Meteorol.*, accepted.
Black, P.G. et al., 2007: Air-Sea Exchange in Hurricanes: Synthesis of Observations from the Coupled Boundary Layer Air-Sea Transfer Experiment. *Bulletin of American Meteorological Society*, **88**, 357-374.
Brutsaert, W., 1982: Evaporation into the Atmosphere. D. Reidel.
Brutsaert, W., 1979: Heat and mass transfer to and from surfaces with dense vegetation or similar permeable roughness. *Boundary-Layer Meteorology*, **16**, 365-388.
Brutsaert, W. and M. Sugita, 1996: Sensible heat transfer parameterization for surfaces with anisothermal dense vegetation. *J. of Atmo. Sci.*, **53**, 209-216.
Cione, J.J., P.G. Blasck and S.H. Houston, 2000: Surface Observation in hurricane environment. *Mon. Wea. Rev.*, **128**, 1550-1561.
DeCosmo, J., K. Katsaros, S.D. Smith, R.J. Anderson, W.A. Oost, K. Bumke and H. Chadwick, 1996: Air-sea exchange of water vapor and sensible heat: The humidity exchange coefficient over the sea (HEXOS) results. *J. of Geophys. Res.*, **101**, 12001-12016.
Donelan, M.A., 1990: Air Sea Interaction. *In:* The Sea: Ocean Engineering Science (Volume 9). E.D. Goldberg et al. (eds). Wiley-Interscience, 239-292.
Drennan, William M., Jun A. Zhang, Jeffrey R. French, Cyril McCormick and Peter G. Black, 2007: Turbulent Fluxes in the Hurricane Boundary Layer. Part II: Latent Heat Flux. *J. Atmos. Sci.*, **64**, 1103-1115.
Emanuel, K.A., 1995: Sensitivity of tropical cyclones to surface exchange coefficients and a revised steady-state model incorporating eye dynamics. *J. Atmos. Sci.*, **52**, 3969-3976.
Fairall, C.W., J.E. Hare, A.A. Grachev, E.F. Bradley and J.B. Edson, 2001: Preliminary results from the ETL open ocean air-sea flux database. 11th Conference on

Interaction of the Sea and Atmosphere. American Meteorological Society, San Diego, California.

Fairall, C.W., J.D. Kepert and G.J. Holland, 1994: The effect of sea spray on surface energy transports over the ocean. *Global Atmos. Ocean Syst.*, **2**, 121-142.

Fairall, C.W., E.F. Bradley, D.P. Rogers, J.B. Edson and G.S. Young, 1996: Bulk parameterization of air-sea fluxes for TOGA COARE. *J. Geophys. Res.*, **101**, 3747-3764.

Fairall, C.W., M.L. Banner, W.L. Peirson, W. Asher, and R.P. Morison, 2009: Investigation of the physical scaling of sea spray spume droplet production. *J. Geophys. Res.*, **114**, C10001, doi:10.1029/2008JC004918.

French, Jeffrey R., William M. Drennan, Jun A. Zhang and Peter G. Black, 2007: Turbulent Fluxes in the Hurricane Boundary Layer. Part I: Momentum Flux. *J. Atmos. Sci.*, **64**, 1089-1102.

Garratt, J.R., 1992: The atmospheric boundary layer. Cambridge University Press, Cambridge.

Gill, A.E., 1982: Atmospheric-Ocean Dynamics. Academic Press.

Gray, W.M., E. Ruprecht and R. Phelps, 1975: Relative humidity in tropical weather systems. *Mon. Wea. Rev.*, **103**, 685-690.

Houze, R.A., Jr., 2010: Clouds in tropical cyclones. *Mon. Wea. Rev.*, **138**, 293-344.

Jarosz, E., D.A. Mitchell, D.W. Wang and W.J. Teague, 2007: Bottom-up determination of air-sea momentum exchange under a major tropical cyclone. *Science*, **315**, 1707-1709.

Janssen, P.A.E.M., 1991: The quasi-linear theory of wind wave generation applied to wave forecasting. *J. of Phys. Oceanogr.*, **21**, 1631-1642.

Kader, B.A. and A.M. Yaglom, 1972: Heat and mass transfer loss for fully turbulent wall flows. *International Journal of Heat and Mass Transfer*, **15**, 2329-2351.

Kepert, J.D., C.W. Fairall and J.-W. Bao, 1999: Modeling the interaction between the atmospheric boundary layer and evaporating sea spray droplets. *In:* Air–Sea Exchange: Physics, Chemistry and Dynamics. G.L. Geernaer (ed.). Kluwer. 363-407.

Kraus, E.B. and J.A. Businger, 1994: Atmosphere-Ocean Interaction. Oxford University Press.

Large, W.G. and S. Pond, 1982: Sensible and latent heat flux measurements over the ocean. *J. of Phys. Oceanogr.*, **12**, 464-482.

Lykossov, V., 2001: Atmospheric and oceanic boundary layer physics. *In:* Wind stress over the Ocean, I.S.F. Jones and Y. Toba (eds). Cambridge University Press, 58-81.

Makin, V.K., 2005: A note on the drag of the sea surface at hurricane winds. *Bound.-Layer Meteorol.*, **115**, 169-176.

Makin, V.K. and C. Mastenbroek, 1996: Impact of waves on air-sea exchange of sensible heat and momentum. *Boundary-Layer Meteorology*, **79**, 279-300.

Moon, I., I. Ginis, T. Hara and B. Thomas, 2007: Physics-based parameterization of air-sea momentum flux at high wind speeds and its impact on hurricane intensity predictions. *Mon. Wea. Rev.*, **135**, 2869-2878.

Mueller, J.A. and F. Veron, 2009: A Lagrangian stochastic model for heavy particle dispersion in the atmospheric marine boundary layer. *Bound.-Layer Meteorol.*, **130**, 229-247.

Persing, J. and M.T. Montgomery, 2003: Hurricane superintensity. *J. Atmos. Sci.*, **60**, 2349-2371.

Powell, M.D., 1990a: Boundary layer structure and dynamics in outer hurricane rainbands. Part I: Mesoscale rainfall and kinematic structure. *Mon. Wea. Rev.*, **118**, 891-917.

Powell, M.D., 1990b: Boundary layer structure and dynamics in outer hurricane rainbands. Part II: Downdraft modification and mixed layer recovery. *Mon. Wea. Rev.*, **118**, 918-938.

Powell, M.D., P.J. Vickery and T.A. Reinhold, 2003: Reduced drag coefficient for high wind speeds in tropical cyclones. *Nature*, **422**, 279-283.

Rotunno, R. and K.A. Emanuel, 1987: An air-sea interaction theory for tropical cyclones. Part II: An evolutionary study using a hydrostatic axisymmetric numerical model. *J. Atmos. Sci.*, **44**, 543-561.

Skamarock, W.C., J.B. Klemp, J. Dudhia, D.O. Gill, D.M. Barker, W. Wang and J.G. Powers, 2008: A description of the Advanced Research WRF Version 3. NCAR Tech Notes-475+STR.

Smith, S.D., 1988: Coefficients for sea surface wind stress, heat flux, and wind profiles as a function of wind speed and temperature. *J. of Geophys. Res.*, **93**, 15467-15472.

Smith, R.K., M.T. Montgomery and S.V. Nguyen, 2009: Tropical cyclone spin-up revisited. *Q. J. R. Meteorol. Soc.*, **135**, 1321-1335.

Stull, R.B., 1997: Reply. *J. Atmos. Sci.*, **54**, 579.

Wang, Y., 1995: On an inverse balance equation in sigma-coordinates for model initialization. *Mon. Wea. Rev.*, **123**, 482-488.

Wang, Y., J.D. Kepert and G. Holland, 2001: On the effect of sea spray evaporation on tropical cyclone boundary layer structure and intensity. *Mon. Wea. Rev.*, **129**, 2481-2500.

Zhang, J.A., P.G. Black, J.R. French and W.M. Drennan, 2008: First direct measurements of enthalpy flux in the hurricane boundary layer: The CBLAST results. *Geophys. Res. Lett.*, 35:L14813, doi:10.1029/2008GL034374.

Zhang, J.A., W.M. Drennan, P.G. Black and J.R. French, 2009: Turbulence structure of the hurricane boundary layer between the outer rainbands. *J. Atmos. Sci.*, **66**, 2455-2467.

Zilitinkevich, S.S., A.A. Grachev and C.W. Fairall, 2001: Scaling reasoning and field data on the sea surface roughness lengths for scalars. *J. of Atmos. Sci.*, **58**, 320-325.

7

Present State of Knowledge of Electrification and Lightning within Tropical Cyclones and Their Relationships to Microphysics and Storm Intensity

Alexandre O. Fierro
Cooperative Institute for Mesoscale Meteorological Studies
University of Oklahoma and NOAA/OAR/National Severe Storms Laboratory, Norman, Oklahoma
afierro@ou.edu

1. Lightning Morphology within Tropical Cyclones and Relationships with Storm Intensity

Over the last three decades, several observational works have supported the existence of a systematic relationship between lightning activity, intensity changes and structural changes within tropical cyclones (TCs). This section summarizes the salient findings of these works.

Samsury and Orville (1994) analyzed the lightning evolution of a weak tropical storm and a major hurricane within an 18 h window that included the periods before and after the storm made landfall. They found that, regardless of cloud-to-ground (CG) flash polarity, lightning primarily occurred within convective regions often associated with the largest reflectivites. In contrast to major hurricanes, tropical storms showed a significant increase in CG flash rate after making landfall by as much as a factor of three, rates which still remain far below those observed within continental systems (e.g., MacGorman et al., 1989). Molinari et al. (1999) confirmed that weaker TCs and tropical storms tended to increase their total flash rate prior to landfall while no significant increase was observed for major hurricanes. Most importantly, they found that three of their hurricane cases showed clear evidence of lightning outbreaks within the eyewall as the latter underwent significant changes in intensity and/

or in structure, such as an eyewall replacement cycle or eyewall contraction (Willoughby, 1982, 1990a). Similar results were found by Abarca et al. (2011) using lightning data collected by the World Wide Lightning Location Network (WWLLN, Jacobson et al., 2006; Lay et al., 2007). Figure 1 shows examples of such eyewall lightning outbreaks in Hurricanes Katrina and Rita (2005) that intensified rapidly before making landfall in the southern coast of the United States (Fierro et al. 2011).

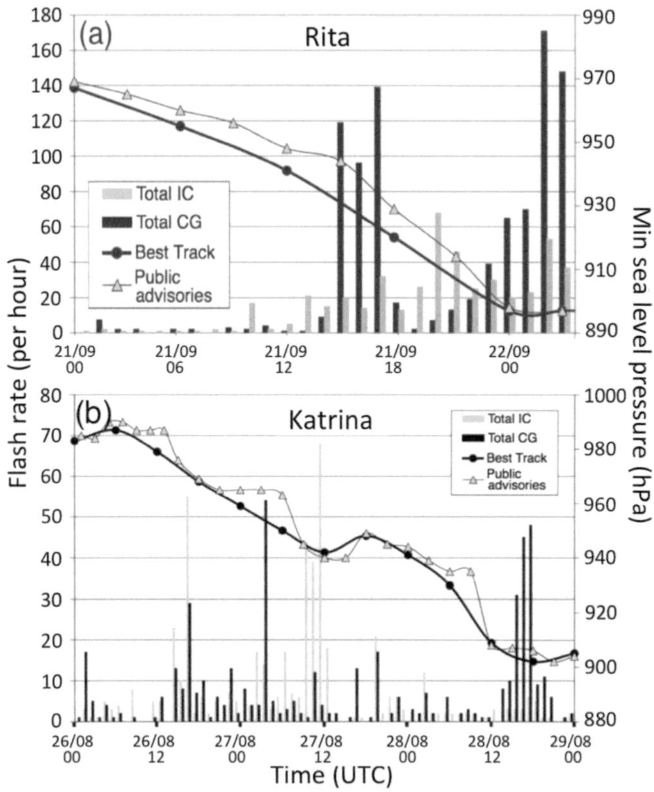

Fig. 1. Time series of intracloud (IC: light gray) and cloud-to-ground (CG: black) lightning hourly flash rate for Hurricane Rita (a), Katrina (b) overlaid with a scatter plot of the NHC best- track (black dots) and NHC public advisories estimated minimum central pressure (triangles). Adapted from Fierro et al. (2011).

Using optical transient detector observations within several TCs over the Atlantic, Cecil and Zipser (1999) showed that the greater flash rates in the eyewall were preferentially associated with strong TCs or weak tropical storms. Cecil and Zipser (1999) also found that a temporal lag existed between the production of ice scattering signatures (proportional to convective intensity and lightning production) and TC strengthening. Based on this result, they argued that lightning activity could be used as a reliable forecast tool as any changes in its patterns were likely associated with intensity change of the TC and even a change in track. Similar eyewall flash trends and spatial patterns were also observed within other rapidly intensifying TCs over the Pacific (Solorzano et al., 2008; Pan et al., 2014), and for TCs worldwide (Price et al., 2009). Using a comprehensive CG lightning dataset of TCs over the north Atlantic and east Pacific basin, DeMaria et al. (2012) suggested that CG lightning activity in the rainbands might be a better surrogate for the imminent occurrence of a rapid intensification cycle. Their study also found that inner core lightning in these two basins was rather correlated with the end of the intensification cycle of a TC.

Similar lightning outbreaks relationships (i.e., with intensification) were also documented within developing tropical systems. Using a very low frequency (VLF) a long-range lightning network, Chronis et al. (2006) and Lepert and Petersen (2010) showed that from all the African easterly waves (AEW) that evolved from the African continent to the Atlantic ocean, the ones exhibiting the strongest total lightning activity were more likely to intensify and become TCs. Lightning activity peaks in late afternoon (Carnegie curve) therefore, the waves exiting the African continent during this period of the day had more chance to become TCs over ocean as they were likely composed of convective elements having collectively stronger updrafts.

Using data from an experimental long-range lightning detection network from Vaisala, Demetriades and Holle (2005) showed that for nine Atlantic hurricanes (such as Lili and Isidore, 2002, Isabel and Fabian, 2003 and Charley, 2004), an eyewall lightning outbreak occurred in 11 of the 12 eyewall cycles present in the study. They also found that eyewall lightning outbreaks were generally occurring during the early intensification stage of the TC. Rodgers et al. (2000) also showed that for intensifying TCs, the closer the lightning is to the storm's centre, the more likely the TC will intensify, since these lightning flash bursts are likely associated with explosive convective bursts or convective band cycles (e.g., Lyons et al., 1989; Black et al., 1993; Simpson et al., 1998) or with supercell(s) growth (Lyons and Keen, 1994). Such sudden convective development was hypothesized to arise from advection of a statically unstable layer into the lower troposphere above an unusually moist boundary layer (BL) with sea surface temperatures (SSTs) of at least 28°C, leading to a sudden local increase in convective available potential energy (CAPE) in the eyewall.

Further supporting the linkage between eyewall lightning and TC intensity change, Kelley et al. (2004) suggested that extremely deep eyewall clouds (most likely to produce lightning) observed via TRMM in the eyewall were coincident with a 70% likelihood of storm intensification. On the other hand, strong mature TCs remaining in a quasi-steady state (e.g., Hurricane Isabel, 2003) exhibited small lightning flash rate in the eyewall (Molinari et al. 1999; Demetriades and Holle, 2005). Willoughby (1990) showed that for weak TC or tropical depressions, convective eruption near the centre could cause the intensification to falter and the cyclone path to become discontinuous. Hence for the case of weaker tropical systems, the lightning outbreaks accompanying strong convective bursts might precede the temporal weakening of the system and/or an imminent change in its track (as in Lei and Chan, 1999).

Due to the relative high frequency of these events, an important scientific question would be to determine if typical lightning exhibits preferential spatial patterns within TCs. Towards this goal, Orville and Coyne (1999) investigated the CG lightning flash characteristics and patterns within 34 TCs between 1986 and 1996. They found that as the TC became more organized, weak lightning activity was present in the eyewall, while the minimum (maximum) activity was observed in the inner band stratiform (outer band) regions. In a more recent study, Cecil et al. (2002b) used aircraft observation of 45 TCs during the TRMM field experiment to investigate lightning activity and reflectivity patterns within the eyewall, inner and outer rainbands. They found that the total lightning flash rates in the eyewall and, particularly within the outer rainbands, were generally much higher than those within the inner band region mainly composed of stratiform precipitation and relatively weak updrafts. Higher lightning flash rates in the rainbands were also documented for Hurricane Andrew (1992) (Molinari et al., 1994) and Hurricane Lili (1996) (Morales et al., 1996). A significant result from this study was that lightning frequency measured within the TC 300 km and 50 km radius tended to indicate a state of TC intensification or temporary weakening.

Using the National Lightning Detection Network (NLDN, Cummins et al., 1998) lightning data derived from nine Atlantic hurricanes, Molinari et al. (1999) was able to establish the existence of three distinct convective regimes in mature hurricanes. These are also used by Cecil et al. (2002a, b) and are:

(i) The eyewall, having similar attributes to that of deep, weakly electrified, monsoonal convection.
(ii) The region outside the eyewall (≈ 90 km radius) resembling the trailing stratiform region of mesoscale convective systems (MCSs) region with higher +CG flash rates relative to the two other regions of the TC mentioned here with a +CG to –CG ratio exceeding 25%, which is still lower than for MCSs (e.g., Holle et al., 1994).
(iii) The outer bands (≈ 250 km radius) characterized by the vast majority of the total ground flashes.

Examples of the three aforementioned patterns are shown in Fig. 2. In the presence of ambient vertical wind shear, a TC producing lightning mostly left of the vertical wind shear vector was indicative of a storm that was successfully resisting the wind shear and, hence, that was deepening. The opposite was true for lightning activity occurring to the right of the wind shear vector (Corbosiero and Molinari, 2002; Molinari et al., 2006; Molinari and Vollaro, 2010).

Most of those pioneer studies on eyewall lightning activity, however, focus largely on CG activity as their higher peak current relative to IC flashes are more readily detected by state-of-the-art very low frequency (VLF)

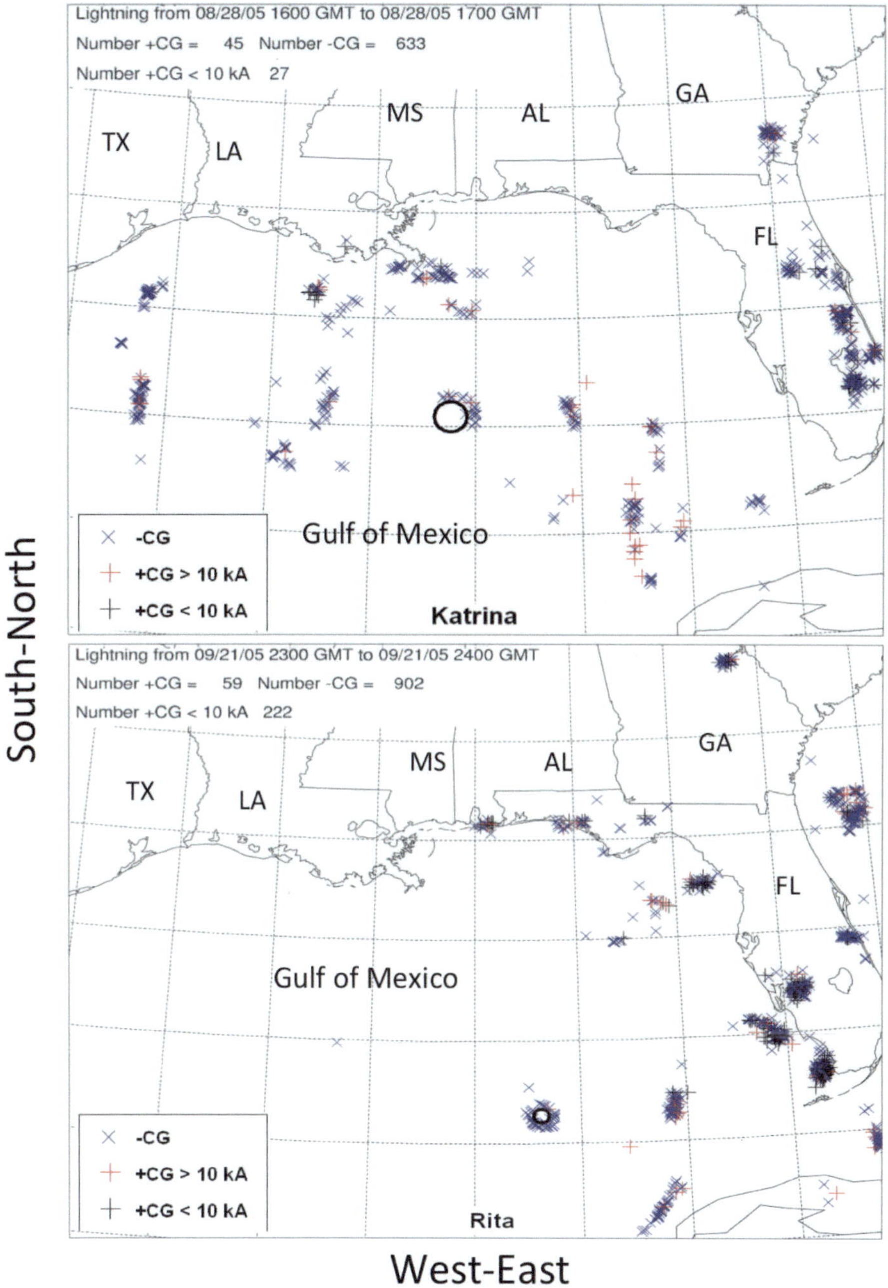

Fig. 2. Maps of CG flashes detected by the National Lightning Detection Network (NLDN) for one hour time period for (top) Hurricane Katrina and (bottom) Rita, on 28 Aug 2005 and 21 Sep 2005, respectively. The approximate eye size is depicted by a thick black circle and the legends for the symbols are shown at the bottom left of each panel. Adapted from Fierro et al. (2007).

lightning networks. Recently, Fierro et al. (2011) carried out the first study showing a three-dimensional view of the total lightning activity (i.e., IC + CG) during the rapid intensification (RI) cycle of three major TCs, namely Katrina, Rita and Charley (Fig. 3 for Rita). They correlated intensification periods with well-defined IC discharge bursts (Fig. 3a, b) associated with eyewall hot towers rotating around the eye (Fig. 3b, c) and a concomitant increase in IC discharge heights (Fig. 3a).

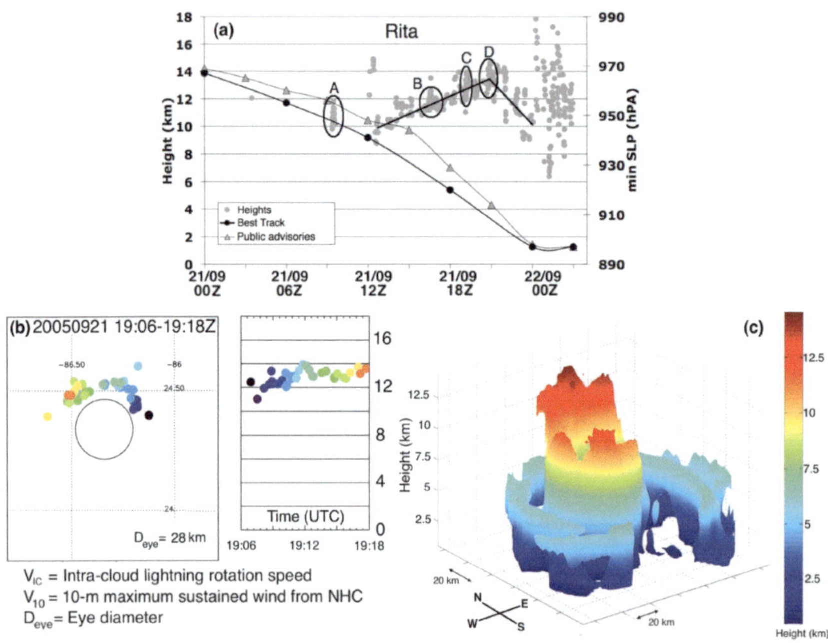

Fig. 3. (a) Time series of minimum surface pressure overlaid with observed IC flash discharge heights. The convective bursts associated with a local maximum in IC flash rate are highlighted by black ovals. (b) Horizontal projection and time-height plot of the IC flashes burst labelled D in (a). (c) Side-view three-dimensional renderings of the 30-dBZ isosurface from in situ aircraft tail Doppler radar data at 1915 UTC 21 Sep for Hurricane Rita. The height above sea level was overlaid onto the isosurface with legends shown on the right. Cardinal axes were added for clarity. (Data courtesy: NOAA/HRD.) Figures adapted from Fierro et al. (2011).

The observations from Fierro et al. (2011), however, only included the most intense IC discharges (or narrow bipolar events) with peak currents on the order of 10 kA (Smith et al., 1999), which only make up about 1-5% of the total IC activity produced by a TC (Shao et al., 2006). The detection of order of magnitude smaller (1 kA amplitude) flashes would require the use of very-high-frequency (hereafter, VHF) sensors similar to the ones mounted in the Oklahoma Lightning Mapping Array and developed at New Mexico Tech (e.g., Rison et al., 1999; MacGorman et al., 2008). The strong relationship between total lightning flash rates with updraft intensification (in contrast with the lack of relationship of CG flash rates, Reap and MacGorman, 1989; Wiens et al., 2005) suggests that systems mapping all lightning would be preferable for monitoring the convection within the eyewall (hot towers) and, hence, the potential for intensification of TCs. Despite this limitation, their study found that high peak current IC discharges were key in helping to distinguish the difference in convective regimes occurring in the eyewall of a TC during or prior to RI periods.

Based on studies focusing on continental systems (e.g., MacGorman et al., 1989), it is likely that if IC flash rates were detected with a better efficiency (i.e., 50% or greater), the observed peak in total lightning flash rate (IC + CG) within TCs would be larger by at least a factor two. This is because IC flashes tend to generally outnumber CG flashes (MacGorman et al., 1989; Rutledge et al., 1992; Boccippio et al., 2001; Schultz et al., 2011). Several investigators such as Mackerras et al. (1998) and Rutledge et al. (1992) showed that the IC to CG ratio exhibited relatively high values in the deep tropics and comparatively small values in the midlatitudes, which they hypothesized could be attributed to deeper cloud tops (higher mean tropopauses). This would cause the volume of cloud above the main negative charge region of the mixed phase region in tropical convection to be two to three times larger than in midlatitude storms, which combined with an exponentially decreasing breakeven electric field with height would be favourable for significant IC production. Rutledge et al. (1992) and Williams and Rutledge (1990) stressed, however, that cloud top height could be misleading and showed that instead, IC lightning flashes were most likely to occur in storms where the 30 dBZ echo top height exceeded the melting level, which is indicative of an active mixed-phase region for the non-inductive charging mechanism involving collisions between riming graupel and ice crystals in the presence of supercooled water (e.g., Takahashi, 1978). On their part, Williams and Stanfill (2002) argued that the larger total flash rates in continental storm (i.e., compared to oceanic storms) were caused by a larger Bowen ratio (sensible heat over latent heat flux) and by more buoyant boundary layer parcels which were able to convert convective available potential energy more efficiently to kinetic energy within the storm's updraft.

2. Lightning, Convection and Microphysics within Tropical Cyclones

In the last 30 years, many studies have focused their attention on maritime tropical convection. These studies do provide a good insight on the kinematical and microphysical conditions leading to significant electrification, mainly within the TC outer rainbands. In TC eyewalls, however, the strong tangential winds at low-to-midlevels and the outwardly tilted nature of the updrafts have to be accounted for. Nonetheless, the necessary ingredients for electrification within TC eyewalls remain the same as for the outer bands.

Takahashi (2003) showed that in maritime convection, the ice crystals and graupel concentration, and hence lightning, was half that of oceanic systems in Eastern Asia. Takahashi (2003), therefore, suggested that different precipitation processes should be expected between oceanic and continental maritime convection. Petersen et al. (1999) also analyzed the electric activity within oceanic convective systems during the Tropical Ocean Global Atmosphere Coupled Ocean-Atmosphere Response Experiment (TOGA COARE) field programme and concluded that lightning occurrence in oceanic convection was preferentially observed in deep, vertically developed cells showing reflectivity values $\geq$30 dBZ above the –10°C level (near $z = 7$ km) for periods exceeding 10 min. Consistent with this, Bedrick and Burgett (1999) showed that the onset of lightning within Supertyphoon Keith (1997) typically occurred after the 30 dBZ contour reached its maximum altitude within the mixed region (at about the –10°C level). Petersen et al. (1999) also showed that, akin to continental convection (Carey and Rutledge, 1998), CG occurrence was highly correlated with the progressive descent of the 30 dBZ reflectivity contour. On the same line, Cecil et al. (2002b) found that the larger flash rates observed within the rainbands of Hurricane Andrew (1992) and Lilly (1996) was coincident with larger reflectivities above the freezing level.

Petersen et al. (1999) findings for tropical convection also supports Simpson et al. (1998)'s, who suggested that in the case of Tropical Storm Oliver (1993), strong updrafts above the melting level were crucial for significant electrification to develop. On the same lines, Black and Hallett (1990) argued that vertical velocities exceeding 12 m s^{-1} near the melting level were associated with substantial electrical activity in the cases of Hurricanes Norbert (1984), Emily (1987) and Hugo (1989). Williams and Rutledge (1990) provided a detailed observational study of the electrical activity of deep convection near Darwin, Australia, during the Down Under Doppler and Electricity Experiment (DUNDEE) field programme. They also found that convective strength (proportional to CAPE) was correlated to intense lightning activity within these cells. Williams and Rutledge (1990) hypothesized that the low lightning rates commonly observed in TCs resulted from low CAPE values. This lower instability within TC centres is consistent with the presence of a warm core at upper levels, where magnitude increases as the TC intensifies (e.g. Jorgensen et al., 1985). Petersen et al. (1995) further proposed that for tropical MCSs, updraft velocities reaching 6 m s^{-1} near the 6 km level might be a threshold required for significant electrification, because this threshold equals that of the terminal fall speed of millimeter sized supercooled water drops.

At present, it is widely accepted that the relatively small electrical activity within TCs is chiefly caused by updrafts in the mixed phase region that are too weak to allow sufficient production and lofting of graupel particles and supercooled water necessary for sufficiently large electric fields to develop via the non-inductive charging process (Takahashi, 1978; Saunders et al., 1991;

Saunders and Peck, 1998). Rogers et al. (2007) showed that the majority of vertical motions within TC eyewalls were weak (i.e., $|w| < 2$ m s^{-1}) with 1-2% reaching 6 m s^{-1}. This finding was consistent with Black et al. (1996) who found that 70% of the vertical velocities in seven Atlantic TCs ranged between –2 and 2 m s^{-1}. Those two studies also showed that the broadest vertical velocity distribution was found in the eyewall with about 5% exceeding 5 m s^{-1}. Rogers et al. (2007) found that modal values of vertical motions were negative in the lowest 2 km and positive above 9 km due to the reduction of liquid water loading caused by hydrometeor loss. Black et al. (1996) found that within TCs, updrafts generally outnumbered downdrafts by at least a factor of two and that updrafts are wider and stronger than downdrafts. They further showed that, unlike updrafts, downdrafts did not exhibit any apparent radial sloping.

Despite the above, it is not uncommon to find updrafts speeds exceeding 10 m s^{-1} within TC eyewalls, particularly when the storm is undergoing rapid intensification. Black et al. (1994) reported updraft/downdraft speeds reaching 24/19 m s^{-1}, respectively, in Hurricane Emily (1987) during its deepening phase. These unusually strong updrafts were associated with the presence of large supercooled water droplets (i.e., with a diameter ranging between 2-3 mm) at altitudes of about 12 km AGL (Black et al., 2003). Such occurrences are not rare; for instance, Bansemer et al. (2002) found high particle concentrations of about 1000 L^{-1} at temperatures as cold as –57°C and large particles (7 mm) at altitudes up to ~13 km in Hurricane Humberto (2001). Similarly, *in situ* aircraft microphysical data measured in Hurricane Bonnie (1998) suggested that supercooled water may exist at temperatures as low as –40°C and that these particles were likely advected rapidly there, preventing instantaneous freezing from occurring (Black and Heymsfield, 2002). Guimond et al. (2010) also reported similar updraft magnitudes (~20 m s^{-1}) during the rapid intensification of Hurricane Dennis (2005).

Consistent with overall weak updrafts, TC eyewalls generally contain little amount of supercooled water. Black (1984) showed that in the case of Hurricane Allen (1980) and Irene (1981), little amount of liquid water drops/droplets were found within updrafts above the freezing level. They found that only updrafts wider than 1 km having vertical velocities greater than 5 m s^{-1} contained predominantly liquid water drops and droplets. Black and Hallett (1986) reported, however, that *not all* updrafts having vertical velocities greater than 5 m s^{-1} contained appreciable amount of liquid water droplets above the freezing level. Supporting the lack of liquid water within TC updrafts above the freezing level, Black and Hallett (1986) showed that TC convection remains largely glaciated at temperatures as warm as –5°C. The sharp decrease in reflectivity with height observed within the eyewall and the outer rainbands further suggest that very little supercooled liquid water is found above the –10°C level. Jorgensen et al. (1985) hypothesized that the rapid falloff of radar reflectivities and cloud water content above the freezing level were attributed to warm-cloud coalescence processes. The relative weak mean updrafts caused an early fallout of the majority of the drops before they reached the freezing level. Moreover, the smaller concentration of cloud condensation nuclei (CCN) over oceans results in a droplet spectrum having fewer smaller droplets (of radius <10 μm) and greater amount of larger droplets (of radius >20 μm) owing to fewer droplets competing for the same amount of available moisture (e.g., Rogers and Yau, 1989).

Consequently, little amount of smaller and lighter droplets are lofted above the freezing level in maritime clouds, which reduces/prevents the formation of graupel pellets within the mixed-phase region. Moreover, the few large graupel pellets often observed within the larger updrafts tend to accrete cloud water quickly leading to the progressive glaciation of the cloud within the mixed phase region. Black and Hallett (1999) found that in the outer region of TCs eyewalls, supercooled water was rapidly depleted due to the erosion of abundant amounts of ice present in downdrafts (scavenging). In other words, they suggested that seeding by ice particles generated upstream of the radial flow limited the amount of supercooled water within the eyewall. Consistent with this, observations made by May and Rajopadhyaya (1996) within a tropical squall line also revealed that mature cells were dominated by the effect of glaciation and that only updrafts characterized by vertical velocities ≥10 m s^{-1} (which are relatively rare within TCs) contained appreciable amounts of supercooled water ~2 km above the freezing level. Strong updrafts were usually found nearby high reflectivity cores (30-40 dBZ echoes), while the opposite was not always true (Black and Hallett, 1986). Moreover, they found that these reflectivity cores primarily contained graupel and rounded ice particles but not liquid water, in line with Willis and Heymsfield (1989).

During the NOAA and NASA Intensity Forecasting Experiment (IFEX, July 2005) aircraft field experiment campaign, *in situ* measurements showed that in the vicinity of Tropical Storm Dennis' ragged "eyewall", updraft speeds at flight levels (between 4.2 and 5.8 km AGL) were reaching 6 m s^{-1} and that liquid (ice) particle concentration there were about 6 g kg^{-1} (4 g kg^{-1}). When the aircraft was at 5.5 km AGL (–4°C) they found a mixture of ice crystals and rimed particles with some having a diameter exceeding 1.6 mm. Houze et al. (1992) found that the ice particles at the 6 km level in Hurricane Norbert (1984) tended to be relatively large in both the eyewall and the outer rainbands. In the eyewall most of these particles were graupel, while in the outer stratiform region the dominant particle type was aggregate in moderate concentration (consistent with Black, 1984; Black and Hallett, 1986; Willis and Heymsfield, 1989). Houze et al. (1992) also hypothesized that the smaller and more abundant (100-300 L^{-1}) particles present between the eyewall and the outer stratiform region might have been produced by splintering of the graupel particle within eyewall updrafts. Graupel particles were spatially coincident with the radius of maximum winds because, in contrast to the lighter aggregates, these are not as susceptible to azimuthal advection owing to their relatively larger terminal fall speed.

Black and Hallett (1999) suggested that charge separation in TCs depends upon the relative mix of supercooled water in the eyewall updrafts and the ice particles that have been ejected from the eyewall inside the inner band stratiform region. They proposed that the electrical activity is favoured in a confined "mixed-phase region" between these updrafts and downdrafts near the storm's core (Fig. 4).

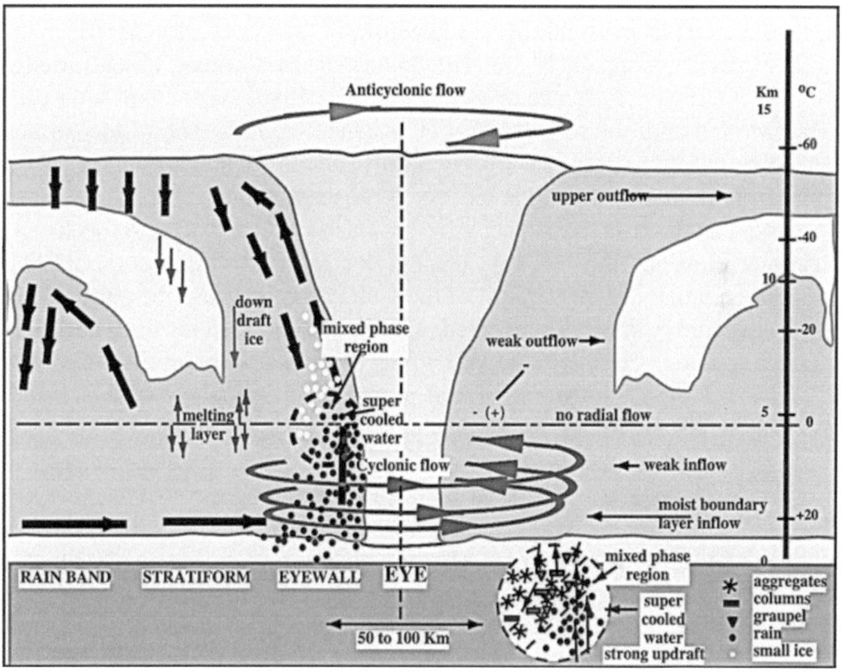

Fig. 4. Schematic of a mature TC cross-section across the eyewall of precipitation processes, cloud particle distribution, charge polarity and flow structure. All symbols are explicitly described in the figure. Adapted from Black and Hallett, 1999.

Using trajectory analysis for the case of Hurricane Emily (1987), Willis and Heymsfield (1989) hypothesized that the ice particles growing and descending through the TC stratiform region were detrained by large reflectivity cores back into the lower levels of the eyewall. The high reflectivity values in the eyewall resulted from the fallout of graupel, raindrops and frozen drops close to their source. They also posited that the graupel confined within convective region was active in the nucleation (through rime splintering) of abundant supply of ice crystals, which initiated the aggregation process outside the convective region. Furthermore, Black and Hallet (1986) argued that most of the ice present in the eyewall updrafts originate from the "ice multiplication" mechanism and is then redistributed throughout the storm by the upper and midlevel anticyclones.

Using various data sources (e.g., TRMM, or CAMEX-3, Kakar, 2006) Zipser and Jiang (2003) found that, in general, eyewall regions tended to have much more precipitation ice and liquid water mass than the outer and inner rainbands. Nevertheless, liquid precipitation mass was still larger than ice precipitation mass within the eyewall.

The strong horizontal winds in the eyewall advects/spreads charged particles over a broader area, thus preventing these particles from residing sufficiently long enough for significant charge separation to occur in a given layer in the eyewall. This process thus might also explain why the vertical and horizontal electric fields observed within TC eyewalls are generally much weaker than in continental MCSs (Goodman and MacGorman, 1986; Rutledge and MacGorman, 1988; MacGorman et al., 1989) and even supercells (MacGorman and Nielsen, 1991). For instance Goodman and MacGorman (1986) showed that a single MCS could generate as much as 25% of the annual total lightning flashes at any site it passes through.

Based on the positive and negative CG flash (+CG and –CG, respectively) locations close to the eyewall (within 80 km), Molinari et al. (1994) hypothesized that the outward tilt of the updraft would cause the upper levels positive charge region to be advected radially outward towards the inner band, producing more +CG there. In other words this would support the possibility of a normal dipole outwardly tilted along the sloping eyewall. Samsury and Orville (1994) found, however, that in the case of TC Jerry (1989) and Hugo (1989), +CG flashes were not necessarily found within the inner band stratiform region. Also, the tilted dipole hypothesis is not widely accepted because for instance, recent modelling work from Mansell et al. (2005) showed that for +CG flashes to occur a negative charge region must exist beneath the lowest region of positive charge. In the eyewall, Black and Hallett (1995) showed that in the deep isolated convective cells of Hurricane Claudette (1991) graupel was carrying most of the negative charges at midlevels (~ –5°C).

3. Modelling of Tropical Cyclone Electrification

During the past 20 years, modelling work primarily focused on the large-scale processes controlling TC dynamics, data assimilation, track prediction, dynamics and thermodynamics, structure and structural changes, all of which led to significant progress in operational hurricane forecasts (e.g., Chapter 3 of this book and references therein; Marks and Shay, 1998; Fierro et al., 2009; Hendricks et al., 2011; DeMaria et al., 2014). In comparison, there have been only three modelling studies dedicated at exploring explicitly the electrical processes within TCs and their relationship(s) to intensity changes (namely, Fierro et al., 2007; Fierro and Reisner, 2011; Fierro et al., 2015). Thus, still little is known about this physical aspect of the TC, though a better understanding of its nature would help improve short-term forecasts of their potential for intensity changes, both at sea and at landfall.

Although idealized (i.e., no terrain, constant Coriolis force and Rankine vortex initialization), the first study (Fierro et al., 2007) made use of a sophisticated three-dimensional stochastic branched lightning model (Mansell et al., 2002, 2005) with small ions (Chiu et al., 1978) coupled with a bulk microphysics scheme featuring 12 hydrometeor categories, 10 of which were in the ice phase (Straka and Mansell, 2005). The availability of a wide range of ice species in this model allowed for a broad spectrum of fall velocities, and hence a more realistic in-cloud charging structures representation (e.g., Saunders and Peck, 1998).

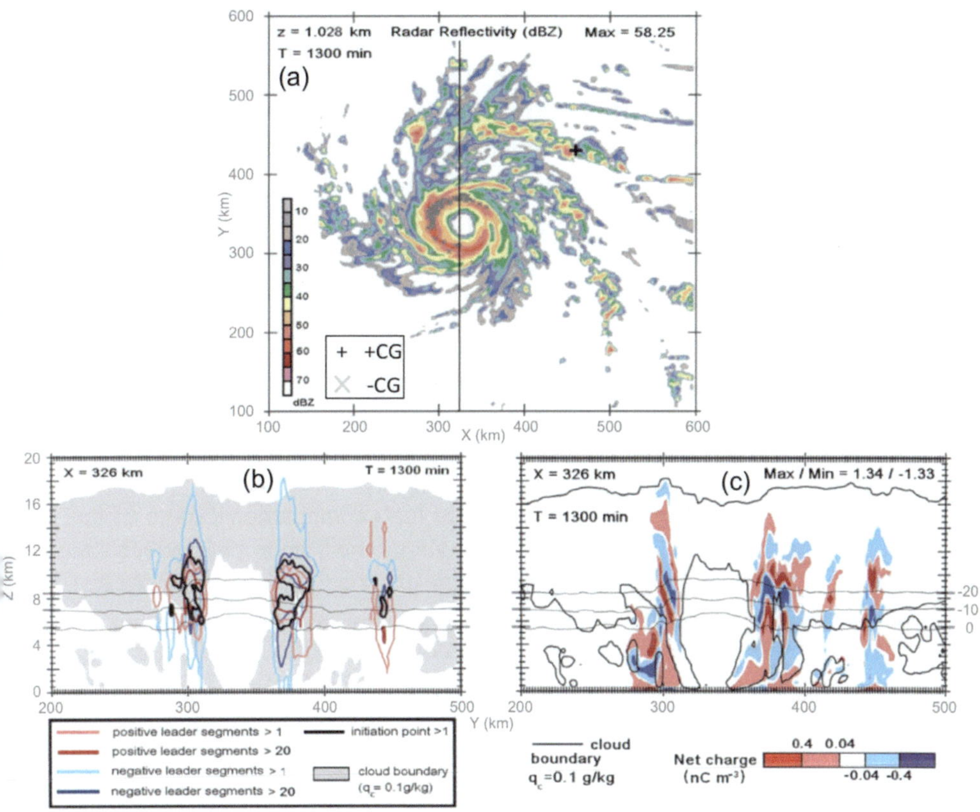

Fig. 5. (a) Horizontal cross section of simulated radar reflectivity (dBZ) from the SAM model. Locations of CG lightning strikes are shown by a grey cross for –CG flashes and by a black "+" symbol for +CG flashes. The flash locations were plotted for a 30 minutes interval up to the cross section time. A black line shows the location of the vertical cross section of (b) and (c). (b) Horizontal cross-sections across the eyewall of cloud mixing ratio >0.1 g kg^{-1} (grey shaded area) and positive (negative) leader segments in red (blue) contours (See legends for details). Total flashes are plotted for a period of 30 min prior to the cross section time. (c) Same as in (b) for net total amount of net charge in nC m^{-3} with legends for colours and shadings shown at the bottom. Adapted from Fierro et al. (2007).

The model was able to generate a realistic TC with a lightning pattern (Fig. 5a) consistent with those reported by past investigators (e.g., Orville and Coyne, 1999; Cecil and Zipser, 2002; Cecil et al., 2002). Similar to observations, the eyewall of the simulated TC was characterized by the presence of –CG flashes (Fig. 5a, b) associated with a simulated normal tripole charge structure (Fig. 5c, Williams et al. 1989). Detailed analysis of the eyewall in-cloud charging revealed that negative (positive) non-inductive charging of small graupel pellets (ice crystals, Saunders and Peck, 1998) dominated at mid (upper) levels while polarization charging (Ziegler et al., 1991) enhanced the lowest charge regions during the fallout of graupel/small hail pellets along with *in situ* charging of cloud water.

The second study from Fierro and Reisner (2011) presented the first real case modelling study (i.e., with terrain and mesoscale initialization) of TC electrification in an attempt to relate the occurrence of intensification phases to the evolution of total lightning activity (Fierro et al., 2011). Further improving upon Fierro et al. (2007), this study made use of observed radar and lightning data during the initialization procedure, which produced a simulation of Rita in good agreement with the observed storm in terms of track and overall convective structure. Consistent with the observations of Fierro et al. (2011) shown in Fig. 3, the model (Reisner et al., 2005) was able to reproduce rapidly rotating electrically active hot towers during intensification phases (Fig. 6), which were hypothesized to arise from transient eyewall barotropic instability modes coupled with local buoyancy enhancements. Despite those improvements, and contingent to the high computational costs of the simulations, the model made use of a single moment microphysics scheme with three ice categories, a simple diagnostic electric field relationship (based on Maxwell's equations) and a simplistic (1-D) discharge model, allowing no possible distinction between flash type and flash polarity.

Using explicit electrification and lightning parameterizations implemented within the Weather Research and Forecasting model (Fierro et al., 2013) coupled with a hybrid three dimensional discharge model (MacGorman et al., 2001). Fierro et al. (2015) performed a high resolution 350 m simulation of the electrification processes within a hurricane. The simulation was analyzed in conjunction with available total lightning observations to further augment our general understanding of some of the key cloud-scale electrification processes within TCs. The general environment and trends of Hurricane Isaac (2012), whose lightning activity was observed by the Earth Networks Total Lightning Network, were utilized to produce a reasonable TC simulation. In line with previous modelling work (Fierro et al., 2007; Fierro and Reisner, 2011), the gross vertical charge structure in the eyewall resembled a normal tripole or a positive dipole, depending on the location (Fig. 7b). The negative charge at middle levels and positive at upper levels arose primarily from non-inductive charging between graupel and ice crystals/snow. As some graupel melted into rain, the main midlevel negative charge region extended down to

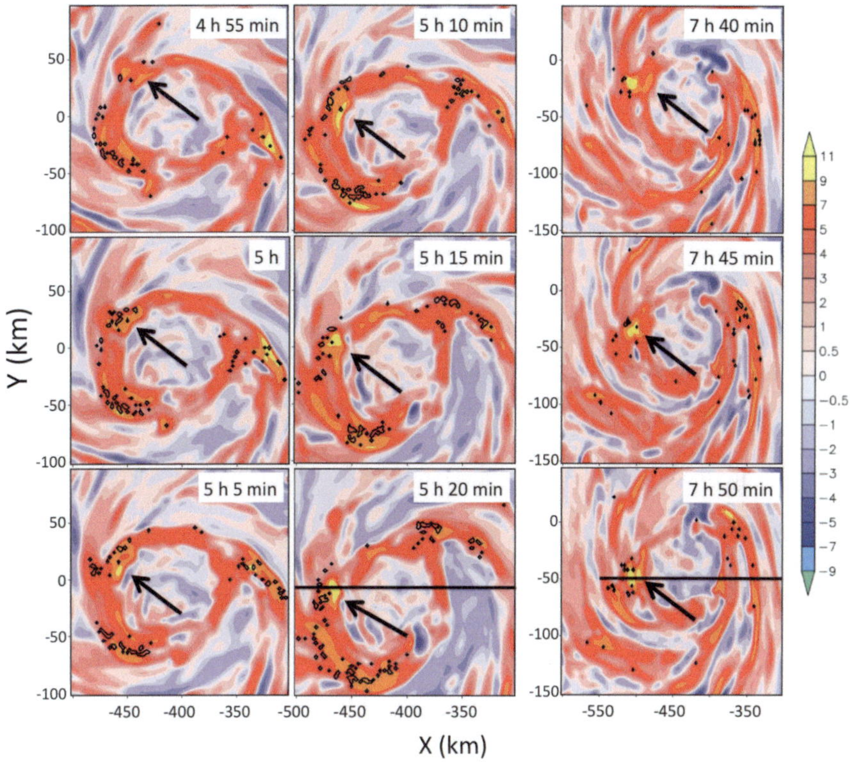

Fig. 6. Horizontal cross sections of 7–9-km layer-averaged vertical velocities (m s^{-1}) at selected times of the 2-km Rita simulation with the HIGRAD model (time shown in the top-right corner of each frame) overlaid with horizontal projections of accumulated 50-s lightning discharges zoomed over the eyewall. Two examples of convective events revolving around the eyewall with the primary circulation are shown. The black arrow indicates the location of the convective event of interest. Figures adapted from Fierro and Reisner (2011).

the surface in some places (Figs 7a, b). The large anvil cloud was comprised of positively charged snow, which in turn resulted in a radially extensive negative screening layer to form on the lighter ice crystals above it (Figs 7a, b). One salient results of this modelling study was that, similar to continental storms and tropical convection, the simulated lightning activity in the eyewall was best correlated with ice water path ($r > 0.7$) and graupel+hail path ($r \approx 0.7$) followed by composite reflectivity at temperatures smaller than −10°C ($r \approx 0.55$). Consistent with observations in continental storms (e.g., Wiens et al., 2005), the relative maxima in eyewall updraft volume and graupel volume were well correlated in time with relative maxima in eyewall total lightning rates. Furthermore, maxima in total eyewall lightning rates were systematically coincident with the end of an intensification phase, in line with DeMaria et al. (2012).

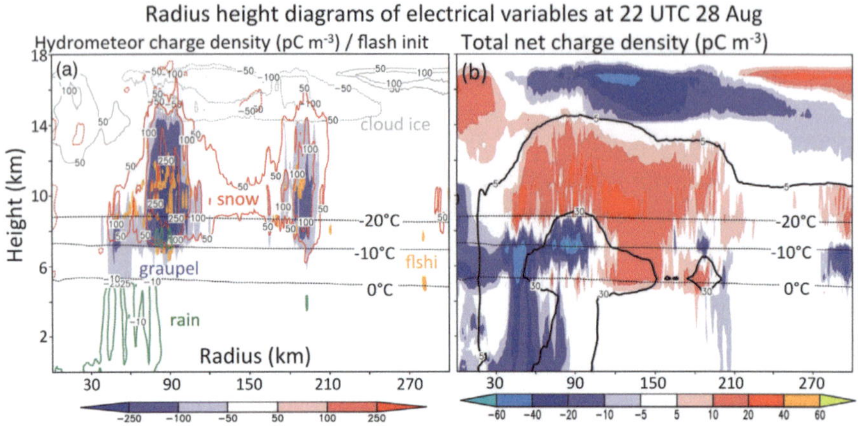

Fig. 7. Storm-centered, radius-height diagrams of azimuthally-averaged radar reflectivity fields for selected simulated electrical variables at 2200 UTC on 28 Aug. 2012. (a) Shows the charge density (in pC m^{-3}) on rain (green), snow (red), cloud ice (grey dashed line) and graupel (blue shading) overlain with flash initiation locations (contour of 1) in orange shading. For convenience, the colour coding of each of the hydrometeor species in (a) also are indicated by text in their corresponding colour. Legends for colours and shadings are shown at the bottom of each corresponding panel. (b) Shows the total net charge density (pC m^{-3}) with the black lines highlighting the 5 dBZ (surrogate for cloud boundary) and 30 dBZ contours (surrogate for significant hydrometeor mass). Figure adapted from Fierro et al. (2015).

Modelling work in this relatively unexplored topic is still warranted to augment our understanding of the physical linkages between lightning activity and intensity changes of a TC, behaviour reported and well-recognized by several investigators (e.g., Molinari et al., 1994, 1999; Kelley et al., 2004; Price et al., 2009; Fierro et al., 2011; DeMaria et al., 2012; Pan et al., 2014). Recent individual and collaborative TC research for about ten years led to original contributions on the understanding of the lightning and electrification within TCs (Fierro et al., 2007, 2011; Fierro and Reisner, 2011; Fierro et al., 2015), tropical (Fierro et al., 2008) and continental (e.g., Mansell et al., 2005; Fierro et al., 2006) convective systems. Since the occurrence of lightning is inherently tied to the evolution of the cloud microphysics, particularly graupel and liquid water content (e.g., Takahashi, 1978; Saunders and Peck, 1998), emphasis should also be directed towards reproducing a reasonable and realistic representation of the hydrometeor fields within the key regions of the TC.

The community is, however, well aware of the inherent difficulty in comparing observed and modelled hydrometeor categories in a TC regardless of hydrometeor categories as very limited TC microphysical data have been collected up to date (McFarquhar and Black, 2004; Black and Hallett, 1999).

McFarquhar and Black (2004) found that while the drop size distribution within TCs are overall well represented by exponential functions (Marshall and Palmer, 1948), there exist noteworthy differences in how the intercept and slope parameters describing these distributions vary with mass content. Current microphysical codes generally assume that the intercepts generally increase with mass content. McFarquhar and Black (2004) found, however, that outside TCs the intercepts decreased as a function of mass content. Further complicating the problem, notable differences were reported between the characteristics of the size distributions in convective regions and the stratiform regions, especially for graupel (the microphysical variable showing the best correlation with lightning). Nonetheless, checking that hydrometeors form in 'expected' regions based on basic microphysical and kinematic aspects of TCs and other storm types remains a reasonable way to verify microphysics and to partially circumvent the limitations arising from limited available observations. For example, is graupel getting more dense in heavy riming regions where cloud drops are larger? Are hailstones formed only in very limited regions of the TC, perhaps in the deep convective towers in the outer bands and eyewall towers? Radar reflectivity patterns in observed TCs are useful for comparison with those of any simulated TCs to ensure that the reflectivity solutions are within the bounds of observations (e.g., Fierro et al., 2011; Fierro and Reisner, 2011).

Also, observations of key electrical variables such as space charge evolution, charge moments and, most importantly, low-sferic IC flashes are not available or cannot be readily observed. Thus, future studies dedicated at explicitly simulating lightning and electrification within TCs using retrospective lightning observations could help augment our understanding of the underlying functional relationship(s) between lightning activity and storm intensity.

REFERENCES

Abarca, S.F., K.L. Corbosiero and D. Vollaro, 2011: The World Wide Lightning Location Network and Convective Activity in Tropical Cyclones. *Mon. Wea. Rev.*, **139**, 175–191.

Bansemer, A.R., A.J. Heymsfield and P.T. Willis, 2002: In situ measurements of particle size distributions in Hurricane Humberto. Preprints, 25th Conference on Hurricanes and Tropical Meteorology, American Meteorological Society, San Diego, CA.

Bedrick, M. and W. Burgett, 1999: A tropical oceanic cloud-to-ground lightning study. Preprints, 23rd Conference on Hurricanes and Tropical Meteorology, American Meteorological Society, Dallas, TX, 243–245.

Black, R.A., 1984: Distribution of particles types above 6.0 km in two Atlantic hurricanes. Preprints, 16th Conference on Hurricanes and Tropical Meteorology, American Meteorological Society, San Diego, CA, 537–541.

Black, R.A. and J. Hallett, 1986: Observations of the distribution of ice in hurricanes. *J. Atmos. Sci.*, **43**, 802–822.

Black, P.G., R.A. Black, J. Hallett and W.A. Lyons, 1986: Electrical activity of the hurricane. Preprints, 23d Conf. on Radar Meteorology, American Meteorological Society, Snowmass, CO, J277–J280.

Black, R.A. and J. Hallett, 1990: Electric field and microphysical measurements in vigorous hurricane eyewalls. Preprints, Conference on Cloud Physics, American Meteorological Society, San Francisco, CA, 662–665.

Black, R.A., J. Hallett and C.P.R. Saunders, 1993: Aircraft studies of precipitation and electrification in hurricanes. Preprints, 17th Conference on Severe Local Storms and Conference on Atmospheric Electricity, American Meteorological Society, St. Louis, MO, J20–J25.

Black, R.A., H.B. Bluestein and M.L. Black, 1994: Unusually strong vertical motions in a caribbean hurricane. *Mon. Wea. Rev.*, **122**, 2722–2739.

Black, R.A., J. Hallett and P. Willis, 1995: Aircraft measurements of ice and electrical evolution in regions of vertical shear in convective cloud. 21st General Assembly of IUGG, Boulder, Colorado.

Black, M.L., R.W. Burpee and F.D. Marks, 1996: Vertical motion characteristics of tropical cyclones determined with airborne doppler radial velocities. *J. Atmos. Sci.*, **53**, 1887–1909.

Black, R.A. and J. Hallett, 1999: Electrification of the hurricane. *J. Atmos Sci*, **56**, 2004–2028.

Black, R.A., G.M. Heymsfield and J. Hallett, 2003: Extra large particle image at 12 km in a hurricane eyewall: Evidence of high altitude supercooled water ? *Geophys. Res. Lett.*, **30**, No 21, 2124, doi:10.1029/2003GL017864.

Blackwell, K.G., 2000: The evolution of hurricane Danny (1997) at landfall: Doppler-observed eyewall replacement, vortex contraction/intensification and low-level wind maxima. *Mon. Wea. Rev*., **128**, 4002–4016.

Boccippio, D.J., Kenneth L. Cummins, H.J. Christian and S.J. Goodman, 2001: Combined Satellite- and Surface-Based Estimation of the Intracloud–Cloud-to-Ground Lightning Ratio over the Continental United States. *Mon. Wea. Rev.*, **129**, 108–122.

Braun, S.A., 2002: A Cloud-Resolving Simulation of Hurricane Bob (1991): Storm structure and eyewall buoyancy. *Mon. Wea. Rev*., **130,** 1573–1592.

Carey, L.D. and S.A. Rutledge, 1998: Electrical and multi-parameter radar observations of a severe hailstorm. *J. Geophys. Res.*, **103(D12)**, 13,979–14,000.

Cecil, D.J. and E.J. Zipser, 1999: Relationships between Tropical Cyclone Intensity and Satellite-Based Indicators of Inner Core Convection: 85-GHz Ice-Scattering Signature and Lightning. *Mon. Wea. Rev.*, **127**, 103–123.

Cecil, D.J., E.J. Zipser and S.W. Nesbitt, 2002a: Reflectivity, ice scattering, and lightning characteristics of hurricane eyewalls and rainbands. Part I: Quantitative description. *Mon. Wea. Rev*., **130**, 769–784.

Cecil, D.J. and E.J. Zipser 2002b: Reflectivity, ice scattering, and lightning characteristics of hurricane eyewalls and rainbands. Part II: Intercomparison of observations. *Mon. Wea. Rev*., **130**, 785–801.

Chiu, C.-S., 1978: Numerical study of cloud electrification in an axisymmetric, time-dependent cloud model. *J. Geophys. Res.*, **81**, 5025–5049.

Chronis, T. and E. Anagnostou, 2006: Evaluation of a Long-Range Lightning Detection Network with Receivers in Europe and Africa. *IEEE Transaction on Geosciences and Remote Sensing*, **44**, 1504–1510.

Corbosiero, K.L. and J. Molinari, 2002: The effects of vertical wind shear on the distribution of convection in tropical cyclones. *Mon. Wea. Rev.*, **130**, 2110–2123.

Cummins, K.L., M.J. Murphy, E.A. Bardo, W.L. Hiscox, R.B. Pyle and A.E. Pifer, 1998: A Combined TOA/MDF Technology Upgrade of the U.S. National Lightning Detection Network. *J. Geophys. Res.*, **103(D8)**, 9035–9044, doi:10.1029/98JD00153.

DeMaria, M., C.R. Sampson, J.A. Knaff and K.D. Musgrave, 2014: Is tropical cyclone intensity guidance improving? *Bull. Amer. Meteor. Soc.*, **95**, 387–398.

DeMaria, M., R. DeMaria, J. Knaff, and D. Molenar, 2011: Tropical Cyclone Lightning and Rapid Intensity Change. *Mon. Wea. Rev.*, **140**, 1828–1842.

DeMaria, M., R.T. DeMaria, J.A. Knaff and D. Molenar, 2012: Tropical cyclone lightning and rapid intensity change. *Mon. Wea. Rev.*, **140**, 1828–1842.

Demetriades, N.W.S. and R.L. Holle, 2005: Long-range lightning applications for hurricane intensity and precipitation nowcasting. Preprints, Conference on Meteorological Application of Lightning Data, American Meteorological Society, San Diego, CA.

Demetriades, N.W.S. and R.L. Holle, 2006: Long-range lightning nowcasting applications for tropical cyclones. Conference on Meteorological Application of Lightning Data, American Meteorological Society, Atlanta, GA.

Fierro, A.O., M.S. Gilmore, E.R. Mansell, L.J. Wicker, J.M. Straka and E.N. Rasmussen, 2006: Electrification and lightning in an idealized boundary-crossing supercell simulation of 2 June 1995. *Mon. Wea. Rev.*, **134**, 3149–3172.

Fierro, A.O., L.M. Leslie, E.R. Mansell, J.M. Straka, D.R. MacGorman and C. Ziegler, 2007: A high resolution simulation of the microphysics and electrification in an idealized hurricane-like vortex. *Meteorology and Atmospheric Physics*, **98**, 13–33.

Fierro, A.O. and J.M. Reisner, 2011: High Resolution Simulation of the Electrification and Lightning of Hurricane Rita during the period of Rapid Intensification. *J. Atmos. Sci.*, **68**, 477–494.

Fierro, A.O., X.-M. Shao, J.M. Reisner, J.D. Harlin and T. Hamlin, 2011a: Evolution of Eyewall Convective Events as Indicated by Intra-Cloud and Cloud-to-Ground Lightning activity during the Rapid Intensification of Hurricanes Rita and Katrina. *Mon. Wea. Rev.*, **139**, 1492–1504.

Fierro, A.O., E.R. Mansell, C.L. Ziegler and D.R. MacGorman, 2013: The implementation of an explicit charging and discharge lightning scheme within the WRF-ARW model: Benchmark simulations of a continental squall line, a tropical cyclone and a winter storm. *Mon. Wea. Rev.*, **141**, 2390–2415.

Fierro, A.O., E.R. Mansell, C.L. Ziegler and D.R. MacGorman, 2015: Explicitly simulated electrification and lightning within a tropical cyclone based on the environment of Hurricane Isaac (2012). *Mon. Wea. Rev.*, In Press.

Frank, W.M. and E.A. Ritchie, 1999: Environmental effects on hurricane core structure. Preprints, 23rd Conference on Hurricanes and Tropical Meteorology, American Meteorological Society, Dallas, TX, 680-681.

Gardiner, B., D. Lamb, R.L. Pitter, J. Hallet and C.P.R. Saunders, 1985: Measurements of initial potential gradient and particles charges in a Montana thunderstorm. *J. Geophys. Res.*, **90**, 6079–6086.

Goodman, S.J. and D.R. MacGorman, 1986: Cloud-to-ground lightning activity in mesoscale convective complexes. *Mon. Wea. Rev.*, **114**, 2320–2328.
Guimond, S.R., G.M. Heymsfield and F.J. Turk, 2010: Multiscale observations of Hurricane Dennis (2005): The effects of hot towers on rapid intensification. *J. Atmos. Sci.*, **67**, 633–654.
Hendricks, Eric A., Melinda S. Peng, Xuyang Ge and Tim Li, 2011: Performance of a dynamic initialization scheme in the Coupled Ocean–Atmosphere Mesoscale Prediction System for Tropical Cyclones (COAMPS-TC). *Wea. Forecasting*, **26**, 650–663.
Jacobson, A.R., R. Holzworth, J. Harlin, R. Dowden and E. Lay, 2006: Performance assessment of the World Wide Lightning Location Network (WWLLN) using the Los Alamos Sferic Array (LASA) as ground truth. *J. Atmos. Oceanic Technol.*, **23**, 1082–1092.
Johnson, R.L. and S.J. Goodman, 1984: Atmospheric electrical activity associated with hurricane Alicia. Preprints, 7th International Conference on Atmospheric Electricity, American Meteorological Society, Albany, NY, 295–298.
Jorgensen, D.P., E.J. Zipser and M.A. LeMone, 1985: Vertical motions in intense hurricanes. *J. Atmos. Sci.*, **42**, 839–856.
Jorgensen, D. and M. LeMone, 1989: Vertical velocity characteristics of oceanic convection. *J. Atmos. Sci.*, **46**, 621–640.
Kakar, R., M. Goodman, R. Hood and A. Guillory, 2006: Overview of the Convection and Moisture Experiment. *J. Atmos. Sci.*, **63**, 5–18.
Kelley, O.A., J. Stout and J.B. Halverson, 2004: Tall precipitation cells in tropical cyclone eyewalls are associated with tropical cyclone intensification. *Geophys. Res. Lett.*, **31**, L24112- L24115.
Lay, E.H., A.R. Jacobson, R.H. Holzworth, C.J. Rodger and R.L. Dowden, 2007: Local time variation in land/ocean lightning count rates as measured by the World Wide Lightning Location Network. *J. Geophys. Res.*, **112**, D13111, doi:10.1029/2006JD007944.
Lei, Y.M. and J.C. Chan, 1999: Relationship between asymmetric convection and tropical cyclone motion. 23rd Conference on Hurricanes and Tropical Meteorology, American Meteorological Society, Dallas, TX, 305-307.
LeMone, M. and E. Zipser, 1980: Cumulonimbus vertical velocity events in GATE. Part I: Diameter, intensity and mass flux. *J. Atmos. Sci.*, **37**, 2444–2457.
Leppert, K.D. and W.A. Petersen, 2010: Electrically Active Hot Towers in African Easterly Waves prior to Tropical Cyclogenesis. *Mon. Wea. Rev.*, **138**, 663–687.
Lin, Y.-L., R.D. Farley and H.D. Orville, 1983: Bulk parameterization of the snow field in a cloud model. *J. Climate Appl. Meteor.*, **22**, 1065–1092.
Lyons, W.A., M.G. Venne, P.G. Black and R.C. Gentry, 1989: Hurricane lightning: A new diagnostic tool for tropical storm forecasting? Preprints, 18th Conf. on Hurricanes and Tropical Meteorology, San Diego, American Meteorological Society, 113-114.
Lyons, W.A. and C.S. Keen, 1994: Observations of lightning in convective supercells within tropical storms and hurricanes. *Mon. Wea. Rev.*, **122**, 1897–1916.
MacGorman, D.R., J.M. Straka and C.L. Ziegler, 2001: A lightning parameterization for numerical cloud models. *J. Appl. Meteor.*, **40**, 459–478.
MacGorman, D.R., D.W. Burgess, V. Mazur, W.D. Rust, W.L. Taylor and B.C. Johnson, 1989: Lightning rates relative to tornadic storm evolution on 22 May 1981. *J. Atmos. Sci.*, **46**, 221–250.

MacGorman, D.R. and W.D. Rust, 1998: The Electrical Nature of Storms. Oxford University Press, New York.

Mackerras, D., M. Darveniza, R.E. Orville, E.R. Williams and S. Goodman, 1998: Global lightning total, cloud and ground flash estimates. *J. Geophys. Res.*, **103**, D16, August 27, 19791– 19809.

Mansell, E.R., D.R. MacGorman, C.L. Ziegler and J.M. Straka, 2002: Simulated three-dimensional branched lightning in a numerical thunderstorm model. *J. Geophys. Res.*, **107**, No. D9, 10.1029/2000JD000244.

Mansell, E.R., D.R. MacGorman, C.L. Ziegler and J.M. Straka, 2005: Charge structure and lightning sensitivity in a simulated multicell storm. *J. Geophys. Res.*, **110**, D12101, doi:10.1029/2004JD005287.

Mansell, E.R., C.L. Ziegler and C. Bruning, 2010: Simulated Electrification of a Small Thunderstorm with Two-Moment Bulk Microphysics. *J. Atmos. Sci.*, **67**, 171–194.

Marks, F.D. and L.K. Shay, 1998: Landfalling tropical cyclones: Forecast problems and associated research opportunities. *Bull. Amer. Meteor. Soc.*, **79**, 305–323.

Marshall, J. and W. Palmer, 1948: The distribution of raindrops with size. *J. of Meteor.*, **5**, 165–166.

May, P.T. and D.K. Rajopadhyaya, 1996: Wind profiler observations of vertical motion and precipitation microphysics of a tropical squall line. *Mon. Wea. Rev.*, **124**, 807–815.

McFarquhar, G.M. and A.J. Heymsfield, 1996: Microphysical characteristics of three anvils sampled during the central equatorial pacific experiment. *J. Atmos. Sci.*, **53**, 2401–2423.

McFarquhar, G.M. and R.A. Black, 2004: Observations of particle size and phase in tropical cyclones: Implications for mesoscale modeling of microphysical processes. *J. Atmos. Sci.*, **61**, 422–439.

Molinari, J., P.K. Moore, V.P. Idone, R.W. Henderson and A.B. Saljoughy, 1994: Cloud-to-ground lightning in Hurricane Andrew. *J. Geophys. Res.-Atmos.*, **99**, 16665–16676.

Molinari, J., P. Moore and V. Idone, 1999: Convective structure of hurricanes as revealed by lightning locations. *Mon. Wea. Rev.*, **127**, 520–534.

Molinari, J., D. Vollaro and K.L. Corbosiero, 2004: Tropical Cyclone Formation in a Sheared Environment: A Case Study. *J. Atmos. Sci.*, **61**, 2493–2509.

Molinari, J., N.W.S. Demetriades, R.L. Holle and D. Vallaro, 2006: Applications of long-range lightning data to hurricane formation and intensification. Preprints, Conf. on Meteor. Appl. of Lightning Data, American Meteorological Society, Atlanta, GA.

Molinari, J. and D. Vollaro, 2010: Rapid intensification of sheared tropical storm. *Mon. Wea. Rev.*, **138**, 3869–3885.

Morales, C.A., J.S. Kriz, E.B. Rodgers and J.A. Weinman, 1997: The evolution of sferics around Hurricane Lilli-1996. Preprints, 22nd Conf. on Hurricanes and Tropical Meteorology, Fort Collins, CO, Amer. Meteor. Soc., 127–128.

Orville, R.E. and J.M. Coyne, 1999: Cloud-to-ground lightning in tropical cyclones (1986-1996). Preprints, 23rd Conference on Hurricanes and Tropical Meteorology, American Meteorological Society, Dallas, TX, 194.

Orville, R.E., R.W. Henderson and L.F. Bosart, 1988: Bipole patterns revealed by lightning locations in mesoscale storm systems. *Geophys. Res. Lett.*, **15**, 129–132.

Pan, L., X. Qie and D. Wang, 2014: Lightning activity and its relation to the intensity of typhoons over the Northwest Pacific Ocean. *Adv. Atmos. Sci.*, **31**, 581–592, doi: 10.1007/s00376-013-3115-y.

Petersen, W.A., R. Cifelli, S.A. Rutledge and B.F. Smull, 1995: Cloud-to-ground lightning and the related kinematic structures of two tropical oceanic MCS's: Contrasting cases. Preprints, Conference on Cloud Physics, American Meteorological Society, Dallas, TX.

Petersen, W.A., R.C. Cifelli, S.A. Rutledge, B.S. Ferrier and B.F. Smull, 1999: Shipborne dual-doppler operations and observations during TOGA COARE. *Bull. Amer. Meteor. Soc.*, **80**, 81–97.

Price, C., M. Asfur and Y. Yair, 2009: Maximum hurricane intensity preceded by increase in lightning frequency. *Nat. Geosci.*, **2**, 329–332, doi:10.1038/NGEO477.

Reap, R.M. and D.R. MacGorman, 1989: Cloud-to-ground lightning: Climatological characteristics and relationships to model fields, radar observations, and severe local storms. *Mon. Wea. Rev.*, **117**, 518–535.

Rison, W., R.J. Thomas, P.R. Krehbiel, T. Hamlin and J. Harlin, 1999: A GPS-based three-dimensional lightning mapping system: Initial observations in central New Mexico. *Geophys. Res. Lett.*, **26**, 3573–3576.

Rodgers, E., J. Weinman, H. Pierce and W. Olson, 2000: Tropical cyclone lightning distribution and its relationship to convection and intensity change. Preprints, 24th Conference on Hurricanes and Tropical Meteorology, American Meteorological Society, Ft. Lauderdale, FL, 537–541.

Rogers, D.P., J.W. Telford and S.K. Chai, 1985: Entrainment and the temporal development of the microphysics of convective clouds. *J. Atmos. Sci.*, **42,** 1846–1858.

Rogers, R.F., 2003: The relationship between vertical shear, storm motion, and rainfall patterns produced from numerical simulations of tropical cyclones. Preprints, 10th Conference on Mesoscale Processes, American Meteorological Society, Portland, OR.

Rogers, R., M. Black, F. Marks, K. Valde and S.S. Chen, 2006: A comparison of tropical cyclone hydrometeor profiles from TRMM, airborne radar, and high resolution simulations. Preprints, 27th Conference on Hurricanes and Tropical Meteorology, American Meteorological Society, Monterey, CA.

Rogers, R.F., M.L. Black, S.S. Chen and R.A. Black, 2007: An Evaluation of Microphysics Fields from Mesoscale Model Simulations of Tropical Cyclones. Part I: Comparisons with Observations. *J. Atmos. Sci.*, **64**, 1811–1834.

Rogers, R.R. and M.K. Yau, 1989: A short course in cloud physics. 3rd edition. Pergamon Press, New York.

Rutledge, S.A. and MacGorman, D.R. 1988: Cloud-to-ground lightning activity in the 10–11 June 1985 mesoscale convective system observed during the Oklahoma–Kansas Pre-storm Project. *Mon. Wea. Rev.*, **116**, 1393–1408.

Rutledge, S.A., E.R. Williams and T.D. Keenan, 1992: The Down Under Doppler and Electricity Experiment (DUNDEE): Overview and preliminary results. *Bull. Amer. Meteor. Soc.*, **73**, 3-16.

Rutledge, S.A. and W.A. Petersen, 1994: Vertical radar reflectivity structure and cloud-to-ground lightning in the stratiform region of MCS's: Further evidence for in-situ charging in the stratiform region. *Mon. Wea. Rev.*, **122**, 1760–1776.

Samsury, C.E. and R.E. Orville, 1994: Cloud-to-ground lightning in tropical cyclones: A study of hurricanes Hugo (1989) and Jerry (1989). *Mon. Wea. Rev.,* **122**, 1887–1896.

Saunders, C.P.R. and L.S. Peck, 1998: Laboratory studies of the influence of the rime accretion rate on charge transfer during crystal-graupel collisions. *J. Geophys. Res.*, **103**, 13949–13956.

Saunders, C.P.R., L.S. Peck, G.G. Aguirre Varela, E.E. Avila and N.E. Castellano, 2001: A laboratory study of the influence of water vapour on the charge transfer during collisions between ice crystal and graupel. *J. Atmos. Res*., **58**, 187–203.

Shao, X.M., J. Harlin, M. Stock, M. Stanley, A. Regan, K. Wiens, T. Hamlin, M. Pongratz, D. Suszcynsky and T. Light, 2005: Katrina and Rita were lit up with lightning. *EOS*, **86(42)**.

Schultz, C.J., W.A. Petersen and L.D. Carey, 2011: Lightning and Severe Weather: A Comparison between Total and Cloud-to-Ground Lightning Trends. *Wea. Forecasting*, **26**, 744–755.

Simpson, J. and J.B. Halverson, 1998: Cloud physics and electrification in forming Tropical Cyclone Oliver, 1993. Preprints, Conference on Cloud Physics, American Meteorological Society, Everett, WA, 185–188.

Smith, D.A. et al., 1999: A distinct class of isolated intracloud lightning discharges and their associated radio emissions. *J. Geophys. Res.*, **104**, 4189–4212.

Solorzano, N.N., J.N. Thomas and R.H. Holzworth, 2008: Global studies of tropical cyclones using the World Wide Lightning Location Network. Preprints, Third Conf. on the Meteorological Applications of Lightning Data, New Orleans, LA, Amer. Meteor. Soc., 1.4. [Available online at http://wwlln.net/publications/Solorzano_AMS_29FEB2008.pdf.]

Squires, K. and S. Businger, 2008: The Morphology of Eyewall Lightning Outbreaks in Two Category 5 Hurricanes. *Mon. Wea. Rev.*, **136**, 1706–1726.

Straka, J.M. and E.R. Mansell, 2005: A bulk microphysics parameterization with. multiple ice precipitation categories. *J. Appl. Meteor*., **44**, 445–466.

Suszcynsky, D.M., A.R. Jacobson, J. Linford, T.E. Light and A. Musfeldt, 2005: VHF lightning detection and storm tracking from GPS orbit. Preprint, Conference on Meteorological Applications of Lightning Data, American Meteorological Society, San Diego, CA.

Takahashi, T., 1978: Riming electrification as a charge generation mechanism in thunderstorms. *J. Atmos. Sci*., **35**, 1536–1548.

Takahashi, T., 2003: Lightning and in-cloud ice phases in the East Asian Mansoon. Preprints, 12th International Conference on Atmospheric Electricity, American Meteorological Society, Versailles, France, 115–118.

Wiens, K.C., S.A. Rutledge and S.A. Tessendorf, 2005: The 29 June 2000 Supercell observed during steps. Part II: Lightning and charge structure. *J. Atmos. Sci.*, **62**, 4151–4177.

Williams, E.R., 1989: The tripole structure of thunderstorms. *J Geophys Res*, **94**, 13151–13167.

Williams, E.R. and S.A. Rutledge, 1990: Studies of electrification and lightning in deep tropical precipitation systems. Conference on Cloud Physics, American Meteorological Society, San Francisco, CA, 290–294.

Williams, E.R. and S. Stanfill, 2002: The physical origin of the land-ocean contrast in lightning activity. *Comptes Rendus—Physique*, **3**, 1277–1292.

Willis, P.T. and A.J. Heymsfield, 1989: Trajectories of hydrometeors in hurricane Emily. Preprints, 16th Conference on Hurricanes and Tropical Meteorology. American Meteorological Society, San Diego, CA, 192–196.
Willoughby, H.E., J.A. Clos and M.G. Shoreibah, 1982: Concentric eye walls, secondary wind maxima, and the evolution of the hurricane vortex. *J. Atmos. Sci.*, **39**, 395–411.
Willoughby, H.E., 1990: Temporal changes of the primary circulation in tropical cyclones. *J. Atmos. Sci.*, **47**, 242–264.
Zhu, H. and R.K. Smith, 2002: The importance of three physical processes in a three-dimensional tropical cyclone model. *J. Atmos. Sci.*, **59**, 1825–1840.
Ziegler, C.L., D.R. MacGorman, J.E. Dye and P.S. Ray, 1991: A model evaluation of non-inductive graupel-ice charging in the early electrification of a mountain thunderstorm. J. Geophys. Res., **96**, 12833–12855.
Zipser, E.J. and H. Jiang, 2003: Variability of ice and liquid precipitation contents and shape of radar reflectivity profiles in tropical cyclones. Preprints, 31st International Conference on Radar Meteorology, American Meteorology Society, Seattle, WA, 375–379.

8

The Role of Land Surface Processes on Tropical Cyclones: Introduction to Land Surface Models

Dev Niyogi[1], Subashini Subramanian[1] and Krishna K. Osuri[2]
[1]Purdue University, West Lafayette, Indiana, USA
[2]School of Earth, Ocean and Climate Sciences, IIT Bhubaneswar, India
dniyogi@purdue.edu

1. Introduction

The role of land surface processes in land falling tropical cyclones is an area of emerging interest. Tropical cyclones are formed as organized convection over warm water (typically 26.5°C, Gray, 1968) packing tremendous amounts of energy. Tropical cyclones have a typical size of 200-2000 km with a life span of about one to two weeks. The cyclone and its environment are interlinked. There are a number of environmental factors that are important for sustaining and intensifying a tropical cyclone including low humidity, cooler sea surface temperature (SST), or higher tropopause temperatures, dry air intrusion from land masses, and large vertical wind shear (Gray, 1968; McBride and Zehr, 1981). However a number of environmental conditions can change the evolution of a landfalling storm.

As tropical cyclones approach land, the situation becomes more complex. There are a multitude of inland heterogeneities such as soils, landuse, and topography that can affect the boundary layer and mesoscale processes which can affect tropical cyclone evolution. Additionally, landfalling storms clearly have major socio-economic impacts which are dependent not just on wind and intensity but also on the rainfall/flood potential. As stated in Emanuel et al. (2006), rainfall, storm water flooding, storm surges, and damaging winds associated with tropical cyclones and depressions over land continue to be the greatest threat to loss of life and property in regions prone to them. While cyclone track prediction has improved considerably over the past few decades due to advances in numerical weather prediction, enhanced atmospheric observations, and data assimilation, the skill in prediction of TC intensity has not similarly improved (Marks and Shay, 1998). Track prediction often depends on the large-scale environment, with intensity being a function of multi-scale interactions that include processes that are active at inner-core/cloud scale, mesoscale, sub-synoptic, and synoptic (large) scales. Some possible scenarios in cyclone evolution as it approaches land/coastal areas are shown in Fig. 1.

Tropical cyclones typically decay as they approach land by encountering unfavourable synoptic conditions such as depletion of its energy source and friction across the surface distorting inflow which leads to weakening of the eye (Kaplan and DeMaria, 1995; Emanuel, 2000). Some storms can also transition into an extratropical cyclone and develop a cold core. In rare instances, the storms can redevelop or reintensify after making landfall by sustaining the necessary convection to maintain tropical cyclone characteristics and in each of these scenarios, there is a distinct threat of rainfall and inundation over land. Heterogeneities in the landscape features (e.g., soil moisture, surface roughness, albedo, vegetated land cover) can create mesoscale boundaries that can impact the associated convection (Pielke, 2001; Emanuel et al., 2004). Whether the storm decays quickly after landfall or sustains its intensity over land, the land surface and its interaction with the tropical cyclone and atmospheric environment is an important component of a landfalling tropical cyclone research efforts. There are many studies highlighting the need for

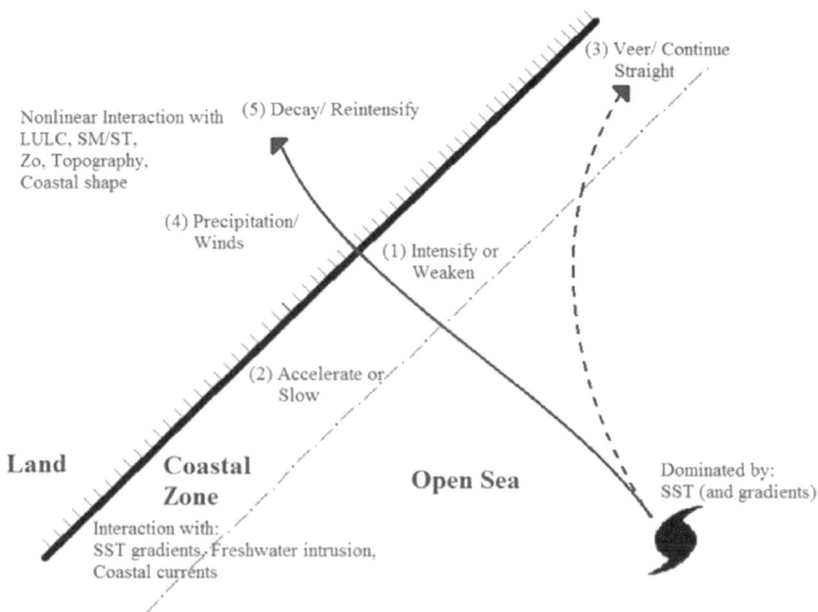

Fig. 1. Example of possible outcomes and factors in the evolution of a tropical cyclone as it approaches land.

accurate representation of land-atmosphere interactions in models that affect atmospheric weather predictions and this continues to be a critical issue as well. Numerical models lack the ability to accurately model converging overland tracks even though great strides have been made in ocean basin track forecasts (Marks and Shay, 1998).

Numerical models continue to be important forecasting tools employed by the operational community to determine track movement and landfall location. However, uncertainty still exists since atmospheric models are not developed enough to capture and simulate the whole range of land-atmosphere interactions particularly at finer scales. Schade and Emanuel (1999) identified potential areas of improvement needed to enhance the overall quality of tropical cyclone prediction including the storm's initial intensity and structure, the thermodynamic state of the storm's environment, and energy exchange between the boundary layer and the storm. Recent modelling work by Emanuel et al. (2008) over Australia and Chang et al. (2009) over India suggests that the antecedent land surface conditions can alter post-landfall storm structure because of the heat and moisture source provided from the wet, warm land masses. Numerous studies have contributed to the understanding of the role that large scale conditions play in the evolution of tropical cyclones (Zehnder, 1991; Breigel and Frank, 1997; Frank and Roundy, 2006).

Recent works have shown the importance of landscape processes in weather and climate (Werth and Avissar, 2005; Feddema, 2005; Alpert et al., 2006). Studies have shown that underlying surface characteristics such as terrain, land use, soil temperature, moisture, albedo, and surface roughness have a great influence on convective systems that pass over areas with surface heterogeneities (e.g., Pielke, 2001). Different land surface conditions result in different boundary layer structures and mesoscale atmospheric circulations. Numerous studies have acknowledged the influence of surface conditions on baroclinic zones, drylines, low-level jet development, capping inversions, and convective storms (Mahfouf et al., 1987; Pielke et al., 1991; Avissar and Liu, 1996; Chase et al., 1999). Although a few studies suggest realistic representation of landscape heterogeneities have no bearing on boundary layer characteristics (Zhong and Doran, 1998; Doran and Zhong, 2000), observational and modelling work by Weaver and Avissar (2001) document that landscape heterogeneity does produce organised cumulus convection using the same model applied by Doran and Zhong. In summary, over the last decade, accurate representation of land surface feedbacks has become a critical component of state-of-the-art NWP models.

A broad objective of this chapter is to summarise some of the recent scientific progress made on land surface processes and landfalling TCs. Beginning with global surface energy budgets, a brief primer on the atmospheric boundary layer is included with an introduction to land surface schemes used in weather models.

2. Surface Energy Budget

Weather and climate in general is fueled by solar radiation and controlled by the amount and distribution of energy radiation. Incoming radiation is scattered, reflected, and absorbed by clouds, atmospheric gases and aerosols. The radiation transmitted by the atmosphere may either be absorbed or reflected back by the Earth's surface. The land surface partitions shortwave radiation into sensible heat flux, latent heat flux and ground heat flux to be stored by the surface before being reflected back as longwave radiation (see Fig. 2). Energy may be stored in various forms and converted to different types, giving rise to a broad variety of weather or turbulent phenomena in the atmosphere. The important interaction between the land surface and atmosphere is the exchange of radiation, sensible heat, latent heat, and momentum fluxes. Land-surface feedback mainly impacts the surface energy and water budgets from the boundary layer perspective.

The energy budget is not the only component driving the Earth's system. The global water cycle also plays an important role in transporting energy and moisture to the atmosphere from the surface. The total radiation absorbed by the land surface is balanced by emissions of thermal, infrared radiation to the atmosphere, latent heat associated with evaporation and transpiration, and sensible heat and diffusion of energy into soil. At the idealized land surface (e.g., if the land surface is flat, bare, and opaque to radiation), the surface net radiation can be expressed as the sum of sensible, latent, and ground heat fluxes. However in reality, the surface is covered by vegetation, buildings, and water, which are opaque to radiation and has a significant heat capacity.

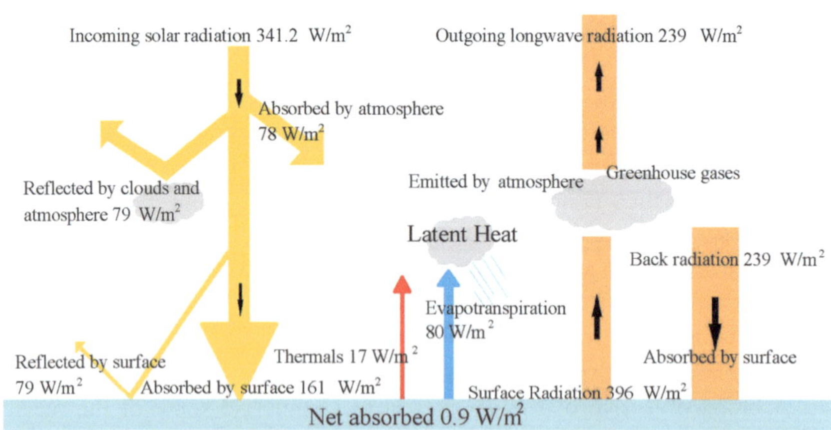

Fig. 2. The global annual mean Earth's energy budget. The broad arrows indicate the schematic flow of energy in proportion to their importance. (Modified from Trenberth et al., 2009)

At the idealized surface, the net downward radiative flux (shortwave and longwave),

R_n = Sensible heat flux (H_S) +Latent heat flux (H_L) + Ground heat flux (H_G)

or

$$R_n = H_S + H_L + H_G$$

The Bowen ratio, $B = H_S/H_L$

Under real surface conditions, the land surface with plant canopy or other features can store some energy. In this case, it is more appropriate to define an interfacial layer which includes such features. $W(t)$ is the energy stored within this layer per unit horizontal area. The revised layer energy budget is:

$$R_n = H_S + H_L + H_G + dW/dt$$

The sensible heat flux, H_S, describes the heat flux transport from or to the Earth's surface. The latent heat flux, H_L, describes the vertical transport of moisture (water vapour) and the heat required for evaporation at the land surface. Radiation and energy fluxes are considered positive if they transport energy away from the Earth's surface (into the atmosphere or into the ground), otherwise they are considered negative. All the terms in the above equation show strong variations in response to the diurnal cycle of heating and cooling at the surface. The turbulent (latent and sensible) fluxes and the ground heat flux are considered positive during daytime and negative during evening/night. During daytime, the land surface receives energy ($R_n > 0$) and is partitioned into the above mentioned three fluxes out of which the sensible and latent heat fluxes are directed upward (into the atmosphere) and the ground heat flux into the surface. The partitioning of the energy fluxes depends on many factors such as land surface characteristics (soil texture, type, vegetation, soil moisture, etc.), geographical location (latitude), season, thermal properties (albedo and emissivity), time of day and of course, local weather (Noilhan and Planton, 1989). For example, in an irrigating field, due to an abundance of soil moisture, the latent heat flux increases due to evaporative cooling of the wet surface and sometimes may exceed the net radiation. The sensible and ground heat fluxes may become negative under such conditions. In contrast, the latent heat flux will be less and sensible heat flux will be greater in a desert region. During the evening/night, the surface loses energy as outgoing radiation which is compensated by the heat from the atmosphere/soil during the formation of dew as latent heat of condensation. Thus, these terms become negative. Comparing the magnitudes of latent and sensible heat fluxes, the energy balance terms are larger during the day and smaller during night time. However, ground heat flux can be assumed typically 10% of the net radiation. Over land, there is large diurnal variation in the surface energy budget and associated surface variables such as humidity and temperature (see Fig. 3). Over large oceans, the large heat capacity and absorption of solar radiation over large depths combine to reduce the near-surface diurnal temperature

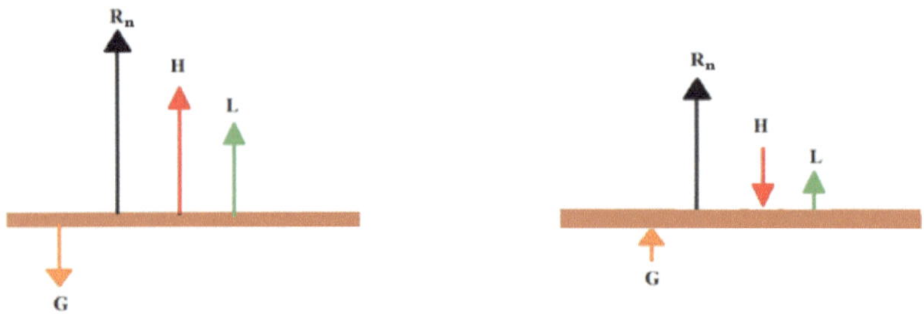

Fig. 3. Schematic representation of typical surface energy budgets for daytime (left) and night time (right). (Adapted from Arya, 1988)

variability, so H_S and H_L varies little. The surface skin temperature of a tropical ocean can vary diurnally by up to 3°C in sunny light wind conditions (Barale, 2010) while over a land surface the temperature variations can be a order of magnitude higher.

2.1 Examples of Energy Balance over Different Landscapes

The energy budget measured over a dry shrubland for a typical cloud free day in January and April is shown in Fig. 4. In this case, latent heat fluxes

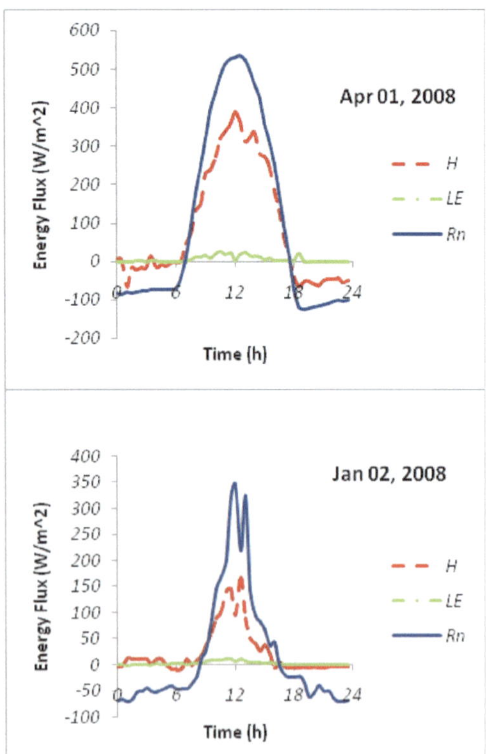

Fig. 4. Observed diurnal energy balance over Sevilleta Desert Shrubland in New Mexico USA Jan 02, 2008 (top) and Apr 01, 2008 (bottom). (Data source: Ameriflux)

are negligible. During the day, as copious solar radiation is absorbed at the surface, the ground heats up rapidly. Initially, most of the heat is conducted into the soil, but as the surface absorbs more heat, H_S dominates and the heat is primarily transferred to the air and increasing the air temperature. This is promoted by extreme differences (up to 28°C) between the ground temperature and the 2m air temperature. At night, surface radiative cooling is balanced by an upward ground heat flux. Since the nocturnal boundary layer is relatively stable, the turbulent heat flux H_S is negligible.

The energy budget of a corn and soybean field for a day in January and April is shown in Fig. 5. During the daytime, surface radiative heating is balanced mainly by latent heat flux due to evapotranspiration, i.e., evaporation from the soil surface and transpiration by the plant leaves. H_L can be so large that the surface gets cooler than the air during early morning and late afternoon and the heat flux is downward. For a field, heat storage is usually small. At night,

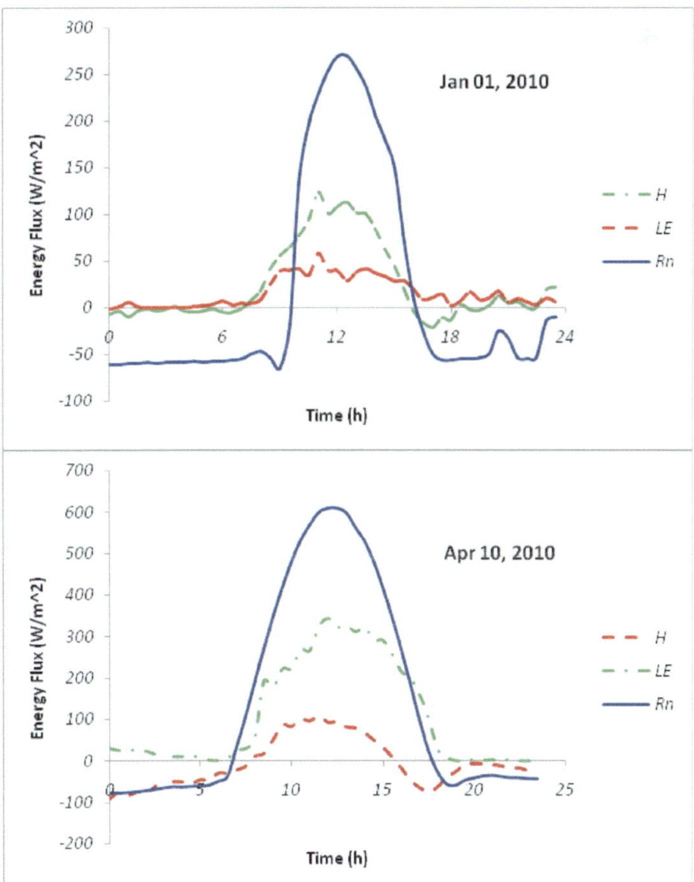

Fig. 5. Observed diurnal energy budget of an agricultural field in Oklahoma, USA, on Jan 01, 2010 (top) and Apr 10, 2010 (bottom). (Data source: Ameriflux)

all terms become much smaller and radiative cooling is mainly balanced by ground heat flux.

Another example is for an evergreen needleleaf forest (Fig. 6). Here latent and sensible heat fluxes are comparable during the day. The storage and ground heat flux become important and for the deeper soil layer, the storage term dominates. At night, the release of heat from the tree canopy and condensation (dew) balance the radiative energy loss.

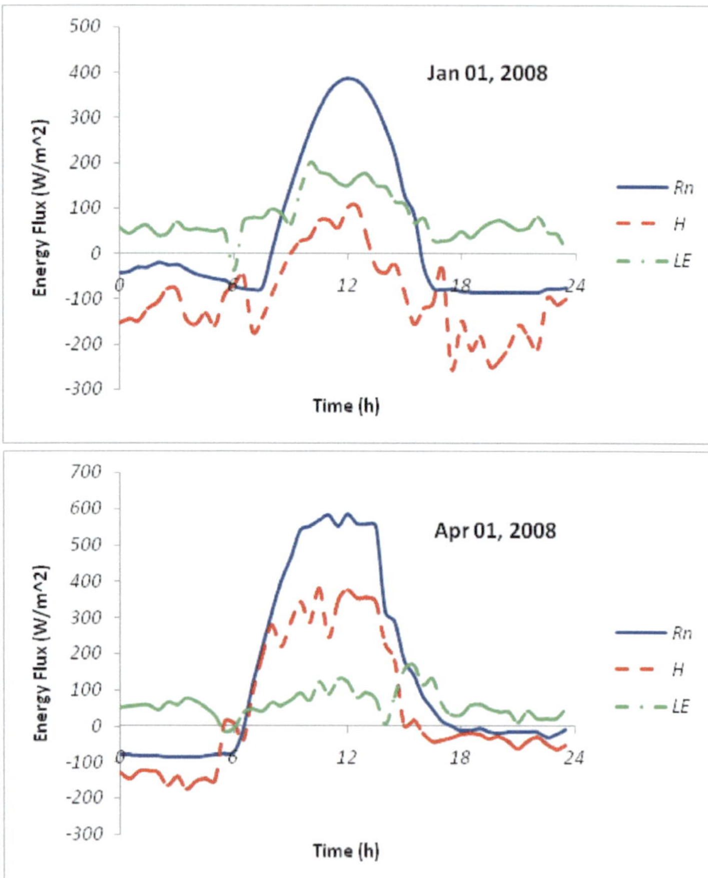

Fig. 6. Observed diurnal energy budget of a needleleaf forest in Niwot Ridge, Colorado, USA, on Jan 01, 2008 (top) and Apr 01, 2008 (bottom). (Data source: Ameriflux)

2.1.1 Net Radiation at the Surface

The net radiation R_n is considered as the difference between downwelling and upwelling shortwave plus longwave radiative fluxes. The net shortwave flux depends on the incident solar radiation $R_{S\downarrow}$ and on surface albedo a_s. The net longwave flux depends upon the downwelling longwave radiation $R_{L\downarrow}$, the surface emissivity ε_s, and the radiating temperature T_s:

$$R_n = R_{S\downarrow} - R_{S\uparrow} + R_{L\downarrow} - R_{L\uparrow} = (1 - a_s)R_{S\downarrow} + R_{L\downarrow} - \{(1 - \varepsilon_s)R_{L\downarrow} + \varepsilon_s\sigma_s T_s^4\}$$
$$= (1 - a_s)R_{S\downarrow} + \varepsilon_s(R_{L\downarrow} - \sigma_s T_s^4)$$

Thus, the surface characteristics critically influence R_n. Albedos are quite diverse, while emissivities are usually near, but not equal to, 1.

2.1.2 Soil Temperatures and Heat Flux

The surface or skin temperature is important for calculating the radiative balance of surface and for predicting moisture processes. The skin or surface temperature can be quite different than the air temperature, which is conventionally measured at 2 m. If there is a plant canopy or surface inhomogeneity, there is no single uniquely definable surface temperature. Radiatively, an apparent surface temperature can be determined from the upward longwave energy flux if the emissivity is known. Large diurnal variations in skin temperature are achieved for bare, dry surfaces for clear calm conditions. Under such conditions, mid-day skin temperature may even reach 50-60°C under some extreme conditions, while early morning skin temperatures can drop to 10-20°C. The surface temperature is related to the temperature profile in the sub-surface medium, as illustrated in Fig. 7. In a solid medium, the sub-surface temperature profile is governed by heat conduction. Deeper in the soil, the diurnal temperature cycle variability decreases and lags the skin temperature cycle.

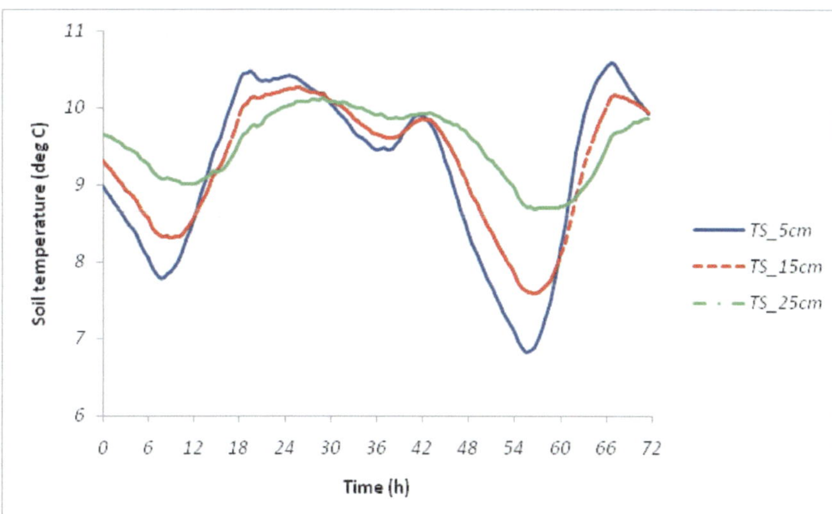

Fig. 7. Observed diurnal course of subsurface soil temperatures at various depths in ARM SGP Main Ameriflux site (Vegetated Cropland) in Oklahoma USA, on Apr 01-03, 2008 (data source: Ameriflux).

3. Boundary Layer

The land surface affects the surface energy balance, which in turn affects the eddies and the turbulent energy exchange within the atmosphere. The atmospheric layer in the vicinity of the land surface is affected by the surface energetics and has detectable turbulence exchanges. This layer can be considered as the atmospheric boundary layer and consists of the lower troposphere. The chief energy source for the Earth's water cycle and the driving force behind boundary layer processes is incoming solar radiation. During daytime and under clear sky conditions, shortwave radiation heats up the Earth's surface creating rising and sinking parcels/plumes of warm air (also called eddies). This causes rapid expansion of the boundary layer height as the surface warms during the day and reaches a maximum height a few hours after solar noon. The height of a typical daytime boundary layer is between 1-2 km. During nighttime, the land cools at a faster rate than the air in the atmosphere resulting in lowering of turbulence consequently causing a collapse in boundary layer height (between 100-200 m). This is also the time when weak mixing occurs. The turbulence is considered to be present only within the boundary layer. Alternately, boundary layer height is also defined as the height of the largest eddy present in the layer.

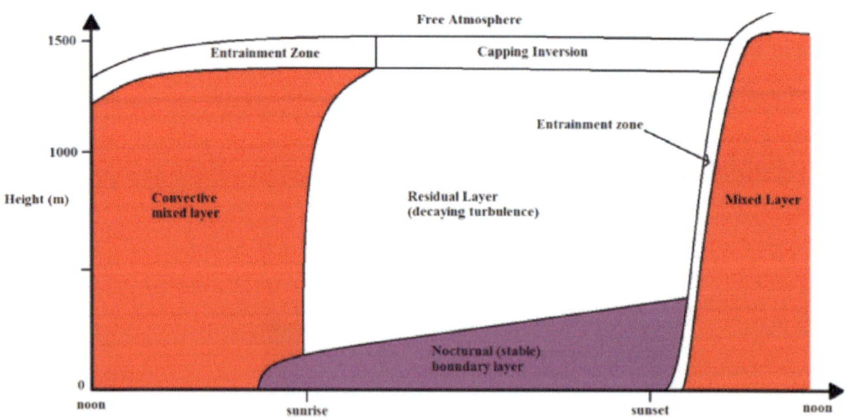

Fig. 8. ABL evolution during a diurnal cycle. (Adapted from Stull, 1988)

Over the oceans, the diurnal variation in temperature is small. The boundary layer does not evolve much during the day and largely depends on synoptic conditions (high pressure or low pressure). This is also due to the fact that water has a large thermal capacity and absorbs most of the incoming radiation. Large-scale conditions, to an extent, also affect the boundary layer structure over land but evolution of boundary layer processes is largely influenced by diurnal variations in ground temperature. The major components of the ABL evolution are a surface layer that is over laid by (i) the mixed layer, (ii) the residual layer, and (iii) the stable boundary layer. The surface layer is typically considered 10% of the depth of the entire boundary layer and has roughly constant flux values or no flux divergence. The layer above the surface layer or the constant flux layer is dominated by vertical exchanges and mixing. There is also a thin layer called the micro layer or the interfacial layer which is the lowest few centimetres of the air just above the ground where molecular transport is more significant than turbulent transport.

The mixed layer is convectively driven and occurs due to heating near the land surface and radiative cooling at the top. Warm air rises, mixes with the atmosphere, and cooling occurs at the top sending the cold air parcels back to the surface aiding the turbulent mixing. Heat, momentum and water vapour are well mixed leading to scalar quantities such as temperature and humidity is almost constant (see Fig. 9).

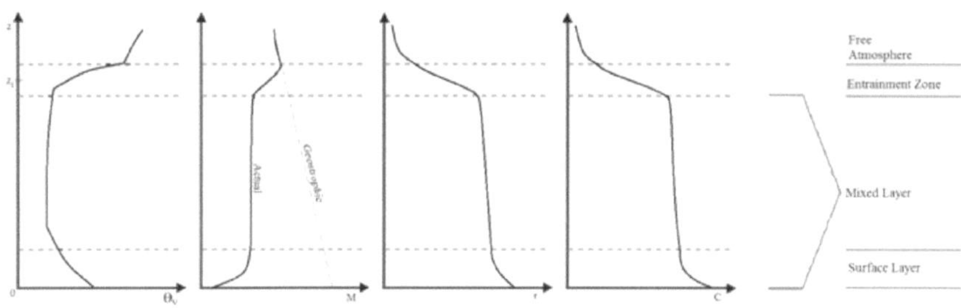

Fig. 9. Typical daytime profiles of mean virtual potential temperature θ_v, wind speed M (where $M^2 = u^2 + v^2$), water vapour mixing ratio r, and pollutant concentration C. (Redrawn from Stull, 1988)

Just before sunset, due to a decrease in incoming radiation, eddy thermals cease to form and turbulence weakens in the formerly well-mixed layer. The resulting layer is called a residual layer (see Fig. 8) and is called so because the initial mean state variables and concentration variables such as humidity resemble the recently decayed mixed layer. Sometimes, over a period of a few days, additional moisture can accumulate in the residual layer through mixing during the day and storage during the night which can lead to formation of clouds in areas where they may normally not form.

As night progresses, the lower part of the residual layer which is in contact with the ground is transformed into a stable boundary layer characterized by very low turbulence and highly stable air. A stable boundary layer has a poorly defined top in contrast to the daytime mixed layer and blends smoothly into the residual layer. The profile of variables for daytime and night time are given in Fig. 10.

In classical fluid dynamics, a boundary layer is the layer in a nearly inviscid fluid next to a surface in which frictional drag associated with that surface is significant (term introduced by Prandtl, 1905). Such boundary layers can be laminar or turbulent, and are often only millimetres thick. In

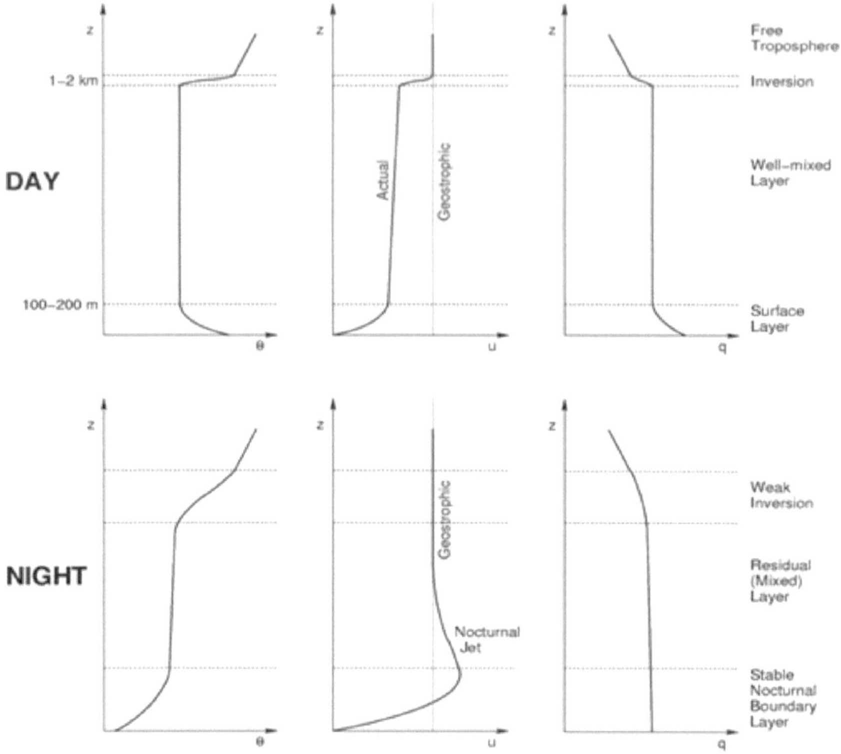

Fig. 10. Typical day and nighttime profiles for potential temperature θ, mean winds *u*, and specific humidity *q*. The boundary layer and mixed layer are marked.

atmospheric models, a similar definition is useful. The atmospheric boundary layer (ABL, also called Planetary BL) is the layer of fluid directly above the Earth's surface in which significant fluxes of momentum, heat and/or moisture are carried by turbulent motions whose horizontal and vertical scales are in the order of the boundary layer depth, and whose circulation timescale is a few hours or less. A similar definition works for the ocean. The complexity of this definition is due to several complications compared to classical aerodynamics.

(i) Surface heat exchange can lead to thermal convection
(ii) Moisture and its effects on convection
(iii) Earth's rotation
(iv) Complex land surface characteristics and topography.

In broad terms, the atmospheric boundary layer provides the surface and mixed layer which in turn provide the link and coupling for the land surface and the atmosphere. The BL is assumed to encompass surface-driven buoyancy, shear and convection.

4. Air-surface Exchange

The interactions between the atmosphere and the land surface are important across a wide variety of space and timescales. Energy from the Sun is converted at the land surface into sources of heat and moisture for the lowest part of the atmosphere (ABL), and the land acts to slow down the surface wind leading to atmospheric turbulence. The ABL is strongly modulated by exchanges of heat, moisture and momentum with the underlying land surface. These exchanges also affect the variations of wind, temperature and humidity in the ABL. As a result, the characterization of land surfaces so as to predict these fluxes (thermal and moisture fluxes, radiative characteristics) accurately becomes important for broader prediction rather than just the boundary layer depth.

Compared to the scale of the atmosphere, the ABL is a shallow layer, albeit an important one. Most of the small-scale processes often occur within this layer, particularly in the surface layer (the lowest 10% of the ABL). These processes are responsible for exchanges of most of the energy and moisture between the surface and the atmosphere which are considered to be important for the evolution of local and large-scale weather phenomena. In addition, the friction at the surface and ABL is primarily responsible for the low-level convergence and divergence of wind and moisture. Thus, an accurate representation of land surface processes and ABL exchange processes becomes an important part of the state-of-the-art weather and climate models.

Representing land surface processes in a numerical weather prediction model involves linking the model's atmospheric boundary layer scheme with its surface schemes and determining the structure of the lowest part of the boundary layer, or the surface layer. A prerequisite for improving the representation of these processes in models is building a good physical representation of the land surface. Observational studies play a prominent role in this research, but detailed process models, such as the Large Eddy Simulations, are also used.

A Summary of ABL

The boundary layer exhibits dynamically distinct sublayers.

(i) Interfacial sublayer – where molecular viscosity/diffusivity dominates vertical fluxes.

(ii) Inertial layer – where turbulent fluid motions dominate vertical fluxes, but the dominant scales of motion are still much less than the boundary layer depth. This is the layer where most surface wind measurements are made.

- Layers (i) + (ii) comprise the surface layer. Coriolis turning of the wind with height is not evident within the surface layer.

(Contd.)

(*Contd.*)

(iii) Outer layer – turbulent fluid motions with scales of motion comparable to the boundary layer depth ('large eddies').

- At the top of the outer layer, the BL is often capped by an entrainment where turbulent BL eddies are entraining non-turbulent, free-atmospheric air. This entrainment zone is often associated with a stable layer or inversion.
- For boundary layers topped by shallow cumulus, the outer layer is subdivided further into sub-cloud, transition, cumulus and inversion layer.

5. Land Surface Models

The total radiation absorbed by the landsurface is balanced by emissions of thermal, infrared radiation to the atmosphere, latent heat loss associated with evaporation and transpiration, and sensible heat losses and diffusion of energy into the soil. The basic task of any land surface model is to simulate the partitioning of net radiation at the land surface into corresponding energy fluxes, when provided with the relevant information on land surface and meteorological forcing. Land covers 30% of the Earth's surface and as stated earlier, the variability of weather above land is greater than the one above oceans. As part of the water cycle, land provides a link (through surface water and ground water) between atmospheric and hydrological models (Boone et al., 2004).

Land surface models (LSMs) based on a solution of the energy balance at the land surface have been subject of ongoing research. Since the late 1980's, a large number of advanced LSMs containing sophisticated parameterizations of vegetation and root zone have been developed.

A widely used approach to land surface modeling is to consider the turbulent energy exchange between the land surface and the land-atmosphere as an electrical analogue. Penman (1948) assumed that the only resistance between the wet surface and the surrounding atmosphere was the so-called atmospheric resistance. The atmospheric resistance expresses the ability of the air to transport a given quantity away from the surface. In unstable conditions, such as those occurring when the surface is strongly heated, buoyancy will enhance the vertical motions allowing a faster exchange, and hence lowering the resistance. Under stable conditions, for example on a clear night with light winds, vertical motion is dampened by the stable stratification of air near the surface, leading to higher resistance. Penman's model is illustrated in Fig. 11.

One of the earliest land surface models using the Penman approach was the so-called 'bucket' model (Manabe, 1969). The bucket model assumes the soil surface holds and evaporates moisture at the same rate as a wet surface. Excess water that the bucket cannot hold was termed as run-off. The bucket

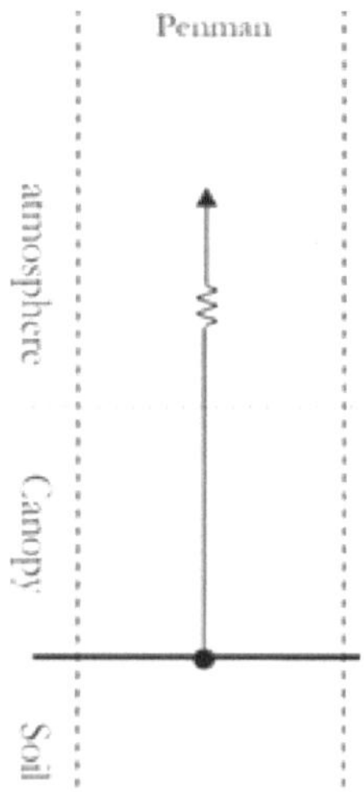

Fig. 11: Penman's electrical analogue.

model does not take into account vegetation or canopy and was the first generation LSM.

However, the land surface acts as free water surface only during and immediately after a precipitation event when the foliage is wet. At all other times, evapotranspiration has two important controls. First, evaporation from bare soil is greatly reduced when the uppermost soil layer dries out, and equally important, the plant retards the transpiration rate because of resistance of the stomata to molecular diffusion of moisture. Recognizing the influence of surface controls on the total evapotranspiration, Monteith (1965) further developed the Penman equation to take the land surface controls into consideration by introducing an additional surface resistance. This resistance depends on the vegetation type, soil moisture conditions and local climate.

Models of the Penman-Monteith type are often referred to as 1-layer, or big-leaf models because they do not distinguish between soil evaporation and transpiration, but treat the land surface as one functionally homogeneous surface. Their simplicity and physically-sound basis has led to the wide application of big-leaf models. In densely vegetated canopies, such 'big-leaf' models have proved sufficient to describe evapotranspiration (Monteith and Unsworth, 1990).

Deardorff (1978) modified the land surface energy consideration thus incorporating a prognostic temperature feedback with vegetation. These second generation land surface models included the interaction of vegetation impacts on energy, water and momentum budgets. These models included more than two soil layers and also considered multilayer soil water interactions.

A common model now in use is the Noah LSM which is an enhancement and is fundamentally based upon Deardorff (1978), Dickinson's (1983) BATS model, and the SiB model (Sellers, 1986). An advancement introduced in these models over the bucket model was to explicitly consider temporal changes in soil moisture (in addition to soil temperature). Some of these models also included snow effects and more realistic land surface interactions. These models outperformed the first generation bucket model improving the modelling of surface-atmosphere interaction on the time scale of days as shown by Beljaars et al. (1996) and Viterbo et al. (1999). The corresponding electrical analog is shown in Fig. 12.

The third generation of LSMs added more complexity by including a semi-empirical representation of vegetation conductance (ET) and considered plant physiology based on how leaf photosynthesis interactions are believed to function (Fig. 13). The evapotranspiration process is considered along with leaf and canopy photosynthesis and conductance. These models have been labeled as BATS2, and SiB2 (e.g., Sellers et al., 1992, 1996; Bonan, 1995).

The level of details being considered as 'adequate' in a LSM appears to depend on the applications and regions being simulated. For instance, for climate studies there have been an early adaptation of more detailed photosynthesis and carbon feedbacks along with increased complexity in the land representation. For the weather forecast community, there has been a general necessity for simplicity and computational efficiency. As a result, while great advances are made in land surface modelling to realistically capture surface variables, the NWP community has continued the use of either the "Bucket"/Slab model (enhanced first generation model) or the Noah LSM (second generation, Noilhan and Planton, 1989) in NWP models.

The slab model that is still operationally used in many regions for its simplicity or for the lack of observations to verify a detailed model, uses the

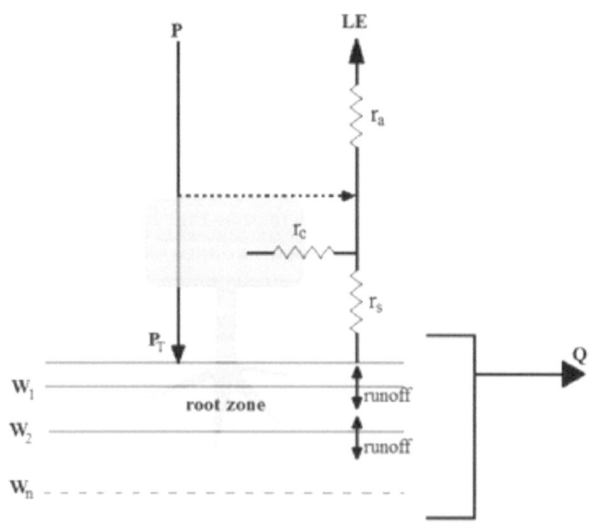

Fig. 12. Illustration of a second-generation land surface model. P and P_T are the total precipitation above the canopy and through-fall respectively, r_a is the aerodynamic resistance, r_s is the soil resistance, r_c is the canopy resistance, and Q is the collective runoff. (Adapted from Gascoin, 2009)

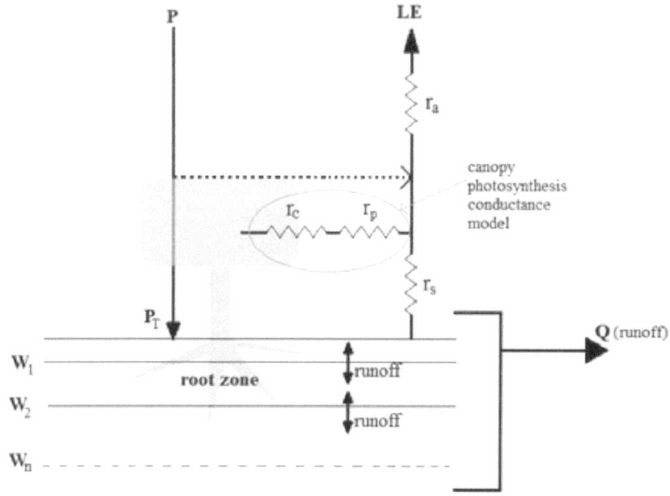

Fig. 13. Illustration of a third-generation land surface model. P and P_T are the total precipitation above the canopy and throughfall respectively, r_a is the aerodynamic resistance, r_s is the soil resistance, r_c is the canopy resistance in series with the resistance from the leaf stomata/photosynthesis, and Q is the collective runoff. (Adapted from Pitman, 2003; Bonan, 2008)

modified force-restore method in the calculation of soil temperature and the soil moisture and heat capacity of soil is set as a seasonally varying function of land-use type (supplied to the model as a look-up table) without explicitly considering the role of vegetation. Soil layers are typically 1, 2, 4, 8 and 16 cm thick, and the temperature below these layers is fixed as the daily average.

The Noah land model that is typically adopted for daily weather forecasts has four soil layers at 10 cm, 30 cm, 60 cm, and 100 cm, respectively. The root zone extends through the first three or all four layers depending on the vegetation class. Desert and shrubland has three layers while forest and crop land has four soil layers. Vertical diffusion, latent heating from soil freezing/ thawing, and heat fluxes at the top and bottom of the soil column are modelled in the soil thermodynamic equation which governs the temperature and in turn the sensible fluxes. Vertical diffusion, hydraulic conductivity (including bottom drainage), surface infiltration of water, and surface evaporation and transpiration by canopy roots are all responsible for moisture changes in the soil column. The surface evaporation and surface infiltration of precipitation are influenced by the vegetation canopy.

Weather forecast models continue to be second generation LSMs which have Penman-Monteith equation as the base framework (e.g., Chen and Dudhia, 2001; Ek et al., 2003). Niyogi et al. (2009) showed that photosynthesis-based gas exchange models (GEM) integrated within a prognostic soil moisture and soil temperature model similar to that developed by Noilhan and Plankton (1989) can be efficiently coupled to weather forecast models. The core component of GEM is the surface resistance scheme and uses the

The governing equations in Noah LSM (Chen et al., 1996) are

Soil Moisture

$$\frac{\partial \theta}{\partial t}=\frac{\partial}{\partial z}\left(D\frac{\partial \theta}{\partial z}\right)+\frac{\partial K}{\partial z}+F_{\theta}$$

(Richard's Equation for soil water movement)

- *D*, *K* functions (soil texture, soil moisture)
- F_{θ} represents sources (rainfall) and sinks (evaporation)

Soil Temperature

$$C(\theta)\frac{\partial T}{\partial t}=\frac{\partial}{\partial}\left(K_{t}(\theta)\frac{\partial T}{\partial z}\right)$$

- C, K_t functions (soil texture, soil moisture)
- Soil temperature information used to compute ground heat flux

relative humidity approach (Ball-Berry model, Ball et al., 1987). Evaluation of this model suggests that it performs better than the Jarvis-based approach (Niyogi et al., 2009; Charusombat, 2012). A more comprehensive review of the evolution and developments in land surface modelling can be found in Pitman (2003).

Land Surface Models within Tropical Cyclone/Hurricane Models

There are at least three categories of LSMs being used in tropical cyclone modelling studies. The first is a simple diagnostic model described in Emanuel et al. (2008). Here, the soil model consists of a series of one-dimensional columns set along the cyclone track. There are a number of soil layers that extend from the surface to 2 m below the surface. Each column integrates a set of two equations of the form (Flerchinger and Saxton, 1989; Chen et al., 1997)

$$\frac{\partial T_{soil}}{\partial t}=\frac{\partial}{\partial z}\left(\kappa\frac{\partial T_{soil}}{\partial z}\right)-I\frac{\partial T_{soil}}{\partial z} \qquad (1)$$

T_{soil} is the soil temperature
κ is thermal diffusivity and

$$I=\frac{\text{hydraulic conductivity}\times\text{soil porosity}}{\text{soil heat capacity}\times\text{density}}$$

This is the measure of the downward flux of water through the soil.The first term on the right hand side of the equation represents the thermal diffusion through the soil and the second term represents the heat transport by hydraulic conductivity through the soil. The above layer is solved for all layers except the topmost layer. The top soil layer interacts with the atmosphere above and is represented by the following equation which when solved is the upper boundary condition for the solution of equation (1).

$$\delta z C_z \rho_z \frac{\partial T_1}{\partial z} = C_s \rho_s \kappa \frac{(T_2 - T_1)}{\delta z} - C_k \rho_\alpha |V| \left(k_0^* - k_b\right) - \rho_l P C_l \left(T_1 - T_{rain}\right) + Q_{rad} \tag{2}$$

where δz is the depth of the uppermost soil layer, T_1 and T_2 are the temperature of first (top) and the second soil layer, C_s is the heat capacity per unit mass of the soil and ρ_s is its density. C_k is the surface enthalpy transfer coefficient and ρ_a is the surface air density, $|V|$ is the 10 m wind speed, k_0^* and k_b are the enthalpy of air in equilibrium with the soil surface and the actual enthalpy of air at 10 m respectively, ρ_1 is the density of liquid water; C_1 is the heat capacity per unit mass of liquid water, P is the rate of precipitation (m/s), T_{rain} is the temperature of rain as it reaches the ground, and Q_{rad} is the radiative heating of soil.

The above two equations conserve energy and efficiently represent a surface boundary for a landfalling hurricane system that evolved over highly porous, sandy soil at high surface temperatures (Emanuel et al., 2008).

The second category of model – the slab model (Tuleya, 1994) – is the most common model and has been used in operational models such as WRF, Hurricane WRF (HWRF) and GFDL (Geophysical Fluid Dynamic Laboratory) model. The slab model follows Deardorff (1978) to assume a surface energy balance equation of the form

$$\sigma T_L^4 + H + LE - (S + F\downarrow) = G \tag{3}$$

$$H = \rho c_p C_e V (T_L - \theta_{va}) \tag{4}$$

$$LE = (WET)\rho L C_e V [R_s(T_L) - R_a] \tag{5}$$

where G is the net ground heat flux, H is the sensible heat flux, LE is the evaporative flux, σT_L^4 is the emission from the Earth's surface assuming an emmisivity value of 1, $(S + F\downarrow)$ is the net downward surface flux, C_e is the drag coefficient of heat and moisture calculated from the Monin Obukhov framework similar to Kurihara and Tuleya (1974), V is the low-level wind speed (normally 10 m winds), θ_{va} is the virtual potential air temperature just above the surface, WET is the wetness coefficient of the land surface, R_s and R_a are the mixing ratios of saturated land surface and the low-level air, L is the latent heat of condensation, ρ is the density of near-surface air, and c_p is the specific heat of air. The land surface temperature can then be computed using the tendency equation

$$\frac{\partial T_L}{\partial t} = \frac{-\sigma T_L^4 - H - LE + (S + F\downarrow)}{\rho_s C_s d} - c\left(T_L - T_{gref}\right) \tag{6}$$

where $\rho_s c_s$ is the soil heat capacity and d is the damping depth, and $c = 2\pi/\tau$ (τ is the period of forcing). The last term in (6) is used to "force-restore" the model's surface temperature to a reference value to avoid spurious values. This sub-surface scheme can adequately represent and simulate the thermal evolution (temperature) of a more complicated multi-level land model and was used as an operational land surface model until recently in the US. Efforts are currently underway to adapt the Noah land model (prognostically predict both surface temperature and moisture) within the hurricane WRF. The Noah model has a long history of development beginning with the adoption of the OSU (Oregon State University) land model into NCEP's Eta model as the Noah model and its continuing upgrades. A brief history of Noah's land model evolution is listed in Table 1. This is only illustrative and not a comprehensive summary.

Table 1. History of evolution in development of Noah land surface model

Date	*Development/Upgrade description*	*Reference(s)*
	Original OSU land model	
	Potential evaporation	Mahrt and Ek (1984)
	Surface fluxes, soil hydraulics and soil thermodynamics	Mahrt and Pan (1984) Pan and Mahrt (1987)
	Noah LSM implementation	
Jan 1996	OSU land model introduced into Eta model	Chen et al. (1996)
	Surface runoff and infiltration	Schaake et al. (1996)
Jul 1996	ISLSCP vegetation greenness changes	
Feb 1997	NESDIS vegetation greenness	Gutman and Ignatov (1998)
	Bare soil evaporation changes	Betts et al. (1997)
	Snow melt changes	Betts et al. (1997)
	Thermal roughness length changes	Chen et al. (1997)
Feb 1998	Increase from 2- to 4-soil layers	
	Self-cycling Eta-EDAS soil moisture and temperature	Ramsay (1998)
Jun 1998	NESDIS snow cover and sea ice analysis	
	Noah LSM Upgrades	
Jul 2001	Frozen soil physics	Koren et al. (1999)
	Snow pack physics upgrade	Koren et al. (1999)
	Max. snow albedo climatology	Robinson and Kukla (1985)
	Shallow snow thermal conductivity	Lunardini (1981)
	Bare soil evaporation refinement	
	Bare soil thermal conductivity changes	Peters-Lidard et al. (1998)
	Vegetation-reduced soil thermal conductivity	Peters-Lidard et al. (1997)
	Transpiration refinements	
Feb 2002	Patchy shallow snow thermal conductivity	
Nov 2003	Improvements to cold season processes	Ek et al. (2003)

Another enhanced version of Noah LSM is the Noah LSM with multi-parameterization options (Noah-MP) which is customizable based on the application for which the land model is being run. This development improves biophysical realism based on Ball-Berry photosynthesis surface resistance with a dynamic vegetation model that allocates carbon to various parts of vegetation (leaf, stem, wood and root) and soil carbon pools (fast and slow). A multi-layer snowpack for improved cold season processes and unconfined aquifer model to accurately represent surface runoff are some of the parameterization options available through this model enhancement. This advanced version of Noah LSM improves on deficiencies identified in treating interactions among vegetation, soil, hydrology, snow and long-term soil state evolution in the current Noah LSM (Niu et al., 2011; Yang et al., 2011). There are sub models that incorporate urban land surface (Chen et al., 2011), crop land and agricultural feedback (Liu et al., 2014).

6. Challenges in Land Surface Modelling

Without a doubt, LSMs have benefited from combined efforts across multiple scientific disciplines and have proved to perform well at different spatial and temporal scales.

The Noah land surface model requires key input such as landuse/land cover (vegetation type), soil texture, initial soil moisture and temperature, slope and other secondary variables that are functions of the above primary variables (Chen et al., 2007). Variations in soil moisture may be caused by factors such as precipitation, soil texture, drainage, irrigation, flooding and shallow groundwater. While most LSMs applied in distributed frameworks include a detailed description of vegetation and root zone, the interactions between groundwater, root zone and surface water, as well as the lateral surface and subsurface flows are normally neglected, and consequently these models will have larger uncertainty in results in areas where such interactions are important. Typically, only three of the land components (soil, snow and vegetation) are explicitly treated while lakes and land ice are neglected (Yang, 2004). Vegetation is treated as 'big leaf' and scaling linearly from the size of a normal leaf to grids measuring 10 km and 100 km can pose problems when dealing with land-surface heterogeneities at smaller scales. The increasing quantity as well as improved quality and resolution of land surface and near-surface climate data obtained from remote sensing in conjunction with the ability to assimilate remotely-sensed data and in-situ data (e.g., NDVI and LAI) into a gridded form has significantly improved the performance of land surface models. From the operational perspective, optimizing the efficiency of LSMs in capturing the multitude of processes and yet retaining simplicity is the continued quest that the land community embarks itself on. The challenges in representing LSMs are even more dominant as post landfall hazards continue causing increasing socio-economic losses in the coastal zones (e.g. Hurricane Sandy (2012) over the New York City).

Acknowledgements

The authors gratefully acknowledge the financial support given by the US National Science Foundation (NSF CAREER AGS-0847472), the Indian National Center for Ocean Information Services (INCOIS), MoES and the Earth System Science Organization, Ministry of Earth Sciences, Government of India (Grant no./Project no MM/SERP/CNRS/2013/INT-10/002) to conduct this research under Monsoon Mission.

REFERENCES

Alpert, P., D. Niyogi, R.A. Pielke Sr., J.L. Eastman, Y.K. Xue and S. Raman, 2006: Evidence for carbon dioxide and moisture interactions from the leaf cell up to global scales: Perspective on human-caused climate change. *Global and Planetary Change*, **54(1–2)**, 202-208, ISSN 0921-8181.

Avissar, R. and Y. Liu, 1996: Three-dimensional numerical study of shallow convective clouds and precipitation induced by land surface forcing. *Journal of Geophysical Research: Atmospheres (1984–2012)*, **101(D3)**, 7499-7518.

Ball, J.T., I.E. Woodrow and J.A. Berry, 1987: A model predicting stomatal conductance and its contribution to the control of photosynthesis under different environmental conditions. *In:* Progress in photosynthesis research. Springer Netherlands.

Barale, V., 2010: Oceanography from Space: Revisited. Springer. p. 263. ISBN 978-90-481-8680-8.

Beljaars, A.C., P. Viterbo, M.J. Miller and A.K. Betts, 1996: The anomalous rainfall over the United States during July 1993: Sensitivity to land surface parameterization and soil moisture anomalies. *Monthly Weather Review*, **124**, 362-383.

Betts, A.K., F. Chen, K. Mitchell and Z. Janjic, 1997: Assessment of land surface and boundary layer models in two operational versions of the NCEP Eta Model using FIFE data. Mon. *Wea. Rev.*, **125**, 2896-2915.

Bonan, G.B., 1995: Land-atmosphere CO_2 exchange simulated by a land surface process model coupled to an atmospheric general circulation model. *Journal of Geophysical Research: Atmospheres (1984–2012)*, **100(D2)**, 2817-2831.

Bonan, G.B., 2008: Ecological Climatology: Concepts and Applications. 2nd edition. Cambridge University Press, Cambridge.

Boone, A. et al., 2004: The Rhone-aggregation land surface scheme intercomparison project: An overview. *Journal of Climate*, **17,** 187-208.

Briegel, L.M. and W.M. Frank, 1997: Large-scale influences on tropical cyclogenesis in the western North Pacific. *Mon. Wea. Rev.*, **125**, 1397-1413.

Chang, H.-I., 2009: Effect of land-atmosphere interactions on mesoscale convection and precipitation over the Indian monsoon region. Ph.D., Purdue University.

Charusombat, U. et al., 2012: Noah-GEM and Land Data Assimilation System (LDAS) based downscaling of global reanalysis surface fields: Evaluations using observations from a CarboEurope agricultural site. *Computers and Electronics in Agriculture*, **86**, 55-74.

Chase, T.N., R.A. Pielke, T.G. Kittel, J.S. Baron and T.J. Stohlgren, 1999: Potential impacts on Colorado Rocky Mountain weather due to land use changes on the adjacent Great Plains. *Journal of Geophysical Research: Atmospheres (1984–2012)*, **104(D14)**, 16673-16690.

Chen, F., K. Mitchell, J. Schaake, Y. Xue, H.L. Pan, V. Koren, Q.Y. Duan, K. Ek and A. Betts, 1996: Modeling of Land-Surface Evaporation by Four Schemes and Comparison with FIFE Observations. *J. Geophys. Res.*, **101**, 7251-7268.

Chen, F., Z. Janjić and K. Mitchell, 1997: Impact of atmospheric surface-layer parameterizations in the new land-surface scheme of the NCEP mesoscale Eta model. *Boundary-Layer Meteorology*, **85(3)**, 391-421.

Chen, F. and J. Dudhia, 2001: Coupling an advanced land surface-hydrology model with the Penn State-NCAR MM5 modeling system. Part I: Model implementation and sensitivity. *Monthly Weather Review*, **129**, 569-585.

Chen, F., Z. Janjić and K.E. Mitchell, 2007: Impact of atmospheric surface-layer parameterizations in the new land-surface scheme of the NCEP mesoscale Eta model. *Boundary Layer Meteorol.*, **85**, 391-421.

Chen, F., H. Kusaka, R. Bornstein, J. Ching, C.S.B. Grimmond, S. Grossman-Clarke, T. Loridan, K.W. Manning, A. Martilli, S. Miao and D. Sailor, 2011: The integrated WRF/urban modelling system: Development, evaluation, and applications to urban environmental problems. *International Journal of Climatology*, **31(2)**, 273-288.

Deardorff, J., 1978: Efficient prediction of ground surface temperature and moisture, with inclusion of a layer of vegetation. *Journal of Geophysical Research: Oceans (1978–2012)*, **83**, 1889-1903.

Dickinson, R.E., 1983: Land surface processes and climate-surface albedos and energy balance. *Adv. Geophys.*, **25**, 305-353.

Doran, J.C. and S. Zhong, 2000: A study of the effects of sub-grid-scale land use differences on atmospheric stability in prestorm environments. *Journal of Geophysical Research: Atmospheres (1984–2012)*, **105(D7)**, 9381-9392.

Ek, M. et al., 2003: Implementation of Noah land surface model advances in the National Centers for Environmental Prediction operational mesoscale Eta model. *Journal of Geophysical Research: Atmospheres (1984–2012)*, **108**.

Emanuel, K., 2000: A Statistical Analysis of Tropical Cyclone Intensity. *Mon. Wea. Rev.*, **128**, 1139-1152.

Emanuel, K., C. DesAutels, C. Holloway and R. Korty, 2004: Environmental control of tropical cyclone intensity. *Journal of the Atmospheric Sciences*, **61**, 843-858.

Emanuel, K., S. Ravela, E. Vivant and C. Risi, 2006: A Statistical Deterministic Approach to Hurricane Risk Assessment. *Bull. Amer. Meteor. Soc.*, **87**, 299-314.

Emanuel, K., J. Callaghan and P. Otto, 2008: A Hypothesis for the Redevelopment of Warm-Core Cyclones over Northern Australia. *Monthly Weather Review*, **136**, 3863-3872.

Feddema, J.J., K.W. Oleson, G.B. Bonan, L.O. Mearns, L.E. Buja, G.A. Meehl and W.M. Washington, 2005: The importance of land-cover change in simulating future climates. *Science*, **310(5754)**, 1674-1678.

Frank, W.M. and P.E. Roundy, 2006: The role of tropical waves in tropical cyclogenesis. *Mon. Wea. Rev.*, **134**, 2397-2417.

Flerchinger, G.N. and K.E. Saxton, 1989: Simultaneous heat and water model of a freezing snow-residue-soil system. I: Theory and development. *Trans. ASAE*, **32(2)**, 565-571.

Gascoin, S., 2009: Etude des paramétrisations hydrologiques d'un modèle de surface continentale: Importance des aquifères et des premiers centimètres du sol. Ph.D. thesis, Université Pierre et Marie Curie – Paris VI.

Gray, W.M., 1968: Global view of the origin of tropical disturbances and storms. *Mon. Wea. Rev.*, **96**, 669-700.

Gutman, G. and A. Ignatov, 1998: The derivation of green vegetation fraction from NOAA/AVHRR data for use in numerical weather prediction models. *Int. J. Remote Sens.*, **19**, 1533-1543.

Kaplan, J. and M. DeMaria, 1995: A simple empirical model for predicting the decay of tropical cyclone winds after landfall. *Journal of Applied Meteorology*, **34(11)**, 2499-2512.

Koren, V., J.C. Schaake, K.E. Mitchell, Q.Y. Duan, F. Chen and J. Baker, 1999: A parameterization of snowpack and frozen ground intended for NCEP weather and climate models. *J. Geophys. Res.*, **104**, 19,569-19,585.

Kurihara, Y. and R.E. Tuleya, 1974: Structure of a tropical cyclone developed in a three-dimensional numerical simulation model. *Journal of the Atmospheric Sciences*, **31(4)**, 893-919.

Liu, X., F. Chen, M. Barlage, G.S. Zhou and D. Niyogi, 2015: Noah-MP-Crop: Enhancing cropland representation in the community land surface modeling systems. 2015 Fall Meeting, AGU, San Francisco, CA, Dec 2015.

Lunardini, V.J., 1981: Heat Transfer in Cold Climates. *Van Nostrand Reinhold,* New York.

Mahfouf, J-F., E. Richard and P. Mascart, 1987: The influence of soil and vegetation on the development of mesoscale circulations. *J. Climate Appl. Meteor.*, **26**, 1483-1495.

Mahrt, L. and M. Ek, 1984: The influence of atmospheric stability on potential evaporation. *J. Climate Appl. Meteor.*, **23**, 222-234.

Mahrt, L. and H. Pan, 1984: A two-layer model of soil hydrology. *Boundary-Layer Meteorology*, **29(1)**, 1-20.

Manabe, S., 1969: Climate and the ocean circulation 1. *Monthly Weather Review*, **97**, 739-774.

Marks, F.D. and L.K. Shay, 1998: Landfalling Tropical Cyclones: Forecast Problems and Associated Research Opportunities. *Bull. Amer. Meteor. Soc.*, **79**, 305-323.

McBride, J.L. and R. Zehr, 1981: Observational analysis of tropical cyclone formation. Part II: Comparison of non-developing versus developing systems. *J. Atmos. Sci.*, **38**, 1132-1151.

Monteith, J., 1965: Evaporation and environment. *Symp. Soc. Exp. Biol.*, **4**.

Monteith, J. and M. Unsworth, 1990: Principles of environmental physics. E. Arnold. London; New York. Distributed in the USA by Routledge. Chapman and Hall.

Niu, G.Y. et al., 2011: The community Noah land surface model with multiparameterization options (Noah-MP): 1. Model description and evaluation with local-scale measurements. *Journal of Geophysical Research: Atmospheres (1984–2012)*, **116**.

Niyogi, D., K. Alapaty, S. Raman and F. Chen, 2009: Development and evaluation of a coupled photosynthesis-based gas exchange evapotranspiration model (GEM) for mesoscale weather forecasting applications. *Journal of Applied Meteorology and Climatology*, **48**, 349-368.

Noilhan, J. and S. Planton, 1989: A Simple Parameterization of Land Surface Processes for Meteorological Models. *Monthly Weather Review*, **117**, 536-549.

Pan, H.L. and L. Mahrt, 1987: Interaction between soil hydrology and boundary-layer development. *Boundary-Layer Meteorology*, **38(1-2)**, 185-202.

Penman, H.L. 1948: Natural evaporation from open water, bare soil and grass. Proceedings of the Royal Society of London. Series A. *Mathematical and Physical Sciences*, **193(1032)**, 120-145.

Peters-Lidard, C.D., M.S. Zion and E.F. Wood, 1997: A soil-vegetation-atmosphere transfer scheme for modeling spatially variable water and energy balance processes. *J. Geophys. Res.*, **102**, 4303-4324.

Peters-Lidard, C.D., E. Blackburn, X. Liang and E.F. Wood, 1998: The effect of soil thermal conductivity parameterization on surface energy fluxes and temperatures. *J. Atmos. Sci.*, **55**, 1209-1224.

Pielke, R.A., 2001: Influence of the spatial distribution of vegetation and soils on the prediction of cumulus convective rainfall. *Reviews of Geophysics*, **39**, 151-177.

Pielke, R.A., G.A. Dalu, J.S. Snook, T.J. Lee and T.G.F. Kittel, 1991: Nonlinear Influence of Mesoscale Land Use on Weather and Climate. *J. Climate*, **4**, 1053-1069.

Pitman, A., 2003: The evolution of, and revolution in, land surface schemes designed for climate models. *International Journal of Climatology*, **23**, 479-510.

Prandtl, L., 1905: On fluid motions with very small friction. Int. Math. -Kongr., Heidelberg.

Ramsay, B.H., 1998: The interactive multi-sensor snow and ice mapping system. *Hydrol. Processes*, **12**, 1537-1546.

Robinson, D.A. and G. Kukla, 1985: Maximum surface albedo of seasonally snow covered lands in the Northern Hemisphere. *J. Clim. Appl. Meteorol.*, **24**, 402-411.

Schade, L.R. and K.A. Emanuel, 1999: The ocean's effect on the intensity of tropical cyclones: Results from a simple coupled atmosphere-ocean model. *Journal of the Atmospheric Sciences*, **56(4)**, 642-651.

Schaake, J.C., V.I. Koren, Q.Y. Duan, K. Mitchell and F. Chen, 1996: A simple water balance model (SWB) for estimating runoff at different spatial and temporal scales. *J. Geophys. Res.*, **101**, 7461-7475.

Sellers, P.J., Y. Mintz, Y.C. Sud and A. Dalcher, 1986: A Simple Biosphere Model (SIB) for Use within General Circulation Models. *J. Atmos. Sci.*, **43**, 505-531.

Sellers, P.J., J.A. Berry, G.J. Collatz, C.B. Field and F.G. Hall, 1992: Canopy reflectance, photosynthesis, and transpiration. III. A reanalysis using improved leaf models and a new canopy integration scheme. **42**, 187-216.

Sellers, P.J., D.A. Randall, G.J. Collatz, J.A. Berry, C.B. Field, D.A. Dazlich, C. Zhang, G.D. Collelo and L. Bounoua, 1996: A Revised Land Surface Parameterization (SiB2) for Atmospheric GCMS. Part I: Model Formulation. *J. Climate*, **9**, 676-705.

Stull, R.B., 1988: An introduction to boundary layer meteorology (Vol. 13). Springer.

Trenberth, K.E., J.T. Fasullo and J. Kiehl, 2009: Earth's Global Energy Budget. *Bulletin of the American Meteorological Society*, **90**, 311-323.

Tuleya, R.E., 1994: Tropical Storm Development and Decay – Sensitivity to Surface Boundary Conditions. *Monthly Weather Review*, **122**, 291-304.

Viterbo, P. and A.K. Betts, 1999: Impact on ECMWF forecasts of changes to the albedo of the boreal forests in the presence of snow. *Journal of Geophysical Research: Atmospheres (1984–2012)*, **104**, 27803-27810.

Weaver, C.P. and R. Avissar, 2001: Atmospheric disturbances caused by human modification of the landscape. *Bulletin of the American Meteorological Society*, **82(2)**, 269-281.

Werth, D. and R. Avissar, 2005: The local and global effects of Southeast Asian deforestation. *Geophys. Res. Lett.*, **32**, L20702, doi:10.1029/2005GL022970.

Yang, Z.-L., 2004: Modeling land surface processes in short-term weather and climate studies. Vol. 3, World Scientific Series on Meteorology of East Asia.

Yang, Z.L. et al., 2011: The community Noah land surface model with multiparameterization options (Noah-MP): 2. Evaluation over global river basins. *Journal of Geophysical Research: Atmospheres (1984–2012)*, **116(D12)**.

Zehnder, J.A., 1991: The interaction of planetary-scale tropical easterly waves with topography: A mechanism for the initiation of tropical cyclones. *J. Atmos. Sci.*, **48**, 1217-1230.

Zhong, S. and J.C. Doran, 1998: An evaluation of the importance of surface flux variability on GCM-scale boundary-layer characteristics using realistic meteorological and surface forcing. *Journal of Climate*, **11(11)**, 2774-2788.

9

The Role of Land Surface Processes on Extreme Weather Events: Land Data Assimilation System

Dev Niyogi[1], Krishna K. Osuri[1, 2], S. Subramanian[1] and U.C. Mohanty[2]

[1]Purdue University, West Lafayette, Indiana, USA

[2]School of Earth, Ocean and Climate Sciences, IIT Bhubaneswar, India

dniyogi@purdue.edu

1. Land Surface and PBL Processes

Synoptic conditions are the main drivers for a tropical cyclone (TC) but equally important and not as well studied, is the role that boundary layer interactions play in the life cycle and the intensity of the storm (Bister and Emanuel, 1998; Emanuel, 2004; Hong et al., 2000; Persing et al., 2002; Tuleya, 1994; Tuleya and Kurihara, 1978). After landfall, most tropical systems are deprived of their surface heat source that fuels the storm or the storm system encounters a high shear environment that leads to the eventual decay of the warm core system (Rhome and Raman, 2006). However, some tropical storm systems have re-intensified inland, away from any warm water source. Examples include TS Erin (2007) and TC Abigail (2001) and Hurricane Fran (1996).

Numerical models are valuable tools in capturing the micro and macro scale interactions, the planetary boundary layer (PBL) have with the environment and studying the two-way feedback between landfalling TC and land surface characteristics. Numerous studies suggest that numerical models predict a TC track reasonably well, but most often fails to accurately capture the intensity and evolution of a storm post landfall. Evans et al. (2011) and Kellner et al. (2012) simulated the re-intensification of tropical storm (TS) Erin (2007) over Oklahoma using WRF-ARW and concluded that intensity of the storm and the vortex was dependent on development and maintenance of PBL. They also found that the storm was sensitive to soil moisture (SM) changes and the high soil moisture condition over Oklahoma before the landfall contributed to the re-intensification of storm.

In their study of land falling typhoon Sepat (2007), Zhang et al. (2011) showed that both latent and sensible heat fluxes (LHF and SHF) helped sustain intensity over land along with the spiral structure of rain bands. In addition, they also found that SM initialization impacted storm prediction over land. This conclusion continues to emerge from studies such as Bozeman et al. (2012) and many others. Kishtawal et al. (2012) used observational datasets and analytical techniques to study the factors influencing the post landfall characteristics of tropical storms over the Atlantic Ocean basin. Their results identify surface roughness and soil heat capacity as two dominant parameters. Model simulations were unable to capture these feedbacks thus highlighting the need to improve the land surface representation for improved post landfall TC characterization. Emanuel et al. (2008) studied the TC Abigail (2001) over Western Australia which underwent a significant period of re-intensification twice over land and concluded that surface and boundary layer interactions play an important role in the life cycle of a storm in intensifying the TC. Over the Indian monsoon region (IMR), it has been observed that monsoon depressions over land sustain themselves longer post landfall, if the 7 day antecedent soil moisture availability was high (Chang et al., 2009; Kishtawal et al., 2013).

Bister and Emanuel (1998) concluded that the bulk of dissipative heating occurs near the atmospheric boundary layer and positively influences the storm centric wind speed maxima. Emanuel (1998) developed an asymmetric hurricane model in which the maximum winds increase with C_k and decrease C_D similar to the Malkus-Riehl and Ooyama models and the hurricane intensity depended on the ratio of transfer coefficients. Braun and Tao (2000) studied a combination of boundary layer schemes with identical values of C_k/C_D, and concluded that the intensity is not only related to the magnitude of C_k/C_D but also on 'the wind speed dependence of roughness parameter z_0'. These results call for a better representation of the surface processes that govern intensity predictions of hurricane and their dependence (or lack thereof) on exchange coefficients. Montgomery et al. (2010) conducted numerical experiments in which the intensification rate and vortex intensity increases with additional surface drag until a certain threshold value is attained after which intensification rate decreases. This emphasises the importance of surface friction in increasing radial inflow into the hurricane boundary layer and thereby creating a zone of convergence for surface fluxes to fuel the cyclone.

The turbulence mixing processes within the hurricane boundary layer plays an important role in regulating the radial and vertical distribution of enthalpy and momentum (from the sea surface to the atmosphere and vice-versa) and, consequently, the TC intensity changes (Gopalakrishnan et al., 2013; Emanuel, 1986, 1995). Smith (1968) concluded that inadequate knowledge of the turbulence structure prevented a more realistic understanding of the boundary layer. To date, even with the best forecast models for TC prediction operating at the highest resolution; it is challenging to accurately represent the turbulent processes in the boundary layer. The transfer and diffusion of fluxes from the land surface to the atmosphere and vice versa are dependent on four key parameters, namely, surface exchange coefficients for momentum (C_d), moisture and heat (C_k) and eddy diffusivity for momentum (K_m), and moisture and heat (K_h). However, there is a large degree of uncertainty in the estimates of these parameters. Consequently, the structure and intensification of TCs as simulated by high-resolution mesoscale models are sensitive to the surface and planetary boundary layer (PBL) parameterizations used in these models (e.g., Braun and Tao, 2000; Nolan et al., 2009a,b; Montgomery et al., 2010; Smith and Thomsen, 2010).

Gopalakrishnan et al. (2013) used an idealized framework of the HWRF system to gain additional understanding of the variability in TC structure and intensity with respect to vertical diffusion. Reduction of eddy diffusivities, K_m (and K_h), to 25% of its original value produced reasonable diffusion coefficients consistent with observations for all wind speeds ranging from about 10 to about 60 m s^{-1} in the HWRF surface parameterization scheme (GFS scheme). Reduction of K_m had a significant influence on the structure, size, and evolution of the vortex. With K_m set to 25% of its original value, the depth of the inflow layer was reduced by 50%, more consistent with the observations. The cross-isobaric flow (inflow) was noted to be stronger with the decrease in the inflow depth. Stronger inflow not only increased the spin of the storm, enhancing the generalized Coriolis term in the equations of motion for tangential velocity, but also resulted in enhanced equivalent potential temperature in the boundary layer, a stronger and warmer core, and, subsequently, a stronger storm.

Other studies that have highlighted the role of boundary layer feedbacks are Anthes and Chang (1978), Emanuel (2003), Smith and Thomsen (2010) and Wang and Wu (2004). Pielke (2001) has shown that underlying surface characteristics such as terrain, land use, soil temperature (ST) and SM, albedo, and surface roughness have great influence on convective systems that pass over areas with surface heterogeneities. Numerous studies have acknowledged the influence of surface conditions on baroclinic zones, drylines, development of low-level jets, capping inversions and convective storms. Hence accurate representation of land surface is becoming increasingly important in this world of rapidly changing landscapes.

As highlighted in Subramanian et al. (2014), TCs are impacted by surface enthalpy fluxes. TCs behave like large heat engines and in the absence of moisture fluxes, they usually decay rapidly. But there are storms that are known to have passed over extremely dry and hot conditions and sustained or even re-intensified after landfall (e.g. TC Abigail, Emanuel et al. (2008)). TC Abigail made landfall on the southern coast of the Gulf of Carpentaria in

Australia around 1200 UTC 26 February 2001 with an estimated maximum sustained 10-min mean wind speed of 33 m/s (Fig. 1). The cyclone weakened while moving in a general westward direction but later surprisingly re-intensified over land. Emanuel et al. (2008) conducted idealized simulations and concluded that the hot, sandy soils of northern Australia may store enough heat and, after they are wetted by the first rains of an oncoming cyclone, may be able to diffuse the stored heat upward rapidly enough to sustain warm-core storms of marginal hurricane intensity. This finding is of interest but emphasizes that underlying soil must be quite hot and have a large enough heat conductivity when wet to support some storms. The simulations also demonstrate the storm intensification may also depend on antecedent moisture availability as well as the translational speed of storm.

One of the major challenges that operational community faces is to estimate the post landfall precipitation and flood threat associated with a cyclone. To predict the rainfall patterns accurately, track forecast accuracy becomes a primary factor. Bozeman et al. (2012) conducted a series of simulation on TS Fay (2008) using the HWRF modeling systems with two land surface parameterizations – the simple GFDL slab model and the more detailed noah land surface model (Fig. 2). The study focused on TS Fay because the storm did not develop a typical TC eye-like structure until after landfall over South Florida (Stuart and Beven, 2009). The Bozemann et al. (2012) study summarized three features that contribute to TC forecasts. First, that land surface parameterization is of importance to the storm forecast track but did not significantly impact the intensity. The improved track resulted in rainfall distributions that correctly reflect observations from past hurricane

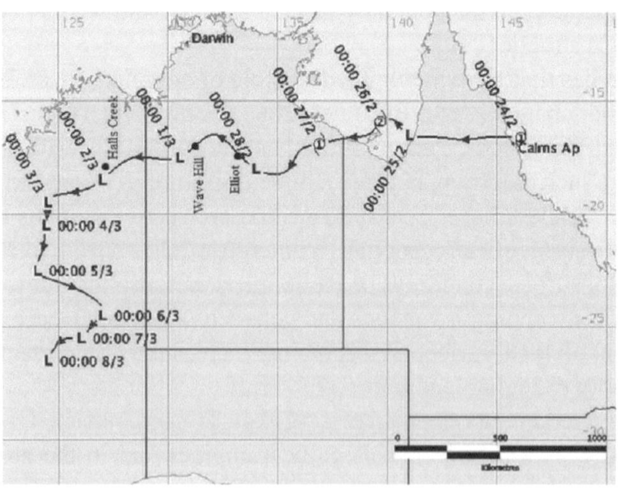

Fig. 1. Track of Abigail following landfall on the southern Gulf of Carpentaria in Australia's coastline from 1100 UTC 27 Feb 2001 (271100 UTC on map) to 00 UTC 8 Mar 2001. Figure obtained and modified from http://www.bom.gov.au/cyclone/history/abigail.shtml.

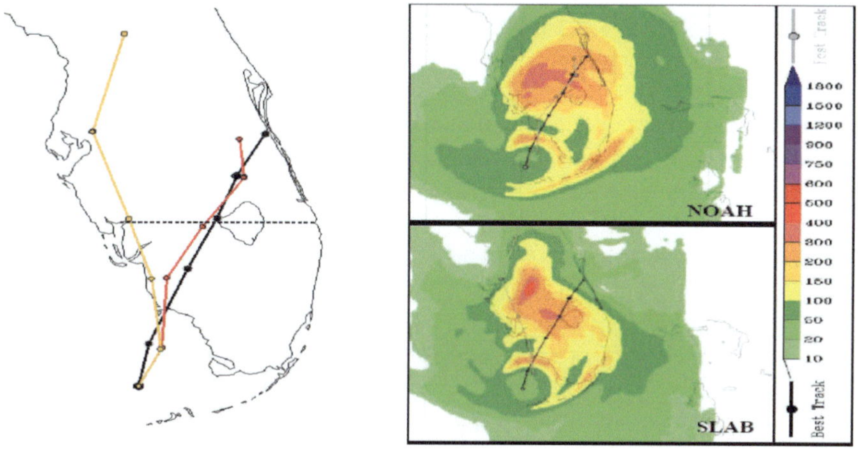

Fig. 2. Left: Dotted line indicates the location of the cross section for the GFDL Slab (S) and Noah (N) LSM tracks. Right: Noah (top) and GFDL Slab (bottom) LSM accumulated rainfall (mm) from 00Z Aug 19 until 06Z Aug 20. (Reprinted with permission from Bozeman et al., 2012)

studies that the core of heavy rain is predominantly in the narrow swath closest to the storm centre (Lonfat, 2004; Marchok et al., 2007; Rodgers et al., 2009). In this sense, the Noah LSM seems to have better forecast performance over the GFDL Slab model (Fig. 2). Secondly, initial/boundary conditions and PBL processes are shown to be important for TC development and intensity forecasts post landfall. Additionally, surface heterogeneity played a small but detectable role in track prediction. The study also noted that the presence of a lake does in fact cause a drop in central pressure, but does not have a large contribution to TC intensity prediction.

A number of studies now make a case for a more realistic land surface representation in numerical weather prediction (NWP) models (e.g Holt et al., 2006; Niyogi et al., 2006; Trier et al., 2004; Kellner et al., 2012). Studies (Tuleya 1994; Kellner et al., 2012) have shown that land surface heterogeneity and feedback could influence convective structure of tropical systems and sustain landfalling storms. Chang et al. (2009) isolated the effect of antecedent soil moisture, storms' rainfall and latent heat release, to provide evidence that warm, wet antecedent soil conditions can lead to more intense and sustained post-landfall TCs. The validity of these results and findings are further strengthened by the analysis of 183 landfalling monsoon TCs by Kishtawal et al. (2012). The hypothesis that, antecedent soil moisture conditions affect post landfall storm sustenance seems to be viable because of similar results found in a global model used over the Indian region (Dastoor and Krishnamurti, 1991) and Northern Australia (Emanuel et al., 2008). While the impact of land surface heterogeneity on convection has been the topic of research for several years (Pielke, 2001), how land surface heterogeneity and antecedent soil moisture conditions affect synoptic scale systems remains to be investigated.

2. Improving Land State Representation Using Land Data Assimilation System

Considerable uncertainty exists in model simulations because of the lack of accuracy in land surface input parameters. Among these, soil moisture and soil temperature (SM/ST) are perhaps the most crucial for BL modelling. The land models coupled with vegetative canopy sub-models (e.g., Noilhan and Planton, 1989; Chen and Dudhia, 2001) and rainfall have become more common to provide case-specific soil moisture profiles at high resolution (e.g., Chen et al., 1996, 2007; Chen and Dudhia, 2001). The rainfall significantly changes the underlying land surface characteristics, creating relatively wet and dry regions. The climatological SM/ST may not be representative of these local rainfall changes and this in turn can result in poor simulation of boundary layer and cloud convection processes and resulting precipitation. Since direct ground-based SM measurements are not readily available or regionally representative, assimilation methods have been developed to create high resolution SM fields by providing realistic forcing of rainfall, near-surface (2 m above the ground level) air temperature and moisture, winds and remotely sensed radiances. Considerable progress has been made recently with such stand-alone approaches for hydrological balance and flux simulation using land data assimilation systems such as the North American land data assimilation system (NLDAS) and the high-resolution land data assimilation system (HRLDAS; Mitchell et al., 2004; Chen et al., 2007).

The application of HRLDAS is evolving over the Indian monsoon region. The community HRLDAS (Chen et al., 1996; Ek et al., 2003, hereafter LDAS) is based on Noah LSM, which is the part of the Weather Research and Forecasting (WRF) system. It is a complex model and has various processes to represent the surface energy and water balance for a variety of land surface and soil conditions at user-specified resolutions. Its dynamical framework calculates temporal changes in surface energy fluxes (e.g., sensible, latent, and ground heat fluxes), net radiation, SM, ST, soil water content, and surface runoff. The LDAS requires a number of observed/analyzed atmospheric forcing and initial land surface conditions and is shown in Table 1. The atmospheric forcing data can be typically obtained from reanalyses products such as the NASA MERRA (Rienecker et al., 2011). The MERRA reanalyses have a resolution of $0.5°\times0.67°$ at sub-daily temporal resolution, and these fields can be bi-linearly interpolated in space to maintain the same resolution in the zonal and meridional directions (i.e., $0.5°\times0.5°$). The reanalyses initial-guess atmospheric fields are prepared based on a large number of atmospheric conventional and non-conventional observations (WCRP Report; Bosilovich et al., 2012). There is some limitation in precipitation estimates over the land (Rienecker et al., 2009) for these reanalyses products. Therefore, the 3-hourly rain rate from other data sources such as TRMMV7 estimates can be used as rainfall forcing after interpolating linearly to an hourly rain rate. The initial data input to local LDAS is typically obtained from the Global Data Assimilation System (GDAS) analysis of the National Centers for Environmental Prediction (NCEP).

The LDAS output includes spatiotemporal fields of SM, ST and surface energy balance components such as latent, sensible and ground heat fluxes (details in Table 2). Both LDAS and the atmospheric model are typically executed on the same grid structure and domain to aid incorporation of the surface fields into the coupled mesoscale model initial conditions.

Table 1. Input fields required to run the LDAS

S. No	*Atmospheric forcing*	*Initializing fields*	*Parametric fields*
1	2 m temperature [K]	Soil moisture [m^3 m^{-3}]	Land use and land cover [13 categories]
2	2 m specific humidity [g kg^{-1}],	Soil temperature [K]	Vegetation type [24 categories]
3	10 m winds [m/s],	Canopy water content [kg/m^{-2}]	Soil type/texture [19 categories]
4	Surface pressure [hPa],	Skin temperature [K]	Surface albedo and max. snow albedo
5	Downward short wave radiation [W m^{-2}] at surface	Water equivalent of accumulated snow depth [m]	Leaf area index
6	Downward long wave radiation [W m^{-2}] at surface		Green veg. fraction
7	Surface total rainfall rate [mm/hr]		Model elevation [m]

Table 2. Output fields from the LDAS

S. No.	*Output variable*	*Corresponding level/depths*
1	Soil moisture [m3 m^{-3}]	0–10 cm
2	Soil temperature [K]	10–40 cm, 40–100 cm and 100–200 cm depths
3	Skin temperature [K]	Surface
4	Skin moisture	Surface
5	Latent heat flux [W m^{-2}]	Surface
6	Sensible heat flux [W m^{-2}]	Surface
7	Ground heat flux [W m^{-2}]	Surface
8	Heat exchange coefficient [W m^{-2} K^{-1}]	Surface

3. Impact of Land Data on Monsoon Depressions and Numerical Experiments

In this and the following sections, numerical experiments are carried out using the WRF model with a single 4 km horizontal resolution domain. A similar domain was used to set up a LDAS and the corresponding SM/ST fields were generated. The first experiment, CNTL uses the initial and boundary conditions from the FiNaL analyses (FNL) of National Centers for Environmental Prediction (NCEP) in 6-hour intervals. In the second experiment (LDAS), the SM/ST profiles with depth and the top layer canopy water content of the CNTL initial condition (FNL analysis) is updated with those obtained from the LDAS. Eight cases are studied from 2007 to 2011 and full details such as, case name, initial date of integration, and forecast length are given in Table 3. The other land surface variables such as vegetation, soil type, land use, and land type are obtained from United States Geological Survey (USGS) data and kept the same for both experiments. The model uses the Mellor Yamada Janjic (MYJ) planetary boundary layer (PBL) scheme, Monin-Obukhov Janjic surface-layer option, RRTM longwave, Dudhia shortwave radiation, and WSM6 microphysics with explicit convection.

Table 3. Number of cases and its model initialization time and forecast length

Cases	*Monsoon depression*	*Forecast length*
Case 1	2007-06-29_00	36 hours
Case 2	2007-06-21_00	48 hours
Case 3	2007-07-04_00	96 hours
Case 4	2007-08-05_00	66 hours
Case 5	2008-06-16_00	48 hours
Case 6	2008-09-16_18	66 hours
Case 7	2009-07-20_00	30 hours
Case 8	2011-06-18_12	96 hours

The top layer volumetric SM (the volume of soil water per unit volume of soil in m^3 m^{-3}) from CNTL, LDAS and the difference (LDAS–CNTL) for case 1 and case 8 are shown in Fig. 3. The closed circle and square symbol represents the location of Ranchi and Kharagpur field stations, respectively. The CNTL shows a smoother SM field due to its coarser resolution initial data; while the LDAS-based SM exhibits strong horizontal heterogeneity based on soil type. These horizontal heterogeneities are of interest, because number of studies (e.g. Klink, 1992), have shown that land surface discontinuities at smaller scales are important for triggering atmospheric circulations. Due to the lack of SM observations, 24-hr accumulated rainfall was used to explain SM variations. The previous day 24-hour accumulated rainfall (Fig. 3g) from TRMM-observed analysis of case 1 shows that there was a heavy rain of the order of 20 mm over the region highlighted in the figure. This heavy rainfall region is in good agreement with that of higher values of SM from the LDAS experiment (Fig. 3b). As seen in Fig. 3g, there was no rainfall over the northern parts of the domain (above 22° N). The observed rainfall variation with latitude is well captured in the LDAS experiment with the northern parts of the domain showing minimum SM compared to the region in the elliptic. However, this is missed and wrongly simulated in the CNTL run (Fig. 3a) inferring that the region of no rainfall has more SM (northern parts of domain) and less SM in the highlighted region. This can be seen in the difference plot (Fig. 3c) suggesting that LDAS could provide more realistic spatial distribution of SM close to the observed rainfall distribution. Similar features are also noticed in case 8 where the LDAS-based initial SM field (Fig. 3e) is in good agreement with the previous day rainfall (Fig. 3h). The region highlighted in Fig. 3h shows the region of heavy rainfall. The impact of the previous day's rainfall on SM is not represented in the CNTL/FNL-based initial SM (Fig. 3d). As in case 1, the CNTL run overestimated SM over the northern parts of the domain in case 8.

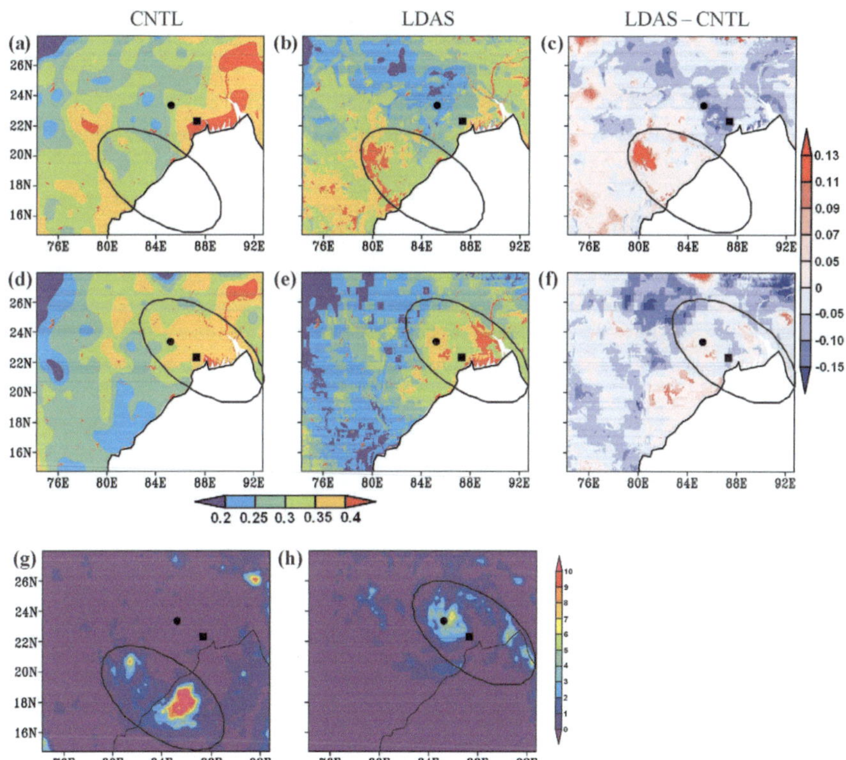

Fig. 3. Initial top layer SM for case 1 (a) CNTL, (b) LDAS and (c) LDAS–CNTL difference. (d-f) are same as (a-c) but for case 8. (g) and (h) are the previous 24-hour accumulated rainfall (mm) corresponding to case 1 and case 8 initial condition respectively. The rectangular and circle symbols show location of Kharagpur (22.31° N, 87.31° E) and Ranchi (23.35° N, 85.23° E) stations respectively.

The surface fields simulated from both experiments corresponding to case 8 are validated with the available observations at Ranchi (first column) and Kharagpur (second column) stations and shown in Fig. 4. The simulated top layer (top 10 cm) SM from the CNTL and LDAS runs for case 8 showed distinctly different patterns. The CNTL run showed an excess of SM throughout the integration period as the initial SM is very high to begin with compared to the LDAS and observed SM. The correlation and RMSE of CNTL-based SM is 0.59 and 0.09 m^3/m^3 respectively. The LDAS-based SM reproduced

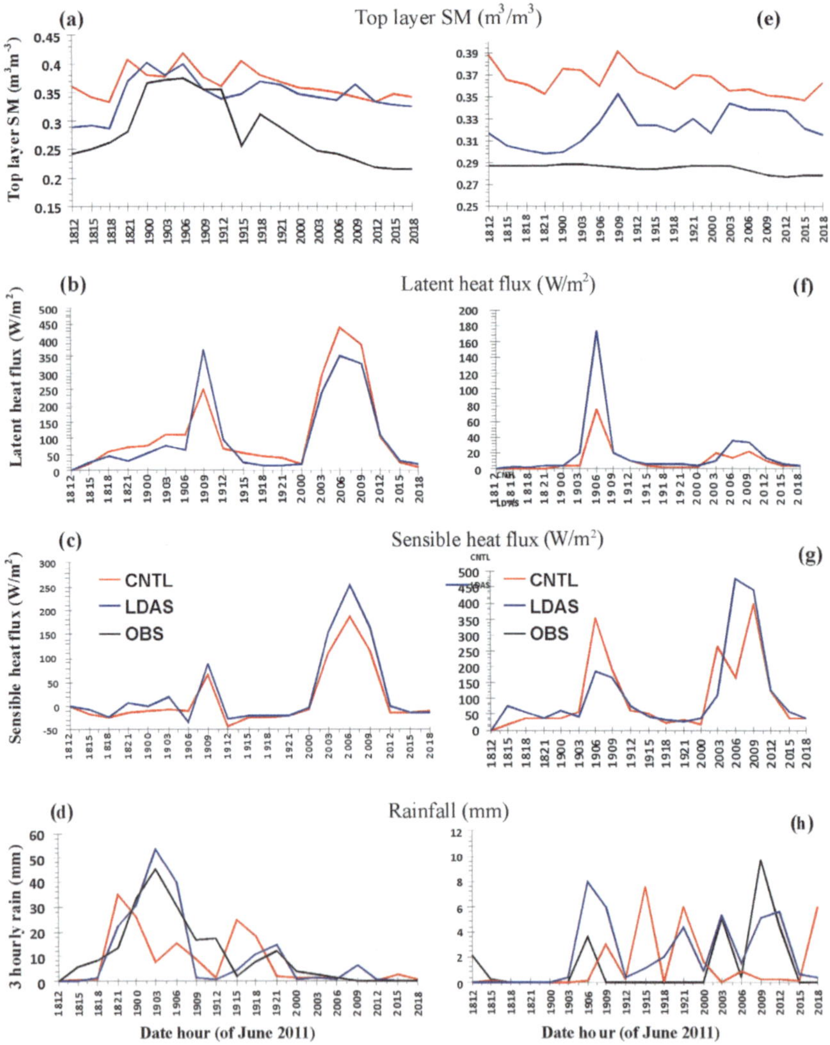

Fig. 4. Verification of model simulated surface parameters at Ranchi station for Case 8 (a) top layer soil moisture (SM, $m^3\ m^{-3}$), (b) latent heat flux (W/m^2), (c) sensible heat flux (W/m^2), (d) 3 hourly accumulated rain (mm), (e-h) are same as (a-d) but for Kharagpur station.

the observed SM pattern with some overestimation during the first 24 hours and correlation of 0.65 and RMSE of 0.006 m^3/m^3. Surface flux observations are not available for this station to validate model results and the model-simulated latent heat fluxes at Ranchi from both CNTL and LDAS exhibit notable differences (Fig. 4b). Warner (2004), for example, highlights that land surface conditions conditions such as soil moisture, impact the partitioning of the available radiative energy at the surface into latent and sensible heat fluxes. The CNTL-based latent heat flux is greater with the overestimation of top layer SM (Fig. 4a).

The surface latent and sensible heat fluxes are estimated by the difference between surface moisture and temperature and air moisture and temperature at the first model level (Flerchinger et al., 1998). Chang et al. (2009) demonstrated that the stronger gradient of heat fluxes at the surface (e.g., warmer, wetter land surface) can help to sustain the boundary layer instability and improve rainfall simulation associated with MDs over the IMR.

Model-simulated tracks of the MD cases (case 1 and case 8) are shown in Fig. 5 for the CNTL and LDAS experiments along with the observed path. It is clear that the overall track prediction by LDAS is reasonably good compared to CNTL. As prescribed, the two experiments have the same initial position for the MD, which is about 57 km from the observed position. In case 1 (Fig. 5a), both experiments simulate the similar tracks over the BoB. Once the system crosses the land, the two runs predict different tracks in response to the the land surface conditions. With the realistic representation of heterogeneity of land surface values (SM and ST) in the LDAS experiment, appreciable improvement is noticed in track prediction. The CNTL-simulated system moved to higher northward latitudes having more vector displacement errors or track errors with respect to the observed track. The track errors from the CNTL run are approximately 60, 98, 55, 140, 170, and 95 km for the 6-36 hour forecast in six-hour intervals, while the same errors in the LDAS run are about 48, 49, 45, 75, 30, and 80 km. For case 1, LDAS simulated better intensity evolution. Observations indicated that the system weakened after crossing the land; however CNTL showed the system intensifying even after landfall in terms of both minimum sea level pressure (MSLP) and higher wind speeds. The LDAS-simulated system is relatively weak, the observed pattern and the error is less compared to CNTL. In case 8 also, the CNTL simulated the storm motion to the higher latitudes due to the availability of higher SM content. Because of improved SM distribution, the LDAS run captured the realistic movement of the system close to the observed. Osuri et al. (2013) showed that the WRF model has a rightward bias in track simulation for the Indian monsoon region, and in the case of westward moving systems, WRF simulated systems move towards higher latitudes. The corresponding track error is shown in Fig. 5d.

The time-longitude cross section of 3-hour rain rate obtained from CNTL and LDAS runs along with TRMM rainfall are also analyzed (not shown).

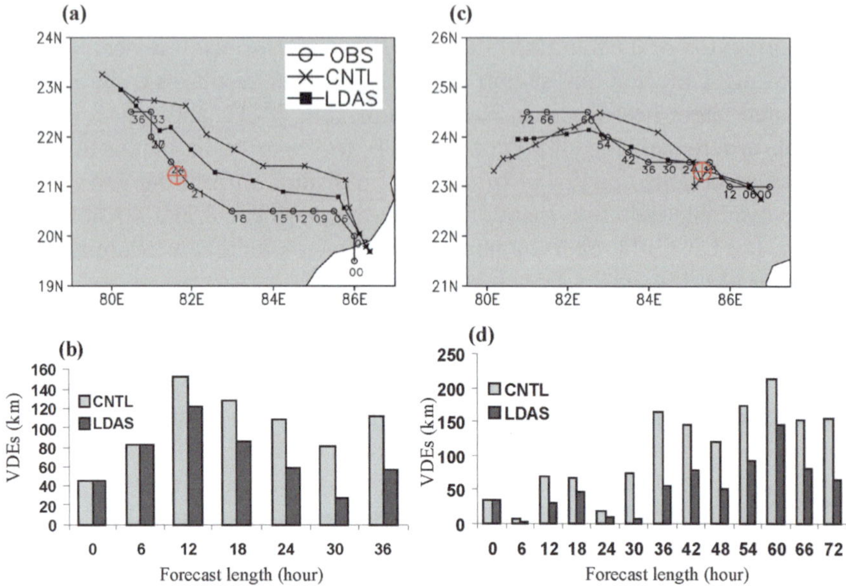

Fig. 5. Model simulated (a) tracks and (b) track errors for case 1 from both the experiments. (c-d) are same (a-b) but for case 8.

Although the CNTL and LDAS show similar rainfall distribution, there is a clear distinction in the timing and amount of rainfall between them. The inclusion of improved high resolution land surface conditions enhanced the boundary layer characteristics and overall, LDAS based model runs could reproduce the observed rainfall with smaller errors.

4. Impact of Improved Land Fields on Mesoscale Convection over Indian Region

In this section, we provide the impact of LDAS-based soil moisture and temperature profiles as land surface conditions in the WRF model and highlight the positive impact on the simulations for the location and timing of convection initiation, as well as evolution of severe thunderstorms and associated mesoscale convection over the Indian monsoon region. We selected three severe thunderstorm cases that occurred on 16 May (Case 1), 22 May (Case 2), and 27 May (Case 3) over the eastern and northeastern parts of India, and Bangladesh during the field phase of the Severe Thunderstorm Observations and Regional Modeling (STORM) program. All three cases are initialized at 12 UTC of the previous day of the thunderstorm event and the forecasted for 36 hours in one hour intervals. Two numerical experiments were conducted, with and without land surface data assimilation using WRF model as outlined in the previous section. The offline LDAS and WRF are executed on the same grid structure/domain to aid incorporation of soil fields initialized from LDAS when assimilating into WRF initial conditions. Except for the assimilation of initial the SM/ST fields, there were no other differences in both the experiments in terms of model configuration and physics. The model-simulated rainfall is also compared with the global climate prediction centre morphing method (CMORPH) rainfall product at 8 km resolution at one hour intervals. The model-simulated reflectivity fields are also compared with the observed reflectivity datasets from Kolkata (22.57° N, 88.35° E) Doppler weather radar (DWR).

Figure 6 shows the time evolution of near-surface meteorological parameters such as surface air temperature, specific humidity, and 10 m wind at three meteorological stations. The CNTL run shows warmer and dry atmosphere during the afternoon than the observed and the LDAS run. The mean bias and RMSE of the CNTL run for surface temperature during the afternoon is about 6 K while it is much lowered and is 2 K in the LDAS run. LDAS shows high atmospheric humidity throughout the simulation. The evolution of 10 m wind is also well captured in the LDAS run. Although both experiments overpredicted during the first 12-hour forecast, the errors are relatively less in the LDAS based WRF runs.

The model-simulated reflectivity from both experiments during the thunderstorm activity are presented for case 3. Figure 7 shows observed reflectivity from Kolkata DWR station from 11 to 14 UTC on 27 May 2007. Results clearly indicate that better representation of high resolution SM/ST leads to improved simulation of severe convection (Fig. 7c, f, i, l). The

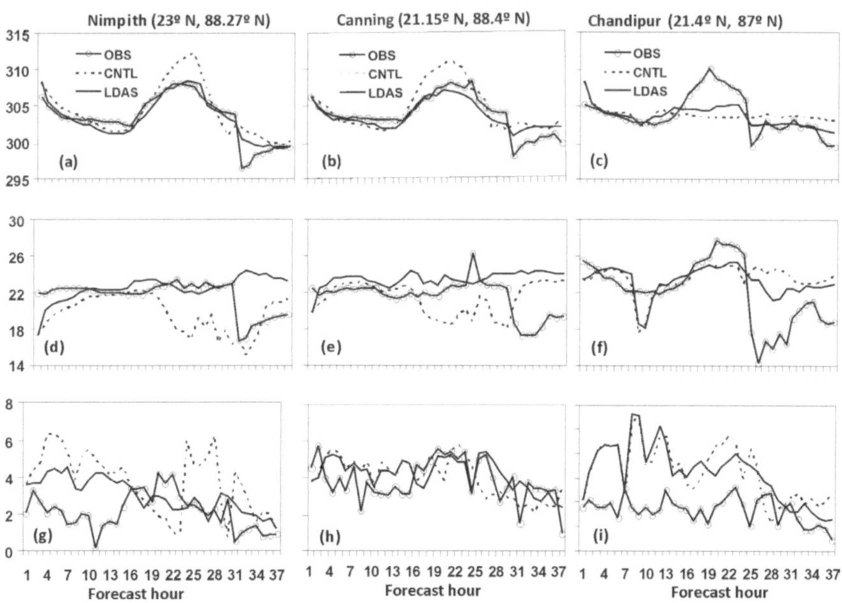

Fig. 6. Same as Fig. 4 but for 2-m air temperature in K (a-c), 2-m specific humidity in g kg^{-1} (d-f) and 10 m wind speed in m s^{-1} (g-i).

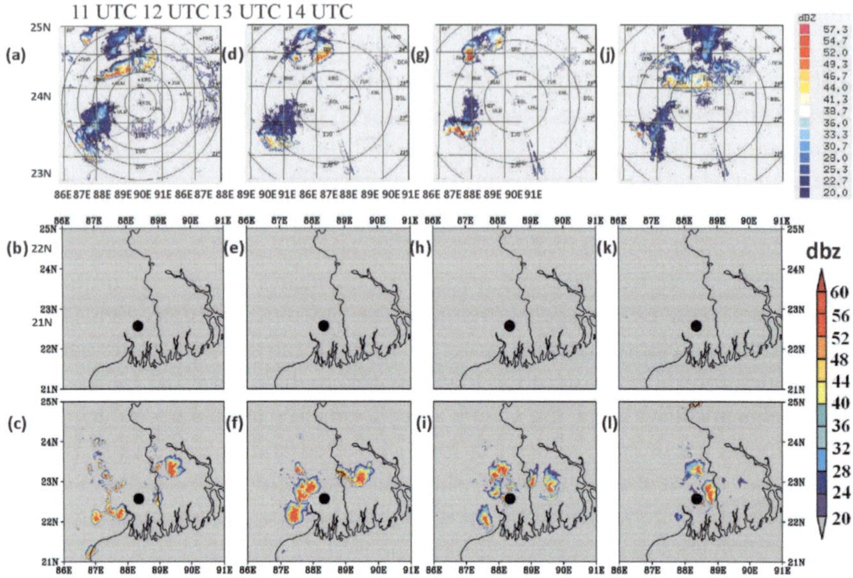

Fig. 7. Reflectivity of Case 3 from (a) observed Kolkata DWR, (b) CNTL and (c) LDAS at 11 UTC of 27 May 2007. (d – f), (g – i) and (j – l) are same as (a – c) but at 12 UTC, 13 UTC and 14 UTC of the same day respectively. The black dot in model figures indicates the location of Kolkata DWR.

CNTL experiment fails to predict the convection over the GWB (Fig. 7b, e, h, k) while the LDAS run predicts strong echoes with a small southward displacement compared to observed echoes. The LDAS run predicted the three bands around the Kolkata DWR station as also noted in the observations. The LDAS run also captured the east-west orientation of echoes at 13 UTC and its movement from the northwest to southeast. Moving in the southeastward direction, the storm was over Kolkata at 14 UTC in the LDAS which shows a high degree of agreement with the observations. The time of initiation of thunderstorm activity is also better simulated in the LDAS experiment (9 UTC 27 May 2007).

Both the experiments capture hourly rainfall distribution corresponding to the simulated reflectivity echoes (not shown). In case 3, the rainfall intensity is significantly improved and closer to the observed rainfall in spite of southward displacement of rainfall distribution. The spatial distribution of low rainfall is improved in LDAS. The CNTL run fails to capture the convection and thus could not predict the rainfall distribution. The observed rainfall over the head of Bangladesh from 13 UTC to 16 UTC is reproduced accurately in LDAS but missed in the CNTL run. Encouraging results have been seen in other two cases also. In spite of the spatial error in simulating convection in the LDAS experiments, the run could capture the intensity of convection and associated rainfall well. Another notable feature is that the LDAS experiment could simulate even the short term heavy rains in agreement with the observations.

5. Future Directions

There exists an emerging need for coupling techniques or ad-hoc land assimilation systems in the atmospheric models to minimize errors. Towards this requirement, both direct and indirect surface and satellite data-based data assimilation methods for coupled land surface–atmosphere models are available and need to be evaluated. Examples include studies by Houser and Shuttleworth (1998), Alapaty et al. (2008), Reichle et al. (2001a, b, 2008), Crow et al. (2008). These studies take reference observations and modify the surface fields to nudge the observed values or derive soil moisture fields using a LDAS like approach. There is a large body of literature now emerging related to developing spatiotemporal soil moisture/temperature fields that can serve as an input to the land models for NWP models.

There are number of satellite products that are available, and integrated within a LDAS-like framework for generating SM/ST fields. As an example, TRMM/TMPA and CMORPH precipitation products can be used for rainfall input as well as verification (both distribution and intensity) along with in situ observations. MODIS aboard Aqua and Terra is widely used to construct a realistic land surface condition using the NDVI and green vegetation faction data along with temperature fields. The Advanced Microwave Scanning Radiometer – EOS (AMSR-E) aboard Aqua could be used for the soil moisture fields for some past studies (the satellite was not functional after Oct 2011), and a newly launched Soil Moisture Active/ Passive (SMAP) sensor now provides gridded SM/ST fields (typically at 36 km grid, which is being reprocessed for a 9 km resolution product). Potential soil moisture products could also be derived from NASA Aquarius and INSAT 3D data products which provide rainfall estimates.

A number of studies have demonstrated the potential of assimilating satellite derived soil moisture into LSMs (e.g., Walker and Houser, 2001; Reichle et al., 2001a, b; Crow and Wood, 2003). However, the direct assimilation of satellite-retrieved soil moisture data is restricted due to the large systematic differences between satellite-retrieved soil moisture and in situ observations, mismatch in spatial scales of data and LSMs, etc. There are studies demonstrating how to overcome these problems using bias removal and downscaling of SM data retrieved from satellites. A brief survey on assimilation of SM and ST into LSM is available in Alavi et al. (2009).

6. Summary

In this chapter, a summary for different experiments that test the hypothesis that the simulation of localized or synoptic scale convection over the Indian region can be improved by consideration of improved land state including of enhanced SM/ST fields as surface initial conditions to mesoscale model is tested.

The results from different cases provide evidence that the proper representation of land surface characteristics (like dry and wet soil boundaries through soil moisture and temperature fields) are important drivers for mesoscale convection even over the Indian region. These spatial heterogeneities can interact with synoptic-scale processes and are important in triggering local-scale convection. Enhanced representation of pre-storm land state conditions can improve model performance in the simulation of thunderstorm convection and rainfall over the pre-monsoon environment.

The LDAS-based initial values of SM and ST profiles consistently helped improve the forecast fields of surface moisture fluxes and near-surface atmospheric conditions such as temperature, moisture, and wind fields. This in turn improved the mass flux in the boundary layer through stronger convective updrafts. Accurate simulation of the moisture fields in the boundary layer over the thunderstorm region helped improve severe convection initiation and maintenance in the model. As a result, LDAS based WRF run could reproduce the observed features of severe convection as assessed by strong reflectivity echoes. The location, timing, and movement of thunderstorms have also improved compared to the runs with climatological SM/ST fields (CNTL). Further, LDAS coupled system also simulated more realistic rainfall distribution and intensity. Improvements in the land models, land surface representation, surface field initialization, and the coupling between the land - atmosphere holds tremendous potential for enhancing tropical cyclone predictions.

Acknowledgements

The authors gratefully acknowledge the financial support given by the US National Science Foundation (NSF CAREER AGS-0847472), the Indian National Center for Ocean Information Services (INCOIS), MoES and the Earth System Science Organization, Ministry of Earth Sciences, Government of India (Grant no. / Project no MM/SERP/CNRS/2013/INT-10/002) to conduct this research under Monsoon Mission.

REFERENCES

Alapaty, K., Dev Niyogi, Fei Chen, Patrick Pyle, Anantharman Chandrasekar and Nelson Seaman, 2008: Development of the Flux-Adjusting Surface Data Assimilation System for Mesoscale Models. *J. Appl. Meteor. Climatol.*, **47**, 2331–2350. doi: http://dx.doi.org/10.1175/2008JAMC1831.1.

Alavi, N., J.S. Warland and A.A. Berg, 2009: Assimilation of Soil Moisture and Temperature Data into Land Surface Models: A Survey, Data Assimilation for Atmospheric, Oceanic and Hydrologic Applications. DOI 10.1007/978-3-540-71056-1 22.

Anthes, R.A. and S.W. Chang, 1978: Response of the hurricane boundary layer to changes of sea surface temperature in a numerical model. *J Atmos Sci*, **35**, 1240–1255.

Bister, M. and K.A. Emanuel, 1998: Dissipative heating and hurricane intensity. *Meteor. Atmos. Phys.*, **65**, 233–240.

Bosilovich, M. et al., 2012: Report of the 4th World Climate Research Programme International Conference on Reanalyses. Silver Spring, Maryland, USA, 7-11 May 2012.

Bozeman, L.M., D. Niyogi, S. Gopalakrishnan, F.D. Marks Jr., X. Zhang and V. Tallapragada, 2012: An HWRF-based Ensemble Assessment of the Land Surface Feedback on the Post–Landfall Intensification of Tropical Storm Fay (2008). *Natural Hazards*, DOI 10.1007/s11069-011-9841-5.

Braun, S.A. and W.-K. Tao, 2000: Sensitivity of high-resolution simulations of Hurricane Bob (1991) to planetary boundary-layer parameterizations. *Mon. Weather Rev.*, **128**, 3941–3961.

Chang, H., D. Niyogi, A. Kumar, C. Kishtawal, J. Dudhia, F. Chen, U.C. Mohanty and M. Shepherd, 2009a: Possible relation between land surface feedback and the post-landfall structure of monsoon depression. *Geophys. Res. Lett.*, **36**, L15826, doi:10.1029/2009GL037781.

Chen, F. and J. Dudhia, 2001: Coupling an Advanced Land Surface–Hydrology Model with the Penn State–NCAR MM5 Modeling System. Part I: Model Implementation and Sensitivity. *Mon. Wea. Rev.*, **129**, 569–585. doi: http://dx.doi.org/10.1175/1520-0493(2001)129<0569:

Chen, F. et al., 1996: Modeling of land-surface evaporation by four schemes and comparison with FIFE observations. *J. Geophys. Res.*, **101(D3)**, 7251–7268.

Chen, F. et al., 2007: Description and evaluation of the characteristics of the NCAR High-Resolution Land Data Assimilation System. *J. Appl. Meteor. Climatol.*, **46**, 694–713.

Crow, W.T. and E.F. Wood, 2003: The assimilation of remotely sensed soil brightness temperature imagery into a land surface model using Ensemble Kalman filtering: A case study based on ESTAR measurements during SGP97. *Adv Water Resour.*, **26**, 137–149.

Crow, W.T., W.P. Kustas and J.H. Prueger, 2008: Monitoring root-zone soil moisture through the assimilation of a thermal remote sensing-based soil moisture proxy into a water balance model. *Remote Sensing of Environment*, **112(4)**, 1268–1281.

Dastoor, A. and T.N. Krishnamurti, 1991: The landfall and structure of a tropical cyclone: The sensitivity of model predictions to soil moisture parameterizations. *Bound Layer Meteorol*, **55**, 345–380.

Ek, M.B., K.E. Mitchell, Y. Lin, E. Rogers, P. Grunmann, V. Koren, G. Gayno and J.D. Tarpley, 2003: Implementation of Noah land surface model advances in the National Centers for Environmental Prediction operational mesoscale Eta Model. *J. Geophys. Res.*, **108**, 8851, doi:10.1029/2002JD003296.

Emanuel, K., J. Callaghan and P. Otto, 2008: A hypothesis for the redevelopment of warm-core cyclones over northern Australia. *Mon Weather Rev*, **136**, 3863–3872. doi:10.1175/2008MWR2409.1.

Emanuel, K.A., 2003: Tropical cyclones. *Annu. Rev. Earth Planet. Sci.*, **31**, 75–104.

Emanuel, K.A., 1986: An air-sea interaction theory for tropical cyclones. Part I. *J. Atmos. Sci.*, **42**, 1062–1071.

Emanuel, K.A., 1988: The maximum intensity of hurricanes. *J. Atmos. Sci.*, **45**, 1143–1155.

Emanuel, K.A., 1995: Sensitivity of tropical cyclones to surface exchange coefficients and a revised steady-state model incorporating eye dynamics. *J. Atmos. Sci.*, **52**, 3969–3976.

Emanuel, K.A., 2004: Tropical cyclone energetics and structure. Atmospheric Turbulence and Mesoscale Meteorology. E. Federovich, R. Rotunno and B. Stevens (eds). Cambridge University Press.

Evans, C., R.S. Schumacher and T.J. Galarneau Jr., 2011: Sensitivity in the Overland Reintensification of Tropical Cyclone Erin (2007) to Near-Surface Soil Moisture Characteristics. *Mon. Wea. Rev.*, **139**, 3848–3870.

Gopalakrishnan, G., F. Marks, J.A. Zhang, X. Zhang, J. Bao and V. Tallapragada, 2013: A study of the impacts of vertical diffusion on the structure and intensity of the tropical cyclones using the high resolution HWRF system. *J. Atmos. Sci.*, **70**, 524–541.

Holt, T., D. Niyogi, F. Chen, K. Manning, M. LeMone and A. Qureshi, 2006: Effect of land–atmosphere interactions on the IHOP 24–25 May 2002 convection case. *Mon. Wea. Rev.*, **134**, 113–133.

Hong, X., S.W. Chang, S. Raman, L.K. Shay and R. Hodur, 2000: The interaction of hurricane Opal (1995) and a warm core ring in the Gulf of Mexico. *Mon. Wea. Rev.*, **128**, 1347–1365.

Kellner, O., D. Niyogi, M. Lei and A. Kumar, 2012: The role of anomalous soil moisture on the inland reintensification of Tropical Storm Erin (2007), *Natural Hazards*, **63(3)**, 1573–1600.

Kimball, S.K., 2008: Structure and Evolution of Rainfall in Numerically Simulated Landfalling Hurricanes. *Mon. Wea. Rev.*, **136**, 3822–3847. doi: http://dx.doi.org/10.1175/2008MWR2304.

Kishtawal, C.M., D. Niyogi, A. Kumar, M.L. Bozeman and O. Kellner, 2012: Sensitivity of inland decay of North Atlantic tropical cyclones to soil parameters. *Natural Hazards*, **63(3)**, 1527–1542.

Lonfat, M., F.D. Marks and S. Chen, 2004: Precipitation distribution in tropical cyclones using the Tropical Rainfall Measuring Mission (TRMM) microwave imager: A global perspective. *Mon. Wea. Rev.*, **132**, 1645–1660.

Marchok, T., R. Rogers and R. Tuleya, 2007: Validation schemes for tropical cyclone quantitative precipitation forecasts: Evaluation of operational models for U.S. landfalling cases. *Weather Forecast*, **22**, 726–746.

Mitchell, K.E. et al., 2004: The multi-institution North American Land Data Assimilation System (NLDAS): Utilizing multiple GCIP products and partners in a continental distributed hydrological modeling system. *J. Geophys. Res.*, **109**, D07S90, doi:10.1029/2003JD003823.

Montgomery, M.T., R.K. Smith and S.V. Nguyen, 2010: Sensitivity of tropical cyclone models to the surface exchange coefficients. *Q. J. R. Meteorol. Soc.* In press.

Niyogi, D., T. Holt, S. Zhong, P.C. Pyle and J. Basara, 2006: Urban and land surface effects on the 30 July 2003 mesoscale convective system event observed in the Southern Great Plains. *J. Geophys. Res.*, **111**, D19107, doi:10.1029/2005JD006746.

Noilhan and Planton, 1989: A simple parametrization of land surface processes for meteorological models. *Mon Wea Rev.*, **117**, 536–549, doi: http://dx.doi.org/10.1175/1520-0493(1989)117<0536:ASPOLS>2.0.CO;2.

Nolan, D.S., J.A. Zhang and D.P. Stern, 2009a: Evaluation of planetary boundary layer parameterizations in tropical cyclones by comparison of in-situ observations and high resolution simulations of Hurricane Isabel (2003). Part I: Initialization, maximum winds, and the outer core boundary layer. *Monthly Weather Review*, **137(11)**, 3651–3674.

Nolan, S.D., J.A. Zhang and D.P. Stern, 2009b: Evaluation of planetary boundary layer parameterizations in tropical cyclones by comparison of in-situ observations and high resolution simulations of Hurricane Isabel (2003). Part II: Inner-core boundary layer and eyewall structure. *Monthly Weather Review*, **137(11)**, 3675–3698.

Osuri, K.K., U.C. Mohanty, A. Routray, M. Mohapatra and D. Niyogi, 2013: Real-Time Track Prediction of Tropical Cyclones over the North Indian Ocean Using the ARW Model. *Journal of Applied Meteorology and Climatology*, **52**, 2476–2492. doi: http://dx.doi.org/10.1175/JAMC-D-12-0313.1.

Persing, J., M.T. Montgomery and R.E. Tuleya, 2002: Environmental interactions in the GFDL hurricane model for Hurricane Opal. *Mon. Wea. Rev.*, **130**, 298–317.

Pielke Sr., R.A., 2001: Influence of the spatial distribution of vegetation and soils on the prediction of cumulus convective rainfall. *Rev. Geophys.*, **39**, 151–177.

Reichle, R.H. et al., 2001a: Variational data assimilation of microwave radio brightness observations for land surface hydrology applications. *IEEE Trans Geosci Remote Sens*, **39**, 1708–1718.

Reichle, R.H. et al., 2001b: Downscaling of radio brightness measurements for soil moisture estimation: A four-dimensional variational data assimilation approach. *Water Resour Res*, **37**, 2353–2364.

Reichle, R.H., W.T. Crow, R.D. Koster, H.O. Sharif and S.P.P. Mahanama, 2008: Contribution of soil moisture retrievals to land data assimilation products. *Geophysical Research Letters*, **35(1)**.

Rhome, J.R. and S. Raman, 2006: Environmental influences on tropical cyclone structure and intensity: A review. *Indian Journal of Marine Sciences*, **35**, 61–74.

Rienecker, M.M., M.J. Suarez, R. Gelaro, R. Todling, J. Bacmeister, E. Liu, M.G. Bosilovich, S.D. Schubert, L. Takacs, G.-K. Kim, S. Bloom, J. Chen, D. Collins, A. Conaty, A. da Silva et al. (2011), MERRA: NASA's Modern-Era Retrospective Analysis for Research and Applications. *J. Climate*, **24**, 3624–3648.

Rodgers, R., F. Marks and T. Marchok, 2009: Tropical cyclone rainfall. *In:* Encyclopedia of hydrological sciences.

Smith, R.K. and G.L. Thomsen, 2010: Dependence of tropical-cyclone intensification on the boundary layer representation in a numerical model. *Q J Roy Meteorol Soc*, **136**, 1671–1685. doi:10.1002/qj.687.

Smith, R.K., 1968: The surface boundary layer of a hurricane. *Tellus*, **20**, 473–483.

Subramanian, S., S. Gopalakrishnan, G.R. Halliwell Jr. and D. Niyogi, 2014: Idealized Study of Land Surface Impacts on Tropical Cyclone Intensity Predictions Using the HWRF Modeling System. 15A.8A presented at 31st Conference on Hurricanes and Tropical Meteorology, San Diego, Calif., March 30-April 04.

Stuart, S.R. and J.L.I.I. Beven, 2009: Tropical cyclone report, Tropical Storm Fay, 15–26 August 2008. National Hurricane Center, Miami.

Trier, S.B., F. Chen and K.W. Manning, 2004: A study of convection initiation in a mesoscale model using high-resolution land surface initial conditions. *Mon. Wea. Rev.*, **132**, 2954–2976.

Tuleya, R.E., 1994: Tropical storm development and decay: Sensitivity to surface boundary conditions. *Mon. Wea. Rev.*, **122**, 291–304.
Tuleya, R.E. and Y. Kurihara, 1978: A numerical simulation of the landfall of tropical cyclones. *J. Atmos. Sci.*, **35**, 242–257.
Walker, J.P. and P.R. Houser, 2001: A methodology for initializing soil moisture in a global climate model: Assimilation of near-surface soil moisture observations. *J Geophys Res-Atmos*, **106**, 11761–11774.
Wang, Y. and C.C. Wu, 2004: Current understanding of tropical cyclone structure and intensity changes—A review. *Meteorol Atmos Phys*, **87(4)**, 257–278. doi: 10.1007/s00703-003-0055-6.
Warner, T.T., 2004: Desert Meteorology. Cambridge University Press, New York.
Zhang, J.A., R.F. Rogers, D.S. Nolan and F.D. Marks, 2011a: On the characteristic height scales of the hurricane boundary layer. *Mon. Wea. Rev.*, **139**, 2523–2535.

10

Ocean Component of the HWRF Coupled Model and Model Evaluation

Richard M. Yablonsky
AIR Worldwide, 131 Dartmouth Street, Boston, MA 02116
richard.m.yablonsky@gmail.com

1. Introduction

Tropical cyclones (TCs) moving above a warm ocean represent an extreme case of air-sea interaction. Strong TC winds near the ocean surface can cool the sea surface temperature (SST) by several degrees Celsius and generate ocean currents of about 1-2 m/s. SST cooling under the hurricane core results in reduction of the enthalpy from the ocean, which limits the TC intensity. SST changes under a hurricane are primarily due to wind-forced turbulent mixing and entrainment of the cooler, deeper waters into the warmer upper ocean mixed layer, but for slow-moving storms in particular, upwelling and horizontal advection can also contribute to SST cooling under the storm core (Yablonsky and Ginis, 2009, 2013). Upwelling under the storm core is a result of mass conservation when the strong cyclonic wind stress generates diverging upper ocean currents away from the storm centre.

Many modelling studies have highlighted the importance of TC–ocean model coupling for simulating TC intensity, including idealized case studies (Ginis et al., 1989; Bender et al., 1993; Hodur, 1997; Schade and Emanuel, 1999; Bao et al., 2000; Chan et al., 2001; Wu et al., 2005; Wu et al., 2007; Liu et al., 2011; Yablonsky and Ginis, 2013), as well as real case studies in the North Atlantic (Bender and Ginis, 2000; Hong et al., 2000; Emanuel et al., 2004; Chen et al., 2007; Wu et al., 2007; Sanabia et al., 2013; Kim et al., 2014; Yablonsky et al., 2015a, b), western North Pacific (Lin et al., 2005; Wu et al., 2007), and South Pacific Oceans (Sandery et al., 2010; Jullien et al., 2014). Uncoupled TC models with a static SST field are restricted by their inability to account for TC-induced SST cooling during TC model integration, which can cause an overestimate of the surface heat and moisture fluxes in the storm core and contribute to a high-intensity bias. Hence, the ocean component of the coupled model's primary objective is to create a realistic SST field, particularly in the storm core. This SST field is subsequently used to calculate the enthalpy fluxes from the ocean to atmosphere.

Hurricane Weather Research Forecast Model (HWRF), a coupled atmosphere-ocean model, was developed at National Oceanic and Atmospheric Administration (NOAA) and has been in operation in the U.S. since 2007. The HWRF is a regional dynamical model, which is mostly used for TC forecast guidance. The atmospheric component of the HWRF is coupled to a fully three-dimensional version of Princeton Ocean model (POM). This version of POM was called POM for Tropical Cyclones (POM-TC) until 2014 (Yablonsky et al., 2015a), at which time POM-TC was replaced with a new version of POM that has message passing interface capabilities and is called MPIPOM-TC (Yablonsky et al., 2015b). A brief history and description of the POM model development for TCs, along with some numerical studies, are discussed in this chapter.

2. History of Ocean Component of Coupled HWRF Model

The three-dimensional primitive equation numerical ocean model known as POM was originally developed at Princeton in the late 1970s (Mellor, 2004). In 1994, a version of POM available at the time was transferred to the University of Rhode Island (URI) for the purpose of coupling to the GFDL hurricane model. At this point, POM code changes were made specifically to address the problem of the ocean's response to hurricane wind forcing in order to create a more realistic SST field for input into the hurricane model. Initial testing showed hurricane intensity forecast improvements when ocean coupling was included (Bender and Ginis, 2000). Since operational implementation of the coupled GFDL–POM model at NOAA in 2001, additional changes to POM were made at URI and subsequently implemented in the operational GFDL model, including improved ocean initialization (Falkovich et al., 2005; Bender et al., 2007; Yablonsky and Ginis, 2008). This POM version, later called POM-TC, was then coupled to the atmospheric component of the HWRF before operational implementation of HWRF at NOAA in 2007.

In addition to the URI-based changes to the POM for hurricane applications, community-based improvements to POM were made between 1994 and 2012, including the addition of the Message Passage Interface (MPI), which enables POM to run in parallel on multiple processors, as in the Stony Brook Parallel Ocean Model (sbPOM; Jordi and Wang, 2012) and the Advanced Taiwan Ocean Prediction (ATOP) system (Oey et al., 2013). Hence, URI merged the 2012 versions of the sbPOM and POM-TC, and additional changes were made to accommodate flexible initialization options and a relocatable grid, culminating in the MPIPOM–TC, which was coupled operationally to HWRF in 2014 (Yablonsky et al., 2015b). A block diagram showing the brief history of POM through the development of MPIPOM–TC is given in Fig. 1 (Brief description is given in Annexure 1).

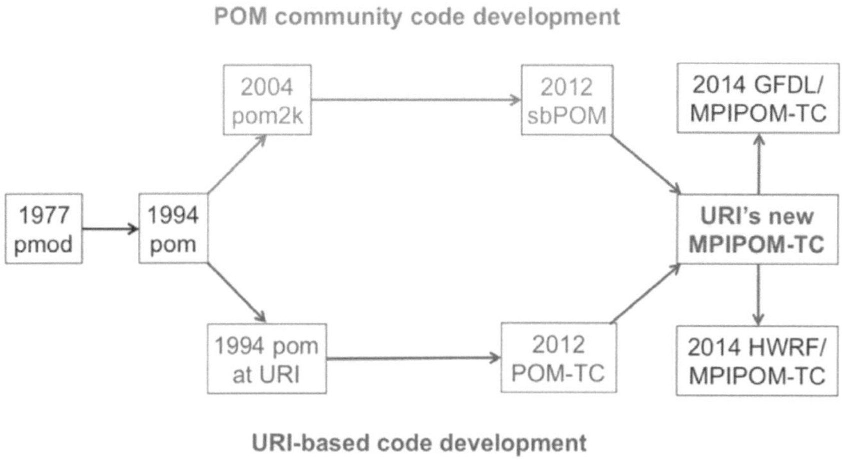

Fig. 1. Block diagram of the history of POM through MPIPOM-TC development from the late 1970's to 2014.

2.1 Configuration of POM-TC

While no longer used operationally in the HWRF due to its replacement with MPIPOM-TC in 2014, it is instructive to examine POM-TC and its evaluation for pedagogical purposes. As discussed in Yablonsky et al., (2015a), the horizontal POM-TC grid uses curvilinear orthogonal coordinates, and Arakawa C grid is used for the spatial differencing of the POM-TC variables. Hence, some model variables are calculated at a horizontally shifted location from other model variables. The North Atlantic Ocean consists of two overlapping grids, both bounded by 10°N latitude to the south and 47.5°N to the north. The first grid covers the Gulf of Mexico, Caribbean, and West Atlantic region (the "United" region), bounded by 98.5°W longitude to the west and 50°W longitude to the east.The second grid covers the east Atlantic region, which is bounded by 60°W longitude to the west and 30°W longitude to the east. Prior to MPIPOM-TC implementation in 2014, HWRF used the current and 72-h forward-extrapolated storm track to choose which POM-TC North Atlantic grid to use for coupling. Both grids are configured with a horizontal grid spacing of ~18 km, which may be too coarse to resolve some details of the ocean response to the TC forcing, but it was chosen because POM-TC runs on a single processor, and the model needed to be able to run sufficiently fast to comply with NOAA's operational requirements.

POM-TC uses a terrain-following sigma coordinate system. There are 23 vertical levels, the depth of which is scaled on the basis of the bathymetry of that location. For the maximum allowable bathymetric depth of 5500 m, the

23 half-sigma vertical levels are located at depths of 5, 15, 25, 35, 45, 55, 65, 77.5, 92.5, 110, 135, 175, 250, 375, 550, 775, 1100, 1550, 2100, 2800, 3700, 4850, and 5500 m. POM-TC has a free surface and a split time step. The internal mode is three dimensional and uses a 9-minute time step during the integration. It uses an explicit scheme for the horizontal differencing and an implicit scheme for the vertical time differencing, eliminating time constraints for the vertical coordinate and permitting high resolution in the surface boundary layer. The Mellor–Yamada level 2.5 turbulence closure model is used for the turbulence parameterization. The configuration and evaluation of MPIPOM-TC is briefly discussed in Annexures 2 and 3 respectively.

2.2 POM–TC Initialization

POM-TC is initialized with realistic three-dimensional temperature and salinity fields from the Generalized Digital Environmental Model (GDEM) monthly ocean temperature and salinity climatology (Teagueet al., 1990) and then integrated without TC model coupling in order to generate realistic ocean currents and incorporate the pre-existing hurricane-generated cold wake (by applying prescribed, observed TC wind stress). The GDEM climatology is modified diagnostically before model integration by interpolating it in time to the POM-TC initialization date (using two months of GDEM), horizontally interpolating it onto the POM-TC United grid, incorporating a land–sea mask and bathymetry data, and employing a feature-based modeling procedure (in the west Atlantic region but not the east Atlantic region) that incorporates historical and near-real-time observations of prominent ocean fronts and eddies (Falkovich et al., 2005; Yablonsky and Ginis, 2008).

Major oceanic fronts and eddies in the western North Atlantic Ocean—namely, the Gulf Stream, the Loop Current, and eddies associated with the Loop Current—are poorly represented by the GDEM climatology's temperature and salinity fields. Historical observations gathered from various field experiments can be used to define the spatial structure of these fronts and eddies (Falkovich et al., 2005, section 3), and then cross-frontal "sharpening" of the GDEM temperature and salinity fields can be performed to obtain more realistic fields by increasing the horizontal density gradients across the fronts. These modified fields yield stronger geotropically-adjusted ocean currents along the front than would be obtained directly from GDEM. In order to improve the initialization of the Loop Current and its eddies, algorithms were incorporated into the feature-based modelling procedure based on guidance from near-real-time observations, such as satellite altimetry (Yablonsky and Ginis, 2008, section 2).

After the aforementioned diagnostic modifications to the GDEM climatology (including the feature-based modifications in the west Atlantic "United" region), the temperature fields are modified by assimilating the real-time daily SST data that is used in NOAA's operational Global Forecast system analysis (Reynolds and Smith, 1994) at the beginning of what is referred to as ocean spin up "phase 1". Although this SST product has relatively coarse resolution, it is chosen instead of other higher-resolution SST products to ensure consistency near the air–sea interface between the ocean initialization and the atmospheric initialization, the latter of which also uses the NOAA's operational Global Forecast System analysis. Further details of the SST assimilation procedure can be found in Yablonsky and Ginis (2008, section 2). Finally, the three-dimensional temperature and salinity fields are interpolated from the GDEM z levels onto the POM-TC vertical sigma levels, and the density is calculated using the modified United Nations Educational, Scientific, and Cultural Organization equation of state (Mellor, 1991), ending the diagnostic portion of the ocean initialization. Ocean spin up during phase 1 requires a 48-h integration, during which dynamic adjustment of the density field and generation of dynamically consistent currents are obtained while the SST is held constant.

After phase 1, the obtained output is used to initialize ocean spin up of "phase 2". During phase 2, which consists of 72 h of integration, the coldwake at the ocean surface and the currents produced by the hurricane prior to the beginning of the coupled model forecast are generated using the observed hurricane surface wind distribution obtained from NOAA's National Hurricane Center along the storm track. The hurricane surface wind distribution is based on interpolation of the radial wind profiles derived from the storm message file (Bender and Ginis, 2000), also known as the TC vitals (Trahan and Sparkling, 2012). These message files contain the storm propagation speed and direction, position, central and environmental pressure, radius of outermost closed isobar, maximum wind speed, and radius of maximum wind and radii of 17, 26, and 33 ms^{-1} winds in the northeast, southeast, southwest, and northwest quadrants of the storm, when available. Once phase 2 is complete, the phase 2 output was used to initialize the POM-TC component of the coupled HWRF from 2007 to 2013.

2.3 POM-TC Coupling to the HWRF Atmosphere (2007-2013 HWRF Operational Implementation)

The atmosphere–ocean coupler operates as an independent interface between the HWRF atmospheric component and the POM-TC. At every internal 9-min ocean time step, which is synchronized with an atmospheric time step, the instantaneous SST is passed to the HWRF atmosphere. The time-averaged total heat and momentum fluxes at the air–sea interface (over all atmospheric time steps between the previous internal ocean time step and the current one) are passed from the atmosphere to the ocean. In the coupler, freshwater fluxes are not exchanged across the air–sea interface, although Jourdain et al. (2013) suggested that fresh water flux in the form of heavy rainfall may have a non-negligible impact on the ocean response to TCs. The coupler serves as a hub for communications between the HWRF atmosphere and POM-

TC; it performs the interpolation of the surface fluxes from the fixed and moving HWRF atmospheric grids to the POM-TC grid, and it performs the interpolation of the SST from the POM-TC grid to the two outermost HWRF atmospheric grids. A generalized bilinear interpolation for nonrectangular quadrilateral grid cells is used; only sea-point values of the surface fields are employed for the interpolation.

3. Results

3.1 Retrospective 2013 Version HWRF–POM-TC Model Forecast Analysis

The operational coupled HWRF system is upgraded annually in order to improve the track, intensity and the storm size. There has been a considerable improvement in the skill of the HWRF system since 2007. The highest skill through 2013 (the final year POM-TC was used before its replacement with MPIPOM-TC) was achieved by the 2013 operational version (Tallapragada et al., 2013) based on analysis of a statistically robust set of retrospective forecasts from 2010-12 Atlantic hurricane season. Figure 2 shows the improvements in the 2013 operational HWRF version as compared with the 2012 operational HWRF version.

3.2 Retrospective 2013 Version HWRF-POM-TC Model Ocean Analysis

The performance of the POM-TC ocean component is analyzed through forecasts obtained from hurricanes Earl (2010), Igor (2010), Irene (2011), Isaac (2012), and Leslie (2012) in which the accuracy of the hurricane track, intensity, and size were sufficient, relative to the National Hurricane Center best track, to analyze the POM-TC SST (and subsurface ocean temperature for the Irene case) using in situ and remotely sensed observations (Yablonsky et al., 2015a). When comparing the POM-TC model simulated cooling with SST observations, two challenges exist: (1) an observational SST data source must be available that indicates the location, magnitude, and spatial extent of the cooling during or shortly after the passage of the hurricane; and (2) the HWRF-forecasted track, propagation speed, intensity, and size must be sufficiently accurate with the observed storm properties to prevent a significant bias of the POM-TC SST response relative to the observations. To address the first challenge, the satellite derived 3-day average SST from the Tropical Rainfall Measuring Mission Microwave Imager (TMI; Gentemann et al., 2004, 2010) is used. The daily TMI-SST contains many discontinuities in the data; hence, 3-day averaged TMI SST is used at an expense of higher temporal resolution. The second challenge is addressed by selecting specific forecast cycles in which the storm properties are well forecasted.

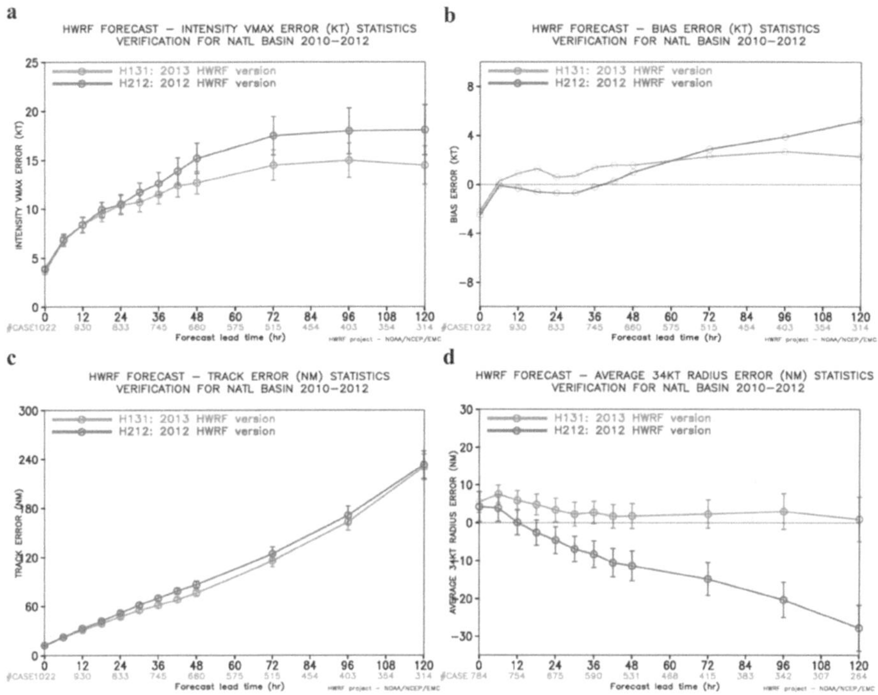

Fig. 2. Average TC intensity (a) error and (b) bias (kt; 1 kt = 0.51 ms^{-1}), as well as (c) track error and (d) average 17 ms^{-1} (34 kt) wind radius bias (nm), as a function of forecast lead time (h) from a homogeneous comparison of the 2013 operational HWRF–POM-TC version (H131, red) to the 2012 operational HWRF–POM-TC version (H212, blue) for all 2010–12 Atlantic TCs, which includes 1022, 930, 833, 745, 660, 575, 515, 454, 403, 354, and 314 forecasts at lead times of 0, 12, 24, 36, 48, 60, 72, 84, 96, 108, and 120 h, respectively.

3.3 Hurricane Earl

The HWRF coupled model forecast of Hurricane Earl, initialized at 0600 UTC 30 August 2010, is shown in Fig. 3. In order to zoom in on the cold wake produced by the model, the first 12h and the final 6h of the forecast track is cut off in Fig. 3. The HWRF cold wake SST at forecast hour 120 (Fig. 3a), and hence the HWRF cold wake SST anomaly at forecast hour 120 relative to forecast hour 0 (Fig. 3c), is up to ~2°C warmer than the observed cold wake SST from the three-day averaged (9/2-9/4) TMI SST product (Fig. 3b, d). In addition, the spatial extent of the HWRF cold wake is only about half as wide as in the TMI in many locations, particularly during the first 84 h of the forecast. Examining the HWRF temperature at 77.5-m depth (Fig. 3e) reveals that Earl propagated along the boundary between areas of higher ocean heat content to the south and west and lower ocean heat content to the north and east, the detailed structure of which could impact the magnitude of the cold wake. In addition, the HWRF intensity was generally ~10 ms^{-1} weaker than observed during the first 84 h of the forecast (Fig. 3f), which likely contributed to the undercooling in the HWRF forecast relative to observations. However, the HWRF storm size, as indicated by the radius of the 17 ms^{-1} wind, is within 50 km of the observed storm size during the first 96 h of the forecast (Fig. 3f), so storm size errors do not explain the under forecasted spatial extent of the cold wake.

Two buoys located close to Earl's track provide an opportunity to compare the temporal evolution of the HWRF SST to observations at point locations. At NOAA/NDBC buoy 41001, located east of Earl's centre at forecast hour ~96-102, the HWRF SST is ~1°C warmer than the buoy SST at the initial time. Assuming the buoy SST is accurate, this result indicates that the blended, coarse-resolution, Global Forecast System SST analysis, which is assimilated into the HWRF/POM-TC initial condition, is ~1°C too warm at the buoy location. The magnitude of the HWRF SST cooling from forecast hour ~96 to ~108 is similar to the buoy SST cooling, with a 3-6 h phase lag in the HWRF SST relative to the buoy SST (Fig. 3g), due to the slower HWRF storm propagation speed relative to the best track (Fig. 3a-e). At NOAA/NDBC buoy 41046, located northeast of Earl's centre at forecast hour ~42, the HWRF SST cooling is underestimated by ~1°C during storm passage (forecast hour ~36-48), but the more significant HWRF SST cooling underestimation occurs later during subsequent upwelling cycles (forecast hour 60 and beyond) (Fig. 3h). Since Earl was propagating rather quickly (~6-7 ms^{-1}), this "post-storm" underestimation of SST cooling is unlikely to have played a significant role in Earl's HWRF forecast intensity bias. Note that in both Figs 3g and 3h, the SSTs from nearby HWRF ocean model grid points are shown in addition to the SST from the closest model grid point to illustrate errors that may be introduced by insufficient ocean model horizontal resolution.

The underforecasted magnitude of Earl's cold wake may be attributable to number of other factors, including insufficient ocean model resolution, limitations in the parameterization of the upper ocean mixing, underestimation of the heat and momentum fluxes at the sea surface, and/or misrepresentation of the initial ocean temperature (and perhaps salinity) structure. For example, the erroneously warm Global Forecast System SST assimilated into the ocean model may have led to an unrealistically stable upper ocean temperature profile, which could have reduced the entrainment rate at the base of the upper ocean mixed layer. Also, Vincent et al. (2012a) have shown that the magnitude of the SST cooling in periphery of the storm is particularly sensitive to the air-sea heat flux.

3.4 Hurricane Igor

The HWRF 120-h coupled model forecast of Hurricane Igor, initialized at 00 UTC on 15 September 2010, is shown in Fig. 4. Similar to Earl, the Igor

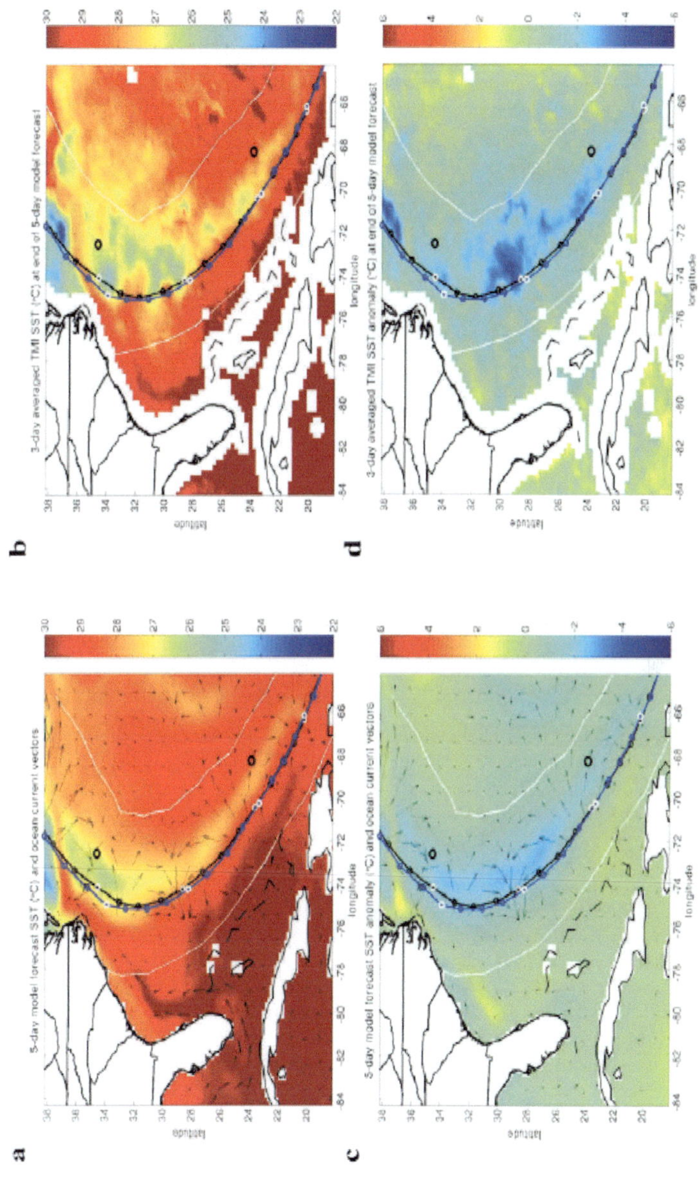

(Contd.)

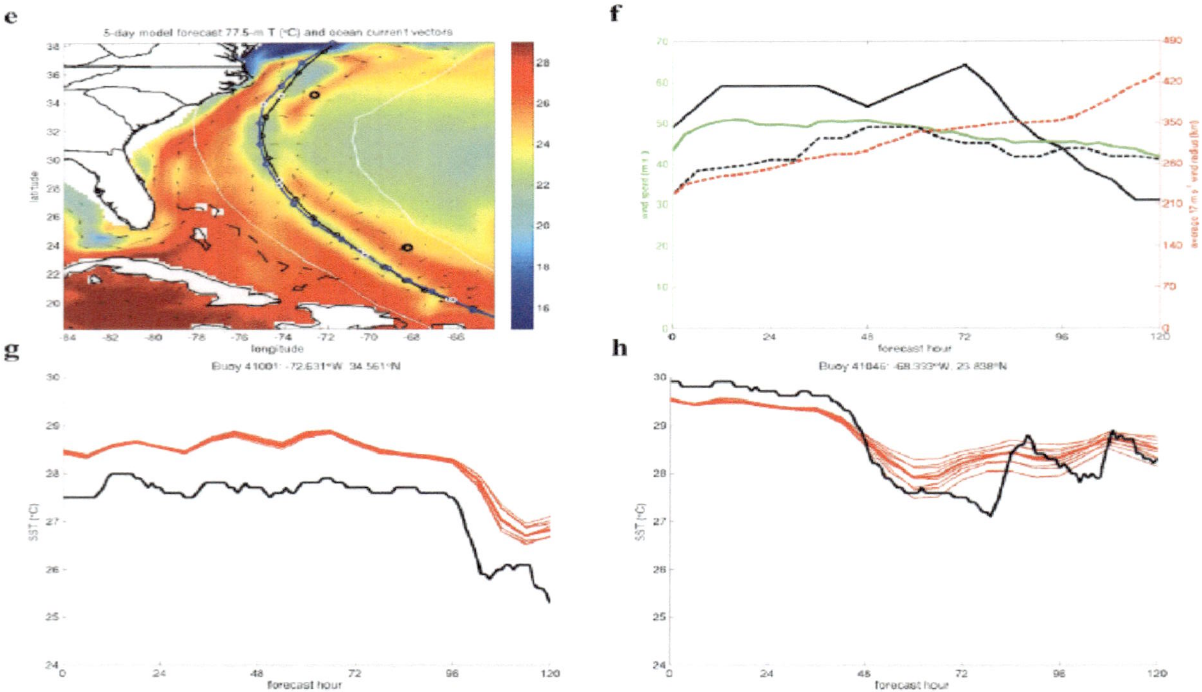

Fig. 3. HWRF-POM-TC forecast of Hurricane Earl, initialized 06 UTC 30 Aug. 2010: (a) 120-h model SST and current with model track (blue with white circles every 24 h), model 17 ms^{-1} wind swath (white), best track (black), and two buoys (black rings); (b) 9/2-9/4 TMI SST with best track 17 ms^{-1} wind swath (white); (c) 120-h model SST anomaly; (d) 9/2-9/4 TMI SST anomaly from 8/28-8/30; (e) 120-h model 77.5-m T; (f) maximum wind from model (green) and observed (black), and 17 ms^{-1} wind radius from model (red dashed) and observed (black dashed); (g) buoy 41001 SST (black) with collocated (thick red) and nearby (thin red) model SST; and (h) as in (g), but buoy 41046.

HWRF cold wake SST at forecast hour 120 (Fig. 4a, c) is up to 2-3°C warmer than the observed cold wake SST from the TMI SST (Fig. 4b, d), and the spatial extent of the cold wake is only about half as wide as in the TMI along the second half of the forecast track. Examining the HWRF temperature at 77.5-m depth (Fig. 4e) reveals that Igor propagated northward across a boundary of higher ocean heat content to the south and lower ocean heat content to the north, impacting the magnitude of the cold wake. The HWRF forecasted intensity was within 5-10 ms^{-1} of the best track, and HWRF storm size was within 50 km of the best track observations (Fig. 4f), so intensity and size errors do not account for the underestimation of the magnitude or spatial extent of Igor's SST cooling. As in Earl, this underestimation could be due to errors in the initial ocean condition and/or shortcomings in the ocean model. The particularly large differences between the model and observed cooling at the end of the forecast, however, may also be impacted by the coarse temporal resolution in the 3-day averaged TMI SST product.

At NOAA/NDBC buoy 41044, located near Igor's centre at HWRF forecast hour ~42 (observed hour ~48), the magnitude of the HWRF SST cooling is generally underestimated by ~0.5-1°C relative to the buoy SST cooling, with a phase lag of ~6 h in the buoy SST relative to the HWRF SST (Fig. 4g) due to the faster HWRF storm propagation speed relative to the best track (Fig. 4a-e). The fast HWRF storm propagation speed may at least partially explain the underestimated HWRF SST cooling. Also, the large spread at nearby grid points indicates the strong sensitivity of the buoy location relative to the storm track. At NOAA/NDBC buoy 41049, located northeast of Igor's centre at HWRF forecast hour ~78 (observed hour ~90), the HWRF SST cooling is again generally underestimated by ~0.5-1°C relative to the buoy SST cooling, with a phase lag of ~12 h in the buoy SST relative to the HWRF SST due to the faster HWRF storm propagation speed relative to the best track (Fig. 4h).

3.5 Hurricane Irene

The HWRF 120-h coupled model forecast of Hurricane Irene, initialized at 12 UTC on 23 August 2011, is shown in Fig. 5. More dramatic than either Earl or Igor, the Irene HWRF cold wake SST at forecast hour 120 (Fig. 5a, c) is up to 3-4°C warmer than the observed cold wake SST from the three-day averaged (8/26-8/28) TMI SST product (Fig. 5b, d). Examining the HWRF temperature at 77.5-m depth (Fig. 5e) reveals that Irene was east of the Gulf Stream until it reached ~32-33°N, at which point Irene's centre crossed the Gulf Stream ~6-12 h before making landfall along the eastern North Carolina coast. Along the second half of the forecast track, where the difference between the model and observed SST is largest (Fig. 5a-d), the HWRF intensity is up to ~10 m s^{-1} *higher* than the best track, and the HWRF storm size is generally within 50 km of the best track (Fig. 5f), so again, neither intensity nor size errors explain the

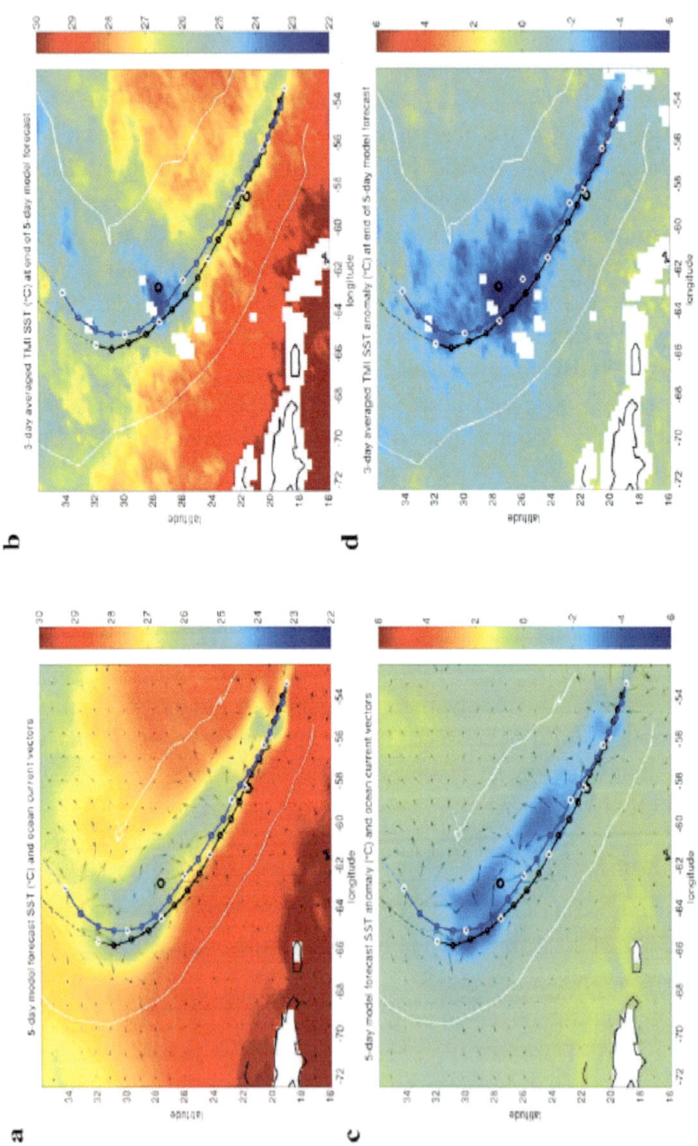
a
5-day model forecast SST (°C) and ocean current vectors
b
3-day averaged TMI SST (°C) at end of 5-day model forecast
c
5-day model forecast SST anomaly (°C) and ocean current vectors
d
3-day averaged TMI SST anomaly (°C) at end of 5-day model forecast
latitude
longitude

(Contd.)

(**Fig. 4.** *Contd.*)

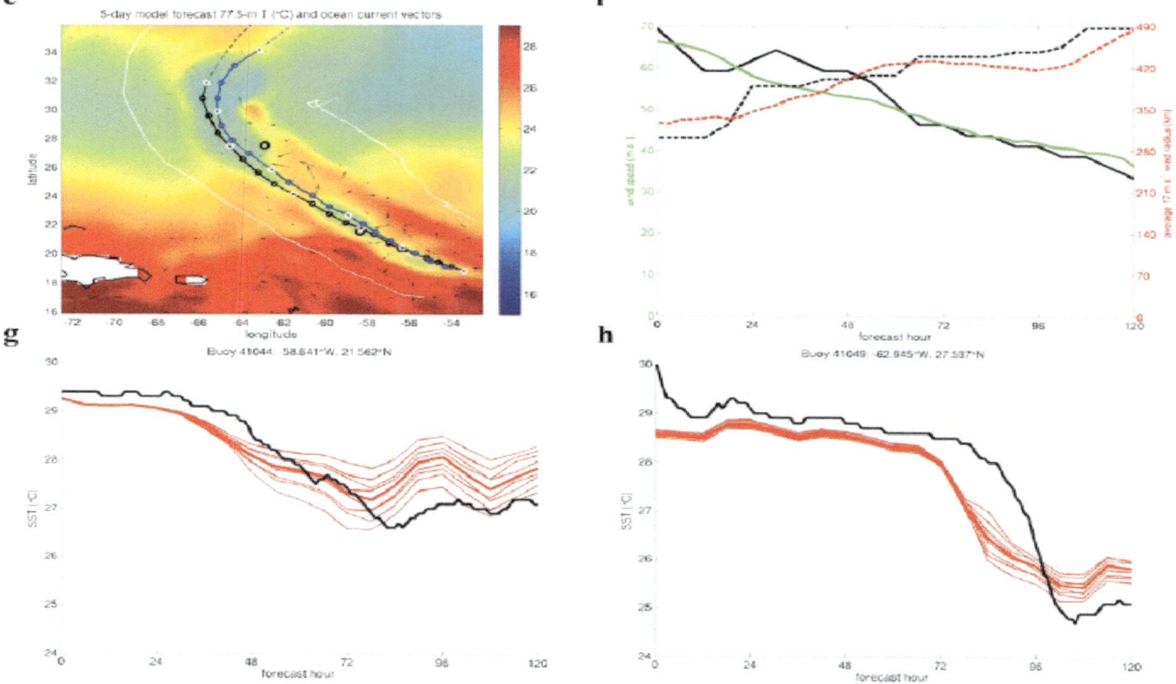

Fig. 4. HWRF-POM-TC forecast of Hurricane Igor, initialized 00 UTC 15 Sept. 2010: (a) 120-h model SST and current with model track (blue with white circles every 24 h), model 17 ms^{-1} wind swath (white), best track (black), and two buoys (black rings); (b) 9/18-9/20 TMI SST with best track 17 ms^{-1} wind swath (white); (c) 120-h model SST anomaly; (d) 9/18-9/20 TMI SST anomaly from 9/13-9/15; (e) 120-h model 77.5-m T; (f) as in Fig. 3f; (g) buoy 41044 SST (black) with collocated (thick red) and nearby (thin red) model SST; and (h) as in (g), but buoy 41049.

SST cooling errors. In Section 3.6, the cold wake during this Irene forecast is examined in more detail by comparing the subsurface temperature evolution against profiles from Airborne eXpendable BathyThermograph (AXBT) instruments and considering the bathymetry (Fig. 5g-h; Figs 6 and 7). As in the previous cases, however, it is possible that ocean to atmosphere heat flux is systematically underestimated in the model, the diagnosis of which would require observations of air-sea fluxes that are not currently available.

3.6 Subsurface Ocean Temperature Analysis (Hurricane Irene)

Fourteen quality-controlled AXBT instruments (Sanabia et al., 2013), dropped in the vicinity of Hurricane Irene during two flights [flight one from ~20 UTC on 8/26 through ~03 UTC on 8/27 (Fig. 6) and flight two from ~15 UTC through ~17 UTC on 8/27 (Fig. 7)], provide a unique opportunity to examine snapshots of the HWRF upper ocean thermal structure at point locations with and without Irene's wind forcing (Fig. 5g). This area is characterized not only by the presence of the Gulf Stream, which impacts the vertical structure of the ocean, but also by the complex local bathymetry (Fig. 5h), which can impact the upper ocean response to Irene's wind forcing. In an attempt to correct for the HWRF storm translation speed being ~6 h too slow during 8/26 and 8/27, a 6-h temporal offset is applied when comparing HWRF and AXBT vertical temperature profiles (Figs 6 and 7). Finally, the nearest (in space) Navy Coupled Ocean Data Assimilation (NCODA) daily temperature profiles are provided as an additional, quasi-independent data source for model analysis (Cummings, 2005; Cummings and Smedstad, 2013), although some of the Irene AXBT profiles were assimilated into NCODA on subsequent days (Sanabia et al., 2013); so not all of the NCODA analyses are independent of the AXBTs. It should be noted that the NCODA product has 1/6° horizontal grid spacing, and the vertical levels in the upper 200 m are defined at 0, 2.5, 7.5, 12.5, 17.5, 25, 32.5, 40, 50, 62.5, 75, 100, 125, 150, and 200 m depth.

3.6.1 Southernmost Temperature Profiles (AXBTs 1, 3, 6, and 7)

The four southernmost AXBTs are numbers 1 (Fig. 6a), 3 (Fig. 6c), 6 (Fig. 6f), and 7 (Fig. 6g), with 1 and 6 (3 and 7) dropped in the left-rear (right rear) quadrant of Irene ~150-250 km away from the storm centre between ~20 UTC on 8/26 and ~02 UTC on 8/27. From before Irene's approach to the time of the AXBT drop, the HWRF upper ocean mixed layer (OML) at AXBT 1 (3) (6) (7) cooled from ~29.7°C to ~29.2°C (from ~28.6°C to ~27.6°C) (from ~29.5°C to ~29.2°C) (from ~28.7°C to ~27.7°C) and deepened from ~25 m to ~60 m (from ~20 m to ~50 m) (from ~25 m to ~60 m) (from ~20 m to ~50 m), including the 6 h offset to account for the HWRF storm translation speed error. By comparison, AXBT 1 (3) (6) (7) indicated an OML temperature of ~28.5°C (~26.1°C) (~28.7°C) (~26.9°C) and depth of ~60 m (~50 m) (~60 m) (~40 m). Below the OML, the HWRF and AXBT vertical temperature gradients

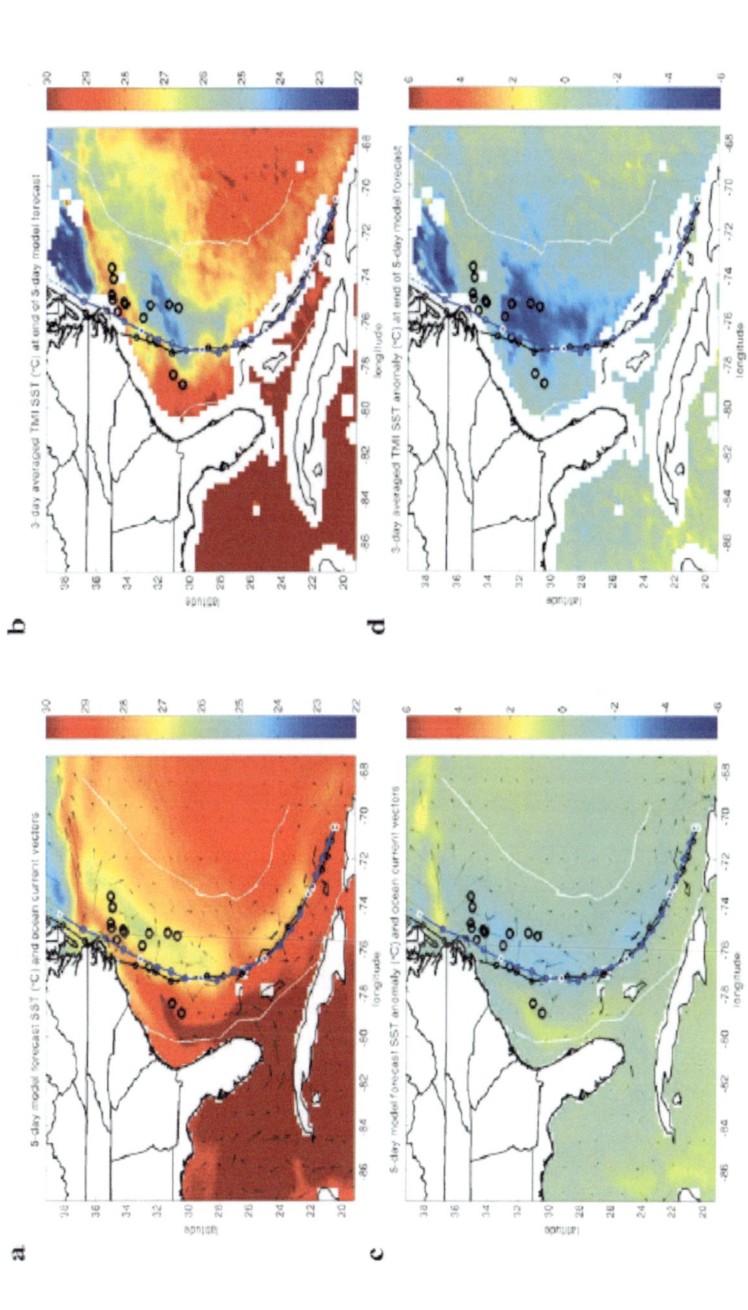
a
b
c
d
5-day model forecast SST (°C) and ocean current vectors
3-day averaged TMI SST (°C) at end of 5-day model forecast
5-day model forecast SST anomaly (°C) and ocean current vectors
3-day averaged TMI SST anomaly (°C) at end of 5-day model forecast
longitude
latitude

(Contd.)

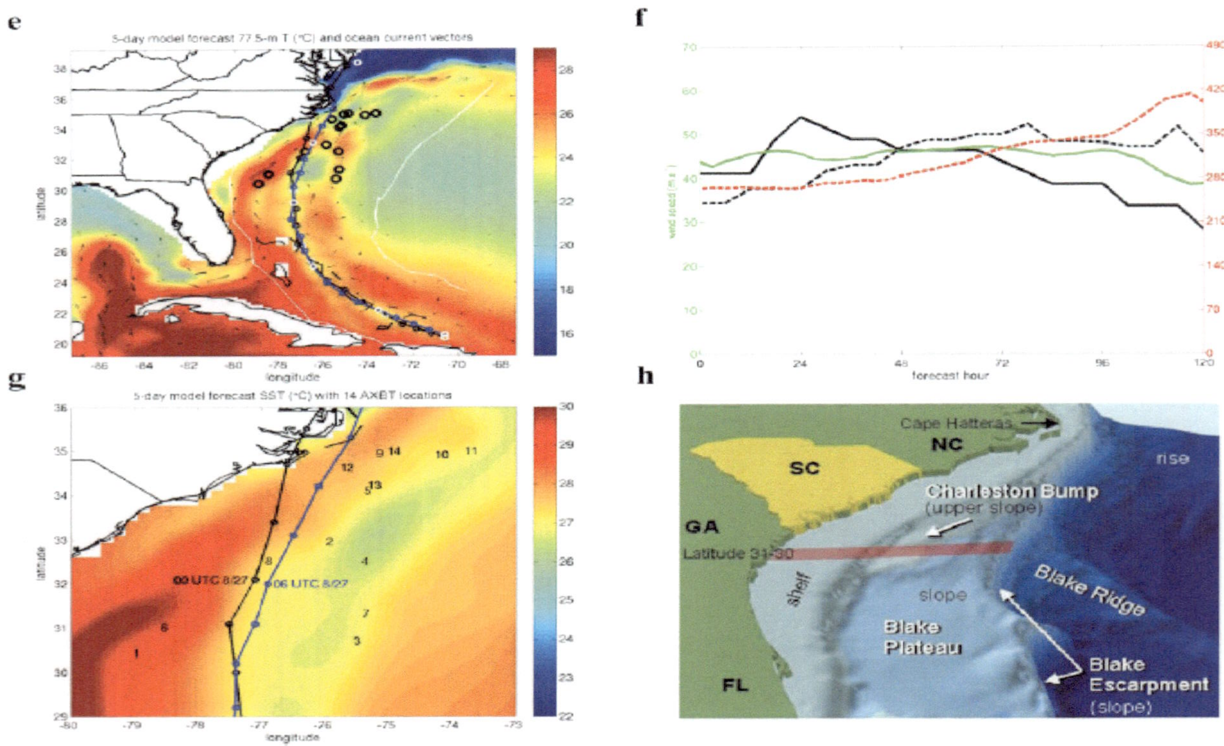

Fig. 5. HWRF-POM-TC 120-h forecast of Hurricane Irene, initialized 12 UTC 23 Aug. 2011: (a) 120-h model SST and current with model track (blue with white circles every 24 h), model 17 ms^{-1} wind swath (white), best track (black), and 14 AXBTs (black rings); (b) 8/26-8/28 TMI SST with best track 17 ms^{-1} wind swath (white); (c) 120-h model SST anomaly; (d) 8/26-8/28 TMI SST anomaly from 8/21-8/23; (e) 120-h model 77.5-m T; (f) as in Fig. 3f; (g) 120-h model SST zoomed in on 8/26-8/28 with 6-h model/best tracks and AXBTs numbered 1-14; and (h) diagram of the continental shelf and slope of the southeastern U.S. (Courtesy: NOAA).

are generally similar in the upper thermocline (down to ~200 m depth), but the anomalously warm HWRF temperature (relative to the AXBTs) continues downward from the OML to the upper thermocline). This warm bias could be due to any combination of insufficient ocean model resolution, inadequate physics (e.g. mixing), or misrepresented initial temperature (and/or salinity) stratification. Storm intensity and size were not a factor here because the model storm was *stronger* and similar in size to the observed storm during this time (Fig. 5f). NCODA's temperature profile, pre-storm, in-storm, and post-storm, generally fall within the range of HWRF and the AXBT, although the NCODA OML is shallow (~20-30 m) even after Irene's forcing is applied, indicating that much of Irene's impact on the upper ocean was probably not captured by NCODA at these times and locations.

3.6.2 Impact of the Gulf Stream Location (AXBTs 2, 5, and 13)

AXBTs 2 and 5 were dropped to the northeast (i.e. ahead) of the Irene near what appears to be the edge of the Gulf Stream at 2100UTC 26 August and 0000UTC 27 August, respectively (Figs 6b and 6e). AXBTs 2 and 5 are reasonable estimates of the pre-storm ocean condition. The pre-storm HWRF, AXBT, and NCODA profiles all agree rather well, except the AXBT has a deeper OML (~55-60 m) than the HWRF and NCODA (~35 m), and at the AXBT 2 location, the NCODA vertical temperature gradient is not as sharp as HWRF and the AXBT near the base of the OML but sharper than HWRF and the AXBT below ~100 m depth, while at the AXBT 5 location, the HWRF and NCODA vertical temperature gradient is sharper than the AXBT near the base of the OML but not as sharp as the AXBT below ~140 m depth. Hence, the AXBTs indicate that these locations may be closer to the Gulf Stream than HWRF (or NCODA) suggests; alternatively, the background ocean stratification associated with the near Gulf Stream environment may also be misrepresented in the HWRF initial ocean condition. Post-storm, HWRF indicates ~2°C of OML cooling, while NCODA indicates only ~1°C of OML cooling, but the timing of these AXBTs (i.e. pre-storm) precludes their usefulness as an analysis tool for this purpose.

AXBT 13 (Fig. 7e) was dropped about 200 km to the southeast of Irene in the near-storm cold wake on 1700 UTC 27 August. AXBT 13, which is approximately collocated with AXBT 5 (in space, not time), provides a unique opportunity to compare the evolution of the upper ocean thermal structure shortly before (AXBT 5) and after (AXBT 13) Irene's wind forcing is applied from two independent in situ observations. Comparing the profiles from AXBT 13 (Fig. 7e) to AXBT 5 (Fig. 6e), it is clear that the main differences in the upper thermocline are due to the AXBT profile indicating that the location is in the Gulf Stream, while HWRF and NCODA indicate that the location is east of the Gulf Stream. The HWRF OML cooling agrees well with the AXBT 13 OML temperature at the time of the AXBT 13 drop (Fig. 7e).

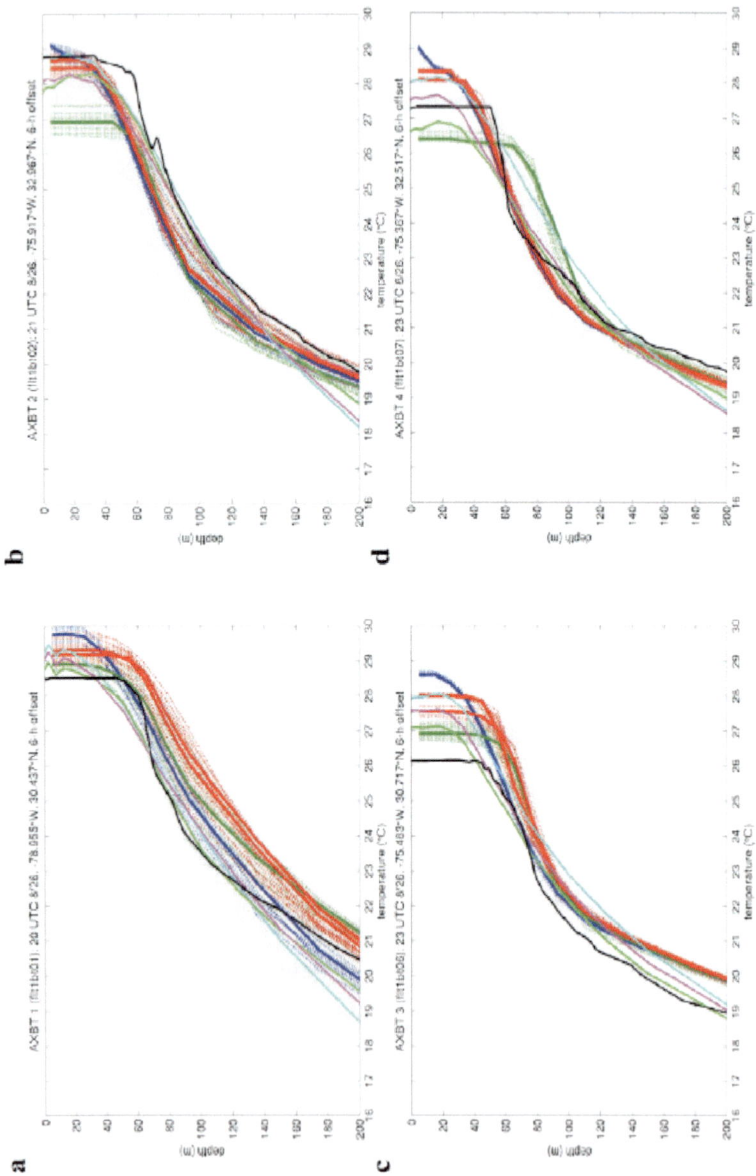
a
AXBT 1 (flt1b01): 20 UTC 8/26, -78.055°W, 30.437°N, 6-h offset
b
AXBT 2 (flt1b02): 21 UTC 8/26, -75.917°W, 32.967°N, 6-h offset
c
AXBT 3 (flt1b06): 23 UTC 8/26, -75.483°W, 30.717°N, 6-h offset
d
AXBT 4 (flt1b07): 23 UTC 8/26, -75.367°W, 32.517°N, 6-h offset
depth (m)
temperature (°C)

(Contd.)

(**Fig. 6.** *Contd.*)

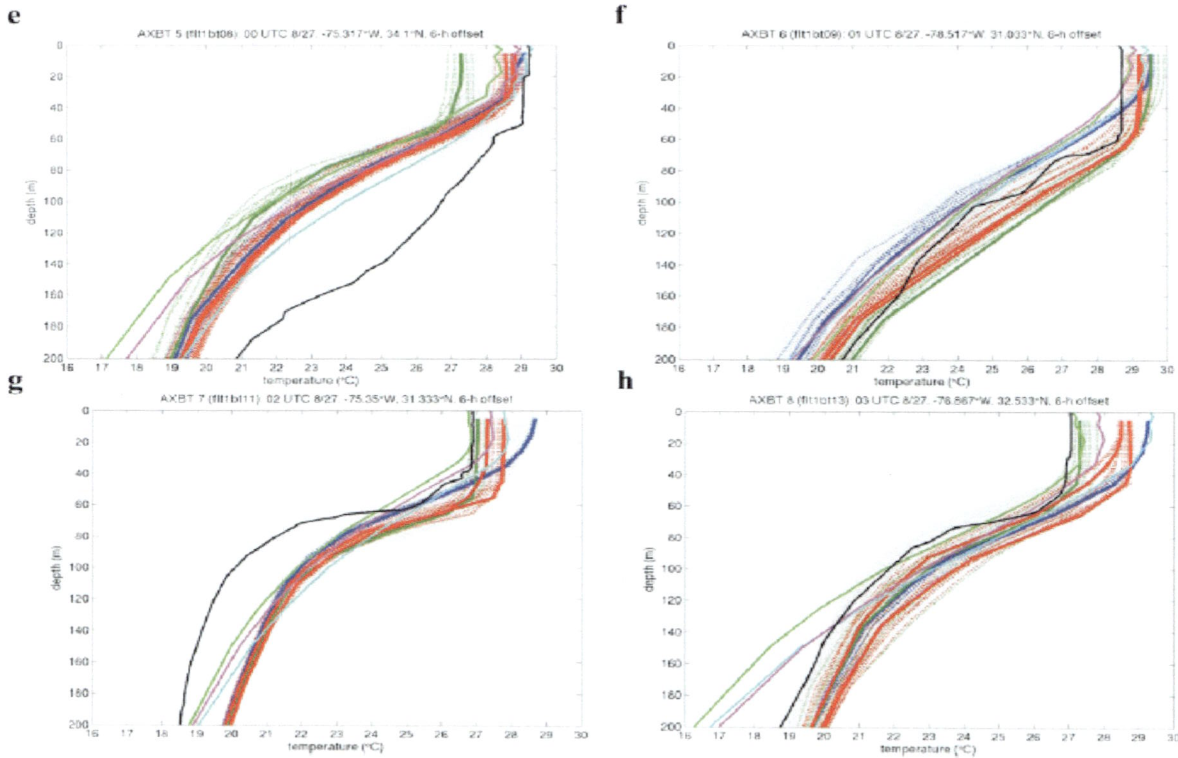

Fig. 6. Vertical temperature profiles during HWRF-POM-TC 120-h forecast of Hurricane Irene, initialized 12 UTC 23 Aug. 2011, at AXBT (a)-(h) 1-8. AXBT number (and Navy flight/BT number), time, position, and offset are given in panel titles. Solid profiles are: AXBT (black); all HWRF/POM-TC 6-h times, with nearby grid points (light gray); HWRF/POM-TC at 26/00Z (dark blue); HWRF/POM-TC at 6-h times bounding AXBT time (red); HWRF/POM-TC at 28/12Z (dark green); NCODA on 26th (light blue); NCODA on 27th (magenta); NCODA on 28th (light green). Dotted profiles are at nearby POM-TC grid points, valid during same-coloured solid POM-TC profiles.

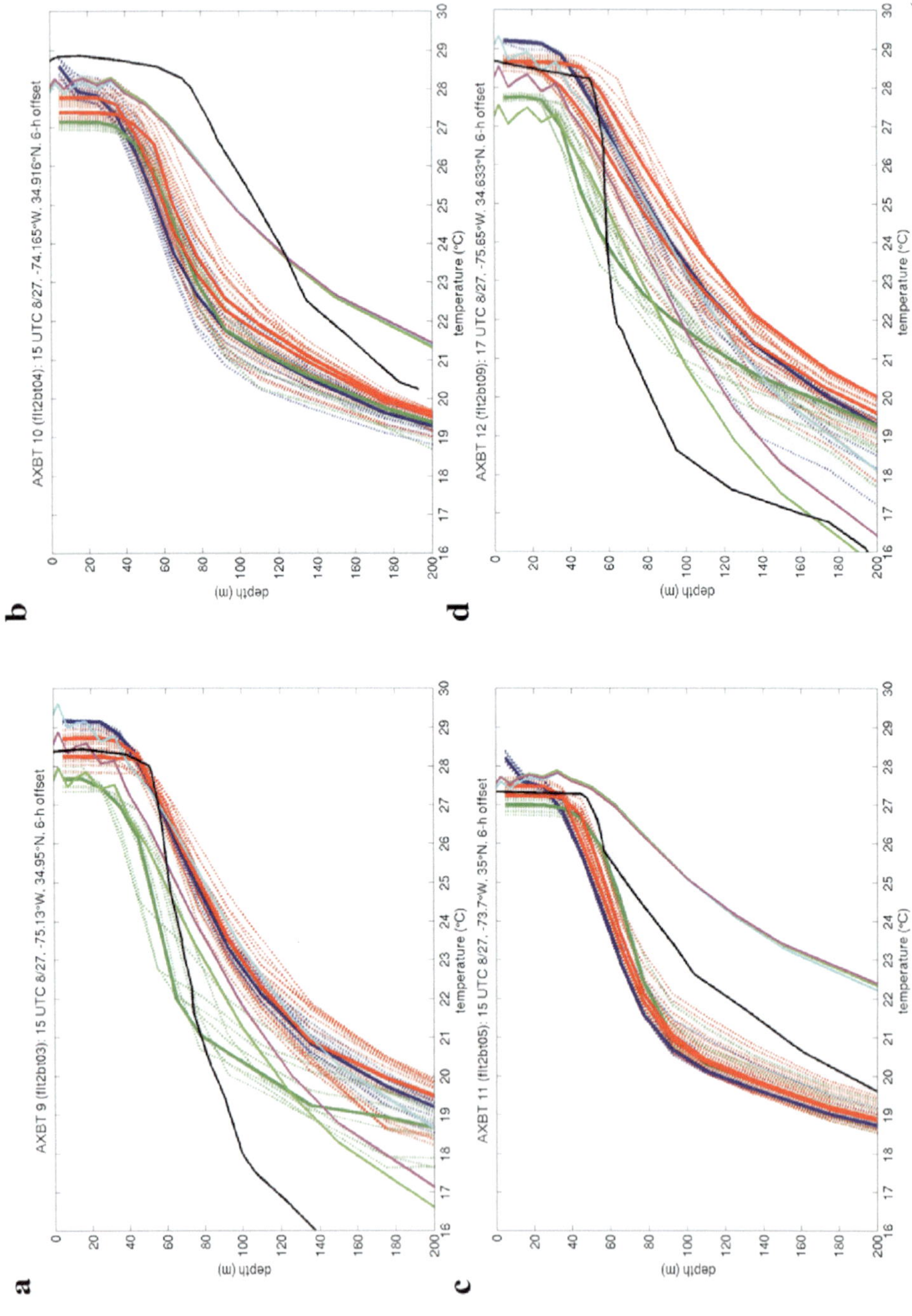
a
AXBT 9 (flt2bt03): 15 UTC 8/27. -75.13°W. 34.95°N. 6-h offset
b
AXBT 10 (flt2bt04): 15 UTC 8/27. -74.165°W. 34.916°N. 6-h offset
c
AXBT 11 (flt2bt05): 15 UTC 8/27. -73.7°W. 35°N. 6-h offset
d
AXBT 12 (flt2bt09): 17 UTC 8/27. -75.65°W. 34.633°N. 6-h offset
depth (m)
temperature (°C)

(Contd.)

(**Fig. 7.** *Contd.*)

e

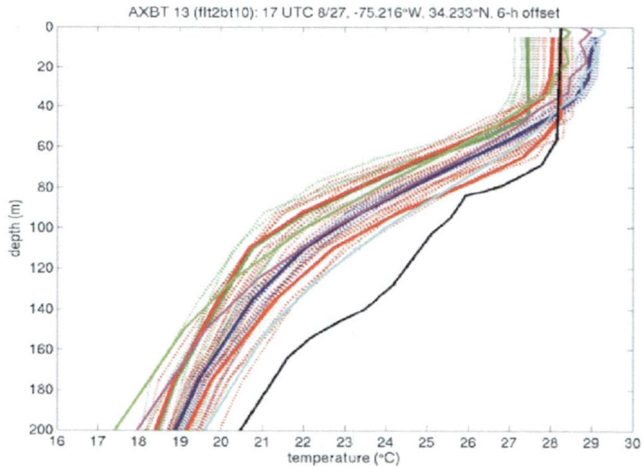

f

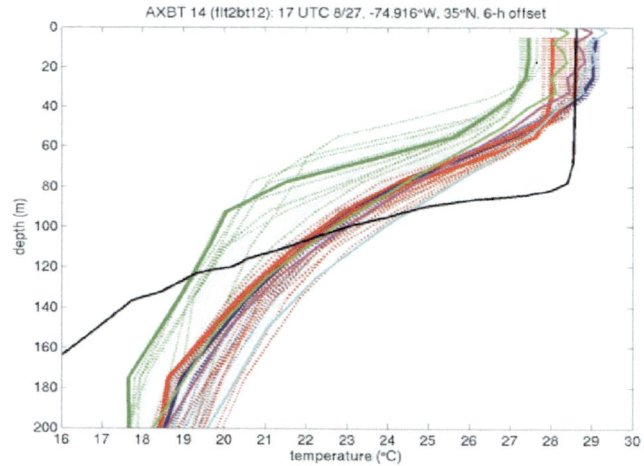

Fig. 7. As in Fig. 6, but at AXBT (a)–(f) 9–14.

3.6.3 Pre-storm Upper-ocean Cooling (AXBTs 4 and 8)

AXBT 4 (Fig. 6d) was dropped about 150 to 200 km to the east-northeast of Irene at 2300 UTC 26 August, in advance of the storm's closest approach. At the same location, from before Irene's approach to the time of the AXBT drop, the HWRF OML cooled from 29°C to 28.1°C and deepened from 0 to 35 m, whereas the AXBT 4 showed an OML temperature of 27.3°C and a depth of 50 m, colder than HWRF. The AXBT-indicated cooling ahead of Irene's closest approach (relative to HWRF) may have important implications for Irene's subsequent intensity change, and in fact Irene weakened more than indicated by HWRF prior to landfall in North Carolina (Fig. 5f), but part of this intensity difference may be due to errors in HWRF storm track and translation speed (Fig. 5g). The underestimated cooling may be due to insufficient mixing in HWRF or advection of the cold wake by the Gulf Stream, similar to the scenario described for a warm ocean eddy in Yablonsky and Ginis (2013).

AXBT 8 (Fig. 6h) was dropped near the centre of Irene at 0300 UTC 27 August. At that location, from before Irene's approach to the time of the AXBT drop, the HWRF OML cooled from ~29.2°C to ~28.6°C and deepened from ~25 m to ~30-45 m. AXBT 8 indicated an OML temperature of ~27°C and depth of ~55 m, consistent with the post-storm HWRF (and NCODA) OML temperatures (~27.2°C). Hence, the observed ocean temperature under the storm core is more representative of the cold wake than it is of the partially-cooled ocean regime that might be expected directly under the storm core; this result would support Gulf Stream induced along-track temperature advection. For both AXBTs 4 and 8, the HWRF and AXBT vertical temperature gradients are similar in the upper thermocline, and NCODA's vertical temperature gradient is sharper than both HWRF and the AXBT below 100-150m depth.

3.6.4 Region of Strong Upwelling (AXBT 9)

AXBT 9 (Fig. 7a) was dropped about 150 km to the east of the Irene at 1500 UTC 27 August, at approximate time of the storm's closest approach. At that location, from before Irene's approach to the time of the AXBT drop, the HWRF OML cooled from ~29.2°C to ~28.5°C and deepened from ~30 m to ~45 m. AXBT 9 indicated an OML temperature of ~28.4°C and depth of ~55 m, similar to HWRF. In the upper thermocline below the OML, the AXBT (but not HWRF) has a sharp vertical temperature gradient, leading to a temperature of 18°C at 100 m depth (vs. 23°C in HWRF), which suggests significant upwelling occurred at this location. The upwelling could be due to the strong wind blowing parallel to the coast from the south. HWRF also indicates strong upwelling in the post-storm profile, but the HWRF temperature never dropped below 19°C in the upper 140 m, while the AXBT profile indicates a 140-m temperature less than 16°C. NCODA's temperature profiles fall between the HWRF and the AXBT. The offshore HWRF track bias might be a reason for the weaker coastal upwelling.

3.6.5 Region of Strong Downwelling (AXBT 10)

AXBT 10 (Fig. 7b) was dropped about 250-300 km to the east of Irene at 1500 UTC 27 August. At these locations, from before the Irene's approach to the time of the AXBT 10 drop, the HWRF OML cooled from 28.5 to 27.5° C and deepened from 0 to 40 m. AXBT 10 indicated an OML temperature of 28.8 °C and a depth of 70 m, significantly warmer than the HWRF. AXBT 10 has a gradual vertical temperature gradient, unlike AXBT 9, leading to a temperature of 26 °C at 100 m depth (compared to 22 °C in HWRF). The results from the AXBT 10 suggest that significant downwelling occurred at that location. HWRF profiles do not indicate downwelling, which may be explained by the offshore track bias.

3.7 AXBT Summary

The Irene AXBTs support the earlier finding that the HWRF upper-ocean mixed layer does not cool as much as observed. However, the close proximity of the Gulf Stream and complex bathymetry make it difficult to draw strong conclusions about the performance of the POM –TC component of the coupled HWRF system. A more indepth analysis of the physical mechanisms responsible for the details of the observed upper ocean AXBT profiles needs to be undertaken before strong conclusions can be drawn.

3.7.1 Hurricane Isaac

The HWRF 48-h coupled model forecast of Hurricane Isaac, initialized at 1200 UTC on 27 August 2012, is shown in Fig. 8. The magnitude of the Isaac HWRF cold wake at forecast hour 48 (Fig. 8a, c) is underestimated relative to the TMI SST (Fig. 8b, d) along the first half of the forecast track, but it is overestimated along the second half of the forecast track. The cold wake differences along the first half of the track may be influenced by subtle differences between the HWRF representation (Fig. 8e, g) and the observed characteristics [based on the sea surface height (SSH) derived from satellite altimetry] (Fig. 8h) of the warm and cold ocean eddies north of the Loop Current in the vicinity of Isaac's track. The overestimation along the second half of the track may be a limitation of the temporal averaging technique used to generate the TMI SST, particularly close to the coast. In addition, the HWRF intensity is ~5-10 ms^{-1} higher than the best track shortly before landfall, although the HWRF storm size is up to ~50 km smaller than the best track at this time (Fig. 8f).

3.7.2 Hurricane Leslie

The HWRF 120-h coupled model forecast of Hurricane Leslie, initialized at 00 UTC on 03 September 2012, is shown in Fig. 9. Unlike Earl, Igor, Irene and Isaac, Leslie was a very slow moving storm during a large part of its lifecycle, allowing for intense upwelling and SST cooling underneath storm core,

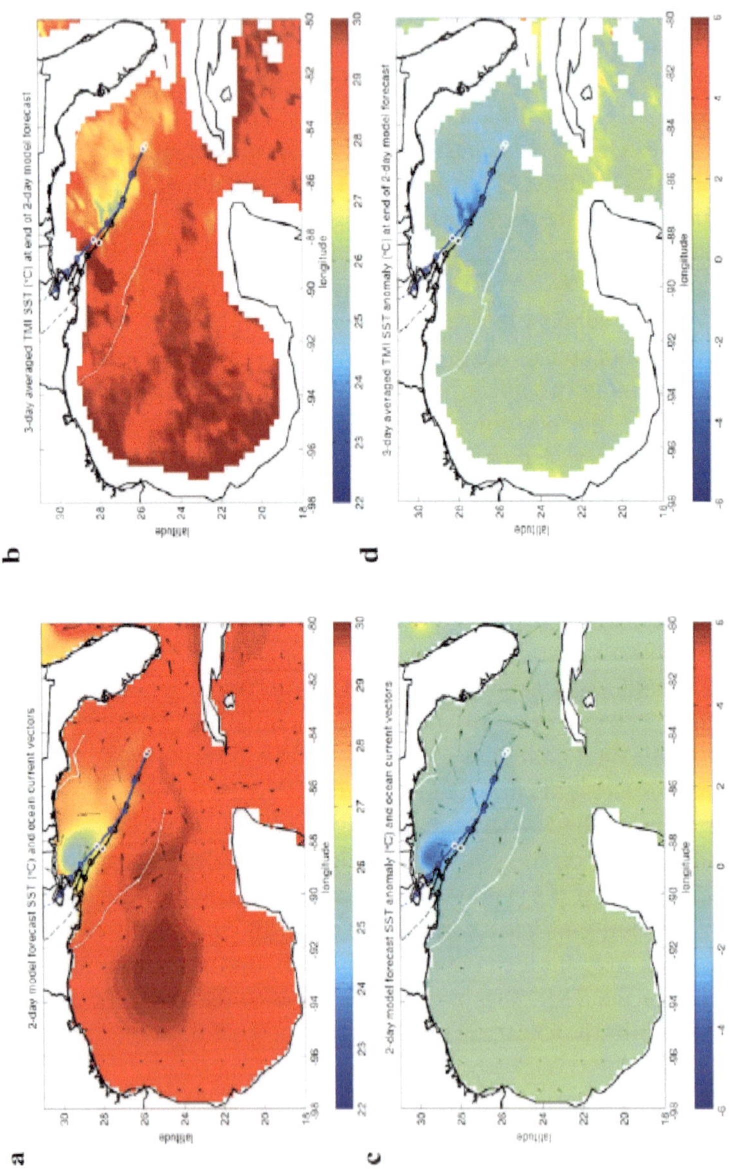
a
b
c
d
2-day model forecast SST (°C) and ocean current vectors
3-day averaged TMI SST (°C) at end of 2-day model forecast
2-day model forecast SST anomaly (°C) and ocean current vectors
3-day averaged TMI SST anomaly (°C) at end of 2-day model forecast
latitude
longitude

(Contd.)

(**Fig. 8.** *Contd.*)

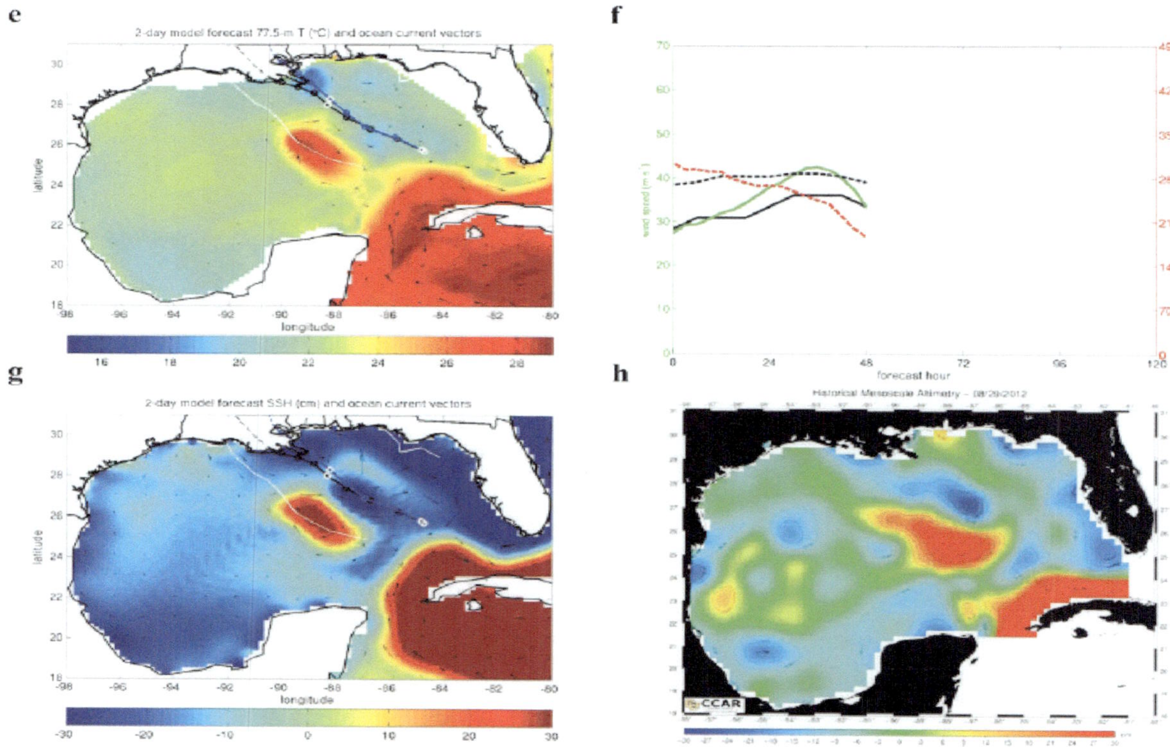

Fig. 8. HWRF/POM-TC 48-h forecast of Hurricane Isaac, initialized 1200 UTC 27 Aug. 2012: (a) 48-h model SST and current with model track (blue with white circles every 24 h), model 17 ms^{-1} wind swath (white), and best track (black); (b) 8/27-8/29 TMI SST with best track 17 ms^{-1} wind swath (white); (c) 48-h model SST anomaly; (d) 8/27-8/29 TMI SST anomaly from 8/25-8/27; (e) 48-h model 77.5-m T; (f) as in Fig. 3f; (g) 48-h model SSH; and (h) 8/29 satellite-derived SSH.

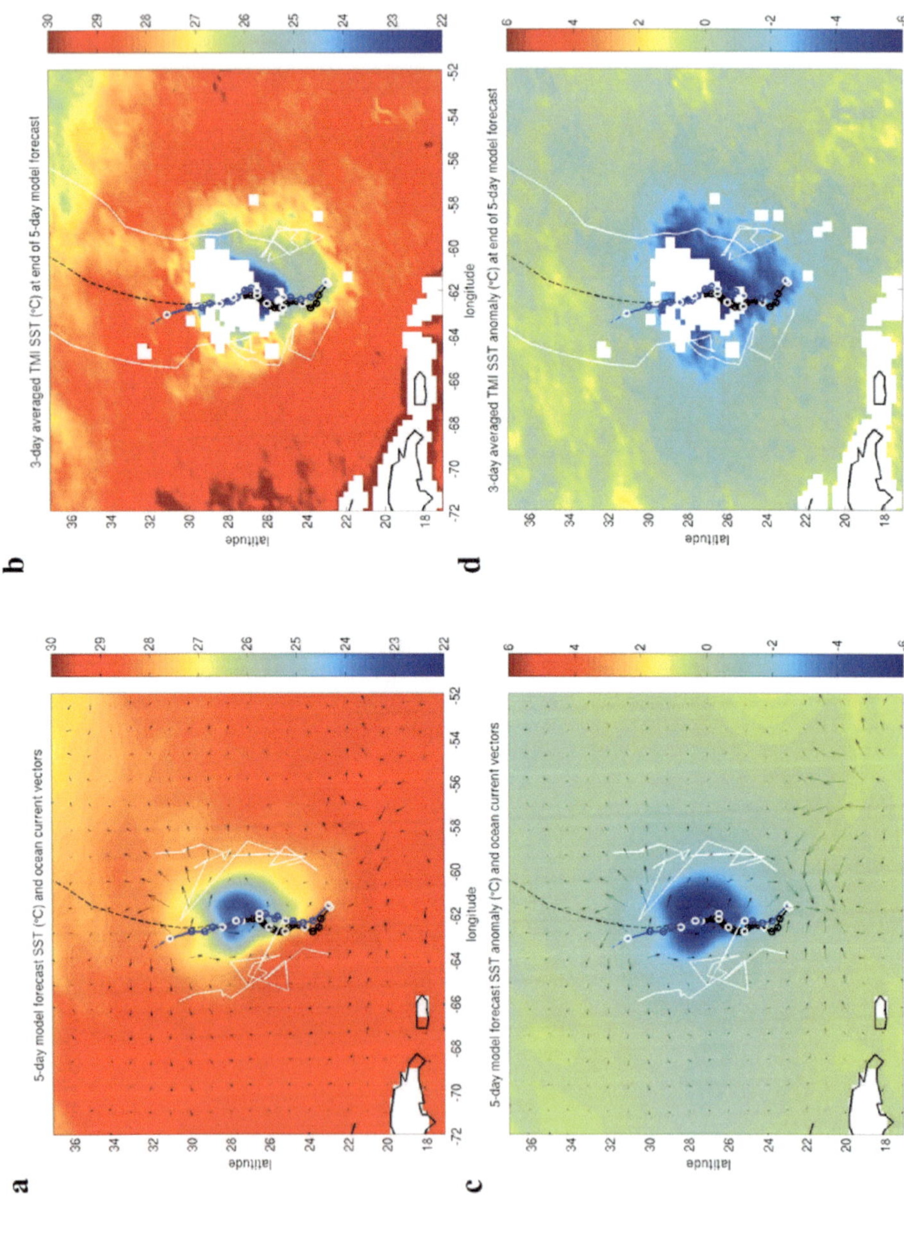
a
5-day model forecast SST (°C) and ocean current vectors
b
3-day averaged TMI SST (°C) at end of 5-day model forecast
c
5-day model forecast SST anomaly (°C) and ocean current vectors
d
3-day averaged TMI SST anomaly (°C) at end of 5-day model forecast
latitude
longitude

(Contd.)

e

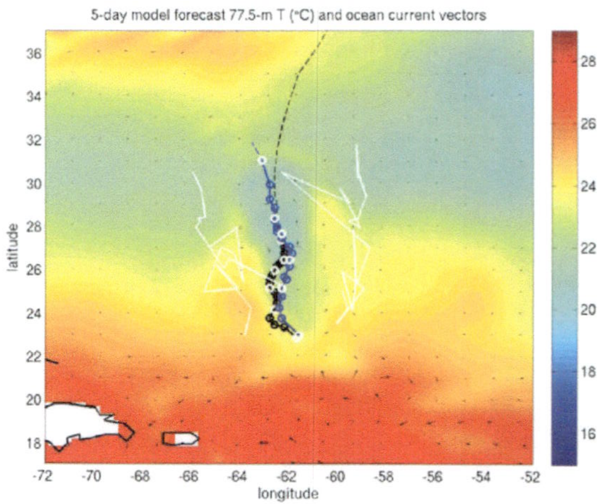

f

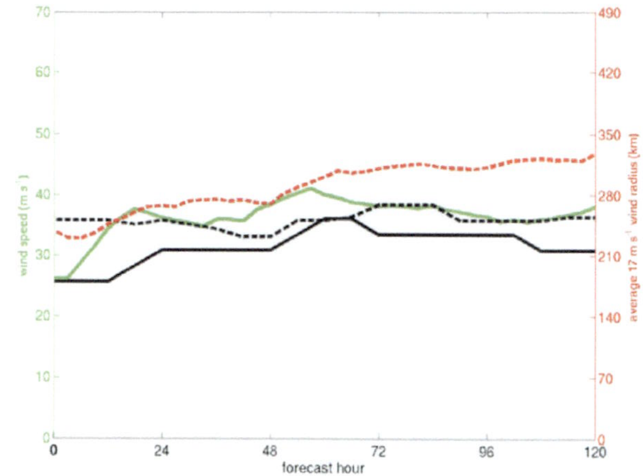

Fig. 9. HWRF-POM-TC forecast of Hurricane Leslie, initialized 00 UTC 03 Sept. 2012: (a) 120-h model SST and current with model track (blue with white circles every 24 h), model 17 ms^{-1} wind swath (white), and best track (black); (b) 9/6-9/8 TMI SST with best track 17 ms^{-1} wind swath (white); (c) 120-h model SST anomaly; (d) 9/6-9/8 TMI SST anomaly from 9/1-9/3; (e) 120-h model 77.5-m T; and (f) as in Fig. 3f.

which limited the storm's intensity. Nonetheless, like the other storms, the magnitude and spatial extent of the Leslie HWRF cold wake SST at forecast hour 120 (Fig. 9a, c) is not as large as the TMI SST (Fig. 9b, d). Examining the HWRF temperature at 77.5-m depth (Fig. 9e) reveals that Leslie propagated northward across a boundary of higher ocean heat content to the south and lower ocean heat content to the north, the structure of which could impact the magnitude of the cold wake. The HWRF intensity forecast is generally 5-10 ms^{-1} higher, and the HWRF size forecast is up to 50-100 km larger, than the best track observations (Fig. 9f), perhaps due to the underestimation of the SST cooling throughout the forecast, which may be explained by the faster HWRF storm propagation speed relative to the best track (Fig. 9a-e).

4. Summary

The selected cases of hurricanes Earl (2010), Igor (2010), Irene (2011), Isaac (2012) and Leslie (2012) reveal that the HWRF coupled model forecasts generally capture the TC-induced cold wake but tend to underestimate the magnitude and spatial extent of the SST cooling relative to available TMI and buoy SST observations.

Annexure 1

Description of MPIPOM-TC

The MPIPOM-TC Ocean modelling system has been widely used for both research and operational applications since 2014. With the availability of the powerful computing resources at NOAA, URI created MPIPOM-TC by merging the 2012 version of the sbPOM and POM-TC and making additional changes to allow for flexible initialization options and a relocatable grid (Yablonsky et al., 2015b).

MPIPOM-TC is an end-to-end ocean modelling system, with both research and operational applications that includes generation of an initial ocean condition, a procedure for prescribed hurricane wind forcing, and/or two-way coupling between the ocean and a prognostic hurricane model (e.g. GFDL or HWRF). The MPIPOM-TC code is built on the parallelized sbPOM framework (Jordi and Wang, 2012), including the main driver (pom.f), problem setup (initialize.f), code advancement in time (advance.f), boundary conditions and forcing (bounds_forcing.f), basic dynamics (solver.f), parallel tasks for communications between processors (parallel_mpi.f), parallel I/O (io_pnetcdf.F), model domain size and decomposition (pom.h), and name list (pom.nml). Key differences from sbPOM, many of which are derived from POM-TC, are summarized in Table 1.

Table 1. Differences between the sbPOM code and the MPIPOM-TC code

File name	Changes from sbPOM to MPIPOM-TC
pom.f	Only used if uncoupled from hurricane model; if coupled, drive_ocean.f used.
initialize.f	Options added for f-plane, spatial smoothing and desmoothing (Bender et al. 1993), 1-D (vertical only) dynamics on an A-grid (vs. 3-D C-grid), wave-dependent turbulence, and non-zero initial currents (geostrophic or prescribed) and/or sea surface height; alternative values for input constants.
advance.f	Prescribed surface forcing only applied when uncoupled from the hurricane model; added missing MPI exchanges; omitted lateral viscosity, mode interaction, external mode, vertical velocity, horizontal advection, and horizontal diffusion for 1-D only dynamics; smoothing and desmoothing of temperature, salinity, and currents optionally applied; temperature growth eliminated; model fails if SST is unreasonably high.
bounds_forcing.f	Maximum depth and the number of vertical sigma levels are not specified here; new boundary conditions added for 1-D only dynamics on an A-grid.
solver.f	Missing MPI exchanges added; A-grid used for 1-D only dynamics; wave breaking energy disabled; wave-dependent turbulence optionally applied.
parallel_mpi.f	None, except for two minor changes when coupled to the hurricane model.
io_pnetcdf.F	I/O parameters added; print times and output directory changed; syntax fixed.
pom.h	Many new parameters added (see comments in code for details).
pom.nml	Ten name list parameters added to control new and enhanced capabilities.

Beyond the changes summarized in Table 1, MPIPOM-TC includes separate files to provide specific capabilities during hurricane-model-coupled and/or uncoupled simulations, including prescribed hurricane wind forcing (hurricane_wind.f) using NOAA's Tropical Cyclone Vitals Database (Trahan and Sparling, 2012) or the International Best Track Archive for Climate Stewardship (Knapp et al., 2010), spatial smoothing and desmoothing (smoothing.f), wave-dependent turbulence (calcpwave.f), and initially geostrophic currents (geovel.f). For coupled simulations only, eight additional files are required: atm_ocm.f, cmp.comm.f, make_om_grid.f, make_om_msk.f, mp_arr_pros_om.f, mpi_more.f, ocm_atm.f, and pom.comm.f. Like sbPOM, MPIPOM-TC is scalable with a wide variety of domain decompositions (Jordi and Wang, 2012).

Annexure 2

Model Initialization and Configuration (MPIPOM-TC)

MPIPOM-TC includes flexible, plug-and-play, Fortran-based initial condition modules. In the 2014 operational GFDL and HWRF models, the initial condition module in the Atlantic Ocean (fbtr) involves feature-based modifications to the U.S. Navy's Generalized Digital Environmental Model (GDEM) monthly temperature (*T*) and salinity (*S*) climatology (Yablonsky and Ginis, 2008; Teague et al., 1990), followed by assimilation of a daily SST product (Reynolds and Smith, 1994); the initial condition module in the eastern North Pacific Ocean (gdm3) uses GDEMv3 (Carnes, 2009), assimilated with daily SST (Reynolds and Smith, 1994). Initial condition modules have also been developed using the stand-alone Navy Ocean Data Assimilation (NCODA) daily *T* and *S* fields (ncda; Cummings 2005; Cummings and Smedstad, 2013) and versions of the HYbrid Coordinate Ocean Model (HYCOM) that use NCODA (e.g. hycu; Chassignet et al., 2009). All of these ocean products are available in the public domain for real time tropical cyclone forecasting. An idealized initial condition module (idel) with user-specified *T* and *S*, flat bathymetry, and no land is available too.

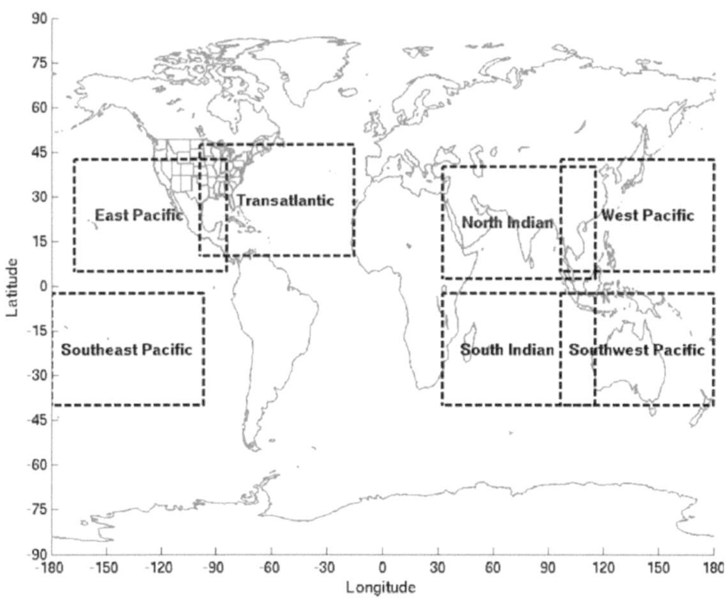

Fig. 10. World ocean domains used in the MPIPOM-TC.

To simulate tropical cyclones worldwide, each MPIPOM-TC initial condition module is relocatable to regions around the world. Currently, these regions include the Transatlantic, East Pacific, West Pacific, North Indian, South Indian, Southwest Pacific, Southeast Pacific, and South Atlantic domains (Fig. 10), as well as the idealized domain. Domain overlap helps to prevent loss of ocean coupling. To avoid domain-specific code (pom.h), all worldwide domains are set to the same size: 869 (449) longitudinal (latitudinal) grid points, covering 83.2° (37.5°) of longitude (latitude) and yielding a horizontal grid spacing of ~9 km. Horizontal domain decomposition is 3 × 3, yielding 291 (151) local grid points on each of the nine processors. There are 23 full sigma levels from the sea surface to the sea floor, with higher vertical resolution in the mixed layer and upper thermocline.

Annexure 3

Model Evaluation (MPIPOM-TC)

MPIPOM-TC was developed to facilitate further improvements to both operational and research versions of hurricane forecast models (e.g. GFDL, HWRF, and the U.S. Navy's version of the GFDL model [GFDN]). Initial operational MPIPOM-TC implementation in GFDL and HWRF was completed in 2014 for the Transatlantic and East Pacific basins after (1) rigorous evaluation of numerous hurricane track and intensity forecasts using either POM-TC or MPIPOM-TC (e.g. Fig. 11), (2) scaling and performance tests to ensure that the forecast completes within the operational time window, and (3) comparison of individual forecasts to both remotely-sensed and in situ ocean observations (e.g. Fig. 11; Yablonsky et al., 2015a). Worldwide MPIPOM-TC operational implementation in GFDL, HWRF and GFDN is planned after further testing and evaluation. The MPIPOM-TC framework described here will allow for more rapid improvements and transition of research to operations (R2O) in the future, including: development and evaluation of alternative initialization modules; physics improvements (especially to the turbulence closure model to properly account for wave-dependent turbulence); three-way coupling between MPIPOM-TC, a hurricane model, and a wave model (e.g. WAVEWATCH III, Tolman, 2014); and high-resolution nesting, especially at the coastline to forecast storm surge. While these new capabilities may improve forecasts of hurricanes and their impacts, R2O must consider a balance between the best model configuration and the computational efficiency required to produce timely forecasts with available resources. Nonetheless, advanced MPIPOM-TC configurations that cannot be immediately transitioned to operations may be used for research and further model development, including future R2O. In addition, simplified MPIPOM-TC configuration options (e.g. 1D only dynamics and idealized initial conditions) can isolate specific physical processes to guide future model improvements.

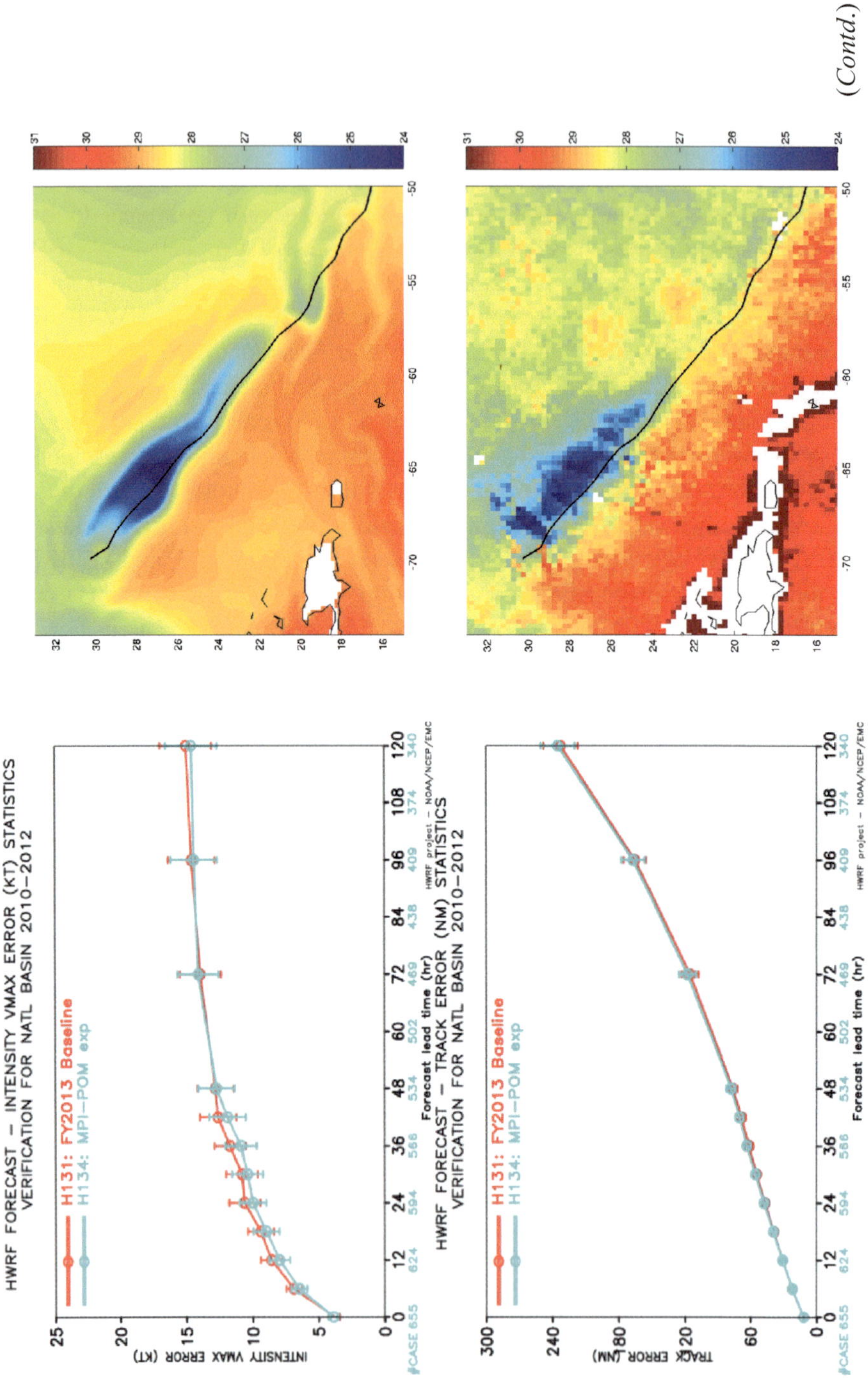
HWRF FORECAST - INTENSITY VMAX ERROR (KT) STATISTICS
VERIFICATION FOR NATL BASIN 2010-2012
H131: FY2013 Baseline
H134: MPI-POM exp
INTENSITY VMAX ERROR (KT)
Forecast lead time (hr)
HWRF FORECAST - TRACK ERROR (NM) STATISTICS
VERIFICATION FOR NATL BASIN 2010-2012
H131: FY2013 Baseline
H134: MPI-POM exp
TRACK ERROR (NM)
Forecast lead time (hr)
HWRF project - NOAA/NCEP/EMC

(Contd.)

(**Fig. 11.** *Contd.*)

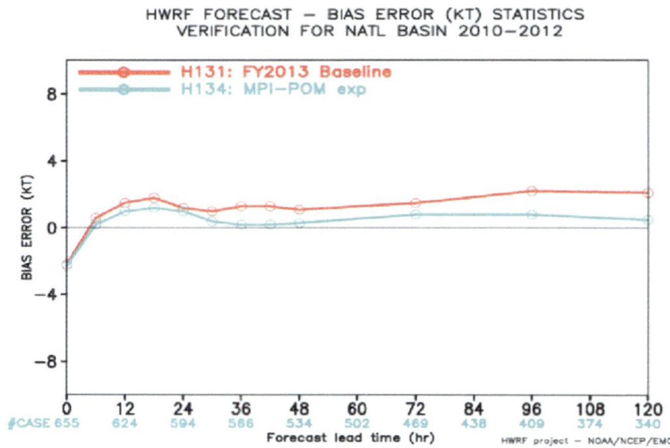

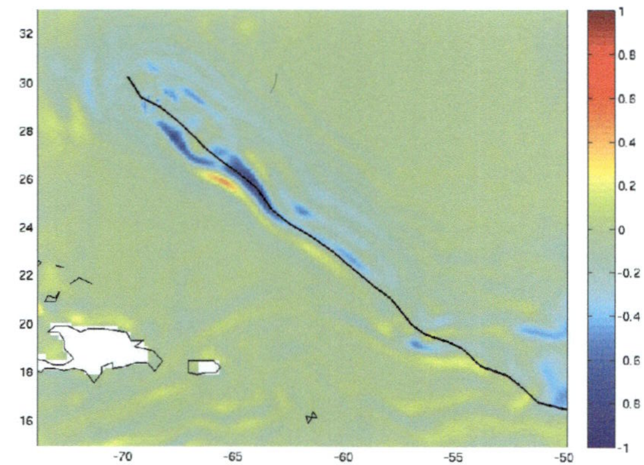

Fig. 11. HWRF intensity error statistics (kt, upper left), track error statistics (nm, centre left), and intensity bias statistics (kt, lower right) using 655 test cases from the 2010-2012 Atlantic hurricane seasons, run with either the 2013 baseline coupled with POM-TC (H131, red), as in Yablonsky et al. (2015), or the 2013 baseline coupled with MPIPOM-TC (H134, blue); courtesy of NOAA's Environmental Modeling Center (EMC). HWRF coupled with MPIPOM-TC reduces the positive intensity bias compared to HWRF coupled with POM-TC; track differences are small. MPIPOM-TC SST (°C) cold wake (upper right), TRMM Microwave Imager 3-day averaged observed SST (centre right), and MPIPOM-TC-POM-TC SST cold wake difference (lower right), generated by Hurricane Katia's forcing through 00 UTC 08 September 2011. MPIPOM-TC cooling compares well with observations and is generally greater than POM-TC, consistent with a reduced positive intensity bias.

REFERENCES

Bao, J.-W., J.M. Wilczak, J.-K. Choi and L.H. Kantha, 2000: Numerical simulations of air-sea interaction under high wind conditions using a coupled model: A study of hurricane development. *Mon. Wea. Rev.*, **128**, 2190-2210.

Bender, M.A., I. Ginis and Y. Kurihara, 1993: Numerical simulations of tropical cyclone-ocean interaction with a high-resolution coupled model. *J. Geophys. Res.*, **98**, 23,245-23,263.

Bender, M.A. and I. Ginis, 2000: Real case simulation of hurricane-ocean interaction using a high-resolution coupled model: Effects on hurricane intensity. *Mon. Wea. Rev.*, **128**, 917-946.

Bender, M.A., I. Ginis, R. Tuleya, B. Thomas and T. Marchok, 2007: The operational GFDL Coupled Hurricane-Ocean Prediction System and a summary of its performance. *Mon. Wea. Rev.*, **135**, 3965-3989.

Carnes, M. R., 2009: Description and evaluation of GDEM-V 3.0. Naval Research Laboratory, Stennis Space Center, MS 39529, 21 pp. [NRL/MR/7330--09-9165].

Chan, J.C.L., Y. Duan and L.K. Shay, 2001: Tropical cyclone intensity change from a simple ocean-atmosphere coupled model. *J. Atmos. Sci.*, **58**, 154-172.

Chassignet, E.P., H.E. Hurlburt, E.J. Metzger, O.M. Smedstad, J. Cummings, G.R. Halliwell, R. Bleck, R. Baraille, A.J. Wallcraft, C. Lozano, H.L. Tolman, A. Srinivasan, S. Hankin, P. Cornillon, R. Weisberg, A. Barth, R. He, F. Werner and J. Wilkin, 2009: US GODAE: Global ocean prediction with the HYbrid Coordinate Ocean Model (HYCOM). *Oceanogr.*, **22**, 64-75.

Chen, S.S., J.F. Price, W. Zhao, M.A. Donelan and E.J. Walsh, 2007: The CBLAST-Hurricane Program and the next-generation fully coupled atmosphere-wave-ocean models for hurricane research and prediction. *Bull. Amer. Meteor. Soc.*, **88**, 311-317.

Cummings, J.A., 2005: Operational multivariate ocean data assimilation. *Quart. J. Royal Meteor. Soc.*, **131**, 3583-3604.

Cummings, J.A. and O.M. Smedstad, 2013: Variational data assimilation for the global ocean. *In:* Data Assimilation for Atmospheric, Oceanic and Hydrologic Applications, Vol. II. S. Park and L. Xu (Eds). Springer.

Emanuel, K., C. DesAutels, C. Holloway and R. Korty, 2004: Environmental control of tropical cyclone intensity. *J. Atmos. Sci.*, **61**, 843-858.

Falkovich, A., I. Ginis and S. Lord, 2005: Ocean data assimilation and initialization procedure for the Coupled GFDL/URI Hurricane Prediction System. *J. Atmos. Oceanic Technol.*, **22**, 1918-1932.

Gentemann, C.L., T. Meissner and F.J. Wentz, 2010: Accuracy of satellite sea surface temperatures at 7 and 11 GHz. *IEEE Transactions on Geoscience and Remote Sensing*, **39**, 1009-1018.

Gentemann, C.L., F.J. Wentz, C.A. Mears and D.K. Smith, 2004: In situ validation of Tropical Rainfall Measuring Mission microwave sea surface temperatures. *J. Geophys. Res.*, **109**, C04021.

Ginis, I., Kh. Zh. Dikinov and A.P. Khain, 1989: A three-dimensional model of the atmosphere and the ocean in the zone of a typhoon. *Dokl. Akad. Nauk. USSR*, **307**, 333-337.

Hodur, R.M., 1997: The Naval Research Laboratory's Coupled Ocean/Atmosphere Mesoscale Prediction System (COAMPS). *Mon. Wea. Rev.*, **125**, 1414-1430.

Hong, X., S.W. Chang, S. Raman, L.K. Shay and R. Hodur, 2000: The interaction between and Hurricane Opal (1995) and a warm core ring in the Gulf of Mexico. *Mon. Wea. Rev.*, **128**, 1347-1365.

Jourdain, N.C., M. Lengaigne, J. Vialard, G. Madec, C.E. Menkes, E.M. Vincent, S. Jullien and B. Barnier, 2013: Observation-based estimates of surface cooling inhibition by heavy rainfall under tropical cyclones. *J. Phys. Oceanogr.*, **43**, 205-221.

Jordi, A. and D.-P. Wang, 2012: sbPOM: A parallel implementation of Princeton Ocean Model. *Environ. Modelling & Softw.*, **38**, 59-61.

Jullien, S., P. Marchesiello, C.E. Menkes, J. Lefèvre, N.C. Jourdain, G. Samson and M. Lengaigne, 2014: Ocean feedback to tropical cyclones: Climatology and processes. *Climate Dyn.*, doi: 10.1007/s00382-014-2096-6.

Kim, H.-S., C. Lozano, V. Tallapragada, D. Iredell, D. Sheinin, H.L. Tolman, V.M. Gerald and J. Sims, 2014: Performance of ocean simulations in the coupled HWRF-HYCOM model. *J. Atmos. Oceanic Technol.*, **31**, 545-559.

Knapp, K.R., M.C. Kruk, D.H. Levinson, H.J. Diamond and C.J. Neumann, 2010: The International Best Track Archive for Climate Stewardship (IBTrACS): Unifying tropical cyclone best track data. *Bull. Amer. Meteor. Soc.*, **91**, 363-376.

Lin, I.-I., C.-C.Wu, K.A. Emanuel, I.-H. Lee, C.-R. Wu and I.-F. Pun, 2005: The interaction of Super typhoon Maemi, 2003: With a warm ocean eddy. *Mon. Wea. Rev.*, **133**, 2635-2649.

Liu, B., H. Liu, L. Xie, C. Guan and D. Zhao, 2011: A coupled atmosphere-wave-ocean modeling system: Simulation of the intensity of an idealized tropical cyclone. *Mon. Wea. Rev.*, **139**, 132-152.

Mellor, G.L., 1991: An equation of state for numerical models of oceans and estuaries. *J. Atmos. Oceanic Technol.*, **8**, 609-611.

Mellor, G.L., 2004: Users guide for a three-dimensional, primitive equation, numerical ocean model (June 2004 version). Prog. in Atmos. and Ocean. Sci., Princeton University.

Mellor, G.L. and T. Yamada, 1982: Development of a turbulence closure model for geophysical fluid problems. *Rev. Geophys. Space Phys.*, **20**, 851-875.

Oey, L., Y.-L. Chang, Y.-C. Lin, M.-C. Chang, F. Xu and H.-F. Lu, 2013: ATOP – The advanced Taiwan ocean prediction system based on the mpiPOM. Part 1: Model descriptions, analyses and results. *Terr. Atmos. Ocean. Sci.*, **24**, 137-158.

Reynolds, R.W. and T.M. Smith, 1994: Improved global sea surface temperature analyses using optimum interpolation. *J. Climate*, **7**, 929-948.

Sanabia, E.R., B.S. Barrett, P.G. Black, S. Chen and J.A. Cummings, 2013: Real-time upper-ocean temperature observations from aircraft during operational hurricane reconnaissance missions: AXBT demonstration project year one results. *Wea. Forecasting*, **28**, 1404-1422.

Sandery, P.A., G.B. Brassington, A. Craig and T. Pugh, 2010: Impacts of ocean-atmosphere coupling on tropical cyclone intensity change and ocean prediction in the Australian region. *Mon. Wea. Rev.*, **138**, 2074-2091.

Schade, L.R. and K.A. Emanuel, 1999: The ocean's effect on the intensity of tropical cyclones: Results from a simple coupled atmosphere-ocean model. *J. Atmos. Sci.*, **56**, 642-651.

Tallapragada, V., L. Bernardet, S. Gopalakrishnan, Y. Kwon, Q. Liu, T. Marchok, D. Sheinin, M. Tong, S. Trahan, R. Tuleya, R. Yablonsky and X. Zhang, 2013:

Hurricane Weather Research and Forecasting (HWRF) Model: 2013 scientific documentation. Developmental Testbed Center. Available from http://www.dtcenter.org/HurrWRF/users/docs/.

Teague, W.J, M.J. Carron and P.J. Hogan, 1990: A comparison between the Generalized Digital Environmental Model and Levitus climatologies. *J. Geophys. Res.*, **95**, 7167-7183.

Tolman, H.L., 2014: User manual and system documentation of WAVEWATCH III version 4.18. Tech. Note 316, NOAA/NWS/NCEP/EMC/MMAB, 282 pp. + Appendices.

Trahan, S. and L. Sparling, 2012: An analysis of NCEP tropical cyclone vitals and potential effects on forecasting models. *Wea. Forecasting*, **27**, 744-756.

Vincent, E.M., M. Lengaigne, G. Madec, J. Vialard, G. Samson, N.C. Jourdain, C.E. Menkes, and S. Jullien, 2012a: Processes setting the characteristics of sea surface cooling induced by tropical cyclones. *J. Geophys. Res.*, **117**, C02020, doi: 10.1029/2011JC007396.

Wu, L., B. Wang and S.A. Braun, 2005: Impacts of air-sea interaction on tropical cyclone track and intensity. *Mon. Wea. Rev.*, **133**, 3299-3314.

Wu, C.-C., C.-Y. Lee and I.-I. Lin, 2007: The effect of the ocean eddy on tropical cyclone intensity. *J. Atmos. Sci.*, **64**, 3562-3578.

Yablonsky, R.M. and I. Ginis, 2008: Improving the ocean initialization of coupled hurricane-ocean models using feature-based data assimilation. *Mon. Wea. Rev.*, **136**, 2592-2607.

Yablonsky, R.M. and I. Ginis, 2009: Limitation of one-dimensional ocean models for coupled hurricane-ocean model forecasts. *Mon. Wea. Rev.*, **137**, 4410-4419.

Yablonsky, R.M. and I. Ginis, 2013: Impact of a warm ocean eddy's circulation on hurricane-induced sea surface cooling with implications for hurricane intensity. *Mon. Wea. Rev.*, **141**, 997-1021.

Yablonsky, R.M., I. Ginis, B. Thomas, V. Tallapragada, D. Sheinin and L. Bernardet, 2015a: Description and analysis of the ocean component of NOAA's operational Hurricane Weather Research and Forecasting (HWRF) Model. *J. Atmos. Oceanic Technol.*, **32**, 144-163.

Yablonsky, R.M., I. Ginis and B. Thomas, 2015b: Ocean modeling with flexible initialization for improved coupled tropical cyclone-ocean model prediction. *Environ. Modelling & Softw.*, **67**, 26-30.

Part 3

Advanced Assimilation and Vortex Initialization Techniques

11

Introduction to Data Assimilation Techniques and Ensemble Kalman Filter

A. Routray[1], Krishna K. Osuri[2]
S. Pattanayak[2] and U.C. Mohanty[2]

[1]NCMRWF, A-50, Institutional Area, Sector-62, Noida, India
[2]School of Earth, Ocean and Climate Sciences
IIT Bhubaneswar – 751 007, India
ashishroutray.iitd@gmail.com

1. Introduction

We are receiving lot of observations day to day from different platforms such as upper air, surface stations and remote sensing instruments. The difficulty is to use these data, which are sometimes conflicting, to find a best estimate of the state of the earth system which will be used for diverse applications. But we need to ensure that a time sequence of these estimated states is consistent with any known equations that govern the evolution of the system. The method to achieve this goal is known as data assimilation (Mathieu and O'Neill, 2008).

Numerical weather prediction (NWP) is an initial value problem. Provided an estimate of the atmospheric state, in terms of the variables of the NWP model, the model simulates the atmospheric state at later times. Determination of the starting state for an NWP model is called data analysis or data assimilation. The resulting state we call the analysis, estimation of which is a non-trivial problem for a number of reasons. On the other hand the quality of the analysis is crucial for the NWP forecast skill, and it is to a large degree due to improvements of the data assimilation systems and addition of satellite data that the forecast skill has improved significantly during recent years.

Data assimilation (DA) is known as the process of creating the best estimate of the initial state for numerical weather prediction (NWP) models through combining all sources of information, including the first guess from previous short-term model forecasts and observations, along with the associated uncertainties in each source of information. Assimilation is an analysis which combines time-distributed observations and a dynamic model. An analysis is an approximation of the true state of a physical system at a given time. An analysis can be useful in itself as a comprehensive and self-consistent diagnostic of the physical system. It can also be used as input data to another operation, notably as the initial state for a numerical weather forecast for instance. It can provide a reference against which to check the quality of observations.

Many assimilation techniques have been developed for meteorology and oceanography. They differ in their numerical cost, their optimality, and in their suitability for real-time data assimilation (Bouttier and Courtier, 1999). The main techniques such as variational analysis (3/4 DVAR), ensemble Kalman filter (EnKF) and hybrid data assimilation as well as the performance skill of the techniques in the weather forecasting are presented in this chapter.

2. Purpose of Data Assimilation

The most common application for data assimilation is numerical weather prediction, for which as accurate as possible initial conditions are needed. Data assimilation is an important tool for model validation. By reducing the error due to the initial conditions, the resulting errors are primarily caused by the model itself. Furthermore, data assimilation is used to create accurate and continuous research data sets (reanalyses). For instance, atmospheric data assimilation is the analytical process of estimating the entire state of the atmosphere from a set of observations. This is considered to be a crucial element of weather forecasting. No forecast model can be wholly correct, and the assimilation procedure aims to equip the numerical model with accurate initial conditions, thus encouraging the model to advance in a realistic direction. It is not just forecasting which benefits. The process is applied to the creation of accurate and continuous research data sets, and also to the diagnosis of model errors. The model generates a fully consistent four-dimensional (4-D) picture of the real ocean/atmosphere. But in the real world, we have to cope with imperfect models (in particular systematic deficiencies), imperfect data (measurement errors, methodological uncertainties, sampling problems, etc.) and often poor data coverage.

Data assimilation is the technique whereby observational data are combined with data from a numerical forecast model to produce an "optimal" estimate (according to some criterions) of the evolving state of the system, referred to as the "analysis." An analysis is the production of an accurate image of the true state of the atmosphere at a given time, represented in a model as a collection of numbers. An analysis can be useful in itself as a comprehensive and self consistent diagnostic of the atmosphere. It can also be used as input data to another operation, notably as the initial state for a numerical weather forecast, or as a data retrieval to be used as a pseudo-observation. It can provide a reference against which to check the quality of observations. The model brings consistency to the observational data, and interpolates or extrapolates data into observational data void regions in space and time.

There are two basic approaches to data assimilation: sequential assimilation, which only considers observation made in the past until the time of analysis, which is the case of real-time assimilation systems and non-sequential, where observation from the future can be used, for instance in a reanalysis exercise. Another distinction can be made between methods that are intermittent or continuous in time. The four basic types of assimilation are depicted schematically in Fig. 1. In Figs 1a and 1c, the bottom row of

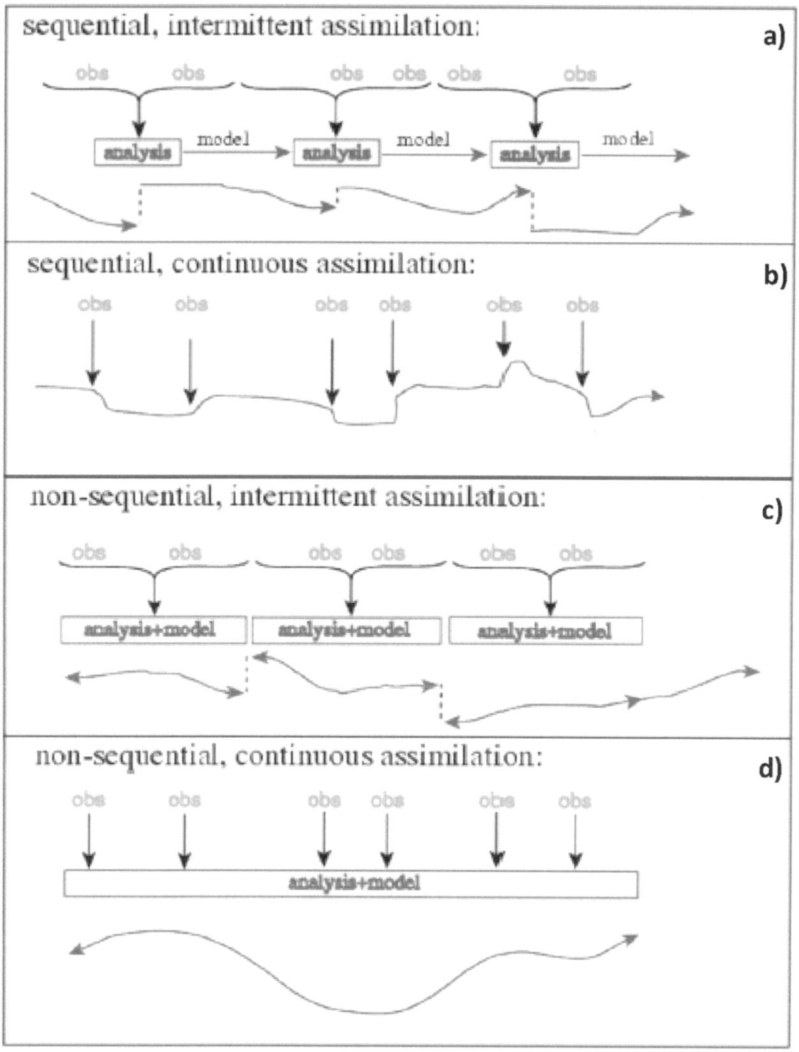

Fig. 1. Representation of the four basic strategies for data assimilation, as a function of time (from Bouttier and Courtier, 1999).

arrows designate the assimilated states. The way the time distribution of observations is processed to produce a time sequence of assimilated states (the lower curve in each panel) can be sequential and/or continuous. In an intermittent method, observations can be processed in small batches, which is usually technically convenient (intermittent data assimilation is used in most global operational systems, typically with a 6-h cycle performed four times a day). In a continuous method, observation batches over longer periods are considered, and the correction to the analyzed state is smooth in time, which is physically more realistic.

3. Quality Control (QC) of Meteorological Observations

All observing systems have their limitations, problems and failures, resulting in the reported measurements being sometimes incorrect. Such data must be identified and rejected by the data assimilation system in order to avoid corruption of the analysis. Due to the amount of data handled this is done by automatic routines, both in the form of preprocessing and during the data assimilation stage. In recent years, the QC systems have become more sophisticated and many observations that would have been thrown out in the past are now corrected, resulting in an improvement in the initial conditions and improved the forecast skill of the numerical models. The newer QC systems allow for a continuous weighting of the observation suspected of having a gross error, rather than yes or no decision to toss it out.

The QC is based on a comparison between observations and some kind of expected value (which could be based on climatology, an average of nearby observations or first guess). The difference between the expected value and the reported observation is called as residual. If the residual is very large (from the prescribed estimated values), then the observation may be considered as erroneous. The importance of performing QC on meteorological observations was recognized since the beginning of NWP. The principle of checking meteorological observations is based upon the redundancy of information within the system. The QC procedure are basically statistical in nature, where the rejection of an erroneous observation is based on a compromise between the risk of accepting erroneous values and the risk of rejecting correct values. The QC has become much more exhaustive in the last decade in connection with the utilization of global data from various sources.

There are different types of QC checks one after another before analysis:

Gross error check (climatological limits): Compare each observation with a prescribed range or climatological mean (maximum five standard deviations), if it is more than the prescribed climatological mean values, the observation was tossed out.

Horizontal consistency/Buddy check: If the observation survived from gross check test, it was then compared to the average of nearby observations and again tossed out if outside a reasonable range.

Internal consistency check: This test is based on the verification of physics or climatologic consistency of each observed parameter. Meteorological relationships among observations measured at a single station. For example, a dew-point temperature observation must not exceed the temperature observation made at the same station. If it does, both the dew-point and temperature observation are flagged as failing their internal consistency check.

Time consistency check: Compare the difference between successive measurements. If the difference exceeds an allowed value, distinct for each parameter, both observations are flagged as suspect. Therefore, this test checks the excessive rate of change of two consecutive values.

Hydrostatic check (vertically layers, $\Delta p = -g.\rho.\Delta z$): It is a most powerful check, it takes advantages of the redundancy between temperature and height information reported by RS/RW observations. The hydrostatic residual is the difference between the values of the thickness of a layer between mandatory level heights, calculated using the reported height and the thickness calculated independently from the reported virtual temperatures.

Incremental check (deviation from first guess): The residual is the increment between the reported observation and first guess (06 hr forecast).

Complex QC (CQC)

Gandin (1988) introduced the idea of complex quality control (where the word "complex" means that it uses several tests simultaneously rather than in series). The basic idea is to estimate several independent residuals and then apply a decision making algorithm based on the information provided by all the independent residuals, rather than performing decision either in a sequential order or as a single OI quality control check using all the information at once. The power of this method lies in the fact that several independent checks can support each other and reduce the level of uncertainty. Furthermore, if the residuals are large and agree reasonably well with each other, they provide the basis for a correction of the observation. The details of the CQC can be found from Gaddin (1988) and Kalnay (2003).

4. Definition and Notations Used in Data Assimilation

The dimension of the model state is n and the dimension of the observation vector is p. We define the following variables:

$\mathbf{x}_t$ true model state (dimension n);

$\mathbf{x}_b$ background model state (dimension n);

$\mathbf{x}_a$ analysis model state (dimension n);

x inputs of the model (dimension n);

$\mathbf{y}_{obs}$ vector of observations (dimension p);

y vector of modelled observations, i.e., the outputs of the model (dimension p);

H observation operator (from dimension n to p);

B covariance matrix of the background errors ($\mathbf{x}_b - \mathbf{x}_t$) (dimension $n \times n$);

R covariance matrix of observation errors ($y_{obs} - \mathbf{H}[\mathbf{x}_t]$) (dimension $p \times p$); and

A covariance matrix of the analysis errors ($\mathbf{x}_a - \mathbf{x}_t$) (dimension $n \times n$).

Derivation of Cost Function *J*

From definition of the joint probability density function of non-independent random variables, the likelihood of the true state x, given the background field x_b is:

$$L_B(x \| x_b) = p_B(x_b|x) = \frac{1}{(2\pi)^{n/2}\,|B|^{1/2}} \exp\left[-\frac{1}{2}(x_b - x)^T B^{-1}(x_b - x)\right] \tag{1}$$

From the definition of joint probability density function of 'n' non-independent random variables. Similarly, the likelihood of the true state x, given the new observation y_0 is given by

$$L_R(x \| y_0) = p_R(y_b|x) = \frac{1}{(2\pi)^{p/2}\,|R|^{1/2}} \exp\left[-\frac{1}{2}(y_0 - H(x))^T R^{-1}(y_0 - H(x))\right] \tag{2}$$

Since the background and new observations are independent, the most likely value of x, given the two independent vector field x_b and y_0, is one that maximizes the joint probability density function i.e.

$$\max_x \; L\,(x \| x_b, y_0) = p_B\,(x_b|x)\; p_R\,(y_0|x)$$
$$= \frac{1}{(2\pi)^{(p+n)/2}\;|B|^{1/2}\;|R|^{1/2}} \exp$$
$$\left[-\frac{1}{2}(x_b - x)^T B^{-1}(x_b - x) - \frac{1}{2}(y_0 - H(x))^T R^{-1}(y_0 - H(x))\right]$$

Taking log on both sides

$$\max_x \; \ln L\,(x \| x_b, y_0) = \max\left[\ln\left(\frac{1}{(2\pi)^{p+n/2}\,|B|^{1/2}|R|^{1/2}}\right)\frac{1}{2}(x_b - x)^T B^{-1}(x_b - x)\right.$$
$$\left.\frac{1}{2}(y_0 - H(x))^T R^{-1}(y_0 - H(x))\right]$$

$$= \max_x\left[Const.\frac{1}{2}\left\{(x_b - x)^T B^{-1}(x_b - x) + (y_0 - H(x))^T R^{-1}(y_0 - H(x))\right\}\right] \tag{3}$$

Thus, maximum likelihood of L is obtained if we get a minimum of the term $\frac{1}{2}\left[(x_b - x)^T B^{-1}(x_b - x) + (y_0 - H(x))^T R^{-1}(y_0 - H(x))\right]$ in Eq. (3). Thus, the cost function to minimize is

$$J(x)=\frac{1}{2}\left[(x_b-x)^T B^{-1}(x_b-x)+(y_0-H(x))^T R^{-1}(y_0-H(x))\right] \quad (4)$$

5. Minimization Algorithms

It is well known, that the high dimensional minimization problem is a major aspect in the numerical data assimilation. To solve the problem of minimization requires efficient algorithms. A few of minimization algorithms are presented here and other minimization algorithms in details can be found from Bishop (1995), Saporta (2005) and Tarantola (2005).

5.1 Cost Function

Aim is to improve the analysis x_a by taking into account in a conjoint way the two sources of available information, the model and the observations.

There are two ways of combining observations with model output:

1. From the point of view of the observations y_{obs}, to interpolate between the observational data, which are available in time and space subject to the constraints provided by the model.

2. From the point of view of the models H, to reduce the uncertainties on the inputs x of the model, under the constraint of the measurements, to refine the outputs y or in the inputs x.

To define the cost function, an easy approach is to measure the distance between the observation y_{obs} (the true measurement) and $H(x)$, i.e.:

$$J(x) = \| y_{obs} - H(x) \|^2, J \text{ is called the } \textit{cost function.}$$

In a vectorial case, it is logical to balance each component by the confidence in the measurement. We can have access to an "a priori" estimation of the desired value, called x_b i.e. *background* information. To assure a compromise between the information given by the observations and the information given by this background value, the cost function is now defined by:

$$J(x) = \alpha x \| y_{obs} - H(x) \|^2 + \beta x \| x - x_b \|^2$$

with $\|.\|$ a norm to determine, and α and β are the weight giving the confidence in the observations and in the background information. These parameters can be determined empirically or analytically knowing the background and observation errors. The definition of the cost function is a major task, and need to know the processes of the problem. It can be difficult to define the different terms of the cost function and the norm. Figure 2 presents the impact of each term of the cost function. The cost function terms have both a quadratic form and tend to "pull" the analysis x_a toward the background x_b and

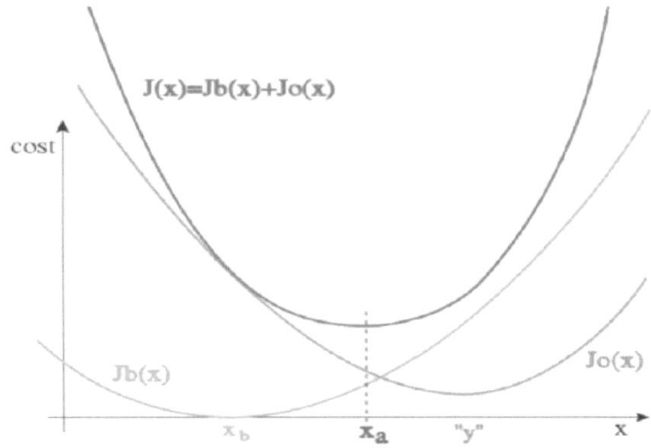

Fig. 2. Schematic representation of a cost function. The term $J_0(x)$ represents $||y - H(x)||^2$ and $J_b(x)$, $||x - x_b||^2$ (from Bouttier and Courtier, 1999).

the observation y, respectively. In this case, x_a is a value of x that makes $J(x)$ as small as possible (given the computational constraints). Finally, the general form of the cost function taking into account the errors covariance matrix is defined as:

$$J(x) = (y_{\text{obs}} - H[x])^{\mathrm{T}} R^{-1}(y_{\text{obs}} - H[x]) + (x - x_b)B^{-1}(x - x_b)$$

5.2 Newton Algorithm

It is known that a minimum of the cost function J, the gradient, ∇J, is zero. Newton's method consists of iteratively approaching the minimum of J using the following formula:

$$X_{k+1} = X_k - H^{-1}{}_k . \nabla J_k$$

where H_k is the Hessian matrix (Tarantola, 2005), which is the square matrix of second-order partial derivatives of a function, and ∇J_k is the gradient of the cost function at the point x_k, k being the iterative step. Note that H_k in this formula is not the same H used in the previous discussion. $H_k^{-1}.\nabla J_k$ is known as the *Newton direction* or the *Newton step* and form the basis for a variety of optimization strategies. The term ∇J_k can be replaced by $-H_k \cdot p_k$, with p_k the descent step, which is usually a fixed value. There are several difficulties with this approach. First, the exact evaluation of the Hessian is, in general, computationally demanding. Second, the Hessian must be inverted and is also computationally demanding. Third, the Newton step may move toward a maximum or a saddle point rather than a minimum. Finally, the step size $(H_k^{-1}.\nabla J_k)$ may be sufficiently large that it takes us outside the range of validity. In this case, the algorithm could become unstable.

5.3 Quasi-Newton Algorithm

Alternative to the Newton algorithm is the *Quasi-Newton* or *variable metric* methods (Bishop, 1995; Tarantola, 2005 for both techniques). The principle of those methods is to use the variations of the ∇J, which has to be minimized, to obtain information on the Hessian. The goal of these methods is, instead of calculating the Hessian directly, and then evaluating its inverse, to build up an approximation to the inverse Hessian over a number of steps.

6. Different Methods of Assimilation

6.1 The Optimal Least-Squares Estimator

A local minimum of J is given by $\nabla J = 0$ and is obtained, usually, when $x = x_a$, the analysis state of the model. The optimal least-squares estimator or *analysis* (Talagrand, 1997) is defined by interpolation equations:

$$x_a = x_b + K(y - H[x_b]) \quad (5)$$

$$K = BH^{T}(HBH^{T} + R)^{-1} \quad (6)$$

where the linear operator K is called the gain or the weight matrix of the analysis and corresponds to the confidence in the observations and in the background. Equation (5) says that the analysis is obtained by adding to the background field the product of the optimal weight matrix and the innovation (what gives the conjoint information observation/background). Equation (6) says that the optimal weight matrix is given by the background error covariance in the observation space (BH^{T}) multiplied by the inverse of the total error covariance.

6.2 Variational Approach—3DVAR

The standard formulation of variational methods was derived from first principles by Lorenc using Bayesian probabilities and assuming Gaussian error distributions (Lorenc, 1986; Lorenc et al., 2000). In the variational method, direct minimization of a cost function determines the analysis (Ide et al., 1997). The cost function that consists of two terms (1) J^b, which measures the distance between the resulting analysis and a background field (or a prior) and (2) J^o, which measures the distance between the analysis and observations.

The cost function is defined as the (weighted) distance between the true state (X) and the background (X_b) plus the (weighted) distance to observations (Y_0).

$$J(X) = \frac{1}{2}\left\{(X - X_b)^T B^{-1}(X - X_b) + [Y_0 - H(X)]^T R^{-1}[Y_0 - H(X)]\right\} \quad (7)$$

[We got this cost function $J(x)$ from Equation (4)]

The minimum of the $J(X)$ is attained for $X = X_a$ such that

$$\frac{\partial J}{\partial X} = \nabla_X J(X_a) = 0 \quad (8)$$

Let the analysis be close to the true state of the atmosphere then we can write

$$X = [X_b + (X - X_b)] \quad (9)$$

Again we can linearize the observation operator $[Y_0 - H(X)]$ by using the Eq. (9)

$$[Y_0 - H(X)] = Y_0 - H\,[X_b + (X - X_b)]$$
$$= [Y_0 - H(X_b)] - H(X - X_b) \quad (10)$$

We can use Eq. (10) in Eq. (7).

Then

$$2J(X) = (X - X_b)^T B^{-1} (X - X_b) + [\{Y_0 - H(X_b)\} - H\,(X - X_b)]^T$$
$$R^{-1}[\{Y_0 - H(X_b) - H\,(X - X_b)\}]$$
$$= (X - X_b)^T B^{-1} (X - X_b) + [\{Y_0 - H(X_b)\}^T - H^T (X - X_b)^T]$$
$$R^{-1}[\{Y_0 - H(X_b)\} - H\,(X - X_b)]$$
$$= (X - X_b)^T B^{-1} (X - X_b) + \{Y_0 - H(X_b)\}^T R^{-1} \{Y_0 - H(X_b)\}$$
$$- \{Y_0 - H\,(X_b)\}^T R^{-1} H(X - X_b) - (X - X_b)^T H^T R^{-1} \{Y_0 - H(X_b)\}$$
$$+ (X - X_b)^T H^T R^{-1} H(X - X_b) \quad (11)$$

The cost function is a quadratic function of the analysis increments $(x - x_b)$. Thus, we need to compute the gradient of J in Eq. (11) w.r.t. x [or w.r.t. $(x - x_b)$] [See the footnote]. Equation (11) can be written after combining the 1st terms [2nd term of Eq. (11) is independent of x and $(x - x_b)$].

Thus, the gradient of the cost function J in Eq. (11) w.r.t. x [or w.r.t. $(x - x_b)$] is

$$2\nabla J\,(x) = 2\,B^{-1} (x - x_b) + 0 - \nabla\,[\{y_0 - H(x_b)\}^T R^{-1} H(x - x_b)]$$
$$-\nabla[(x - x_b)^T H^T R^{-1} \{y_0 - H(x_b)\}] + 2\,H^T R^{-1} H\,(x - x_b)$$

As $\nabla\,[\frac{1}{2} x^T Ax] = Ax$
So $\nabla\,[(x - x_b)^T B^{-1} (x - x_b)] = 2\,B^{-1} (x - x_b)$
Similarly for last term in the Eq.

$$= 2\,B^{-1} (x - x_b) + 2\,H^T R^{-1}\,H(x - x_b)$$
$$-\nabla\,[((\{y_0 - H(x_b)\}^T R^{-1}\,H)^T)^T (x - x_b)]$$
$$-\,H^T R^{-1}[y_0 - H(x_b)]$$

$\because\ (A^T)^T = A$ (For 3rd term)
As $\nabla\,(x^T d) = d$
So $(x - x_b)^T H^T R^{-1}(y_0 - H(x_b)) = H^T R^{-1}(y_0 - H(x_b))$

$$= 2\,B^{-1} (x - x_b) + 2\,H^T R^{-1}\,H(x - x_b) - (\{y_0 - H(x_b)\}^T R^{-1}\,H)^T$$
$$-\,H^T R^{-1}[y_0 - H(x_b)]$$

As $\nabla\,(d^T x) = d$ (For 3rd term)

Since J in Eq. (11) is a quadratic function of the analysis increment $(x - x_b)$, for a given generalized quadratic function F

$$F(x) = \tfrac{1}{2} x^T Ax + d^T x + c, \quad (A)$$

we have

$$\nabla F(x) = Ax + d \quad (B)$$

and $$\nabla(d^T x) = \nabla(x^T d) = d. \quad (C)$$

$$= 2\,B^{-1}(x - x_b) + 2\,H^T R^{-1} H(x - x_b) - (R^{-1} H)^T (\{y_0 - H(x_b)\}^T)^T$$
$$- H^T R^{-1}[y_0 - H(x_b)]$$

As $(AB)^T = B^T A^T$ (For 3rd term)

$$= 2\,B^{-1}(x - x_b) + 2\,H^T R^{-1} H(x - x_b) - H^T R^{-1}(y_0 - H(x_b)) - H^T R^{-1}$$
$$[y_0 - H(x_b)]$$

$(A^T)^T = A$ and
R is symmetric $\Rightarrow (R^{-1})^T = R^{-1}$(For 3rd term)

$$\Rightarrow 2\nabla J(x) = 2\,B^{-1}(x - x_b) + 2\,H^T R^{-1} H(x - x_b) - 2\,H^T R^{-1}[y_0 - H(x_b)]$$

$$\Rightarrow \nabla J(x) = B^{-1}(x - x_b) + H^T R^{-1} H(x - x_b) - H^T R^{-1}[y_0 - H(x_b)] \tag{12}$$

Since $\nabla J(x_a) = 0$ to ensure that J is minimum, from Eq. (12) (Put $x \approx x_a$)

$$B^{-1}(x_a - x_b) + H^T R^{-1} H(x_a - x_b) - H^T R^{-1}[y_0 - H(x_b)] = 0$$

$$\Rightarrow B^{-1}(x_a - x_b) + H^T R^{-1} H(x_a - x_b) = H^T R^{-1}[y_0 - H(x_b)]$$

$$\Rightarrow [B^{-1} + H^T R^{-1} H](x_a - x_b) = H^T R^{-1}[y_0 - H(x_b)]$$

$$\Rightarrow x_a - x_b = [B^{-1} + H^T R^{-1} H]^{-1} H^T R^{-1}[y_0 - H(x_b)]$$

$$\Rightarrow x_a = x_b + [B^{-1} + H^T R^{-1} H]^{-1} H^T R^{-1}[y_0 - H(x_b)] \tag{13}$$

This Eq. (13) is the solution of the 3-dimensional variational (3DVAR) analysis problem.

The 2nd term of Eq. (13) $[B^{-1} + H^T R^{-1} H^{-1}]\, H^T R^{-1}$ is called *weight matrix* (W). It looks similar to the OI result.

The solution, i.e., the analysis of the model x_a, is sought iteratively by performing several evaluations of the cost function and its gradient in order to approach the minimum using a suitable descent algorithm (Discussed in above section). We look for the analysis as an approximate solution to the equivalent minimization problem defined by the cost function J. The approximation lies in the fact that only a small number of iterations are performed. The minimization can be stopped by limiting artificially the number of iterations or by requiring that the norm of the gradient $\|\nabla J(x)\|$ decreases by a predefined amount during the minimization. The process is shown in Fig. 3.

It is a formal solution, the computation x_a requires the inversion of a huge matrix, which is impractical.

In practical 3DVAR, we do not invert a huge matrix. We find the minimum of $J(X)$ by computing the cost function for a range of values of X and using an optimization technique. The location of the minimum depends greatly on the nature of the function J.

Note that the control variable i.e. the variable with respect to which we are minimizing the cost function J is used in the minimization process to finally

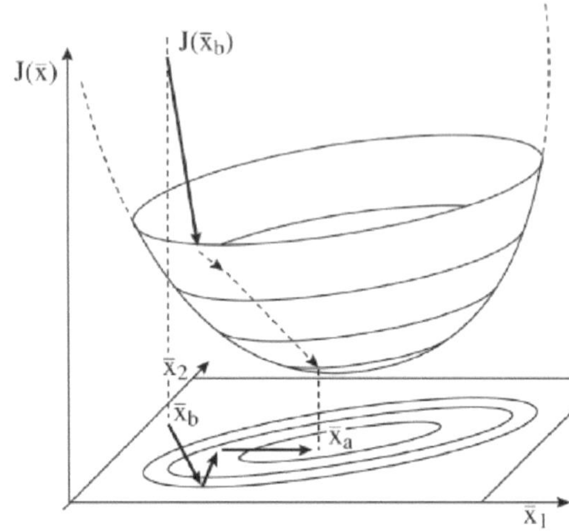

Fig. 3. Minimization of a quadratic cost function $J(x)$. The gradient of the cost function and the cost function itself are supplied to a minimization algorithm, which works to change x to obtain a smaller value of the cost function (from Bouttier and Courtier, 1999).

obtain the analysis. Although the two statistical methods, 3DVAR and OI, have been shown to formally solve the same problem, but there are important differences in the method of solution.

In practice, OI requires the introduction of a number of approximations and local solution of the analysis, grid point by grid point or small volume by small volume (Lorence, 1981). This in turn requires the use of a "radius of influence" and selection of only the stations closet to the grid point or volume being analyzed. The background error covariance matrix also has to be locally approximated.

Despite their formal equivalence, 3DVAR have several important advantages with respect to OI, because the cost function is minimized using global minimization algorithms and as a result it makes unnecessary many of the simplifying approximations required by OI (Parrish and Derber, 1992; Courtier et al., 1987).

6.2.1 Error Covariances

A correct specification of observation and background error covariances is crucial to the quality of the analysis, because they determine to what extent the background fields will be corrected to match the observations. The essential parameters are the variances, but the correlations are also very important because they specify how the observed information will be used in model space if there is a mismatch between the resolution of the model and the density of the observations.

Observation error variances

The reported atmospheric observations used in data assimilation are not perfect, they contain several kinds of errors, including instrumental errors and representativeness errors. The representativeness errors i,e. actually correct observations may reflect the presence of a subgrid-scale atmospheric phenomenon that cannot be resolved by the model or the analysis. The representativeness errors indicate that the observation is not representative of the area averaged measurement required by the model grid.

As explained before, they should also include the variance of representativeness errors which is not negligible when analysing phenomena. It is wrong to leave observation biases as a contribution to the observation error variances because it will produce biases in the analysis increments; whenever observation biases can be identified, they should be removed from the observed value or from the background fields, depending on whether one thinks they are caused by problems in the model or in the observation procedure.

Background error variances

They are usually estimates of the error variances in the forecast used to produce x_b. In the Kalman filter they are estimated automatically using the tangent-linear model, so they do not need to be specified (although this means that the problem is moved to the specification of the model error and the tuning of approximated algorithms that are less costly than the complete Kalman filter). A crude estimate can be obtained by taking an arbitrary fraction of climatological variance of the fields themselves. If the analysis is of good quality (i.e. if there are a lot of observations) a better average estimate is provided by the variance of the differences between the forecast and a verifying analysis. If the observations can be assumed to be uncorrelated, much better averaged background error variances can be obtained by using the observational method. However, in a system like the atmosphere the actual background errors are expected to depend a lot on the weather situation, and ideally the background errors should be flow-dependent. This can be achieved by the Kalman filter, by 4DVAR to some extent, or by some empirical laws of error growth based on physical grounds. If background error variances are badly specified, it will lead to too large or too small analysis increments. In least-squares analysis algorithms, only the relative magnitude of the background and observation error variances is important. However, the absolute values may be important if they are used to make quality-control decisions on the observations (it is usually desirable to accept more easily the observations with a large background departure if the background error is likely to be large).

The role of the Background error statistics (B) are:

1. To spread out the information from the observations.

2. To provide statistically consistent increments at the neighbouring gridpoints and levels of the model.
3. To ensure that observations of one model variable (e.g. temperature) produce dynamically consistent increments in the other model variables (e.g. vorticity and divergence).

There are two major problems to be addressed in specifying B:

1. We want to describe the statistics of the errors in the background. However, we don't know what the errors in the background are, since we don't know the true state of the atmosphere.
2. The B matrix is enormous ($\sim 10^7 \times 10^7$). We are forced to simplify it just to fit it into the computer. Even if we could fit it into the computer, we don't have enough statistical information to determine all its elements.

Three approaches to estimating the background error statistics:

1. The Hollingsworth and Lönnberg (1986) method: Differences between observations and the background are a combination of background and observation error. The method tries to partition this error into background errors and observation errors by assuming that the observation errors are spatially uncorrelated.
2. The NMC method (Parrish and Derber, 1992): This method assumes that the spatial correlations of background error are similar to the correlations of differences between 48h and 24h forecasts (or 24 hrs and 12 hrs for regional scale modelling) verifying at the same time.
3. The Analysis-Ensemble method (Fisher, 2003): This method runs the analysis system several times for the same period with randomly-perturbed observations. Differences between background fields for different runs provide a surrogate for a sample of background error.

6.2.2 Estimating Background Errors—Prospective and Constraint of Methods

Innovation statistics

- The only direct method for diagnosing background error statistics.
- Provides statistics of background error in observation space.
- Statistics are not global, and do not cover all model levels.
- Requires a good uniform observing network.
- Statistics are biased towards data-dense areas.

Forecast differences

- Generate global statistics of model variables at all levels.
- Inexpensive and easy.
- Statistics are a mixture of analysis and background error.
- Not good in data-sparse regions.

Ensembles of analyses

- Assumes statistics of observation error are well known.

- Diagnoses the statistics of the actual analysis system.
- Danger of feedback (noisy analysis system gives noisy statistics, which gives noisier system).

6.2.3 Comparison of 3DVAR and OI

- In 3DVAR there is no data selection; all available data are used simultaneously. This avoids jumpiness in the boundaries between regions that have selected different observations.
- The background error covariance matrix for 3DVAR can be defined with a more general, global approach, rather than the local approximations used in OI.
- It is possible to add constraints to the cost function without increasing the cost of the minimization. These can be used to control spurious noise.
- For example, we may require the analysis increments to approximately satisfy the linear global balance equation.
- With the implementation of 3DVAR at NCEP, it became unnecessary to perform a separate initialization step in the analysis cycle.
- It is possible to incorporate nonlinear relationships between observed variables and model variables in the H operator. This is harder to do in the OI approach.
- 3DVAR has allowed three-dimensional variational assimilation of radiances.
- The quality control of the observations becomes easier and more reliable when it is made in the space of the observations than in the space of the retrievals.

6.3 Variational Approach—4DVAR

The 3DVAR finds the best estimation of the state of the process (e.g., atmosphere or ocean) at a given time with the help of a forecast obtained with a model and observations of the process at this given time. However, the observations are often distributed in time. So, we need to do several analyses in sequence. For instance, in meteorology, we desire not only to realize a real-time state of the atmosphere but we also wish to do successive analyses during the evolution of the meteorological conditions.

In contrast to OI and 3DVAR, the 4DVAR method includes the dynamic evolution of the model in the assimilation. The essential difference between 3-D and 4-D VAR is illustrated in Fig. 4. There are not many differences between 3-D and 4-D VAR methods, the 4DVAR is a simple generalization of 3DVAR for observations that are distributed in time. The equations are the same, provided the observation operators are generalized to include a forecast model that will allow a comparison between the model state and the observations at the appropriate time. The main difference consists in including observations at different times in the definition of the cost function $J(x)$:

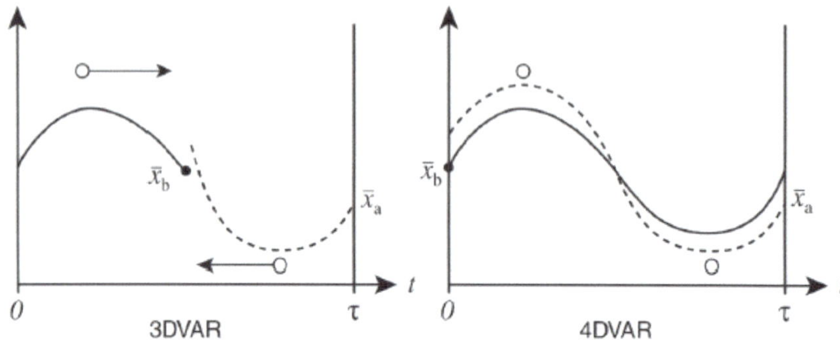

Fig. 4. The difference between 3-D and 4-D VAR. 4DVAR is a time/model consistent interpolator of departures (from Hòlm, 2003).

$$J(x) = (x - x^b)^T B^{-1}(x - x^b) + \sum_{i=0}^{n}(y_i - H_i[x_i])^T R_i^{-1}(y_i - H_i[x_i])$$

where the index *i* is any given observation time.

The principle of the 4DVAR assimilation technique is given in Fig. 5. The goal of 4DVAR assimilation method is to find the initial condition such that its forecast best fits the observations within the assimilation interval. The evaluation of the cost function *J*(*x*) and its gradient ∇*J*(*x*) requires one direct model integration from times 0 to *n* and one suitably modified *adjoint integration* mode of transposes of the tangent linear model time-stepping operators *Mi*.

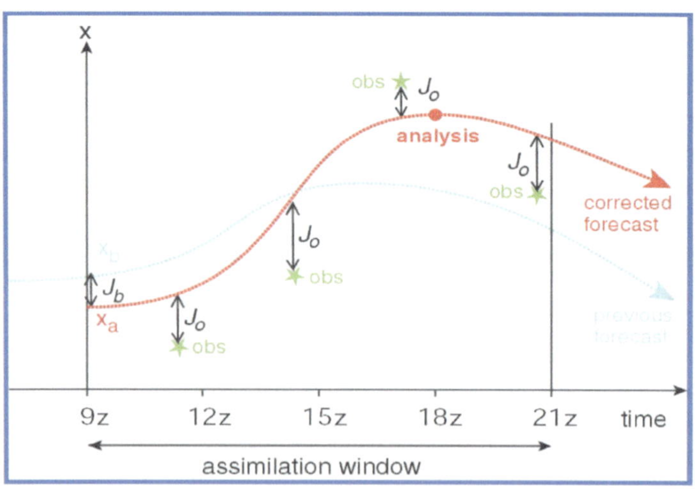

Fig. 5. Example of 4DVAR intermittent assimilation in a numerical forecasting system. Every 6 h, a 4DVAR is performed to assimilate the most recent observations, using a segment of the previous forecast as background. This updates the initial model trajectory for the subsequent forecast (from Bouttier and Courtier, 1999).

The 4DVAR technique has a number of advantages over 3DVAR schemes including:

The ability to use observations at the time of their measurement or in predetermined time bins.

Implicit definition of flow-dependent forecast error covariances, of vital importance in the optimal use of observations in fast-developing weather systems.

The ability to use a forecast model as a constraint enhancing the dynamic balance of the final analysis.

6.4 Kalman Filter (KF)

The Kalman filter is a set of mathematical equations that provides an efficient computational (recursive) means to estimate the state of a process, in a way that minimizes the mean of the squared error. The filter is very powerful in several aspects: it supports estimations of past, present, and even future states, and it can do so even when the precise nature of the modelled system is unknown (Welch and Bishop, 2004). Rudolf Emil Kalman defined and published in 1960 a recursive solution to the discrete signal, linear filtering problem.

6.4.1 The Basic Idea of a Kalman Filter

Noisy data in → Hopefully less noisy data out

The applications of a Kalman filter are numerous such as

- Tracking objects, fitting Bezier patches to point data
- Economics
- Navigation
- Many computer vision applications viz Stabilizing depth measurements, feature tracking
    - Cluster tracking
    - Fusing data from radar, laser scanner and stereo-cameras for depth and velocity measurement and many more.

In a very general sense, "noise" is an unwanted contribution to a measured signal, and there are studies on various kinds of noise related to a defined context (acoustic noise, electronic noise, environmental noise, and so forth).

Noise is here typically a high-frequency random perturbation of measured pixel values, caused by electronic noise of participating sensors (such as camera or scanner), or by transmission or digitization processes.

The direct extension of the OI is the KF. The idea is to do successive OIs between each pair of observation times so as to evolve the model state and the error covariance using the model. Consider the situation in which we have several observations distributed in time. After having realized an analysis from the observations $y_{\text{obs}}(t_i)$ and a forecast $x_f(t_i)$ given by the model, obtained an estimate, $x_a(t_i)$, of the state of the process. Integrate the model from this state until the time of the following observations $y_{\text{obs}}(t_i + 1)$, we obtain $x_f(t_i + 1)$. If the errors associated with the observations $y_{\text{obs}}(t_i + 1)$ and with the forecast $x_f(t_i + 1)$ are uncorrelated, then we can use equations (5) and (6) to do a second analysis.

In this context, the background information at time $t + 1$ is replaced by a forecast value given by the model from t to $t + 1$, from the best estimate of the state at t, namely, the analysis $x_a(t)$. The notations are the same as defined in section 3, except that:

The background and analysis error covariance matrix B and A are, respectively, replaced by P_f and P_a to denote the fact that the background is now a forecast. The background state x_b is a forecast denoted by x_f.

The time index of each quantity is denoted by the suffix 'i'. The model forecast operator from dates i to $i + 1$ is denoted by $M_{i \to i+1}$. Concerning the hypotheses, we assume the independence of the forecast, of the model and analysis errors, and consider that the forecast operator is linear (Bouttier and Courtier, 1999).

Under the specified hypotheses, the optimal way (in the least squares sense) to assimilate sequentially the observations is given by the *KF algorithm*. This direct extension of the formula of the OI between t_i and t_{i+1} leads to the following framework:

1. Step of the forecast of the state:

$$x_f(t_{i+1}) = M_{ti \to ti+1}\, x_a(t_i) \tag{14}$$

2. Error covariance forecast:

$$P_f(t_{i+1}) = M_{ti \to ti+1}\, P_a\, M^T_{ti \to ti+1} + Q_{ti} \tag{15}$$

with Q_{ti}, an error covariance matrix.

3. State analysis at t_{i+1}:

$$x_a(t_{i+1}) = x_f(t_{i+1}) + K(t_{i+1})[y(t_{i+1}) - H(t_{i+1})x_f(t_{i+1})] \tag{16}$$

with the Kalman gain computation

$$K(t_{i+1}) = P_f(t_{i+1})H^T_{ti+1}\,[H(t_{i+1})P_f(t_{i+1})H^T(t_{i+1}) + R(t_{i+1})]^{-1} \tag{17}$$

4. Error covariance analysis at t_{i+1}:

$$P_a(t_{i+1}) = [I - K(t_{i+1})H(t_{i+1})]P_f(t_{i+1}) \tag{18}$$

5. The analysis are the sequences of $x_a(t_i)$.

Figure 6 summarizes the process discussed above from step 1 to 5.

We can interpret the Kalman filter as solving the generalized inverse problem at each time step of the model, so that the assimilation windows shrinks to $[t, t + 1]$. If one accepts the linear hypothesis, this algorithm fulfills all the properties for being an "ideal" assimilation algorithm, namely, to use all the available information in order to produce the most accurate possible description of the process, together with the uncertainty resulting from the uncertainties on the various sources of information. The application of KF to assimilation of meteorological and oceanographical observations has

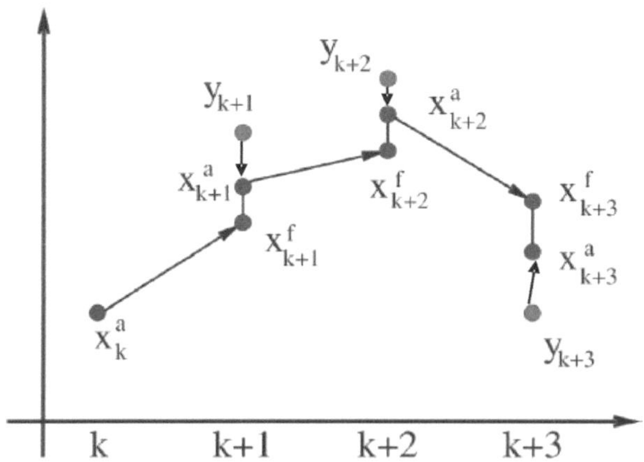

Fig. 6. Sequential method: Successive steps of forecast and analysis (from Bocquet, 2006).

produced convincing results as to the capability of the method for effectively extracting the information contained in the observations and the model (Talagrand, 1997).

6.4.2 Limitations of the KF

- From a practical point of view, the major difficulty in applying the method is the computational cost.
- The main disadvantage of the sequential method is that a given observation has an impact only on the future evolution of the physical process and is not used to correct the last states. This is due to the main characteristic of the sequential method, i.e., the one-way pass.
- Probability distributions associated with nonlinear systems are generally non-Gaussian.
- Estimation algorithms often cannot be presented in closed form.
- Integration of the covariance matrix involves computation of MP^aM^T which requires two-fold matrix multiplication (very costly).
- For accurate atmospheric science applications, state size is ~$O(10^6\text{-}10^8)$ so that storage of the covariance matrices is nearly impossible.

6.4.3 Extended KF (EKF)

The Kalman filter algorithm can be generalized to non-linear H and M operators, although it means that neither the optimality of the analysis nor the equivalence with 4DVAR hold in that case. If H is non-linear, H can be defined as its tangent linear in the vicinity of x_b, as discussed in a previous section. Similarly, if M is non-linear, which is the case of most meteorological and oceanographical models, M can be defined as the tangent linear forecast model in the vicinity of x_a, i.e. we assume that for any likely initial state $x\ (i)$ (notably $x\ t\ (i)$),

$$M_{i\to i+1}[x(i)] - M_{i\to i+1}[x_a(i)] \approx M_{i\to i+1}[x(i) - x_a(i)]$$

and the realism of this hypothesis must be appreciated using physical arguments, as already discussed about the observation operator and 4DVAR. If H and/or M are non-linear, the algorithm written above is called the *Extended Kalman Filter* (*EKF*). Note that the linearization M interacts with the model errors in a possibly complicated way, as can be seen from the proof of Eq. (15) above. If non-linearities are important, it may be necessary to include empirical correction terms in the equation, or to use a more general stochastic prediction method such as an *ensemble prediction* (*or Monte Carlo*) method, which yields an algorithm known as the *Ensemble Kalman Filter* (*EnKF*).

6.4.4 Discussion on the KF Algorithm

The input to the algorithms is: the definition of the model and the observation operator, the initial condition for (x, P) when the recurrence of the filter is started, the sequence of observations y, and the sequence of model and observation error covariance matrices (Q, R). The output is the sequence of estimates (x_a, P_a) of the model state and its error covariance matrix. The organization of the KF assimilation looks like a coupled stream of estimations of model states and error covariances (Fig. 7). The variational form of the least-squares analysis can be used in the analysis step of the Kalman filter, instead of the explicit equations written above.

The numerical cost of the KF or EKF is that of the analysis itself, plus the estimation of the analysis error covariances, discussed in a specific section, plus the covariance forecast equation (15) which requires n forecasts of the tangent linear model (n being the dimension of the model state) to build the operator M. The storage cost itself is significant, since each P matrix is $n \times n$ (only a half can be stored since they are symmetric) and in (Eq. 18) the KH matrix must be evaluated and stored too (unless the variational form is used, in which case n evaluations of the gradient of the cost function must be performed to build the Hessian which must then be inverted). It means that the cost of the KF is much larger than 4DVAR, even with small models. The algorithm should rather be regarded as a reference in the design of more approximate assimilation algorithms which are being developed now-a-days. It is still not clear what is the best way to approximate the KF, and the answer will probably be application-dependent.

There are many similarities between 4DVAR and the EKF and it is important to understand the fundamental differences between them (Ghil, 1989; Lacarra and Talagrand, 1988; Errico et al., 1993):

- 4DVAR can be run for assimilation in a realistic NWP framework because it is computationally much cheaper than the KF or EKF.
- 4DVAR is more optimal than the (linear or extended) KF inside the time interval for optimization because it uses all the observations at once, i.e. it is not sequential, it is a smoother.

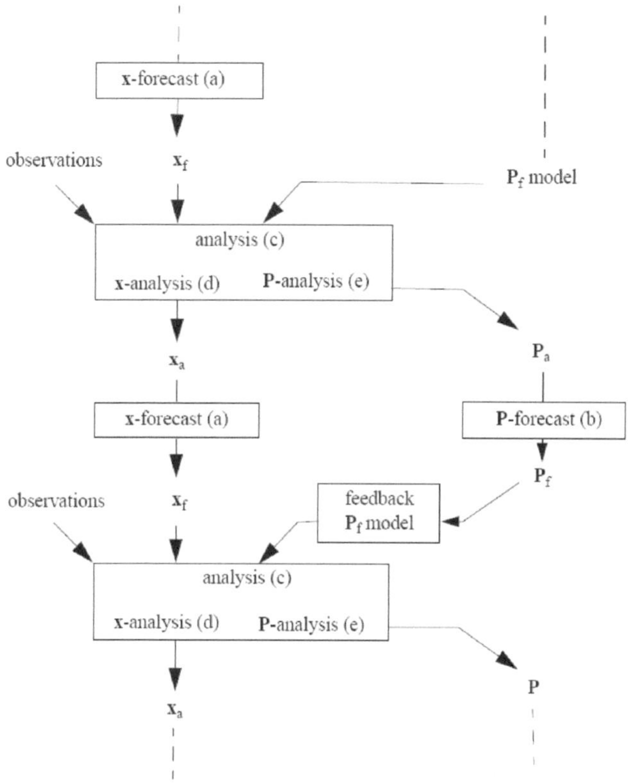

Fig. 7. Schematic for computation of KF or EKF assimilation.

- Unlike the EKF, 4DVAR relies on the hypothesis that the model is perfect (i.e. $Q = 0$).
- 4DVAR can only be run for a finite time interval, especially if the dynamical model is non-linear, whereas the EKF can in principle be run forever.
- 4DVAR itself does not provide an estimate of P_f; a specific procedure to estimate the quality of the analysis must be applied, which costs as much as running the equivalent EKF.

6.5 Ensemble KF (EnKF)

The ensemble Kalman filter (EnKF) is an alternative data assimilation method first proposed for geophysical applications by Evensen (1994). It's based on the original linear, recursive Kalman filter that produces the unbiased minimum variance estimate, in a least-square sense, under the assumption of unbiased noise processes. Unlike 3DVAR, it uses ensemble forecasts to estimate flow-dependent background error covariance. As in the standard Kalman filter, the update equation can be formulated as

$$x_a = x_f + K(y - Hx_f)$$

where x_f represents the prior estimate or first guess, x_a is the posterior estimate or analysis, y is the observation vector, H is the observation operator that returns observed variables given the state, and K is the Kalman gain matrix defined as

$$K = P_f H^T (HP_f H^T + R)^{-1}$$

where P_f and R represent the background and observational error covariance, respectively. In EnKF, the flow-dependent P_f is estimated through an ensemble of short-range forecasts. The analysis looks like

$$x_a = x_f + P_f H^T (HP_f H^T + R)^{-1} (y - Hx_f)$$

Observations are taken sequentially with the assumption of uncorrelated observation errors. EnKF is equivalent to variational methods for linear systems with Gaussian error distributions and an infinite ensemble size. However, it offers significant advantages over variational methods. It uses short-term ensemble forecasts to estimate more realistic, flow-dependent forecast (background) error covariance that doesn't rely on specified physical balances. Such balances might not hold true for smaller scale systems with active moist convection. EnKF provides not only the best state estimation, it can also seamlessly couple the associated flow-dependent uncertainty with ensemble forecasting. Nevertheless, operational NWP centres around the globe continue to predominantly use variational data assimilation techniques, although many have begun to incorporate flow dependencies in their background error estimates. Figure 8 shows schematic of EnKF data assimilation.

Ensemble forecast provides better, flow-dependent estimate of background error covariance: $P_f = N_e^{-1} \Sigma\, (x_i^f - \underline{x})(x_i^f - \underline{x})^T$

If observations are independent these can be assimilated sequentially, the inversion $(HP_f H^T + R)^{-1}$ is a scalar inversion, which has low computational overhead.

The ensemble Kalman filter (EnKF) is an approximate KF that obtains its covariance information from a set of parallel forecasts that constitute an ensemble. This has several advantages:

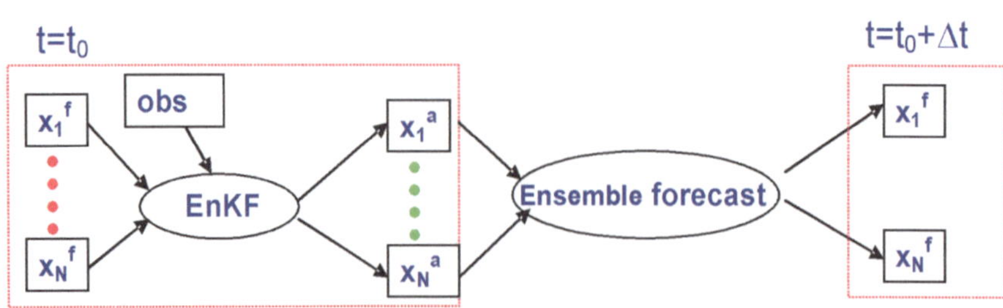

Fig. 8. Schematic of EnKF data assimilation.

- The forecast model itself can be fully nonlinear, avoiding the need to construct approximate *linearized* models.
- The ensemble mean itself is the minimum-variance best estimate according to Bayesian perspective.
- The covariance matrix is *sampled* from the available ensemble members, circumventing the necessity of expensive matrix operations.
- The posterior uncertainty is consistent with the mean analysis and can be directly used for ensemble forecasting.
- It has similarities with important sampling and particle filtering but uses much fewer samples for computational feasibility.

The EnKF has various applications, details as given below:

- Geophysical systems in general
  - provide continuous gridded analysis from heterogeneous observations (and dynamical models)
  - infer the unobserved quantities
- Numerical weather and climate prediction
  - generate initial conditions for numerical weather prediction models
  - information about prior and posterior analysis uncertainty (from ensembles)
  - explicit treatment of model error through parameter estimation (by augmenting the state with uncertain model parameters)
- Designing of observing systems
  - estimating the importance of specific observations through OSE's or OSSE's
  - adaptive observations, controllable observing systems
- Others: probabilistic evaluation of system dynamics and source attributions.

REFERENCES

Bishop, C.M. (ed.), (1995): Neural Networks for Pattern Recognition. Oxford Univ. Press.

Bocquet, M., 2006: Introduction aux principes et méthodes del'assimilation de données en géophysique. EnSTA ParisTech and Ecole des Ponts Paris Tech lectures notes.

Bouttier, F. and P. Courtier, 1999: Data assimilation concepts and methods. Training Course notes of ECMWF.

Courtier, P., E. Andersson, W. Heckley, J. Pailleux, D. Vasiljevic, M. Hamrud, A. Hollingsworth, F. Rabier and M. Fisher, 1998: The ECMWF implementation of three-dimensional variational assimilation (3D-Var). Part 1: Formulation. *Quart. J. Roy. Meteor. Soc.*, **124**, 1783-1807.

Errico, R., T. Vukicevic and K. Raeder, 1993: Examination of the accuracy of a tangent linear model. *Tellus*, **45A**, 462-477.

Evensen, G., 1994: Sequential data assimilation with a nonlinear quasi-geostrophic model using Monte Carlo methods to forecast error statistics. *J. Geophys. Res.*, **99**, 10,143-10,162.

Ghil, M., 1989: Meteorological Data Assimilation for Oceanographers. Part I: Description and Theoretical Framework. *Dyn. of Atmos. Oceans*, **13**, 171-218.

Ide, K., P. Courtier, M. Ghil and A.C. Lorenc, 1997: Unified notation for data assimilation: Operational sequential and variational. *J. Meteor. Soc. Japan*, **75**, 180-189.

Kalman, R.E., 1960: A New Approach to Linear Filtering and Prediction Problems. Transaction of the ASME—*Journal of Basic Engineering*, 35-45.

Lacarra, J.-F. and O. Talagrand, 1988: Short-Range Evolution of Small Perturbations in a Barotropic Model. *Tellus*, **40A**, 81-95.

Lorenc, A.C., 1981: A global three-dimensional multivariate statistical interpolation scheme. *Mon. Weather Rev.*, **109**, 701-721.

Lorenc, A.C., 1986: Analysis methods for numerical weather prediction. *Q.J.R. Meteorol. Soc.*, **112**, 1177-1194.

Lorenc, A.C., S.P. Ballard, R.S. Bell, N.B. Ingleby, P.L.F. Andrews, D.M. Barker, J.R. Bray, A.M. Clayton, T. Dalby, D. Li, T.J. Payne and F.W. Saunders, 2000: The Met. Office global three-dimensional variational data assimilation scheme. *Quart. J. Roy. Meteor. Soc.*, **126**, 2991-3012.

Mathieu, P.-P. and A. O'Neill, 2008: Data assimilation: From photon counts to Earth System forecasts. *Remote Sens. Environ.*, **112**, 1258-1267.

Parrish, D.F. and J.C. Derber, 1992: The National Meteorological Center's spectral statistical interpolation analysis system. *Mon. Wea. Rev.*, **120**, 1747-1763.

Saporta, A. (ed.), 2005: Probability, Data Analysis and Statistics. Technip, Paris.

Talagrand, O., 1997: Assimilation of observations: An introduction. *J. Meteorol. Soc. Jpn.*, **75**, 191-209.

Tarantola, A. (ed.), 2005: Inverse Problem Theory and Model Parameter Estimation. SIAM.

Welch, G. and G. Bishop, 2004: An Introduction to the Kalman Filter. University of North Carolina (UNC), Chapel Hill, North Carolina. Technical Report (TR 95-041).

12

Data Assimilation and Predictability of Tropical Cyclones

Fuqing Zhang
Department of Meteorology, The Pennsylvania State University
University Park, Pennsylvania, USA
fzhang@psu.edu

1. Introduction

Hurricanes are one of deadliest and costliest natural hazards, with total losses topping $100 billion for the first time in 2005 (Pielke et al., 2008). Accurate predictions of hurricanes, therefore, have enormous economic value, and demand is increasing for more accurate forecasts with longer lead times and more precise warnings to minimize losses due to hurricane preparation and evacuation as well as to destruction. Over the past decade, significant progress has been made in short-range (up to five days) track forecasts of tropical cyclones. The current day average 48-h forecast position is as accurate as a 24-h track forecast 10 yr ago (Franklin, 2004).

Unfortunately, today's intensity predictions continue to have significant error. There is virtually no improvement in our ability to predict hurricane intensity in terms of minimum sea level pressure, maximum wind speed, or amount of precipitation (Houze et al., 2007) and we, therefore, have very limited skill in predicting tropical cyclone formation, rapid intensification, fluctuation, or decay (Elsberry et al., 2007). The discrepancy between track and intensity forecast accuracy of tropical cyclones may be due to the fact that track is largely dependent on the large-scale environment, which has become more and more predictable with rapid advances in numerical weather prediction models and better observing systems. Meanwhile, intensity is largely determined by far less predictable internal dynamics that are only modulated by the larger-scale environment (Holland, 1997; Willoughby, 1999; Emanuel, 1999). Upscale growth of moist convection, such as in the form of vertical hot towers (VHTs), may play a critical role in internal dynamics (Hendricks et al., 2004; Krishnamurti et al., 2005; Montgomery et al., 2006). Therefore, limited predictability of moist convection may also ultimately limit the predictability of tropical cyclones, as is the case for extra-tropical cyclones (Zhang et al., 2002, 2003, 2007) or continental warm season mesoscale convective systems (Zhang et al., 2006; Hawblitzel et al., 2007; Bei and Zhang, 2007).

Understanding the source of uncertainty in tropical cyclone intensity forecasts continues to be a significant concern, especially given their recent lack of improvement (e.g., Franklin, 2005; Elsberry et al., 2007). Predictions of tropical cyclone formation, rapid intensification, and decay remain particularly problematic (Houze et al., 2007), yet our knowledge on how error grows in such forecasts remains very limited (Van Sang et al., 2008). A better understanding of this problem can potentially lead to improved forecasts and is the motivation of the current study. The ultimate goals is to improve tropical cyclone track and intensity prediction through further development of the regional-scale, cloud-resolving ensemble-based data assimilation and prediction system capable of efficiently assimilating in situ and remotely sensed observations. Thus, the primary objective is to examine the dynamics and predictability, as well as observation targeting and observability, of tropical cyclones and thus to enhance the ensemble analysis and prediction system through inter-comparison and potential coupling with other limited-area operational models.

2. Why Lack of Progress in Intensity Forecast?

Tropical cyclone track is mostly determined by larger-scale environment whose forecast improves with better observations, better models, higher resolution and faster computers. Figure 1 demonstrates the evolution of the annually averaged position errors over the Atlantic basin for the official forecasts provided by the National Hurricane Centre (NHC). It is clearly noticed that the track forecast is being improved greatly and steadily over the past couple of decades with the current-day averaged 48-h forecasts error as small as the 24-h error 15 years ago. However, limited predictability is noticed with the intensity forecast. Figure 2 represents the official intensity forecast of tropical cyclones over Atlantic. There is virtually no improvement in the NHC official forecast for intensity for the past few decades. In essence, despite the advancement in new modelling and data assimilation approaches, tropical cyclone forecast, especially in terms of intensity, is still a difficult and challenging problem worldwide.

This is at least partially due to the fact that tropical cyclone intensity is strongly dependent on internal dynamics and moist convection which are smaller in scales, more chaotic, less understood and intrinsically less predictable. Thus the limitations in the current operational hurricane prediction models are the inaccurate model, insufficient resolution, no inner-core observations, and inefficient data assimilation methods.

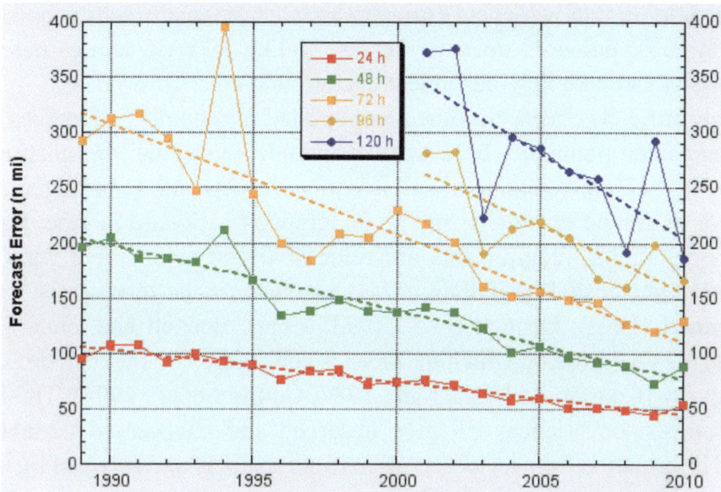

Fig. 1. Annual average track errors as provided by NHC over Atlantic basin.

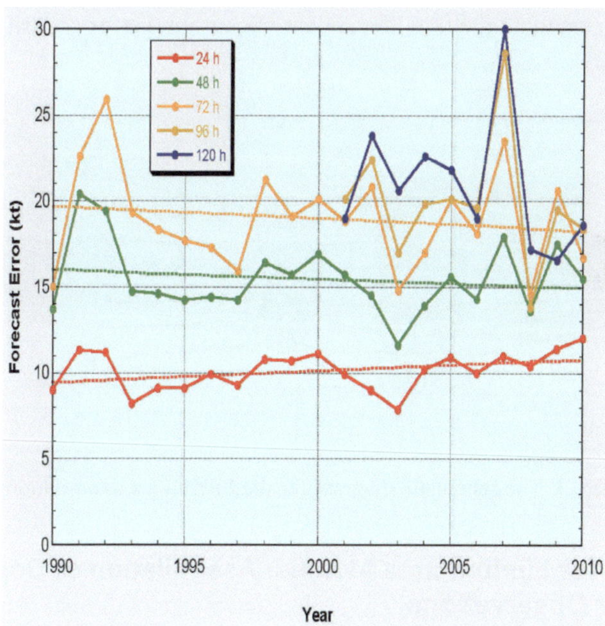

Fig. 2. The official intensity forecast as provided by NHC over Atlantic basin.

3. Ensemble Performance and Predictability

The ensemble Kalman filter (EnKF) is an alternative data assimilation method first proposed for geophysical applications by Geir Evensen (Evensen, 2003). It is based on the original linear, recursive Kalman filter that produces the unbiased minimum variance estimate, in a least-square sense, under the assumption of unbiased noise processes. Unlike 3DVAR, it uses ensemble forecasts to estimate flow-dependent background error covariance.

The EnKF has been examined and applied in a number of studies since and has gained popularity because of its simple conceptual formulation and relative ease of implementation, e.g., it requires no derivation of a tangent linear operator or adjoint equations, and no integrations backward in time. Further, the computational requirements are affordable and comparable with other popular sophisticated assimilation methods such as the representer method by Bennett (1992), Bennett et al. (1993, 1996), Bennett and Chua (1994) and the 4DVAR method which has been much studied by the meteorological community (Courtier and Talagrand, 1987; Courtier et al., 1994). The EnKF uses forecast covariances between observed and unobserved variables to spread information from the observations both spatially and between variables. These covariances are estimated from an ensemble of prior forecasts initialized when observations were last available. Figure 3 demonstrates the flow-chart for the cycling data assimilation system of the EnKF that uses the short-term ensemble to estimate the flow-dependent background error covariance.

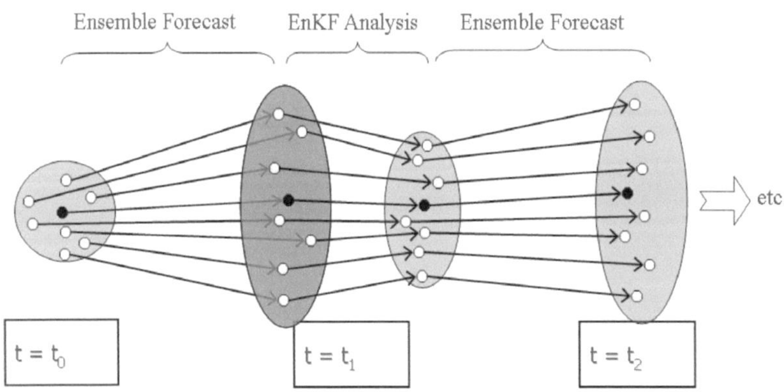

Fig. 3. The schematic diagram for the EnKF data assimilation.

3.1 EnKF for Limited-area Models: Assimilation of Doppler Radar Observations

The EnKF uses an ensemble of short-range forecasts to estimate the flow-dependent background error covariances (Fig. 3) required in data assimilation. The feasibility of the EnKF for convective-scale data assimilation has been demonstrated in perfect-model experiments (Snyder and Zhang, 2003; Zhang et al., 2004; Zhang et al., 2006) for mesoscale and regional scale data assimilation experiments. The performance of EnKF for limited area models using the Doppler radar observations of supercells is discussed in this section.

Radial-velocity and reflectivity observations from various radars are our primary tools in probing the structure and dynamics of convective-scale phenomena. How best to estimate the atmospheric state at this scale given limited radar observations has been the pursuit of many researchers since meteorological radars were first deployed. Such estimation from single-Doppler observations relied primarily on various retrieval algorithms (e.g., Qiu and Xu, 1992; Laroche and Zawadzki, 1994; Shapiro et al., 1995; Xu et al., 2001; Weygandt et al., 2002) that typically combine the radial velocity observations with the continuity equation to first estimate the wind fields and then derive (or retrieve) the thermodynamic variables through other physical constraints and assumptions. Three-dimensional and four-dimensional variational data assimilation methods, namely 3DVAR and 4DVAR, have also shown significant potential in convective-scale data assimilation using both simulated and real observations (Sun and Crook, 1997, 1998; Gao et al., 1999).

Snyder and Zhang (2003) explored for the first time the use of an ensemble Kalman filter (EnKF) to assimilate single-Doppler radar observations in a cloud-scale model. The Kalman filter is designed to use all available information in order to produce the most accurate possible description of the state of the flow. It also provides the uncertainty in the state of the flow resulting from the uncertainties in various sources of information. The environmental sounding used in Snyder and Zhang (2003) is presented in Fig. 4. It is based on the Oklahoma City, Oklahoma, sounding from 0000 UTC

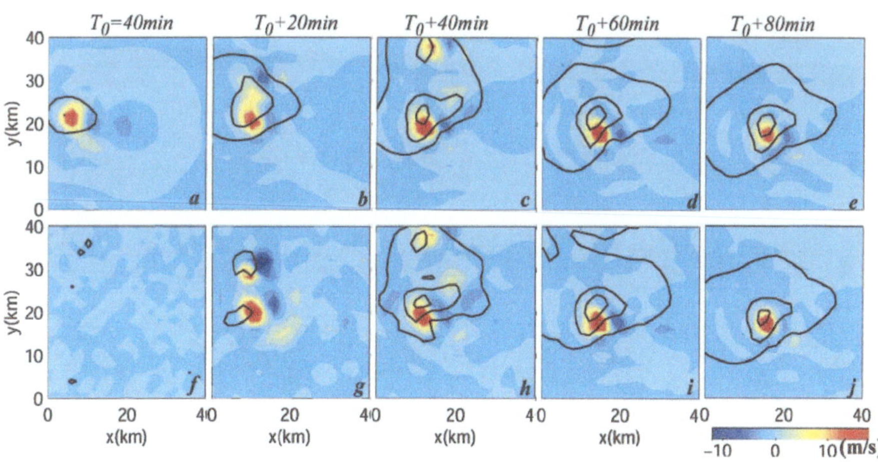

Fig. 4. The 5-km-AGL vertical velocity (coloured) and surface perturbation liquid-water potential temperature (black lines, every 2 K) for (a)–(e) the reference simulation and (f)–(j) the EnKF analysis in the CNTL experiment. Shades of red and blue indicate upward and downward motion, respectively, with gradations of colour every 2 m s^{-1}. Fields are shown at t_0, t_0 + 20, t_0 + 40, t_0 + 60 and t_0 + 80 min, labelled on top as T_0 = 40 min, etc. (t_0 = 40 min).

25 July 1997, but 7 m s^{-1} is subtracted from the zonal wind at all levels in order to minimize the movement of the right-moving supercell through the domain.

In the control EnKF experiment described below, the assimilation cycle starts at time t_0 = 40 min after the initialization of the warm bubble. Figures 4a–e show the vertical velocity (w) at 5 km above ground level (AGL) at t_0 and every 20 min after t_0 in the reference simulation. The corresponding 5-km-AGL rainwater mixing ratio (q_r) is shaded in Figs 5a–e. Also shown in these figures are the 5-km-AGL relative vorticity (contours in Fig. 5) and the surface liquid-water potential temperature (contours in Fig. 4), which are the primary indicators of the mid-level vortex and low-level cold pool structure of the supercell, respectively.

These quantities from the reference simulations are used to evaluate the EnKF assimilation results showing in the bottom panels. The warm bubble initiates a convective cell, which strengthens over the first 20 min and begins to rain at 30 min. As is typical for such a sounding (not shown), the initial cell forms a downdraft and then splits into a strong primary cell that moves to the right of the environmental shear and a weaker, secondary cell moving to the left of the shear. Here, the initial storm splits around 60 min after the initialization of the warm bubble. As can be seen from the bottom panels of Figs 4 and 5, even initialized with pure random noise in the initial ensemble, the EnKF control analysis assimilating the Doppler radial velocity only where there is significant reflectivity (dBZ > 5) every five minutes will quick drive the analysis towards the reference truth simulation, in as little as 20 minutes for vertical motions and voracity, slightly later for rainwater and the surface cold pool (Snyder and Zhang, 2003). Further sensitivity experiments in Zhang

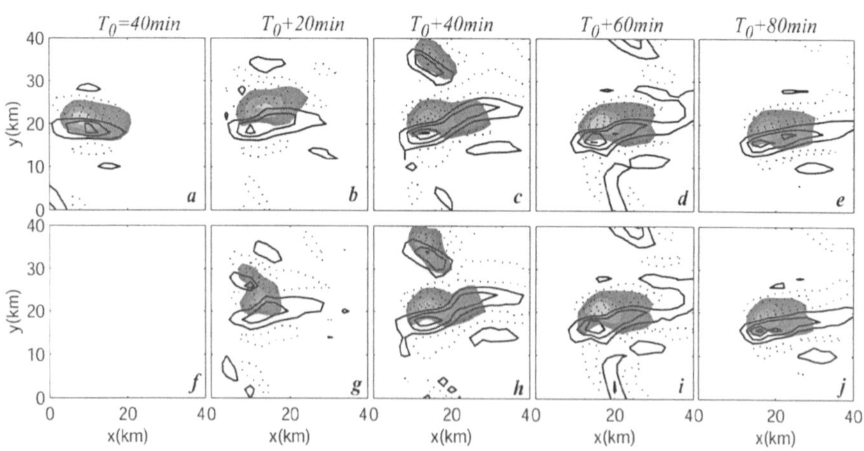

Fig. 5. The 5-km-AGL rainwater mixing ratio (shaded, every 2 g kg^{-1}) and relative vorticity (every 0.001 s^{-1}; solid lines are positive and dotted lines are negative) in (a)–(e) the reference simulation and (f)–(j) the EnKF analysis for the CNTL experiment. Fields are shown at t_0, $t_0 + 20$, $t_0 + 40$, $t_0 + 60$ and $t_0 + 80$ min, labelled on top as T_0 = 40 min, etc. (t_0 = 40 min).

et al. (2004) demonstrated that the EnKF analysis is robust under different realistic data coverage, frequency, observation error, as well as the quality of the initial guess.

3.2 Hurricane Initialization and Prediction Using Doppler Radar Observation with EnKF

3.2.1 Assimilation of W88D Doppler Radar Radial Velocity for Humberto '07 (Zhang et al., 2009)

The assimilation of Doppler radar radial velocity observations directly through the EnKF is demonstrated for cloud-resolving hurricane analysis and prediction, both deterministically and probabilistically. Hurricane Humberto (2007), the first landfalling hurricane in the United States since the end of the 2005 busy hurricane season and the most rapidly intensifying near-landfall storm in U.S. history is examined. Humberto strengthened from a 40 mile per hour (mph) depression at 1200 UTC 12 September 2007 to a 92-mph hurricane at 0700 UTC 13 September, a 52 mph increase in surface wind speed in only 19 h. The storm caused extensive damage along the southeast Texas coast and was poorly predicted by operational models and forecasters. The real-time forecast by the operational Global Forecast System (GFS) running at the National Centres for Environmental Prediction (NCEP) failed to capture the intensification and genesis of the storm (Fig. 6). The Weather Research and Forecasting (WRF) model also failed in post-event, 4.5-km, cloud-resolving simulations that were initialized with the GFS analyses (as in the control ensemble analysis and forecast detailed in subsequent sections) with lead times every 6 h from 6 to 48 h.

At different stages of formation and intensification, Humberto was within range of coastal Weather Surveillance Radars-1988 Dopplers (WSR-88Ds) at

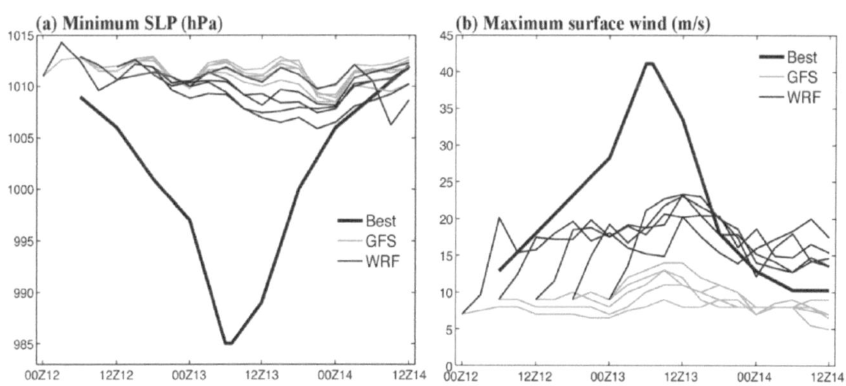

Fig. 6. Time evolution of (a) minSLP and (b) maxWSP forecasts by operational GFS forecasts starting every 6 h from 0000 UTC 12 Sep to 0000 UTC 12 Sep 2007 and by 4.5-km WRF forecasts starting from the operational GFS analyses in comparison to the NHC best-track estimate.

Corpus Christi (KCRP) and Houston–Galveston (KHGX) in Texas and Lake Charles (KLCH) in Louisiana. These radars provided valuable convective-scale observations of the storm but were not assimilated by real-time NCEP operational models. With large volumes of radar observations that are recorded at a much higher resolution than the forecast model grid spacing for the EnKF data assimilation, significant data thinning and quality control of observations become necessary. The process of combining multiple observations into one high-accuracy "super" observation (SO) is often referred to as "superobbing." An SO for radar radial velocity is developed in Zhang et al. (2009) following some of the related approaches in earlier studies (Lindskog et al., 2000, 2004; Alpert and Kumar, 2007).

Figure 7 shows the time evolution of the minimum sea level pressure (minSLP) and maximum surface wind speed (maxWSP) estimated from the posterior EnKF mean analysis field (gray; hereafter "the EnKF analysis") as well as the average (red) of the maximum– minimum values estimated from each ensemble member's posterior (green) in comparison to the NHC best-track analyses (black). Minimum SLP in the EnKF analysis (gray) agrees well with the average of the members' minimum values (red) due to collocation of the centres of most members. The analysis mean and some ensemble members also compare favourably with the NHC best-track analysis. However, due to the strong spatial and temporal variability of the maximum surface wind speed, maxWSP in the EnKF analysis (gray) is significantly smaller than that of the average (red) of the members' maxima. Yet, the average of each member's maxWSP (red) matches well with the NHC best-track estimate (black) but larger errors are seen after landfall. Hence, it is more appropriate to use the

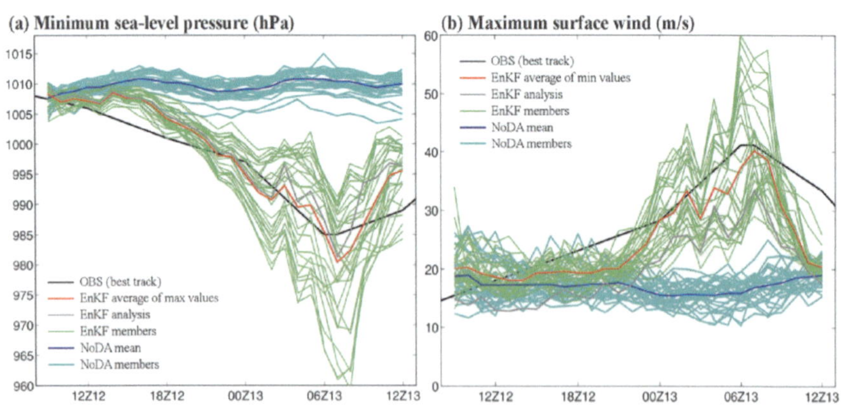

Fig. 7. Time evolutions of (a) minSLP and (b) maxWSP estimated from the EnKF analysis and NoDA ensemble forecast. Thin green (cyan) lines represent the maxWSP and minSLP estimated from each ensemble member of the EnKF analysis (NoDA forecast) with thick red (blue) lines representing the average of maxWSP and minSLP in each member. The gray line indicates the maxWSP–minSLP estimated from the EnKF analysis mean while the black curve is the NHC best-track estimate.

averages of maxWSP and minSLP from each member for verifying ensemble forecasts in terms of extreme values.

There is large ensemble spread of both minSLP and maxWSP among the analysis members (Fig. 7). Standard deviation of minSLP (maxWSP) increases from 1 to 2 hPa (1–3 m s^{-1}) during the first few assimilation cycles to 10.2 hPa (8.3 m s^{-1}) near the peak intensity time at 0800 UTC 13 September but drops quickly to 5.4 hPa (4.0 m s^{-1}) at 0900 UTC 13 September after the storm in most members makes its landfall (not shown). Likewise, the minSLP (maxWSP) at the peak intensity time varies from 992 (27 m s^{-1}) to 960 hPa (59 m s^{-1}). In terms of the corresponding intensity categories, this represents a range from a strong tropical storm to a major hurricane. Large disparities between ensemble members demonstrate significant EnKF analysis uncertainties during and after the rapid intensification of Humberto (Zhang et al., 2009).

Figure 8 shows the simulated cyclone position, minSLP, and maxWSP in deterministic WRF forecasts initialized with EnKF analyses at different times. Despite a slight delay in peak intensity compared to the best-track estimates, all forecasts initialized with EnKF analyses simulate significant tropical cyclone formation and intensification if WSR-88D radial velocity observations are assimilated for 9 h or longer.

On average, continuous assimilation through time and assimilating more observations produces both better analyses of the initial storm and better deterministic track and intensity forecasts, which is especially evident for forecasts initialized before 1800 UTC 12 September. These WRF forecasts from the EnKF analyses are in strong contrast to the nearly complete forecast failure of the WRF forecasts (with the same model configuration) cold started from the GFS analyses at all lead times (Fig. 6). This signifies the importance and potential of assimilating radar observations in improving cloud-resolving tropical cyclone initialization and prediction at all lead times (Zhang et al., 2009).

3.2.2 Assimilation Airborne Doppler Winds with WRF-EnKF (Weng and Zhang, 2012)

Through a Weather Research and Forecasting model (WRF)-based ensemble Kalman filter (EnKF) data assimilation system, the impact of assimilating airborne radar observations for the convection-permitting analysis and prediction of Hurricane Katrina (2005) is examined in this section.

In this study, the impact of assimilating airborne Doppler radar radial velocity observations for convection-permitting analysis and prediction of hurricanes will be explored in detail. The selected case study is for Hurricane Katrina (2005) (Knabb et al., 2006), which developed from a tropical depression formed near Turks and Caicos around 1800 UTC 22 August 2005. It reached tropical storm strength (>17.5 m s^{-1} maximum sustained winds) around 1200 UTC 24 August and attained hurricane strength (>35 m s^{-1} winds)

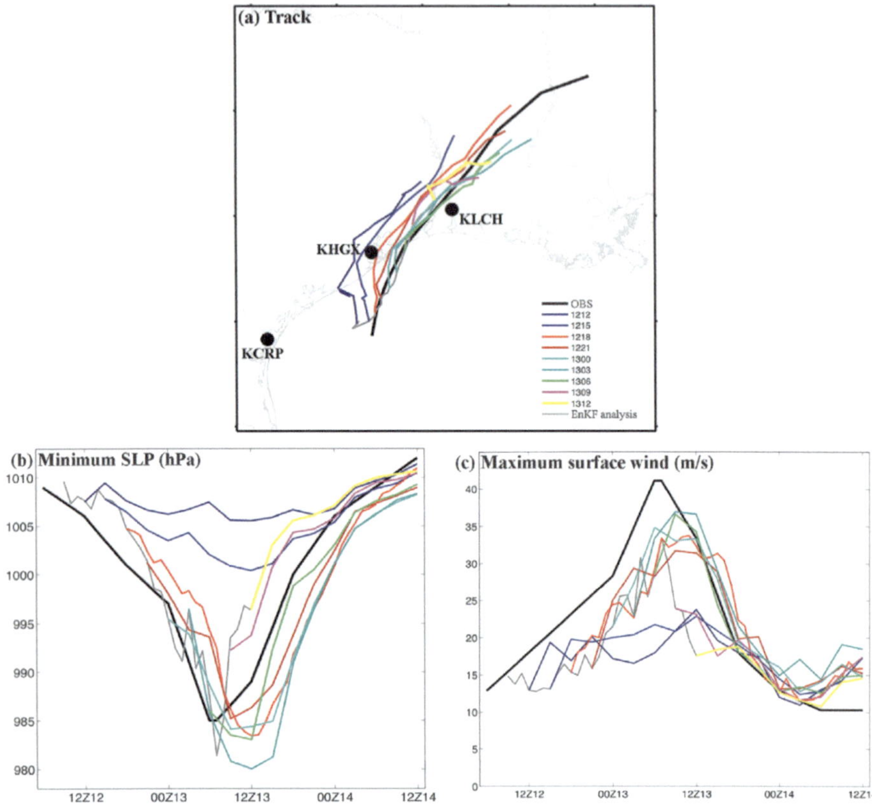

Fig. 8. The simulated (a) positions, (b) minSLP, and (c) maxWSP of Humberto in the deterministic WRF forecasts (colour curves) initialized with the EnKF analyses every 3 h from 1200 UTC 12 Sep to 1200 UTC 13 Sep 2007 in comparison with the NHC best-track estimate (black).

just before making its first landfall on the southeastern coast of Florida during the afternoon of 25 August. The storm weakened slightly to a tropical storm immediately following landfall, but it quickly regained hurricane intensity after crossing into the southeastern Gulf of Mexico. Katrina intensified to a category-3 hurricane on the Saffir–Simpson scale (>50 m s^{-1} winds) by 1200 UTC 27 August, and intensified further to category 5 (>72.5 m s^{-1} winds) by 1200 UTC 28 August. The storm reached its peak intensity on the afternoon of 28 August with maximum sustained winds of 75 m s^{-1} and a central pressure of 902 hPa, which was the fourth lowest pressure ever recorded by 2010 for an Atlantic tropical cyclone. While inner-core structural changes weakened the hurricane to a strong category-3 storm with about 920-hPa minimum central pressure as it neared shore, winds of this intensity were still powerful enough (about 57 m s^{-1}) to cause significant structural damage (Knabb et al. 2006). Katrina's extraordinary size contributed to devastating storm surge of exceeding 10-m peaks in several locations as Katrina made landfall in southeast Louisiana and finally along the Mississippi coast (Fritz et al., 2007). In addition, 1464 lives from Louisiana were lost according to the report of Louisiana Department of Health and Hospitals (http://www.dhh.louisiana.gov/offices/page.asp?ID=192&Detail=5248), and $81 billion in property damage was estimated (Pielke et al., 2008).

The NCEP GFS operational analysis at 0000 UTC 25 August and its forecast were used to generate the initial and boundary conditions respectively. A 60-member ensemble initiated by adding perturbations derived from an application of the WRF 3DVAR default background error statistics to the GFS analysis was integrated to 1430 UTC 25 August. Half of the SOs of each leg were chosen randomly and assimilated by the EnKF at 1430, 1530, 1630, 1700, 1900 and 2000 UTC over all domains with the successive covariance localization (SCL) method. No other observation was assimilated in this study whose primary objective is to examine the impact of assimilating airborne radar observations. The ensemble mean before assimilation will be called "prior," while the mean after assimilation will be called "posterior." A deterministic WRF forecast initialized with the GFS analysis at 0000 UTC 25 August without assimilating observations will be called "NoDA." The forecasts are initialized at 2000 UTC 25 August with fixed model domains, but the two inner domains become movable and follow the vortex centre starting at 0000 UTC 26 August. The NCEP GFS operational forecast initialized at 0000 UTC 25 August is used as the boundary conditions. A deterministic forecast (denoted as "EnKF_DF") is initialized from the EnKF mean (EnKF updates the ensemble mean first, then updates the perturbations to each member), while an ensemble of forecasts (denoted as "EnKF_EF") is initialized with the EnKF analysis perturbations. The NHC official forecast (OFCL) and model forecast (GFDL) available at 0000 UTC 26 is used as baselines in this study. The WRF forecast initialized with the GFS operational analysis at 0000 UTC 25 (NoDA) is also used to examine the impacts of hurricane forecasts with airborne radar data assimilation.

To make a comparison between OFCL and GFDL operational forecasts, we first evaluate the deterministic hurricane track and intensity forecasts of EnKF_DF. As shown in Fig. 9a, hurricane tracks from OFCL and GFDL operational forecasts are located to the right of the NHC best track. Hurricanes from these forecasts made landfall at locations close to Panama City, Florida, while EnKF_DF closely followed the NHC best track, despite a 6-h delay. Despite an initial deviation to the left, the landfall position error without considering the lag from EnKF_DF was only 30 km. The forecasts without EnKF assimilation of observations, NoDA, are in the middle of the operational forecasts and the EnKF deterministic forecast. From this result, it is encouraging that the track forecast for EnKF_DF, especially the landfall location, shows significant improvements over the control WRF forecast without EnKF assimilation (NoDA).

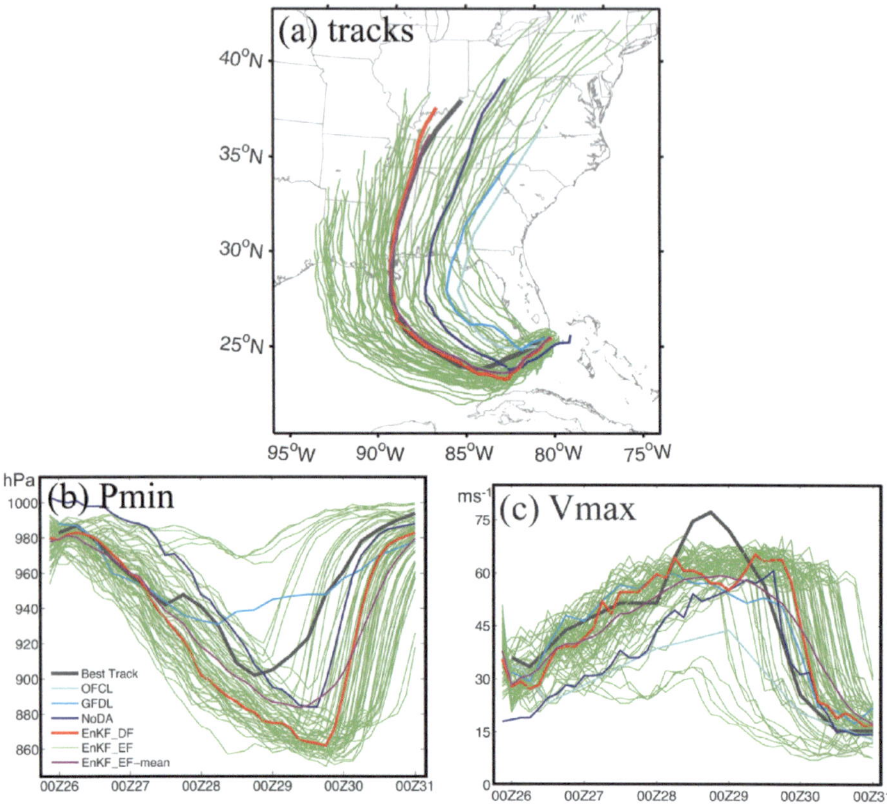

Fig. 9. (a) Hurricane track, (b) minimum sea level pressure, and (c) maximum 10-m wind speed forecasts for the best track (black), OFCL (cyan), GFDL (sky blue), NoDA (blue), EnKF_DF (red), EnKF_EF (green), and EnKF_EF-mean (magenta).

For intensity forecasts in terms of minimal sea level pressure ("Pmin," Fig. 3b) and 10-m maximum wind speed ("Vmax," Fig. 3c), both OFCL and GFDL are weaker than the NHC best track. At 2000 UTC 25 August (initial time of EnKF forecasts) the Pmin and Vmax for EnKF_DF were much closer to the NHC best track than NoDA. After 1800 UTC 27 August, Pmin for EnKF_DF was lower than that of NoDA and the NHC best track, but Vmax of EnKF_DF intensified quickly and reached its peak intensity at 1800 UTC 28 August (Weng and Zhang, 2012).

Shortly later, the observed Katrina became weaker while the WRF simulations with and without data assimilation continued to develop before making landfall. The simulated storm movement trails the observations, a result similar to previous studies using a mesoscale (Davis et al., 2008) or global (Shen et al., 2006) model.

Another advantage of the EnKF over variational data assimilation methods is that it naturally provides an ensemble of initial conditions. Ensemble track forecasts from the EnKF perturbations are also shown in Fig. 9 along with the deterministic forecasts. The large ensemble forecast spread indicates the large amount of uncertainty associated with the prediction of Hurricane Katrina at this initialization time. Ensemble means for track and intensity (EnKF_EF-mean) show the same improvement as the deterministic forecast EnKF_DF. It is worth noting that despite overall improved forecast by assimilating radar observations, the track forecasts of both EnKF_DF and ensemble mean have a substantial southward error compared to the NHC best track. This southward bias may be caused by the initial adjustment in the model after all observations are assimilated, the impacts of sampling/model error, and the inhomogeneity due to interactions with land at the earlier forecast hours.

3.2.3 Verification with independent in-flight measurement (Weng and Zhang, 2012)

In this subsection we will compare the inner-core structures between the WRF forecast without data assimilation (NoDA), the short-term ensemble mean forecast before the EnKF analysis (named as prior) and after the EnKF analysis (named as posterior), and then verify each experiment with independent observations from the aircraft in-flight measurements and the (Stepped Frequency Microwave Radiometer) SFMR-retrieved winds.

Figure 10 shows the comparison between the wind speeds derived from the lowest model level (~35 m above sea level) of NoDA, prior, posterior, and the SFMR-retrieved surface wind speeds in a storm-relative coordinate. The SFMR data were retrieved from the NOAA P-3 flight mission on 25 August and organized into six stages based on the legs of the flight (taken at the same time as the airborne Doppler measurements), and the model-derived wind speeds are calculated at the same corresponding time of six legs of SOs (the SFMR retrieved surface winds were time tagged using a similar method that was used for the airborne Doppler velocity SOs). The wind speed of NoDA (gray dashed curves in Fig. 10) fails to match the SFMR-derived measurements (gray curves in Fig. 10) for all six legs, with the intensity of the hurricane vortex during this period being significantly weaker. The prior valid at 1430 UTC before assimilating the first leg of airborne radar observations (gray dash–dotted curve in Fig. 10a) is similar to that of NoDA, both of which are much weaker than that estimated by the SFMR derived measurement in the first leg. The posterior from the EnKF analysis valid at 1430 UTC, after assimilation of the first leg of radar observations, reproduces the observed wind structure along the flight track across the inner core, despite still weaker in intensity (gray dash–dotted vs. black solid curves in Fig. 10a).

The ensemble forecast mean initialized with the prior at 1530 UTC before assimilation of the second-leg radar observations preserves much of the observed inner-core structure (according to the SFMR estimates), but the errors in the hurricane intensity remain quite large (gray dash–dotted curve in Fig. 10b). However, the priors at the later assimilation times apparently benefit from the improved posteriors during the previous assimilation cycles

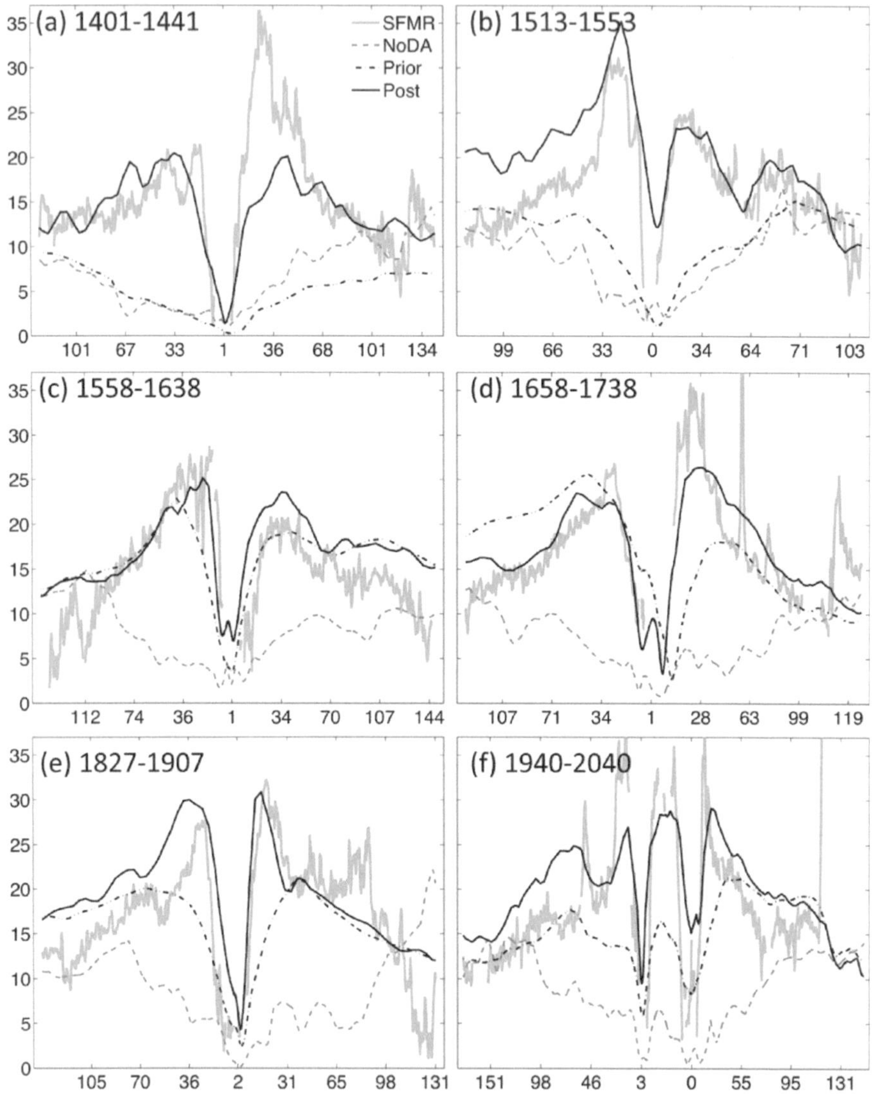

Fig. 10. Surface wind speed retrieved by SFMR (gray solid) and the lowest model level wind speed derived from NoDA (gray dashed), prior (gray dash–dotted), and posterior (solid) in storm-relative space (a)–(f) for the six legs. The *x* axis is the horizontal distance (km) between the flight and the storm centre of the NHC operational analyses; the *y* axis is the wind speed (m s^{-1}).

(gray dash–dotted and black solid curves in Figs 10 c, e, f) in terms of the vortex structure.

Verification of the flight-level temperature (Fig. 11) with in situ measurements on the airplane is also encouraging. Even though the analyses (posterior) of temperature cannot completely reproduce the sharp rise of temperature inside the eyewall (except for maybe after the last assimilation cycle), the temperature structure is improved by continuous assimilation of inner-core radar observations with the EnKF. Unlike the improvements for the wind and temperature fields, the assimilation of radar radial velocity observations appears to have a much smaller impact on the moisture field (Fig. 12). This is possibly due to weaker correlations between winds and

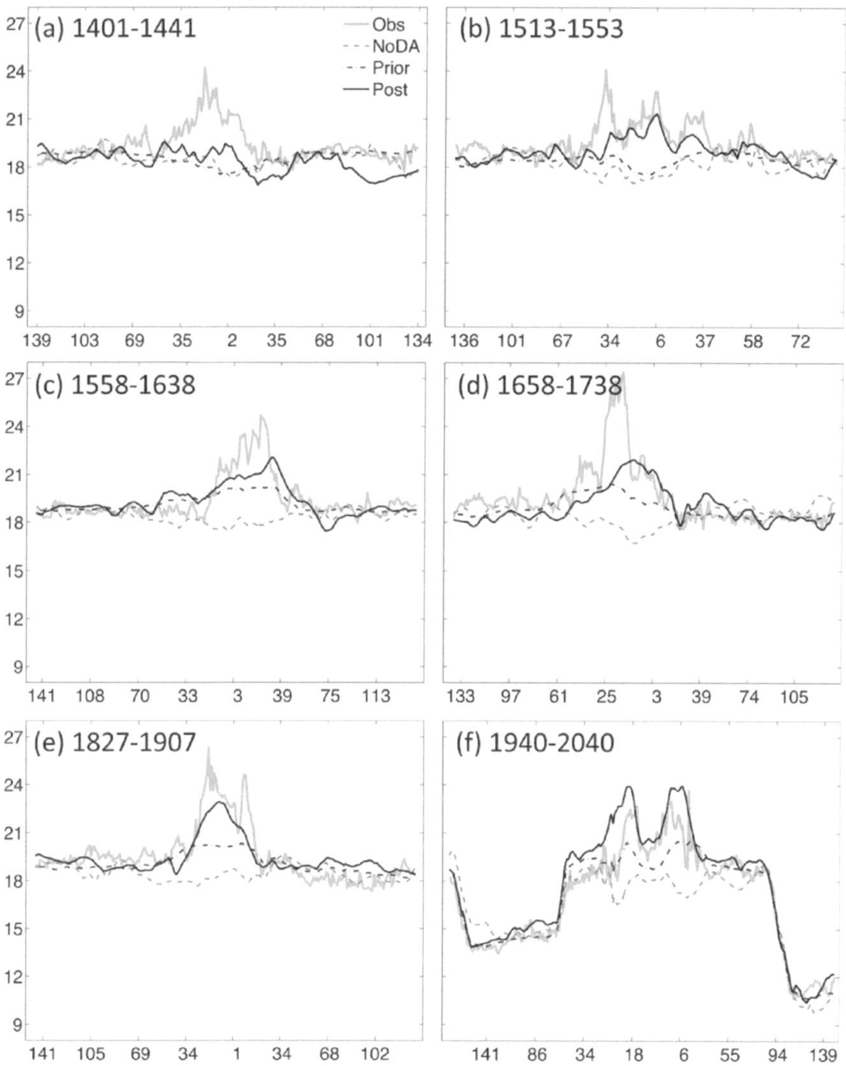

Fig. 11. Flight-level temperature retrieved by SFMR (gray solid) and the lowest model level wind speed derived from NoDA (gray dashed), prior (gray dash–dotted), and posterior (solid) in storm-relative space (a)–(f) for the six legs. The x axis is the horizontal distance (km) between the flight and the storm centre of the NHC operational analyses; the y axis is the temperature (°C).

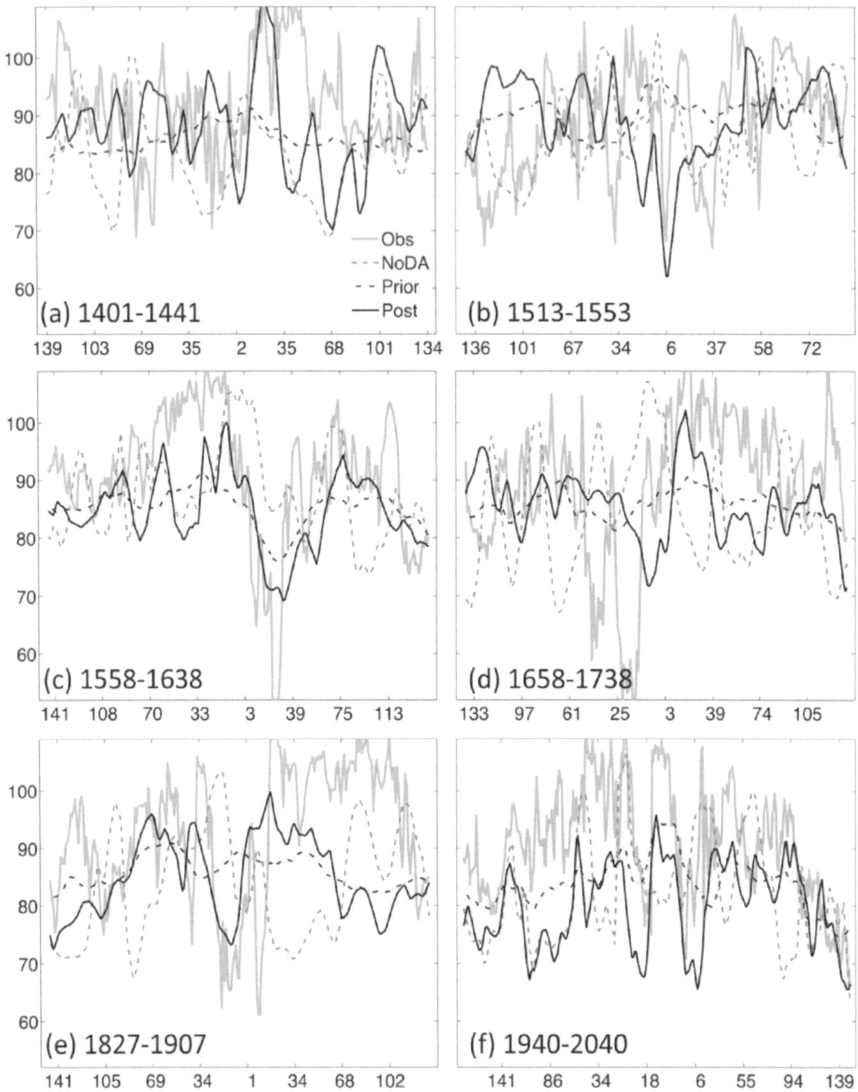

Fig. 12. Flight-level relative-humidity retrieved by SFMR (gray solid) and the lowest model level wind speed derived from NoDA (gray dashed), prior (gray dash–dotted), and posterior (solid) in storm-relative space (a)–(f) for the six legs. The *x* axis is the horizontal distance (km) between the flight and the storm centre of the NHC operational analyses; the *y* axis is the relative-humidity (%).

moisture than between winds and winds or between winds and temperature. The moisture field has relatively more spectral power for both total and error energy in smaller scales which are less predictable and more random resulting in smaller correlations with the Doppler winds (also refer to Zhang et al., 2006).

To further evaluate the performance of the EnKF analyses, the structure and magnitude of analysis increments for different variables sampled in two-dimensional (both horizontal and vertical) planes are examined. Figure 13 shows the horizontal winds at the 3-km height from the dual-Doppler radar wind analyses retrieved from the NOAA P-3 airborne radar observations (first column), compared with the EnKF prior (second column), posterior (third column), and their difference (analysis increment, last column) at the six assimilation times (from the first row to the last row). During this period, the NoDA forecast presents a very weak vortex (not shown). In the first cycle (1430 UTC 25 August 2005, the first row in Fig. 13), the winds for the prior (Fig. 13b) also present a very weak vortex, and the maximum wind speed is less than 7 m s^{-1}. After the first assimilation cycle, the EnKF posterior (Fig. 13c) shows a stronger cyclonic vortex at 3 km with a maximum of ~23 m s^{-1} in both the northwest and the southeast quadrants. Wind speeds in the centre are weaker than 7 m s^{-1} and the centre of the circulation is very close to what is shown in the dual-Doppler radar analyses (Fig. 13a). Although still weaker than the dual-Doppler analysis, the EnKF posterior wind structure and amplitude are much closer to the observations than the prior. The EnKF analysis increment (Fig. 13d) shows that the EnKF assimilation increases the wind speed around the observed vortex centre and the maximum increment is up to 15 m s^{-1}.

The cyclonic vortex of the prior (second column in Fig. 13) becomes increasingly close to the observed vortex centre and size (first column) after two assimilation cycles in both centre position and size, but still contains some biases in structure and intensity. This is shown at 1730 (fourth row in Fig. 13), 1900 (fifth row), and 2000 UTC (last row), where the areas of maximum wind speed are displaced from locations indicated by the dual-Doppler analyses. Starting from the second assimilation cycle (second row), the posterior (third column) improves the vortex intensity and the distribution of maximum wind speed. The analysis increment (last column) shows that at the first assimilation cycle, the positive increment spans most of the display domain (Fig. 13d). Starting from the third cycle, the area of positive increment becomes much smaller with large increments only around the vortex centre. Nevertheless, the EnKF assimilation continuously increases the vortex intensity and improves the depiction of the vortex structure.

Vertical cross sections of horizontal wind speeds along the flight tracks of each leg marked in the first column of Fig. 13 are presented in Fig. 14. A clear storm eye and a strong wind band around the eye are shown in the wind speed field derived by HRD from the dual-Doppler radar analysis (Fig. 14a) for all six legs. Similar to the result of the horizontal wind analysis at 3 km, the NoDA forecast does not simulate the storm structures for any of the six legs (not shown).

The prior EnKF (at 1430 UTC) also does not show the storm structure before assimilating the first-leg radar data (Fig. 14b). After the first-leg

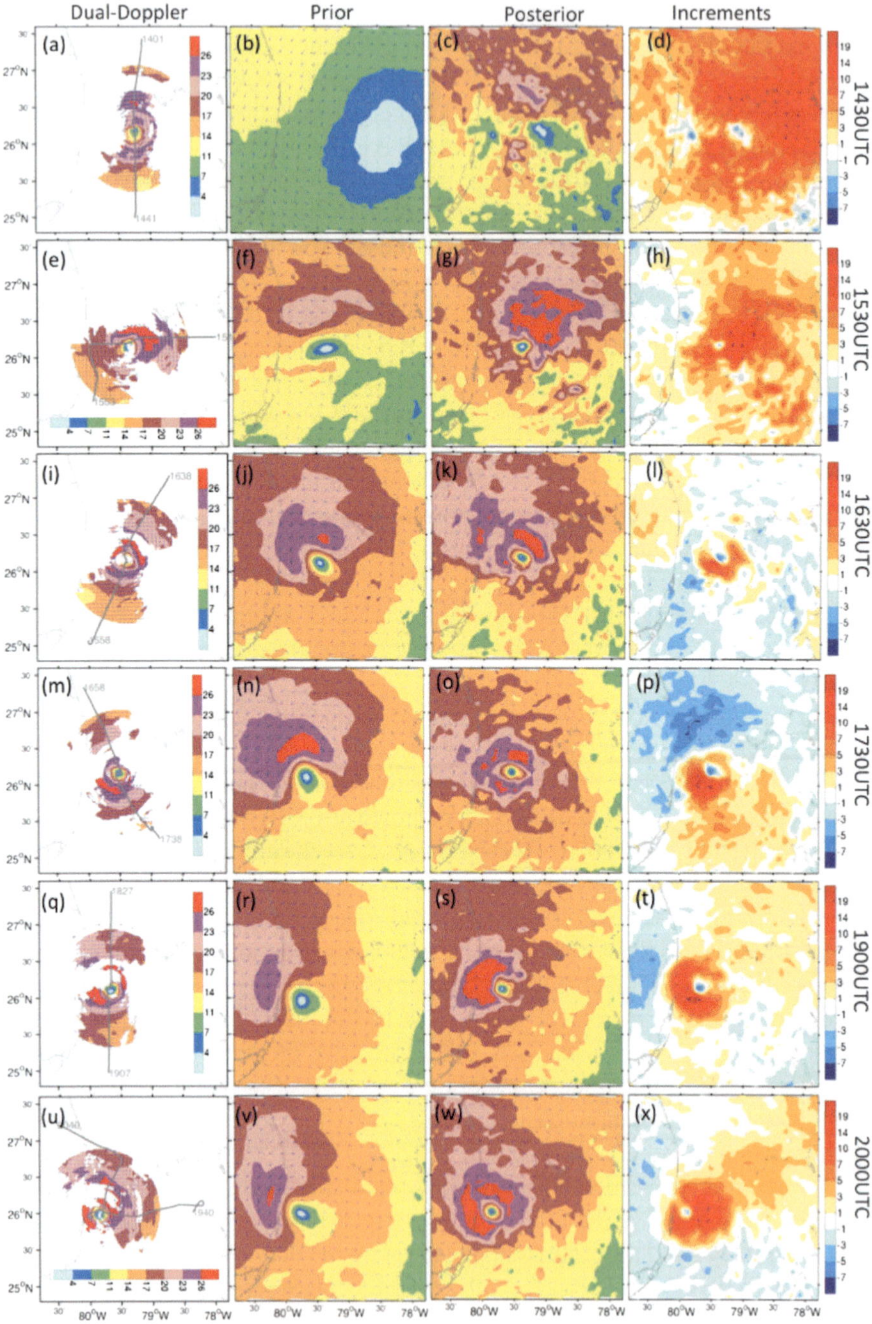

Fig. 13. Winds and wind speed at 3-km height. (from left to right) Dual-Doppler radar analyses, prior, posterior, and EnKF increments (posterior minus prior). (from top to bottom) For 1430, 1530, 1630, 1730, 1900 and 2000 UTC. The gray lines in the 1st column are tracks of the NOAA P-3 flight. The circled shaded areas in the last column indicate the increments are less than 1 m s^{-1}.

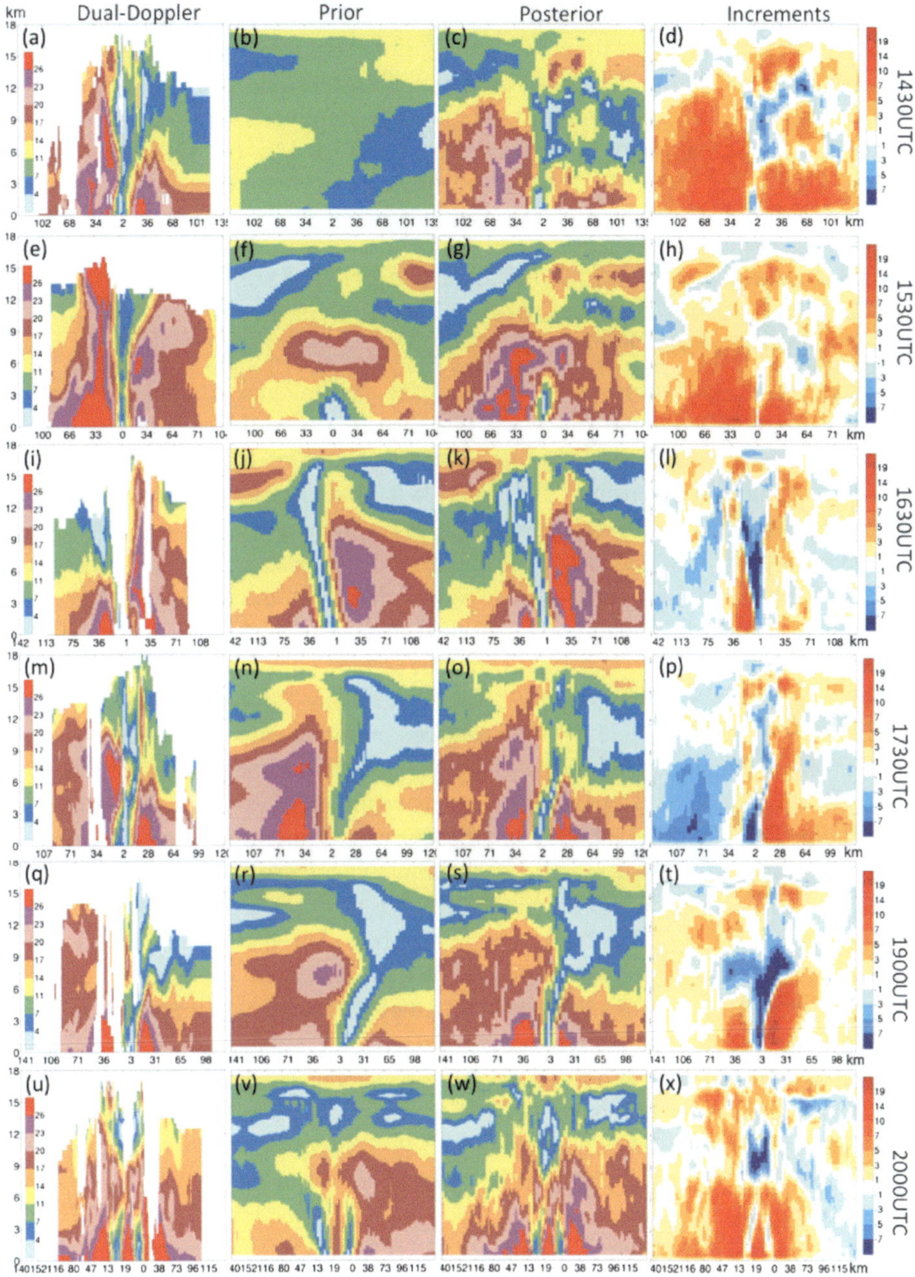

Fig. 14. Track-height wind speed sections along the leg tracks shown in the first column of Fig. 13. Other information is the same as that in Fig. 13.

airborne radar observations are assimilated, the storm centre position and the storm structure (Fig. 14c) are considerably closer to the dual-Doppler analyses (Fig. 14a); however, the storm intensity of posterior (Fig. 13c) is still weaker than that estimated by the dual-Doppler analysis (Fig. 14a). At the second assimilation time, the EnKF prior (Fig. 14f) shows a very weak vortex under 3 km in the same location as the analyses, while the posterior (Fig. 14g) strengthens the storm and builds the storm vortex up to 6 km. From the third assimilation cycle onward, the prior (Figs 14 j, n, r, v) catches the storm's location and structure well, and the EnKF posterior (Figs 14 k, o, s, w) continues to converge towards the dual-Doppler analyses (Figs 14 i, m, q, u).

3.3 Summary on Performance of the EnKF Assimilation of Inner-core Airborne Doppler Radar Observations (Zhang et al., 2011)

Zhang et al. (2011) presents the summary performance of the hurricane forecasts initialized with this WRF-EnKF system using inner-core airborne Doppler radar observations for all 61 applicable cases over the 2008-2010 Atlantic hurricane seasons. The same WRF-based ensemble analysis and prediction system as in Zhang et al. (2009) and Weng and Zhang (2012) is capable of assimilating hurricane inner-core observations using EnKF as described in previous sections. The WRF model has three two-way-nested domains "D1 to D3" with horizontal grid spacing of 40.5, 13.5, 4.5 and 1.5 km, covering a total area of 10206 km × 6561 km, 3402 km × 3402 km, 1134 km × 1134 km, and 378 km × 378 km, respectively. D1 covers a fix area as shown in Fig. 8a while D2-D3 are centered on the initial storm location and these latter domains are movable following the centre of the tropical cyclones.

Zhang et al. (2011) focusses exclusively on the added values of assimilating airborne Doppler observations from the NOAA P-3 aircrafts (Gamache et al., 1995; Marks, 2003). There were 30 ensemble members used in the real-time WRF-EnKF system while the hindcast EnKF analyses and forecasts for all the 61 cases during 2008-2010 used 60 members. The ensemble is initialized with perturbations derived from the WRF variational background error covariance (Barker et al., 2004) based on the NOAA Global Forecast System (GFS) operational analysis at 9-12 h prior to the assimilation of the first available or expected airborne Doppler observations, while the updated GFS operational forecast closest to the airborne radar observation time is used for boundary conditions for the subsequent 126-h forecasts. Model physics and generation of the initial and boundary perturbations for the ensemble are the same as in Weng and Zhang (2012). For example, if the airborne observations are to be taken between 2130 and 2330 UTC, the WRF ensemble will be initialized at 1200 UTC based on the 1200-UTC GFS analysis, and the cycled EnKF assimilation will be performed at 2200 and 2300 UTC, respectively. The EnKF analysis at 2300 UTC will then be time-shifted and regarded as initial condition of 0000 UTC of the next day for the subsequent 126-h single deterministic WRF forecast, which uses the next-day 0000-UTC GFS run as the boundary conditions.

The raw airborne Doppler velocity observations first pass through a rigorous data thinning and quality control procedure (Joint Hurricane Testbed final report 2005)[1] that was developed and implemented on the aircraft radar computer workstation (Griffins et al., 1992). Since the airborne radar observations are not used routinely in any current NOAA operational models in real-time, and since the volume of radar data is too big to be transferred from the reconnaissance aircraft to the ground in a timely manner, we developed a data thinning and additional quality control procedure implemented on the aircraft computers to allow us to summarize the massive amount of airborne Doppler radar observations into a few thousand 'super-observations' (SOs) per leg (straight flight path through the storm centre) that can be transferred to the ground computers in real-time. The detailed procedure to generate SOs for the airborne Doppler observations can be found in Weng and Zhang (2012), which is similar to that developed for the ground-based Doppler radars in Zhang et al. (2009).

There are a total of 61 applicable NOAA P-3 airborne Doppler radar missions that sampled a total of 14 tropical cyclones over the Atlantic basin that include 26 missions in 2008 for five storms (Hurricanes Dolly, Gustav, Ike, Paloma and Tropical Storm Fay), 10 missions in 2009 for three storms (Tropical Storms Ana and Danny, and Hurricane Bill) and 25 missions in 2010 for six storms (Hurricanes Alex, Earl, Karl, Richard, Tomas and Tropical Depression Two). These are primarily storms that posed a potential threat to the US.

Using the same WRF and EnKF configurations in the real-time system demonstrated for Hurricane Ike (2008) except for doubling the ensemble size to 60 members, we performed the 4.5-km WRF-EnKF analysis for each of these 61 applicable airborne Doppler missions during the 2008-2010 Atlantic hurricane seasons. The deterministic forecast in each case is integrated for 126 h initialized with the EnKF analysis (time-shifted to the closest 0000, 0600, 1200 or 1800 UTC) that assimilates the airborne Doppler radar observations in D1-D3. Figure 15(a-b) shows the overall performance of the WRF-EnKF system in terms of mean absolute forecast errors of track and maximum 10-m wind speed (verified against NHC best track estimates) with the two operational regional-scale dynamic models (HWRF and GFDL) as well as the NHC official forecasts issued at similar times (as well as at 6 h later). The homogeneous comparison was applied which means that these errors are averages over the same number of forecasts for each model or guidance at each forecast time. Following a similar procedure termed as "variable interpolator" for the "late models" used operationally at NHC[1], a simple case-dependent bias correction is used in producing Fig. 15b that modifies the maximum wind

[1] Online information is available at http://www.nhc.noaa.gov/modelsummary.shtml.

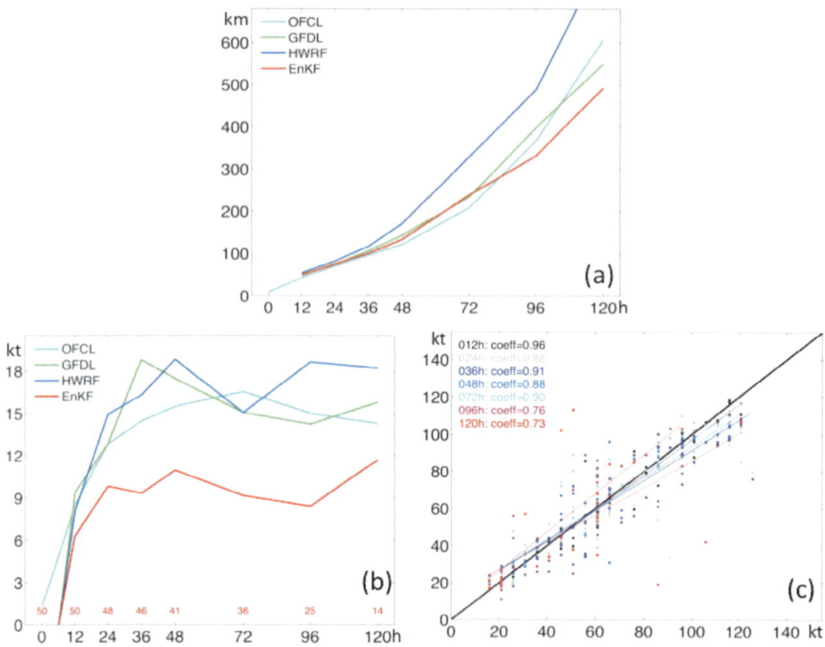

Fig. 15. Mean absolute forecast error averaged over 50 samples homogenized by all 61 airborne Doppler missions during 2008-2010 for the NHC official forecast ('OFCL', solid cyan), the 6-h later official forecast ('OFCL_+6h, dash solid), the GFDL hurricane model ('GFDL', green), HWRF ('HWRF', dark blue), and the WRF deterministic forecast in 4.5-km resolution initialized with EnKF analysis ('EnKF', red) after making the samples homogeneous for all five forecasts for (a) the track position error (km) and (b) the 10-m maximum wind speed error (knots) with simple bias correction for all forecasts except for OFCL. (c) Scatters of NHC Best-track (Y-axis) and forecasts (X-axis) Vmax (knots) at 12, 24, 36, 48, 72, 96 and 120 h lead-time for the WRF deterministic forecast in 4.5-km resolution initialized with EnKF analysis.

speed analysis and forecasts at 6-30 h lead times from 0 h by subtracting the difference between each model forecast and the NHC best track estimate from the intensity forecast scaled linearly as (36-t)/30, where t is the forecast lead time from 6 to 120 h. Likely due to the use of a coarser GFS analysis at approximately 9 to 12 h to initialize the initial ensemble for each case before assimilating the first airborne Doppler data, the intensity of the WRF forecasts initialized with the EnKF analysis have a low bias for the first 30 h (thus benefit from the bias correction).

The WRF forecasts initialized with the EnKF assimilating high-resolution airborne Doppler radar observations substantially outperform both operational forecast models (HWRF and GFDL) as well as the NHC official intensity forecast at nearly all forecast lead times (Fig. 15b), despite producing comparable or slightly smaller track forecast errors (Fig. 15a). With the simple bias correction, the WRF-EnKF produced 20-40% smaller intensity forecast errors than the NHC official forecast issued at the similar times for lead times from 24-120 h. Moreover, the correlations of the EnKF-initialized intensity forecasts with the NHC best-track estimates are high at all forecast lead times (Fig. 15c). Although not shown, we also performed the 1.5-km deterministic forecasts for all cases initialized from the 4.5-km EnKF analysis; the averaged performance of the 1.5-km WRF forecasts is comparable to the 4.5-km forecasts for both track and intensity. In the meantime, the WRF forecast initialized directly from the GFS analysis with the same model configuration but without the EnKF assimilation of airborne observations also outperformed considerably the operational models of GFDL and HWRF for this same set of cases, though the forecast with the EnKF assimilation of the inner-core observations was still more skillful than without (with 10-20% smaller intensity forecast up to 72 h, not shown). Most recently Zhang and Weng (2015) extended the study of Zhang et al. (2011) to include 100 +P3 missions with higher resolution (3 - km) EnKF analyses and forecasts that achieved similar promising performance.

3.4 Hurricane Predictability under Different Shear (Zhang and Tao, 2013)

In this section, the effect of vertical wind shear and system-scale flow asymmetry on the predictability of tropical cyclone (TC) intensity during different stages of the TC life cycle (i.e., initial formation, rapid intensification, and quasi-steady mature phase) is discussed. A series of ensemble experiments are performed with varying magnitudes of vertical wind shear, each initialized with an idealized weak TC-like vortex, with small-scale, small-amplitude random perturbations added to the initial conditions. It is found that the environmental shear can significantly affect the intrinsic predictability of tropical cyclones, especially during the formation and rapid intensification stage. The larger the vertical wind shear, the larger the uncertainty in the intensity forecast, primarily due to the difference in the timing of rapid intensification. A recent study by Zhang et al. (2011) showed that the hurricane intensity forecast accuracy can be improved by as much as 30-40% through convection-permitting ensemble simulations that assimilate airborne Doppler radar observations within the inner-core area with an advanced data assimilation method. The recent study of Zhang and Sippel (2009) showed that small initial condition errors can lead to very different intensity forecasts due to the presence of moist convection. This suggests that at least for some storms the intensity forecasts may be intrinsically limited. This effect of moist convection in limiting predictability has also been seen for the mesoscale structure of mid-latitude extratropical cyclones (Zhang et al., 2002, 2003, 2007). The limited predictability of tropical cyclones has also been examined in the recent studies of Sippel and Zhang (2008, 2010) and Nguyen et al. (2008).

The WRF-ARW model is used in Zhang and Tao (2013) with three nested domains, with domain sizes of 4320 km by 4320 km (D1), 1440 km by 1440 km (D2) and 720 km by 720 km (D3), with horizontal grid spacings of 18, 6 and 2 km, respectively with two movable (D2, D3) domains. There are four sets (no shear, 1 m s^{-1}, 3 m s^{-1} and 5 m s^{-1}) of ensemble experiments, each with 20 members. Different moisture perturbations with the magnitude randomly selected from a uniform distribution of (–0.5, 0.5) g/kg are applied to the water vapour mixing ratio throughout the innermost domain below 950 hPa. The vertical profiles of the environmental flow for all ensemble experiments are shown in Fig. 16.

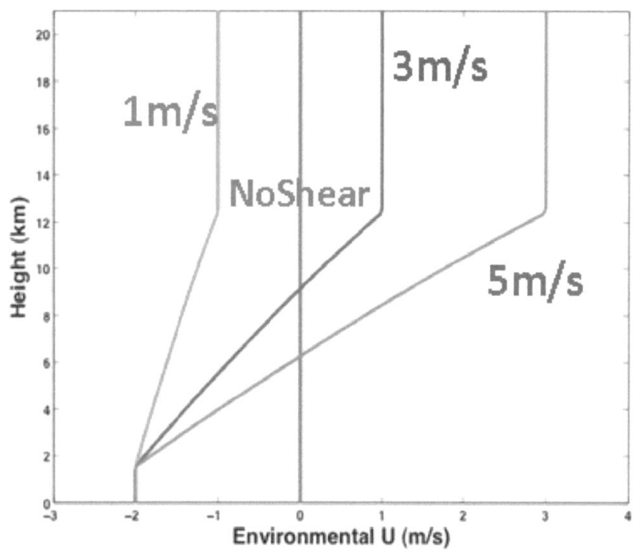

Fig. 16. Vertical profiles of environmental flow for SH0 (0 m/s vertical wind shear), SH1 (1 m/s), SH3 (3 m/s) and SH5 (5 m/s).

There is no environmental mean flow or vertical wind shear in ensemble experiment with "no shear (SH0)". The other three experiments (SH1, SH3 and SH5) have westerly vertical wind shear of 1, 3 and 5 m s^{-1}, respectively, but all have the same surface mean easterly wind of 2 m s^{-1}. Figure 17 shows the time evolution of the tropical cyclone intensity in terms of the 10-m maximum wind speed along with the mean and standard deviation (ensemble spread) from all ensemble experiments under different environmental flow conditions (with and without vertical wind shear). We can see increasingly larger intensity forecast divergence among ensemble members for experiments with increasingly stronger vertical wind shear.

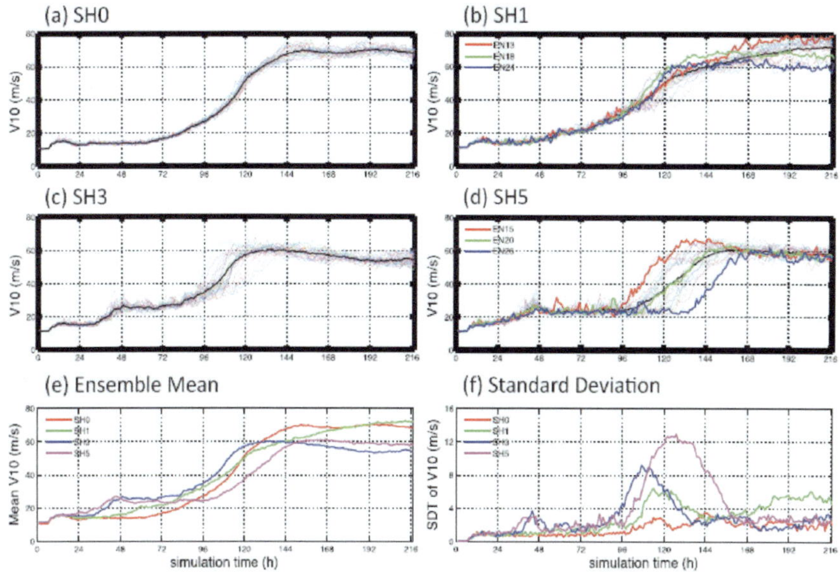

Fig. 17. Time evolution of the tropical cyclone intensity in terms of the 10-m maximum wind speed for all ensemble members of (a) SH0, (b) SH1, (c) SH3 and (d) SH5, along with (e) the ensemble mean and (f) standard deviation (ensemble spread) from all ensemble experiments. The thick black lines in the top two rows denote the ensemble mean while EN13, EN18, EN24 of SH1 in (b) and EN15, EN20, EN26 of SH5 in (d) are denoted with thick coloured lines.

To further understand how and when these three members start to differ systematically, and what results in the loss of predictability in experiment SH5, we examine the time evolution of the direction and magnitude of the local vertical wind shear and the storm centre tilt in Fig 18. The increasing of tilt and shear magnitude if occurred at the beginning, then start the decrease of tilt and shear magnitude and after tilting angle reaches 90° (N), the tilt vector swirls faster. When tilt and shear magnitude reach their minima, TC rapid intensification stage starts.

4. Conclusions

Based on the light of the results and discussions, following few conclusions can be placed as follows.

Despite large improvement in the track forecast of tropical cyclones, there is almost no improvement in the intensity forecast at all lead times. The intensity may be intrinsically less predictable than the track forecast, but also because of deficiencies in the forecast models or in the initial conditions that are used to produce the forecasts. However, the advance data assimilation approach like the Ensemble Kalman filter (EnKF) shows its potential for the improved intensity forecast.

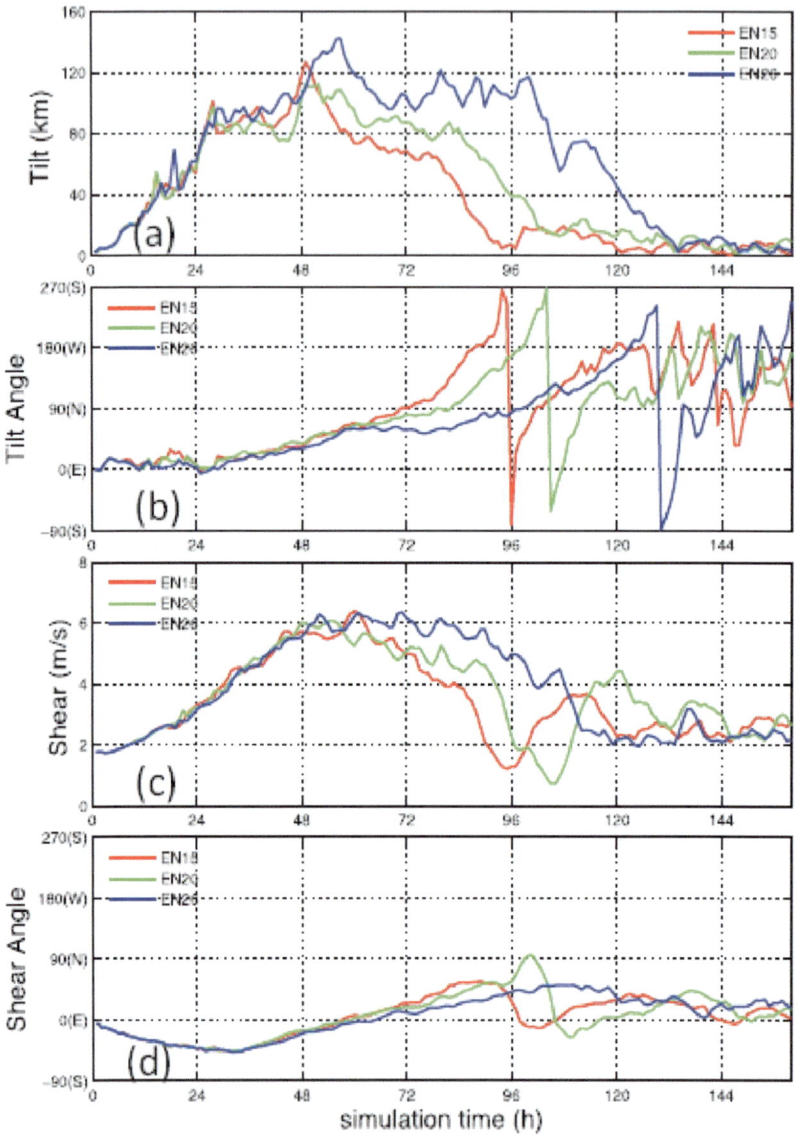

Fig. 18. Time evolution of the direction and magnitude of the storm centre tilt and the local vertical wind shear. (a) tilt magnitude; (b) tilt angle; (c) shear magnitude; and (d) shear angle for ensemble members EN15 (red), EN20 (green), and EN26 (blue) in SH5.

The uses of Doppler radar observations is demonstrated for cloud-resolving hurricane analysis, initialization, and prediction with EnKF for Hurricane Humberto (2007), the first landfalling hurricane in the United States since the end of the 2005 hurricane season and the most rapidly intensifying near-landfall storm in U.S. history. It is found that the EnKF analysis, after assimilating radial velocity observations from three WSR-88Ds along the Gulf coast, closely represents the best-track position and intensity of Humberto. Deterministic forecasts initialized from the EnKF analysis, despite having considerable variability with different lead times, are also capable of predicting the rapid formation and intensification of the hurricane. These forecasts are superior to operational forecasts, simulations without radar data assimilation, and forecasts initialized with the assimilation of the same observations with a three-dimensional variational method implemented with the same forecast model. Moreover, ensemble forecasts initialized with EnKF analysis perturbations before the rapid intensification show large spread among ensemble members (Zhang et al., 2009).

The assimilation of airborne Doppler radar observations within the inner-core area through convection permitting ensemble simulation enhances the hurricane intensity forecast accuracy for about 30-40% (Zhang et al., 2011). The assimilation of airborne Doppler radar observations for the convection-permitting analysis and prediction of Hurricane Katrina (2005) shows promising results. Verifications against independent radar, in situ, and remotely-sensed measurements show that the EnKF analyses with assimilation of airborne radar observations successfully depict the inner-core structure of the hurricane vortex at the analysis time in terms of both dynamic (wind) and thermodynamic (temperature) fields. Despite an initial trajectory error, the forecasts initialized with the EnKF analyses of airborne radar observations improved the hurricane track forecasts at lead times beyond 12 h compared to NOAA's operational forecasts and the forecast without assimilating airborne radar observations. The error of the landfall forecast was about 30 km with a 6-h delay. Besides the improved deterministic analysis and forecasts, the ensemble forecasts initiated from the EnKF analyses also indicate large forecast uncertainty at this lead-time for the hurricane track and intensity forecasts (Weng and Zhang, 2012).

It is found that the predictability of tropical cyclone intensity can be significantly influenced by the environmental vertical wind shear, especially during the formation and rapid intensification stage. The larger the vertical wind shear, the larger the uncertainty in the intensity forecast, primarily due to the difference in the timing of rapid intensification. In the presence of vertical wind shear, small-scale, small-amplitude random initial noise may lead to changes in the onset and ending of rapid intensification by as much as 1-2 days, through the randomness and chaotic nature of moist convection (Zhang and Tao, 2013).

Upscale error growth from differences in moist convection first alters the tilt amplitude and angle of the incipient tropical storms, which later leads to significant differences in the timing of precession and vortex alignment. During the precession process, both the vertical tilt of the storm and the effective (local) vertical wind shear are considerably decreased after the tilt angle reaches 90 degrees to the left of the environmental shear. The tropical cyclones rapidly intensify immediately after the tilt and the effective local shear reach their respective minima. The time at which this occurs is different in each member, with the differences among members becoming larger with increasing vertical wind shear, consistent with the recent study of Rappin and Nolan (2012).

The predictability of tropical cyclone intensity during the formation and rapid intensification stages is intrinsically more limited in the presence of stronger vertical wind shear (Zhang and Tao, 2013). Readers interested in the dynamics and impacts of vertical wind shear on the predictability of tropical cyclones can refer to the follow up studies in Tao and Zhang (2014, 2015).

REFERENCES

Alpert, J.C. and V.K. Kumar, 2007: Radial wind super-obs from the WSR-88D radars in the NCEP operational assimilation system. *Mon. Wea. Rev.*, **135**, 1090-1108.

Bei, N. and F. Zhang, 2007: Mesoscale predictability of the torrential rainfall along the mei-yu front of China. *Quart. J. Roy. Meteor. Soc.*, **133**, 83-99.

Bennett, A.F., 1992: Inverse methods in physical oceanography. Cambridge University Press, Cambridge.

Bennett, A.F. and B.S. Chua, 1994: Open-ocean modelling as an inverse problem: The primitive equations. *Mon Weather Rev*, **122**, 1326-1336.

Bennett, A.F., B.S. Chua and L.M. Leslie, 1996: Generalized inversion of a global numerical weather prediction model. *Meteorol Atmos Phys*, **60**, 165-178.

Bennett, A.F., L.M. Leslie, C.R. Hagelberg and P.E. Powers, 1993: Tropical cyclone prediction using a barotropic model initialized by a generalized inverse method. *Mon Weather Rev*, **121**, 1714-1729.

Chen, S.S., J.F. Price, W. Zhao, M.A. Donelan and E.J. Walsh, 2007: The CBLAST-Hurricane Program and the next generation fully coupled atmosphere–wave–ocean models for hurricane research and prediction. *Bull. Amer. Meteor. Soc.*, **88**, 311-317.

Courtier, P. and O. Talagrand, 1987: Variational assimilation of meteorological observations with the adjoint vorticity equation II: Numerical results. *Q J R Meteorol Soc*, **113**, 1329-1347.

Courtier, P., Thepaut and A. Hollingsworth, 1994: A strategy for operational implementation of 4DVAR, using an incremental approach. *Q J R Meteorol Soc*, 120, 1367-1387.

Davis, C.A. et al., 2008: Prediction of landfalling hurricanes with the advanced hurricane WRF model. *Mon. Wea. Rev.*, **136**, 1990-2005.

Elsberry, R.L., T.D.B. Lambert and M.A. Boothe, 2007: Accuracy of Atlantic and eastern North Pacific tropical cyclone intensity forecast guidance. *Wea. Forecasting*, **22**, 747-762.

Emanuel, K.A., 1999: Thermodynamic control of hurricane intensity. *Nature*, **401**, 665-669.

Evensen, G., 2003: The ensemble Kalman filter: Theoretical formulation and practical implementation. *Ocean Dyn.*, **53**, 343-367.

Franklin, J.L., 2004: National Hurricane Centre verification report. Proc. 57th Interdepartmental Hurricane Conf., Miami, FL, NOAA. Available online at http://www.nhc.noaa.gov/verification/pdfs/Verification_2008.pdf.

Franklin, J.L., 2005: 2004 National Hurricane Centre forecast verification report. NHC. Available online at http://www.nhc.boulder.noaa.gov/verification/pdfs/Verification_2004.pdf.

Gao, J., M. Xue, A. Shapiro and K.K. Droegemeier, 1999: A variational method for the analysis of three-dimensional wind fields from two Doppler radars. *Mon. Wea. Rev.*, **127**, 2128-2142.

Hawblitzel, D., F. Zhang, Z. Meng and C.A. Davis, 2007: Probabilistic evaluation of the dynamics and predictability of mesoscale convective vortex event of 10–13 June 2003. *Mon. Wea. Rev.*, **135**, 1544-1563.

Hendricks, E.A., M.T. Montgomery and C.A. Davis, 2004: The role of "vortical" hot towers in the formation of Tropical Cyclone Diana (1984). *J. Atmos. Sci.*, **61**, 1209-1232.

Holland, G.J., 1997: The maximum potential intensity of tropical cyclones. *J. Atmos. Sci.*, **54**, 2519-2541.

Houze, R.A., S.S. Chen, B.F. Smull, W.-C. Lee and M.M. Bell, 2007: Hurricane intensity and eyewall replacement. *Science*, **315**, 1235-1238.

Knabb, R.D., J.R. Rhome and D.P. Brown, 2006: Tropical cyclone report: Hurricane Katrina: 23–30 August 2005. National Hurricane Centre, Miami, FL. Available online at http://www.nhc.noaa.gov/pdf/TCR-AL122005_Katrina.pdf.

Krishnamurti, T.N., S. Pattnaik, L. Stefanova, T.S.V. Kumar, B.P. Mackey, A.J. O'Shay and R.J. Pasch, 2005: The hurricane intensity issue. *Mon. Wea. Rev.*, **133**, 1886-1912.

Laroche, S. and I. Zawadzki, 1994: A variational analysis method for the retrieval of three-dimensional wind field from single-Doppler radar data. *J. Atmos. Sci.*, **51**, 2664-2682.

Lindskog, M., H. Järvinen and D.B. Michelson, 2000: Assimilation of radar radial winds in the HIRLAM 3D-Var. *Phys. Chem. Earth*, **25B**, 1243-1249.

Lindskog, M., K. Salonen, H. Jarvinen and D.B. Michelson, 2004: Doppler radar wind data assimilation with HIRLAM 3DVAR. *Mon. Wea. Rev.*, **132**, 1081-1092.

Montgomery, M.T., M.E. Nicholls, T.A. Cram and A.B. Saunders, 2006: A vortical hot tower route to tropical cyclogenesis. *J. Atmos. Sci.*, **63**, 355-386.

Pielke, R.A., Jr., J. Gratz, C.W. Landsea, D. Collins, M.A. Saunders and R. Musulin, 2008: Normalized hurricane damages in the United States. *Nat. Hazards Rev.*, **9**, 29-42.

Qiu, C. and Q. Xu, 1992: A simple adjoint method of wind analysis for single-Doppler data. *J. Atmos. Oceanic Technol.*, **9**, 588-598.

Rappin, E.D. and D.S. Nolan, 2012: The effect of vertical shear orientation on tropical cyclogenesis. *Q. J. Roy. Meteorol. Soc.*, DOI: 10.1002/qj.977.

Shapiro, A., S. Ellis and J. Shaw, 1995: Single-Doppler velocity retrievals with Phoenix II data: Clear air and microburst wind retrievals in the planetary boundary layer. *J. Atmos. Sci.*, **52**, 1265-1287.

Shen, B.W., R. Atlas, O. Reale, S.-J. Lin, J.-D. Chern, J. Chang, C. Henze and J.-L. Li, 2006: Hurricane forecasts with a global mesoscale-resolving model: Preliminary results with Hurricane Katrina (2005). *Geophys. Res. Lett.*, **33**, L13813, doi:10.1029/2006GL026143.

Snyder, C. and F. Zhang, 2003: Assimilation of simulated Doppler radar observations with an ensemble Kalman filter. *Mon. Wea. Rev.*, **131**, 1663-1677.

Sun, J. and N.A. Crook, 1997: Dynamical and microphysical retrieval from Doppler radar observations using a cloud model and its adjoint. Part I: Model development and simulated data experiments. *J. Atmos. Sci.*, **54**, 1642-1661.

Talagrand, O. and P. Courtier, 1987: Variational assimilation of meteorological observations with the adjoint vorticity equation. I: Theory. *Q J R Meteorol Soc*, **113**, 1311-1328.

Tao, D. and F. Zhang, 2014: Effect of environmental shear, sea-surface temperature and ambient moisture on the formation and predictability of tropical cyclones: An ensemble-mean perspective. *Journal of Advances in Modeling Earth Systems (JAMES)*, DOI: 10.1002/2014MS000314.

Tao, D. and F. Zhang, 2015: Effects of vertical wind shear, on the predictability of tropical cyclones: Pratical versus intrinsic limit *J. Adv. Model. Earth Syst.*, DOI: 10.1002/2015MS00474.

Van Sang, N., R.K. Smith and M.T. Montgomery, 2008: Tropical cyclone intensification and predictability in three dimensions. *Quart. J. Roy. Meteor. Soc.*, **134**, 563-582.

Weygandt, S.S., A. Shapiro and K.K. Droegemeier, 2002: Retrieval of model initial fields from single-Doppler observations of a supercell thunderstorm. Part I: Single-Doppler velocity retrieval. *Mon. Wea. Rev.*, **130**, 433-453.

Willoughby, H.E., 1999: Hurricane heat engines. *Nature*, **401**, 649-650.

Xu, Q., H.D. Gu and S. Yang, 2001: Simple adjoint method for three-dimensional wind retrievals from single-Doppler data. *Quart. J. Roy. Meteor. Soc.*, **127**, 1053-1067.

Zhang, F., A. Odins and J.W. Nielsen-Gammon, 2006: Mesoscale predictability of an extreme warm-season rainfall event. *Wea. Forecasting*, **21**, 149-166.

Zhang, F. and C. Snyder, 2007: Ensemble-based data assimilation. *Bull. Amer. Meteor. Soc.*, **88**, 565-568.

Zhang, F. and J.A. Sippel, 2009: Effects of moist convection on hurricane predictability. *J. Atmos. Sci.*, **66**, 1944-1961.

Zhang, F., C. Snyder and R. Rotunno, 2002: Mesoscale predictability of the "surprise" snowstorm of 24–25 January 2000. *Mon. Wea. Rev.*, **130**, 1617-1632.

Zhang, F., C. Snyder and R. Rotunno, 2003: Effects of moist convection on mesoscale predictability. *J. Atmos. Sci.*, **60**, 1173-1185.

Zhang, F., N. Bei, R. Rotunno, C. Snyder and C.C. Epifanio, 2007: Mesoscale predictability of moist baroclinic waves: Convection-permitting experiments and multistage error growth dynamics. *J. Atmos. Sci.*, **64**, 3579-3594.

Zhang, F. and Y. Weng, 2015: Predicting Hurricane, Intensity and Associated Hazards: A Five year Real-Time Forecast Experiment with Assimilation of Airborne Doppler Radar Observations. Bulletin of the American Meteorological Society, **96**, 25-32.

Zhang, F., Z. Meng and A. Aksoy, 2006: Test of an ensemble Kalman filter for mesoscale and regional-scale data assimilation. Part I: Perfect model experiments. *Mon. Wea. Rev.*, **134**, 722-736.

Zhang, F., Z. Meng, J.F. Gamache and F.D. Marks, 2011: Performance of convection-permitting hurricane initialization and prediction during 2008-2010 with ensemble data assimilation of innercore airborne Doppler radar observations. *Geophys. Res. Lett.*, **38**, L15810, doi:10.1029/2011GL048469.

13

Data Assimilation: Comparison and Hybridization between Ensemble and Variational Methods

Fuqing Zhang[1] and Ashish Routray[2]

[1]Department of Meteorology, The Pennsylvania State University
University Park, Pennsylvania, USA

[2]NCMRWF, A-50, Institutional Area, Sector-62, Noida, India

fzhang@psu.edu

1. Introduction

The accuracy of numerical weather prediction (NWP) depends critically on the qualities of the initial conditions and the forecast model. The initial conditions of an NWP model usually come from data assimilation (DA), a procedure that aims to estimate the state and uncertainty of the atmosphere as accurately as possible by combining all available atmospheric information. In other words, DA is known as the process of creating the best estimate of the initial state for NWP models through combining all sources of information, including the first guess from previous short-term model forecasts and observations, along with the associated uncertainties in each source of information.

Variational and ensemble-based approaches are two of the most advanced and widely used data assimilation techniques. Operational NWP models have predominantly used three- or four-dimensional variational (3DVAR or 4DVAR) data assimilation methods over the past couple of decades (Lewis and Derber, 1985; Parrish and Derber, 1992; Courtier et al., 1994). Ensemble-based data assimilation methods, in various forms of ensemble Kalman filters (EnKFs) first proposed by Evensen (1994), which use a short-term ensemble to estimate the flow-dependent background error covariance, are becoming a popular alternative to the variational approaches.

2. Basic Algorithms of 3DVAR, 4DVAR and EnKF

2.1 Variational Algorithms

Mathematically, the variational method is aimed at obtaining a balanced state analysis subjective to both dynamical and statistical constrains by minimizing a cost function J:

$$J = J_b + J_o + J_c = \frac{1}{2}(x_0 - x^b)^T \mathbf{B}^{-1}(x_0 - x^b) + \frac{1}{2}\sum_{k=0}^{K}[H_k M_k(x_0) - y_k^o]^T \mathbf{O}^{-1}[H_k M_k(x_0) - y_k^o] + J_c , \quad (1)$$

where J_b, J_o and J_c are the background, observation and penalty terms respectively, and the subscript k denotes an analysis time during the assimilation window. In the background term J_b, x_0 is the analysis at the initial time, x^b is the first guess field and **B** is the background error covariance. In the observation term J_o, H_k is the nonlinear observation operator, M_k is the nonlinear model, **O** is the observational error covariance, and x_k and y_k^0 are the analysis and observation states distributed at time k during the assimilation window. In J_c, a digital filter is introduced to remove high-frequency waves in the analysis state.

The fundamental difference between 4DVAR and 3DVAR is that the 4DVAR usually performs minimization of the cost function J over different times (through the use of tangent linear and adjoint models over a trajectory) while the 3DVAR only has $k = 0$, meaning the analysis occurs for a fixed time. The subsequent advantages of 4DVAR over 3DVAR are addressed in Huang et al. (2009) and include: (i) the ability to deal with asynchronous observations; (ii) implicit definition of flow-dependent background error covariance, which has vital importance for fast-developing weather systems; and (iii) the ability to use a forecast model as a constraint, thus enhancing the dynamic balance of the final analysis.

For most practical applications of the variational analysis, an incremental approach (Coutier et al., 1994) is used to resolve the minimization, as Equation (1) can be rewritten as:

$$J = \frac{1}{2}\delta x_0{}^T \mathbf{B}^{-1}\delta x_0 + \frac{1}{2}\sum_{k=0}^{K}(\mathbf{H}_k \mathbf{M}_k \delta x_0 - d_k)^T \mathbf{O}^{-1}(\mathbf{H}_k \mathbf{M}_k \delta x_0 - d_k) + J_c, \quad (2)$$

where δx_0 is the analysis increment at the initial time from the first guess, $\mathbf{H}_k$ and $\mathbf{M}_k$ are the tangent linear versions of the observation operator and forecast model, and d_k is the innovation vector in the observation space at time k, i.e., $d_k = y_k^0 - H_k M_k(x_0)$. The incremental approach allows the minimization to occur within inner loops with the simplified and/or low-dimensional linear models, which is more computationally efficient for operational uses. And multiple outer loops may help the variational analysis partly overcome the impacts from the nonlinearity problems for complex physical parameterizations (Lorenc, 2003b).

In order to reduce the condition numbers and accelerate the minimization, preconditioning of the background term in the cost function is conducted with a control variable v (Lorenc et al., 2000; Barker et al., 2004):

$$\delta x_0 = \mathbf{U}v, \tag{3}$$

where the preconditioning matrix $\mathbf{U}$ is defined as $\mathbf{B} = \mathbf{U}\mathbf{U}^T$. Then the incremental formulation of the cost function (Equation 2) is adapted as:

$$J = \frac{1}{2}v^T v + \frac{1}{2}\sum_{k=0}^{K}(\mathbf{H}_k\mathbf{M}_k\mathbf{U}_k v - d_k)^T\mathbf{O}^{-1}(\mathbf{H}_k\mathbf{M}_k\mathbf{U}_k v - d_k) + J_c\ , \tag{4}$$

and the gradient of the Equation (4) with respect to the control variable v is then:

$$\nabla_v J = v + \mathbf{U}^T\sum_{k=0}^{K}\mathbf{M}_k^T\mathbf{H}_k^T\mathbf{O}^{-1}(\mathbf{H}_k\mathbf{M}_k\mathbf{U}_k v - d_k) + \nabla_v J_c\ , \tag{5}$$

where $\mathbf{H}_k^T$. denotes the adjoint observation operator at time k (the transpose of $\mathbf{H}_k$), and $\mathbf{M}_k^T$ is the adjoint model that propagates the analysis residual at each observation time k backward to the beginning of the assimilation window (i.e., $k = 0$).

In reality, the background error covariance should evolve the flow-dependence with multivariate and multiscale correlations within the assimilation time, but the $\mathbf{B}$ in most variational algorithms is estimated based on the climatological statistics, typically using the so-called the "NMC method" (Parrish and Derber, 1992). The NMC-based $\mathbf{B}$ assumes that the background error statistics can be approximated by the averaged forecast differences, e.g., a month-long average of the differences between 24-h and 12-h hour forecasts valid at the same time (Barker et al., 2004):

$$\mathbf{B} = \overline{(x^b - x^t)(x^b - x^t)^T} = \overline{\varepsilon\varepsilon^T} \approx \overline{[x^f(24h) - x^f(12h)][x^f(24h) - x^f(12h)]^T}, \tag{6}$$

x^b, x^t and x^f represent the background, truth and forecast state vectors, respectively, and the estimated background errors are calculated from two forecasts with lead times of 24 and 12 h.

2.2 EnKF Algorithm

EnKF is based on the original linear, recursive estimator that produces the unbiased minimum variance estimate, in a least square sense, under the assumption of unbiased noise processes. Unlike variational approaches, it uses ensemble forecasts to estimate the flow-dependent background error covariance rather than the NMC method. As in the extended Kalman filter, there are two basic steps: the forecast step and the analysis step.

In the forecast step, the state vector (x) and the associated flow uncertainty ($\mathbf{P}^f$) evolve in the time period from $k - 1$ to k with the forward nonlinear (M) and linear ($\mathbf{M}$) models:

$$x_k^f = M(x_{k-1}^a) \tag{7}$$

$$\mathbf{P}_k^f = \mathbf{M}\mathbf{P}_{k-1}^a\mathbf{M}^T + \mathbf{Q}, \tag{8}$$

where the superscript *f* and *a* denote the prior (forecast) and posterior (analysis) quantities, **Q** represents the model error covariance matrix. Note, the flow uncertainty term $\mathbf{P}^f$ is referred as the same matrix as the **B** in variational approaches.

In the analysis step of time *k*:

$$x_k^a = x_k^f + \mathbf{K}_k d_k \tag{9}$$

$$\mathbf{P}_k^a = (\mathbf{I} - \mathbf{K}_k\mathbf{H}_k)\mathbf{P}_k^f(\mathbf{I} - \mathbf{K}_k\mathbf{H}_k)^T + \mathbf{K}_k\mathbf{O}\mathbf{K}_k^T$$
$$= (\mathbf{I} - \mathbf{K}_k\mathbf{H}_k)\mathbf{P}_k^f \tag{10}$$

$$\mathbf{K}_k = \mathbf{P}_k^f\mathbf{H}_k^T(\mathbf{H}_k\mathbf{P}_k^f\mathbf{H}_k^T + \mathbf{O})^{-1}, \tag{11}$$

where d_k is the innovation vector and **K** is the Kalman gain, the curial factor determining how to balance the weights between the observation and the background state, and spread their impacts with spatial and multivariate correlations to get a minimum variance analysis. However, the extended Kalman filter has difficulty for resolving realistic atmospheric data assimilation issues, because of the large dimension (usually 10^6~10^8) and nonlinearity of the NWP models. Likewise, the tangent linear and adjoint models are also needed as in 4DVAR.

Evensen (1994) firstly used a group of short-term forecasts to approximate the background error covariance matrix, as:

$$\mathbf{P}^f \approx \overline{x'^f(x'^f)^T} = \frac{1}{N-1}\sum_{i=1}^{N}(x_i^f - \overline{x^f})(x_i^f - \overline{x^f}), \tag{12}$$

in which $x'^f = x^f - \overline{x^f}$ is the ensemble perturbation (departures from the mean), *N* is the total number of ensemble forecasts, i.e., ensemble size, $\overline{x^f}$ is the ensemble mean, and x_i^f denotes the ith ensemble member. The EnKF is equivalent to the extended Kalman filter if the error growth is linear and the ensemble-size is near to infinite.

The advantages of EnKF are obvious, including: (i) the flow-dependent information can be explicitly resolved by the $\mathbf{P}^f$ matrix with relatively low dimensions; (ii) the nonlinear model and observation operators are allowed; (iii) the tangent linear and adjoint models are no longer needed, which are complicated to code; (iv) it can provide ensemble analysis and flow uncertainty estimates, rather than the deterministic analysis in variational methods; and (v) the ensemble forecasts can be easily initialized with EnKF analyses.

The issues involved in EnKF are also considerable due to the ensemble approximation:

(i) The analysis error is usually underestimated in EnKF, because the Equation (10) is revised for its ensemble form as:

$$\mathbf{P}^a \approx \overline{x'^a (x'^a)^T} = (\mathbf{I} - \mathbf{KH})\mathbf{P}^f(\mathbf{I} - \mathbf{KH})^T , \tag{13}$$

where the x'^a is the updated ensemble perturbation, and the **K** and $\mathbf{P}^f$ are both in the manner of ensemble approximations. Note, the positive term $\mathbf{KOK}^T$ (Equation 10) is ignored in Equation (13); therefore, the analysis uncertainty would be underestimated, the background estimations would be wrongly weighted in the analysis step and then result in the shrunk ensemble spread (i.e., variance of the ensemble) in following cycles, which is known as "filter divergence" as the analysis shifting away from the truth. Besides, the model error treatment **Q** could be another source for the underestimated background error covariance during the forecast step. To avoid such problem, several derivatives of the ensemble-based Kalman filter are investigated. E.g., the EnKF with sequential perturbed observations (Houtkamer and Mitch, 2001), the "double" EnKF with parallel sampling groups (Houtkamer and Mitchell, 1998), the ensemble square root filter without perturbed observations (EnSRF, Whitaker and Hamill, 2002), the ensemble adjustment Kalman filter (EAKF, Anderson, 2001) and the ensemble transform Kalman filter (ETKF, Bishop et al., 2001) both with singular-vector decomposition (SVD). On the other hand, the manipulated inflation for the background error covariance is also a practical realization and widely used in previous studies (Anderson and Anderson, 1999; Hamill et al., 2001; Zhang et al., 2004; Anderson, 2007; Whitaker et al., 2008).

(ii) A limited ensemble size could incur large sampling errors in the background error covariance, which may lead to both the spurious correlations and the underestimated background errors (Whitaker and Hamill, 2002). The localization method with a "Schur product" (Gaspari and Cohn, 1999) is considered as an effective way to reduce the sampling errors. It multiples the ensemble-based background covariance element by element with a distance-dependent Gaussian-shaped correlation function that varies from one at observation point to zero at specific radius of influence.

Since the first application in atmospheric sciences (Houtekamer and Mitchell, 1998), the EnKF has been widely examined with different models at different scales (e.g., Hamill and Snyder, 2000; Anderson, 2001; Whitaker and Hamill, 2002; Mitchell et al., 2002; Snyder and Zhang, 2003; Zhang and Anderson, 2003; Zhang et al., 2004, 2006, 2009a; Aksoy et al., 2005, 2006a, b; Houtekamer et al., 2005, 2009; Tong and Xue, 2005; Dirren et al., 2007; Meng and Zhang, 2007, 2008a,b; Whitaker et al., 2008; Torn and Hakim, 2008a, 2009a; Buehner et al., 2010a,b). There are several recent review articles on the EnKF including, Evensen (2003, 2007), Hamill (2006) and Ehrendorfer (2007).

3. Advantages and Limitations of Different Data Assimilation Methods

Recently, systematic inter-comparisons between the EnKF and 3DVAR/4DVAR have been conducted in various NWP models. For example, Houtekamer et al. (2005) compared the EnKF with 3DVAR for the Canadian global model, and Whitaker et al. (2008) did so for the National Centers for Environmental Prediction's (NCEP's) Global Forecast System (GFS). Both studies showed convincingly the advantages of the EnKF over 3DVAR in global models. For limited-area models, Meng and Zhang (2008a,b) also showed that the EnKF consistently outperformed the 3DVAR within the Weather Research and Forecasting (WRF) model framework at the regional scales. More recently, Yang et al. (2009) compared the EnKF with 3D/4DVAR using a quasi-geostrophic model. Buehner et al. (2010a, b) compared the EnKF with 4DVAR in an operational scenario of Canadian global model while Miyoshi et al. (2010) did so for the Japan Meteorological Agency's (JMA's) operational global model; both studies showed that the EnKF has comparable performance to 4DVAR in global models. For limited-area models, Caya et al. (2005) compared the performance of the EnKF with 4DVAR using synthetic observations for a supercell event in an elastic cloud model and found that the EnKF and 4DVAR performed comparably. Moreover, both Buehner et al. (2010a, b) and Caya et al. (2005) found that the EnKF initialized models will have better forecast performance at a later stage of their forecasts after a few assimilation cycles even though 4DVAR may fit the observations better at the analysis time and requires less than spin-up time from a cold start.

From these aforementioned comparisons, the advantages and limitations of different data assimilations methods are summarized below.

3.1 Advantages and Limitations of Variational Assimilation System

- Observations can easily be assimilated directly without the need for prior retrieval. This results in a consistent treatment of all observations and, as the observation errors are less correlated (with each other and the background errors), practical simplifications to the analysis algorithm.
- The VAR solution is found using all observations simultaneously, unlike the OI technique for which a data selection into artificial sub-domains is required.
- Asynoptic data can be assimilated near its validity time. This is implicit to 4DVAR but can also be achieved using a "rapidly-updating" 3DVAR technique.
- Balance (e.g. weak geostrophy, hydrostatic) constraints can be built into the preconditioning of the cost-function minimization. In 4DVAR, use is also made of the implicit balance of the forecast model.

Although the variational data assimilation system has some advantages, it has some limitations too. The variational analysis is based on a number of assumptions. Firstly, given both imperfect observations and prior (e.g. background) information as inputs to the assimilation system, the quality of the output analysis depends crucially on the accuracy of prescribed errors. Secondly, although the variational method allows for the inclusion of linearized dynamical/physical processes, in reality real errors in the NWP system may be highly nonlinear. This limits the usefulness of variational data assimilation in highly nonlinear regimes e.g. the convective scale or in the tropics. The 3DVAR system is lack of flow-dependent background uncertainty. The 4DVAR searches the model trajectory with best fit to the temporally distributed observations and the background fit, but it is computationally expensive and requires high code maintenance.

3.2 Advantages and Limitations of Ensemble Assimilation System

The ensemble Kalman filter (EnKF) is an *approximate* Kalman filter that obtains its covariance information from a set of parallel forecasts that constitute an ensemble. This has several advantages:

- The background error covariance is flow dependent, which reflects the error of the day.
- The forecast model itself can be fully nonlinear, avoiding the need to construct approximate *linearized* models and so also the observation operator.
- The ensemble mean itself is the minimum-variance best estimate according to Bayesian perspective.
- It provides not only the best estimation of the state, but also the associated flow-dependent uncertainty; therefore it can be seamlessly coupled with ensemble forecasting and easier to account the model error.
- There is no need to code a tangent linear or adjoint model.
- The covariance matrix is *sampled* from the available ensemble members, circumventing the necessity of expensive matrix operations.
- The posterior uncertainty is consistent with the mean analysis and can be directly used for ensemble forecasting.
- The ensemble members can be run simultaneously, making it easy to parallelize (Evensen, 2003; Hamill, 2006; Zhang and Snyder, 2007).
- It has similarities to and is capable of more sampling and particle filtering but uses much fewer samples for computational feasibility.
- It has flow-dependent background error covariance with moderate computational cost and low code maintenance.

The EnKF also has certain limitations.

- It is highly dependent on the quality of ensembles and the first-guess.

- Probability distributions associated with nonlinear systems are generally non-Gaussian.
- Estimation algorithms often cannot be presented in closed form.
- Integration of the covariance matrix involves computation of **MP**a**M**T which requires two-fold matrix multiplication (very costly).
- For typical atmospheric science applications, state size is ~O(106-108) so that storage of the covariance matrices is nearly impossible.

4. Inter-comparison of 3DVAR, 4DVAR and EnKF

The inter-comparison of 3D/4D variational with EnKF data assimilation system for the regional scale models is discussed in this section.

4.1 Experimental Design

The regional Advanced Research Weather Research and Forecasting (ARW-WRF) model (Skamarock et al., 2005) version 3.1.1 is employed as the platform to investigate all DA approaches in a realistic scenario. All experiments are conducted over a single domain covering the contiguous United States and surrounding areas (Fig. 1), which has 71 × 51 horizontal mesh grids with 90-km spacing and 27 vertical levels up to 50 hPa.

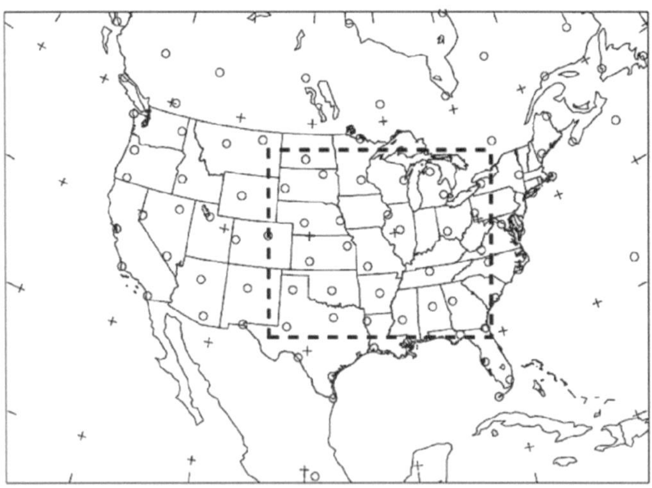

Fig. 1. Model domain configuration. The circles denote the locations of the radiosonde observations used for assimilation and verification. The dashed box shows the subset of the domain selected for verification statistics.

The Grell–Devenyi cumulus scheme, WRF single-moment six-class graupel microphysics scheme, and Yonsei State University (YSU) boundary layer scheme are used for all deterministic forecasts, whereas multi-physics parameterizations are considered for ensemble forecasts in the EnKF [as in Meng and Zhang (2008a, b)]. The first forecast cycle of this month-long experiment is initialized at 0000 UTC 1 June 2003, where the NCEP global final analysis (FNL) data are used to create the initial and lateral boundary conditions (IC and LBC). In subsequent cycles, the lateral boundary conditions are interpolated from the FNL analyses, whereas the initial conditions are updated and propagated from previous DA analyses.

The WRF variational data assimilation system (WRFVar) version 3 (Huang et al., 2009) and the WRF EnKF system of Meng and Zhang (2008a, b) are used for all DA experiments. The WRF-Var has both 3DVAR and 4DVAR capabilities; the newly released 4DVAR component is developed as an extension from the previous WRF 3DVAR system (Barker et al., 2004). The background error covariance in 3DVAR/4DVAR experiments is static and prescribed by the National Meteorological Center (NMC) method (Parrish and Derber, 1992), which assumes homogeneous and isotropic correlations for a set of independent control variables derived from the forecast difference between the 24- and 12-h lead time forecasts of the preceding month for the given domain (i.e., May 2003 in this case). The variance scale parameter is carefully tuned to 3.0 rather than the default value of 1.0 to give the best performance of the 3DVAR among various experiments tested (not shown; this also improves the performance of 4DVAR as compared to the default). The EnKF adopts settings similar to those in Meng and Zhang (2008a, b), that is, an ensemble of 40 members with diverse physics parameterization schemes, perturbed lateral boundary conditions from FNL analyses, a relaxation coefficient of 0.8 (Zhang et al., 2004), and the covariance localizations of Gaspari and Cohn (1999) using a radius of influence of 1800 km for radiosondes and profilers and 600 km for other observations. The initial ensemble perturbations are randomly generated at 0000 UTC 1 June 2003 through the CV5 option (Barker et al., 2004) of the WRF-Var system, as in Meng and Zhang (2008a, b).

4.2 Results and Discussion

The performance of all three DA systems over the one month period is examined in this section. The rootmean-square error (RMSE) of horizontal winds (U, V), temperature T, and the mixing ratio of water vapour Q are calculated between model forecasts and radiosonde observations over a subset of the model domain (dashed box in Fig. 1). The same area is used for verification of forecast performance as in Meng and Zhang (2008b). There are a total of 59 individual 72-h WRF deterministic forecasts, initialized respectively from the variational and EnKF mean analyses at 0000 and 1200 UTC each day, used in all the verification statistics shown below. We first examine the vertical distribution of the mean RMSE of the 12-h WRF forecasts of U, V, T and Q averaged over the entire month (Fig. 2). At a first glance, the largest RMSE of U, V and T is around 200 hPa near the tropopause while the largest error of Q is in the mid to lower troposphere. Comparison of vertical profile of the mean 12-h forecast error from each DA experiment shows that the EnKF performs comparably to 4DVar for the horizontal winds and temperature fields while the EnKF has noticeably smaller RMSE than 4DVAR for moisture throughout the troposphere, especially at mid levels. The comparable performance of the EnKF versus 4DVAR for the wind and temperature fields and the advantage of the EnKF with regard to moisture are fairly persistent throughout the month, as can be seen from the time series of vertically averaged RMSE shown in Fig. 3. Both the EnKF and 4DVAR have substantially smaller RMSEs than does 3DVAR for all forecast variables at nearly all times (Figs 2 and 3).

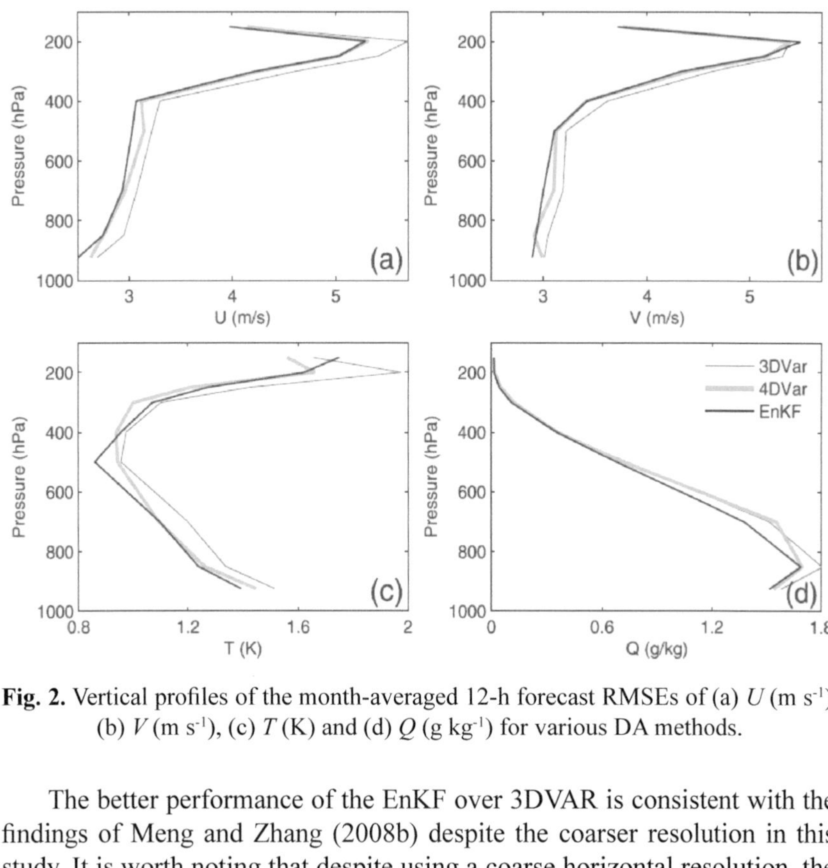

Fig. 2. Vertical profiles of the month-averaged 12-h forecast RMSEs of (a) U (m s^{-1}), (b) V (m s^{-1}), (c) T (K) and (d) Q (g kg^{-1}) for various DA methods.

The better performance of the EnKF over 3DVAR is consistent with the findings of Meng and Zhang (2008b) despite the coarser resolution in this study. It is worth noting that despite using a coarse horizontal resolution, the RMSEs for both the EnKF and 3DVAR in the current study are similar in magnitude to those in Meng and Zhang (2008b), suggesting that the difference between variational and ensemble methods discussed herein is not sensitive to model resolution.

Figure 4 displays the vertical profiles of the RMS differences between the analysis of each DA scheme and the sounding observations averaged over all 59 analyses. Note that these differences are no longer called "errors" but are considered as a measure of the fit of the analysis to observations, since all verifying observations are assimilated by each scheme and thus no

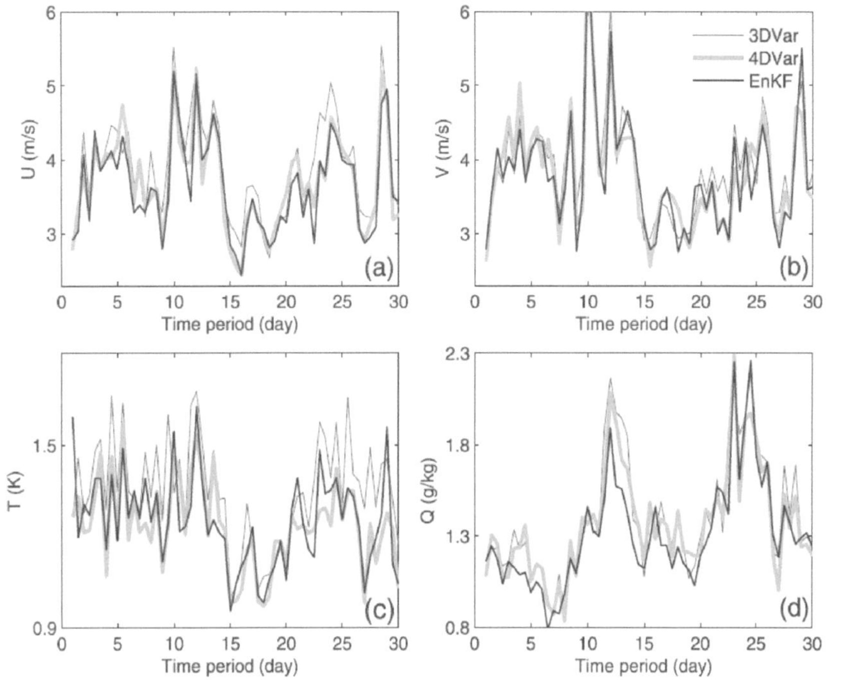

Fig. 3. Time evolution of the vertically averaged 12-h forecast RMSE of (a) *U* (m s^{-1}), (b) *V* (m s^{-1}), (c) *T* (K) and (d) *Q* (g kg^{-1}).

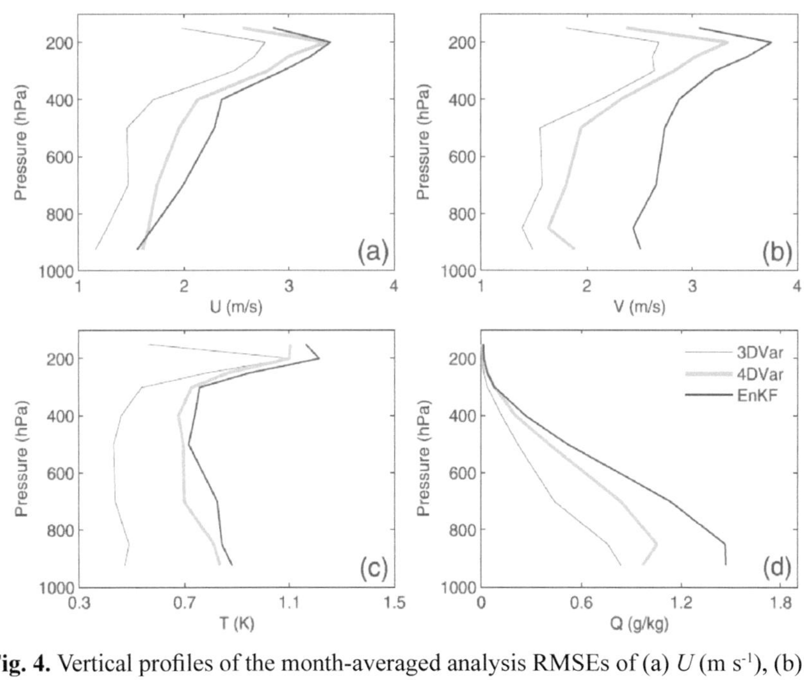

Fig. 4. Vertical profiles of the month-averaged analysis RMSEs of (a) *U* (m s^{-1}), (b) *V* (m s^{-1}), (c) *T* (K) and (d) *Q* (g kg^{-1}) for various DA methods.

longer are considered to be independent verifications at the analysis time. In sharp contrast to Fig. 2 of the 12-h forecast error, Fig. 4 shows that the 3DVAR scheme best fits the observations, followed by 4DVAR, while the EnKF analysis has the largest difference between the radiosonde observations throughout the vertical domain. This result confirms that an analysis with a closer fit to observations does not necessarily lead to a better forecast.

Figure 5 summarizes the domain-averaged (in both vertical and horizontal directions) RMSEs further averaged over all 59 WRF forecasts of the month for each DA experiment at forecast lead times from 0 to 72 h evaluated every 12 h. Beyond what have been shown in Figs 2 and 3, it is found that 4DVAR has consistently smaller error than 3DVAR from 12- to 48-h forecast lead times for horizontal winds and temperature fields, but their forecast error amplitude becomes comparable afterward (at 60 and 72 h). The differences in the RMSE of the moisture field between 4DVAR and 3DVAR are relatively small, since few moist processes are considered in the linear models of 4DVAR (Huang et al., 2009). This implies that the 4DVAR may not resolve more flow-dependent moist information through adjoint minimization than that in the static background error covariance as in 3DVAR. On the other hand, the advantage of the EnKF over both 3DVAR and 4DVAR becomes more evident after the 36-h forecast time for all forecast fields, while the EnKF moisture forecast field is superior to those of both 3DVAR and 4DVAR at all lead times despite fitting less closely to the observations at the analysis time.

This result is consistent with the time dependence of the relative performance of 4DVAR and EnKF in previous works (Caya et al., 2005;

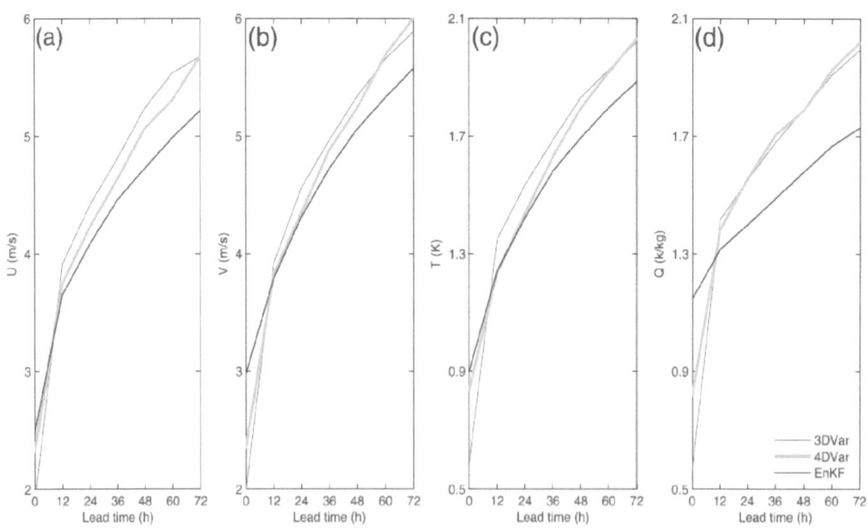

Fig. 5. Domain-averaged RMSEs further averaged over all 59 WRF forecasts of the month for each DA experiment at forecast lead times from 0 to 72 h evaluated every 12 h for (a) U (m s^{-1}), (b) V (m s^{-1}), (c) T (K) and (d) Q (g kg^{-1}).

Buehner et al., 2010 a,b). It is rather remarkable that the 72-h forecast error of the EnKF is comparable in magnitude to the 48-h error of 3DVAR/4DVAR, a gain of nearly 1-day lead time in forecast accuracy. Figure 6 shows the time evolution of the vertically averaged 72-h RMSEs from all DA experiments, which displays an apparent fluctuation between different initialization times. The 72-h RMSE of the 4DVAR-initialized forecasts is comparable to that of the 3DVAR over all months while the advantage of the EnKF over the two variational schemes is most evident for a few episodes (e.g., 10–13 June) that feature the passage of strong mesoscale convective systems (Davis et al., 2004; Hawblitzel et al., 2007).

The clear advantage of the EnKF over 3DVAR and 4DVAR in the moisture field in all verifications signifies the benefit of using a flow-dependent background error covariance, since the balance used to estimate the background error covariance inWRF-Var (the same for both 3DVAR and 4DVAR) does not include the multivariate correlations for moisture field. The lack of multivariate balance may be one of the key factors impacting the DA performance. The EnKF also benefits from using a multi-physics ensemble mean for an improved forecast prior to the analysis (Meng and Zhang, 2008a).

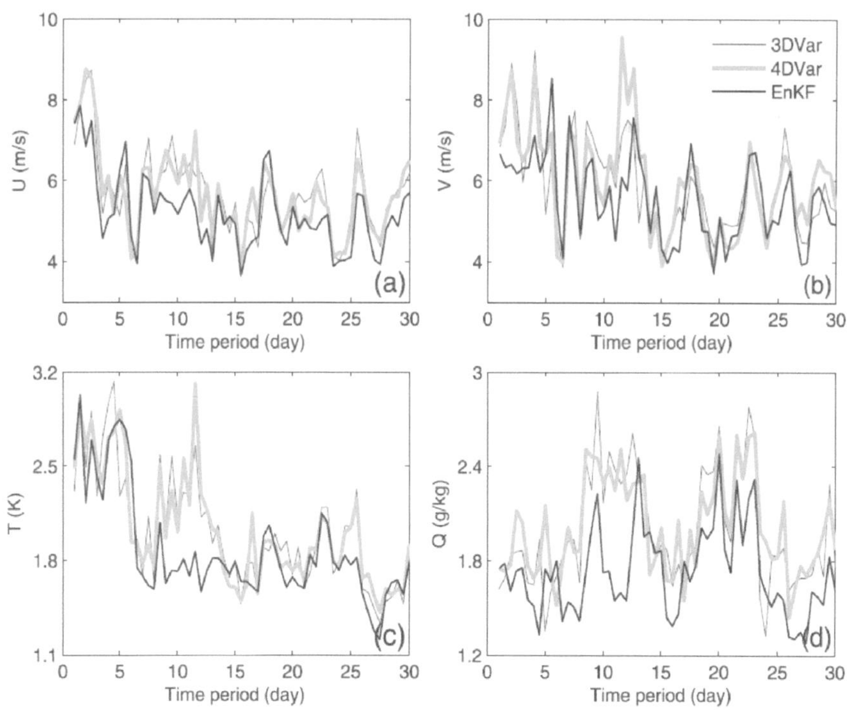

Fig. 6. Time evolution of the vertically averaged 72-h forecast RMSE of (a) U (m s^{-1}), (b) V (m s^{-1}), (c) T (K) and (d) Q (g kg^{-1}).

5. Coupling and Hybrid of EnKF with 3DVAR/4DVAR

Given the disadvantage of using static, mostly isotropic background covariance, there are increasing efforts to introduce the flow-dependent error statistics estimated from short-term ensemble forecasts into a variational data assimilation technique. Results show that 3DVAR may benefit from using a homogeneous background error covariance calculated from an ensemble at either the initial time only or evolving with time (Buehner, 2005; Meng and Zhang, 2008a).

Lorenc (2003) proposed an extended control variable method using an ensemble-based background error covariance in the background term of a variational analysis scheme. Another approach to ingest an ensemble background error covariance into a variational method is by directly combining the two error covariances (Hamill and Snyder, 2000). The extended control variable and direct covariance combination approaches have been proven to be theoretically equivalent by Wang et al. (2007). As an extension of Lorenc (2003), Liu et al. (2008, 2009) designed an ensemble-based, four-dimensional variational algorithm for both a 1D shallow-water model and WRF by calculating the gradient of the cost function using an ensemble background covariance in the observation space. This avoids the use of a tangent linear and adjoint model. This method quickly converges to the true solution and can produce results that are comparable to 4DVAR, but at far less computational cost in its minimization.

5.1 Hybrid Formulation within WRF-Var

Mathematically, the variational method is aimed at obtaining a balanced state analysis subject to both dynamical and statistical constraints by minimizing a cost function J. The hybrid formulation within WRF-Var framework can be represented as,

$$J_b = J_{b1} + J_{b2} = \frac{1}{2}\delta x^T \left[(1-\beta)\mathbf{B} + \beta \mathbf{P}^f \circ \mathbf{C}\right]^{-1} \delta x \tag{14}$$

The background term is adapted with the weighted ensemble covariance. Using the Alpha-control variables in the cost function,

$$J_b = \frac{1}{1-\beta} J_{b1} + \frac{1}{\beta} J_{b2} = \frac{1}{1-\beta}\left(\frac{1}{2} v^T v\right) + \frac{1}{\beta}\left[\frac{1}{2}\alpha^T \begin{pmatrix} \mathbf{C} & & 0 \\ & \cdots & \\ 0 & & \mathbf{C} \end{pmatrix} \alpha\right] \tag{15}$$

The background term in cost function is then calculated by traditional preconditioning control variable v (Barker et al., 2004) with the ensemble-based Alpha-control variable (Lorenc, 2003), which is also localized with Schur product, as in Wang et al. (2009).

In the E3DVAR, 3DVAR and EnKF run separately in parallel but also with two-way exchanges. There is a "coupler" between them that contains three major variable exchanges. The schematic diagram of E3DVAR is presented in Fig. 7.

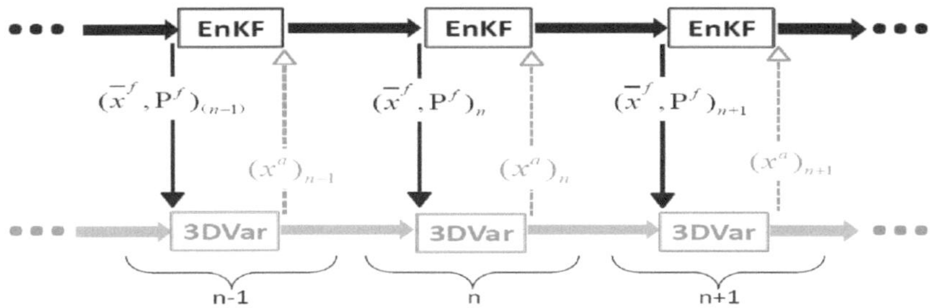

Fig. 7. Schematic diagram of E3DVAR that couples 3DVAR with EnKF.

The EnKF provides ensemble-based background error covariance ($\mathbf{P}^f$) for 3DVAR. EnKF provides the prior ensemble mean ($\overline{x}^f$) as the first guess for 3DVAR. 3DVAR provides deterministic analysis ($\overline{x}^a$) to replace the posterior ensemble mean for the next ensemble forecast.

Similarly, in E4DVAR, 4DVAR and EnKF run separately in parallel but also with two-way exchanges. There is a "coupler" between them that contains three major variable exchanges (Fig. 8): (i) introduce ensemble-based background error covariance $\mathbf{P}^f$ into the 4DVAR cost function, (ii) use prior ensemble mean $\overline{x}^f$ as the first guess for each 4DVAR cycle, and (iii) replace posterior ensemble mean by the 4DVAR analysis x_0^a for next ensemble forecast (Zhang et al., 2009).

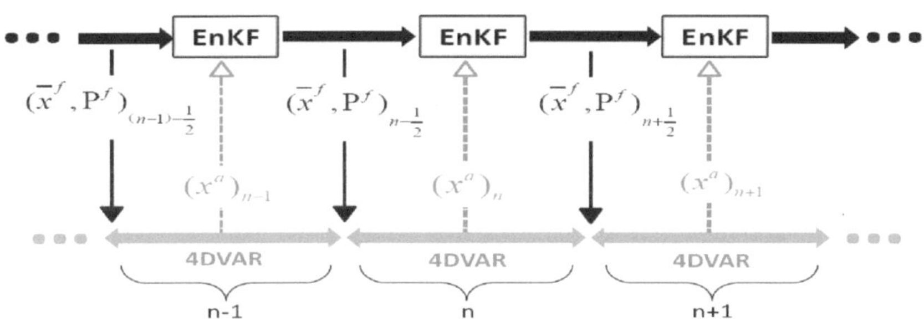

Fig. 8. Schematic diagram of E4DVAR that couples 4DVAR with EnKF.

5.2 Comparison of Different Assimilation Methods

Single-observation experiments were conducted to demonstrate the capability of each DA method on resolving the structure of flow-dependent background error covariance. The resulting analysis increments of temperature (shaded), horizontal winds (vector), and moisture (contour) from various data assimilation experiments are illustrated in Fig. 9.

In the 3DVAR (Fig. 9a), the computed analysis increments have an isotropic structure, centered at the observation location and completely independent of the background flow, because the NMC-based background error covariance in 3DVAR is obtained using averaged error statistics (over the previous month in

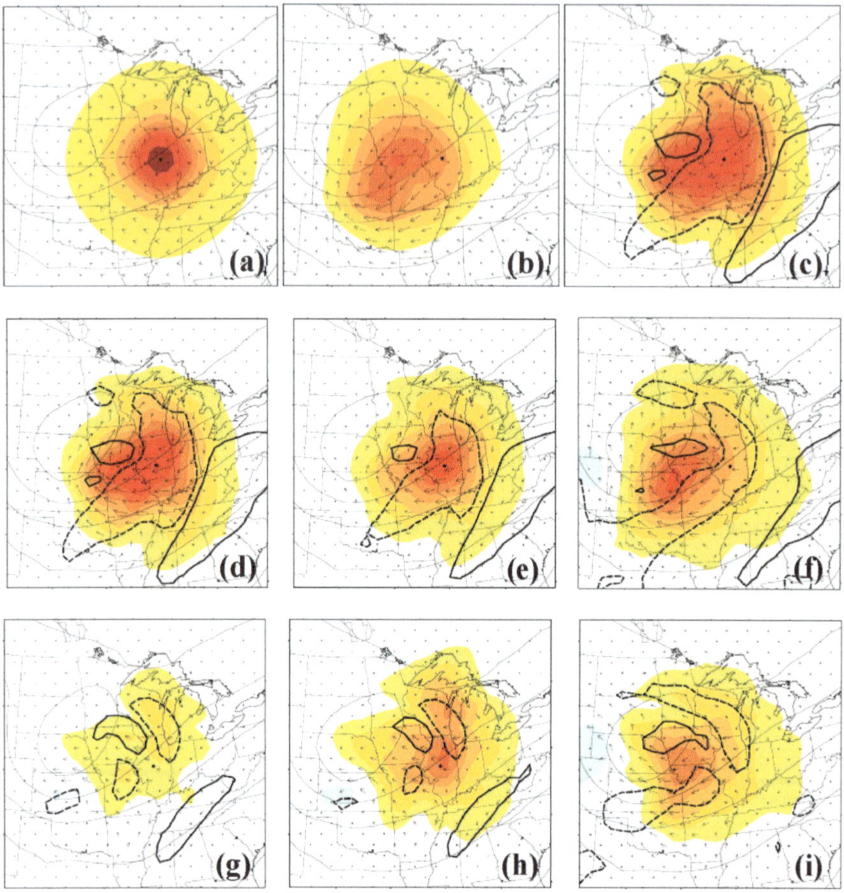

Fig. 9. The DA analysis increments of temperature (shaded, units in K), winds (vectors), and moisture (thick contours, units in g kg^{-1}, solid for positive and dashed for negative) for 1-K innovation at 500 hPa over Illinois (black dot) for (a) 3DVAR, (b) 4DVAR, (c) EnKF with 40 members, (d) EnKF with 40 members, (e) E3DVAR with 40 members, (f) E4DVAR with 40 members, (g) EnKF with 10 members, (h) E3DVAR with 10 members and (i) E4DVAR with 10 members. Thin contours are background geopotential height (unit in m).

this case), which assumes homogenous and isotropic correlations. In contrast, flow-dependent forecast uncertainty and correlations are taken into account during the 4DVAR assimilation (Fig. 9b), with the largest temperature increments shifted upstream and the wind increments located southwest along the background geopotential height contours. There are virtually no moisture increments in the 3DVAR analysis as no multivariate correlations exist between moisture and other variables in the NMC-based background error covariance of the current WRF-Var, though such correlations could play an important role in severe weather systems. A similar situation is shown in the 4DVAR experiment, but with some weak increments to moisture variables that were gained through the linear model iterations. However, with only very simple physics included in the linearization process of 4DVAR (Huang et al., 2009), the moisture increments were an order of magnitude smaller than those from EnKF and thus were not plotted in Fig. 3b. Compared to the 4DVAR, even stronger flow-dependent analysis increments result from the EnKF with 40 ensemble members (Fig. 9c), including stronger temperature increments and northeasterly wind increments.

The coupling of 3DVAR and 4DVAR with EnKF namely E3DVAR (Fig. 9e) and E4DVAR (Fig. 9f), respectively well matches the analysis increment pattern and may be an equivalent solution of the time-evolving error covariance, which can be implicitly modelled through a finite-time window or explicitly estimated by a group of ensemble members (Lorenc, 2003b). One major advantage of using ensemble-based background error covariance in the EnKF is the estimation of multivariate correlations between humidity and other state variables. The representation of cross covariances between temperature and moisture could be highly impacted by the physical parameterizations (Berre, 2000), as accounted for by the model errors. However, if only a small ensemble size is used, sampling errors may contaminate the background uncertainty estimate, even after applying a tighter localization (e.g., Fig. 9g). This would result in a less reliable estimate of forecast covariance, followed by inappropriate analysis increments. The analysis increments from EnKF, E3DVAR and E4DVAR with 10 ensemble members are presented in Figs 9(g-i), respectively. The weighting factor β can be used to control the relative weights of NMC and ensemble-based background error covariance in the E3DVAR and E4DVAR systems.

This shows the advantages of E3DVAR over the standalone 3DVAR and EnKF methods, and E4DVAR hybrid system over 4DVAR and EnKF (Zhang and Zhang, 2012). The performance of E3DVAR and E4DVAR are described as follows.

Figure 10 compares 12-h forecast RMSE for all control experiments, including simulations run from the two coupled systems, E3DVAR and E4DVAR, and the uncoupled component systems, 3DVAR, 4DVAR and EnKF. All forecasts are verified using sounding observations, and errors are averaged over all fifty-nine WRF forecasts for the same month used in the previously discussed sensitivity experiments. Figure 11 shows the corresponding 0- to 72-h time evolution of the domain-averaged RMSE for each control experiment, evaluated every 12 h. The configurations of the control 4DVAR and E4DVAR experiments follow a similar setup as the 3DVAR and E3DVAR cases, except the 4DVAR option in the WRFDA system (Huang et al., 2009) is used. As first observed in Zhang et al. (2011), the WRF versions of the EnKF and 4DVAR systems perform similarly, except that the EnKF has substantially smaller error in the moisture analysis and slightly lower error for winds and temperature. The EnKF and 4DVAR systems perform considerably better than 3DVAR, but are both less accurate than E3DVAR (Figs 10-11). E4DVAR

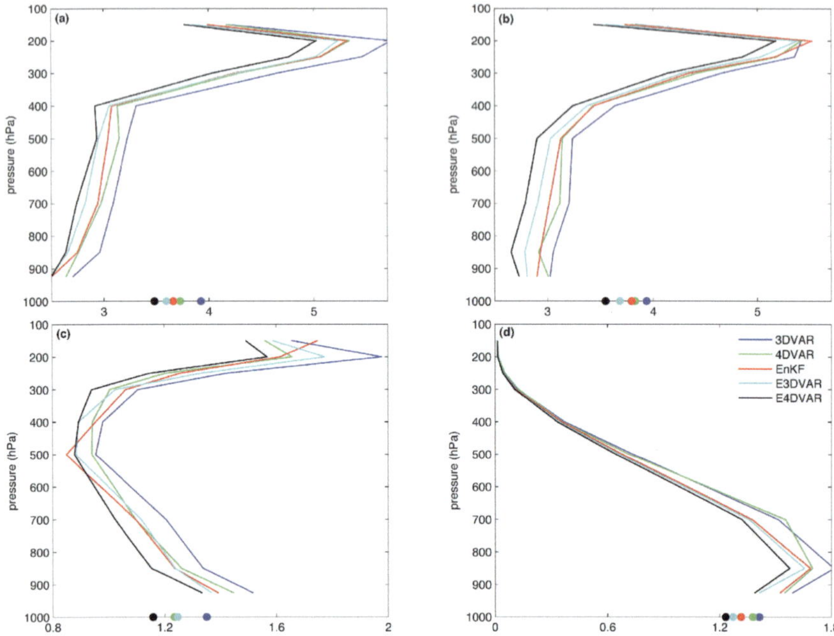

Fig. 10. Vertical profiles of the month-averaged 12-h forecast RMSE of (a) U (m/s), (b) V (m/s), (c) T (K) and (d) Q (g/kg) for various DA methods.

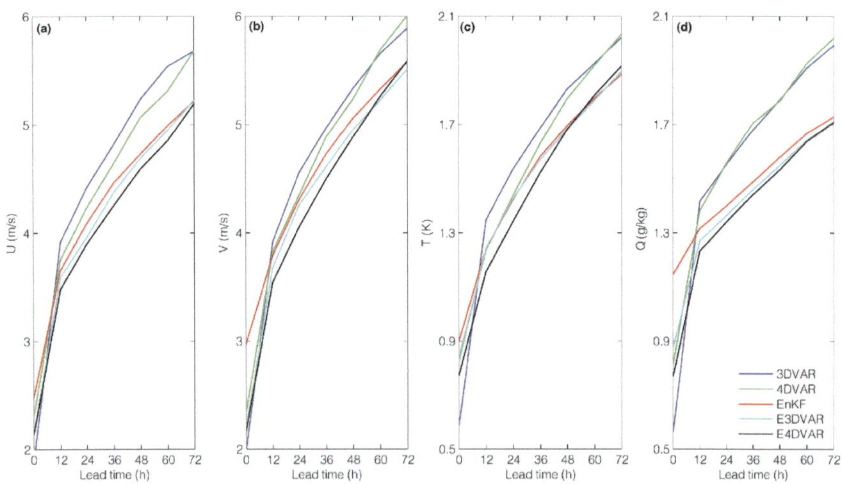

Fig. 11. Domain-averaged RMSE further averaged over all 59 WRF forecasts of the month for each control DA experiments at forecast lead times from 0 to 72 h evaluated every 12 h for (a) U (m/s), (b) V (m/s), (c) T (K) and (d) Q (g/kg).

shows further improvements over E3DVAR by reducing 12-h forecast error at nearly all levels and variables (Fig. 10).

The benefits of E4DVAR over E3DVAR persist for the 48-h forecast lead times, but the two methods converge to a similar level of forecast accuracy as the EnKF by 60 to 72 h (Fig. 11). It is also worth noting that the error amplitudes of all three ensemble-based data assimilation methods (EnKF, E3DVAR and E4DVAR) are quantitatively similar at the 72 h forecast time, all of which are considerably smaller than those of the two variational based methods (3DVAR and 4DVAR) that use static background error covariance (Fig. 11).

The ensemble forecasts provide information regarding the background flow field that goes unresolved by the NMC-based covariance; therefore, the variational data assimilation cases experience large peaks in forecast error during days in which the uncertainty 23 in the flow field deviates from the climatological statistics used to derive the static covariance matrix. The probabilistic methods may also be benefiting from an improved first-guess estimate at each data assimilation cycle that is provided from an ensemble mean.

Based on the month-long statistical results, E4DVAR significantly outperformed all the other DA methods for a 48-h lead time, whose root mean square errors kept in a lower level than the others during the whole month. The monthly mean results also show that E4DVAR had the lowest forecast errors for both dynamical and thermal variables, followed by E3DVAR, EnKF, 4DVAR and 3DVAR in sequence. However, it may be noticed that the E4DVAR is superior to E3DVAR, though it is highly computationally expensive than E3DVAR.

Conclusions

After a brief introduction of the algorithms for EnKF and 3/4DVAR, this chapter explores the inter-comparison and coupling of ensemble kalman filter (EnKF) with variational data assimilation system. The comparison of EnKF system with both the three-dimensional and four-dimensional variational data assimilation (3DVAR and 4DVAR) methods of the Weather Research and Forecasting (WRF) model is presented. The performances of these methods are evaluated through verifying the 12- to 72-h forecasts initialized twice daily from the analysis of each method against the standard sounding observations. It is found that 4DVAR has consistently smaller error than that of 3DVAR for winds and temperature at all forecast lead times except at 60 and 72 h when their forecast errors become comparable in amplitude, while the two schemes have similar performance in moisture at all lead times. The forecast error of the EnKF is comparable to that of the 4DVAR at 12–36-h lead times, both of which are substantially smaller than that of the 3DVAR, despite the fact that 3DVAR fits the sounding observations much more closely at the analysis time. The advantage of the EnKF becomes even more evident at 48–72-h lead times; the 72-h forecast error of the EnKF is comparable in magnitude to the 48-h error of 3DVAR/4DVAR.

The coupling of the EnKF with the variational method seems to be a promising endeavour. The successful implementation of a hybrid data assimilation approach that has two-way coupling of the EnKF with 3DVAR and 4DVAR methods in a limited area weather prediction model, is complementary to recent studies with a global model as in Buehner et al. (2010 a,b). In coupled E3DVAR system, the ensemble-based background error covariance is introduced into the 3DVAR cost function, and the mean of the EnKF analysis is replaced by the 3DVAR analysis. The strengths and weaknesses of the set of data assimilation schemes are investigated under a realistic operational modelling scenario for a month-long cycling data assimilation experiment. Data assimilation approaches that use flow-dependent, multivariate background error covariance produced superior forecasts than methods that solely rely on static covariance in a setting in which conventional data were assimilated. Data assimilation experiments using a 40-member multiple-physics ensemble show reductions in 12- to 72-h forecast root-mean-square error when simulations are initialized with the E3DVAR system as opposed to a similarly configured EnKF. Additional sensitivity experiments show only a small sensitivity of forecast errors from E3DVAR to the weighting coefficient that is used to balance the amount of ensemble-based and static background covariance. E3DVAR may also perform as well as a standard EnKF that uses a much larger ensemble size, given that the system is carefully tuned to account for sampling error. In addition, the hybrid system is less sensitive to the use of multi-physics ensembles that are typically used to treat model error in ensemble data assimilation systems. Our results also show that additive inflation may lead to further improvements in the ensemble-based approaches, because ensemble forecasts typically underestimate the true background uncertainty.

Furthermore, the E3DVAR system is compared with an E4DVAR hybrid that couples the EnKF with a four-dimensional variational system. E4DVAR leads to further forecast improvements at nearly all model levels and variables for forecast lead times between 12 and 48 h, but converges to the same accuracy as the other ensemble-based methods (EnKF and E3DVAR) for lead times between 60 and 72 h. Forecasts from the three ensemble-based methods have considerably smaller errors than forecasts initialized from the two variational methods (3DVAR and 4DVAR) at later lead times, likely due to the use of flow-dependent forecast uncertainty and an ensemble mean first guess field during cycling.

The computational cost of the E4DVAR will be a concern for its full applications, where the major costs are from the iterative inner loops in the 4DVAR and the sequential analysis in the EnKF. The 4DVAR and EnKF have roughly 45- and 5-min wall-clock time for each analysis cycle, respectively, and such costs may increase dramatically for high-resolution modelling, especially in the 4DVAR component. Several methods to minimize these costs include: (i) improve the efficiency of parallel computing of the related codes;

(ii) use a coarse-resolution inner loop for 4DVAR and EnKF analysis, so-called multi-incremental 4DVAR (Courtier et al., 1994) and dual resolution EnKF (Gao and Xue, 2008); and (iii) introduce a second outer loop in 4DVAR with preconditioning from the prestart, half-window analysis.

REFERENCES

Aksoy, A., F. Zhang and J.W. Nielsen-Gammon, 2006a: Ensemble-based simultaneous state and parameter estimation with MM5. *Geophys. Res. Lett.*, **33**, L12801, doi:10.1029/2006GL026186.

Aksoy, A., F. Zhang and J.W. Nielsen-Gammon, 2006b: Ensemble-based simultaneous state and parameter estimation in a two-dimensional sea-breeze model. *Mon. Wea. Rev.*, **134**, 2951-2970.

Aksoy, A., F. Zhang, J.W. Nielsen-Gammon and C.C. Epifanio, 2005: Ensemble-based data assimilation for thermally forced circulations. *J. Geophys. Res.*, **110**, D16105, doi:10.1029/ 2004JD005718.

Anderson, J.L., 2001: An ensemble adjustment Kalman filter for data assimilation. *Mon. Wea. Rev.*, **129**, 2884-2903.

Barker, D.M., W. Huang, Y.-R. Guo, A. Bourgeois and Q.N. Xiao, 2004: A three-dimensional variational data assimilation system for MM5: Implementation and initial results. *Mon. Wea. Rev.*, **132**, 897-914.

Berre, L., 2000: Estimation of synoptic and mesoscale forecast error covariances in a limited area model. *Mon. Wea. Rev.*, **128**, 644-667.

Buehner, M., 2005: Ensemble-derived stationary and flow-dependent background-error covariances: Evaluation in a quasi-operational NWP setting. *Quart. J. Roy. Meteor. Soc.*, **131**, 1013-1043.

Buehner, M., P.L. Houtekamer, C. Charette, H.L. Mitchell and B. He, 2010a: Intercomparison of variational data assimilation the ensemble Kalman filter for global deterministic NWP. Part I: Description and single-observation experiments. *Mon. Wea. Rev.*, **138**, 1550-1566.

Buehner, M., P.L. Houtekamer, C. Charette, H.L. Mitchell and B. He, 2010b: Intercomparison of variational data assimilation and the ensemble Kalman filter for global deterministic NWP. Part II: One-month experiments with real observations. *Mon. Wea. Rev.*, **138**, 1567-1586.

Caya, A., J. Sun and C. Snyder, 2005: A comparison between the 4DVAR and the ensemble Kalman filter techniques for radar data assimilation. *Mon. Wea. Rev.*, **133**, 3081-3094.

Davis, C. et al., 2004: The Bow-Echo and MCV Experiment: Observations and opportunities. *Bull. Amer. Meteor. Soc.*, **85**, 1075-1093.

Dirren, S., R.D. Torn and G.J. Hakim, 2007: A data assimilation case study using a limited-area ensemble Kalman filter. *Mon. Wea. Rev.*, **135**, 1455-1473.

Ehrendorfer, M., 2007: A review of issues in ensemble-based Kalman filtering. *Meteor. Z.*, **16**, 795-818.

Evensen, G., 1994: Sequential data assimilation with a nonlinear quasi-geostrophic model using Monte Carlo methods to forecast error statistics. *J. Geophys. Res.*, **99**, 10143-10162.

Evensen, G., 2003: The ensemble Kalman filter: Theoretical formulation and practical implementation. *Ocean Dyn.*, **53**, 343-367.
Evensen, G., 2007: Data Assimilation: The Ensemble Kalman Filter. Springer.
Gaspari, G. and S.E. Cohn, 1999: Construction of correlation functions in two and three dimensions. *Quart. J. Roy. Meteor. Soc.*, **125**, 723-757.
Hamill, T.M., 2006: Ensemble-based atmospheric data assimilation. *In:* Predictability of Weather and Climate. T. Palmer and R. Hagedorn (Eds). Cambridge University Press, 124-156.
Hamill, T.M. and C. Snyder, 2000: A hybrid ensemble Kalman filter 3D variational analysis scheme. *Mon. Wea. Rev.*, **128**, 2905-2919.
Hawblitzel, D.P., F. Zhang, Z. Meng and C.A. Davis, 2007: Probabilistic evaluation of the dynamics and predictability of a mesoscale convective vortex of 10–13 June 2003. *Mon. Wea. Rev.*, **135**, 1544-1563.
Houtekamer, P.L. and H.L. Mitchell, 1998: Data assimilation using an ensemble Kalman filter technique. *Mon. Wea. Rev.*, **126**, 796-811.
Houtekamer, P.L., H.L. Mitchell and X. Deng, 2009: Model error representation in an operational ensemble Kalman filter. *Mon. Wea. Rev.*, **137**, 2126-2143.
Houtekamer, P.L., H.L. Mitchell, G. Pellerin, M. Buehner, M. Charron, L. Spacek and B. Hansen, 2005: Atmospheric data assimilation with an ensemble Kalman filter: Results with real observations. *Mon. Wea. Rev.*, **133**, 604-620.
Huang, X.-Y. et al., 2009: Four-dimensional variational data assimilation for WRF: Formulation and preliminary results. *Mon. Wea. Rev.*, **137**, 299-314.
Liu, C., Q. Xiao and B. Wang, 2008: An ensemble-based four-dimensional variational data assimilation scheme. Part I: Technical formulation and preliminary test. *Mon. Wea. Rev.*, **136**, 3363-3373.
Liu, C., Q. Xiao and B. Wang, 2009: An ensemble-based four-dimensional variational data assimilation scheme. Part II: Observing system simulation experiments with Advanced Research WRF (ARW). *Mon. Wea. Rev.*, **137**, 1687-1704.
Lorenc, A.C., 2003: The potential of the ensemble Kalman filter for NWP—A comparison with 4D-Var. *Quart. J. Roy. Meteor. Soc.*, **129**, 3183-3203.
Meng, Z. and F. Zhang, 2007: Tests of an ensemble Kalman filter for mesoscale and regional-scale data assimilation. Part II: Imperfect model experiments. *Mon. Wea. Rev.*, **135**, 1403-1423.
Meng, Z. and F. Zhang, 2008a: Tests of an ensemble Kalman filter for mesoscale and regional-scale data assimilation. Part III: Comparison with 3DVar in a real-data case study. *Mon. Wea. Rev.*, **136**, 522-540.
Meng, Z. and F. Zhang, 2008b: Tests of an ensemble Kalman filter for mesoscale and regional-scale data assimilation. Part IV: Comparison with 3DVar in a month-long experiment. *Mon. Wea. Rev.*, **136**, 3671-3682.
Mitchell, H.L., P.L. Houtekamer and G. Pellerin, 2002: Ensemble size, balance, and model-error representation in an ensemble Kalman filter. *Mon. Wea. Rev.*, **130**, 2791-2808.
Parrish, D.F. and J.C. Derber, 1992: The National Meteorological Center's spectral statistical-interpolation analysis system. *Mon. Wea. Rev.*, **120**, 1747-1763.
Skamarock, W.C., J.B. Klemp, J. Dudhia, D.O. Gill, D.M. Barker, W. Wang and J.G. Powers, 2005: A description of the Advanced Research WRF Version 2. NCAR Tech. Note 4681STR.
Snyder, C. and F. Zhang, 2003: Assimilation of simulated Doppler radar observations with an ensemble Kalman filter. *Mon. Wea. Rev.*, **131**, 1663-1677.

Tong, M. and M. Xue, 2005: Ensemble Kalman filter assimilation of Doppler radar data with a compressible nonhydrostatic model: OSSE experiments. *Mon. Wea. Rev.*, **133**, 1789-1807.

Torn, R.D. and G.J. Hakim, 2008a: Performance characteristics of a pseudo-operational ensemble Kalman filter. *Mon. Wea. Rev.*, **136**, 3947-3963.

Torn, R.D. and G.J. Hakim, 2009a: Ensemble data assimilation applied to RAINEX observations of Hurricane Katrina (2005). *Mon. Wea. Rev.*, **137**, 2817-2829.

Wang, X., C. Snyder and T.M. Hamill, 2007: On the theoretical equivalence of differently proposed ensemble–3DVAR hybrid analysis schemes. *Mon. Wea. Rev.*, **135**, 222-227.

Wang, X., T.M. Hamill, J.S. Whitaker and C.H. Bishop, 2009: A comparison of the hybrid and EnSRF analysis schemes in the presence of model error due to unresolved scales. *Mon. Wea. Rev.*, **137**, 3219-3232.

Whitaker, J.S. and T.M. Hamill, 2002: Ensemble data assimilation without perturbed observations. *Mon. Wea. Rev.*, **130**, 1913-1924.

Whitaker, J.S., T.M. Hamill, X. Wei, Y. Song and Z. Toth, 2008: Ensemble data assimilation with the NCEP global forecast system. *Mon. Wea. Rev.*, **136**, 463-482.

Yang, S.C., E. Kalnay, B. Hunt and N.E. Bowler, 2009: Weight interpolation for efficient data assimilation with the local ensemble transform Kalman filter. *Quart. J. Roy. Meteor. Soc.*, **135**, 251-262.

Zhang, F. and C. Snyder, 2007: Ensemble-based data assimilation. *Bull. Amer. Meteor. Soc.*, **88**, 565-568.

Zhang, F., C. Snyder and J. Sun, 2004: Impacts of initial estimate and observation availability on convective-scale data assimilation with an ensemble Kalman filter. *Mon. Wea. Rev.*, **132**, 1238-1253.

Zhang, F., Y. Weng, J.A. Sippel, Z. Meng and C.H. Bishop, 2009a: Cloud-resolving hurricane initialization and prediction through assimilation of Doppler radar observations with an ensemble Kalman filter. *Mon. Wea. Rev.*, **137**, 2105-2125.

Zhang, F., Z. Meng and A. Aksoy, 2006: Tests of an ensemble Kalman filter for mesoscale and regional-scale data assimilation. Part I: Perfect model experiments. *Mon. Wea. Rev.*, **134**, 722-736.

Zhang, M. and F. Zhang, 2012: E4DVar: Coupling an ensemble Kalman filter with four dimensional variational data assimilation in a limited-area weather prediction model. *Mon. Wea. Rev.*, **140**, 587-600.

Zhang, M., F. Zhang, X. Huang and X. Zhang, 2011: Intercomparison of an ensemble Kalman filter with three- and four-dimensional variational data assimilation methods in a limited-area model over the month of June 2003. *Mon. Wea. Rev.*, **139**, 566-572.

Zhang, S. and J.L. Anderson, 2003: Impact of spatially and temporally varying estimates of error covariance on assimilation in a simple atmospheric model. *Tellus*, **55A**, 126-147.

14

Operational ACCESS-TC: Vortex Specification, 4DVAR Initialization, Verification and Structure Diagnostics

Noel Davidson, Yi Xiao, Harry Weber, Yimin Ma, Charlie Lok, Lawrie Rikus, Jeff Kepert, Peter Steinle, James Fraser and Radhika Kanase[1]

Centre for Australian Weather and Climate Research (CAWCR)
P.O. Box 1289, Melbourne, Australia 3001
[1]Indian Institute of Tropical Meteorology, Pune, India
n.davidson@bom.gov.au

1. Introduction

The Australian Community Climate and Earth System Simulator, ACCESS (Puri et al., 2010, 2012) has been configured for operational and research applications on Tropical Cyclones. ACCESS is an implementation at the Australian Bureau of Meteorology of the full UK Meteorological Office's Numerical Modelling System (Gregory and Rowntree, 1990; Wilson and Ballard, 1999; Webster et al., 2003; Davies et al., 2005; Rawlins et al., 2007 and references therein). The ACCESS-TC system runs at a resolution of 0.11° and 50 levels. The domain is re-locatable and nested in coarser-resolution forecasts. Initialization consists of five cycles of 4DVAR over 24 hours and forecasts to 72 hours are made. Without vortex specification, initial conditions usually contain a weak and misplaced circulation (Zou and Xiao, 2000). This makes it necessary to incorporate a synthetic TC vortex at the observed location in initial conditions and preferably during the data assimilation (e.g., Kurihara et al., 1995; Davidson and Weber, 2000). In ACCESS-TC, based on estimates of central pressure and storm size, vortex specification is used to filter the analysed circulation from the original analysis, construct the inner-core of the storm, merge it with the large scale analysis at outer radii, and relocate it to the observed position. Since TC reconnaissance is not conducted in the Australian Region, significant effort has been devoted to building physically based, synthetic inner-core structures, validated using historical dropsonde archives from the Atlantic and northeast Pacific. The power of the 4DVAR method allows the use of only synthetic mean sea level pressure observations to build the horizontal and vertical structure of storms.

This report summarizes some of the components and characteristics of ACCESS-TC.

2. System Configuration

The system configuration is summarized as follows:

(i) Resolution
0.11° × 50L, on a 300 × 300 re-locatable grid with TC near centre of domain. Option is available for higher resolution forecasts, initialized from and nested within the standard configuration.

(ii) Vortex Specification
(a) Structure based on observed location, central pressure and size (tuned and validated using ~6000 dropsonde observations from the Atlantic).
(b) Only synthetic MSLP observations used in the 4DVAR to
- relocate the storm to observed location,
- define the inner-core circulation, and
- impose steering flow asymmetries consistent with the past motion.

(iii) Initialization using 4DVAR assimilation
Five cycles of 4DVAR over the 24 hours prior to the start of the forecast. Uses all standard observational data in the 24 hours including those available at the initial time of the forecast. The observational data base includes: surface stations, ships, drifting buoys, aircraft, radiosondes in addition to pilot and profiler winds, ATOVS (HIRS, AMSU-A and AMSU-B), AIRS, atmospheric motion-vector winds (AMVs) derived from geostationary satellite data, scatterometer winds (ASCAT) and synthetic MSLP observations from the vortex specification (no upper air synthetic observations).

(iv) Forecast Model: UKMO Unified Model from ACCESS.

3. Vortex Specification and Validation

Our goal is to make skillful track, intensity and structure forecasts. However, because of the lack of observational data to define the location and inner-core structure of most storms, it is necessary to develop techniques for vortex initialization. The vortex specification is based on the idealized profile of Chan and Williams (1987), but tuned on observed structures for the North Atlantic where reconnaissance is frequently used to define vortex structure. The vortex specification uses estimates of storm location, central pressure and large scale analyses to obtain the vortex structure parameter set (latitude, longitude, CP,

Vmax, RMW, R34, ROCI, where CP = Central Pressure, Vmax = Maximum winds, RMW = Radius of Maximum Winds, R34 = Radius of gale force (34 knot) winds and ROCI = Radius of Outer Closed Isobar). From this idealized vortex we extract synthetic mean sea level pressure observations to pass to the 4DVAR assimilation. The density and radial extent of synthetic observations is sufficient to define the maximum wind at the RMW and to allow smooth merging of the vortex at outer radii.

Validation of the idealized structures is presented in Fig. 1 which shows a comparison between values of R34 and RMW from subjective estimates obtained from an extended best track data set (Demuth et al., 2006) and from the vortex specification. RMW, and to a lesser extent R34, are difficult to estimate. Subjective estimates are made within the time constraints of an operational forecast office, do not contain any dynamical constraints (all parameter estimates are independent of one another), and can be variable in quality depending on the skill and experience of the duty forecaster. Scatterometer observations and AMVs are not always available and this makes subjective estimation difficult. The mean absolute differences for RMW and R34 are 15.8 and 40.5 nm. Given the large uncertainties in the subjective estimates of RMW and R34, we think the objective estimates have comparable quality. In the mean, the differences are not large. For individual cases, as illustrated, estimates can be rather different. Mean differences are likely within the observational errors for these parameters and so we suggest that the vortex structures used in ACCESS-TC are the best that can be specified without actual reconnaissance observations.

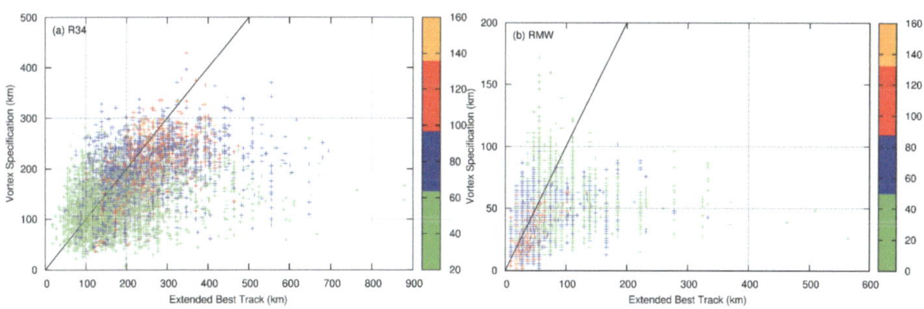

Fig. 1. Comparison of (a) R34 and (b) RMW between the Extended Best Track estimates and the vortex specification calculation based only on estimates of central pressure and radius of outer closed isobar. Units are km. Colour key follows the Saffir-Simpson scale according to Vmax: tropical storm (green), hurricane categories 1-2 (blue), 3-4 (red) and 5 (orange).
Source: Demuth et al. (2006)

4. 4DVAR Initialization

In simple terms, the 4DVAR iterates back and forth over a time-window (in this case six hours) and attempts to fit both the observations and the model trajectory, taking into account (i) the quality of the observations, defined via observational errors, and (ii) the quality of the model forecast, defined via a forecast-error covariance function.

The 4DVAR uses the synthetic observations to:

(a) define the horizontal structure of the inner-core circulation (CP, Vmax, RMW, R34) at the observed location,
(b) build the vertical structure of the circulation from MSLP observations,
(c) construct the secondary circulation without constraining it through the use of synthetic wind observations,
(d) create a balanced TC circulation at the observed location, with correct structure and intensity, and
(e) create a structure which is responsive to environmental influences (e.g., wind shear) without imposing constraints on the vertical-stacking or tilt of the circulation. These are important for vortex dynamics and the development of appropriate cloud asymmetries.

(i) Primary Circulation

The impact of the synthetic observations used by the 4DVAR is shown in Fig. 2. Note the impact of the vortex specification on (a) the initial location of the storm, and (b) prediction of the track. Note also the construction of the 3D structure (500 hPa cyclonic circulation) from the 4DVAR based only on the use of synthetic MSLP observations. 4DVAR uses the synthetic MSLP observations to define the horizontal structure, vertical depth and tilt of system; these are all important for the evolution of the vortex. The technique defines the primary circulation.

(ii) Secondary Circulation

It is also important that the initialization builds the secondary circulation to be in quasi-balance with the primary circulation (see for example, the quasi-balanced constraint of the Sawyer–Eliassen equation). We illustrate the construction by 4DVAR of the ascent branch of the secondary circulation using the vertical motion field within the inner circulation (Fig. 3).

Note that the vertical motion and moisture fields are initialized during the 4DVAR assimilation. Further there is quite reasonable correspondence between the regions of observed active inner rainbands and eyewall convection, and regions of strong and weak ascent obtained after initialization. The 4DVAR has built the vertical motion field which is quite consistent with the observed structure.

We illustrate the construction by 4DVAR of the whole TC circulation using a case study of TC Yasi from the Australian region. Figure 4 shows (i) observed and forecast tracks and intensities for TC Yasi from base time 00UTC 31 January 2011, (ii) cross-sections of zonal and meridional wind and vertical motion at the initial time of the forecast following the application of 4DVAR over 24 hours in five cycles. Note first that the forecast skill for both

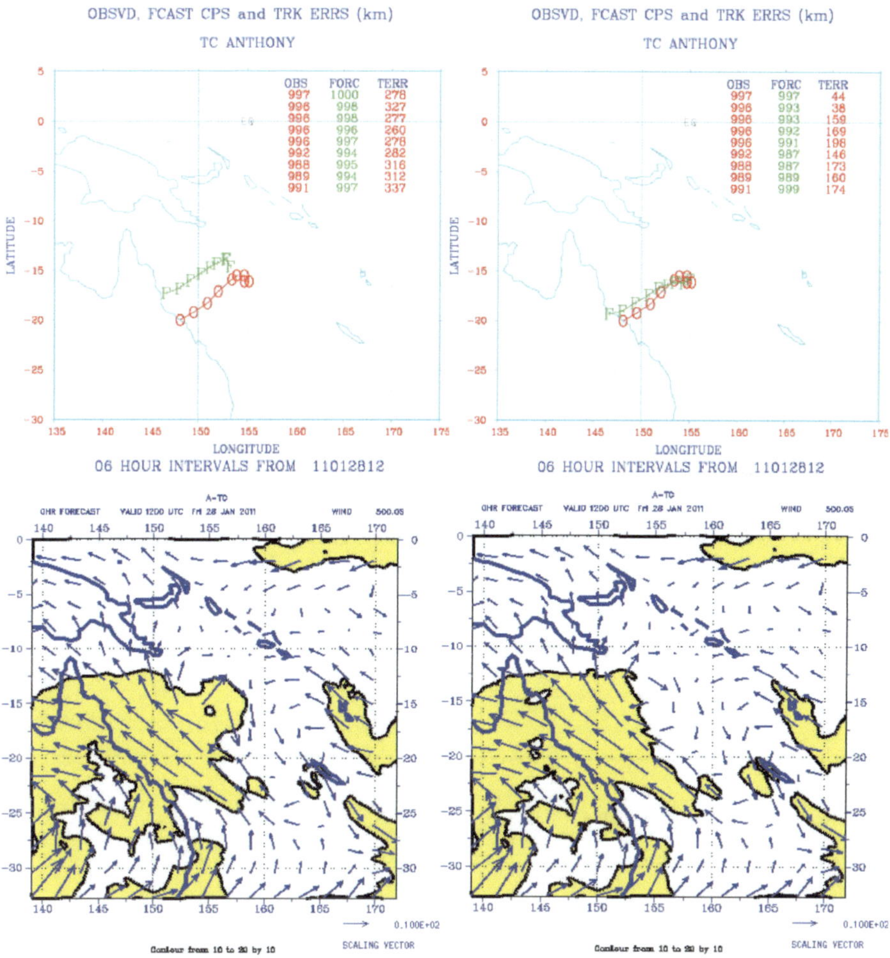

Fig. 2. For TC Anthony: Upper panel: observed and forecast track and intensity without and with vortex specification. Tables show observed (OBS) and forecast (FORC) central pressures (hPa), and track errors (TERR, kms). Lower panel: initial condition for 500 hPa wind without and with vortex specification (synthetic MSLP observations only). Contour interval is 10 ms^{-1}. Winds greater than 10 ms^{-1} are shaded yellow.

track and intensity is rather encouraging. The 60-hour track error near landfall is around 100 km and the forecast has skill in predicting the intensification but does not forecast the rapid intensification just prior to landfall. Initialization of the primary circulation can be seen in the zonal wind cross-section, which reveals an inner-core with a low-level Vmax of approximately 35 ms^{-1} at a radius of maximum winds of approximately 50 km on the equatorward side of the storm. The initialized secondary circulation can be seen in the north-south cross-sections of meridional wind and vertical motion, with (a) radial inflow (outflow) concentrated mostly in the boundary layer (upper troposphere) and

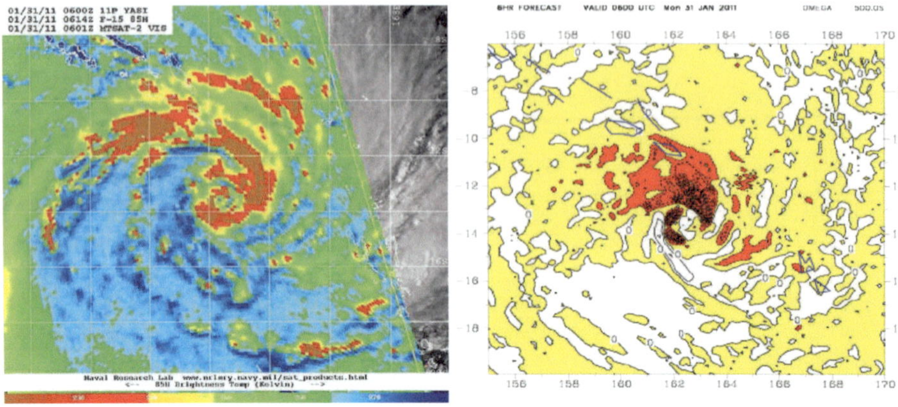

Fig. 3. 85 GHz Imagery (left panel) and ACCESS-TC 500 hPa vertical motion field at $t = 6$ hrs (initialized with 4DVAR) for Yasi from base time 00Z 20110131.

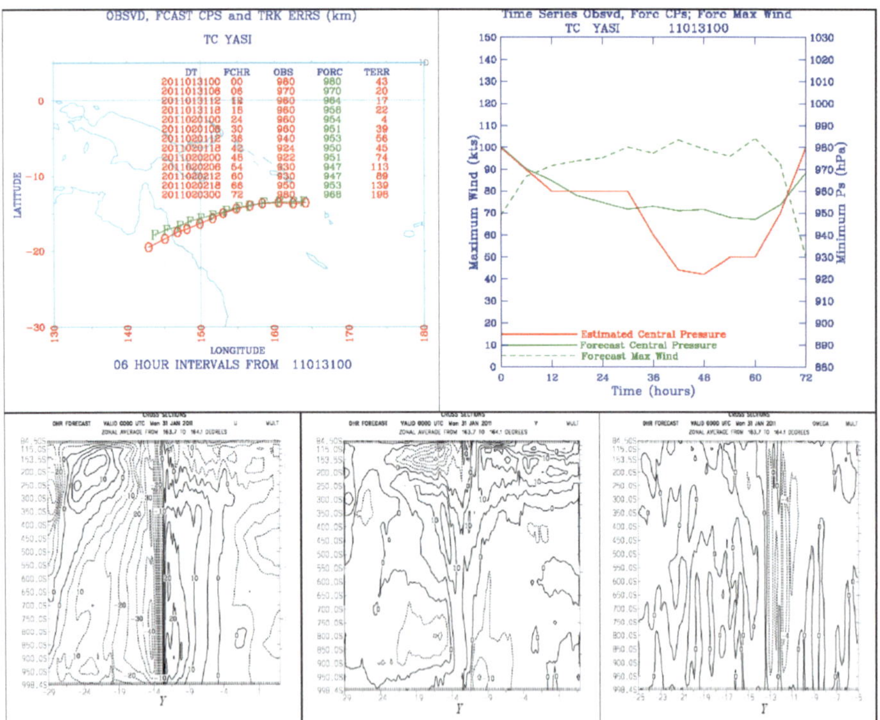

Fig. 4. Top panels: Observed and forecast tracks and intensities for TC Yasi from base time 0000UTC 31 January 2011. Tables show observed (OBS) and forecast (FORC) central pressures (hPa), and track errors (TERR, kms). Lower panels, left to right: north-south cross-sections of zonal wind (west winds are positive, ms^{-1}), meridional wind (south winds are positive, ms^{-1}), and vertical motion (hPa/s, ascent is negative). The 'Y' on the abscissa indicates Yasi's location.

extending to at least 1000 km from the centre of the storm, and (b) ascent focussed at small radii on Yasi's equatorward side. Note that at this time, Yasi was being influenced by southeasterly environmental wind shear and so we might expect convection to be mostly located to the north of the circulation.

5. Verification

Verification statistics for the 2011 tropical cyclone season over the North West Pacific are shown in Fig. 5. The track and intensity forecasts for around 100 forecasts are quite encouraging. So far we have based intensity verification on central pressure. We hope to extend the verification to maximum wind and other vortex parameters (possibly RMW and R34) in the near future.

An example of track and intensity forecasts for one storm, TC Ma-On, is shown in Fig. 6. The track forecasts mostly stay tight on the observed track and the intensity forecasts show some skill but tend to be slow with the intensification and slow with the weakening. These systematic errors are the subject of ongoing experiments. Testing suggests that the errors will be reduced with the next upgrade of ACCESS-TC.

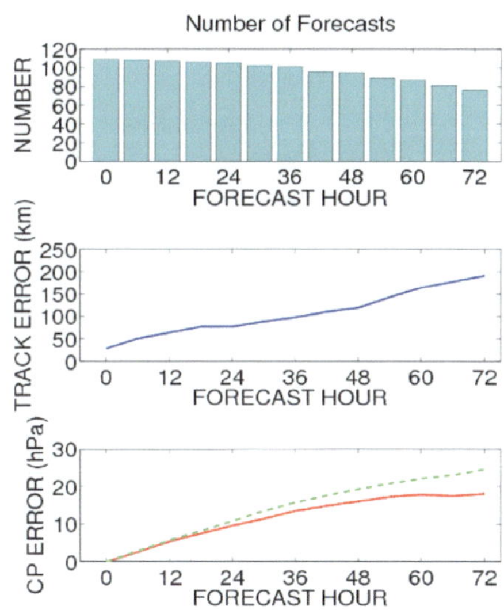

Fig. 5. Verification statistics for ACCESS-TC for the 2011 tropical cyclone season over the North West Pacific. Top panel shows the number of forecasts at each forecast time; middle panel shows track error (km) at each forecast time; lower panel shows mean absolute central pressure error (hPa) at each forecast time. The green curve on the lower panel is the mean absolute central pressure error using persistence as the forecast.

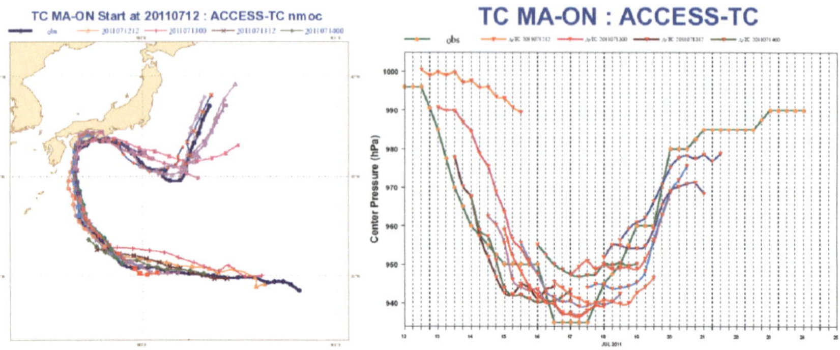

Fig. 6. Observed and forecast tracks and intensities for TC Ma-On using the ACCESS-TC model.

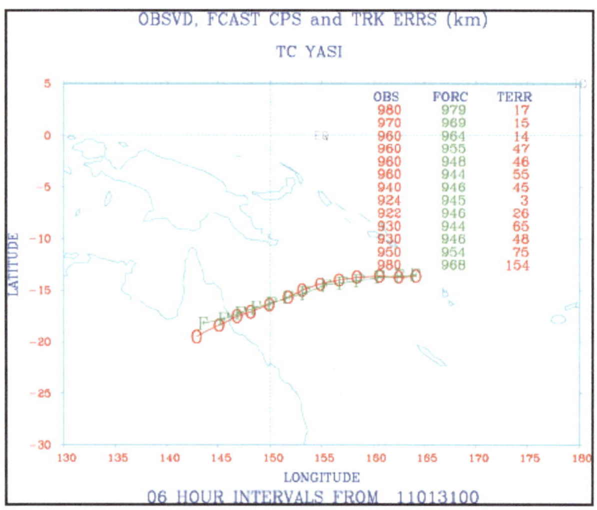

Fig. 7. Estimated and forecast tracks and central pressures for TC Yasi from base time 0000UTC 31 January 2011, at 6-hourly intervals to 72 hours. The red "o"s are observed locations and the green "F"s are corresponding forecast locations. The table shows observed (OBS) and forecast (FORC) central pressures (hPa), and forecast track errors (TERR, kms).

6. Illustrative Example of Forecast Vortex Structure Change: TC Yasi

Historically track and intensity (as measured by central pressure or maximum wind) have been the main metrics used to verify TC forecasts. However understandably, TC structure, as indicated by for example, RMW and/or R34, is becoming increasingly important (Ma et al., 2012; Davidson and Ma, 2012). It is thus interesting to document forecasts of structure change from a sophisticated NWP system like ACCESS-TC for an intensifying, high impact storm at landfall. Figure 7 shows estimated and forecast tracks and central pressures from an ACCESS-TC forecast for TC Yasi from base time 00 UTC 31 January 2011. This forecast was run at a resolution of 0.04° with 50 levels and initialized from the 0.11° resolution forecast as described above for operational ACCESS-TC. The skill of the forecast is rather encouraging. The track error near landfall was less than 50 km, and estimated and forecast central pressures were around 930 and 945 hPa respectively. Figure 8 shows time series of forecast central pressure, and azimuthal-mean values of maximum wind, radius of maximum wind, R64, R50 and R34 from the ACCESS-TC forecast. The time series show the forecast intensification and also interestingly show the inward contraction of Rmax, the growth and expansion of the circulation as indicated by the R34 time series, and the development of R50 and then R64 together with the inward contraction of these radii. During and following intensification the circulation grows in size while the inner circulation contracts. Documentation and understanding of the way in which these structure changes occur will be the focus of ongoing research.

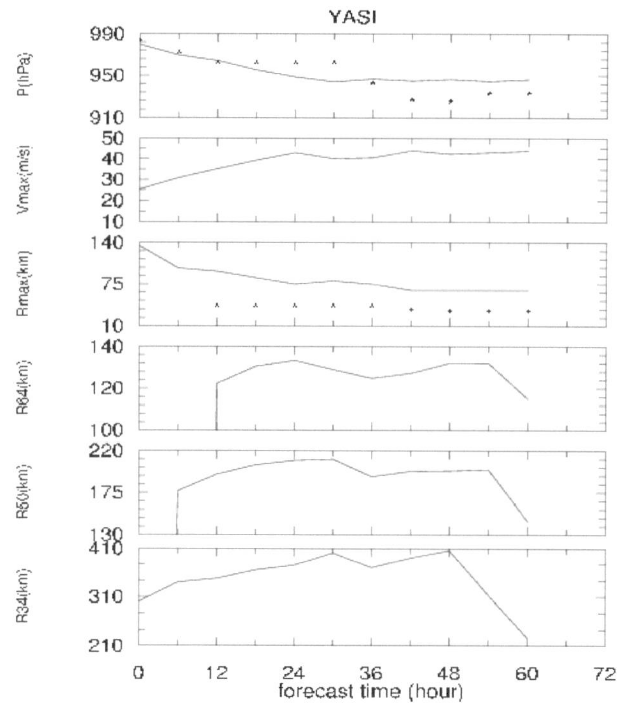

Fig. 8. Time series of forecast central pressure (hPa), and azimuthal-mean values of maximum wind (ms^{-1}), radius of maximum wind, R64, R50 and R34 (kms), for TC Yasi from base time 0000UTC 31 January 2011. The symbols in the top panel are real-time estimates of central pressure. The symbols in the third top panel are spot values of RMW, as opposed to the graphed azimuthal-mean values.

7. Summary and Future Plans

ACCESS-TC has been developed for operations and research. Unique aspects of the initialization and performance are:

- Use of 4DVAR and comprehensive observational data sets to define the LSE of storms.
- Application of a physically based, validated vortex specification to construct the inner-core of an idealized TC circulation.
- Use of 4DVAR assimilation to build a balanced horizontal and vertical structure, and the secondary circulation from only synthetic MSLP observations extracted from the idealized TC.
- Verification of track, intensity and structure for Australian region seasons 2010-11, 2011-12 and NW Pacific seasons 2011, 2012 has been rather encouraging.

Future plans include:

- General system upgrades (use of more satellite data in assimilation, higher resolution, improved physics).
- Real time forecasts from pre-genesis over the north Atlantic (assimilation of inner-core reconnaissance data with 4DVAR).
- Specification, prediction and validation of TC structure (CP, Vmax, RMW, R34, ROCI): Critical for prediction of track, intensity, structure, storm surge and rainfall.
- Experiments with high resolution initialization and prediction:
  - (i) Experiments with Ensemble Prediction and "Hybrid Assimilation" (as implemented by the UKMO).
  - (ii) Experiments with revised and new physics: Diagnostics for TC boundary layer and moist processes.
- Possible enhancements to 4DVAR (inner and outer loops).
- NWP and basic research applications from special experimental data sets: TPARC/TCS08, PREDICT: Genesis and Rapid Intensification

REFERENCES

Chan, J.C. and R.T. Williams, 1987: Analytical and numerical studies of the beta-effect in tropical cyclone motion. Part I: Zero mean flow. *J. Atmos. Sci.*, 44, 1257-1265.

Davidson, N.E. and Y. Ma, 2012: Surface Pressure Profiles, Vortex Structure and Initialization for Hurricane Prediction. Part II: Numerical Simulations of Track, Structure and Intensity. *Meteorol. Atmos. Phys*, **117**, 25-45.

Davidson, N.E. and H.C. Weber, 2000: The BMRC High-Resolution Tropical Cyclone Prediction System: TC-LAPS. *Mon. Wea. Rev.*, **128**, 1245-1265.

Davies, T., M.J.P. Cullen, A.J. Malcolm, M.H. Mawson, A. Staniforth, A.A. White and N. Wood, 2005: A new dynamical core for the Met Office's global and regional modelling of the atmosphere. *Q. J. Roy. Meteorol. Soc.*, **131**, 1759-1782.

Demuth, J., M. DeMaria and J.A. Knaff, 2006: Improvement of advanced microwave sounder unit tropical cyclone intensity and size estimation algorithms. *J. Appl. Meteor.*, **45**, 1573-1581.

Gregory, D. and P.R. Rowntree, 1990: A mass flux convection scheme with representation of cloud ensemble characteristics and stability-dependent closure. *Mon. Wea. Rev.*, **118**, 1483-1506.

Kurihara, Y., M.A. Bender, R.E. Tuleya and R.J. Ross, 1995: Improvements in the GFDL Hurricane Prediction System. *Mon. Wea. Rev.*, **123**, 2791-2801.

Rawlins, F., S.P. Ballard, K.J. Bovis, A.M. Clayton, D. Li, G.W. Inverarity, A.C. Lorenc and T.J. Payne, 2007: The Met Office global four-dimensional variational data assimilation scheme. *Quart. J. Roy. Met. Soc.*, **133**, 347-362.

Ma, Y., M. Kafatos and N.E. Davidson, 2012: Surface Pressure Profiles and Initialization for Hurricane Prediction. Part I: Analysis of Observed and Synthetic Structures. *Meteorol. Atmos. Phys*, **117**, 5-23.

Puri, K. et al., 2013: Implementation of the initial ACCESS Numerical Weather Prediction system. *Aust. Meteor. Oceanograph. J.*, **63**, 265-284.

Puri, K. et al., 2010: Preliminary results from Numerical Weather Prediction implementation of ACCESS. *CAWCR Research Letters*, **5**, 15-22. http://www.cawcr.gov.au/publications/researchletters/CAWCR_Research_Letters_5.pdf.

Webster, S., A.R. Brown, D.R. Cameron and C.P. Jones, 2003: Improvements to the representation of orography in the Met. Office Unified Model. *Q. J. Roy. Meteorol. Soc.*, **129**, 1989-2010.

Wilson, M. and S.P. Ballard, 1999: A microphysically based precipitation scheme for the UK Meteorological Office Unified Model. *Q. J. Roy. Meteorol. Soc.*, **125**, 1607-1636.

Zou, X. and Q. Xiao, 2000: Studies on the Initialization and Simulation of a Mature Hurricane Using a Variational Bogus Data Assimilation Scheme. *J. Atmos. Sci.*, **57**, 836-860.

15

Physical Initialization in Tropical Cyclone Forecasting

T.N. Krishnamurti[1], A. Jaya Kumar[2], Y.E.A. Raj[3] and S.B Thampi[3]

[1]Florida State University, Tallahassee, Fl-32306, USA
[2]National Centre for Medium Range Weather Forecasting (NCMRWF) NOIDA, UP – 201309, India
[3]Regional Meteorological Centre, India Meteorological Department Chennai – 600001
tkrishnamurti@fsu.edu

1. Introduction

To improve the definition of the initial state and forecasts within the tropics is an objective at many numerical weather prediction centres. Physical initialization is an attempt to improve the analysis of the basic state variables by using nonconventional data like rain rates, outgoing longwave radiation (OLR) or condensation heating rates derived from rain rates. The initialization procedure consists of two steps (Krishnamurti et al., 1991, 1994). This presentation covers the handling of physical initialization for cloud resolving non-hydrostatic models:

1. Diagnostic calculation of surface flux, humidity analysis consistent with surface flux, observed rainfall data and net OLR.
2. Newtonian Relaxation process applied during each time step of a model forecast for certain variables.

Details of procedure are as follows:

1.1 Yanai Fluxes

Following Yanai et al. (1973), integral technique to obtain diagnostic measures of surface flux of water vapour and sensible heat using input of observed rainfall rate and vertically integrated radiative heating rates are used.

The surface fluxes of sensible heat (F_S) and moisture (F_L) are expressed as

$$F_S \uparrow = Q_1 - Q_R - LP$$

where Q_1 denotes vertically integrated apparent heat source.

$$Q_1 = \frac{\partial \bar{s}}{\partial t} + \nabla_\sigma . \overline{sv} + \frac{\partial}{\partial \sigma} \bar{s}\bar{\sigma}$$

where $\bar{S}$ is the large scale dry static energy and Q_R denotes the vertically integrated net radiative heating;

$$F_l \uparrow = -Q_2 + LP$$

where Q_2 is the vertically integrated apparent moisture sink and P is the measure of the observed rainfall rate, where

$$Q_2 = -L\left(\frac{\partial \bar{q}}{\partial t} + \nabla_\sigma . \overline{qv} + \frac{\partial}{\partial \sigma} \bar{q}\dot{\sigma}\right)$$

The vertical integeral approach of Yanai fluxes provides information for Q_2 from the deep troposphere. The $\dot{\sigma}$ is calculated using mass continuity equation. Tropical precipitation estimate is based on the mix of raingauge and OLR based regression methods.

1.2 Reverse Similarity Theory

Given the Yanai fluxes of sensible and latent heat as an input to the similarity theory, one solves for the potential temperature and the moisture variable at the top of the constant flux layer. Different sets of similarity equations describe stable ($L > 0$) and unstable ($L < 0$) for the surface layers.

Non-dimensionalized vertical gradients of large-scale quantities such as wind, potential temperature and specific humidity can be expressed as universal function of a non-dimensional height z/L, where z is the height above the earth surface and L is the *Monin-Obukhov* length.

The empirical fits of the boundary layer observation are explained in terms of stability. Stability is expressed as a function of L, Richardson number, Ri. The definition of stability is based on surface heat flux; unstable (heat flux up), stable (heat flux down) or neutral (heat flux = 0) corresponds to $L < 0$, $L > 0$ and $L = 0$, respectively.

$$L = \frac{u^{*2}}{\kappa \beta \theta}$$

Following, similarity fluxes of momentum, heat and moisture respectively by the relations,

$$F_M = \rho C_M (U_2 - U_1)^2,$$
$$F_M = \rho C_p C_H (U_2 - U_1)(\theta_2 - \theta_1),$$
$$F_Q = \rho C_q (U_2 - U_1)(q_2 - q_1) g_w$$

Empirical fits of the boundary layer observations for the unstable case are expressed as

$$\frac{\kappa z}{u^*}\frac{\partial \bar{u}}{\partial z} = \left(1 - 15\frac{z}{L}\right)^{-1/4},$$

$$\frac{\kappa z}{\theta^*}\frac{\partial \bar{\theta}}{\partial z} = 0.74\left(1 - 9\frac{z}{L}\right)^{-1/2}$$

$$\frac{\kappa z}{q^*}\frac{\partial \bar{q}}{\partial z} = 0.74\left(1 - 9\frac{z}{L}\right)^{-1/2}$$

and for the stable case are expressed as,

$$\frac{\kappa z}{u^*}\frac{\partial \bar{u}}{\partial z} = 1.0 + 4.7\frac{z}{L},$$

$$\frac{\kappa z}{\theta^*}\frac{\partial \bar{\theta}}{\partial z} = 0.74 + 4.7\frac{z}{L}$$

$$\frac{\kappa z}{q^*}\frac{\partial \bar{q}}{\partial z} = 0.74 + 4.7\frac{z}{L}$$

Reverse similarity theory solves for θ_2 and q_2 given the surface flux of heat and moisture. For the stable case the exact solution for θ_2 and q_2 have no convergence problem, whereas in unstable case the Newton Rhapson iterations provide convergence to the prescribed Yanai fluxes within the 3 to 4 iterations is expressed schematically as

$$Q2(Yanai) \rightarrow q'w' \rightarrow q^* \rightarrow L \rightarrow (q2, T2, U2)$$

1.3 Reverse Cumulus Parameterization

The vertical distribution of specific humidity is reanalyzed such that the rainfall implied by the cumulus parameterization algorithm closely matches the prescribed observed rainfall rates (observed estimates are interpolated to each time step). The old procedure for the construction of a reverse cumulus parameterization is designed for Kuo's scheme (Krishnamurti et al., 1991). This involves two parameters b (moistening) and n (meso-scale convergence) parameter. The large-scale moisture convergence is defined by $I_L = -\frac{p_s}{g}\int_{\sigma_\tau}^{\sigma_B} \sigma \frac{\partial q}{\partial \sigma} d\sigma$ and the total moisture supply is denoted by $I = I_L\,(1 + n)$.

The total moistening and rainfall rates are expressed as

$$M = I_L\,(1 + \eta)b$$

$$R = I_L\,(1 + \eta)(1 - b).$$

The modified Kuo's scheme makes use of multiple regression to optimize the heating, moistening and rainfall rates. Here M and R are expressed as functions of large-scale variable that have strong control on deep convection.

Thus the evolving large scale variables of a forecast determine the local values of M and R, which in turn determine b and n. Thus a closure of the parameterization is accomplished. In the reverse Kuos, the specific humidity q is modified using the relation

$$q_m = \frac{q^R}{-\frac{p_s}{g}\int_{\sigma_T}^{\sigma_n}\sigma\frac{\partial q}{\partial\sigma}d\sigma} + \frac{\frac{1}{g}\int_{\sigma_T}^{\sigma_n} q d\sigma}{\frac{1}{g}\int_{\sigma_T}^{\sigma_n} d\sigma}\left(1 - \frac{R}{-\frac{p_s}{g}\int_{\sigma_T}^{\sigma_n}\sigma\frac{\partial q}{\partial\sigma}d\sigma}\right)$$

Here q_m is the modified specific humidity, q is the specific humidity prior to modification and R is the observed rainfall rate. The moisture convergence corresponding to the modified specific humidity matches the observed rainfall, given by

$$-\frac{p_s}{g}\int_{\sigma_T}^{\sigma_n}\sigma\frac{\partial q_m}{\partial\sigma}d\sigma = R$$

The *classical physical initialization* (Krishnamurti et al., 1991) utilized a Newtonian nudging of the observed precipitation estimates (inversely) using *reverse algorithms* to adjust the values of the variables that go into the cumulus parameterization and the PBL physics for the nowcasting of rain rates. This procedure appears to be far more robust compared to a variational assimilation of brightness temperatures within *3DVAR* or *4DVAR* since the operational models at coarse resolution do not carry adequate microphysical parameterizations to translate brightness temperatures to rain rates. Typical correlations for the *physical initialization* with respect to *observed rain rates* reach values of the order of 0.8 consistently whereas the correlations for assimilation of brightness temperatures are around 0.3 to 0.4. Figure 1 shows an example of physical initialization for a global model. This illustration conveys the robust aspects of rain rate initialization by this method. The top panel shows precipitation from a mix of satellite and raingauge-based estimates. The middle panel shows the results from physical initialization and the bottom panel shows the results from the same model that does not use physical initialization. The respective domain averaged correlations for rains based on physical initialization and for no physical initializations shown on the left depict respective values of 0.91 and 0.33. The lower value is typical for most operational models so-called Global Data Assimilation System Model Rain. In this sense this nowcasting of rain is impressive. That advantage must transcend to the forecast. Numerous studies (Krishnamurti et al., 2008) have shown that the physical initialization does show a positive impact for the forecasts of rains at days 1, 2 and 3. We shall next address the physical initialization for cloud resolving models.

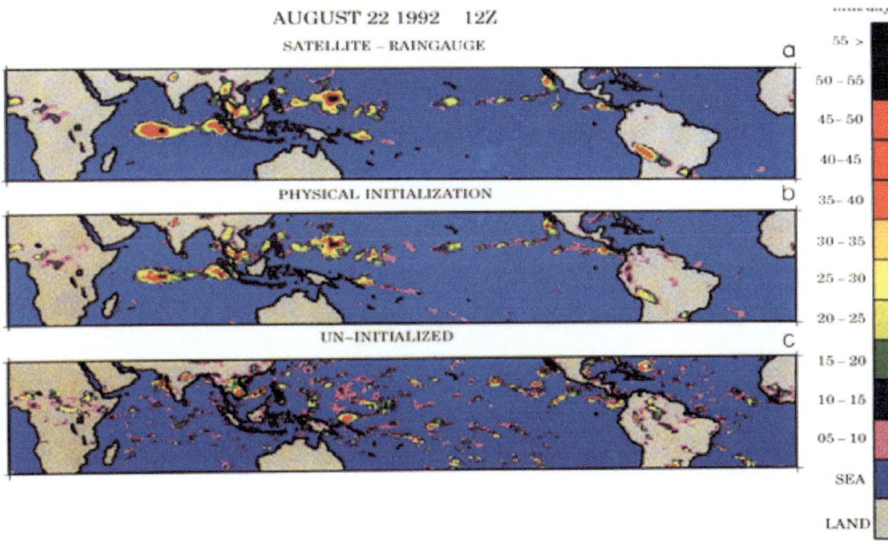

Fig. 1. Observed (based on SSM/I OLR and rain gauge) 24-hr rainfall estimate (mm, day-1) between 1200 UTC 21st August and 1200 UTC 22nd August 1992, compared with (b) physically initialized precipitation (using FSU GSM at T255 resolution) and (c) control run (un-initialized run).

2. Physical Initialization for Cloud Resolving Models

Study calls for the use of cloud resolving mesoscale non-hydrostatic models with microphysics at a horizontal resolution of near 1 km with a direct use of radar-based rainfall rate estimates. This entails the restructuring of the vertical profile of specific humidity in rain areas and the restructuring of the vertical velocity in the rain-free areas. It should be noted that in non-hydrostatic cloud resolving models the largest contribution to rains arise from the disposition of supersaturation of vertically ascending air parcels. There are also rains that can directly be attributed to microphysical processes such as melting of ice, grouple and snow. However make an assumption here that the entire rain in a non-hydrostatic microphysical model arises from the disposition of supersaturation and for making the model rain (each time step) to match a radar-based (time step interpolated) rain we invoke a change in the vertical profile of moisture, to accomplish that.

Matching model rain rates at any horizontal grid point to satellite-based estimates of rain rates (interpolated to that time step) calls for a change of the model's vertical profile of moisture by a factor of $(1 + \varepsilon)$, where ε is defined by the relation:

$$\varepsilon = \frac{\text{Rain Rate Observed}}{\text{Rain Rate Model}} - 1$$

Model total column rain rate:

$$\text{Rain Rate}_{\text{Model}} = -\frac{P_s}{g}\int_0^1 \dot{\sigma}\frac{\partial q}{\partial \sigma}d\sigma$$

where $\dot{\sigma}$ is the vertical velocity, q is the specific humidity and P_s is the surface pressure.

Note that the relation $R_{\text{Model}} = -\frac{\partial q_S}{\partial t}$ is a measure of the rate of condensation, if super saturation is not permitted.

This is approximated by $R_{\text{Model}} = -\frac{\partial q_S}{\partial t}$ and also noted that $q = q_s$, the saturation value.

The model rain rate is being forced towards the satellite-based value with the modification of the moisture by the parameter 1 + ε. However, that does not guarantee that the model will accept that value since all other model variables must come into equilibrium with that change of moisture and the rain rate. Disposition of supersaturation alone does not take into account the inclusion of melting of ice, snow and grouple. This requires other steps of physical initialization (those are not included in this study).

Any changes in the moisture profile can change the total precipitable water. In order to conserve moisture during these computational steps, whenever arbitrary changes are made to the moisture profile, counter-corrections are required for moisture conservation. Total precipitable water prior to changes in moisture profile is:

$$\frac{P_s}{g}\int_0^1 q d\sigma$$

Total precipitable water after the changes in moisture profile is:

$$\frac{P_s}{g}\int_0^1 q_{\text{modified}}$$

In order to accomplish this, moisture at any vertical level is re-modified by the relation:

$$q_{\text{Remodified}} = q_{\text{modified}}\left\{\frac{\frac{P_s}{g}\int_0^1 q\mathrm{d}\sigma}{\frac{P_s}{g}\int_0^1 q_{\text{modified}}\mathrm{d}\sigma}\right\}$$

where q modified on the right hand side of the equation is the result provided at the end of step 1.

Thus two modifications in the moisture profile are made: one to match the model rain rates toward the observed satellite-based rain rate estimates and a second, which is a passive step that adjusts for the conservation of total precipitable water. We noted that this second change of moisture is much smaller than what is called for in the first step which is the disposition of supersaturation, thus the second change from conservation does not alter the precipitation matching very much. This is one of the reasons for the success of physical initialization. These two steps are carried out sequentially during each time step of the physical initialization. The final step invokes an iteration between forecast hour-24 and hour 0 during physical initialization. During this step the moisture profile is updated by physically initialized profiles and the initialization forecast process is repeated roughly five times. This iteration improves the physically initialized rain rates considerably providing correlations (between the satellite-based estimates and the model rain rates) in excess of 0.8, compared to control forecasts with values of around 0.3. The resultant rain rates of the physical initialization process seem to converge nicely in roughly five iterations.

Tropical Cyclone Thane

Cyclone Thane was the strongest tropical cyclone of 2011 within the North Indian Ocean. Thane initially developed as a tropical disturbance and was declared a Depression during December 25, before being declared Cyclonic Storm Thane during the next day. As it was named, Thane started to turn towards the west. Its development slowed down during December 27. Thane became a Very Severe Cyclonic Storm during December 28. Before it approached states of Tamilnadu and Andhra Pradesh, it weakened slightly. Thane then made landfall early on December 30, on the north Tamil Nadu coast between Cuddalore and Pondicherry and rapidly weakened into a depression. The official track of Thane is shown in Fig. 2. Table 1 provides some of the storm pertinent data on time, location and intensity (maximum winds in miles per hour) prior to the landfall of Thane. On December 29th the strongest winds (90 mph) were noted by the India Meteorological Department. A radar reflectivity image and the derived rainfall (using a Z-R relationship) is shown in Figs 3a, b. The radar was ground-based and was located at Chennai. The heavist rainfall rates along the eye wall of Thane were around 22 mm/hour. An animation of the entire radar reflectivity and the radar-based rainfall estimates was prepared for the purposes of physical initialization. That data was fed into the WRF ARW models forecasts from hour 24 to hour 0. The model's vertical profiles of moisture in rain areas and the vertical profiles of downward vertical velocity were adjusted at every time step for the physical initialization. More details on this procedure will be published soon by Krishnamurti. Figure 4 shows the time history of physical initialization and prediction at Cuddalore where very heavy rains were encountered after the landfall of Thane. Time is along the abscissa and the ordinate shows rain rates in the units mm/hour. The solid black line shows the observed radar-based estimates, the dashed line shows the physical initialization based results and the red line is the control

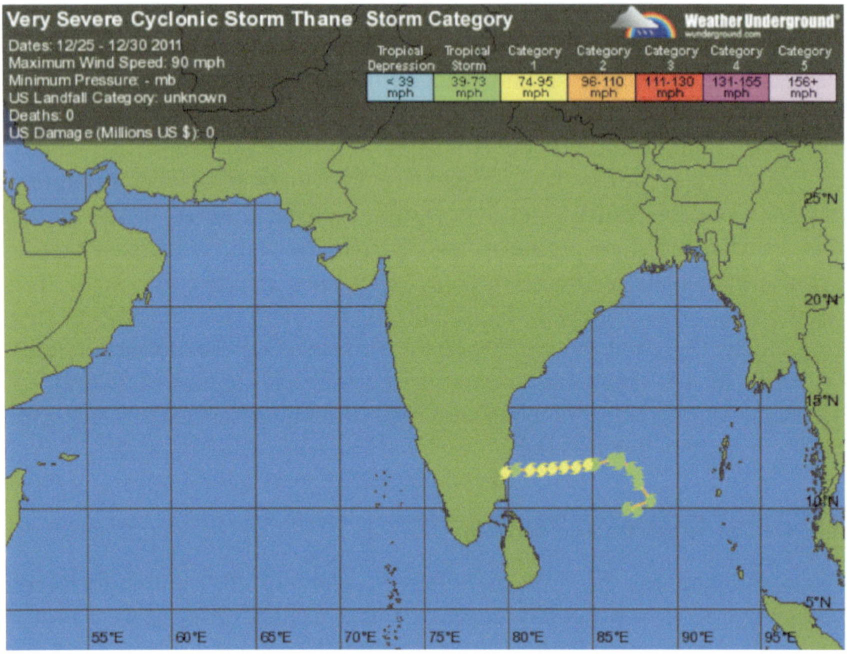

Fig. 2. Official track of cyclone Thane.

Table 1: Data on time, location and intensity (maximum winds in miles per hour) prior to the landfall of Thane

Date	*Time (GMT)*	*Lat*	*Lon*	*Wind (mph)*
12/25	18	9.80	87.70	40
12/26	0	9.90	87.10	45
12/26	6	10.30	88.50	45
12/26	12	11.20	87.80	45
12/26	18	11.60	87.70	50
12/27	0	11.90	87.30	50
12/27	6	12.40	86.70	65
12/27	12	12.20	86.60	70
12/27	18	12.40	86.20	70
12/28	0	12.10	85.20	70
12/28	6	12.10	84.80	75
12/28	12	12.00	84.10	85
12/28	18	12.00	83.40	85
12/29	0	11.90	82.70	85
12/29	6	11.80	82.10	90
12/29	12	11.80	81.40	85
12/29	18	11.80	80.50	70
12/30	0	11.70	79.90	75

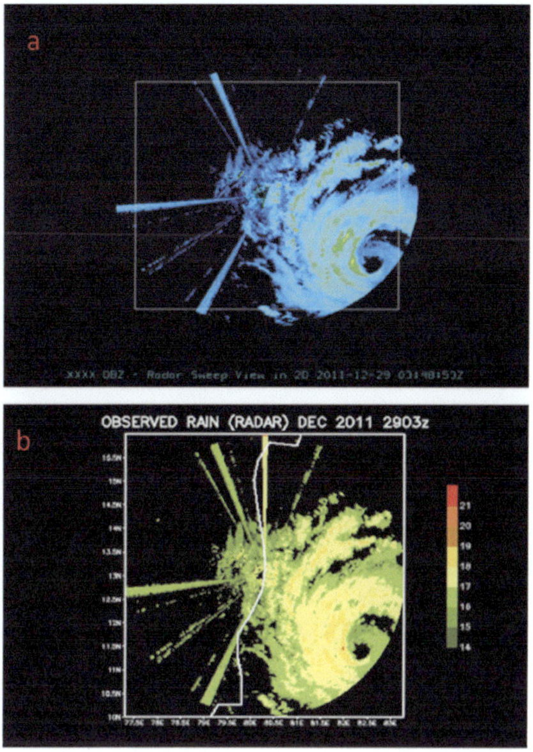

Fig. 3. (a) Radar reflectivity; (b) Observed radar rain calculated using reflectivity.

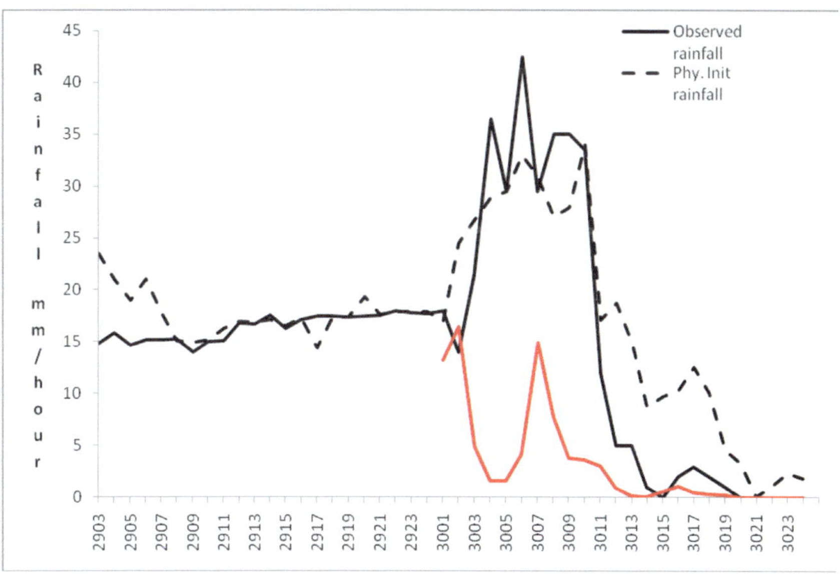

Fig. 4. Time history of physical initialization and prediction at Cuddalore, where the solid black line shows the observed radar-based estimates, the dashed line shows the physical initialization based results and the red line is the control run result without rain rate initialization.

run result that does not invoke a rain rate initialization. The vertical arrow near hour 3000 denotes the time of start of forecast after physical initialization. The impressive aspect of physical initialization based day one of forecasts is the ability of this method to show a very reasonable history of heavy rains of the order of 40 mm/hour at Cuddalore during the passage of Thane after the landfall. This is the kind of research and forecasts that are possible from physical initialization.

REFERENCES

Krishnamurti, T.N., J. Xue, H.S. Bedi, K. Ingles and D. Oosterhof, 1991: Physical initialization for numerical weather prediction over the tropics. *Tellus*, **43AB**, 51-81.

Krishnamurti, T.N., H.S. Bedi, G.D. Rohaly, D. Oosterhof, R.C. Torres, E. Williford and N. Surgi, 1994: Physical Initialization. ECMWF Workshop, Modelling, validation and assimilation of clouds, 31.

Yanai, M., S. Esbensen and J.H. Chu, 1973: Determination of bulk properties of tropical cloud clusters from large scale heat and moisture budgets. *J. Atmos. Sci.*, **30**, 611-627.

Part 4

Monitoring and Prediction of Tropical Cyclones

16

Monitoring and Forecasting of Tropical Cyclones over North Indian Ocean

M. Mohapatra
India Meteorological Department
Mausam Bhavan, Lodi Road, New Delhi – 110 003
mohapatraimd@gmail.com

1. Introduction

Tropical cyclones (TCs) are the most devastating phenomena among all natural disasters, having taken more than half a million lives all over the world in the last five decades. TCs are accompanied by very strong winds, torrential rains and storm surges. The havoc caused by TCs to shipping in the high seas and coastal habitats along the Indian coasts due to above mentioned adverse weather have been known since hundreds of years. The tropical warm north Indian Ocean (NIO), like the tropical North Atlantic, the South Pacific and the NW Pacific, is a breeding ground for the disastrous TC phenomenon. Historically, in terms of loss to human life, the Bay of Bengal TCs have accounted for deaths ranging from a thousand to three hundred thousands. The Bay of Bengal has experienced more than 75% of the total world-wide TCs causing human death of 5000 or more in last 300 years (Dube et al., 2013). For example, the death toll had exceeded 3,00,000 and 1,40,000 in earlier Bangladesh severe cyclones of 1970 and 1991 respectively in which the damages were beyond imagination. Orissa Super Cyclonic Storm (SuCS) of 29-30 October 1999 had caused over 10,000 human deaths and resulted in destruction and damage to over 1.9 million houses in 14 districts of Odisha (Kalsi, 2006; Mohapatra et al., 2002). Recently there was death toll of 1,40,000 in Myanmar due to very severe cyclonic storm, Nargis during April-May, 2008 (Tyagi et al., 2010). There are several such examples of the enormous storm-havoc along not only the Eastern Indian coast but also along the coasts of the other rim countries of Bay of Bengal and Arabian Sea. The district-wise TC hazard proneness for India has been analysed by Mohapatra et al. (2012a). It indicates that entire east and west coast of India are prone to TC activity. However, TC proneness varies from district to district, being maximum along West Bengal and north Odisha coastal districts.

The reduction of TC disasters depends on several factors including hazard analysis, vulnerability analysis, and preparedness & planning, early warning, prevention and mitigation. The early warning is a major component for the south Asian region due to its socio-economic conditions. The early warning component includes skill in monitoring and prediction of TC, effective warning products generation and dissemination, coordination with emergency response units and the public perception about the credibility of the official predictions and warnings.

In the last three decades the prediction skills have substantially improved in the north Atlantic and Pacific Ocean basins as a result of improvement in observations and research campaigns. Though there has been improvement (Mohapatra et al., 2013a, 2013b, 2013c), the potential for further improvement exists in the Indian Ocean basin too provided a special effort is launched in observations and modelling. The purpose of this document is to present the current status of monitoring and prediction aspects of TCs over the NIO by India Meteorological Department (IMD) along with the problems and prospects for the future.

2. IMD's role in TC Monitoring and Forecasting over the NIO

TC forecasting and warning is one of the most important functions of IMD. It was the first service undertaken by the Department as early as in 1865 and thus the service started before the establishment of the department in 1875. IMD at present is the nodal agency in the country to monitor and predict the cyclonic disturbances and issue the warning and advisory bulletins. IMD, New Delhi acts as a Regional Specialised Meteorological Centre (RSMC) for providing TC advisories to the World Meteorological Organisation (WMO)/Economic and Social Cooperation for Asia and the Pacific (ESCAP) Panel member countries, viz., Bangladesh, Myanmar, Thailand, Srilanka, Maldives, Pakistan and Oman. IMD, New Delhi also works as a Tropical Cyclone Advisory Centre (TCAC) for international civil aviation as per the requirement of International Civil Aviation Organisation (ICAO). The area of responsibility of RSMC, New Delhi is shown in Fig.1(RSMC, New Delhi, 2012).

The broad functions of IMD/RSMC, New Delhi with respect to TC forecasting over NIO are as follows:

- Round the clock watch on weather situations over the NIO.

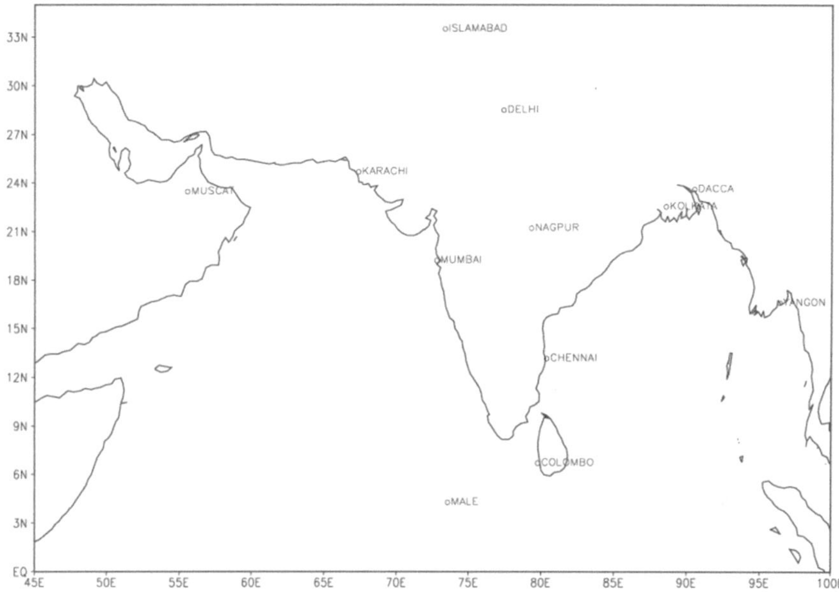

Fig. 1. Area of responsibility of RSMC-TC, New Delhi.

- Analysis and processing of global meteorological data for diagnostic and prediction purposes.
- Detection, tracking and prediction of cyclonic disturbances in the NIO.
- Running of numerical weather prediction (NWP) models for TC track, intensity (pressure and wind), rainfall and storm surge predictions.
- Implementation of the Regional Cyclone Operational Plan of WMO/ESCAP Panel.
- Issue of Tropical Weather Outlook and TC Advisories to the Panel countries.
- Issue of Numbered Cyclone Warning Bulletins to disaster management agencies, All India Radio (AIR), national television (Doordarshan) and other TV Channels, and print media.
- Research on storm surge, rainfall, track and intensity prediction techniques.

IMD continuously expands and strengthens its activities in relation to all the above to ensure most critical meteorological support through observations, analysis, prediction and warning to disaster managers and decision makers not only in the country but also to the NIO rim countries.

3. Climatological Characteristics of TC over the NIO

The low pressure systems over the NIO are classified (Table 1) based on the associated maximum sustained surface wind (MSW) at the surface level

(IMD, 2003; WMO, 2009). The systems with the intensity of depressions and above are considered as cyclonic disturbances. The 'cyclone' is a generic term associated with a low pressure system with MSW of 34 knots or more. It corresponds to the definition of tropical storms over other Ocean basins like Pacific and Atlantic Oceans. Over these basins, it is called as TC/Typhoon when the MSW is 64 knots or more corresponding to very severe cyclonic storm over the NIO.

It is now a well known fact of climatology (IMD, 2008) that about 5 to 6 TCs (MSW of 34 knots or more) occur in a year over the NIO prominently during the pre-monsoon season (March-April-May) and the post-monsoon season (October-November-December). It accounts for about seven percent of the global TCs. The maximum frequency is in the two months of May and November (Fig. 2). The tracks of TCs over the NIO during 1891-2007 (IMD, 2008) are shown in Fig. 3. TCs generally move in a northwesterly direction. However, they may recurve sometimes depending upon the environmental conditions. The TCs crossing different coastal states are shown in Fig. 4. The Bay of Bengal TCs more often strike Odisha-West Bengal coast in October, Andhra coast in November and the Tamil Nadu coast in December. Over 50 percent of the TCs in the Bay of Bengal strike different parts of the east coast of India, 30 percent strike coasts of Bangladesh, Myanmar and Sri Lanka and about 20 percent dissipate over the sea itself. Percentage of TCs dissipating over the Arabian Sea is higher (60%) as the western Arabian Sea is cooler. Maximum landfall occurs over Gujarat coast (18% of total TCs in Arabian Sea) of India followed by Oman coast. Several efforts have been made in the IMD to update climatological records on TCs of the NIO. Recently an electronic atlas has been published for tracks of TCs over the Bay of Bengal and Arabian Sea (IMD, 2008). It is available in cyclone page of IMD's website with free access (www.imd.gov.in). Life period of a TC over the NIO is 5-6 days. It has very severe cyclonic storm (64 knots or more) intensity for 2-3 days as against six days of global average.

Table 1. Classification of cyclonic disturbances over the NIO (since 1999)

Low pressure system	*Maximum sustained surface winds*	*No. of closed isobars at interval of 2 hPa within 5 deg lat/long square*
Low pressure area	< 17 knots	1
Depression	17–27 kts	2
Deep depression	28–33 kts	3
Cyclonic storm	34–47 kts	4-7
Severe cyclonic storm (SCS)	48–63 kts	8-10
Very severe cyclonic storm (VSCS)	64–119 kts	11-39
Super cyclonic storm (SuCS)	120 kts and above	40 or more

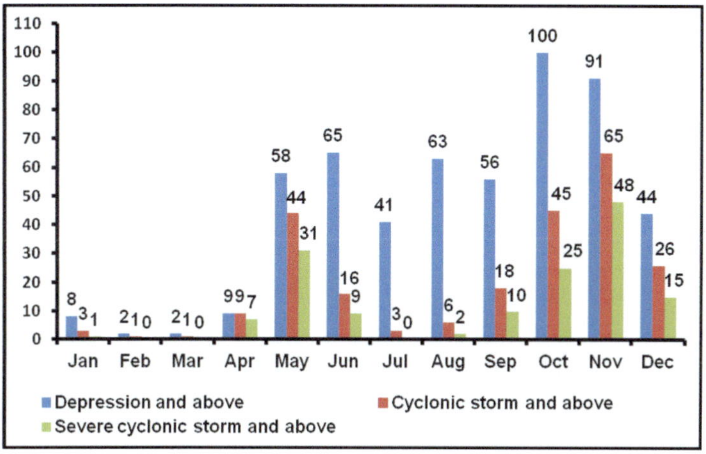

Fig. 2. Monthly frequency of cyclonic disturbances during 1961-2010 over NIO.

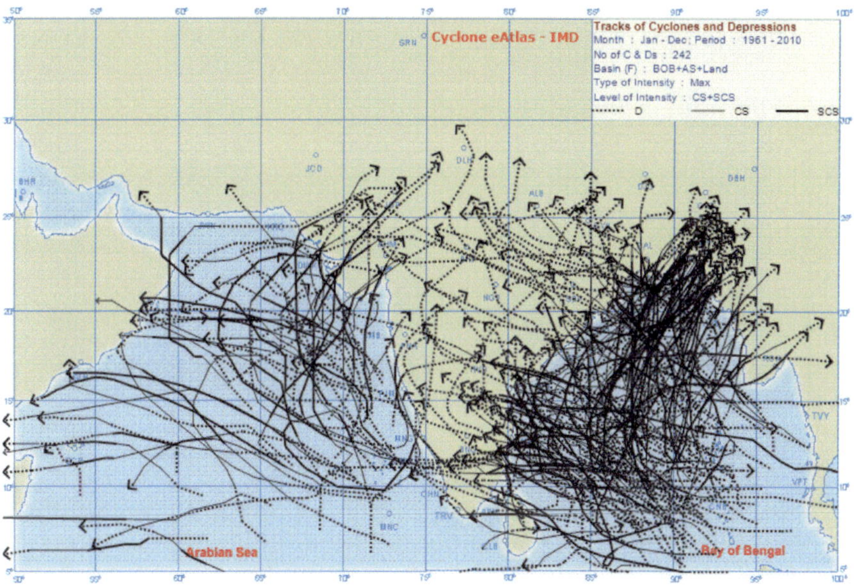

Fig. 3. Tracks of TC over the NIO during 1961-2010.

NIO TCs, like the hurricanes and the typhoons, develop from an incipient low over a period of a few days. Many incipient cloud cluster-scale disturbances and low pressure areas form but only some develop as depressions (sustained wind speed 17 to 33 kt) and only about 50% of the depressions develop into TC intensity (wind speed 34 kt or more) and only less than 25% of TCs further intensify into SCS and VSCS (wind speed over 48 kt or more). While it takes about 2 to 4 days to develop a low pressure area into a depression, the intensification from a depression to VSCS can occur in 24- to 48-hrs.

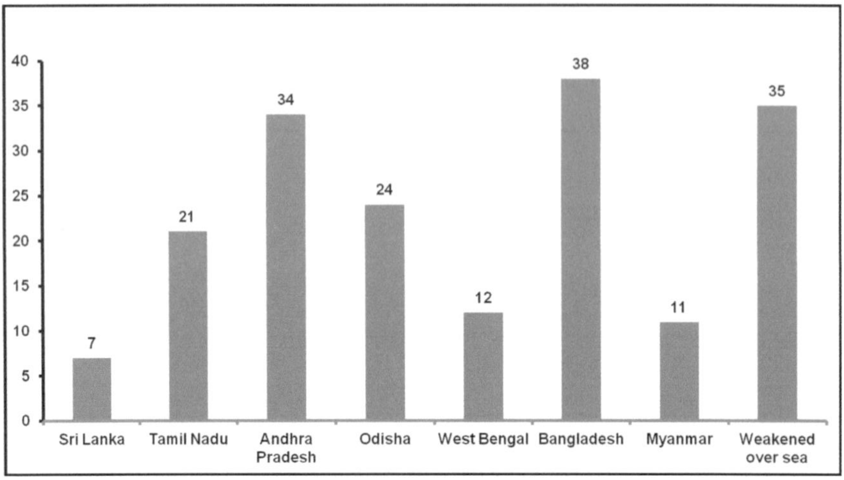

Fig. 4(a). Frequency of TCs over the Bay of Bengal landfalling over different coastal states during 1961-2010.

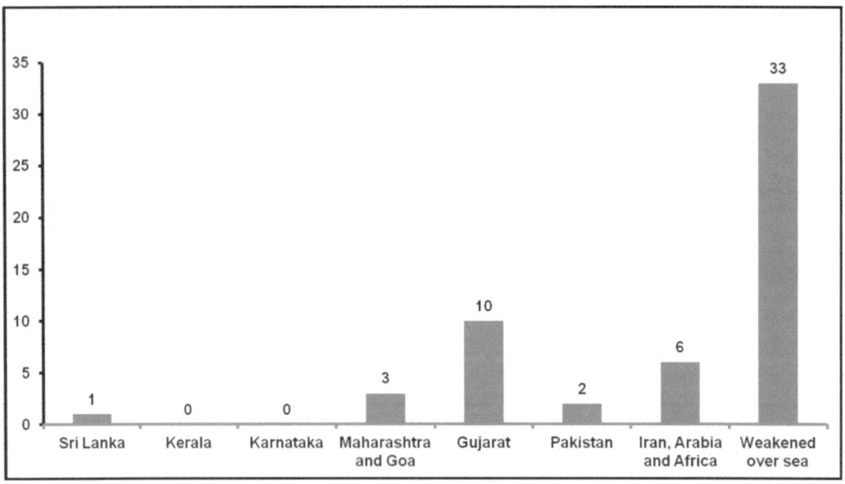

Fig. 4(b). Frequency of TCs over the Arabian Sea landfalling over different coastal states during 1961-2010.

4. Structural and Environmental Characteristics of TCs over NIO

The structural characteristics are similar to those over other Ocean basins. However, due to absence of aircraft reconnaissance, detailed structural characteristics over the NIO are not yet known. Studies made so far in this respect are mainly based on satellite and radar observations (Raghavan, 1990, 1997, 2013; Kalsi, 2002; Bhatia and Sharma, 2013). The INSAT imagery of

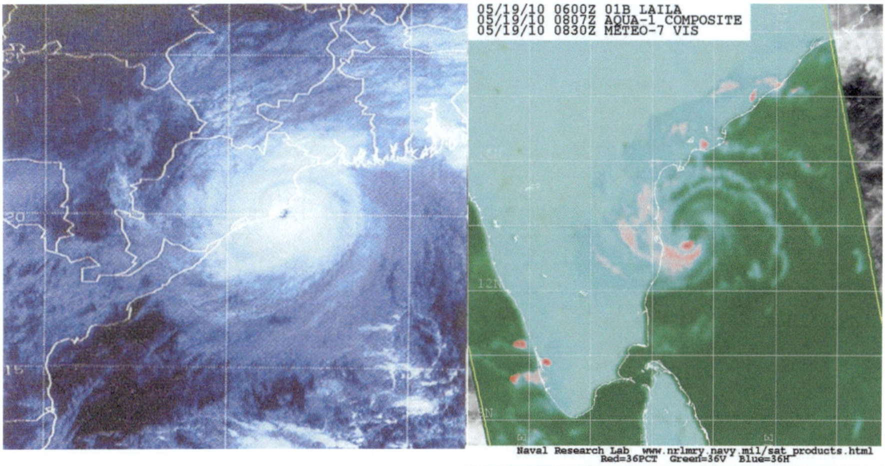

Fig. 5. (a) A typical INSAT satellite imagery of Orissa super cyclone (29 October, 1999), showing different characteristics like eye, eye wall, spiral bands etc. and (b) Microwave imagery of TC, Laila (19 May 2010).

Odisha Super cyclone on 29th October, 1999 indicates satellite view of clouds associated with a TC as shown in Fig. 5. In recent years, the microwave imageries from the polar orbiting satellites have helped further to understand the structure of TC as it can provide the imageries in different levels of the TC (Jha et al., 2013). A radar image of a matured TC consists of eye, eye wall, spiral bands, pre-cyclone squall lines and streamers as shown in Figs 5 and 6. However, in recent years, the Doppler weather radar (DWR) has provided better understanding of structure of TCs, as it provides the associated wind distribution in addition to horizontal and vertical extent of convection in the TC. An example is shown in Fig. 6.

That a TC structure may broadly agree in the NIO and North Atlantic Ocean was found in the study by Colon et al. (1970) with respect to the Arabian Sea and Atlantic systems and other studies based on satellite and radar data. However, there could be subtle differences in their genesis and even perhaps structures as the climatological atmospheric and oceanic conditions differ over NIO from those over North Atlantic.

Differences in climatological features in the two basins manifest as follows:

- The Bay of Bengal is a vast warm pool adjoining the warm pool of the western North Pacific. Several of the precursors of the Bay of Bengal TCs emerge from the neighbouring Pacific basin as remnant of typhoons and re-intensify over the Bay after passing over rather narrow strip of SE Asian countries. The Atlantic hurricanes are born out of the easterly waves coming out of West Africa.

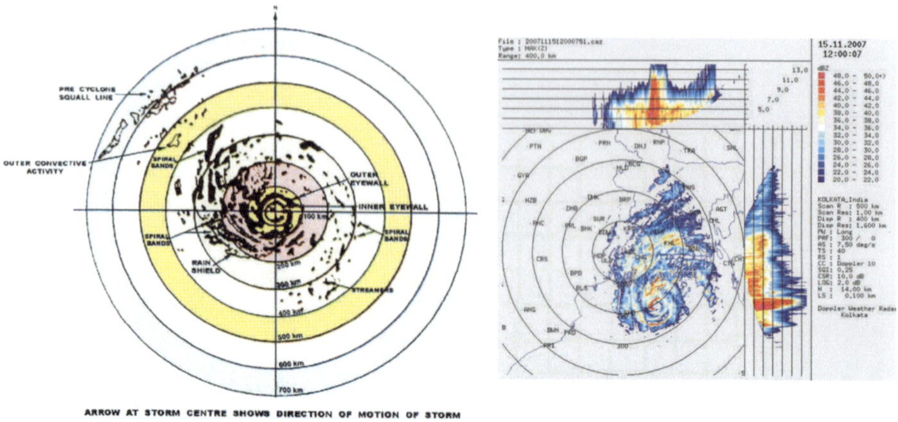

Fig. 6. (a) Composite structure of TC as seen in radar (Raghavan, 1997) and (b) DWR imagery of a TC, SIDR (15 November 2007).

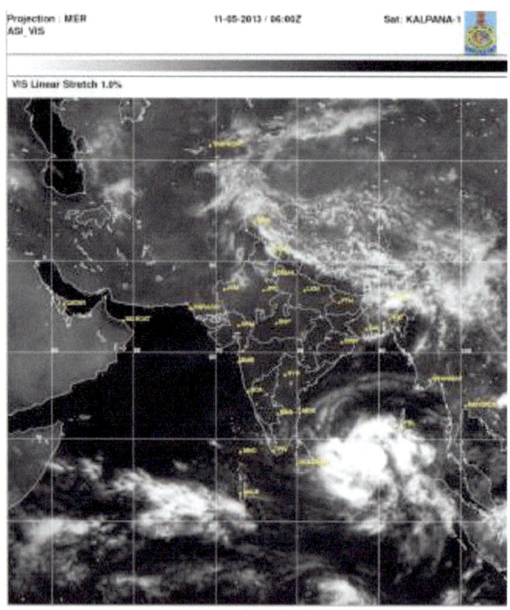

Fig. 6 (c). Kalpana imagery at 0400 UTC of 11 May 2013 indicating genesis of TC (T-2.5) over Bay of Bengal in association with MJO index lying in phase 4 with amplitude of 1.6.

- Bay of Bengal TCs form in the monsoon trough with low-level equatorial (or monsoon) westerlies to their south. Prior to the genesis of a Bay of Bengal cyclone, the westerlies show strengthening. This strengthening may occur under the eastward passing Madden-Julian Oscillation (MJO) (Mohapatra and Adhikary, 2011) in which the envelope of large scale organized convection is moving eastward as a super cluster (2000 km or more) but the intensifying meso-scale cloud cluster (200-500 km) moves westward. The convection on the super cluster scale, under large scale strengthening westerly flow, and meso-scale convection, though still unexplained processes, leads to meso-scale organsation of convection and mergers of meso-scale clusters to result in the genesis of a cyclone. An example in this aspect is shown in Fig. 6(c) in the form of a satellite imagery at the genesis phase of TC Viyaru. The genesis of TC Viyaru took place in association with phase 4 (maritime continents) of MJO with amplitude of 1.6. Thus both phase and amplitude of MJO were favourable for genesis of TC, Viyaru. Venkatesh (2006) has shown that meso-scale merger of vortices takes place in the Bay of Bengal like the observations by Simpson et al. 1997 in the Western Pacific basin. During pre-monsoon and post-monsoon seasons, the inter-tropical convergence zone (ITCZ) and convection play pre-dominant role apart from the remnants from south China Sea. On the contrary hurricanes of the Atlantic basin form in the ITCZ region of North Atlantic through processes involving interactions of the easterly waves with the large scale flow.

- The upper tropospheric environment over the south and central Bay of Bengal is dominated by strong tropical easterly flow in the southern Bay and strong sub-tropical westerly flow north of 18° N during main TC season (October-December). The sub-tropical westerlies are channelized by the Himalayan massif north of 27° N, in the large scale environment of the TCs. On the other hand Atlantic hurricanes are generated in weak easterly flow in the upper troposphere coming out of West Africa but strong sub-tropical westerly flow existing to north of 20°N.
- Bay of Bengal is a narrow basin and an intensifying TC takes only 3-4 days to cover it before land fall. Similarly Arabian Sea is also smaller. Tropical North Atlantic is a much larger basin and TCs continue to strengthen into high intensity hurricanes as they enter the Gulf of Mexico after several days of travel in the ITCZ of North Atlantic.
- The ocean currents in the NIO are quite complex and in the post-monsoon season low salinity warm waters begin to cover the India's eastern coastal waters from October to December through the East India Coastal Current.
- Another interesting part is the dissipation/rapid weakening of systems over NIO. It poses a great challenge to a forecaster as it very often leads to wrong/over-warning. The dissipation/rapid weakening over the sea may happen due to various reasons including colder ocean thermal energy, entrainment of dry and cold continental air into the core of TC and increase in vertical wind shear in the horizontal wind.

The above similarities and differences in the two large scale environments also provide important scientific challenges and motivations for undertaking detailed studies on structure of TCs over the NIO.

5. Monitoring and Forecasting of TCs over the NIO

The entire process of TC monitoring and forecasting (Mohapatra et al., 2013a) is shown in a schematic diagram (Fig. 7). The observational network for TC monitoring consists of land-based surface and upper-air stations, observations from Doppler Weather Radars (DWRs)/Cyclone Detection Radars (CDRs) and data from geo-stationary satellites. In addition, observations from ships and also from buoys are of immense importance for the analysis and forecasting of cyclones. Brief descriptions on various aspects of monitoring are given below.

It is important to correctly determine the location and intensity of the system, as initial error in location and intensity can lead to increase in error in forecast location and intensity (Mohanty et al., 2010; Osuri et al., 2012). Hence, there is a need of dense observational network over the sea and along the coast. Detailed observational network over the NIO is discussed in following sections.

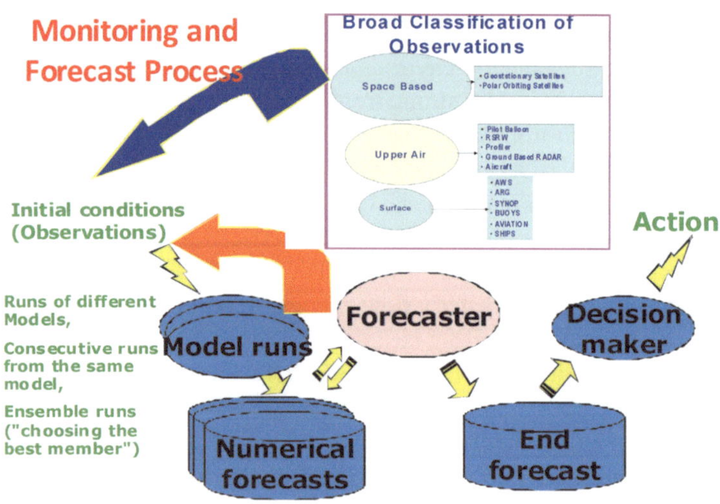

Fig. 7. Monitoring and forecasting process of tropical cyclone (after Mohapatra et al., 2013).

5.1 Monitoring of TCs over NIO

5.1.1 Conventional Observational Network

IMD has a good network of surface observatories satisfying the requirement of WMO. There are 70 departmental manned surface observatories of IMD at present all along the coast (Fig. 8) and over Bay and Arabian Sea islands. There are 21 pilot balloon observatories and 15 radiosonde/radio wind (RS/RW) observatories (Fig. 9). The meteorological data thus collected all over these stations are used on real time basis for CD monitoring. Details of the historical developments of the observational network have been discussed by

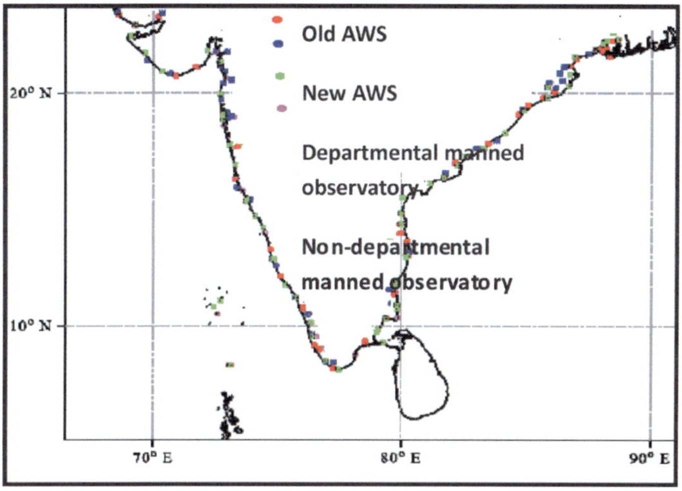

Fig. 8. Coastal surface observatory network by the end of 2010 (after Mohapatra et al., 2012b).

Mohapatra et al. (2012b). The optimum representation of surface observatories was obtained during 1940s and 1950s for the whole east and west coasts except Kerala. Bay islands got due representation during 1952, while Arabian Sea islands and Kerala during 1960s. The present network of pilot balloon and RS/RW observatories in India was achieved during 1971-80 (Fig. 9), though pilot balloon observation started as early as 1915 over Pune and 1919 over Kolkata (Mohapatra et al., 2012b). The upper air meteorological data over the rim countries of NIO are not sufficient and even certain coastal strips lack adequate surface observations (WMO, 2009).

5.1.2 Buoy and Ship Observations

Government of India has established a National Data Buoy Programme (NDBP) at National Institute of Ocean Technology (NIOT), Chennai. Under this programme, twelve moored data buoys were deployed during August '97 to February '98 at a sea depth of 20 m to 4500 m and spread over the Bay of Bengal and Arabian Sea. Since 2002, the number has increased to 25. The data buoys are fitted with sensors to measure air pressure, air temperature, wind speed and direction and sea surface temperature among other parameters. All these instruments have the accuracy as per WMO standard. These buoys have resulted in better monitoring and reduction of location and intensity error in association with ship and satellite observations. However, due to various reasons, only a few buoys are operational in the NIO in recent years (12 numbers by end of 2010).

As per the guidelines issued to Indian voluntary observing fleet (IVOF), synoptic observations are made at the main standard times: 0000, 0600, 1200 and 1800 UTC by the ships. When additional observations are required, they are made at one or more of the intermediate standard times: 0300, 0900,

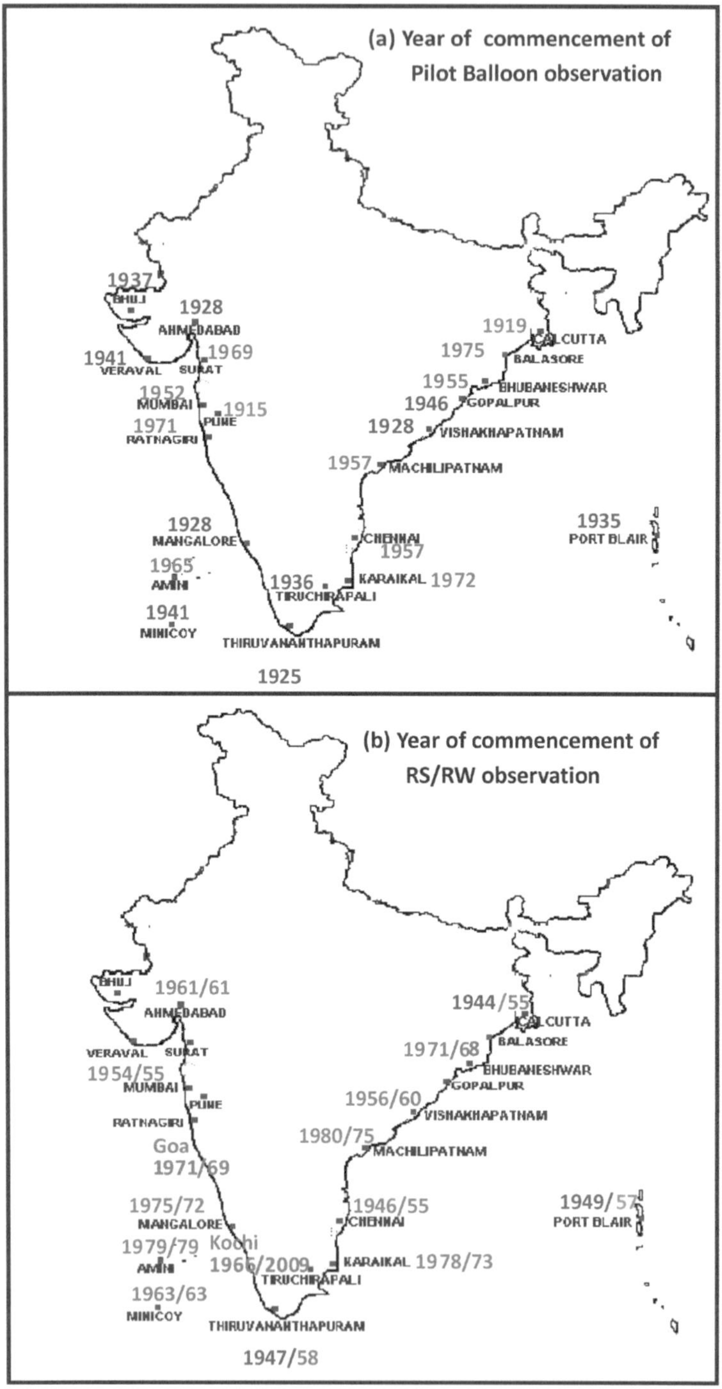

Fig. 9. Year of commencement of (a) pilot balloon and (b) RS/RW observations along the coast of India (after Mohapatra et al., 2012b).

1500 and 2100 UTC. When sudden or dangerous weather developments are encountered, observations are made for immediate transmission without regard to the standard times of observation. Over the NIO, ships are registered under IVOF. The ship observations were quite high during pre-satellite period (before 1960s), it gradually decreased with the advent of polar orbiting satellite in 1960s and reduced further with the introduction of Indian geostationary satellites during 1980s. The IVOF at present consists of 186 ships of Indian and foreign registry.

5.1.3 Satellite-based Observations

India has launched geo-stationary satellite METSAT, later named KALPANA-I, on September 2002 purely for meteorological applications. It is positioned over the equator at 74° E (Mohapatra et al., 2012b). Another geostationary satellite under INSAT series (INSAT-3A) was launched in April 2003 with the meteorological payloads identical to those of INSAT-2E which became operational since May 2003. It is positioned over the equator at 93.5° E. Both these satellites provide imageries in visible (VIS), infrared (IR) and water vapour (WV) channels. In addition INSAT-3A is also equipped with Charged Coupled Device (CCD) cameras capable of providing imageries in VIS, near IR (NIR) and Short-Wave IR (SWIR) channels with greater resolution.

During cyclone situation, data from KALPANA-1 are processed at hourly/half hourly intervals to assess the location and intensity of cyclonic disturbances out at sea. In addition to above, the products like outgoing long wave radiation, quantitative precipitation estimates, sea surface temperatures, cloud motion vectors, water vapour derived wind vector, isotherm analysis on enhanced infrared images are also generated on operational mode and posted on website of IMD.

The sea surface wind as estimated by scatterometer-based satellite is very useful in locating the centre of the TC (Uhlhorn and Black, 2003). This product is in use since inception of Quikscat in 1999. However, it has the limitations as it provides only two observations. It also suffers from rain contamination and inability to measure the wind speed more than 50 knots (28 mps). ASCAT, Windsat and OSCAT winds are used currently for TC monitoring. However, ASCAT has the less swath. The ASCAT ocean surface winds are a 10-metre neutral stability wind. The OSCAT-based surface winds are being used since November 2009 with effect from cyclone Phyan over the Arabian Sea and is similar to Quikscat, which stopped working in 1999. The examples of Windsat, ASCAT and OSCAT winds are shown in Fig. 10. The microwave imageries were introduced for monitoring and guidance in a subjective manner during later part of 2000s. It was used more objectively to locate the system centre since 2010 by utilizing the TC module available in TC forecasting work station at IMD, New Delhi. Microwave imageries are more helpful in predicting the structural characteristics and intensification in short

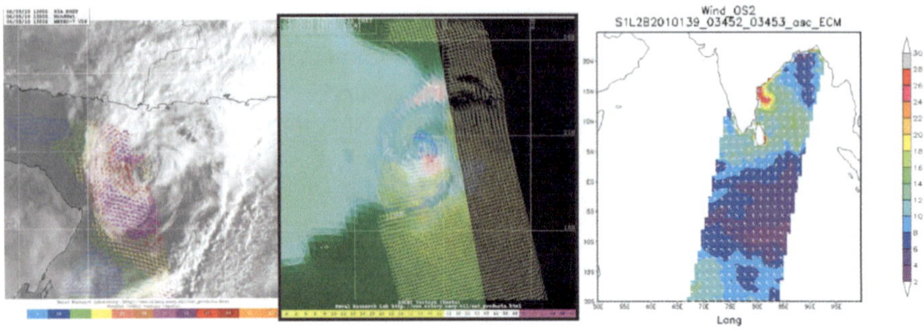

Fig. 10. Examples of utility of Windsat, ASCAT and Oceansat-II winds in TC monitoring.

range, as the characteristics of intensification is first observed in microwave imageries unlike visible and infrared imageries (Jha et al., 2013).

5.1.4 Radar Observation

There are 11 numbers of S-band CDR stations viz. Kolkata, Paradip, Visakhapatnam, Machilipatnam, Chennai, Sri Harikota, Karaikal, Kochi, Goa, Mumbai and Bhuj (Fig.11). Out of these 11 stations, five stations (Chennai, Kolkata, Sriharikota, Visakhapatnam and Machilipatnam) are using DWRs and the remaining stations have conventional S-band radars. DWR provides vital information on radial velocity within a TC which is not available in

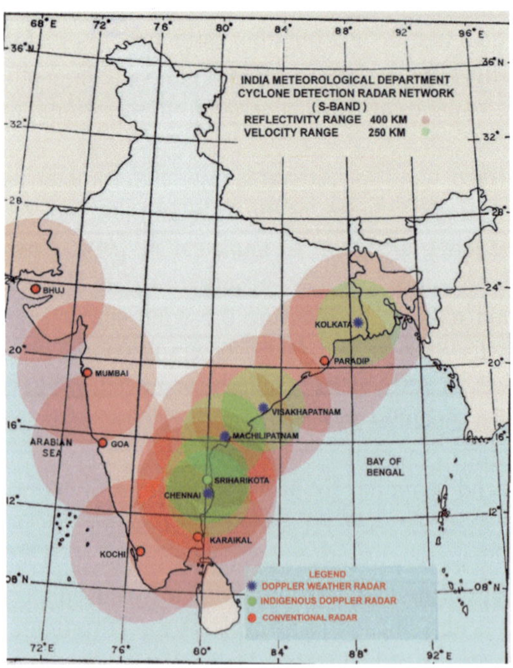

Fig. 11. S-band conventional cyclone detection radar and DWR network.

conventional radar. Conventional radar provides information on reflectivity and range only, whereas a DWR provides velocity and spectral width data along with various meteorological, hydrological and aviation products which are very useful for forecasters in estimating the storm's centre, its intensity and predicting its future movement. The data from radars of neighbouring countries are also used. However, only two coastal DWRs (Khepupara and Cox' Bazar) are available in Bangladesh and no such radars in other neighbouring countries. Detailed review of radar network in India and neighbourhood are given by Mohapatra et al. (2012b).

5.2 TC Analysis Tools and Techniques

Various kinds of analytical procedure are described in Cyclone Manual (IMD, 2003, 2013). A systematic check list is prepared for identification of location and intensity of TC and also for the prediction of track and intensity. The procedure necessarily deals with determination of location and intensity along with other characteristics of the TC like associated MSW, estimated central pressure and pressure drop at the centre, shape and size, radius of outermost closed isobar, point and time of landfall, if any or area of dissipation etc. with the available observations in the storm region.

To ensure the availability of the data and forecast products from various national and international sources at Cyclone Warning Division, IMD, New Delhi, an institutional mechanism has been developed in consultation with all the stake holders. A standard operation procedure (SOP) has been prepared for monitoring and prediction of cyclonic disturbances and issue of warning. It includes the road map and check lists for this purpose.

The TC analysis, prediction and decision-making process is made by blending scientifically based conceptual models, dynamical & statistical models, meteorological datasets, technology and expertise. A weather analysis and forecasting system in a digital environment is used to plot and analyse different weather parameters, satellite, radar and numerical weather prediction (NWP) model products. In this hybrid system synoptic method could be overlaid on NWP models supported by modern graphical and GIS applications to produce high quality analyses and forecast products. The TC Module installed in this forecasting system has the following facilities:

- Analysis of all synoptic, satellite and NWP model products for genesis, intensity and track monitoring and prediction
- Preparation of past and forecast tracks upto 120 hrs
- Depiction of uncertainty in track forecast
- Structure forecasting (forecast of wind in different sectors of TC)

However all the data are not still available in TC module. For better monitoring and prediction, additional help is taken of ftp and websites to collect and analyse:

- Radar data and products from IMD's radar network and neighbouring countries

- Satellite imageries and products from IMD and international centres
- Data, analysis and forecast products from various national and international centres

The automation of the process has increased the efficiency of system, visibility of IMD and utility of warning products (Mohapatra et al., 2013a).

5.3 Monitoring and Prediction of Genesis of TC

The synoptic charts are prepared and analysed every three hours to monitor the TCs over the NIO. As the NIO is a data sparse region and a genesis (formation of depression) takes place in mid-oceanic region, most of the cyclogenesis are determined with the satellite observation. However, there are cases of genesis near the coast within the radar range. The synoptic and radar monitoring determine the genesis of such system. There are cases like meso-cyclone Ogni (28-30 October 2006), which could be detected only with the radar and coastal observations (Hatwar et al., 2008). Similar is the case with monsoon depressions over the northern part of the Bay of Bengal. Gray's Parameters (Gray, 1968) including sea surface temperature and ocean heat content, convective instability, wind shear, low level relative vorticity, coriolis parameter and upper level divergence ate monitored for genesis and these parameters are mostly estimated by satellite technique. With the introduction of weather satellites, Indian meteorologists began to use satellite data in their observational studies on the TC (Mishra and Gupta, 1976; Kalsi, 2002; Kelkar, 1997; Bhatia and Sharma, 2013). These and other studies have built up a storehouse of knowledge on the formation of TCs in the NIO, emphasizing the role of SST, synoptic and convective processes in their formation and evolution of structural changes as the intensification of TCs proceeds. Following steps are followed for monitoring and predicting the cyclogenesis by examining the genesis parameters. These steps are based on Gray's parameters and the findings of recent studies (Gray, 1968; Roy Bhowmik, 2003; Kotal et al., 2009; Kotal and Bhattacharya, 2013).

Step I: SST, depth of 26°C isotherm and ocean thermal energy

- SST exceeding 26°C and a deep thermocline (50 m)
- Analysis of SST and ocean thermal energy from available models and satellite derived products (http: //www.aoml.noaa.gov/phod/cyclone/data/ni.html)
- Depth of the 26°C isotherm. SST, ocean thermal energy, sea height anomaly etc. field provided on a daily basis from TMI measurements (Tropical Rainfall Measuring Mission's Microwave Imager).

Step II: Conditional instability through a deep and moist atmospheric layer

- Area of deep convection on satellite imagery
- Large values of relative humidity in the lower and middle troposphere

- Favourable synoptic scale conditions for development of deep convection (Madden Jullian Oscillation/Outgoing Longwave Radiation): (http://www.bom.gov.au/bmrc/clfor/cfstaff/matw/maproom/OLR_modes/)
- Maps of the MJO real-time filtered OLR anomalies each averaged for a period of seven days.

Step III: Pre-existing disturbance

- Presence of a pre-existing disturbance in the lower atmosphere: analysis of pre-existing lows in the ITCZ (their location, organisation, intensity)
- Animated visible and IR geostationary imagery: estimation of the areas of organising deep convection, their evolution during the last 24 hours; first estimation of associated LLCCs (low level circulation centres); analysis of exposed LLCCs (when out of deep convection).
- Micro-Wave imagery over suspected convective areas (Monterey website, or Synergie Cyclone), to see the low or mid-level improving cyclonic organisation, under Cirrus clouds. (http://www.nrlmry.navy.mil/tc_pages/tc_home.html)
- Ocean surface winds derived from satellite: Oceansat-II (OSCAT): http://manati.orbit.nesdis.noaa.gov/quikscat/ Ocean Surface Winds derived from the SeaWinds Scatterometer aboard the Oceansat-II satellite
- Ascat : http://manati.orbit.nesdis.noaa.gov/ascat/ Ocean Surface Vector Winds derived from the Advanced Scatterometer (ASCAT – 50 km) aboard the EUMETSAT METOP satellite
- Windsat : http://manati.orbit.nesdis.noaa.gov/windsat/ Ocean Surface Winds derived from WindSat/Coriolis Measurements

Step IV: Environmental conditions

- Significant value of planetary vorticity (Coriolis force) ~5 degrees distance from Equator.
- Tropical storm can be observed ~ 2.5° sometimes also
- High relative vorticity at 850 hPa level (http://cimss.ssec.wisc.edu/tropic2/)
- Weak vertical shear of the horizontal winds (http://cimss.ssec.wisc.edu/tropic2/)
- Good upper level outflow (Models analysis/satellite observations)
- Good low level inflow (Models analysis/satellite observations)

Wind shear based on satellite observation is defined as follows:
Wind shear = (150-300) hPa layer mean minus (700-925) hPa layer mean
Wind shear is classified as follows:
5-10 kt: weak (favourable for development)
10-20 kt: moderate (unfavourable for weak system, or neutral for mature cyclone)
more than 20kt: strong (unfavourable)

Step V: NWP and dynamical-statistical model forecasts for genesis

- Examine each NWP model output to find out the initial condition
- Compare the model initial condition with the actual observations
- Find the model with best initial condition and find out its forecast for next three days
- From experience, examine the reliability of the model
- Examine the guidance from dynamical-statistical model
- If many or all the models predict genesis, the forecast confidence for genesis is high. If a few including the most reliable model predict genesis, the confidence is moderate. If less than 50% of the models suggest genesis, the confidence may be given as low. If no model predicts genesis, no genesis should be predicted.

Step VI : Consensus genesis forecast

- Based on synoptic, statistical, dynamical-statistical and NWP models guidance, a consensus forecast is issued for likely genesis during next 3 days

5.4 Determination of location of centre

The location of the centre of the system is determined based on (a) Synoptic position, (b) Satellite position (INSAT/METSAT/microwave position), (c) Radar position and (d) Centre determined by other satellite centres. When the system is far away from the coast and not within the radar range, satellite position gets more weight, though position is modified some times with availability of ship and buoy observations. When the system comes closer to the coast, radar position gets maximum preference followed by the satellite position. When the system is very close to coast or over the land surface, the coastal observations get the highest preference followed by radar and satellite observations. The average confidence level of locating the centre of the system over the NIO is about 50 km (Mohapatra et al., 2012b; Goyal et al., 2013).

5.4.1 Synoptic Technique

In this technique, the centre of the system is determined by considering the centroid of the wind distribution at the surface level. In the pressure field, the location of lowest mean sea level pressure is considered as the centre of the system (IMD, 2003). The direction and speed of the movement of a TC are determined primarily from the three hourly displacement vectors of the centre of the system. The synoptic technique has got serious limitations over the open sea due to non-availability of sufficient observations. When the system comes closer to the coast line, this technique takes precedence to determine the centre. based on observations from CDR/DWR stations as well as coastal observatories. The AWS stations along coast are also very useful as they provide hourly observations on real time basis (Bhatia et al., 2008). The coastal hourly observations help not only in correctly analyzing the location, but also in determining the landfall point and time along with various TC parameters at the time of landfall and hence help in adverse weather warning. To highlight the above facts, these observations in association with the TC Nisha during November, 2008 are shown in Fig. 12.

5.4.2 Satellite Technique

In the initial stage, the centre is determined, from the centre of the low cloud lines (IMD, 2003). Similar is the case in shear pattern, when the convection lies away from the centre. As the system intensifies and acquires the banding pattern, the centre is determined from the banding feature using logarithmic spiral. In the Central Dense Overcast (CDO) pattern, the centre of CDO is the centre of the system. In the eye pattern, the centre determination is easier and accurate as it is same as the centre of the eye of the TC.

5.4.3 Radar Techniques

The *eye* or the centre of the TC can be derived from a continuous and logical sequence of observations. The geometric centre of the echo-free area will be reported as the *eye location*. If the wall cloud is not completely closed, it is still usually possible to derive an eye location with a high degree of confidence by sketching the smallest circle or oval that can be superimposed on the inner edge of the existing portion of the wall cloud. When the wall cloud is not developed fully but a centre of circulation is identifiable, then this feature is observed and reported similar to the eye. When the eye or centre is indistinct or outside the range or the radar beam overshoots the inner eyewall when it does not extend very high, spiral band overlays are used to estimate the location of the centre. Use of logarithmic spirals has been described in detail in the Appendix II of Chapter 5 of instructions for Cyclone Detection Radar

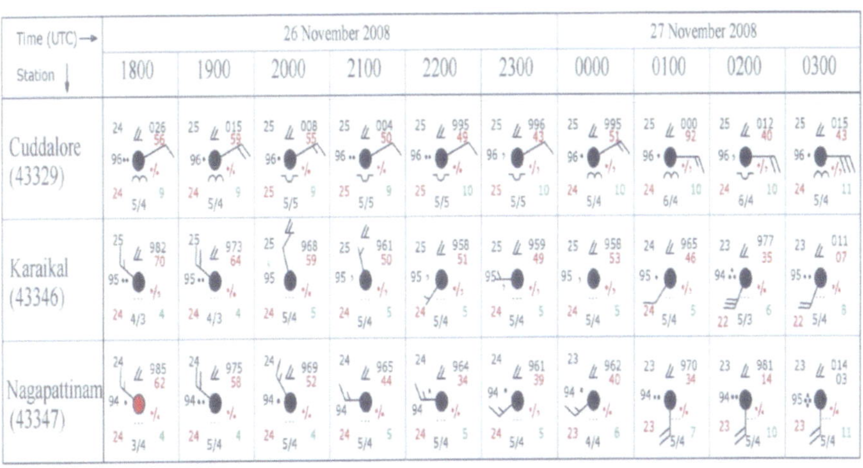

Fig. 12. Hourly coastal observations (260000 to 270300 UTC) in association with cyclonic storm NISHA (25-27 November 2008).

stations in I. Met. D. *Weather Radar Observations Manual* – 1976 (IMD, 1976).

The location estimation error has been about 55 km over the sea areas (standard error of satellite estimation according to Goyal et al. (2013)). The location error of a depression is more than a TC over the open sea area due to similar error in estimation by the satellite. According to Elsberry (2003), the errors in determining the TC centre over the northwest Pacific Ocean can be upto 50 km by satellite fixes, 20-50 km by radar observations and by about 20 km by aircraft reconnaissance. The induction of DWR reduces the error in fixing the centre of the TCs in radar range.

The landfall point estimation error has been 140 km or more prior to 1891 for west coast and more than 105 km for east coast. It reduced to about 100 km by the end of 1940 for both the coasts and to 55 km by the end of 1960. It further reduced to about 25 km by 2010 mainly due to installation of coastal AWS during late 2000s. The landfall point estimation error was more (>100 km) between Machilipatnam-Cudalore and Nagapatinam-Kanya Kumari coasts till 1940. It was about 75 km between Puri and Kakinada and Nellore and Chennai during 1941-2009 and reduced to about 25 km in 2010. The landfall time estimation error may be about half an hour since 1974 with the introduction of coastal hourly observations and CDRs. During 1960-1974, it may be at least one and a half hour with the three hourly observations.

5.5 Determination of Intensity

The intensity estimation takes into consideration (a) satellite, (b) radar and (c) synoptic analysis (IMD, 2003). Like the location of the system, when the system is far away from the coast and not within the radar range, satellite estimated intensity gets maximum weight. When the system comes closer to the coast, radar estimated intensity is considered along with satellite estimated intensity. When the system is very close to coast or over the land surface, the coastal observations are considered alongwith radar and satellite observations for estimating the intensity. As operational practice, a check list is prepared for this purpose to take care of all the available information, which in turn helps in decision making process to determine the above characteristics of TC.

Intensity for a TC is generally reported as the MSW over a specified time period. However, forecast centres report wind speeds using different averaging periods (Harper et al., 2008), which becomes a significant issue when comparing wind speed estimates from different agencies. Averaging periods include 1-, 2-, 3-min, and the WMO 10-min standard. Operationally, the value of MSW is almost never measured. It is generally inferred from satellite using the Dvorak technique. Comparing and converting between different measures of MSW are further complicated by the large variety of methods from which wind speeds are inferred. Such complications include the predominant use of satellite-derived current intensity to wind speed conversions, variation in estimated intensity for the same TC by different satellite centres, estimates from winds measured by radar wind measured at flight level, and new remote sensing techniques such as scatterometry winds, etc. (Mishra and Gupta, 1976; Olander and Velden, 2007; Nakazawa and Hoshino, 2009; Raghavan, 2013).

5.5.1 Intensity of TC by Synoptic Technique

In this technique, the available surface observations are taken into consideration to find out MSW and number of closed isobars at the interval of 2 hPa within a specified region around the system centre (IMD, 2003). The classification of system as discussed in section 3 is used for this purpose.

5.5.2 Intensity of a TC by satellite technique

The intensity classification by satellite technique is based on Dvorak's technique (Dvorak, 1975, 1984; Velden et al., 2006; Velden et al., 1998). The intensity of the tropical system is indicated by a code figure called T Number based on above technique, which is a pattern recognition technique. Another feature of the technique is the Current Intensity number (C.I.) which relates directly to the intensity (in term of wind speed) of the cyclone. The C.I. number may differ from the T number on some occasions to account for certain factors which are not directly related to cloud features. The empirical relationship between C.I. number and the maximum wind speeds (according to Dvorak) are given in Table 2. Column 3 of the table gives the pressure depths (peripheral pressure minus central pressure in hPa) as applicable for Indian Sea area using the relation Vmax = 14.2 × SQRT (Pn–Po) following Mishra and Gupta (1976).

As there is no aircraft reconnaissance in the NIO, Dvorak's technique, which is based on one minute averaging has not been verified. Comparison of satellite-based intensity and the best track estimates of IMD indicate a difference of about 0.5 T (Goyal et al., 2013). However, coastal stations, equipped with cup anemometer and Dynes P.T. anemograph use 3-minute averaging in Indian region. The wind conversion factor has not been applied to the data over the NIO as estimated by IMD. Historically, wind-pressure relationships have been used to relate minimum central pressure based on the estimated MSW for TCs. However, their application has been inconsistent across the forecast centres. Atkinson and Holliday wind-pressure relationship (Atkinson and Holliday, 1977) has been most widely used since its inception in the mid-1970s. Koba et al. (1991) and Nakazawa and Hoshino (2009) discussed the wind-pressure relationship over the west Pacific Ocean. Knaff and Zehr (2007) have developed a wind-pressure relationship that is now widely used. Forecasters at the Australian Bureau of Meteorology have recently adopted a modified version of the Courtney and Knaff (2007) wind-pressure relationship that has shown improved performance at low latitudes.

Recently the microwave imageries and brightness temperatures are also used to determine the intensity. The brightness temperature from the polar orbiting satellites is utilized to estimate the central pressure and MSW. Though this technique has not been validated over the NIO, the algorithm used in north Atlantic basin is being used for the NIO (Jha et al., 2013).

Table 2. Maximum sustained wind speed (MSW) and pressure depth in relation to Dvorak's CI number

C.I. number	*Max. Wind Speed (knots)*	*Pressure depth (in mb)*
1	25	
1.5	25	
2	30	4.5
2.5	35	6.1
3	45	10.0
3.5	55	15.0
4	65	20.9
4.5	77	29.4
5	90	40.2
5.5	102	51.6
6	115	65.6
6.5	127	80.0
7	140	97.2
7.5	155	119.1
8	170	143.3

5.5.3 Radar Technique

This technique can be used to find out the pattern of the TC and associated MSW. Based on this, the intensity can be determined. Further, radar helps in finding out the structure of TC. Details have been discussed by Raghavan (1997, 2013). As satellite- and radar-based wind are not available at surface level, the wind observations from these techniques are converted to 10-metre wind using the suitable conversion technique like those used in case of aircraft reconnaissance technique in Atlantic.

The average error in MSW estimation has reduced over the years. During the pre-satellite era (till 1960), the average error in intensity estimation may be at least one stage in Beaufort scale (5-15 knots or 3-8 mps upto severe cyclonic storm stage). There is no classification of intensity between very severe cyclonic storm and above intensity in Beaufort scale. The error could have reduced gradually during polar satellite era. It could have been T0.5 (05-20 knots or 3-10 mps) with the introduction of Dvorak's classification of intensity since 1974. Based on seven TCs, Mishra and Hem Raj (1975) have shown large difference between wind speed inferred from synoptic data and those derived from satellite technique. According to them, the MSW could be under-estimated by 8-17 knots (5-9 mps) in depression/deep depression stage,

26-28 knots (13-14 mps) in cyclonic storm stage and 37 knots (19 mps) in severe cyclonic storm or higher stage in pre-satellite era.

5.6 TC Track Forecasting

5.6.1 Historical Development of Forecasting Technique

A variety of observational data have been used in India till 1960s to forecast the track and landfall of TCs such as patterns of 24-hr pressure changes along coastal stations, approach of a westerly trough in mid- and upper troposphere etc. Satellite era, since 1960s, added another feature with respect to observations in terms of the direction of major cirrus outflow away from the central region of a TC which indicates future direction of its movement. Development of objective techniques for forecasting tracks of TCs in the NIO began in 1972 (Sikka and Suryanarayna, 1972; Neumann and Mandal, 1978) by using a computer-based half persistence and half climatology technique and adoption of analogue technique by Datta and Gupta (1975). Ramanathan and Bansal (1977) examined the track prediction with a quasi-geostrophic baroclinic model. The detailed review on NWP models for TC prediction is given by Mohanty and Gupta (1997), Gupta (2006) and Sikka (2006). With the regional quasi-Langrangian model in operational use in IMD and the T-80 atmospheric general circulation model in the NCMRWF in 1990s, TC motion is being operationally predicted in India using dynamical models. As the predictions with these models suffer from the inaccuracy of precise location of a TC at the initial time by as much as 200 to 300 km, researchers began to use synthetic vortex (Singh et al., 2005; Chourasia et al., 2013) by replacing the vortex position and structures at the initial time by an idealized vortex based on satellite observations. With this the model errors have reduced. However, data by Prasad et al. (1998), Prasad and Rama Rao (2003) and Gupta (2006) show that the statistical vector error for prediction of TCs in the NIO basin remains high (24-hr. error of 140 km, 48-hr error of 220 km and 72-hr error of over 300 km), with standard deviation approaching 50-70 percent of the respective vector errors. Recent work by Ali et al. (2007) using neural network approach also did not show any significant improvement in 48- and 72-hr track forecast. Kotal and Bhowmik (2011) used multi-model ensemble (MME) technique for track prediction, which showed better performance in track prediction. Multi-model ensemble based on five global and regional models have been developed by IMD for TC track prediction since 2009 using multiple regression technique. There is also facility in TC module of the forecasters' work station to develop MME using equal weightage to individual model tracks available in TC module. Ensemble prediction system (EPS) and super-ensemble have also been developed based on ECMWF, UKMO, JMA and NCEP models under TIGGI project of WMO with the help of JMA in 2011. It helps in providing location specific striking probability.

The challenge for the Indian researchers in the last couple of years has been to reduce the track prediction error and also, if possible, predict the intensity changes in the evolution of TCs. Several groups in India have applied high resolution meso-scale models of U.S. origin with a variety of physical parameterization in isolation and also in combination. In 2005 an exercise was undertaken to evaluate the performance of meso-scale models on the genesis and movement of Orissa Super Cyclone of 29-30 Oct, 1999. Several papers on simulation on this cyclone were published in the special issue of *Mausam* (2006). Sikka and Rao (2007) have found a considerable scatter in the track and intensity predictions on this cyclone as the investigators used different initial conditions (NCMRWF, NCEP, ECMWF), synthetic vortex, parameterization schemes etc. Meso scale WRF models were introduced in IMD and other research agencies during 2009 with spatial resolution of 27 km (Mohanty et al., 2013). It was further improved with local data assimilation and increase in resolution to 9 km in 2010 and 2011. Currently an effort is underway in which high resolution hurricane WRF model with the support from NCEP, USA is being used in track and intensity predictions. However, it is worth stressing that this exercise would not have any aircraft reconnaissance data to improve the environmental flow which is so crucial (as found by US investigators) for improving the track predictions.

5.6.2 Current Methods of Track Forecasting

To summarise, currently following methods are used by IMD for TC track forecasting:

(i) Statistical techniques
- Analogue
- Persistence
- Climatology
- Climatology and persistence (CLIPER) developed by Neuman and Mandal (1978),

(ii) Synoptic techniques – Empirical techniques
(iii) Satellite techniques – Empirical technique
(iv) Radar techniques – Empirical technique
(v) NWP models
- Individual models (Global and regional)
- IMDGFS (574), ARP (Meteo-France), ECMWF, JMA, UKMO, NCEP, WRF (IMD, IITD, IAF), HWRF (IMD), Quasi-Lagrangian model (QLM)
- MME (IMD) and MME based on TC module
- EPS (Strike probability, Location specific probability)

Examples of track prediction by individual models in case of TC Rashmi and EPS product in case of TC Jal are shown in Fig. 13(a) and Fig. 13(b) respectively.

All the NWP models as mentioned above except QLM are late models as the model output is available to forecasters as late as six to twelve hrs. The QLM is a near-real time model and the statistical models like CLIPER

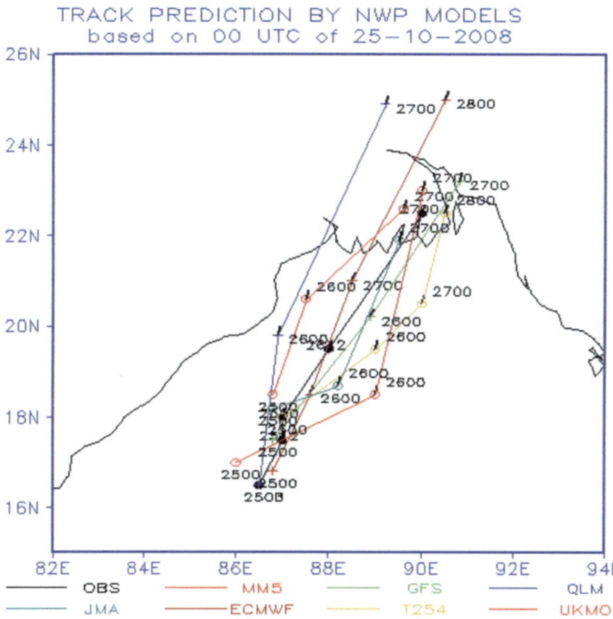

Fig. 13(a). Track prediction of TC Rashmi by NWP models.

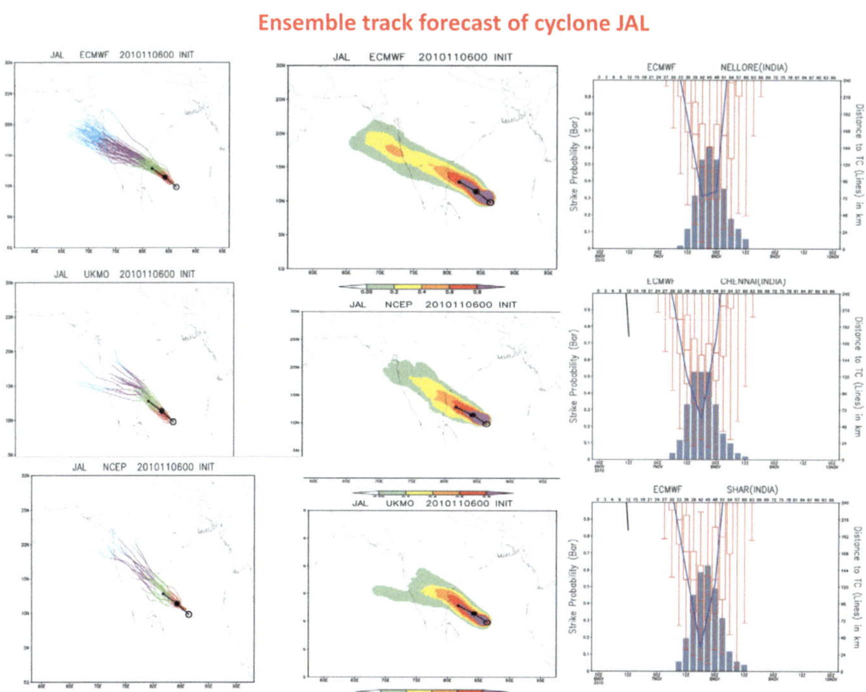

Fig. 13(b). Track prediction of TC Jal by EPS.

and ISRO model are real time models. Similarly all the NWP models as mentioned above do not use synthetic vortex and vortex relocation. While QLM and NCEP use synthetic vortex, global model like ECMWF does not use these processes.

In the synoptic method, prevailing environmental conditions like wind shear, low to upper level wind and other characteristics as mentioned in cyclone manual and check list (IMD, 2003, 2013) are considered. All these fields in the NWP model analyses and forecasts are also considered. The development of characteristic features in satellite and radar observations is also taken into consideration. While the synoptic, statistical and satellite/radar guidances help in short range forecast (upto 12/24 hrs), the NWP guidance is mainly used for 24-120 hr forecasts. Hence, the RSMC forecast tracks result from a manually analyzed forecasting process, which relies on output from several NWP models as discussed above. Consensus forecasts that gather all or part of the numerical forecast tracks and uses synoptic and statistical guidance are utilised to issue official forecast.

5.6.3 Operational Track Forecast Products

Considering recent development in prediction capability, IMD introduced the objective TC track forecast valid for next 24 hrs over the NIO in 2003. It further extended the validity period upto 72 hrs in 2009 (Mohapatra et al., 2013b). The track forecast has been issued by RSMC, New Delhi from deep depression stage onwards since 2009 for 12, 24, 36, 48, 60 and 72 hr forecast periods. It introduced 96 and 120 hr forecasts in 2013. The TC forecast is issued six times a day at the interval of three hours, i.e. based on 00, 03, 06, 09, 12, 15, 18 and 21 UTC observations. The forecasts are issued about three hours after the above mentioned observation time. An example of the product during TC Thane is shown in Fig. 14. It may be mentioned that the RSMC, Tokyo introduced 24 hr objective forecast in 1982, 48 hr forecast in 1988, 72 hr forecast in 1997 and 96 & 120 hr forecast in 2009 for northwest Pacific Ocean (RSMC, Tokyo, 2011). Similarly, National Hurricane Center (NHC) and RSMC, Miami, USA introduced 24 hr objective forecast in 1954, 48 hr forecast in 1961, 72 hr forecast in 1972 and 96 and 120 hr forecast in 2001 for Atlantic Ocean (NHC, 2012). Prior to 2003, the TC forecasts issued by IMD for the NIO were subjective and textual in form.

5.6.4 Cone of Uncertainty in Track Forecast

The "cone of uncertainty (COU)"—also known colloquially as the "cone of death," "cone of probability," and "cone of error"—represents the forecast track of the centre of a TC and the likely error in the forecast track based on predictive skill of past years. Most WMO designated RSMCs for TCs including Miami, Florida; Tokyo, Japan; and Hawaii, USA and TC Warning Centers (TCWCs) currently display COU around their official track forecasts, using a climatological method based on their area of responsibility. For each forecast lead time, an uncertainty circle is built whose radius is taken as a

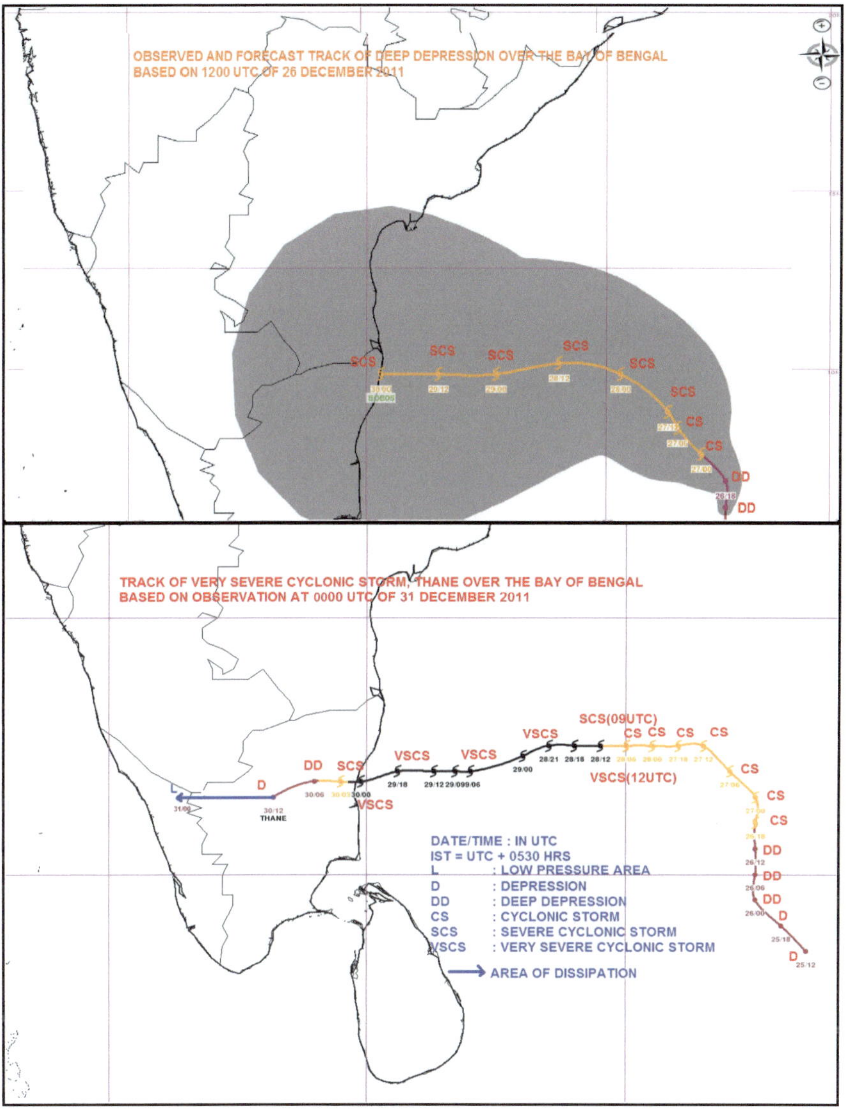

Fig. 14. A typical example of observed and forecast track of depression which later on became the very severe cyclonic storm Thane.

fixed quantile (e.g., 67% for Miami, 70% for Tokyo) of the distribution of direct position error (DPE) computed over several previous seasons. The Joint Typhoon Warning Center (JTWC) in Hawaii produces COU for TCs over Pacific Ocean and Indian Ocean, whose radii are the sum of the climatological average DPE and the predicted 34-knot wind radius. RSMC, La Re Union uses ECMWF EPS guidance to issue COU forecast.

The COU in the forecast of IMD has been introduced with effect from the TC 'WARD' during December, 2009. It is helpful to the decision makers as it indicates the standard forecast errors in the forecast for different periods like 12, 24, 36, 48, 60 and 72 hrs. The radii of circles used to construct the COU are 75, 150, 200, 250, 300 and 350 km respectively for 12, 24, 36, 48, 60 and 72 hrs forecasts based on past average errors of official forecasts for 12 and 24 hrs forecast and of QLM for 36-72 hr forecasts. A typical example of COU forecast showing the uncertainty circles for different forecast periods are shown in Fig.15 following Mohapatra et al. (2012c). It is found that the observed track lies within the forecast COU in about 60% of the cases. It is

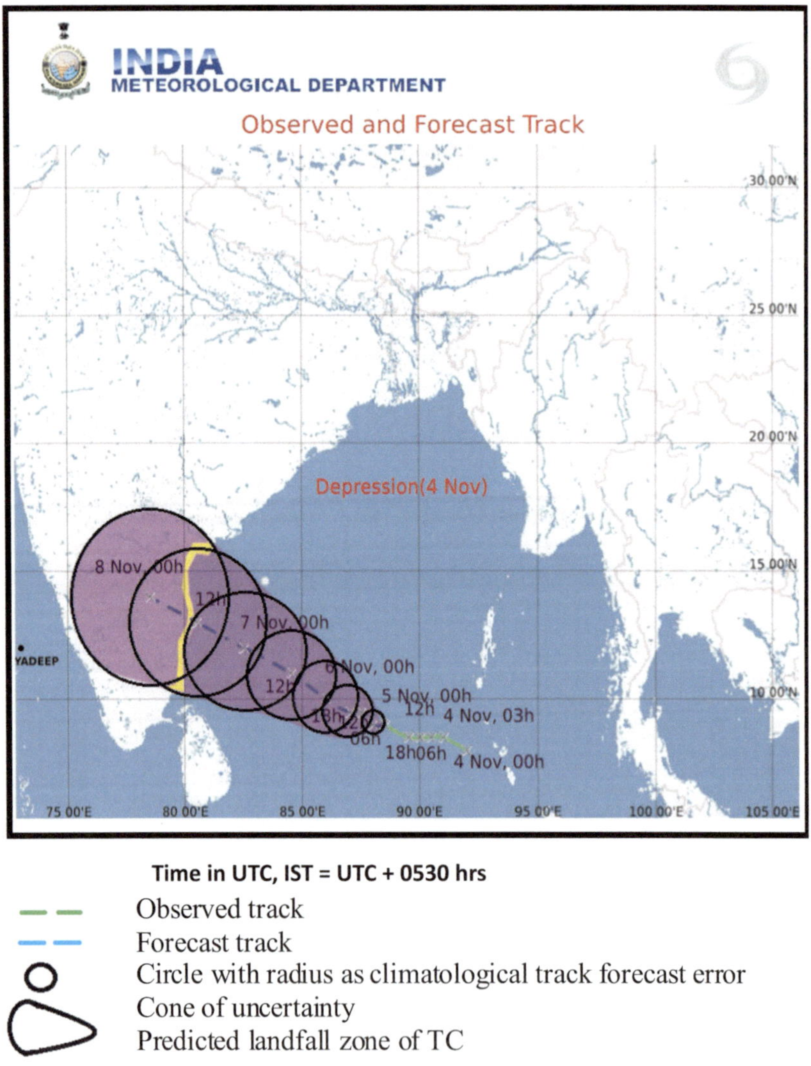

Fig. 15. A typical example of observed and forecast track of depression which later on became the severe cyclonic storm (SCS) Jal (4-8 November 2010) (after Mohapatra et al., 2012).

in agreement with those over other ocean basins as the entire track of the TC remains within the COU roughly 60-70% of the time over the northern Atlantic Ocean and Pacific Oceans.

5.7 Intensity Forecasting

The intensity forecast has been issued by RSMC, New Delhi from deep depression stage onwards since 2009 for 12, 24, 36, 48, 60 and 72 hr forecast periods. It has been issued during CS stage onwards during 2003-2008 and for 12 and 24 hrs forecast periods only. The TC intensity forecast is issued four times a day at the interval of six hours, i.e. based on 00, 06, 12 and 18 UTC observations valid upto 24 hrs during 2003-2008. It has been issued upto 72 hrs since 2009 in every three hourly intervals and upto 120 hrs since 2013. The forecasts are issued about three hours after the above mentioned observation time. Following methods are used by IMD for intensity forecasting of TCs over the NIO:

(i) Statistical Techniques
- Analogue
- Persistence
- Climatology

(ii) Synoptic Technique – Empirical Techniques
(iii) Satellite Technique – Empirical technique
(iv) Radar Technique – Empirical technique
(v) NWP Models
- Individual models (Global and regional)
- IMDGFS (574), ARP (Meteo-France), ECMWF, JMA, UKMO, NCEP
- WRF (IMD, IITD, IAF), HWRF (IMD)

(vi) Dynamical Statistical Model (Statistical cyclone intensity prediction-SCIP, Kotal et al., 2008)

There is no CLIPER model for intensity prediction unlike the Atlantic and Pacific basins. In the synoptic method, prevailing environmental conditions like wind shear, sea surface temperature, ocean thermal energy, low level inflow, upper level outflow etc. as mentioned in the section 5.3 for genesis are considered. All these fields in the NWP model analyses and forecasts are also considered. The development of characteristic features in satellite and radar observations is also taken into consideration for predicting the intensity. While the synoptic, statistical and satellite/radar guidance helps in short range intensity forecast (upto 12/24 hrs), the NWP and dynamical-statistical guidance (Kotal et al., 2008) is mainly used for 24-120 hr forecast. Hence, the RSMC, New Delhi official forecast intensity results from a manually analyzed forecasting process, which relies on output from several models as mentioned above. Consensus forecasts that gather all or part of the numerical forecast intensities and use synoptic and statistical guidance are utilised to issue official forecast.

5.8 Quadrant Wind Radii Forecasting

The TC wind radii forecasts are generated in terms of the radii of 34 kts, 50 kts and 64 kts (1 kt = 0.52 ms^{-1} or 1.85 kmph) winds in four geographical quadrants around the TC. These wind radii represent the maximum radial extent of winds reaching 34 kts, 50 kts and 64 kts in each quadrant (NW, NE, SE and SW). The initial estimation and forecast of the wind radii of TC is rather subjective and strongly dependent on the data availability, climatology and analysis methods. The subjectivity and reliance on climatology is amplified in NIO in the absence of aircraft observations. However, recently with the advent of easily accessible remotely sensed surface and near-surface winds (e.g. Ocean Sat., Special Sensor Microwave Imager (SSMI), low level atmospheric motion vectors and Advanced Microwave Sounder Unit (AMSU) retrieval methods) and advances in real time data analysis capabilities, IMD introduced TC wind radii monitoring and prediction product in Oct., 2010. The initial wind radii estimates have become less subjective due to the tools and products mentioned above. While better initial estimates of wind radii are becoming available, forecasting these wind radii remains a difficult task. It is mainly because of the fact that we do not have any objective wind radii forecast methods and current NWP models fail to produce forecasts that are better than climatology (Knaff et al., 2006, 2007; Knabb et al., 2006). The inputs for wind radii monitoring are obtained from following observations:

- Ship
- Buoy
- Scatterometry winds
- Lower level Atmospheric motion vectors
- Cloud Motion Vectors
- Water vapour based wind vectors
- Special Sensor Microwave Imager (SSMI) data
- Advanced Microwave Sounder Unit (AMSU)
- Latest advances in real time data analysis capabilities
- DWR (when system is within the radar range)
- Coastal wind observations (when the TC is near coast)

The road map for monitoring and prediction of quadrant wind radii is given below.

(a) Date and time of initial condition of forecast
(b) Official location and intensity (T/C.I. No., maximum wind and centre position)
(c) Initial TC wind radii estimation
    - Wind radii based on Oceansat/ASCAT/Windsat wind
    - SSMI based wind radii
    - Wind radii based on lower level atmospheric motion vectors
    - Wind radii by AMSU retrieval method
    - Wind radii based on global and regional NWP model analyses

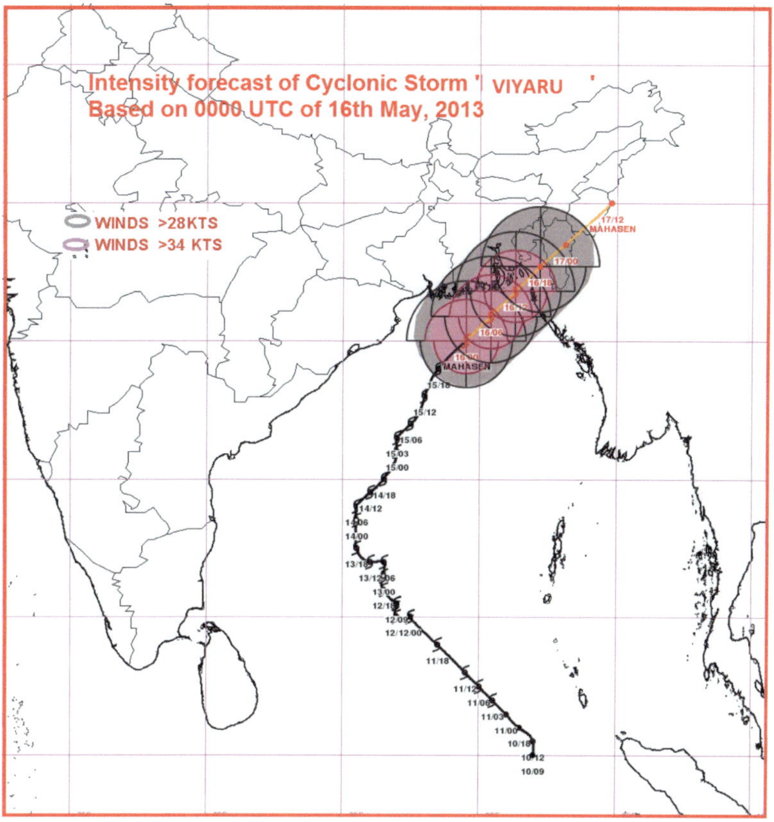

Fig. 16. A typical graphical presentation of quadrant wind forecast during TC Viyaru (10-17 May 2013).

- Wind radii based on DWR wind retrieval
- Value addition based on coastal, ship and buoy observations
- Climatological consideration

(d) Official forecast of TC intensity and track upto 72 hrs.
(e) Persistence forecast based on initial wind radii and past 12 hrs trend.
(f) Climatological forecast of TC wind radii and CLIPER
(g) NWP model forecasts of 10 metre wind radii
   - Select the model most appropriate to initial condition
   - Compare the wind field distribution to the actual wind
   - Calculate the wind radii in four quadrants for the threshold of 34 kts, 50 kts and 64 kts surface wind
   - Make corrections based on actual wind, climatology, intensity of the system

(h) Official TC wind radii forecast in four quadrants for the threshold of 34 kts, 50 kts and 64 kts based on S.N. (b-g)

A typical example of the quadrant wind radii product is shown in Fig. 16.

6. TC Forecast Accuracy over NIO

6.1 Track Forecast Accuracy

Mohapatra et al (2013b) evaluated the TC track forecast issued by IMD during 2003–2011 (nine years) by calculating the direct position error (DPE) and skill in track forecast. The accuracy of TC track forecast has been analysed with respect to basin of formation (Bay of Bengal, Arabian Sea and NIO as a whole), season of formation (pre-monsoon and post-monsoon seasons), intensity of TCs (CS and SCS or higher intensities) and type of track of TCs (climatological/straight moving and recurving/looping type). The average DPE is about 140, 262 and 386 km and skill is about 27%, 39% and 50%, respectively for 24, 48 and 72 hr forecasts over the NIO as a whole during 2009–2011. Though the DPE is higher and skill is less as compared to those in northwest Pacific and north Atlantic Ocean, the rate of decrease (increase) in DPE (skill) is higher over the NIO in recent years. The DPE (skill) over the NIO has decreased (increased) at the rate of about 7.3 km (3%) per year during 2003–2011 for 24 hr forecasts. The average DPE of NHC, USA for the TCs over the northern Atlantic Ocean during 2003-2011 is about 95 km, 150 km and 250 km for 24, 48 and 72 hr forecasts respectively (NHC, 2012). Similarly, the DPE of RSMC, Tokyo for the TCs over the northwest Pacific Ocean is about 113, 208 and 310 km for 24, 48 and 72 hr forecasts respectively during 2003-2011 (RSMC, Tokyo, 2011). The average skill of NHC, USA for the TCs over the northern Atlantic Ocean during 2003-2011 is about 45%, 55% and 54% for 24, 48 and 72 hr forecasts respectively (NHC, 2012). The lower accuracy over the NIO may be due to the fact that the NIO is a data sparse region compared to northwest Pacific and north Atlantic Oceans, as there is no aircraft probing facility in the NIO unlike the northwest Pacific and north Atlantic Oceans. Further, the in-situ data from the meteorological buoys are significantly less over the NIO, with only 12 in 2011 (Mohapatra et al., 2012b). Scarcity of observational data leads to poor initial condition in NWP models. It further leads to increase in forecast error of NWP models, which are the backbone for official track forecast of TCs, especially in 36-120 hr forecast range (Osuri et al., 2012; Mohanty et al., 2010).

6.2 Intensity Forecasting Accuracy

Mohapatra et al. (2013c) evaluated the TC intensity forecast issued by IMD during 2005–2011 (seven years) by calculating the absolute error (AE), root mean square error (RMSE) and skill in intensity forecast in terms of MSW. The accuracy of TC intensity forecast has been analysed by them with respect to basin of formation (Bay of Bengal, Arabian Sea and NIO as a whole), season of formation (pre-monsoon and post-monsoon seasons), intensity of TCs (CS and SCS or higher intensities) and type of track of TCs (climatological/straight moving and recurving/looping type). The study shows that the average AE

(RMSE) in intensity forecast is about 11 (14), 14 (19) and 20 (26) knots, respectively, for 24-, 48- and 72-h forecasts over the NIO as a whole during 2009–2011. The skill of intensity forecast is about 44% (48%), 60% (58%) and 60% (65%) for 24-, 48- and 72-h forecasts during 2009–2011 with respect to AE (RMSE). There is no significant improvement in terms of reduction in AE and RMSE of MSW forecast over the NIO like that over the northwest Pacific and northern Atlantic Oceans during 2005–2011. However, the skill in intensity forecast compared to persistence method has significantly improved by about 6% (10%) and 9% (8%) per year, respectively, for 12- and 24-h forecasts considering the AE (RMSE) during 2005–2011.

7. Problems and Prospects

Over the past several years, there have been large improvements in track forecast skill (Mohapatra et al., 2013b, 2013c) and modest improvements in the intensity skill like other ocean basins. These errors, particularly the intensity errors negatively affect wind radii forecasts. The poor intensity forecast is particularly pronounced when intensity forecast fail to or falsely forecast winds that exceed the 34 kts, 50 kts and 64 kts thresholds. There is also scope for improvement in future, as evident from the recent past cases, due to existing and proposed buoy network over the NIO, ongoing modernization plan of IMD resulting in improved observational system like DWR and AWS along the coast, improved modelling technique, ongoing forecast demonstration project (FDP) on landfalling TCs over the Bay of Bengal. Modernisation of Voluntary Observing Fleet (50 nos) scheme is in progress. However we may need cooperation from other member countries of WMO/ESCAP Panel to install more buoys and share the data with IMD/RSMC, New Delhi.

As a result of increase in requirements of meteorological support, it has become necessary to provide adequate and realistic observations for frequent initialization of NWP models for short to medium range forecasting of TCs (track, intensity and associated adverse weather).

NIO being the data sparse region, the best track estimate is subject to errors. The mean best track error may be considered as 50 km. In the absence of the observations over the NIO, the best track of the TC is mostly estimated with the satellite imagery interpretation with the help of Dvorak's technique. This technique has been applied over the NIO since its inception during 1970s. However, this technique, which has been developed for north Atlantic Ocean basin, needs to be validated for NIO with aircraft reconnaissance. Further automated objective Dvorak's technique has to be validated/developed for NIO to minimize the human error and evaluate its efficiency.

A new geostationary meteorological satellite INSAT-3D is being designed by ISRO. It will have a six channel imager and a nineteen-channel sounder. In addition to satellite imagery in six channels several new derived products will be available after processing the data from INSAT-3D. These products will be useful in NWP and in day-to-day weather forecasting. In addition to ground level products, the vertical profiles of atmospheric temperature, moisture and ozone will be available from the satellite data of INSAT-3D. It will provide 1 km resolution imagery in visible band, 4 km resolutions in IR band and 8 km in water vapour channel. It will provide much improved capabilities to the meteorological community and users. New ground segment facilities to receive and process data from INSAT-3D are being installed. Three NOAA/ MODIS/Metops Satellite data receiving and processing systems have been installed at New Delhi, Chennai and Guwahati for obtaining Satellite derived products that are useful in weather forecasting. The polar orbiting satellite, OCEANSAT-II has been launched by Indian Space Research Organisation during September, 2009. It has the capability to estimate the ocean surface wind vectors. It will be helpful in monitoring of cyclonic disturbances over the NIO. Similarly the Meghatropiques launched in October 2011 and Saral in February, 2013 may provide the data on evolution of tropical weather systems and sea state conditions required for monitoring and prediction of cyclogenesis respectively.

The FDP on landfalling TCs over the Bay of Bengal has been taken up. It will help us in minimizing the error in prediction of TC track and intensity forecasts and hence adverse weather. During pre-pilot phase (15 Oct-30 Nov 2008-12), several national institutions participated for joint observational, communicational and NWP activities resulting in improved forecast and delivery of services (Mohapatra et al., 2013a). With possible manned and unmanned aircraft reconnaissance, it will help in improving TC track and intensity forecasting and hence the adverse weather warning over the NIO region, as demonstrated in Atlantic and Pacific Ocean basins (Burpee et al., 1996; Holland et al., 2001; Aberson, 2003; Aberson et al., 2006; Wu et al., 2007; Chou and Wu, 2008). With the completion of ongoing modernization programme, the error is likely to reduce by about 20% by 2015 and by 40% by 2020 from the base year of 2010.

Thus, IMD continuously expands and strengthens its activities in relation to observing strategies, forecasting techniques, disseminating methods and research relating to different aspects of TCs to ensure most critical meteorological support through observations, analysis, predictions and warnings to disaster managers and decision makers not only in India but also to the NIO rim countries.

Acknowledgement

The author is thankful to Dr. L.S. Rathore, Director General of Meteorology, IMD for his encouragement and support to carry out this work.

REFERENCES

Aberson, S.D., 2003: Targeted observations to improve operational tropical cyclone track forecast guidance. *Mon. Wea. Rev.,* **131**, 1613-1628.

Aberson, S.D., M.L. Black, R.A. Black, R.W. Burpee, J.J. Cione, C.W. Landsea and F.D. Marks, Jr., 2006: Thirty years of tropical cyclone research with the NOAA P-3 aircraft. *Bull. Amer. Met. Soc.*, **87**, 1039-1055.

Ali, M.M., C.M. Kishtwal and S. Jain, 2007: Predicting cyclone tracks in the NIO: An artificial neural network approach. *Geophys. Res. Lett.*, **34**, L 04603

Atkinson, G.D. and C.R. Holliday, 1977: Tropical cyclone minimum sea level pressure/ maximum sustained wind relationship for the Western North Pacific. *Mon. Wea. Rev.*, **105**, 421-427.

Bhatia, R.C. and A.K. Sharma, 2013: Recent advances in observational support from space-based systems for tropical cyclones, *Mausam*, **64**, 97-104.

Bhatia, R.C., Sunit Das, M. Mohapatra and S.K. Roy Bhowmik, 2008: Use of Satellite and AWS data for weather Prediction. *IMD Met. Monogr. Synop. Met.* No. **6/2008**. Published by India Meteorological Department, pp. 109-121.

Burpee, R.W., J.L. Franklin, S.J. Lord, R.E. Tuleya and S.D. Aberson, 1996: The impact of Omega dropwindsondes on operational hurricane track forecast models. *Bull. Amer. Met. Soc.*, **77**, 925-933.

Chou, K.-H. and C.-C. Wu, 2008: Development of the typhoon initialization in a mesoscale model combination of the bogused vortex and the dropwindsonde data in DOTSTAR. *Mon. Wea. Rev.*, **136**, 867-879.

Chourasia, Manjusha, R.G. Ashrit and John P. George, 2013: Impact of cyclone bogusing and regional assimilation on tropical cyclone, track and intensity predictions. *Mausam*, **64**, 135-148.

Colon, J.A., C.R.V. Raman and V. Srinivasan, 1970: On some aspects of tropical cyclone of 20-29 May 1913 over the Arabian Sea. *Ind. J. Met. Hydrol. Geophys*, **21**, 1-22.

Courtney, J. and J.A. Knaff, 2009: Adapting the Knaff and Zehr Wind-Pressure Relationship for operational use in Tropical Cyclone Warning Centres. *Australian Meteorological and Oceanographic Journal*, **58(3)**, 167-179.

Datta, R.K. and R.N. Gupta, 1975: Tropical storms in the Bay of Bengal and calculation of probabilities of its striking east coast by storm analogue technique. *Ind. J. Met. Hydrol. & Geophys.*, **26**, 345-348.

Dube, S.K., A.D. Rao, Jismy Poulose, M. Mohapatra and T.S. Murty, 2014: Storm surge inundation in south Asia under climate change scenarios. *In:* Monitoring and prediction of tropical cyclones in the Indian Ocean and climate change. (Eds) U.C. Mohanty, M. Mohapatra, O.P. Singh, B.K. Bandyopadhyay and L.S. Rathore. Capital Publishing Company, New Delhi.

Dvorak, V.F., 1975: Tropical cyclone intensity analysis and forecasting from satellite imagery. *Mon. Wea. Rev.*, **103**, 420-430.

Dvorak, V.F., 1984: Tropical cyclone intensity analysis using satellite data. Technical Report (NOAA TR NESDIS 11), National Oceanic and Atmospheric Administration, National Environmental Satellite, Data and Information Service.

Goyal, Suman, M. Mohapatra and A.K. Sharma, 2013: Comparison of best track parameters of RSMC, New Delhi with satellite estimates over north Indian Ocean. *Mausam*, **64**, 25-34.

Gray, W.M., 1968: Global view of origin of Tropical disturbance and storms. *Mon. Wea Rev.*, **96**, 669-700.

Gupta, A., 2006: Current status of tropical cyclone track prediction techniques and forecast errors. *Mausam*, **57**, 151-158.

Harper, B.A., J. Kepert and J. Ginger, 2008: Wind speed time averaging conversions for tropical cyclone conditions. Proc. 28th Conf on Hurricanes and Tropical Meteorology, AMS, Orlando.

Hatwar, H.R., V. Subrahmanyam, M. Mohapatra, S.K. Roy Bhowmik, B.K. Bandyopadhay, Ch. Singh and K. Srivastava, 2008: A report on the cyclonic storm "Ogni" 2006, India Meteorological Department, Meteorological Monograph, Cyclone Warning Division.

Holland, G.J., P.J. Webster, J.A. Curry, G. Tyrell, D. Gauntlett, G. Brett, J. Becker, R. Hoag and W. Vaglienti, 2001: The Aerosonde robotic aircraft: A new paradigm for environmental observations. *Bull. Amer. Met. Soc.*, **82**, 889-901.

IMD, 1976: Weather Radar Observations Manual – 1976. IMD, Lodi Road, New Delhi.

IMD, 2003: Cyclone Manual. India Meteorological Department, New Delhi.

IMD, 2008: Track of storm and depressions over the Indian Seas during 1891-2007. *In:* Cyclone e-Atlas of IMD, published by IMD.

IMD, 2013: Cyclone Warning Services: Standard Operation Procedure. Cyclone Warning Division, IMD, New Delhi.

Jha, T.N., M. Mohapatra and B.K. Bandyopadhyay, 2013: Estimation of intensity of tropical cyclone over Bay of Bengal using microwave imagery. *Mausam*, **64**, 105-116.

Kalsi, S.R., 2006: Orissa super cyclone – A synopsis. *Mausam*, **57**, 1-20.

Kalsi, S.R., 2002: Use of satellite imagery for tropical cyclone intensity analysis and forecasting. Met. Monograph, Cyclone Warning Div. No. **1/2002**.

Kelkar, R.R., 1997: Satellite-based monitoring and prediction of tropical cyclone intensity and movement. *Mausam*, **48**, 157-168.

Knabb, R., E. Rappaport, M. Mainelli, J. Franklin, C. Lauer and A. Krautkramer, 2006: Progress toward operational implementation of tropical cyclone wind probability products. [Available online at http://www.ofcm.gov/ihc06/ Presentations/06%20 session6%20Decision-making% 20Products/s6-04Session%206%20(knabb). ppt#492,31,Slide31.]

Knaff, J.A., Mark Demaria, Timothy P. Marchok, James M. Gross and Colin J. Mcadie, 2007: Statistical Tropical Cyclone Wind Radii Prediction Using Climatology and Persistence. *Weather and Forecasting*, **22**, 781-791.

Knaff, J.A., C.P. Guard, J.P. Kossin, T.P. Marchok, T. Smith and N. Surgi, 2006: Operational guidance and skill in forecasting tropical cyclone structure change. Workshop Topic Reports of the Sixth WMO International Workshop on Tropical Cyclones, Tropical Meteorology Research Programme Rep. 72. [Available from World Meteorological Organization, Case Postale 2300, CH-1211 Geneva 2, Switzerland, and online at http://www.wmo.ch/web/arep/arep-home.html.]

Knaff, John A. and Raymond M. Zehr, 2007: Reexamination of Tropical Cyclone Wind–Pressure Relationships. *Wea. Forecasting*, **22**, 71-88.

Koba, H., T. Hagiwara, S. Osano and S. Akashi, 1991: Relationships between CI number and minimum sea level pressure/maximum wind speed of tropical cyclones. *Geophysical Magazine*, **44,** 15-25.

Kotal, S.D. and S.K. Roy Bhowmik, 2011: A multimodel ensemble technique for cyclone track prediction over north Indian Sea. *Geofizika*, **28**, 275-291.

Kotal, S.D., P.K. Kundu and S.K.R. Bhowmik, 2009: An analysis of cyclogenesis parameters for developing and non-developing low pressure systems. *Nat. Hazards*, **50**, 389-402.

Kotal, S.D., S.K. Roy Bhowmik, P.K. Kundu and A.K. Das, 2008: A statistical cyclone intensity prediction model for Bay of Bengal. *J Earth Syst Sci*, **117**, 157-168.

Kotal, S.D. and S.K. Bhattacharya, 2013: Tropical cyclone Genesis Potential Parameter (GPP) and its application over the north Indian Sea. *Mausam*, **64**, 149-170.

Mausam, 2006: Special Issue on Orissa super cyclone of October 1999. *Mausam*, **57**, 1-196.

Mishra, D.K. and G.R. Gupta, 1976: Estimates of maximum wind speed in tropical cyclones occurring in the Indian Seas. *Ind. J. Met. Geophys*, **27**, 285-290.

Mishra, D.K. and H. Raj, 1975: A satellite study of intensities of cyclonic storms in the Bay of Bengal. *Ind. J Meterol Geophys*, **26**, 455-464.

Mohanty, U.C., Krishna K. Osuri, A. Routray, M. Mohapatra and Sujata Pattanayak, 2010: Simulation of Bay of Bengal Tropical Cyclones with WRF model: Impact of initial and boundary condition. *Marine Geodesy*, **33**, 294-314.

Mohanty, U.C., Krishna K. Osuri and Sujata Pattanayak, 2013: A study on high resolution mesoscale modeling systems for simulation of tropical cyclones over the Bay of Bengal. *Mausam*, **64**, 117-134.

Mohanty, U.C. and A. Gupta, 1997: Deterministic methods for prediction of tropical cyclone tracks. *Mausam*, **48**, 257-272.

Mohapatra, M., B.K. Bandyopadhyay and Ajit Tyagi, 2012b: Best track parameters of tropical cyclones over the North Indian Ocean: A review. *Nat. Hazards*, **63**, 1285-1317, doi:10.1007/s11069-011-9935-0.

Mohapatra, M., D.R. Sikka, B.K. Bandyopadhyay and Ajit Tyagi, 2013a: Outcomes and Challenges of Forecast Demonstration Project (FDP) on landfalling cyclones over the Bay of Bengal. *Mausam*, **64**, 1-12.

Mohapatra, M. and S. Adhikary, 2011: Modulation of cyclonic disturbances over the north Indian Ocean by Madden – Julian oscillation. *Mausam*, **62**, 375-390.

Mohapatra, M., D.C. Gupta, N.K. Chanchlani and S.K. Dastidar, 2002: Orissa Super Cyclone, 1999 – A case study. *Journal of Indian Geophysical Union*, **6**, 93-106.

Mohapatra, M., G.S. Mandal, B.K. Bandyopadhyay, Ajit Tyagi and U.C. Mohanty, 2012: Classification of cyclone hazard prone districts of India. *Nat Hazards*, **63**, 1601-1620, DOI 10.1007/s11069-011-9891-8.

Mohapatra, M., B.K. Bandyopadhyay and D.P. Nayak, 2013c: Evaluation of operational tropical cyclone intensity forecasts over north Indian Ocean issued by India Meteorological Department. *Nat Hazards*, **68**, 433-451, DOI 10.1007/s11069-013-0624-z.

Mohapatra, M., D.P. Nayak, R.P. Sharma and B.K. Bandyopadhyay, 2013b: Evaluation of official tropical cyclone track forecast over north Indian Ocean issued by India Meteorological Department. *J. Earth Syst. Sci.*, **122**, 589-601.

Mohapatra, M., D.P. Nayak and B.K. Bandyopadhyay, 2012c: Evaluation of Cone of Uncertainty in Tropical Cyclone Track Forecast over north Indian Ocean issued by India Meteorological Department, *Tropical Cyclone Research and Review*, **2**, 331-339.

Nakazawa, T. and S. Hoshino, 2009: Intercomparison of Dvorak Parameters in the Tropical Cyclone Datasets over the Western North Pacific. *Scientific Online Letters on the Atmosphere*, **5**, 33-36.

Neumann, C.J. and G.S. Mandal 1978: Statistical prediction of tropical storm motion over the Bay of Bengal and Arabian Sea. *Indian J. Meteorol. Hydrol. Geophys.*, **29**, 487-500.

NHC USA, 2012: Forecast verification report. National Oceanic and Atmospheric Administration, http://www.nhc.noaa.gov/verification.

Olander, T.L. and C.S. Velden, 2007: The advanced Dvorak technique: Continued, development of an objective scheme to estimate tropical cyclone intensity using geostationary infrared satellite imagery. *Wea. Forecasting*, **22**, 287-298.

Osuri, Krishna K., U.C. Mohanty, A. Routray and M. Mohapatra, 2012: The impact of satellite-derived wind data assimilation on track, intensity and structure of tropical cyclones over the North Indian Ocean. *Int. J. Rem. Sens.*, **33**, 1627-1652, doi: 10.1080/01431161.2011.596849.

Prasad, K., S.R. Kalsi, Y.V. Rama Rao, S.K. Bhowmik and D. Joardar, 1998: Evaluation of operational cyclone track forecasts by IMD limited area NWP model. *Mausam*, **51**, 91-95.

Prasad, K. and Y.V. Rama Rao, 2003: Cyclone track prediction by quasi-Lagrangian limited area model. *Meteor.Atmos.Phys.*, **83**, 173-185.

Raghavan, S., 1997: Radar observations of tropical cyclone. *Mausam*, **48**, 169-188.

Raghavan, S., 2013: Observational aspects including weather radar for tropical cyclone monitoring. *Mausam*, **64**, 89-96.

Raghavan, S., 1990: Structure of tropical cyclones in the Bay of Bengal. *Mausam*, **41**, 325-328.

Ramanathan, Y. and R.K. Bansal, 1977: On the application of a primitive equation barotropic model for the prediction of storm tracks in the Indian region. *Ind. J. Met. Hydrol & Geophys.*, **28**, 169-176.

Roy Bhowmik, S.K., 2003: An evaluation of cyclone genesis parameters over the Bay of Bengal using model analysis. *Mausam*, **54**, 351-358.

RSMC Tokyo, 2011: Annual Report on activities of the RSMC Tokyo Typhoon Center, http://www.jma.go.jp/rsmc-hppub-eg/annualreport.html.

Sikka, D.R. and R. Suryanarayna, 1972: Forecasting the movement of tropical storms/ depression in the Indian region by computer oriented technique with climatology and persistence. *Ind-J. Met. Geophys.*, **23**, 35-40.

Sikka, D.R., 2006: Major advances in understanding and prediction of tropical cyclones over the NIO. *Mausam*, **57**, 165-196.

Simpson, J., E. Ritchie, G.J. Holland, Halverson and S. Stewart, 1997: Meso-scale interactions in tropical cyclone genesis. *Mon. Wea. Rev.*, **125**, 2643-2661.

Singh, P., P.K. Pal, C.M. Kishtwal and P.C. Joshi, 2005: Impact of bogus Vortex on track and intensity prediction of tropical cyclone. *J. Earth Sys. Sci.*, **114**, 427-436.

Tyagi, Ajit, M. Mohapatra, B.K. Bandyopadhyay and Naresh Kumar, 2010: Characteristics of very severe cyclonic storm "NARGIS" over the Bay of Bengal during 27th April to 3rd May, 2008. *In:* Indian Ocean tropical cyclones and climate change. (Ed.) Yassine Charabi. Springer Publication Ltd, 315-326.

Uhlhorn, E.W. and P.G. Black, 2003: Verification of remotely sensed sea surface winds in hurricanes. *J. Atmos. Ocean. Technol.*, **20**, 99-116.

Velden, C.S. et al., 2006: The Dvorak tropical cyclone intensity estimation technique: A satellite-based method that has endured for over 30 years. *Bull. Amer. Met. Soc.*, **87**, 1195-1210.

Velden, C.S., T.L. Olander and R.M. Zehr, 1998: Development of an objective scheme to estimate tropical cyclone intensity from digital geostationary satellite infrared imagery. *Wea. Forecasting,* **13**, 172-186.

Venkatesh, T.N., 2006: Meso-scale interactions during the genesis and intensification of October, 1999 Orissa super cyclone. *Mausam*, **57**, 31-36.

WMO, 2009: Tropical cyclone operational plan for the Bay of Bengal and the Arabian Sea. Tropical Cyclone Programme Report no. Tcp21 Edition 2009, WMO/TD No.84.

Wu, C.C., K.H. Chou, P.H. Lin, S.D. Aberson, M.S. Peng and T. Nakazawa, 2007: The impact of dropwindsonde data on typhoon track forecasts in DOTSTAR. *Wea. Forecasting*, **22**, 1157-1176.

17

Operational Tropical Cyclone Forecasts Models at IMD and Their Performance

S.K. Roy Bhowmik
India Meteorological Department
New Delhi – 110003, India
skrb.imd@gmail.com

1. Operational NWP System of IMD

There has been a massive up-gradation of weather forecasting capabilities in India under the modernization programme of the Government of India, which covers various components such as atmospheric observation network; strengthening of computing facilities; data integration and product generation; and dissemination of information to an optimum level. It has improved forecasting capabilities for high impact weather events like cyclones, severe thunderstorm, heavy rainfall and floods in a significant manner. IMD now has a network of automatic weather stations, Doppler Weather Radars (DWR), state-of-the-art upper air systems etc. These observations are now being used to run numerical prediction models on High Performance Computing Systems (HPCS). Global Forecast System (GFS T574/L64) was made operational at IMD New Delhi, incorporating Global Statistical Interpolation (GSI) scheme as the global data assimilation for the forecast up to seven days. Mesoscale forecast system WRF (ARW) with 3DVAR data assimilation is being operated daily twice, at 27 km, 9 km and 3 km horizontal resolutions for the forecast up to three days using initial and boundary conditions from the IMD GFS T574. At ten other regional centres, very high resolution mesoscale models (WRF at 3 km resolution) are made operational with the installation of High End Server. Doppler weather and mesoscale dynamical model-based Nowcast system was made operational for the national Capital of Delhi. Polar WRF is implemented to provide day-to-day short range (48 hours) weather forecast for the Maitri region over Antarctica. District level quantitative five days weather forecasts based on Multi-Model Ensemble (MME) system are being generated to support Agro-Meteorological Advisory Service of India. All these NWP products are routinely made available on the IMD web site www.imd.gov.in.

2. Numerical Modelling of Tropical Cyclone

Tropical cyclones are among the most deadly natural hazards as they are associated with very strong winds, heavy rain and storm surge. Various stages of cyclone forecasting are genesis, movement, intensity and decay after landfall. Due to the increasing human habitation near the coasts, accurate and timely forecasting of tropical cyclone has posed a challenging task to the operational forecasters. The synoptic methods have been the mainstay of tropical weather forecasting. Of late, NWP methods have acquired greater skills and are playing increasingly important role in delivering operational real time weather forecasts. However, limitations remain, particularly in the prediction of track and intensity of tropical cyclone. As such there has been an operational requirement to formulate an objective procedure to handle operational cyclone forecasting work in a more efficient and effective manner.

IMD implemented an NWP-based objective method for the operational cyclone forecasting work. The method comprises four forecast components, namely (a) Cyclone genesis potential parameter (GPP), (b) Multi-model ensemble (MME) technique for track prediction, (c) Cyclone intensity prediction (SCIP) model and (d) Predicting decaying intensity after the landfall. The four forecasting components of tropical cyclone are described below.

2.1 Genesis Potential Parameter (GPP)

Every year about one dozen low pressure areas form over the Indian seas, but only a few of them intensify into a cyclonic storm. So from the operational point of view, it is very important to know at the initial stage the potentiality of a low pressure system to intensify into a cyclonic storm.

The process of initiation of a cyclonic disturbance over the sea area is called cyclogenesis. To quantify the cyclogenesis, McBride et al. (1981) proposed a Daily Genesis Potential parameter (DGP) on the basis of model analysis fields over the Atlantic and Pacific Ocean basin. In their study, DGP is defined as the difference of vorticity between 900 hPa and 200 hPa. The study showed that DGP is three times greater for developing systems than that of non-developing systems. An analysis of Cyclone Genesis Parameter for the Bay of Bengal, conducted by Roy Bhowmik (2003), showed that the procedure is capable of providing useful predictive signal. Kotal et al. (2009) extended the work further by defining Genesis Potential Parameter (GPP) as:

$$\mathrm{GPP} = \frac{\xi_{850} \times M \times I}{S} \quad \text{if } \xi_{850} > 0,\ M > 0 \text{ and } I > 0$$

$$= 0 \qquad \text{if } \xi_{850} \le 0,\ M \le 0 \text{ or } I \le 0$$

where ξ_{850} = Low level relative vorticity (at 850 hPa) in $10^{-5}\,s^{-1}$

S = Vertical wind shear between 200 and 850 hPa (knots)

$M = \dfrac{[\text{RH} - 40]}{30}$ = Middle troposphere relative humidity,

where, RH is the mean relative humidity between 700 and 500 hPa and

$I = (T_{850} - T_{500})$ °C = Middle-tropospheric instability (Temperature difference between 850 hPa and 500 hPa).

All the variables are estimated by averaging of all grid points over an area of radius 2.5° around the centre of cyclonic systems.

The study showed that GPP values are 3 to 5 times greater for the developing systems (T.No. > 2.5; maximum wind speed >35 knots) than for non-developing systems (T.No. ≤ 2.5; maximum wind speed ≤ 35 knots) and is useful in differentiating between developing and non-developing systems at their early stages of development. They also showed that GPP values are equal and above 8.0 for developing systems and below 8.0 for non-developing systems in more than 85% of cases. GPP values for developing and non-developing systems as reported by Kotal et al. (2009) are shown in Table 1.

Table 1. Genesis potential parameter (GPP) for developing systems and non-developing systems

GPP ($\times 10^{-5}$) →					
T.No. →	1.0	1.5	2.0	2.5	3.0
Developing	11.1	12.3	13.3	13.5	13.6
Non-developing	3.4	4.2	4.6	2.7	-

Various thermo-dynamical parameters, which are used for real time analyzing Genesis Potential Parameter (GPP) for cyclonic storms, are derived from the operational Global Data Assimilation System (GDAS) of India Meteorological Department (IMD), New Delhi.

2.2 Track: Multi-Model Ensemble (MME) Technique

A Multi-Model Ensemble (MME) technique for predicting the track of tropical cyclones for the Indian seas is developed using cyclone data of 2007-2009 (Kotal and Roy Bhowmik, 2012). The technique is based on the multiple linear regression method using the member models WRF, QLM, IMD-GFS, ECMWF and JMA. The predictors (shown in Table 2) selected for the ensemble technique are forecasts latitude and longitude position at 12-hour interval up to 72-hour of five operational models. In the MME forecasts, model-forecast latitude position and longitude position of the member models are linearly regressed against the observed latitude and longitude position respectively for each forecast time at 12-hours intervals for the forecast up to 72-hour. Multiple linear regression technique is used to generate weights (regression coefficients) for each model for each forecast hour (12hr, 24hr, 36hr, 48hr, 60hr, 72hr). These coefficients are then used as weights for ensemble forecasts.

12-hourly forecast latitude (LAT^f) and longitude (LON^f) positions by multiple linear regression technique are defined as:

$$LAT^f_t = a_0 + a_1 ECMWF_t^{lat} + a_2 IMD_t^{lat} + a_3 JMA_t^{lat} + a_4 MM5_t^{lat} + a_5 QLM_t^{lat}$$

$$LON^f_t = a'_0 + a'_1 ECMWF_t^{lon} + a'_2 IMD_t^{lon} + a'_3 JMA_t^{lon} + a'_4 MM5_t^{lon} + a'_5 QLM_t^{lon}$$

for t = forecast hour 12, 24, 36, 48, 60 and 72.

The dependent variable latitude (LAT^f) is in °N and longitude (LON^f) in °E. The details of model predictors are given in Table 2.

Table 2. Model parameters

S. No.	*Member models*	*Symbol of predictors*	
		Latitude position	*Longitude position*
1.	European Centre for Medium-Range Weather Forecasts (ECMWF),	$ECMWF^{lat}$	$ECMWF^{lon}$
2.	GFS of National Centers for Environmental Prediction (IMD)	IMD^{lat}	IMD^{lon}
3.	Japan Meteorological Agency (JMA)	JMA^{lat}	JMA^{lon}
4.	WRF model	WRF^{lat}	WRF^{lon}
5.	Quasi-Langrangian model (QLM)	QLM^{lat}	QLM^{lon}

2.3 Intensity

In view of limitations of NWP models in the prediction of intensity of tropical cyclones (Elsberry et al., 2007; Houze et al., 2007), Kotal et al. (2008a) developed a statistical-dynamical cyclone intensity prediction (SCIP) model for prediction of intensity at 12-hour interval up to 72 hours to aid operational cyclone forecasting work over the Indian seas. The maximum potential intensity (MPI) of a cyclone during life time of the system for the Bay of Bengal is also investigated (Kotal et al., 2008b).

Roy Bhowmik et al. (2007) proposed a simple empirical model for predicting cyclone intensity over the Bay of Bengal. The study is based on the assumption that tropical cyclone intensifies exponentially, where the intensification factor is determined using past 12 hours intensity changes. A major limitation of this empirical model (Roy Bhowmik et al., 2007) is that it does not include parameters to take into account the physical and dynamical processes involved. The study warranted further investigation in a more general manner incorporating other synoptic and thermodynamical factors, which play important role for intensification of storms. In order to overcome these shortcomings, Kotal et al. (2008a) developed a Statistical Cyclone Intensity Prediction (SCIP) model for the Bay of Bengal for predicting 12 hourly cyclone intensity (up to 72 hours), applying multiple linear regression technique using various dynamical and physical parameters as predictors. The model equation (Kotal et al., 2008a) is given as:

$$dv_t = a_0 + a_1 IC12 + a_2 SMS + a_3 VWS + a_4 D200 + a_5 V850 + a_6 ISL + a_7 SST + a_8 ISI$$

for $t =$ forecast hour 12, 24, 36, 48, 60 and

$dv_t =$ Intensity change during the time interval t.

The details of model predictors are given in Table 3.

Table 3. Model parameters

S.No.	*Predictors*	*Symbol of predictors*	*Unit*
1.	Intensity change during last 12 hours	IC12	Knots
2.	Vorticity at 850 hPa	V850	$\times 10^5\ s^{-1}$
3.	Storm motion speed	SMS	ms^{-1}
4.	Divergence at 200 hPa	D200	$\times 10^5\ s^{-1}$
5.	Initial storm intensity	ISI	Knots
6.	Initial storm latitude position	ISL	°N
7.	Sea surface temperature	SST	°C
8.	Vertical wind shear	VWS	Knots

The Sea Surface Temperature (SST) analysis at 1° latitude-longitude grid interval of NCEP (National Center for Environmental Prediction) is used in the Statistical Cyclone Intensity Prediction (SCIP) model. The other model parameters (vorticity, divergence and vertical wind shear), which are used in the SCIP model, are derived from the operational GDAS of IMD. The SCIP model is implemented in April 2008 and is being used for real time forecasting.

2.4 Decay of Intensity after the Landfall

The forecast of inland wind after the landfall of a cyclone is of great concern to disaster management agencies. To address this problem, Roy Bhowmik et al. (2005) proposed an empirical model for predicting 6-hourly maximum sustained surface winds (intensity) that is valid till the system becomes a weak low pressure area after the landfall over the Indian region. According to the decay equation (Roy Bhowmik et al., 2005), the maximum sustained surface wind speed (MSSW) after the landfall at time t is given by:

$$V_{t+6} = V_b + (V_t - V_b) \times R_1, \text{ for } t = 0 \quad (1)$$

$$= V_b + (V_t - V_b) \times R_2, \text{ for } t = 6, 12, 18 \text{ and } 24 \quad (2)$$

where reduction factors

$$R_1 = \exp(-a_1 \times 6.0) \quad (3)$$

$$\text{and } R_2 = \exp(-a_2 \times 6.0) \quad (4)$$

Decay constant a_1 for the first six hours after the landfall (for $t = 0$ to 6) is given by:

$$a_1 = [\ln \{(V_o - V_b)/(V_6 - V_b)\}]/6 \quad (5)$$

The decay constant a_2 for the remaining 12 hours (for $t = 6$ to 18 hours) is taken as:

$$a_2 = [\ln \{(V_6 - V_b)/(V_{18} - V_b)\}]/12 \quad (6)$$

Regression equation relating R_1 and R_2 is given below:

$$R_2 = 0.982 \times R_1 - 0.081 \quad (7)$$

where, V_0 is the maximum sustained surface wind speed at the time of landfall, V_t is the wind speed at time t after the landfall and V_b is the background wind speed. After landfall, tropical cyclone decays to some background wind speed. The background wind speed V_b and the reduction factors R_1 and R_2 as determined (Table 4) in the decay model (Roy Bhowmik et al., 2005) are used in this study.

The steps suggested by Roy Bhowmik et al. (2005) for the operational forecasting are:

(i) At the time of landfall (at $t = 0$), employ the observed landfall intensity V_0 and the values of R_1, R_2 and V_b, that are obtained based upon the sample average decay rate (Table 4), to make a six-hourly prediction of V_t using equation (1).

(ii) Six hours after the landfall (at $t = 6$), use V_0 and V_6 from observation and V_b from Table 4 to compute actual R_1 from equations (3) and (5). Then get new R_2 from equation (7) and use equation (2) to revise the forecast for 12 hours after the landfall and later times.

(iii) Twelve hours after the landfall (at 12), employ observed V_{12} to make a six-hourly prediction using equation (2).

(iv) Eighteen hours after the landfall, employ observed values of V_0 and V_{18} to calculate actual R_2 from equations (4) and (6) and revise the forecast for 24 hours and beyond using equation (2).

(v) Twenty four hours after the landfall, use observed V_{24} to make a final forecast for V_{30}.

An example of prediction of decay after landfall of cyclone Thane 2011 over the Bay of Bengal is shown in Fig. 1.

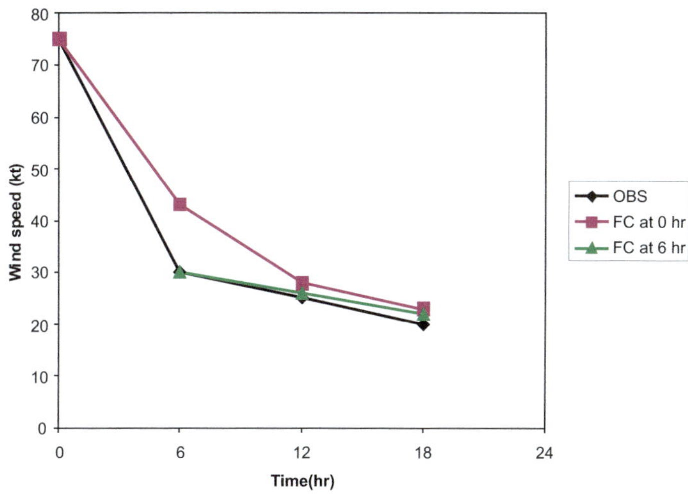

Fig. 1. Prediction of decay after landfall of cyclone Thane 2011 over the Bay of Bengal.

Table 4. Decay parameters of mean curve

MSSW (knots)	$a_1\ (h^{-1})$	$R_1\ (6\ h)^{-1}$	$a_2\ (h^{-1})$	$R_2\ (6\ h)^{-1}$	V_b *(knots)*
< 65	0.099	0.552	0.149	0.408	19.0
≥65	0.154	0.339	0.194	0.311	21.0

3. Performance Evaluation

3.1 Case Study: Cyclone "Jal" (4-7 November 2010)

In order to demonstrate the performance of the objective NWP system, case study of very severe cyclonic storm Jal of Bay of Bengal of November 2010 is illustrated. The observed track of the system is shown in Fig. 2.

Severe cyclonic storm Jal developed over the Bay of Bengal from the remnant of a depression which moved from northwest Pacific Ocean to the Bay of Bengal across southern Thailand. It moved west-northwestwards and intensified upto severe cyclonic storm on 6 November, 2010. The system intensified into a Deep Depression in the early morning of 5 November and into a Cyclonic Storm "Jal" at 0600 UTC of the same day with centre near lat. 9.00N and long. 87.50E, about 900 km east-southeast of Chennai. The cyclonic storm "Jal" over southeast Bay of Bengal continued to move west-northwestwards and intensified further into a severe cyclonic storm in the early hours of 6 November. However as the severe cyclonic storm Jal moved to the southwest Bay of Bengal closer to India coast, it entered into a region of lower ocean thermal energy and moderate to high vertical wind shear in association with the strong easterlies in the upper tropospheric level. The high wind shear led to westward shearing of the convective clouds form the system centre

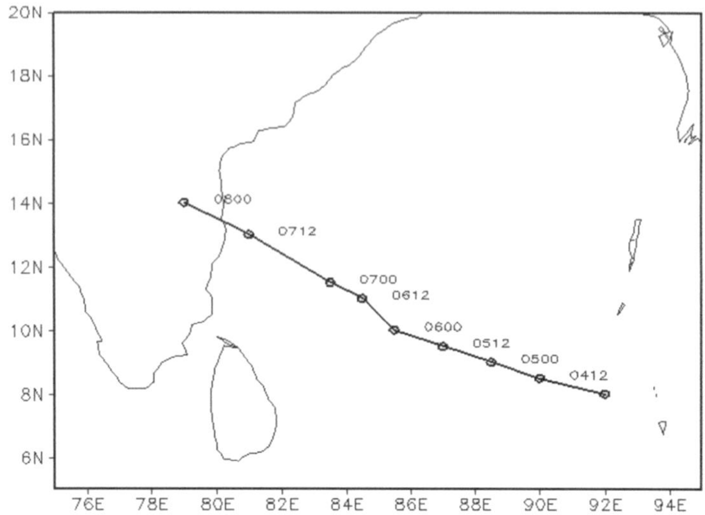

Fig. 2. Observed track of cyclone Jal of November 2010.

and lower ocean thermal energy led to unsustainability of convection over the region. Due to these two factors, the severe cyclonic storm Jal weakened into a cyclonic storm at 0600 UTC of 7 November 2010 over southwest Bay of Bengal with centre near lat. 12.50N and long. 82.50E, about 250 km east-southeast of Chennai. It weakened further into a deep depression and crossed north Tamilnadu–south Andhra Pradesh coast, close to the north of Chennai near 13.30N and 80.30E around 1600 UTC of 7 November 2010. It continued to move west-northwestwards, further weakened into a depression at 0300 UTC and into a well marked low pressure area over Rayalaseema and adjoining south interior Karnataka at 0600 UTC of 8 November 2010.

(a) Genesis

Analysis of Genesis Potential Parameter (GPP) values computed for this cyclone on the basis of real time model analysis fields (IMD GFS) along with the GPP values for Developing Systems and Non-Developing Systems are shown in Table 5. The higher GPP values (>8.0, the threshold value) at early stages of development (T.No. 1.5, 2.0) have clearly indicated that the cyclone "Jal" had enough potential to intensify into a developing system (>35 knots).

(b) Track and Landfall

Figures 3(a-b) display the forecast tracks of the cyclone "Jal" by various operational NWP models (ECMWF, IMD-GFS, NCEP-GFS, JMA, WRF and MME) with the initial conditions of 0000 UTC of 4 and 5 November 2010 respectively. It is encouraging to note that all models showed northwestward movement of cyclonic storm "Jal" towards north Tamilnadu and south Andhra Pradesh coast. The forecast errors of member models based on different initial conditions and the corresponding consensus forecasts (MME) are summarized in Tables 6 (a-c).

Table 5. GPP ($\times 10^{-5}$) for developing system, non-developing system and cyclone "Jal"

Date/Time	*04.11.2010 0000 UTC*	*05.11.2010 0000 UTC*
T.No.	1.5	2.0
Developing	11.1	13.3
Non-developing	3.4	4.6
Cyclone Jal	**13.5**	**17.4**

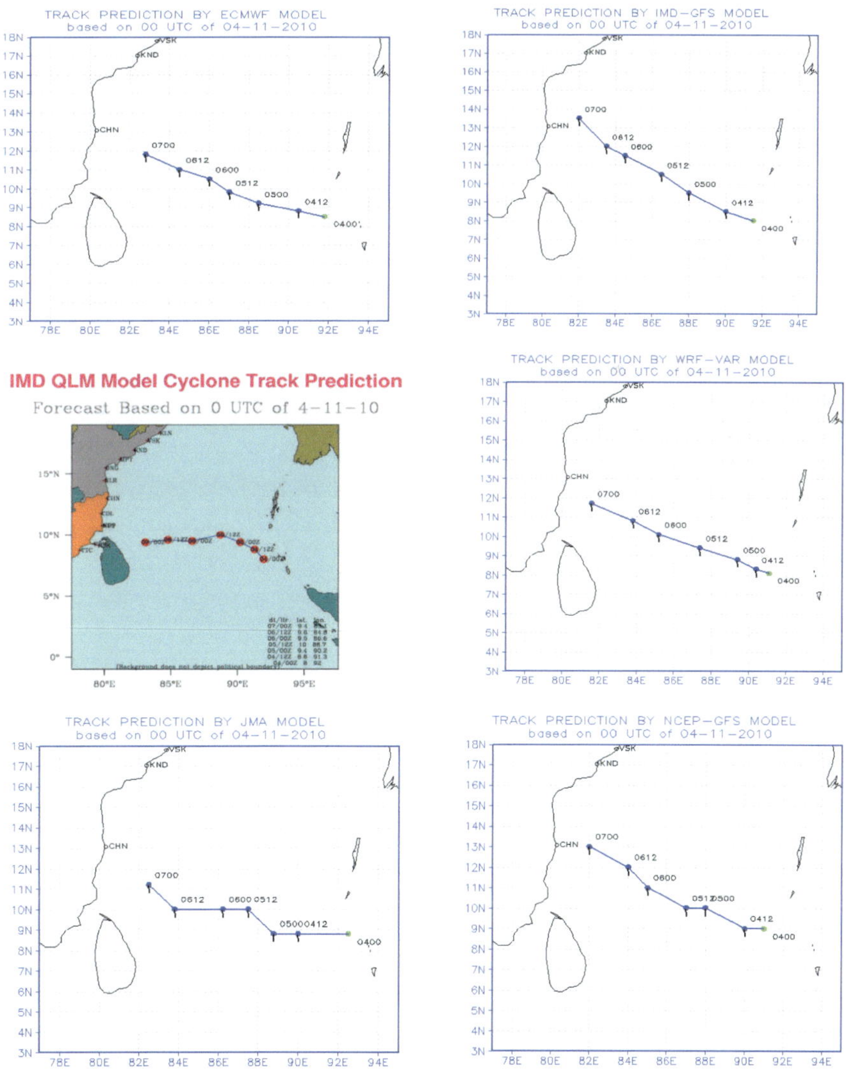

(Contd.)

(Fig. 3(a) *Contd.*)

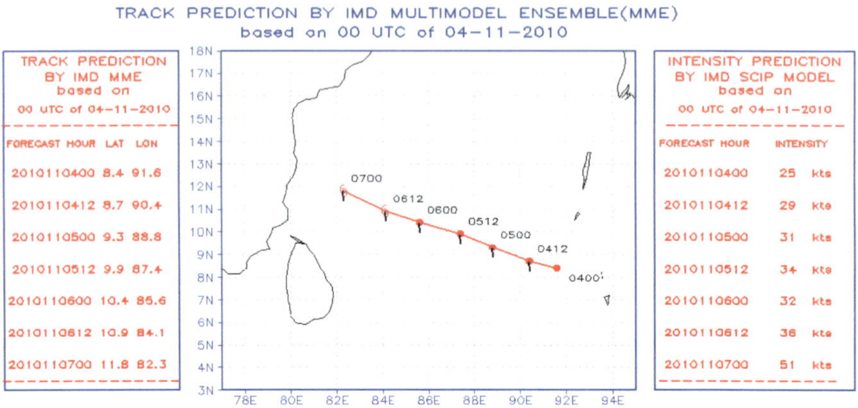

Fig. 3(a). Track forecasts of multi-model ensemble and its member models based on 0000 UTC of 04.11.2010.

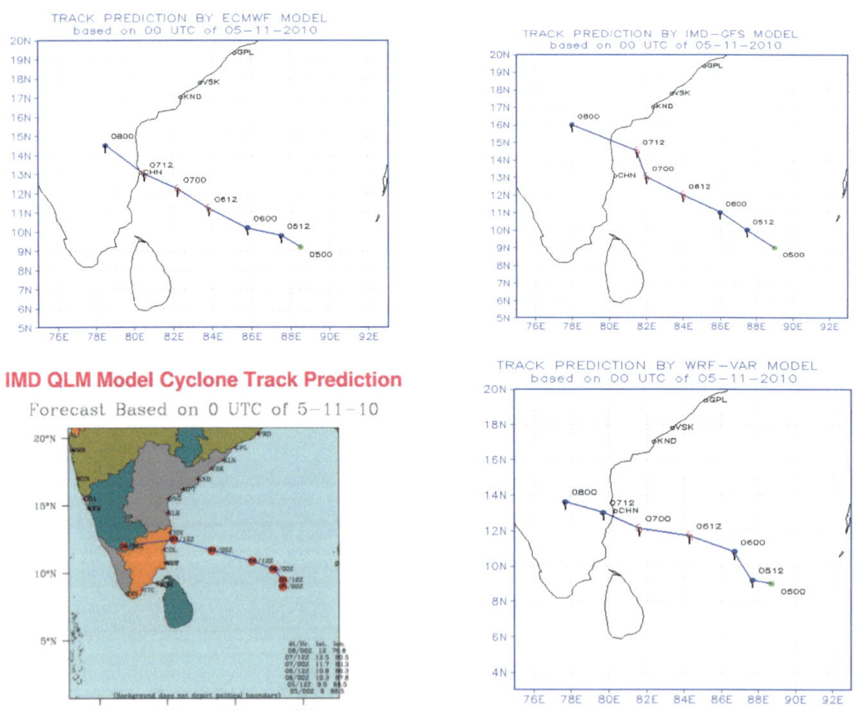

(*Contd.*)

(Fig. 3(b) *Contd.*)

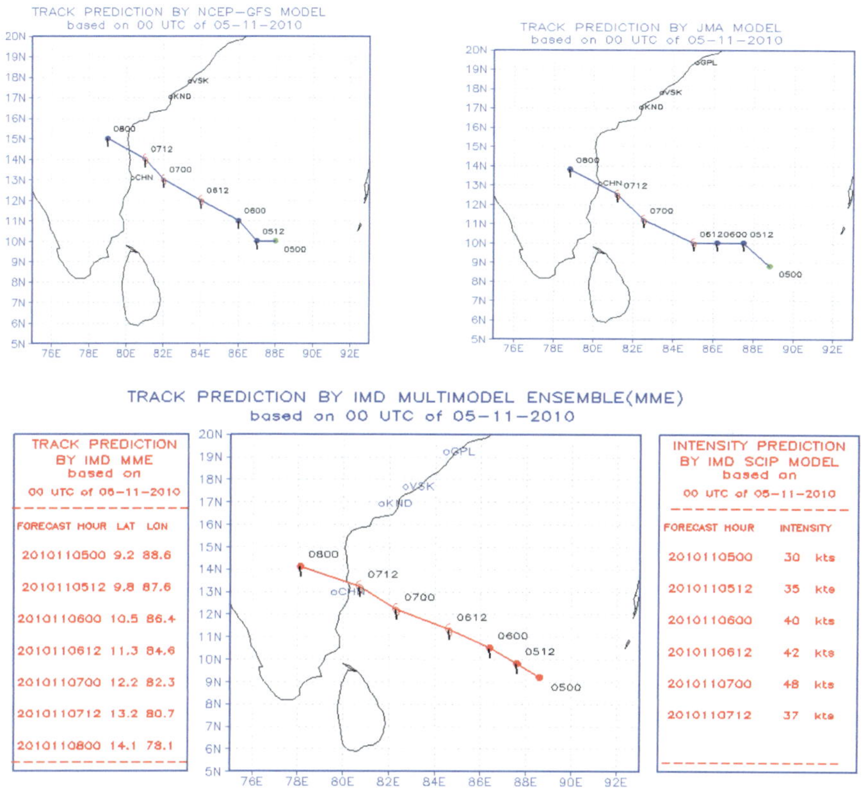

Fig. 3(b). Track forecasts of multi-model ensemble and its member models based on 0000 UTC of 05.11.2010.

Table 6(a). Average track forecast errors (Direct position error in km)

	Lead time →					
Models	12 hr	24 hr	36 hr	48 hr	60 hr	72 hr
ECMWF	46	49	45	120	27	83
GFS (NCEP)	69	141	97	178	117	233
JMA	57	77	87	95	97	114
QLM(IMD)	178	207	206	82	118	237
GFS (IMD)	59	119	124	216	165	275
WRF(IMD)	66	113	50	126	110	208
MME(IMD)	64	78	54	99	42	135

Table 6(b). Landfall forecast error (km) in respect of Tamilnadu-Andhra Coast (13.3°N/80.2°E at 1600 UTC of 07.11.2010)

Lead time (hr) →	*12 (16:00) hr based on 07.11.2010*	*36 (40:00) hr based on 06.11.2010*	*60 (64:00) hr based on 05.11.2010*
Model			
ECMWF	46	56	22
GFS (NCEP)	111	133	133
JMA	33	45	35
QLM (IMD)	291	189	90
GFS (IMD)	189	144	200
WRF (IMD)	111	33	57
MME (IMD)	123	90	11

Table 6(c). Landfall time error (in hour) at Tamilnadu-Andhra (E: Early; D: Delayed)

Lead time (hr) →	*12 (16:00) based on 07.11.2010*	*36 (40:00) based on 06.11.2010*	*60 (64:00) based on 05.11.2010*
Model			
ECMWF	02:30 E	04:00 D	01:00 D
GFS (NCEP)	Exact on L/F Time	04:00 D	04:00 D
JMA	01:30 D	07:00 D	01:30 D
QLM (IMD)	09:00 D	09:30 D	02:30 E
GFS (IMD)	07:30 D	02:00 D	01:30 D
WRF (IMD)	11:00 D	01:00 D	07:00 E
MME (IMD)	05:00 D	04:00 D	01:00 E

(c) Intensity

The 12 hourly intensity forecasts by Statistical intensity Prediction (SCIP) model valid up to 72 hours against the observations are shown in Tables 7(a-e). Results are found to be reasonably good.

Table 7(a). Model (SCIP) performance based on 0000 UTC of 04.11.2010

Forecasts hours →	*00 hr*	*12 hr*	*24 hr*	*36 hr*	*48 hr*	*60 hr*	*72 hr*
Observed (knots)	25	25	30	45	55	60	60
Forecasts (knots)	25	29	31	34	32	36	51
Error (knots)	0	4	1	-11	-23	-24	-9

Table 7(b). Model (SCIP) performance based on 0000 UTC of 05.11.2010

Forecasts hours →	*00 hr*	*12 hr*	*24 hr*	*36 hr*	*48 hr*	*60 hr*	*72 hr*
Observed (knots)	30	45	55	60	60	35	-
Forecasts (knots)	30	35	40	42	48	37	-
Error (knots)	0	-10	-15	-18	-12	2	-

Table 7(c). Model (SCIP) performance based on 0000 UTC of 06.11.2010

Forecasts hours →	*00 hr*	*12 hr*	*24 hr*	*36 hr*	*48 hr*	*60 hr*	*72 hr*
Observed (knots)	55	60	60	35	-	-	-
Forecasts (knots)	55	57	55	55	-	-	-
Error (knots)	0	-3	-5	20	-	-	-

Table 7(d). Model (SCIP) performance based on 0000 UTC of 07.11.2010

Forecasts hours →	*00 hr*	*12 hr*	*24 hr*	*36 hr*	*48 hr*	*60 hr*	*72 hr*
Observed (knots)	60	35	-	-	-	-	-
Forecasts (knots)	55	55	-	-	-	-	-
Error (knots)	-5	20	-	-	-	-	-

Table 7(e). Mean intensity forecast error (Jal)

Forecasts hours →			*12 hr*	*24 hr*	*36 hr*	*48 hr*	*60 hr*	*72 hr*
Jal	00/04.11.10	Error (knots)	4	1	-11	-23	-24	-9
	00/05.11.10	Error (knots)	-10	-15	-18	-12	2	-
	00/06.11.10	Error (knots)	-3	-5	20	-	-	-
	00/07.11.10	Error (knots)	-5	20	-	-	-	-
Mean absolute error (knots)			5-5	10	16	18	13	9

3.2 Forecast Verification Statistics

Cyclone track prediction skill of operational NWP models during 2010 and 2011 are summarized here in tabular form in Tables 8 and 9 respectively.

Table 8. Mean track position errors (km) of NWP models for cyclones during 2010 over the Indian region

Average	*12 hr*	*24 hr*	*36 hr*	*48 hr*	*60 hr*	*72 hr*
ECMWF	54	71	102	170	202	246
NCEP-GFS	158	178	177	236	253	334
JMA	195	96	176	203	232	268
IMD-MM5	118	141	241	350	363	356
IMD-QLM	103	144	167	181	256	311
IMD-MME	72	104	140	205	190	244
IMD-T382	94	124	164	212	246	290
IMD-WRF-VAR	155	137	236	253	234	265

Table 9. Mean track position errors (km) of NWP models for cyclones during 2011 over the Indian region

	Lead time →					
Models	*12 hr*	*24 hr*	*36 hr*	*48 hr*	*60 hr*	*72 hr*
ECMWF	103	77	75	86	116	118
GFS (NCEP)	80	103	77	126	202	114
JMA	119	129	241	179	204	307
QLM (IMD)	76	126	174	295	425	515
MME (IMD)	70	76	95	119	122	164
GFS (IMD)	122	169	202	230	276	332
WRF (IMD)	147	198	245	278	246	293

4. Models in the Experimental Mode

4.1 Hurricane WRF for Indian Seas

The basic version of the model HWRFV (3.2+) which was operational at EMC, NCEP was ported on IBM P-6/575 machine, IMD, New Delhi with nested domain of 27 km and 9 km horizontal resolution and 42 vertical levels with outer domain covering the area of 80° × 80° and inner domain 6° × 6° with centre of the system adjusted to the centre of the observed cyclonic storm.

The model has special features such as vortex initialization, coupled with ocean model to take into account the changes in SST during the model integration, tracker and diagnostic software to provide the graphic and text information on track and intensity prediction for real-time operational requirement. HWRF model was tested to run the model in cycling mode at IMD, New Delhi. In this run only the atmospheric model (HWRF) was tested. The ocean model (POM-TC) and ocean coupler require the customization of ocean model for Indian seas. IMD is expecting to implement the ocean cupping in collaboration with INCOIS, Hyderabad. The model is presently under testing for experimental operational implementation.

4.2 TC Ensemble Forecast for Indian Sea

As per the advice from WMO to provide a guidance of tropical cyclone forecasts in near real-time for the ESCAP/WMO members based on the TIGGE Cyclone XML (CXML) data, under the joint project of World Weather Research Program (WWRP) and Tropical Cyclone Program (TCP). TC homepage was developed by JMA and the same software was transferred to IMD to generate similar page for RSMC, New Delhi. The software was implemented at NWP Division, IMD, New Delhi and ensemble TC products from ECMWF, UKMO, NCEP and JMA will be provided experimentally during the post-monsoon cyclone season 2011 for ESCAP/WMO members of RSMC, New Delhi region.

REFERENCES

Elsberry, R.L., T.D.B. Lambert and M.A. Boothe, 2007: Accuracy of Atlantic and eastern North Pacific tropical cyclone intensity forecast guidance. *Wea. Forecasting*, **22**, 747-762.

Goerss, J.S., 2000: Tropical cyclone track forecasts using an ensemble of dynamical models. *Mon. Wea. Rev.*, **128**, 1187-1193.

Houze, R.A., S.S. Chen, B.F. Smull, W.C. Lee and M.M. Bell, 2007: Hurricane intensity and eyewall replacement. *Science*, **315**, 1235-1238.

Kotal, S.D., Ajit Tyagi and S.K. Roy Bhowmik, 2011: Potential vorticity diagnosis of rapid intensification of very severe cyclone Giri (2010) over the Bay of Bengal. *Natural Hazards*, online DOI 10.1007/s/11069-011-0024-1.

Kotal, S.D. and S.K. Roy Bhowmik, 2011: A multi-model ensemble (MME) technique for cyclone track prediction over the North Indian Sea. *Geofizika*, **28**: 275-291.

Kotal, S.D., P.K. Kundu and S.K. Roy Bhowmik, 2009: Analysis of cyclogenesis parameter for developing and non-developing low pressure systems over the Indian Sea. *Nat. Hazards*, **50**, 389-402.

Kotal, S.D., S.K. Roy Bhowmik, P.K. Kundu and A.K. Das, 2008a: A Statistical Cyclone Intensity Prediction (SCIP) Model for Bay of Bengal. *J. Earth. Sys. Sci.*, **117**, 157-168.

Kotal, S.D., P.K. Kundu and S.K. Roy Bhowmik, 2008b: An analysis of Sea Surface Temperature and Maximum Potential Intensity of Tropical Cyclones over the Bay of Bengal between 1981 and 2000. *Meteorological Applications*, **16**, 169-177.

McBride, J.L. and R.M. Zehr, 1981: Observational analysis of tropical cyclone formation. Part II: Comparison of non-developing versus developing systems. *J. Atmos. Sci.*, **38,** 1132-1151.

Mackey, B.P. and T.N. Krishnamurti, 2001: Ensemble Forecast of a Typhoon Flood Event. *Wea. and Forecasting*, **16**, 399-415.

Roy Bhowmik, S.K. and S.D. Kotal, 2010: A dynamical statistical model for prediction of a tropical cyclone. *Marine Geodesy*, **33**, 412-425.

Roy Bhowmik, S.K., S.D. Kotal and S.R. Kalsi, 2007: Operational tropical cyclone intensity prediction—An empirical technique. *Nat. Hazards*, **41**, 447-455.

Roy Bhowmik, S.K., S.D. Kotal and S.R. Kalsi, 2005: An empirical model for predicting decaying rate of tropical cyclone wind speed after landfall over Indian region. *Journal of Applied Meteorology*, **44**, 179-185.

Roy Bhowmik, S.K., 2003: An evaluation of cyclone genesis parameter over the Bay of Bengal using model analysis. *Mausam*, **54**, 351-358.

Sen Roy, Soma, S.K. Roy Bhowmik, V. Lakshmanan and S.B. Thampi, 2010: Doppler Radar-based Nowcasting of the Bay of Bengal Cyclone Ogni of October 2006. *J. Earth Sci. Sys.*, **119(2)**, 183-199.

Vijaya Kumar, T.S.V., T.N. Krishnamurti, M. Fiorino and M. Nagata, 2003: Multimodel superensemble Forecasting of Tropical Cyclones in the Pacific. *Mon. Wea. Rev.*, **131**, 574-583.

Weber, H.C., 2003: Hurricane Track Prediction Using a Statistical Ensemble of Numerical Models. *Mon. Wea. Rev.*, **131**, 749-770.

Williford, C.E., T.N. Krishnamurti, Correa Torres Ricardo, Cocke, Steven, Zaphiris Christidis and T.S. Vijaya Kumar, 2003: Real-Time Multimodel Superensemble Forecasts of Atlantic Tropical Systems of 1999. *Mon. Wea. Rev.*, **131**, 1878-1894.

18

Tropical Cyclone Research over the North Indian Ocean: Impact of Data and Vortex Initialization in High Resolution Mesoscale Models

U.C. Mohanty, Krishna K. Osuri and Sujata Pattanayak
SEOCS, Indian Institute of Technology Bhubaneswar, Odisha, India
ucmohanty@gmail.com

1. Introduction

The nomenclature of TCs is different in different parts of the world. In the Atlantic and eastern Pacific, they are known as 'hurricane' and in western Pacific as 'typhoon'. In the North Indian Ocean (NIO) region, they are termed as 'tropical cyclone'. Though, considered as a relatively rare phenomenon in any basin, annual frequency of occurrence is quite steady with an average variation of ±7%. Approximately, 80 tropical storm reach cyclonic storm intensity every year that cause an average number of 20,000 deaths and a total economic loss of $6-7 billion (Southern, 1979). Almost all these storms form within 25° latitude on both sides of the equator except in the equatorial region of 5° S to 5° N due to negligible Coriolis force.

The NIO basin is unique in nature than any other cyclone basins of the world, as far as the genesis, period of occurrence are concerned (Mohanty et al., 2010). The NIO comprised of Bay of Bengal (BoB) and Arabian Sea (AS) [BoB and AS together are also known as Indian seas] contributes 5-6 TCs every year (~7% of the world's TCs) (Mohanty, 1994). An observational study on TC activity over a period of 120 yrs (1891-2010) suggest that the BoB and AS contributes about 76% and 24%, respectively from the total number of TCs over NIO (Mohanty et al., 2011). It contributes more than 75% to the total number of TCs over the NIO (Mohanty et al., 2011). The genesis of TCs over this region is highly seasonal with primary maximum in post-monsoon season (mid-September to December) and secondary maximum during pre-monsoon season (April and May), unlike in the other ocean basins which occur around late summer to early fall. It is also found that the post-monsoon storms are more devastating (De Angelis, 1976; Mohanty, 1994). The TCs over the NIO move usually the west, north-west and northward, sometimes, recurve towards north-east or east (Fig. 1). Westward moving TCs are more in number and ~85% of these TCs form during post-monsoon season (October to December), with a peak in October (over 30%) (Fig. 1b). Northward moving TCs form primarily during post-monsoon season, similar as the westward moving systems. The northeastward-moving with recurvature and landing over Myanmar is unique, not only for its fewest number but for the peaks in pre-monsoon season (Fig. 1a), which implicit that connection to summer monsoon onset may be considered. During the pre-monsoon season, the inter tropical convergence zone (ITCZ) is located sufficiently over the open water of the Indian seas which trigger low pressure system and its development into a mature cyclone (Lee et al., 1989). As the storm gets intensified, it passes through several stages. Based on pressure drop and maximum sustainable surface wind, the World Meteorological Organization (WMO) classified the TCs over the NIO broadly into seven categories and is provided in Table 1.

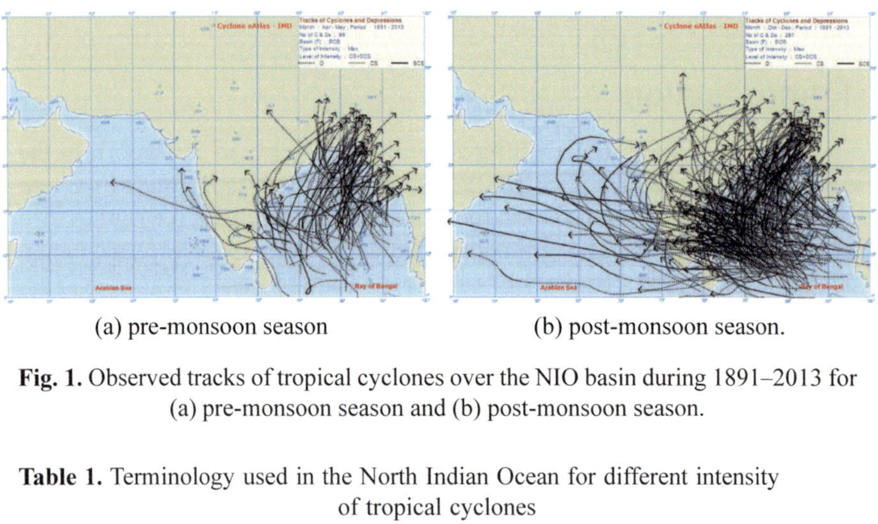

(a) pre-monsoon season (b) post-monsoon season.

Fig. 1. Observed tracks of tropical cyclones over the NIO basin during 1891–2013 for (a) pre-monsoon season and (b) post-monsoon season.

Table 1. Terminology used in the North Indian Ocean for different intensity of tropical cyclones

Sl. No.	*Category of system*	*Pressure drop (hPa)*	*Maximum wind speed knot (kmph)*
1.	Low (L)	1.0	< 17 (< 32)
2.	Depression (D)	1.0-3.0	17-27 (32-50)
3.	Deep depression (DD)	3.0-4.5	28-33 (51-59)
4.	Cyclonic storm (CS)	4.5-8.5	34-47 (60-90)
5.	Severe cyclonic storm (SCS)	8.5-15.5	48-63 (90-119)
6.	Very severe cyclonic storm (VSCS)	15.5-65.6	64-119 (119-220)
7.	Super cyclonic storm (SUCS)	> 65.6	> 119 (>220)

Though relatively moderate in intensity as compared with typhoons and hurricanes, the NIO cyclones that cross surrounding countries are highly devastating (De Angelis, 1976; Mohanty et al., 2011). During the past 312 years (1701-2012), out of 25 recorded cyclones with heavy loss of human lives more than 5000, 19 cases (76%) occurred in Indian subcontinent and it may be noted that the 20th century was worst affected. For example, about 130 lakhs of people affected and 10,000 people died due to the BoB super cyclone that crossed Orissa (India) coast on 29 November 1999. Due to very severe cyclone Sidr (15 November 2007) over the BoB, more than 3360 deaths occurred and about 2.5 millions of acres crop was damaged. VSCS Nargis over the BoB caused more than 138,000 deaths in Myanmar. SCS Aila (25 May 2009) over the BoB was responsible for 270 deaths and thousands of injuries in West Bengal, India and Bangladesh. The SCS Laila (20 May 2010) over the BoB caused a storm surge of 2 to 3 metres inundated the low lying areas leading to six deaths and damage to agriculture and fisheries in the coastal region of Andhra Pradesh and Tamil Nadu. Such disaster is mainly due to shallow bathymetry, nearly funnel shape of the coastline and the long stretch of the low-lying delta region entrenched with a large number of river systems leading to high storm surges and coastal inundations. The densely populated coastal region enhance the higher mortality rate over this region. This damage is mainly due to strong winds, heavy precipitation and the worst of all the cumulative effect of storm surges and astronomical tides that overwhelm miles of shores. Hence, reasonably accurate prediction of track, intensity, structure and associated storm surges of these devasting storms at least in 48 h advance would be highly desirable for planning and implementation of the mitigation measures effectively and hence reduction of loss of life and property considerably.

For disaster warnings and mitigation efforts, forecasting TC tracks is a critical component; track forecast errors over the NIO are high relative to those over the Atlantic and Pacific Oceans (Mohapatra et al., 2013b). Synoptic and statistical methods have limitations in predicting track and intensity beyond 24 h over the NIO (Mohanty and Gupta, 1997; Gupta, 2006). Several mesoscale models at cloud resolving resolutions were introduced for TC predictions over the NIO basin. A sustained investment in research and development of numerical models, observations, and data assimilation techniques for providing improved initial states to these models led to reliable and realistic TC forecast over the NIO basin. India Meteorological Department also uses many global and regional models such as the IMD Global Forecast System (GFS), the National Center for Atmospheric Research (NCAR) version of Advanced Research version of the Weather Research and Forecasting (ARW), and the National Oceanic Atmospheric Organizations (NOAA) Hurricane WRF (HWRF, Gopalakrishnan et al., 2002) system of 2010.

Pattanayak and Mohanty (2008) made a comparative study on the performance of both Mesoscale model version 5 (MM5) and Advanced

Research Weather Research Forecasting (ARW) models in the simulation of TCs over Indian seas and demonstrated the superiority of the ARW model over MM5. ARW model is used extensively worldwide for the simulation of various weather events, such as heavy rainfall (Mohanty et al., 2012; Hong and Lee, 2009), monsoon depressions (Routray et al., 2010) and tropical cyclones (Osuri et al., 2012a; Pattanaik and Ramarao, 2009; Davis et al., 2008). The model is also used to study the land surface processes (Niyogi et al., 2006). The performance of the ARW model in prediction of tropical cyclones (TCs) over the north Indian Ocean (NIO) at different 27, 18 and 9 km resolutions up to 72-hour forecast length is evaluated during the TC seasons of 2007–11. The study highlights the capabilities of the ARW model over the Indian monsoon region and the continued need for operational forecasts from high-resolution models. The 9-km runs improve the track and intensity prediction by about 8%–24% and 15%–40% respectively over the 18-km predictions. The Non-hydrostatic Mesoscale Model of WRF (NMM) is also skillful for prediction of TCs over different basins. Pattanayak and Mohanty (2010) and Mohandas and Ashrit (2011) demonstrated the skillful performance of NMM model in simulation of TCs over the NIO. The atmospheric component of HWRF modelling system was proven for the simulation of track, intensity and vertical structure of the cyclone Mala with reasonable accuracy over the BoB (Pattanayak et al., 2012a).

According to Mohanty et al. (2010), the initial vortex position error in global analyses is about 80-100 km and further contributes to more track forecast errors. The primary and important task is to reduce the errors in initial conditions. It is well known that the performance of high resolution mesoscale models highly depends on the quality of initial conditions. The quality of initial conditions can be improved with the mesoscale data assimilation of high dense observations. Moreover, the global analyses and forecast products are available in coarser resolution to integrate mesoscale models. Because of lack of sufficient conventional observations over the oceans where TCs form and evolve, the global analyses are ill-defined in representing the initial structure and position of the vortex. Several previous studies have demonstrated that the assimilation of sea surface and upper air satellite-derived winds near and around the centre of the storm can substantially improve the initial analyses of TCs and hence the prediction of track, intensity and structure (e.g. Velden et al., 1998; Chen 2007; Pu et al., 2008; Osuri et al., 2012b). Further, the 3DVAR assimilation of DWR reflectivity and radial wind observations clearly showed improvements in the simulation of mesoscale events (Gao et al., 1999; Xiao et al., 2005, 2007). Routray et al. (2010) also showed better simulation of intense convective systems influencing the large-scale Indian monsoonal flow and showed positive impact of DWR data on the prediction of the location, propagation and development of rain bands associated with the monsoon depressions over the BoB. Govindankutty et al. (2010) indicate the positive impact of DWR radial wind along with IMD global telecommunication system (GTS) data in the simulation of a TC.

2. Numerical Modelling Systems and Its Applications

Numerical weather prediction (NWP) is a complex system involving non-linear dynamics and physics of atmosphere, hydrosphere, biosphere and cryosphere. The NWP models, especially the non-hydrostatic meso-scale models, with high spatial resolution have shown great potential to predict severe weather events such as tropical cyclones, severe thunderstorms and heavy rainfall. With increasing resolution, these models are also capable of resolving cloud scale phenomena with explicit representation of cloud microphysics and more realistic representation of orographic features. Much of this success is due to increase in computing resources, developments in numerical techniques, improved understanding of physical processes, and improvements in observing systems, data assimilation and initialization. Assimilation of regional observational at high temporal resolution are being used to generate high resolution mesoscale analysis.

(a) ARW Model

The ARW modelling system was developed for a broad range of applications such as idealized simulations, parameterization research, data assimilation research, real-time NWP etc. WRF is a fully compressible non-hydrostatic system of equations with complete coriolis and curvature terms. Model equations are in the mass-based terrain following co-ordinate system and solved in Arakawa-C grid. Runge-Kutta 2nd and 3rd order time integration technique is used for model integration. Table 1 summarizes details and different options of ARW model. Also the 3-dimensional variational (3DVAR) data assimilation system is embedded within the framework of ARW modelling system. The detailed description of the model equations, physics, dynamics and data assimilation system can be found in Skamarock et al. (2005) and Barker et al. (2004), see also online at http://wrf-model.org.

(b) NMM Model

The WRF-NMM model has been credited by its adoption in the National Centre for Enviromental Prediction (NCEP), as their operational mesoscale modelling system for USA in June 2006. By moving over to WRF-NMM as their operational model, NCEP has been able to produce weather analyses and forecast products based on a resolution of 0.1 Deg. The WRF-NMM is a fully compressible, non-hydrostatic mesoscale model with a hydrostatic option. Its vertical coordinate is a hybrid sigma-pressure coordinate. In horizontal Arakawa E-grid staggering is used. The model uses a forward-backward scheme for horizontally propagating fast waves, implicit scheme for vertically propagating sound waves, Adams-Bashforth scheme for horizontal advection, and Crank-Nicholson scheme for vertical advection. The same time step is used for all terms. The dynamics conserve a number of first and second order quantities including energy and enstrophy. This model supports a variety of capabilities including real-data simulations, full physics options, non-hydrostatic and hydrostatic (runtime option), one-way static nesting and applications ranging from metres to thousands of kilometres. A detailed description of the model is provided by Janjic et al. (2001) and Janjic (2003).

(c) HWRF Model

The Weather Research and Forecast system for hurricane prediction i.e. HWRF system was developed by the Environmental Modeling Center (EMC)/National Centers for Environmental Prediction (NCEP). It is designed to be a flexible, state-of-the-art atmospheric simulation system that is efficient on available single as well as parallel computing platforms. The HWRF model is a fully compressible, non-hydrostatic model with a hydrostatic option (Janjic et al., 2001; Janjic, 2003). The model uses the rotated lat-long projection that minimizes the convergence over meridians over the domain and the Arakawa E-grid staggering for computational efficiency. The terrain following hybrid sigma-pressure vertical coordinate system is used. The dynamics conserve a number of first- and second-order quantities including energy and enstrophy (Janjic, 1984). Forward–backward schemes are used for the horizontally propagating fast-waves, and implicit scheme is used for the vertically propagating sound waves. Adams–Bashforth scheme and Crank–Nicholson scheme are used for horizontally and vertically propagating waves. The Geophysical Fluid Dynamics Laboratory (GFDL) long-wave and short-wave radiation schemes are incorporated in the model. The GFDL long-wave radiation scheme has the multi-band, transmission table look-up scheme with carbon dioxide, ozone and water vapour absorptions (Fels and Schwarzkopf, 1975; Schwarzkopf and Fels, 1985; Schwarzkopf and Fels, 1991) in which cloud microphysics effects are included. The GFDL short-wave radiation scheme has two spectral bands, *k*-distribution scheme with ozone and water vapour as the main absorbing gases (Lacis and Hansen, 1974) and cloud microphysics effects are also included. It has a well-tested NCEP GFS scheme for planetary boundary layer parameterization (Hong and Pan, 1996). The PBL height is determined using an iterative bulk-Richardson approach working from the ground upward, whereupon the profile of the diffusivity coefficient is specified as a cubic function for the PBL height. The diffusivity coefficient values are obtained by matching the surface layer fluxes. A counter-gradient flux parameterization is included. The Simplified Arakawa-Schubert cumulus parameterization scheme is used (Arakawa and Schubert, 1974; Grell, 1993). The moving structure of inner nest along with the automatic tracking system of tropical cyclones and diagnosis of its intensity and structure enhances the importance of the HWRF modelling system.

3. Performance of High Resolution Mesoscale Models for TC Prediction

The application of high resolution mesoscale models and its improved performance with 3-dimensional variational data assimilation system (3DVAR) for prediction of track and intensity of TC over BoB is discussed without focussing on the inter-comparison of their performances.

3.1 ARW Model Performance

The ARW model has been customized for NIO basin in terms of different parameterization schemes, resolutions and domain size. The model configuration including domain and physical parameterization schemes follows the sensitivity study of Osuri et al. (2012a). The ICs for the ARW model are obtained from the analysis and forecast fields of the NCEP GFS. The lateral boundary conditions are updated in 6-h intervals with a fixed sea surface temperature throughout the model integration. The land surface boundary conditions are taken from the U.S. Geological Survey with a horizontal grid spacing of 10 min. A single experimental domain is fixed between 77° and 102° E and between 3° and 28° N over the BoB, and between 48° and 78° E and between 5° and 30° N over the AS, with 51 vertical layers and 27-km, 18-km and 9-km horizontal grid resolution. The integration time step (dt) is chosen as three times of model resolution in seconds with the Kain–Fritsch cumulus parameterization scheme, the slab model for the land surface representation, the WRF single-moment 3-class microphysics scheme, and the Yonsei University planetary boundary layer scheme.

Among different combinations of physical parameterization schemes, the Kain-Fristch (KF) convection scheme with Yonsui-University (YSU) planetary boundary layer (PBL) scheme with WSM-3 microphysics produces better TC predictions over the NIO. It also revealed that the convection strongly affects the cyclone tracks and boundary layer physics influence the intensity predictions strongly. The mean RMSE of intensity clearly depicts that this combination shows least error (13 hPa and 11 m s^{-1}) as compared to others (Osuri et al., 2012a). Further, it is also clear that the KF convection scheme with any PBL scheme is better with minimum RMSE of 14 hPa and 12.5 m s^{-1} compared to other schemes. Similarly, the mean RMSE of intensity with YSU scheme is 18 hPa (14 m s^{-1}). It is seen from the track predictions that this combination predicted the tracks with less vector displacement errors ranging from 70 to 200 km for 12 to 72 hrs forecast length. For all other combinations, it ranges from 75 to 410 km. Similarly, the landfall position error (km) and time error (hour) was reduced with this combination, as the errors vary from 30 to 110 km for different forecast lead times. The time errors are marginally high, but acceptable. In all the cases, only YKF experiment predicts the landfall. While, most of the other combinations fail even to predict the landfall for Gonu and Sidr. This combination is also successful in predicting structure and rainfall.

(i) Impact of Resolution

Though the realistic predictions/forecast of TCs demand horizontal and vertical high resolutions (Gopalakrishnana et al., 2012), the developing countries like India do not have computational facilities that required running the mesoscale models at really high resolutions (≤3 km). There is also a time constraint to provide real-time forecast of TCs to IMD to use for their operational use. Hence, an extensive study has been carried out to assess the impact of resolution for TC predictions in real-time endeavour over NIO basin and individual TC basins (BoB and AS) at different horizontal resolutions (27, 18 and 9 km). Further, the TCs are stratified based on (1) basins of formation, (2) nature of movement i.e. straight-moving and recurving TCs, (3) TC intensity at model initialization, and (4) season of occurrence. The study of this kind may provide a baseline configuration for real-time predictions at reasonable high resolution and bridge to transfer research to operations in India. For this evaluation, 17 cyclones of 2007-2011 seasons were predicted in 12-hr cycle during TC life (resulted in 100 forecast cases) using NCEP GFS analyses and forecast products as initial and boundary condition to the ARW model. Complete details on the stratification of TCs can be found in Table 1 of Osuri et al. (2013). Besides the direct forecast position error, we also calculated the directionality errors (cross-track (CT) and along-track (AT) errors) to understand the role of resolution in improving the TC track. The information on CT and AT errors is very helpful for TC disaster management; because the CT error gives idea on the spread of the tracks and will help in determining the area of evacuation needed in case of a landfalling TC and the AT error will help in determining the time of evacuation. A relatively lower CT error is desirable by disaster managers because it will lower the cost of evacuation. The positive values of CT errors represent right or eastward bias and negative values represent left or westward bias. Similarly, the positive values of AT errors represent the time or speed error in movement of the TC.

The mean forecast track errors (skill with reference to persistence track) of the model at 27-km resolution over the NIO ranges from 113 to 375 km (7%–51%) for a 12–72-h forecast (not shown). It was found that the higher the resolution of the model is, the better are the track and intensity predictions by the model. For the NIO TCs, the forecast errors vary from 106 to 359 km for the 12–72-h forecast at 18 km horizontal resolution, while the mean DPE of the 9-km runs is 115, 204 and 329 km for the same forecast lengths, respectively. It is also noted that the 9-km predictions improved the TC movement by about 12%–28% as compared with 27-km runs and 4%–14% as compared with 18-km predictions. The error statistics in predicting analyses straight-moving TCs (48 forecast cases corresponds to 10 TCs) and recurving TCs (52 forecast cases corresponds to 7 TCs) over the NIO highlight that the model is more skillful (30–67% from 12–72 hour forecast) to predict straight movers than recurving TCs (not shown). However, the prediction of recurving TCs can be improved with use of mesoscale models at high resolution. The possible reasons are discussed for the same in Osuri et al. (2013). Further, the skill of the model is significant in track prediction when initialized at severe cyclonic storm (SCS) stage as compared to the initialization at depression and cyclonic storm (CS) stages (not shown).

The analysis is based on 38, 28 and 34 forecast cases issued with depression, CS and SCS stages at the time of model initialization, respectively. The 99% CI is also significantly less for the SCS initializations (12 km at 12-hour forecast and 90 km for the 72-hour forecast) as compared to that of CS initializations (47 and 187 km respectively). There is no significant difference in track errors up to 36 hours, when predictions are made from the depression and CS stages. However, there is an improvement of about 20-25 km in 48-72 hr forecast times from the CS stage initializations as compared to those at the depression stage (Fig. 1c). Mohapatra et al. (2013a) also demonstrated the above fact by verifying the operational TC forecasts of IMD for the CS and SCS categories. With respect to season of occurrence, the model exhibits less error and 3–15% more gain in skill for post-monsoon TCs than that of pre-monsoon TCs (not shown).

The analyses of mean CT error analyses revealed that the ARW model shows a bias to predict right or eastward movement of TCs over the NIO basin (both BoB and AS). This bias is particularly true for straight-moving TCs. In the case of recurving TCs, the model showed a rightward bias up to the 36-h forecast and thereafter a westward bias. Similar to the CT analyses, the analyses of mean AT errors of the ARW model reveal that track positions are generally behind the observed tracks as seen by the result that AT errors are negative for all forecast lengths. In comparing the magnitudes of AT and CT errors of the ARW model, it is seen that the AT errors are larger for all forecast lengths. The ARW forecasts are in general slower relative to the actual translation speed of the system for all forecast lengths, and they predict a delayed landfall. In other words, the spread of the track relative to the observed track is less. Hence the ARW model is more accurate in predicting TC landfall location than landfall time. The 99% CI of mean forecasts error from the ARW model at all forecast intervals is smaller when compared with those of persistence track errors suggesting that the ARW model forecasts are in general more consistent for all forecast intervals. A comparison of 18- and 9-km predictions showed that the intensity prediction at 9-km resolution was improved by ~15%–40% over the 18-km predictions up to the 72-h forecast. More details of these error statistics can be found in Osuri et al. (2013). Mohapatra et al. (2013a, b) compiled IMD official forecast errors in terms of TC movement and intensity. A comparison of ARW model performance at 9-km resolution for track and intensity prediction with the IMD official errors is shown in Table 2.

Table 2. Comparison of ARW model track and intensity errors with IMD official forecast errors

	Track forecast position errors (km)		*Intensity forecast errors (kts)*	
	ARW model	*IMD official*	*ARW model*	*IMD official*
24 hr	119	130	16	14
48 hr	186	262	12	19
72 hr	327	386	8	26

(ii) Impact of 3-dimensional Data Assimilation

The statistics of error covariances are vital input to the WRF-Var analysis system along with first guess and observations (Lee et al., 2006). Forecast ("first guess" or "background") error covariances (BE) are required for the WRF-Var data assimilation systems. The background error (BE) is the covariance of forecast minus truth (observed) in analysis control variable space. The optimal use of observations and prior (background) information depends on the accuracy of observation and background errors (Barker et al., 2004; Lee et al., 2006). The BE statistics are required over the experimental domain as the analyses are sensitive to BE. In WRF-Var systems, these covariances are typically calculated off-line, and significant tuning is required to optimize performance for a particular application (Ingleby, 2001). Particularly, the impact of the BE statistics is very important in data-sparse areas. Numerous studies (Barker et al., 2004; Jianfeng et al., 2005; Bansal and Rizvi, 1993) have also indicated that superior results are obtained by using the own calculated domain-specific error covariances.

(a) Impact of Background Error Covariance (BE)

The multivariate, three-dimensional nature of 3DVAR's background error covariance (BE) is examined by studying 3DVAR's response to single bogus zonal and meridional winds, temperature and humidity observations. In WRF-Var system, the single observation test is used as a proof-of-concept to determine the performance of a single observation through the established correlations among WRF-Var variables (Wu et al., 2000).

Various single observation tests at different locations (latitude, longitude and sigma level) are performed over the Indian domain and a few of them will be discussed in this sub-section. Background error covariance matrix (B) controls how information is spread out from the single observation, to provide statistically consistent increments at the neighbouring grid points and at different vertical levels of the model. It also ensures that observations of one variable e.g. temperature produce dynamically consistent increments in the other model variables (e.g. vorticity and divergence). The innovation (observation minus background; O–B) of the single temperature and single u-wind is assigned to 1 °K and 1 ms^{-1} respectively.

Figure 2 depicts the WRF-Var analysis increments in response to single temperature observation at (20.1 °N; 80.6 °E; sigma level 25) of the experimental domain. The shape of horizontal temperature increment (Fig. 2c) is confined to about 150 km. The vertical profile (figure not provided) of *T*, *q* and *u*, *v* analysis increment in response to single temperature observation at (20.1 °N; 80.6 °E; sigma level 25) of the domain respectively. Maximum temperature increment (0.18 °K) is seen at the 25 sigma level. The vertical profile of wind increment in response to the temperature observation found maximum value above the 25 sigma level. The magnitude and spread of temperature increment (Fig. 2c) is very small. The magnitude of wind (Fig. 2a and b) response from temperature increment is also very small, however the spatial impact is significant as compared to the temperature increments (Fig. 2c).

Figure 3 represents the analysis increments in response to the single *u*-wind component (1 ms) observation applied at (20.1 °N; 80.6 °E; sigma level 25) of the domain. It is seen that the spread of *u* increment (Fig. 3b) of 0.7 ms^{-1} is observed. The response of *v*-increments (Fig. 3c) supports cyclonic and anti-cyclonic circulation to the north and south of *u*-increment location respectively. Accordingly, cold and warm temperature increments (Fig. 3a) are also found. As seen earlier in Fig. 2c, the scale of temperature response on wind (<*T*, *u*>) was large and the same is noted for wind response on temperature (<*u*, *T*>) in Fig. 3b and c. Results of single observation test with global model's B statistics over Indian region (Bansal and Rizvi, 1993; Das Gupta, 2006) also revealed similar type of structures of temperature and *u*-wind analysis increments.

The temperature increment is created via the hydrostatic balance, mass-wind balance and Richardson equation. The temperature increment (Fig. 3a) response is small, but its length scale is larger than that of the wind increments. Therefore, setting up the mesoscale background error statistics and mesoscale balance is still a challenge.

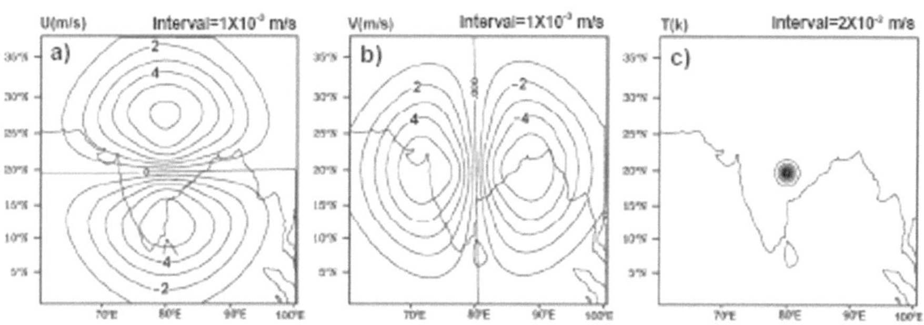

Fig. 2. Response of the analysis increments to a single temperature observation 1 °K at (20.1 °N; 80.6 °E; 25-sigma level) for (a) *U* (ms^{-1}); (b) *V* (ms^{-1}) and (c) *T* (°K).

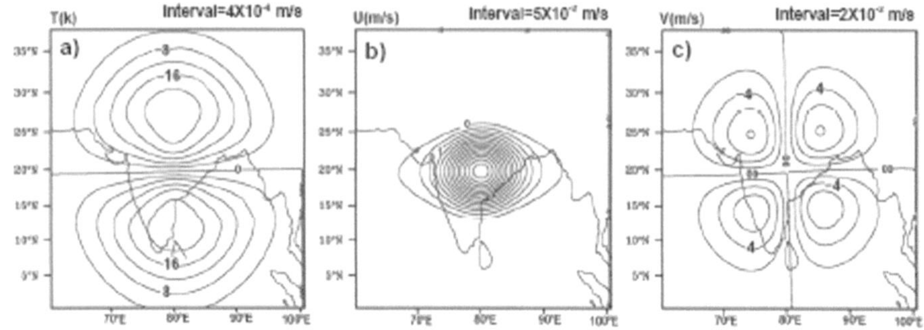

Fig. 3. Response of the analysis increments to a single *u*-wind component observation 1 ms^{-1} at (20.1 °N; 80.6 °E; 25-sigma level) for (a) T (°K); (b) U (ms^{-1}) and (c) V (ms^{-1}).

(b) Impact of Sea Surface Winds on Initialization and Simulation of NIO TCs

The single experimental domain of 9 km horizontal resolution is fixed to 3° N–28° N, 78° E–103° E, centred at 15.5° N and 89.5° E for BoB TCs and 5° N–30° N, 48° E–78° E with centre at 17.5° N and 63° E for AS TCs. The model parameterization schemes are taken from Chaper 3 established to be more suitable for NIO TCs. Details on numerical experiments can be found in Osuri et al. (2012b). A total of six recent TCs with unique in nature are considered. Full details of synoptic situations of these cyclones can be found in Regional Specialized Meteorological Centre reports (RSMC, 2008, 2009, 2010 and 2011). The Nargis is simulated with five different initial conditions (ICs), Gonu with four, Sidr with four, KhaiMuk with three ICs, Aila with four, and Laila with four ICs. Therefore, a total of 24 cases are considered for mean error statistics and details are given in Table 3 including initialization time, forecast length.

With each IC, a set of two numerical experiments are conducted. The first experiment, control simulation (CNTL) is carried out by using the NCEP FNL analyses as initial and boundary conditions (IBCs) to the model. In the second experiment known as 3DVAR, the global FNL analysis is improved by insertion of satellite derived wind observations from QSCAT and wind speed from SSM/I over the sea surface using 3DVAR analysis system. The amount of observations that were assimilated in each initial condition is given in Table 6. A number of quality checks are performed including the removal of the observations that are outside of the time range and domain (horizontal and top), hydrostatic assumptions and vertical consistency checks (Barker et al., 2004). Data distribution and analyses fit to the observations can be seen in Osuri et al. (2012b). It is mentioned that the vortex is well covered by QSCAT winds and the outer area is covered by SSM/I winds.

The initial vortex position and structure is improved significantly in 21 cases out of total 23 cases (Fig. 4). The mean initial vortex position error is

Table 3. Design of the experiments and the number of observations (QSCAT and SSM/I) ingested in each case of initial condition for both the TCs

TCs	*Cases*	*Initial conditions (forecast length in hours)*	*QSCAT*	*SSM/I*
VSCS Nargis (2008)	1	00 UTC of 28 April 2008 (120)	5171	2175
	2	12 UTC of 28 April 2008 (108)	4290	2851
	3	00 UTC of 29 April 2008 (96)	4328	2713
	4	12 UTC of 29 April 2008 (84)	5478	3230
	5	00 UTC of 30 April 2008 (72)	723	2132
SuCS Gonu (2007)	6	00 UTC of 2 June 2007 (132)	5794	2701
	7	12 UTC of 2 June 2007 (120)	5594	3299
	8	00 UTC of 3 June 2007 (108)	5660	3208
	9	12 UTC of 3 June 2007 (96)	5699	3559
VSCS Sidr (2007)	10	00 UTC of 13 Nov 2007 (72)	470	3252
	11	12 UTC of 13 Nov 2007 (60)	727	2871
	12	00 UTC of 14 Nov 2007 (48)	1497	2702
	13	12 UTC of 14 Nov 2007 (36)	3465	4387
CS KhaiMuk (2008)	14	12 UTC of 13 Nov 2008 (60)	1026	2561
	15	00 UTC of 14 Nov 2008 (48)	681	2096
	16	12 UTC of 14 Nov 2008 (36)	2946	1037
SCS Aila (2009)	17	00 UTC of 23 May 2009 (72)	3213	2719
	18	12 UTC of 23 May 2009 (60)	2956	3081
	19	00 UTC of 24 May 2009 (48)	2743	2938
	20	12 UTC of 24 May 2009 (36)	3113	2514
VSCS Laila (2010)	21	00 UTC of 18 May 2010 (60)	-	2711
	22	12 UTC of 18 May 2010 (48)	-	3759
	23	00 UTC of 19 May 2010 (36)	-	3211
	24	12 UTC of 19 May 2010 (24)	-	3575

reduced from 72 km (FNL analyses) to 48 km in 3DVAR analyses (by 34%). The 3DVAR experiment improves the track prediction in all the cases but showed one representative case for each cyclone (Fig. 5). From Fig. 6, the mean 24-hr, 48-hr, 72-hr and 96-hr forecast track errors (in km) are 176, 291, 327 and 425 for CNTL experiment and 132, 228, 176 and 224 for 3DVAR experiment respectively. Hence, there is an improvement of 25%, 22%, 46% and 47% respectively with 3DVAR assimilation experiment. There is an improvement of 10% in MSLP prediction and 21% in 10 m wind prediction with 3DVAR experiment. Out of 24 cases, 3DVAR shows better intensity predictions in 18 cases (in terms of MSLP) and in 19 cases (in terms of 10 m

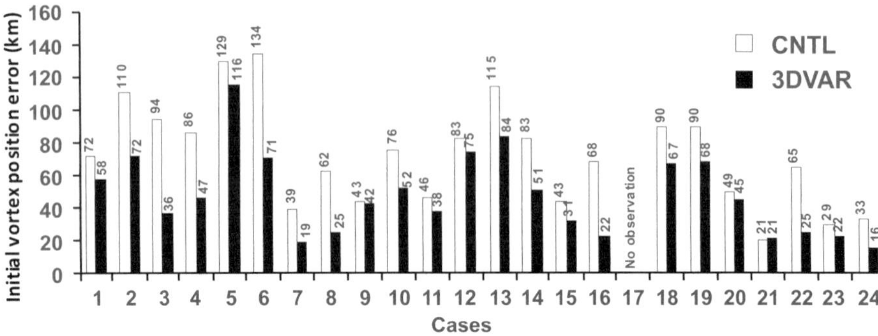

Fig. 4. Initial vortex position errors (km) for each case from CNTL and 3DVAR analysis.

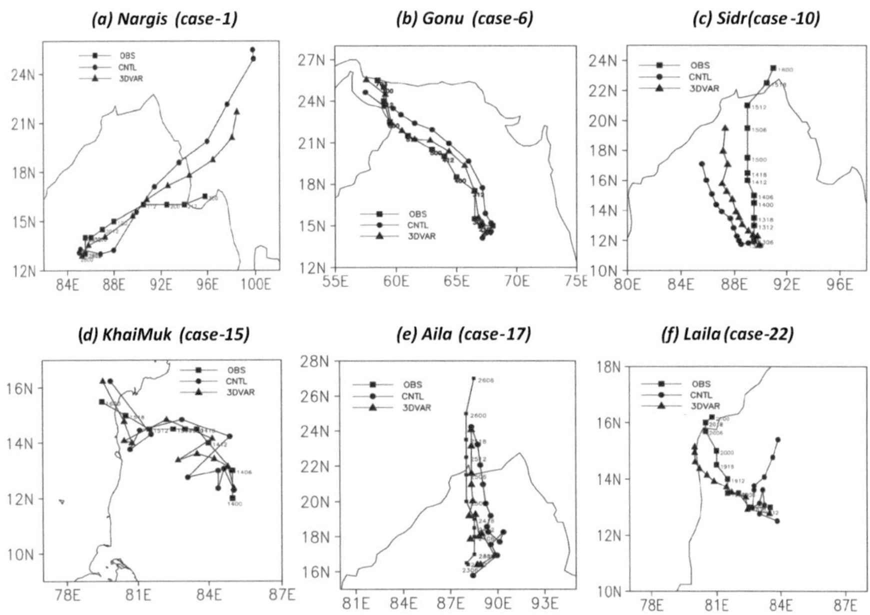

Fig. 5. Model simulated tracks from two experiments along with IMD observed track of (a) Nargis, (b) Gonu, (c) Sidr, (d) KhaiMuk, (e) Aila and (f) Laila.

wind). The 3DVAR experiment predicts the realistic structure of eye, eye wall and outer core of TC. It could also predict the large scale surface latent heat fluxes from the underlying warm ocean due to the strong updrafts in eye wall region. Hence, unlike the CNTL experiment, the 3DVAR retains the warmer core structure with well-defined constant equitable potential temperature profiles extending through the depth of the entire troposphere which ultimately influences the intensity of the TCs also. With the improvement in movement of TCs, 3DVAR predicts landfall position more accurately (Table 4). Out of 24 cases, CNTL fails to predict the landfall in nine cases and 3DVAR in

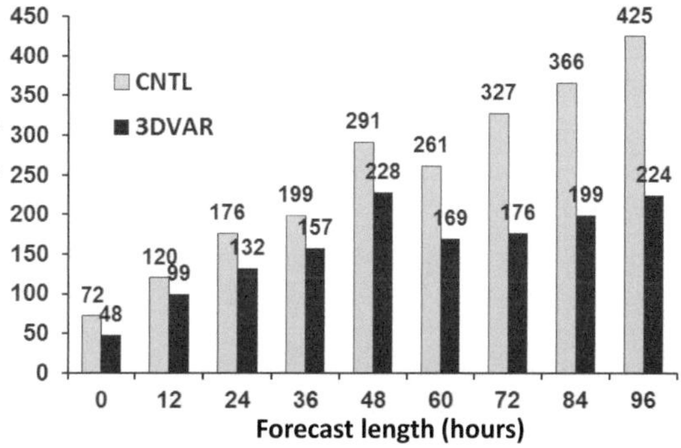

Fig. 6. Mean (of 24 cases) vector displacement errors (km) for each case from CNTL and 3DVAR runs.

four cases. The 3DVAR experiment could improve the landfall prediction in 17 cases (out of 20 landfall cases). The mean landfall error is reduced by 28% after the assimilation satellite derived winds. It is 38% if consider the homogeneous cases of 15 in which both CNTL and 3DVAR made landfall. The improved structure, intensity and track with the assimilation of QSCAT and SSM/I winds improved the spatial distribution of rainfall. In quantitative prediction of rainfall, the 3DVAR performs better up to about 9 cm of rainfall as ETS is greater than 0.2.

3.2 Performance of NMM Model

The performance evaluation of NMM model includes impact of horizontal resolution, assimilation of satellite derived wind and assimilation of both satellite and upper air data model in simulation of TCs over BoB. The optimized physical parameterization schemes follows the sensitivity study of Pattanayak et al. (2012b) are considered. The initial (IC) and boundary conditions (BC) for the NMM model are obtained from NCEP FNL analysis.

(i) Impact of Horizontal Resolution

The results as obtained with different horizontal resolution are presented in this section. For this purpose the very severe cyclonic storm Nargis (2008) is considered. Figures 7 (a), (b) and (c) represent the MSLP (hPa) with 27 km, 9 km and 3 km. The horizontal resolution of 27 km could simulate the MSLP of 1001 hPa, whereas the horizontal resolution of 9 km and 3 km simulates 992 and 994 hPa respectively. The model simulated pressure drop and the percentage improvement relative to 27 km simulation is presented in Table 5. It is noticed that the model simulation with 9 km and 3 km provides almost similar type of result. Figures 7 (d), (e) and (f) represent the wind (m/s) at 850 hPa with 27 km, 9 km and 3 km, respectively.

Figures 8 (a), (b), (c) and (d) represent the TRMM and day-1 forecast from model valid at 03 UTC 29 April 2008 with 27 km, 9 km and 3 km horizontal

Table 4. Peak intensity simulated and landfall errors from both CNTL and 3DVAR runs for all the cases

Cases	Observed intensity	*Peak intensity*				*Landfall errors*			
		MSLP (hPa)		*10 m wind (m s^{-1})*		*Position (km)*		*Time (hours)*	
		CNTL	*3DVAR*	*CNTL*	*3DVAR*	*CNTL*	*3DVAR*	*CNTL*	*3DVAR*
1	Nargis Peak Intensity 962 hPa and 46 m s^{-1} Landfall over Myanmar at 12 UTC 2 May 2008	942	956	48	45	339	205	+33	+24
2		950	942	45	47	174	113	+12	+3
3		966	962	40	42	140	92	+15	+12
4		959	954	42	44	42	31	+12	+6
5		963	957	40	43	77	24	+6	0
6	Gonu Peak Intensity 920 hPa and 65 m s^{-1} Landfall over Oman at 03 UTC of 6 June 2007	941	928	47	51	NL	61	NL	+2
7		945	933	47	51	102	55	-4	+2
8		952	936	42	48	NL	NL	NL	NL
9		995	995	19	19	219	NL	+30	NL
10	Sidr Peak Intensity 944 hPa and 59 m s^{-1} Landfall at over Bangladesh at 18 UTC 15 Nov. 2007	968	960	41	44	NL	NL	NL	NL
11		972	965	37	39	NL	NL	NL	NL
12		996	992	19	21	NL	NL	NL	NL
13		999	997	17	18	117	73	-9	-6
14	KhaiMuk Peak Intensity 994 hPa and 20 m s^{-1} Landfall over India 00 UTC of 16 Nov. 2008	996	995	18	19	83	51	-9	-3
15		1000	998	14	17	60	43	-9	-9
16		1001	998	14	17	65	39	-6	-3

(*Contd.*)

Table 4. (*Contd.*)

17	Aila Peak Intensity	972	970	31	34	110	49	-6	-9
18	968 hPa and 31 m s^{-1}	973	970	28	32	49	45	-6	-9
19	Landfall over India 09 UTC	980	976	23	28	66	76	-6	-9
20	of 25 May 2009	983	983	25	24	95	114	-6	-9
21	Laila Peak Intensity 986 hPa	952	975	45	40	NL	NL	NL	NL
22	and 28 m s^{-1}	962	985	40	34	NL	184	NL	+6
23	Landfall over India 12 UTC	972	990	34	25	NL	108	NL	+15
24	of 20 May 2010	989	996	23	20	NL	40	NL	-3

Observed information is in column 2. +/- sign represents the ahead/delay in time. "NL" represents No Landfall simulated.

Table 5. Model simulated pressure drop at different horizontal resolution

Date & Time	*Pressure drop (hPa) from different horizontal resolutions*			*% of improvement w. r. t. 27 km*		*Observed pressure drop (hPa)*
	27 km	*9 km*	*3 km*	*9 km*	*3 km*	
00 UTC 28 Apr 08	11	12	12	8	8	6
12 UTC 28 Apr 08	11	12	13	8	15	14
00 UTC 29 Apr 08	10	10	10	0	0	14
12 UTC 29 Apr 08	12	12	12	0	0	20
00 UTC 30 Apr 08	8	10	10	20	20	20
00 UTC 01 May 08	8	15	16	47	50	20
00 UTC 02 May 08	9	15	12	40	35	40
00 UTC 03 May 08	10	10	10	0	0	6
% of mean improvement w.r.t. 27 km				24.6	25.6	

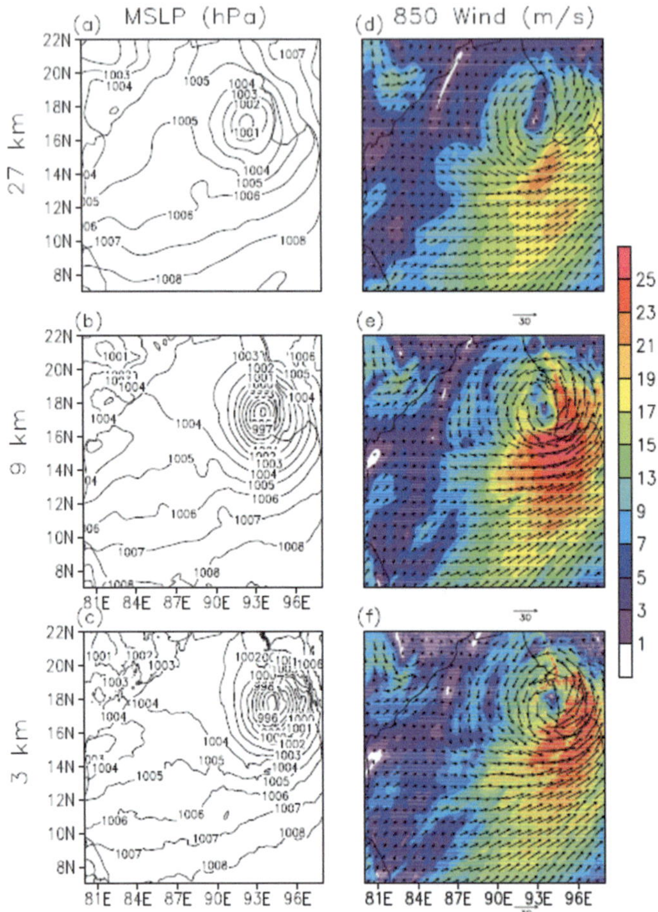

Fig. 7. Day-3 forecast of MSLP (hPa) and wind (m/s) at 850 hPa valid at 00 UTC 01 May 2008. (a, d) with 27 km, (b, e) with 9 km, and (c, f) with 3 km horizontal resolution.

resolutions, respectively. With the higher resolution i.e. from 9 km to 3 km model, the model could simulate higher precipitation. Figures 8 (e), (f), (g) and (h) represent the TRMM and day-3 forecast from model valid at 03 UTC 01 May 2008 with 27 km, 9 km and 3 km horizontal resolutions, respectively. The day-3 forecast enhances the model performance with increasing the horizontal resolution.

The performance of the model enhances in increasing the model horizontal resolution from low (27 km) to mesoscale resolution (<10 km). The horizontal resolution of 3 km marginally improves intensity and structure of the cyclones. However, with the existing computer resources, it increases the computational cost to more than 20 times w.r.t. to the simulation time with 9 km horizontal resolution. Therefore, the horizontal resolution of 9 km is the optimum configuration and used in rest of the experiments.

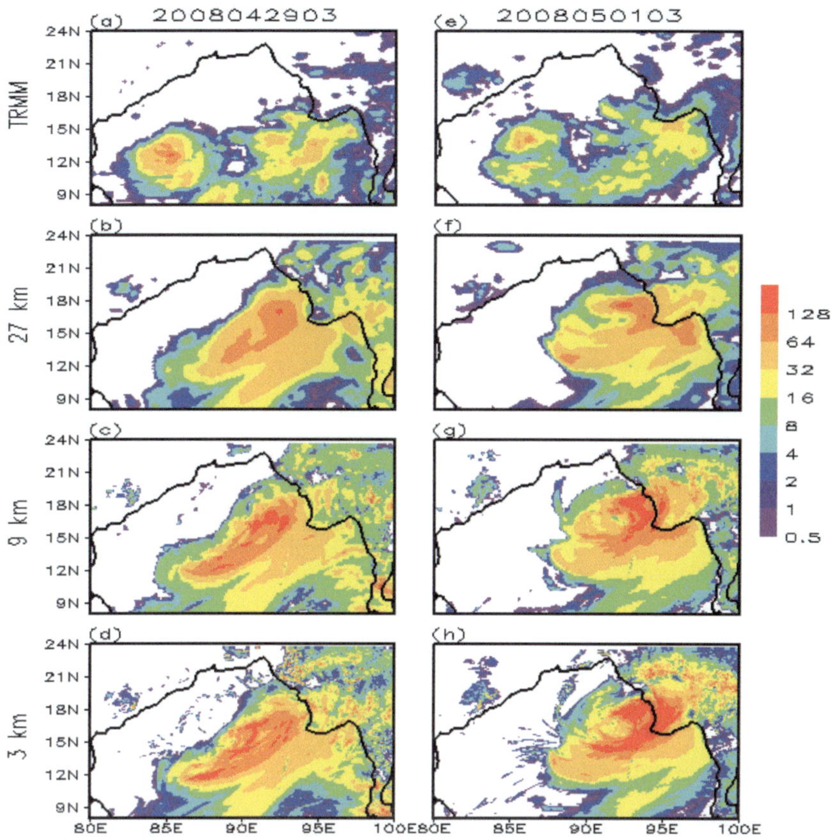

Fig. 8. 24 hrs accumulated precipitation (mm). (a) TRMM precipitation valid at 03 UTC 29 April 2008. (b), (c) and (d) are model simulated 24 hrs accumulated precipitation (mm) with 27 km, 9 km and 3 km horizontal resolutions, respectively. (e-h) are same as (a-d), but valid at 03 UTC 01 May 2008.

(ii) Impact of Satellite Data

The impact of satellite derived wind in simulation of track and intensity of tropical cyclones over BoB is illustrated. The observational datasets include scatterometer wind fields from Qscat, SSM/I and cloud motion vector from Kalpana satellite. The model is integrated in a single domain with the horizontal resolution of 9 km along with 51 levels up to a height of 30 km in the vertical.

Figures 9 (a) and (b) represent model simulated tracks for the cyclone Nargis with different initial conditions from CNTL (IC from FNL analysis) and SAT (IC improved with satellite data, taking FNL as background) experiments, respectively. In both the experiments, the track of the cyclone moves towards Myanmar coast from each initial condition. However, at the landfall point, the wide spread of about 400 km in latitudinal direction is seen in CNTL

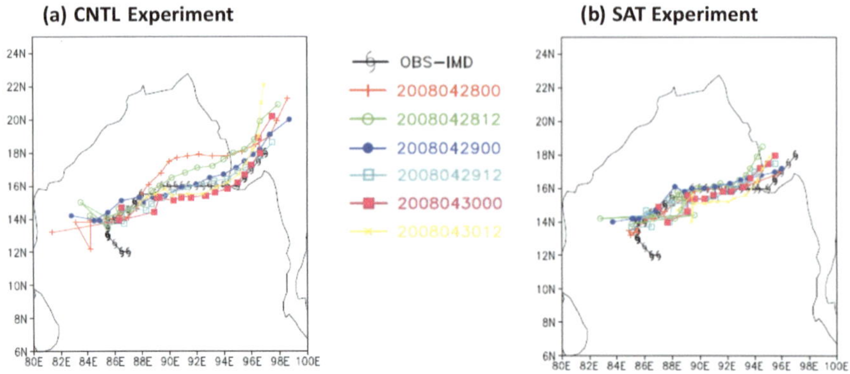

Fig. 9. Tracks with different initial conditions from (a) control simulation (FNL), and (b) satellite data assimilation (SAT) experiments for the cyclone Nargis.

simulation which reduces to about 200 km with SAT experiments. The initial positional error is reduced in assimilation experiments in comparison to the control simulations.

The time evolution of MSLP (hPa) (Figure not shown) as simulated with both the experiments is compared with estimated central pressures (with time) as obtained from IMD. The observed minimum MSLP was 962 hPa valid at 06 UTC 02 May 2008. In CNTL experiment, the model could simulate the minimum MSLP of 972 hPa valid at 06 UTC 02 May 2008. The SAT experiments simulate the MSLP of 968 hPa.

(iii) Combined Effect of Conventional and Satellite Data

In this section the impact of both satellite and conventional data in simulation of landfalling TCs over the BoB is presented. These observations include Synop, Sound, Metar, Pilot, Buoys, Airep, Qscat etc. In this regard the BoB TCs viz., Nargis (2008), Bijli (2009), Jal (2010) and Thane (2011) are simulated at different initial times. A total of 34 initial states (eight initial conditions for Nargis, seven initial conditions for Bijli, nine initial conditions for Jal and ten initial conditions for Thane) are considered. The numerical experiments are broadly categorized into two sets such as (i) control (CNTL) experiment with FNL as the initial and boundary condition and (ii) data assimilation (DA) experiment in which model initial background field is improved through assimilation of additional observations as mentioned above.

Figure 10 represents the model simulated tracks from both CNTL and DA experiments along with best-fit track from IMD. Figures 10 (a) and (b) present the track of the cyclone Nargis with CNTL and DA experiments, respectively. The initial vortex position is improved and close to the actual realization in DA experiment than the CNTL simulation. Also, the subsequent forecasted track is improved in DA experiments. Similar results are noticed in all the other cases. Figures 10 (c) and (d) present the track of the cyclone Bijli with

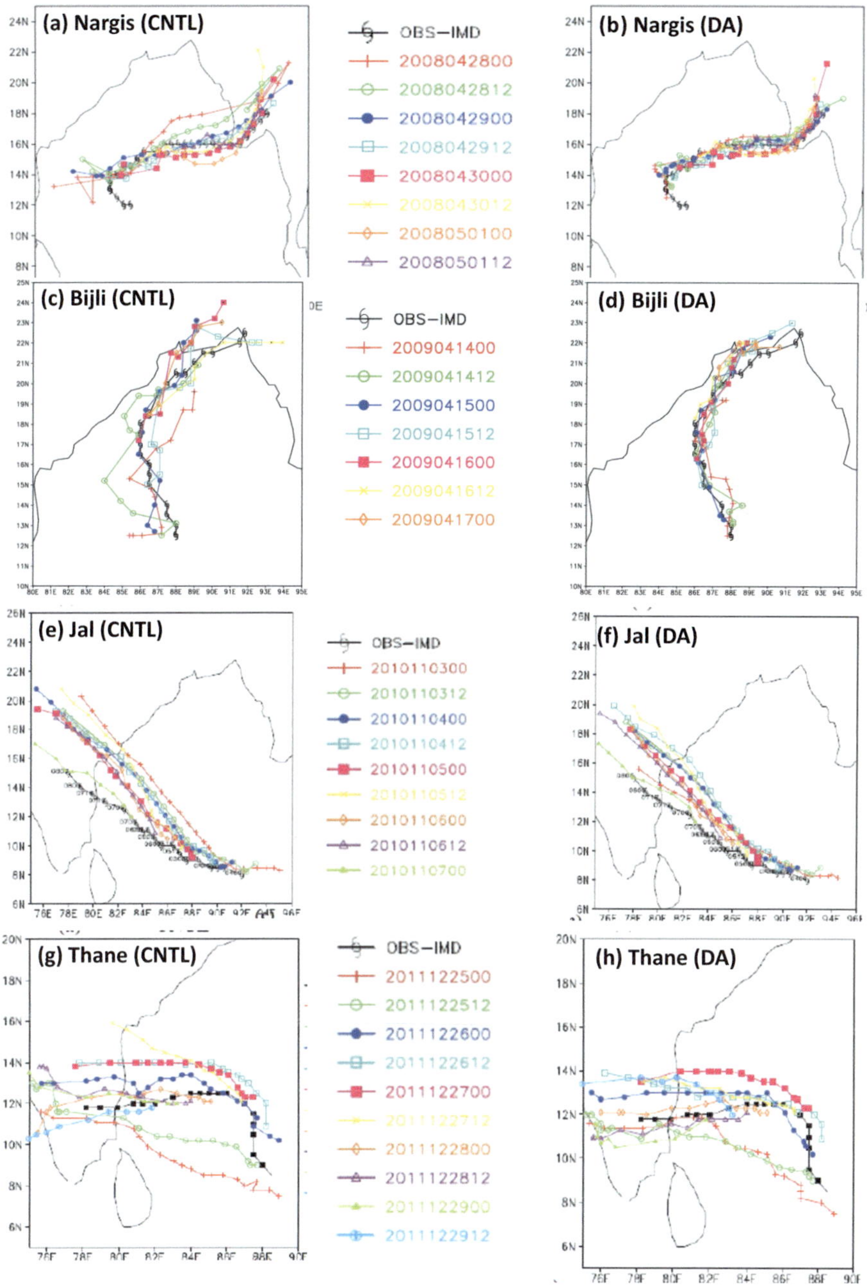

Fig. 10. Model simulated tracks with CNTL and DA experiments along with best-fit track from IMD for (a, b) Nargis, (c, d) Bijli, (e, f) Jal and (g, h) Thane cyclones.

CNTL and DA experiments, respectively. The results show that at the initial time of model integration, the improvement of nearly 50% is achieved in DA experiments than that of the CNTL experiments. The landfall point and time error are also reduced in DA experiments. Figures 10 (e) and (f) present the track of the cyclone Jal with CNTL and DA experiments, respectively. At the initial time, the improvement of about 45 % is noticed in DA experiments, however not much improvement is seen in subsequent forecast hour. Figure 10 (g) and (h) presents the track of the cyclone Thane with CNTL and DA experiments, respectively. The improvement in the range of 10-31% is seen in DA experiments than that of the CNTL simulations. A wide spread of nearly 550 km in landfall location is calculated in CNTL simulations which reduces to 280 km in DA experiment.

The mean VDEs for all the 34 cases are also calculated. The mean improvement of 35%, 23%, 26%, 27%, 28% and 22% is noticed at 00, 24, 48, 72, 96 and 120 hrs respectively. Thus, it may be noted that the track position of the cyclones are improved in initial as well as forecast time of model integration.

The model simulated 24 hrs accumulated precipitation is validated against TRMM (3B42-V6) observations. The precipitation data are presented at the landfall time for each case. Figure 11 presents the spatial distribution of 24 hrs accumulated precipitation as obtained from TRMM and model simulations for cyclones Nargis, Bijli and Jal valid at 12 UTC 02 May, 2008, 03 UTC 18 April 2009 and 16 November, 2010, respectively.

Results show that the model simulations from both CNTL and DA experiments always overestimate compared to TRMM precipitation. However, the spatial distribution of precipitation is well captured by the model. In case of Nargis, wide spread rainfall is observed and model simulates the precipitation of 15 and 22 cm along the Myanmar coast during the landfall of the cyclone. Both the experiments overestimate than the observed precipitation. In case of Bijli, both the experiments well simulate the precipitation pattern, however the maximum precipitation is shifted towards right of the track in CNTL experiments of the cyclone. In case of Jal, the model well simulates the precipitation pattern, however the amount of precipitation is better simulated by DA experiments. Rainfall occurred at most places with heavy to very heavy rainfall at a few places over north Tamil Nadu, Puducherry, coastal Andhra Pradesh, Rayalaseema, south Interior Karnataka and coastal Karnataka.

3.3 Performance of HWRF Model

In this section, the performance of HWRF modelling system for the prediction of TC Giri is investigated as it is a rapid intensified cyclone just before the landfall. The model configuration and the physics employed are given in Table 1. HWRF uses two domains, i.e., the outer static domain with 27 km and a movable 9 km nested domain. The size of the inner moving nest is $6° \times 6°$ and follows the vortex. In this study, bogus vortex initialization, relocation

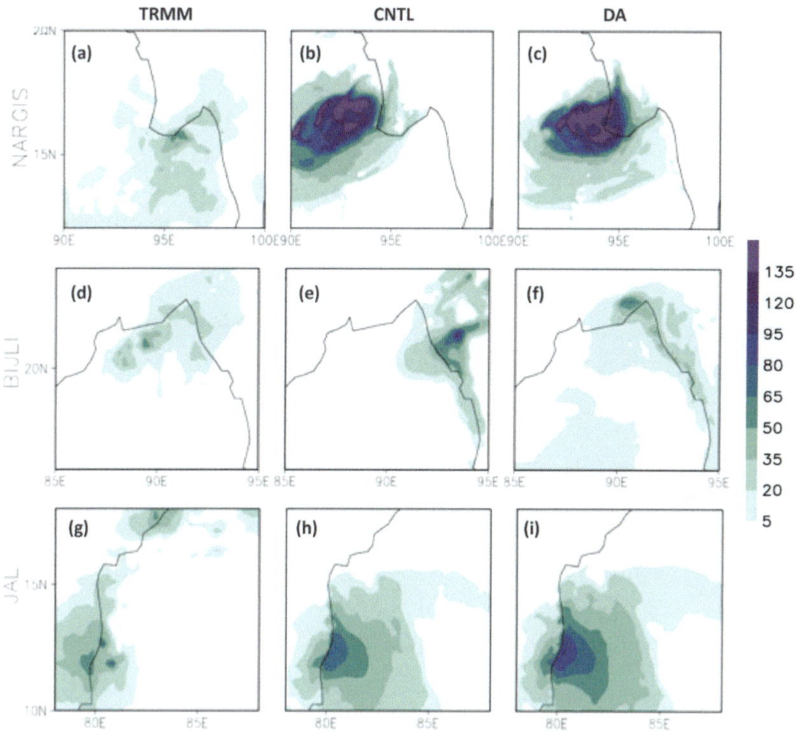

Fig. 11. 24 hrs accumulated precipitation (mm) from TRMM (left panel), model simulation from CNTL (middle panel) and DA (right panel) experiments for Nargis, Bijli and Jal cyclones valid at the landfall time.

and assimilation methods are used. For bogus vortex, TC position and intensity are taken from IMD observations and structure details (like radius of maximum winds and radius of outermost closed isobar) are taken from Joint Typhoon Warning Centre (JTWC). After creating the bogus vortex, depending upon the intensity of convection (shallow, medium or deep) associated with TC, the bogus vortex undergoes intensity and structure corrections in vortex initialization procedure. The vortex, in medium and deep convective TCs, receives the same vortex corrections mentioned above, while the vortex with shallow convection undergoes only two corrections i.e., the vortex top is set to 700 hPa and the warm core structures are removed. TCs with deep convection, Grid Statistical Interpolation (GSI) system is used to assimilate large scale observations (upper-level, surface and satellite radiance data) to further improve the vortex. If the TC is not deep, data assimilation will not be used. The detailed information regarding the procedures can be found from HWRF scientific documentation by Gopalakrishnan et al. (2010). The TC Giri is initialized with HWRF model at three intensity stages i.e., depression (DD), cyclonic storm (CS) and severe cyclonic storm (SCS) stages to predict the track, intensity and structure. According to NCEP TC vitals, Giri has shallow convection at DD stage, medium convection at CS and deep convection at SCS stage.

Figure 12 provides the intensity evolution, track and associated errors for each experiment initialized at above mentioned three stages. The initial position error is zero as in all cases, vortex relocation is carried out. It is seen that the experiment initialized from depression stage could not capture the intensity evolution, track movement even after the insertion of bogus vortex which results with large track errors of about 85 km to 295 km in 6 hours to

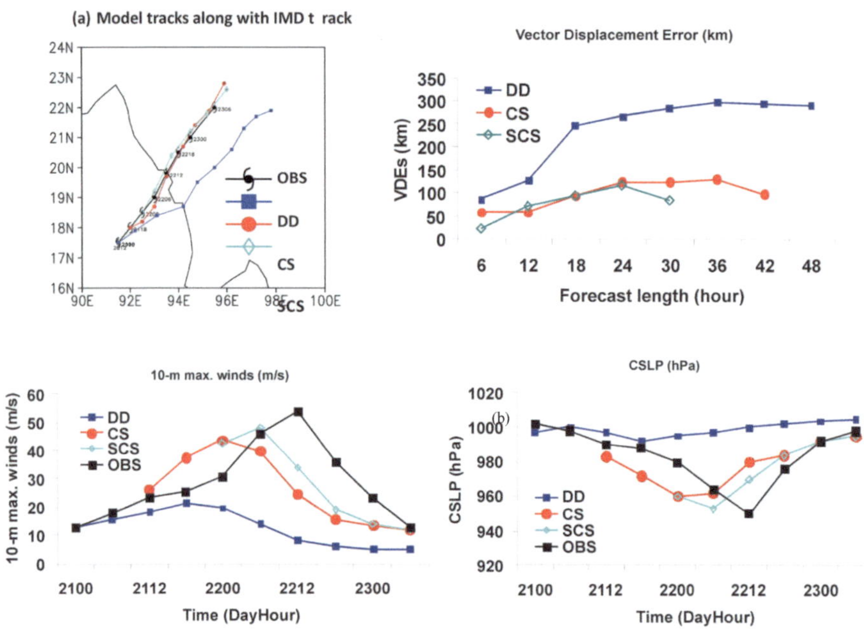

Fig. 12. HWRF predicted (a) tracks and (b) vector displacement error, km and intensity evolution in terms of (c) 10-m maximum winds, m/s and (d) MSLP, hPa, for TC Giri at three stages of intensity. The forecast length decreases by 12 hours from DD to CS, CS and SCS i.e., depression case has 48 hours forecast length, cyclonic storm case has 42 hours and SCS case has 36 hours forecast.

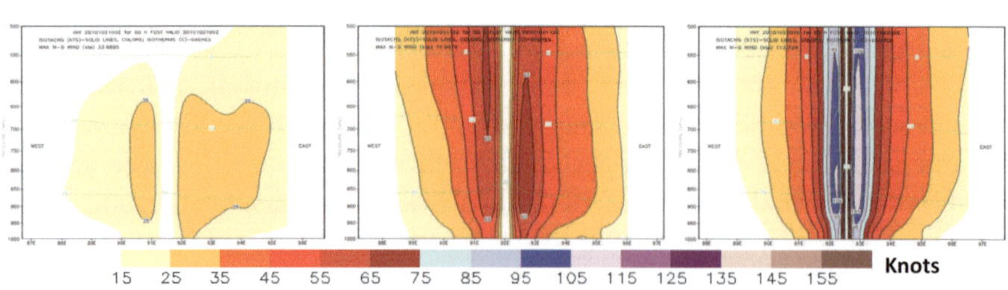

Fig. 13. HWRF initialized vortex at the initial time of each case (a) depression stage, 00UTC 21 Oct, (b) cyclonic storm, 12UTC 21 Oct and (c) severe cyclonic storm, 00 UTC 22 Oct 2010.

48 hours forecast lengths (Figs 12 a, b). The maximum intensity predicted in depression stage experiment is about 992 hPa (22 m/s) against the observed intensity of 950 hPa (54 m/s) (Figs 12 c, d). This may be due to the fact that the TC Giri is a shallow vortex at depression stage and hence vortex is not improved (Fig. 12a) as there are no vortex intensity and size corrections, additional data assimilation, except the two above mentioned corrections. Hence, the intensity evolution is also poor compared to observation and other two experiments.

Figure 13 shows different vortex structures generated through different initialization procedures as mentioned above when TC Giri is at shallow (DD), medium (CS) and deep (SCS) convective stages. However, in the later two cases the results are very impressive having minimum errors in both intensity and track prediction because of improved initial structure. Though, data assimilation is not used in CS stage, it could capture the rapid intensification (wind increased about 18 m/s within 12 hours) before landfall and a maximum intensity of 40 m/s. Similar results are observed in the last case in which TC is initialized at SCS stage and capture the right peak intensity of 48 m/s (Figs 12 c, d). This may be due to the improved vortex initialization procedure at CS and both initialization and assimilation at SCS stages. It could also predict rapid dissipation after the landfall. The track errors are about 100 km in six hours to 48 hours forecast lengths (Figs 12 a, b) from both the CS and SCS initializations. These results show the importance of high resolution mesoscale data assimilation along with improved vortex initialization procedure for more realistic TC forecasts (as seen in CS and SCS stage initializations).

4. Conclusions

The study demonstrates some of the important aspects such as research and development of numerical models, observations, and data assimilation techniques for providing improved initial states in those models for improved prediction of tropical cyclones (TCs) over the North Indian Ocean (NIO). The increments in the official forecast skills of track and intensity is mainly due to more accurate and consistent numerical guidance from different mesoscale and global models. We conducted experiments to address these issues for improved TC predictions. Mohanty et al. (2010) demonstrated that real boundary conditions provide better simulations than the forecast boundary conditions. The importance of physical parameterization schemes for track and intensity predictions of TCs over the NIO basin are studied by Pattanayak et al. (2012b) and Osuri et al. (2012a). Osuri et al. (2012b) demonstrated the usefulness of satellite derived sea surface winds in improving initial states and consequent TC simulations when they are over the open waters. Realizing the increasing severity of TCs over NIO (Mohanty et al., 2011) with warming environment, it is important and challenging task to provide guidance for intensity and track in advance. Therefore, the damage can be minimized considerably.

High resolution mesoscale modelling systems provide better guidance for TC forecast in real-time up to 72 hours. The model at high-resolution (18 and 9 km) yielded an improvement of about 4%–14% and 12%–28% in mean DPE over 27-km runs. High resolution is desirable for improved predictions of recurving TCs. The track errors are much lesser when the models are initialized at severe cyclonic storm stage as compared to that of depression and cyclonic storm stages. The analyses of directional errors reveal that ARW model tracks are slower and biased to rightwards.

The track and intensity error is relatively more in real-time predictions using ARW model when initialized with global analyses and forecast fields without data assimilation. Assimilation of sea surface winds improved the initial vortex position and structure and achieved an improvement of 25%, 22%, 46% and 47% respectively at 24-hr, 48-hr, 72-hr and 96-hrs forecast in track prediction as compared to no-data assimilation run. There is an improvement of 10% in MSLP prediction and 21% in 10-m wind prediction with 3DVAR experiment. Out of 24 cases, 3DVAR shows better intensity predictions in 18 cases (in terms of MSLP) and in 19 cases (in terms of 10 m wind). With the improvement in movement of TCs, 3DVAR predicts landfall position more accurately. Out of 24 cases, CNTL fails to predict the landfall in nine cases and 3DVAR in four cases. The 3DVAR experiment could improve the landfall prediction in 17 cases (out of 20 landfall cases).

Considering NMM model, the track and intensity errors are substantially improved in DA experiments than that of the CNTL experiments. The mean track error in WRF-NMM model based on the 34 cases varies from 91 km to 352 km in CNTL simulations for 00 hr to 120 hrs forecast length. At the same time, in case of DA experiments the mean track error for 00 hr to 120 hrs forecast length ranges from 61 km to 297 km. The mean VDEs are improved by about 35%, 23%, 26%, 27%, 28% and 22% at 00, 24, 48, 72, 96 and 120 hrs respectively. The DA experiments significantly improve the intensity forecast as well in most of the cases.

HWRF model with advanced capabilities like vortex relocation, initialization and data assimilation techniques showed impressive improvement for the prediction of TC Giri over the BoB. It could be able to capture the rapid intensification and weakening of the system. Therefore, it can be a value-added model for the research and operational purposes.

5. Limitations and Future Directions

The simulation of track and intensity of TCs has always been the challenging problem for weather forecasters. There have been considerable improvements in the numerical prediction of TCs in last two decades. Though significant improvement in prediction of track of the cyclonic storms is reported in a number of studies, including the present one, prediction of genesis and intensity of the storms continues to be a challenging task to the modellers. As an extension to the present work, the improvement in model physics, especially the cumulus parameterization and microphysics need to be focused in further study.

The variational method is capable of dealing with different sources of observations that have complicated nonlinear relation with model variables. In the process of converting observations into a specific format that is accepted by assimilation system, there will be some truncations and round-off which leads to errors. For example, instead of assimilating satellite derived products into model initial conditions, it will be better to assimilate the satellite radiance directly which reflects the improved use of observations and becomes increasingly important for enhancing the weather predictions. India is in the process of launching weather satellites to monitor and provide high resolution three dimensional data. It is important to develop a procedure to assimilate the Indian satellite (such as INSAT-3D and Kalapna-1) radiance data and its impact assessment is also of great significance. Realizing the importance of DWRs, the India Meteorological Department (IMD) has been improving the network of DWRs in the coastal as well inland regions to observe the severe weather phenomena. More improvement in TC forecast can be expected by assimilating multiple DWR data. Further, the WRF-3DVAR is capable to assimilate the satellite radiance and multiple DWR data.

A set of initial states that are slightly different from the original analysis, or, a suite of multimodels provide different forecast spreads. A statistical mean of such forecasts might help in providing a realistic state of the atmosphere than a single forecast. Combining these forecasts is known as the ensemble or super ensemble method. In recent days, this is recognized as an important component for 3-4 days TC forecast.

Still, the numerical models are limited in providing the sudden intensity changes such as rapid intensification, weakening over the sea, or re-intensification of the system after landfall. Along with the initial state of the atmosphere, improved forecasts rely on accurate representation of land surface also. The prediction of boundary layer processes, cloud, precipitation and surface meteorological conditions are limited and can be improved by adequate illustration of land surface physics and land state initialization i.e. vegetation characteristics, soil moisture and soil temperature etc. It is realized that among all the land surface parameters, the soil moisture plays prime role in genesis and development of deep convection. Therefore, research is required in addressing the role of such land surface data on the intensity changes associated with the extreme weather events such as tropical cyclones, severe thunderstorms and heavy rainfall events.

The performance of these mesoscale models in the prediction of very heavy rainfall associated with landfalling TCs is satisfactory over the Indian region. However, there is a scope to improve the rainfall prediction over India.

A common problem in rainfall prediction is that the model overestimates the moderate rainfall (<10 cm) and underestimates the heavy to very heavy rainfall (>30 cm). This uncertainty in the model predictions can be reduced with the cyclic assimilation of high resolution observations. Now, IMD is expanding the DWR network in India and rapid update cyclic data assimilation of multi DWR observations could solve the problem to some extent. Rogers (2010) also highlighted the requirement of cloud-scale resolutions (of 1–3 km) for predicting hurricanes and related inland rainfall. Further, providing improved land surface conditions to the high resolution models is another area of research to improve rainfall prediction. The interaction of land surface parameters (such as land use, land cover, roughness length, leaf area index, soil moisture etc.) is also crucial in modulating the rainfall (Niyogi et al., 2006).

Acknowledgement

We owe thanks to Indian National Centre for Ocean Information Services (INCOIS) for the needful financial support to carry out the research on tropical cyclones over NIO and duly acknowledge. Also acknowledge the Indo-US Science and Technology Forum (IUSSTF) support for the conference. The authors also acknowledge to IMD for providing TCs observations. The authors acknowledge NCAR, NCEP and DTC for providing the ARW, NMM and HWRF modelling systems and global analysis and forecast products used in the present study.

REFERENCES

Arakawa, A. and W.H. Schubert, 1974: Interaction of a cumulus cloud ensemble with the large scale environment. Part I. *J. Atmos. Sci.*, **31**, 674-701.

Bansal, R.K. and S.R.H. Rizvi, 1993: A New Approach to the Analysis Problem. *Proceeding of National Symposium:* Advances in Tropical Meteorology, Meteorology and National Development, TROMPET 93, 17-19 March 1993, NCMRWF, New Delhi, (ed.) R. K. Datta. pp. 23-32.

Barker, D.M., W. Huang, Y.-R. Guo, A.J. Bourgeois and Q.N. Xiao, 2004: A three-dimensional variational data assimilation system for use with MM5: Implementation and initial results. *Mon. Weather Rev.*, **132**, 897-914.

Chen, S.H., 2007: The Impact of assimilating SSM/I and QuikSCAT satellite winds on Hurricane Isidore simulation. *Mon. Wea. Rev.*, **135**, 549-566.

Das Gupta, M., 2006: Impact of Certain Meteorological Data on Monsoon Simulation with a Global Analysis-Forecast System. Ph.D. Thesis, Berhampur University.

Davis, C.A. et al., 2008: Prediction of landfalling hurricanes with the Advanced Hurricane WRF model. *Mon. Wea. Rev.*, **136**, 1990-2005.

De Angelis, D., 1976: World of tropical cyclones – North Indian Ocean. *Mar. Weather Log.*, **20**, 191-194.

Fels, S.B. and M.D. Schwarzkopf, 1975: The simplified exchange approximation: A new method for radiative transfer calculations. *J. Atmos. Sci.*, **32**, 1475-1488.

Gao, J., M. Xue, A. Shapiro and K.K. Droegemeier, 1999: A variational method for the analysis of three-dimensional wind fields from two Doppler radars. *Mon. Weather Rev.*, **127**, 2128-2142.

Gopalakrishnan, S.G., D.P. Bacon, N.N. Ahmad, Z. Boybeyi, T.J. Dunn, M.S. Hall, Y. Jin, P.C.S. Lee, R.V. Madala, R. Ananthakrishna Sarma, M.D. Turner and T. Wait, 2002: An operational multi-scale atmospheric model with grid adaptivity for hurricane forecasting. *Mon Wea Rev*, **130(7)**, 1830-1847.

Gopalakrishnan, S.G., Q. Liu, T. Marchok, D. Sheinin, N. Surgi, R. Tuleya, R. Yablonsky and X. Zhang, 2010: Hurricane Weather and Research and Forecasting (HWRF) model scientific documentation. NOAA/NCAR/Development Tech Center, 75 pp. [Available online at http://www.dtcenter.org/HurrWRF/users/docs/scientific_documents/HWRF_final_2-2_cm.pdf.]

Gopalakrishnan, S., S. Goldenberg, T. Quirino, X. Zhang, F. Marks, K.-S. Yeh, R. Atlas and V. Tallapragada, 2012: Toward improving high resolution numerical hurricane forecasting: Influence of model horizontal grid resolution, initialization, and physics. *Wea. Forecasting*, **27**, 647-666.

Govindankutty, M., A. Chandrasekar, A.K. Bohra, J.P. George and M. Das Gupta, 2008: The impact of assimilation of MODIS observations using WRF-VAR for the prediction of a monsoon depression during September 2006. *Open Atmos. Sci. J.*, **2**, 68-78.

Grell, G.A., 1993: Prognostic Evaluation of Assumptions Used by Cumulus Parameterizations. *Mon. Wea. Rev.*, **121**, 764-787.

Gupta, A., 2006: Current status of tropical cyclone track prediction techniques and forecast errors. *Mausam*, **57**, 151-158.

Hong, S.-Y. and H.-L. Pan, 1996: Nonlocal boundary layer vertical diffusion in a medium-range forecast model. *Mon. Wea. Rev.*, **124**, 2322-2339.

Hong, S.-Y. and J.-W. Lee, 2009: Assessment of the WRF model in reproducing a flash-flood heavy rainfall event over Korea. *Atmos. Res.*, **93**, 818-831.

Ingleby, N.B., 2001: The statistical structure of forecast errors and its representation in the Met. Office global 3-D variational data assimilation scheme. *Quart. J. Roy. Meteor. Soc.*, **127**, 209-232.

Janjic, Z.I., 1984: Non-linear advection schemes and energy cascade on semi-staggered grids. *Mon. Wea. Rev.*, **112**, 1234-1245.

Janjic, Z.I., 2001: Nonsingular Implementation of the Mellor-Yamada Level 2.5 Scheme in the NCEP Meso model. NCEP Office Note No. 437.

Janjic, Z.I., 2003a: A Nonhydrostatic Model Based on a New Approach. *Meteorology and Atmospheric Physics*, **82**, 271-285 (Online: http://dx.doi.org/10.1007/s00703-001-0587-6).

Janjic, Z.I., 2003b: The NCEP WRF Core and Further Development of Its Physical Package. 5th International SRNWP Workshop on Non-Hydrostatic Modeling, Bad Orb, Germany, 27-29 October.

Jianfeng, G.U., Qingnong Xiao, Ying-Hwa Kuo, Dale M. Barker, Xue Jishan and M.A. Xiaoxing, 2005: Assimilation and Simulation of Typhoon Rusa (2002) using the WRF System. *Adv. in Atm. Sc.*, **22(3)**, 415-427.

Lacis, A.A. and J.E. Hansen, 1974: A parameterization for the absorption of solar radiation in the earth's atmosphere. *J. Atmos. Sci.*, **31**, 118-133.

Lee, C.S., R. Edson and W.M. Gray, 1989: Some large-scale characteristic associated with tropical cyclone development in the North Indian Ocean during FGGE. *Mon. Wea. Rev.*, **117**, 407-426.

Lee, M.S., Y.H. Kuo, and D.M. Barker, 2006: Incremental Analysis Updates Initialization Technique Applied to 10-km MM5 and MM5 3DVAR. *Mon. Wea. Rev.*, **134**, 1389-1404.

Mohandas, S. and R.G. Ashrit, 2011: Tropical cyclone prediction using different convective parameterization schemes in a mesoscale model. Research Report, NMRF/RR/01/2011.

Mohanty, U.C., A. Routray, Krishna K. Osuri and S. Kiran Prasad, 2012: A Study on Simulation of Heavy Rainfall Events over Indian Region with ARW-3DVAR Modeling System. *Pure Appl. Geophys.*, **169**, 381-399.

Mohanty, U.C., 1994: Tropical cyclones in the Bay of Bengal and deterministic methods for prediction of their trajectories. *Sadhana*, **19**, 567-582.

Mohanty, U.C. and A. Gupta, A., 1997: Deterministic methods for prediction of tropical cyclone tracks. *Mausam*, **48**, 257-272.

Mohanty, U.C., Krishna K. Osuri, A. Routray, M. Mohapatra and S. Pattanayak, 2010: Simulation of Bay of Bengal tropical cyclones with WRF Modeling system: Impact of Initial Value and Boundary Conditions. *Marine Geodesy*, **33**, 294-314.

Mohanty, U.C., Krishna K. Osuri, S. Pattanayak and P. Sinha, 2012: An observational perspective of tropical cyclone activity over Indian seas in a warming environment. *Natural Hazards*, **63**, 1319-1335, DOI 10.1007/s11069-011-9810-z.

Mohanty, U.C., Krishna K. Osuri and S. Pattanayak, 2013: A study on high resolution mesoscale modeling systems for simulation of tropical cyclones over the Bay of Bengal. *Mausam*, **64**, 117-134.

Mohapatra, M., B.K. Bandyopadhyay and D.P. Nayak, 2013a: Evaluation of operational tropical cyclone intensity forecasts over north Indian Ocean issued by India Meteorological Department. *Nat. Hazards*, **68**, 433-451.

Mohapatra, M., D.P. Nayak, R.P. Sharma and B.K. Bandyopadhyay, 2013b: Evaluation of official tropical cyclone track forecast over north Indian Ocean issued by India Meteorological Department. *J. Earth Syst. Sci.*, **122**, 589-601.

Niyogi, D., T. Holt, S. Zhong, P.C. Pyle and J. Basara, 2006: Urban and land surface effects on the 30 July 2003 mesoscale convective system event observed in the southern Great Plains. *J. Geophys. Res.*, **111**, D19107, doi:10.1029/2005JD006746.

Osuri, Krishna K., U.C. Mohanty, A. Routray and M. Mohapatra, 2012b: Impact of Satellite Derived Wind Data Assimilation on track, intensity and structure of tropical cyclones over North Indian Ocean. *International Journal of Remote Sensing*, **33**, 1627-1652, DOI:10.1080/01431161.2011.596849.

Osuri, Krishna K., U.C. Mohanty, A. Routray, Makarand A. Kulkarni and M. Mohapatra, 2012a: Sensitivity of physical parameterization schemes of WRF model for the simulation of Indian seas tropical cyclones. *Natural Hazards*, **63**, 1337-1359, DOI 10.1007/s11069-011-9862-0.

Osuri, Krishna K., U.C. Mohanty, A. Routray, M. Mohapatra and Dev Niyogi, 2013: Real-Time Track Prediction of Tropical Cyclones over the North Indian Ocean Using the ARW Model. *Journal of Applied Meteorology and Climatology*, **52**, 2476-2492.

Pattanaik, D.R. and Y.V. Rama Rao, 2009: Track prediction of very severe cyclone 'Nargis' using high resolution Weather Research Forecasting (WRF) model. *J. Earth Syst. Sci.*, **118**, 309-329.

Pattanayak, S. and U.C. Mohanty, 2008: A comparative study on performance of MM5 and WRF models in simulation of tropical cyclones over Indian seas. *Current Science*, **95**, 923-936.

Pattanayak, S. and U.C. Mohanty, 2010: Simulation of track and intensity of tropical cyclones with WRF-NMM modeling system. WMO Technical Report, WMO/TD No. 1541.

Pattanayak, S., U.C. Mohanty and S.G. Gopalakrishnan, 2012a: Simulation of very severe Cyclone Mala over Bay of Bengal with HWRF modeling system. *Nat. Hazards*, **63**, 1413-1437.

Pattanayak, S., U.C. Mohanty and Krishna K. Osuri (2012b): Impact of parameterization of physical processes on simulation of track and intensity of tropical cyclone Nargis (2008) with WRF-NMM model. *The Scientific World Journal*, DOI: 10.1100/2012/671437.

Pu, Z., X. Li, C. Velden, S. Aberson and W.T. Liu, 2008: Impact of aircraft dropsonde and satellite wind data on the numerical simulation of two landfalling tropical storms during TCSP. *Wea Forecast*, **23**, 62-79.

Routray, A., U.C. Mohanty, S.R.H. Rizvi, D. Niyogi, Krishna K. Osuri and D. Pradhan, 2010: Impact of Doppler weather radar data on simulation of Indian monsoon depressions. *Quart. J. Roy. Meteor. Soc.*, **136**, 1836-1850.

Schwarzkopf, M.D. and S.B. Fels, 1985: Improvements to the algorithm for computing CO_2 transmissivities and cooling rates. *J. Geophys. Res.*, **90**, 541-550.

Schwarzkopf, M.D. and S.B. Fels, 1991: The simplified exchange method revisited: An accurate, rapid method for computations of infrared cooling rates and fluxes. *J. Geophys. Res.*, **96**, 9075-9096.

Skamarock, W.C., J.B. Klemp, J. Dudhia, D.O. Gill, D.M. Barker, W. Wang and J.G. Powers, 2005*: A description of the Advanced Research WRF Version 2. NCAR Technical Note NCAR/TN-408 + STR. Available from* www.wrf-model.org.

Southern, R.L., 1979: Global socio-economic impact of tropical cyclones. *Aust. Meteor. Mag.*, **27**, 175-195.

Velden, C., T. Olander and S. Wanzong, 1998: The impact of multispectral GOES-8 wind information on Atlantic tropical cyclone track forecasts in 1995. Part I: Dataset methodology, description, and case analysis. *Mon. Wea. Rev.*, **126**, 1202-1218.

Wu, B., J. Verlinde and J. Sun, 2000: Dynamical and microphysical retrievals from Doppler radar observations of a deep convective cloud. *J. Atm. Sci.*, **57**, 262-283.

Xiao, Q., Y.-H. Kuo, J. Sun, W.-C. Lee, D.M. Barker and E. Lim, 2007 : An approach of radar reflectivity data assimilation and its assessment with the inland QPF of Typhoon Rusa (2002) at landfall. *J. Appl. Meteorol. Clim.*, **46**, 14-22.

Xiao, Q., Y.-H. Kuo, J. Sun, W.-C. Lee, E. Lim, Y.-R. Guo and D.M. Barker, 2005: Assimilation of Doppler radar observations with a regional 3DVAR system: Impact of Doppler velocities on forecasts of a heavy rainfall case. *J. Appl. Meteorol.*, **44**, 768-788.

19

Superensemble Technique for Tropical Cyclone Prediction

T.N. Krishnamurti[1], S. Pattnaik[2] and M. Mandal[3]

[1]Department of Earth, Ocean and Atmospheric Science
Florida State University, Tallahassee, U.S.A.
[2]School of Earth, Ocean and Climate Sciences, IIT Bhubaneswar, India
[3]Centre for Oceans, Rivers, Atmosphere and Land Sciences
IIT Kharagpur, India
tkrishnamurti@fsu.edu

1. Introduction

The superensemble is a weighted mean that extracts useful information from an ensemble of multimodel forecasts to produce optimal guidance for decision makers. The superensemble has significant advantages over simple forecast averaging: it corrects for known biases in the individual objective forecasts, and it weights them to leverage the fact that objective forecasting tools often have regional skill variations. It makes use of as many as 10 million weights, that is a product of grid points in the horizontal, vertical levels in the model, the number of variables and the number of models.

This technique was first applied in meteorology to improve weather and seasonal climate forecasts (Krishnamurti et al., 1999, 2000b). Then tropical precipitation forecasts (Krishnamurti et al., 2000a) and tracking of tropical cyclones in the Pacific (Kumar et al., 2003) also benefited from the SE. Along the last few years, the method has been further investigated with dynamical linear models, from the Kalman Filter (Shin and Krishnamurti, 2003a) to probabilistic approaches (Shin and Krishnamurti, 2003b) for short- to medium-range precipitation forecasts using satellite products. In oceanography, it has been shown that the prediction of temperature (Logutov and Robinson, 2005; Rixen et al., 2009) or acoustic properties (Rixen and Ferreira-Coelho, 2006) in the water column could also be improved by using multi-model statistics. More recently, Rixen and Ferreira-Coelho (2007) introduced the hyper-ensemble, a SE combining models of different nature, to improve surface drift prediction; a method also tested in Vandenbulcke et al. (2009).

2. Methodology

Predictions from an ensemble of slightly different initial conditions and/or various versions of models using a single base model are often carried out by the weather services. An ensemble mean is defined as the average of all models involved in the ensemble suite. Another type of ensemble mean is the bias-removed ensemble mean, where the bias of each model is removed prior to the execution of an ensemble average. This SE technique entails the partition of a time line into two parts. One part is a 'training' phase, where forecasts by a set of member models are compared to the observed or the analysis fields with the objective of developing statistics (i.e., weights a_i) on the least squares fit of the forecasts to the observations. The second part is the forecast phase where estimates for a_i from the training phase are used to create the superensemble. The performance of the individual models is obtained in the training phase using multiple linear regressions against observed (analysis) fields. The outcome of this regression is the weights assigned to the individual models in the ensemble, which are then passed on to the forecast phase to construct the superensemble forecasts. The temporal model anomalies of the variables are regressed against the observed anomalies when formulating the superensemble forecasts, and the weights are multiplied to the corresponding model anomalies. The constructed forecast is:

$$S = \bar{O} + \sum_{i=1}^{N} a_i (F_i - \bar{F}_i) \tag{1}$$

where $\bar{O}$ is the observed climatology over the training period; a_i is the weight for the ith member in the ensemble; and F_i and $\bar{F}_i$ are the ith model's forecasts and the forecast mean (over the training period) respectively. N is the number of member models. The weights a_i's are obtained by minimizing the error term G, where G is expressed as:

$$G = \sum_{t=1}^{N_{train}} (S'_t - O'_t)^2 \tag{2}$$

Here N_{train} is the number of time samples in the training phase, and S'_t and O'_t are the respective superensemble and observed field anomalies at training time 't'.

This exercise is performed at all model grid points. A fit performed for all model variables at all model grid points at all vertical levels typically yields close to 10^7 regression weights. These spread of weights are fractional, positive or negative. This large number arises from the number of transform grid points, number of vertical levels, number of basic variables and the number of models. Over many such locations we have noted diverse performance characteristics of the member models that arises from differences in horizontal and vertical discretization, treatment of physics, handling of inhomogeneity of land surface, orography, water bodies, surface physics and boundary conditions.

All such peculiarities tend to leave their signature in the error distributions and hence on these weights. These may be thought of as a collective bias correction procedure. The second part of the time line is composed of model predictions. The superensemble approach combines each of these forecasts according to the weights determined during the training phase using the formulation. The prediction '*S*' is referred to as the 'superensemble' forecast. This forecast should be contrasted with the more standard anomaly forecasts known as the bias-removed ensemble mean or ensemble mean forecast

$$E = \bar{O} + \frac{1}{N}\sum_{i=1}^{N}(F_i - \bar{F}_i)$$

Or

$$\hat{E} = \bar{O} + \frac{1}{N}\sum_{i=1}^{N}(F_i - \bar{O}) \tag{3}$$

The skill of the multimodel superensemble method significantly depends on the error covariance matrix, since the weights of each model are computed from a designed covariance matrix. The classical method for the construction of the superensemble utilizes a least square minimization principle within a multiple regression of model output against observed 'analysis' estimates. This entails a matrix inversion that is solved by the Gauss Jordan elimination technique. That matrix can be ill-conditioned and singular depending on the interrelationships of the member models of the superensemble. We have recently designed a singular value decomposition (SVD) method (Wilks, 1995) for the multimodel superensemble that overcomes this problem and removes the ill conditioning of the covariance matrix entirely (Yun et al., 2005). Tests of this method have shown improved skills for weather and seasonal climate forecasts.

3. Global Models

This section illustrates performance of multi-model superensemble prediction of track and intensity of tropical cyclones using a suite of global models. Prediction/simulation of tropical cyclones in the Atlantic Ocean, Pacific Ocean and Bay of Bengal is discussed in the next three sub-sections. The focus is on illustration of superiority of multi-model superensemble prediction over the prediction by individual member models.

3.1 Atlantic Hurricanes

After the introduction of superensemble prediction technique by Krishnamurti et al. (1999), a series of publications by Krishnamurti et al. (2000, 2011), Goerss (2000), Williford et al. (2003) and Franklin (2006) reported performance of multi-model superensemble forecast of Atlantic hurricanes. Some of these results of multi-model superensemble prediction of Atlantic hurricanes using a suite of global models will be presented in this section. It may also be mentioned here that besides multi-model superensemble, several ensemble averages based on combination of good models have been used as models to predict Atlantic hurricanes viz., GUNA (GFDL model, UK Met Office model, NOGAPS and Aviation), GUNS (GFDL, UKMO, NOGAPS) and CONU (using at least two of the five models – NCEP GFS, GFDL, US Navy version of GFDL, NOGAPS and UKMO).

Krishnamurti et al. (2011) demonstrated real time performance of the FSU (Florida State University) superensemble prediction during 2004, 2005 and 2006 hurricane seasons. This superensemble method was based on combination of bias corrected every 6-hour forecast from each of the member models. In this method separate statistical weights were considered for track and intensity forecasts from each of the member models. The member models considered for track prediction included OFCI (A time interpolated dataset of the NHC forecast), UKMI (A time interpolated dataset of the UKMO model forecast), NGPI (A time interpolated dataset of the NOGAPS forecast), GFDI (A time interpolated dataset of the GFDL model forecast), GFSI (A time interpolated dataset of the NCEP GFS forecast) and GUNA. The member models used for intensity prediction included OFCI, UKMI, GFSI, SHF5 (A statistical model of HRD for hurricane intensity forecast) and DSHP (A statistical model of HRD for hurricane intensity forecast with inland decay).

3.1.1 Mean Track Errors

The mean errors in predicting the track of the hurricanes in the entire hurricane seasons for the period 2004-2008 using the FSU superensemble (FSSE) prediction and the member models were computed and compared. The mean track errors (up to 120 hours) for different models including the superensemble for the entire 2004, 2005 and 2006 hurricane seasons is shown in Fig. 1. There were 300 entries for 12-h forecasts, which drops to 129 entries for 120-h forecasts since several models terminated their forecasts prior to day 5 if the tropical cyclone dissipates in the model forecast or in the observations. In 2004 hurricane season, the multi-model superensemble had the smallest errors for the track forecast through day 5 of the forecasts (Fig. 1a). The mean error in day-5 (120 hours) forecasts is reduced errors by as much as 39 km compared to the best model. Among the member models, the forecasts from the GUNA were somewhat superior compared to all other member models. The ensemble mean of the member models was better than member models including the ensemble GUNA (GFDI, UK Met, US Navy model NOGAPS and Aviation the US GFS model) 12–120 h except 72 and 84 h.

The forecast errors of the ensemble mean suffer from the fact that the statistical weights of the poorer models and the best model are all assigned the same weight. The superensemble is better in this regard as it assigns fractional positive or negative weights based on the performance of the model in the training phase at all grid locations for each model separately. The track errors for superensemble during 2005 (Fig. 1c) were the smallest among the models

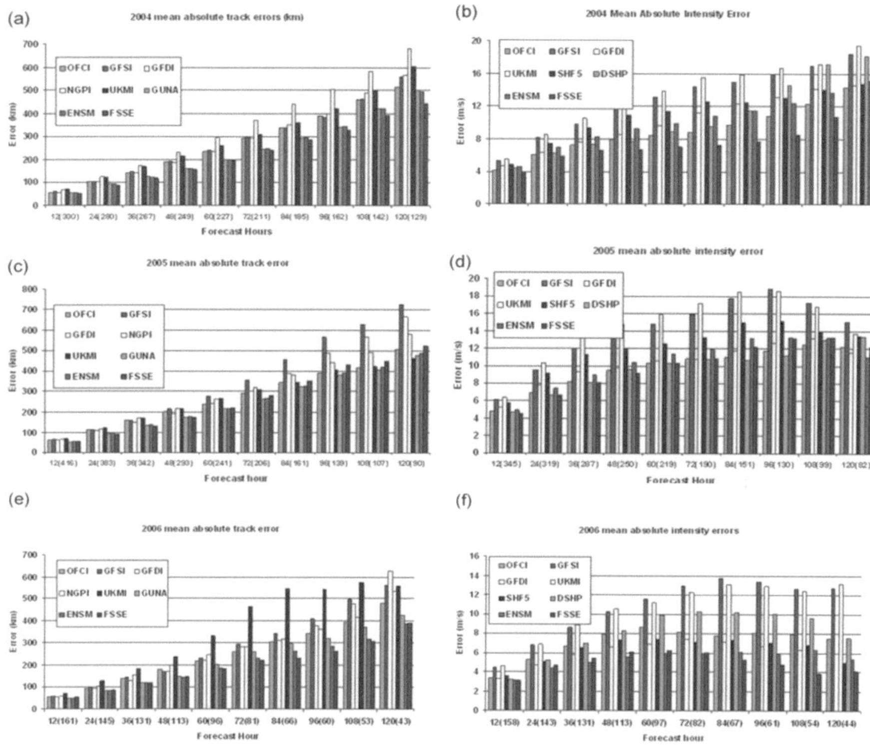

Fig. 1. Homogeneous comparisons of (a) Mean absolute track errors (in km) for 2004 season. (b) Mean absolute intensity errors (in ms^{-1}) for 2004. (c) Mean absolute track errors (in km) for 2005. (d) Mean absolute intensity errors (in ms^{-1}) for 2005. (e) Mean absolute track errors (in km) for 2006. (f) Mean absolute intensity errors (in ms^{-1}) for 2006. The number within the parentheses in the abscissa shows the number of cases for that particular forecast hour. (Taken from Krishnamurti et al., 2011)

in 12–48 h forecasts but were ranked second or third in 72, 96, 108 and 120 h forecasts. GUNA performed better than the FSU superensemble after 60-h forecasts. In 2006, the mean track errors were least in the superensemble forecast except for 12- and 24-h forecasts where GUNA was slightly better. In 2007 and 2008 (figure not shown), the FSU superensemble was able to outperform all the member models and the ensemble mean when the averaged skills over several years were examined. Among the dynamical models GFDI performed better than GFSI, UKMI and NGPI for the 2008 season.

3.1.2 Mean Absolute Intensity Error

The mean absolute intensity errors for the 2004 season are shown in Fig. 1b. This shows that the multi-model superensemble had the lowest errors at all forecast hours. The corresponding intensity forecast errors in different models for the year 2005 are presented in Fig. 1d. The intensity forecast errors in the superensemble forecast during 2005 were lowest at some forecast lead times compared to the member models. For 72–108 h forecasts, DSHP (a statistical model) was better than the superensemble. For 120 h forecast, the intensity error was more in the superensemble than the ensemble mean of the member models. This may be due to the fact that the training period of the 2005 season was affected by model changes that took place during the 2005 hurricane season. The intensity forecast errors for 2006 hurricane season is shown in Fig. 1f. It shows that the intensity errors in the superensemble forecast were lowest for all forecast times except for 120-h forecasts where SHF5 provided the best results. Among the dynamic models GFDI performed reasonably well during the 2006 season. The FSU superensemble was able to correct the model biases and produce a consistent forecast compared to the member models and their ensemble mean.

Figure 2 shows several forecasted tracks of hurricane Charley from several models initialized every 6 h with the positions shown at 12-h intervals. The

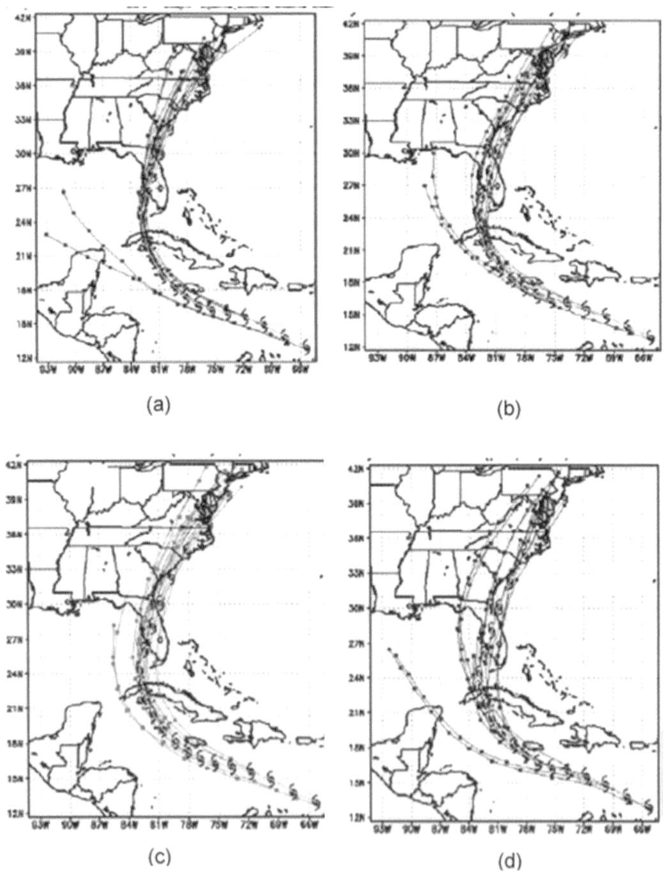

Fig. 2. Successive 6-hourly forecast tracks of hurricane Charley from (a) FSU superensemble, (b) OFCI, (c) GFSI and (d) UKMI models. The observed best track is shown by the yellow line with hurricane symbols representing positions plotted every 12 h.

collection of forecast tracks shown are from the superensemble, the OFCI, the best performing member model (GFSI) and the worst performing member model (UKMI). Baring two initial forecasts, the superensemble forecasts were consistent in predicting the track and landfall position of the storm. Left biases were clearly seen in the tracks obtained from OFCI, GFSI and the UKMI forecasts which were reduced in the superensemble forecast. Figure 3 presents several forecasted tracks of hurricane Frances from several models initialized every 6 h. The forecasted tracks shown are from the superensemble, the OFCI, the best performing member model (GUNA) and the worst performing member model (UKMI). A large right bias in forecasts was noted for several member models. Ensemble averaging, such as those implicit in GUNA, does not remove a consistent right bias in the tracks. The superensemble recognizes such features of forecasts from consistent errors during its training phase and does correct it to a large extent. The superensemble forecast was able to reduce the right bias by removing the individual biases of the member models.

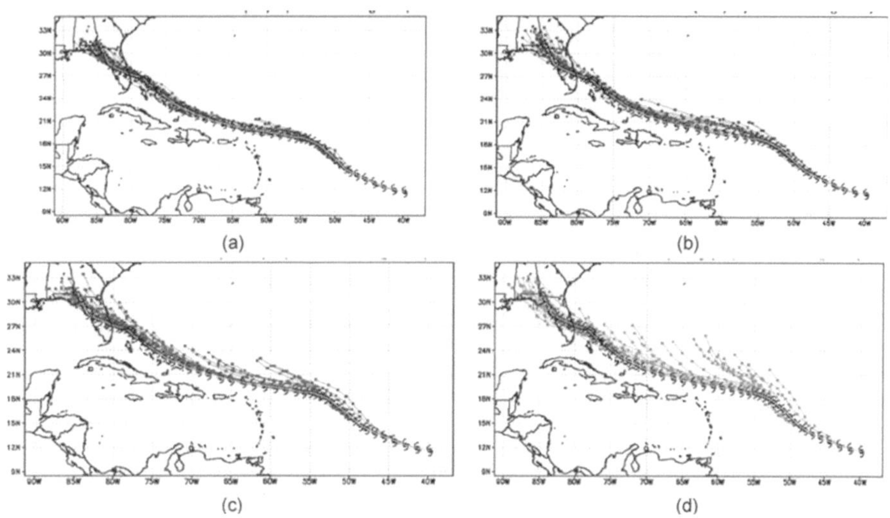

Fig. 3. Successive 6-hourly forecast tracks of hurricane Frances from (a) FSU superensemble, (b) OFCI, (c) GUNA and (d) UKMI models. The observed best track is shown by the yellow line with hurricane symbols representing positions plotted every 12 h.

3.2 Pacific Ocean Typhoons

Kumar et al. (2003) used superensemble technique to provide track and intensity forecast of the tropical cyclones over the Pacific Ocean for the years 1998–2000. Some of the results from this study are used here to demonstrate the superiority of superensemble forecast over the forecast generated by the member models. The superensemble technique used in the above mentioned study is similar to that has been used for Atlantic hurricanes (Williford et al., 2003). Forecasts from the currently available operational model datasets on the tracks and intensities of Pacific storms were in this study. The datasets used here include track and intensity forecasts from the ECMWF, NCEP/EMC (AVN and MRF), the U.S. Navy (NOGAPS), UKMO and JMA (GSM and TYM). The errors in track forecast (in km) for the tropical cyclones over the entire Pacific Ocean from the member models, ensemble mean and superensemble for the years 1998, 1999, and 2000 are shown in Figs 4a–c, respectively. The intensity errors for the same are presented in Figs 5a–c. It can be noted that the skill from the superensemble is consistently higher compared to the member models and the ensemble mean.

The performances of the member models were found to be quite different from year to year. This is due to some major changes in the models. This

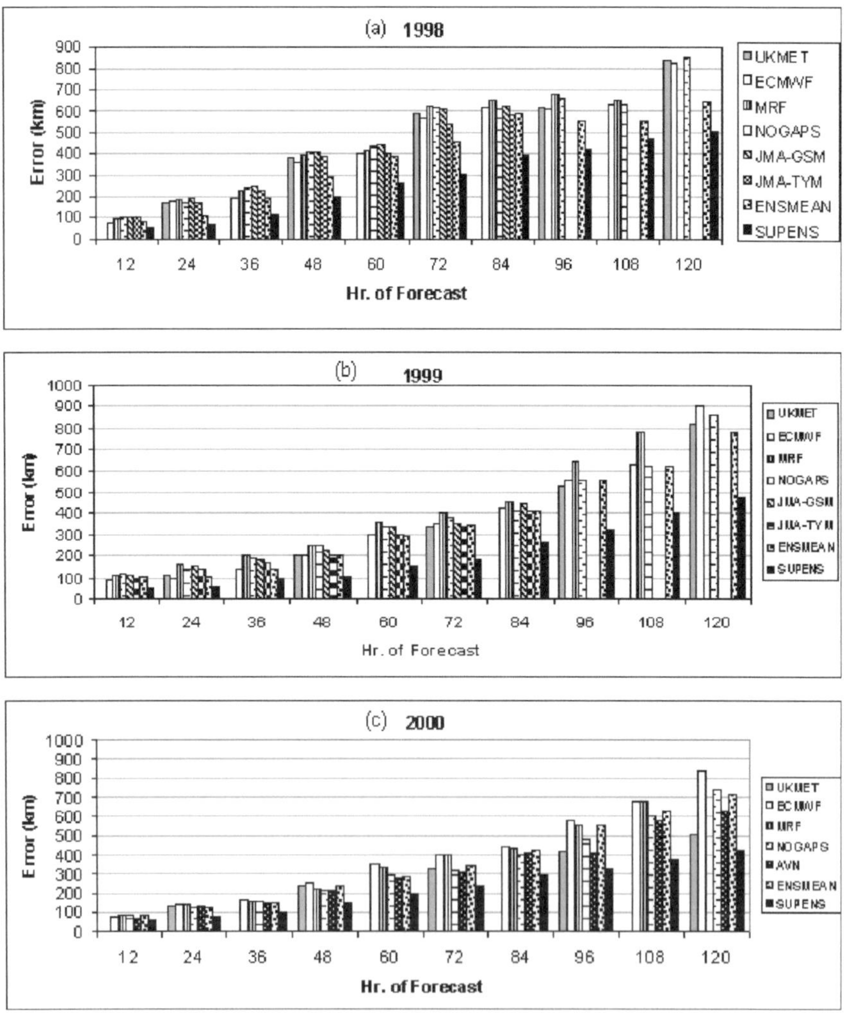

Fig. 4. Mean track errors (km) for tropical cyclones over the Pacific Ocean for (a) 1998, (b) 1999 and (c) 2000.

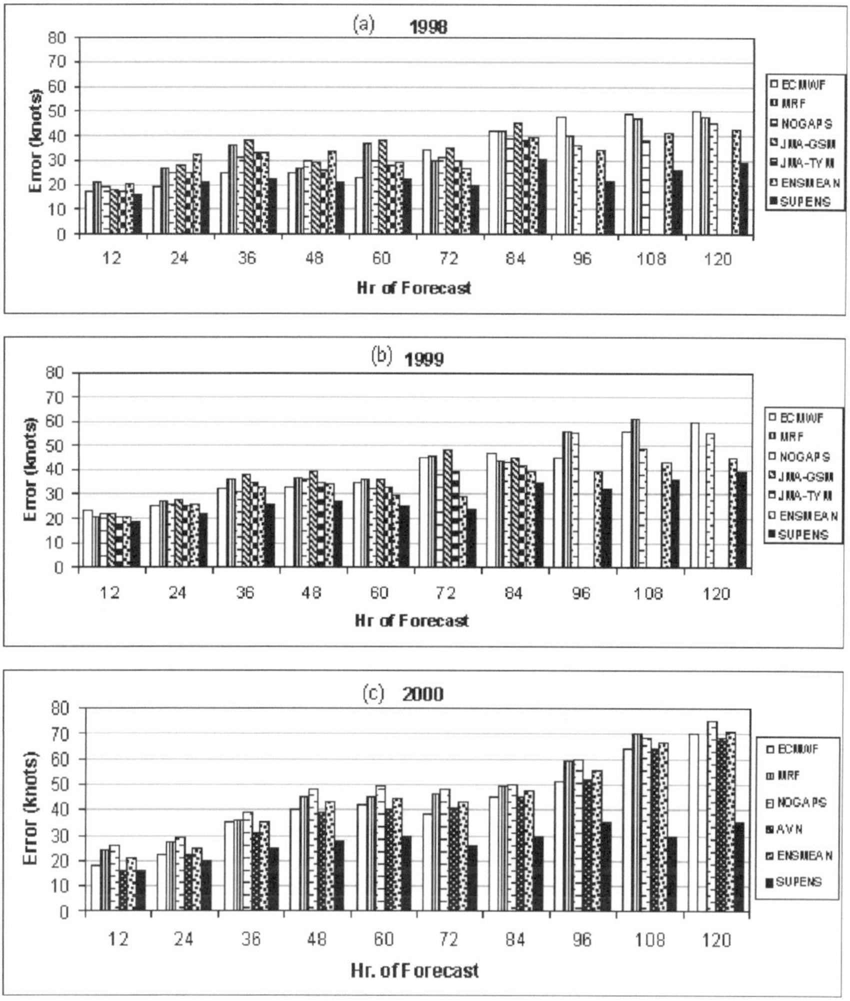

Fig. 5. Mean intensity errors (knots) for tropical cyclones over the Pacific Ocean for (a) 1998, (b) 1999 and (c) 2000.

seems to influence the performance of the superensemble over different years. Models such as ECMWF appear to perform very well after 2–3 days. That is related to an improved definition of the storm environment at scales larger than the tropical cyclone where the better models and the superensemble appear to do better. It is apparent that if no major model changes take place during the training and the forecast phases, the superensemble technique can improve the forecast skill on the order of 60, 140, 160 and 200 km for the tropical cyclone position errors over the best models in 24, 48, 72 and 96 hours forecast respectively. The intensity forecast skills of the superensemble can exceed those of the best models by about 5, 10, 13, and 20 knots, in 24, 48, 72 and 96 hours forecast respectively. The issue of intensity forecasts of the typhoons and hurricanes is a problematic area where further research is clearly warranted. It is however of interest here to note that the superensemble systematically pushes the skill toward a slight improvement over those of the best model and the ensemble mean. These are some of the important results from our study over the whole of the Pacific region. Independent basin statistics have also been generated and the results for each of the basins confirm the superior performance of the superensemble methodology as revealed by its lesser track and intensity errors compared to member models and the ensemble mean.

Results from the superensemble track and intensity forecasts for the west Pacific typhoons for the years 1998–2000 are presented in Figs 6a and 6b. Here the superensemble is constructed using only the multi-model forecasts of typhoons over the west Pacific for each year separately. The results shown in this figure compose independent forecast phases of the superensemble for each year of study. In Fig. 6a the position errors at different times of the

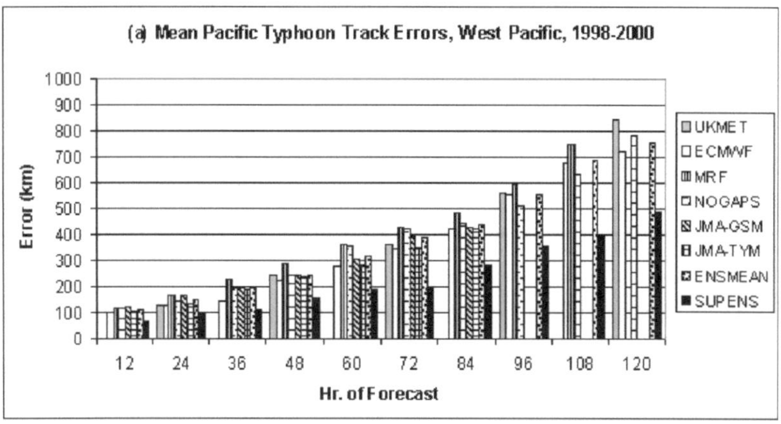

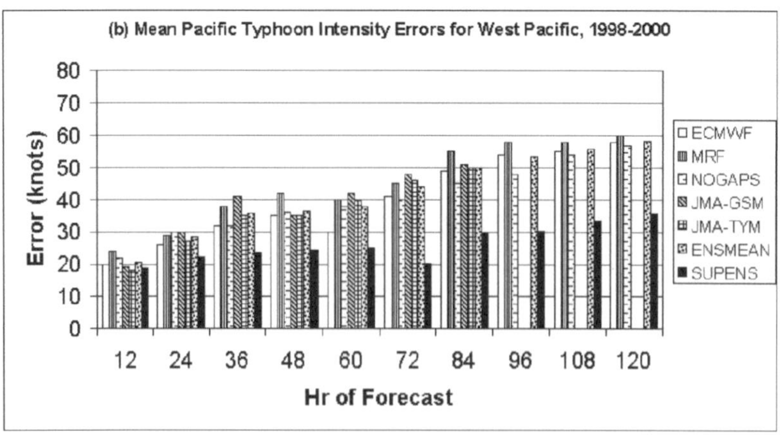

Fig. 6. (a) Mean typhoon track errors (km) for the west Pacific, 1998–2000. (b) Mean typhoon intensity errors (km) for the west Pacific, 1998–2000.

forecast for multi-models, the ensemble mean, and the superensemble are displayed. The intensity errors are shown in Fig. 6b. One should note that the results in Fig. 6 depict the average values for all three years of study, though the superensemble is constructed separately for each year using storms from an earlier part of each year in the training phase and the rest of the season in the forecast phase. The results from tropical cyclone forecasts over the other basins, east Pacific and South Pacific (not shown here), are similar to the west Pacific and the total Pacific. The error trends were similar for member models and the superensemble outperforms them and the ensemble mean for all basins. The magnitudes of error for member models and for the superensemble were slightly higher in the South Pacific. It is clear from these illustrations that there were no basic differences in the results presented here whether they are applied to the entire Pacific Ocean or to a portion thereof.

3.3 The Bay of Bengal Cyclones

There are very limited efforts in superensemble prediction of Bay of Bengal cyclones. Kumar et al (2007) made an attempt to provide superensemble forecast of Bay of Bengal cyclones using a mixture of global, regional and mesoscale models. In this section, some of their results are presented to provide a brief account of the possible improvement in prediction of Bay of Bengal cyclones using superensemble technique. Since no training data base for several past tropical cyclones from a number of models were available for this basin, they first generated such a data base using different versions of the FSU global and regional spectral models and the PSU-NCAR mesoscale model MM5. The model considered for this are: FSU global spectral model with Kuo convection scheme (FSUGK), FSU global spectral model with Arakawa-Schubert convection scheme (FSUGA), FSU regional spectral model with Kuo convection scheme (FSURK), FSU regional spectral model with Arakawa-Schubert convection scheme (FSURA), MM5 model with Grell convection scheme (MM5A) and MM5 model with Kain-Fritsch convection scheme (MM5B). A total of 77 forecast experiments were conducted for as many as 17 tropical cyclones over the Bay of Bengal during 1996–2000. About 70 of these forecasts were used in the training phase. The model coefficients were generated for each variable (latitude, longitude and intensity) at every 12 h up to 72 h of the forecasts. About eight storm days comprising the November 1996 Andhra Pradesh cyclone and October 1999 Orissa super cyclone were subjected to the forecast phase of the superensemble.

The mean track and intensity errors for the forecast phase of the superensemble are presented in Figs 7a and b. It was observed that the superensemble forecast has less than 300 km error in 72-h forecast and is about 60 km less than the best model (FSURA). The intensity errors are also least in the superensemble forecast throughout at the end of 24-, 48- and 72-h compared to the ensemble mean and the member models. The individual model forecasts, their ensemble mean, and the superensemble forecasts for the

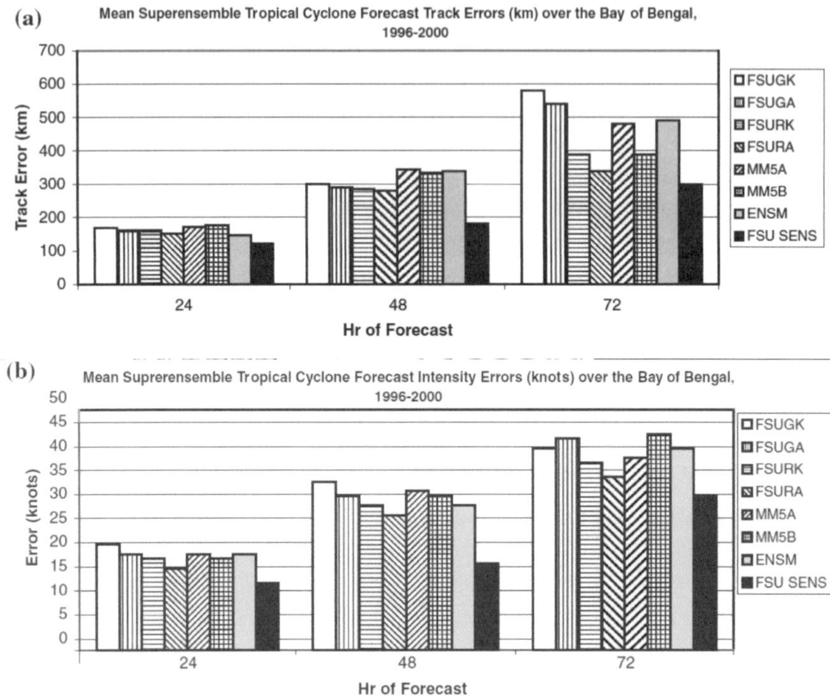

Fig. 7. (a) Mean track error (km) and (b) Mean intensity error (knots) for tropical cyclone forecasts over the Bay of Bengal for the years 1996–2000.

track of the 1996 Andhra Pradesh cyclone and the 1999 Orissa super cyclone are presented in Figs 8a and b, respectively. The superensemble tracks are more close to the observed tracks in both the cases, and possess least track error compared to other member models. The member models have shown a large northward bias for the 1996 Andhra Pradesh cyclone, which was successfully corrected by the superensemble. In case of 1999 Orissa super cyclone the member models have shown both southward and northward bias. It was found that the ensemble mean also have northward bias. Superensemble forecast has shown much better track that closely followed the observed path, though the landfall time and location has not been well predicted with the superensemble technique. Landfall location was about 60 km north of the actual landfall location and was 12 hours advance. The mean track errors in superensemble forecast of the Bay of Bengal cyclones are of the order of 110, 190 and 295 km, respectively, in 24, 48 and 72 h, which are about 20, 95 and 55 km improvement over the best model. The corresponding mean intensity errors are 14, 18 and 32 knots, showing an improvement of about 3, 9 and 4 knots over the best model.

It may be important to note that though the quantitative reduction in intensity error was not possible as a result of the superensemble technique, it was however possible to improve the intensity forecast by improving

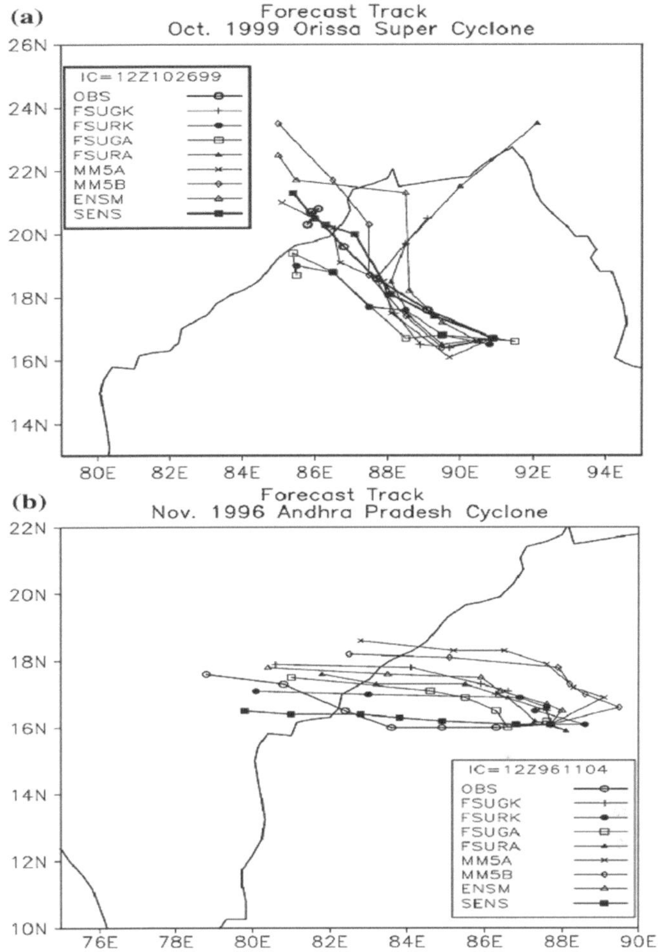

Fig. 8. (a) Multi-model and superensemble forecasts for October 1999 Orissa Super Cyclone. (b) Multi-model and superensemble forecasts for November 1996 Andhra Pradesh Cyclone.

the intensity forecast of the member models. That alone contributed to the improvements for the multimodel superensemble. The magnitudes of the errors in track and intensity forecast may further decrease by expanding the training data base and adding more diverse member models.

4. Mesoscale Models

Application of superensemble technique in regional model for track and intensity forecast of the tropical cyclones are discussed in this section. Here results for experimental forecast for 58 Atlantic storms (Table-1) spread across three years (i.e. 2004-2006) using six models WRF (two versions),

MM5, HWRF, GFDL and DSHP are considered (Krishnamurti et al., 2010). The forecast duration is extended up to 72 hours. Finally skill comparison has been done between mesoscale and global suite of model (GFS, NOGAPS,

Table 1. Name, duration and domains of tropical storms

Year	*Storm name*	*IC for 72 h FCST (00/12 UTC)*	*Domain*
2006	Chris	1 Aug	55W–75W, 10N–30N
	Debby	22 Aug	20W–50W, 5N–35N
	Ernesto	27 Aug	65W–85W, 10N–30N
	Gordon	13 Sep	45W–65W, 15N–40N
	Helene	17 Sep	40W–60W, 15N–35N
	Isaac	30 Sep	50W–70W, 25N–50N
2005	Dennis	7 July	65W–95W, 10N–35N
	Emily	15 July	60W–90W, 5N–30N
	Harvey	3 Aug	50W–70W, 20N–40N
	Irene	13 Aug	60W–80W, 20N–40N
	Katrina	27 Aug	75W–95W, 20N–40N
	Maria	4 Sept	45W–65W, 20N–40N
	Nate	6 Sept	55W–70W, 20N–40N
	Ophelia	8 Sept	65W–85W, 25N–45N
	Philippe	18 Sept	50W–70W, 10N–35N
	Rita	21 Sept	80W–100W, 20N–35N
	Stan	02 Oct	80W–100W, 10N–25N
	Wilma	22 Oct	70W–90W, 15N–35N
2004	Charley	10 Aug	60W–87W, 5N–22N
	Danielle	15 Aug	25W–45W, 10N–35N
	Frances	1 Sept	65W–85W, 15N–40N
	Frances	2 Sept	65W–85W, 15N–40N
	Gaston	28 Aug	65W–85W, 25N–45N
	Hermine	30 Aug	60W–75W, 30N–50N
	Ivan	12 Sept	75W–95W, 15N–35N
	Ivan	13 Sept	75W–95W, 15N–35N
	Jeanne	24 Sept	65W–85W, 20N–35N
	Karl	18 Sept	30W–55W, 10N–30N
	Lisa	25 Sep	35W–55W, 10N–30N

Note: All storm cases are run for 0000 and 1200 UTC.

UKMet) results. The bias corrected ensemble mean (BCEM) is a label that is used to identify the superensemble in this section.

4.1 Intensity and Track Forecast

Figure 9 (a, b) shows the absolute track errors (km) and the absolute errors for intensity (m s^{-1}) for forecasts of all hurricanes of the year 2004, 2005 and 2006 as shown in Table 1 for day 1 forecast (24-hour). This figure also includes the error distributions of each member model, the ensemble mean and the bias corrected ensemble mean of the mesoscale suite. These illustrations also include the models that are included in Table 2 and the mean error for each

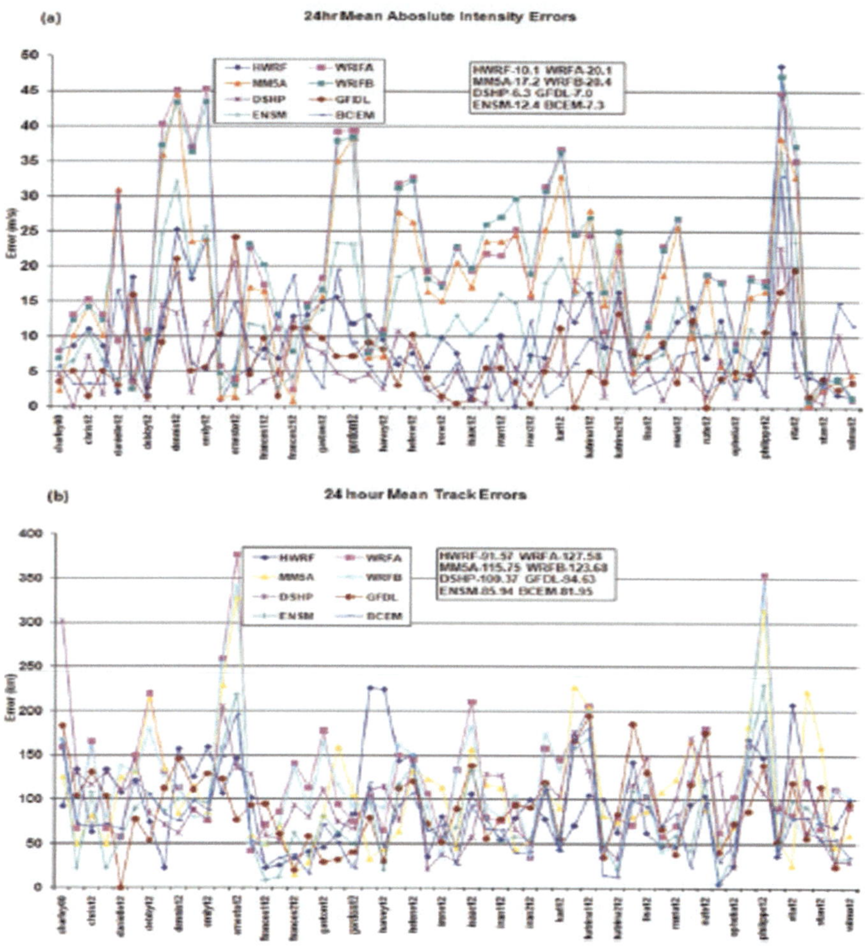

Fig. 9. Twenty-four hour mean absolute errors for member models, ensemble mean and bias corrected ensemble mean for (a) intensity (m s^{-1}) and (b) track (km). In the inset the average errors for 24 h forecast are shown for member models, ensemble mean and bias corrected ensemble mean.The storm names are shown in the x-axis followed by the initial time. The storms which have two initial dates as shown in table is named as <storm_name>2<XX> where XX is the initial time.

Table 2: List of real time forecast models

	Model (Track)	*Model (Intensity)*
1.	HWRF (H212) (M)	HWRF (H212) (M)
2.	COTC (M)	COTC (M)
3.	AHW4 (M)	AHW4 (M)
4.	ARFS (M)	ARFS (M)
5..	SPC3 (S)	SHIPS (S)
6.	FIM9 (L)	SPC3 (S)
7.	UWN8 (M)	UWN8 (M)
8.	GFDL (M)	GFDL (M)
9.	NGPS (L)	FIM9 (L)
10.	ECMF (only for training data from 2011) (L)	NGPS (L)
11.	AVNI (L)	AVNI (L)

HWRF – Hurricane Model Forecast Track/Intensity
COTC – U.S. Navy COAMPS-TC Model Forecast
SPC3 – Statistical Prediction of Intensity Forecast (6 members)
FIM9 – Finite-Volume Icosahedral Model Forecast
GFDL – GFDL Hurricane Model Forecast Track/Intensity
NGPS – U.S. Navy NOGAPS Model Forecast
ECMF – ECMWF Model Forecast
AVNI – Previous GFS Forecast Interpolated Ahead 6 Hours
AHW4 – SUNY Advanced Hurricane WRF Forecast Track/Intensity
SHIPS – Statistical Hurricane Intensity Prediction Scheme
UNW8 – University of Wisconsin Non-hydrostatic Research Model
ARFS – FSU WRF Forecast Track/Intensity (Thomas and Anu: not since 2011)

model. It is found that the bias corrected ensemble mean (BCEM) clearly provides one of the better forecasts compared to the member models in a consistent manner. Figure 9(a, b) indicates that most models do well for some storms and do poorly for others. In addition this study also suggest that there is a strong initial state dependence that leads to model errors The clustering of the model errors is an important issue for ensemble forecasts. As most of these mesoscale models use GFS model to provide their initial conditions, therefore clustering of forecast tracks can occur, this may partly be attributed to the use of same initial conditions.

It is also noted large errors in the intensity forecast (35-45 ms^{-1}) at 24 hours for storms Dennis, Gaston, Karl of 2004 and Philipe of 2005. The intensity errors for the same storms for DSHP's is in the range of 5-15 ms^{-1} and those of biased corrected ensemble mean were between 10-15 ms^{-1}. It is also found that the large contributions of errors are generated from WRFA and WRFB models. The range of errors in HWRF model is generally lower than those other models in the suite. The intensity errors for these storms at 48 hours are in the range of 30 ms^{-1} for these storms (Figure not shown). Similarly, large intensity errors are noted during 72 hours forecast (Figure not shown). More research efforts are required to understand the behaviour of a specific model which forecasts a specific storm.

The differences in track errors among model for 24 (Fig. 9) shows a few well-marked spikes. For 24 h forecasts Emily of 2004 and Ophelia (2005) carried large errors of the order of 350–375 km. By 48 h as many as six forecasts had errors in excess of 400 km (Figure not shown). There were seven forecasts by hour 72 that had errors in excess of 600 km. These forecast errors describe the nature of the performance of the mesoscale suite of models that were used. It was noted that a consistent pattern of errors (from storm to storm) is reduced by the bias correction proposed here. The bias corrected ensemble mean reduces the track errors significantly.

4.2 Mean Errors (Intensity and Track)

Figure 10 illustrates the mean errors for intensity and track forecast for the storms (Table 1). The bias corrected ensemble mean carries the least errors from 12 to 72 hour forecast. It has less track error compared to any individual model as well as ensemble mean forecasts. Again HWRF provides

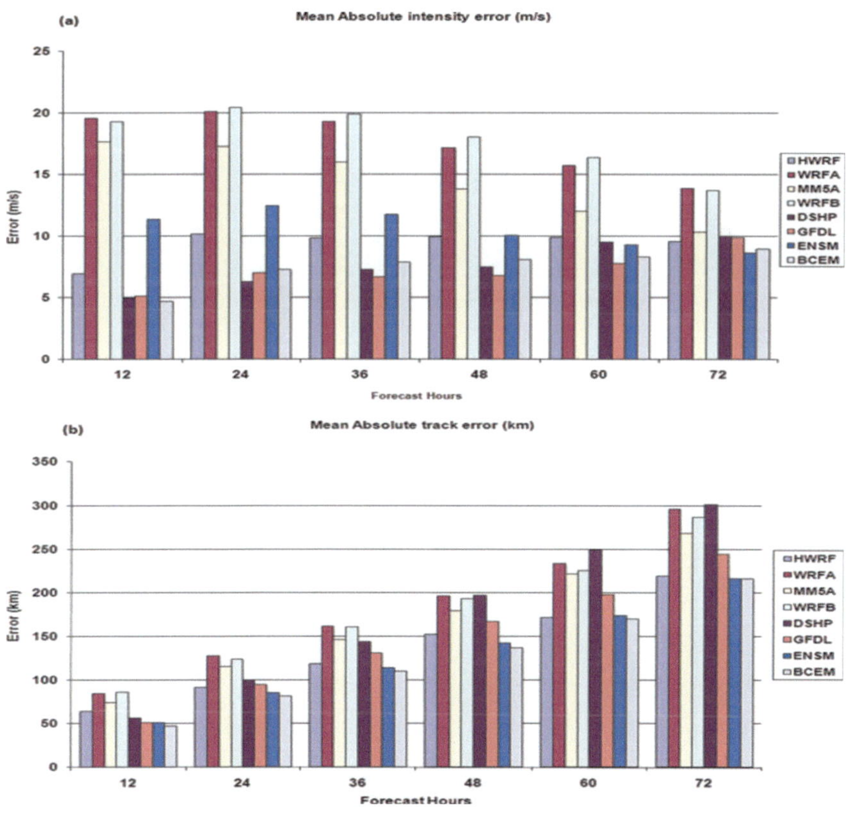

Fig. 10. Mean absolute errors for (a) intensity (in m s^{-1}) and (b) track (in km) for member models, ensemble mean and bias corrected ensemble mean of the mesoscale suite of models.

some of the best track forecasts among the member mesoscale models. The improvement in the bias corrected ensemble forecast is roughly 10-20 kms. For intensity forecast biased corrected ensemble mean compares close to the best models (DSHP and GFDL). The DSHP, GFDL, ensemble mean and bias corrected ensemble mean provide best guidance for intensity and the differences in errors of intensity forecasts for these models were within 1-2 ms^{-1}.

Figure 11 shows the summary of track errors for all of the hurricanes of the year 2004, 2005 and 2006. The bias corrected ensemble mean shown as the last bar for these forecasts from 12 to 72 h carries the least error among all of the mesoscale member models and the ensemble mean for track forecasts. The HWRF provides some of the best track forecasts among the member models of the mesoscale suite. Overall the bias corrected ensemble mean improves the forecast performance for tracks by roughly 10–20 km.

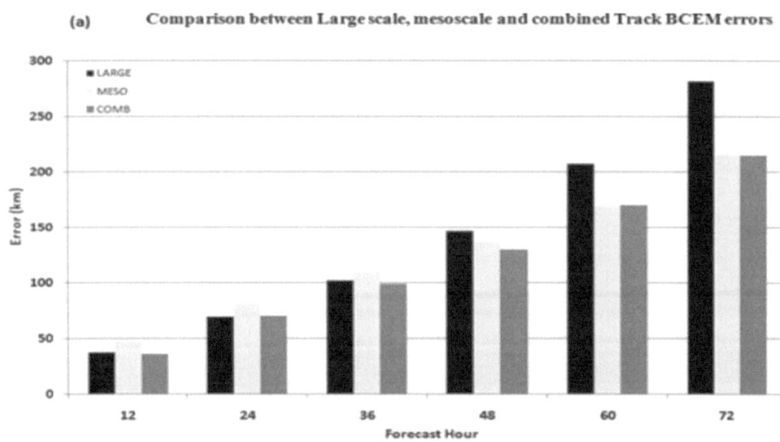

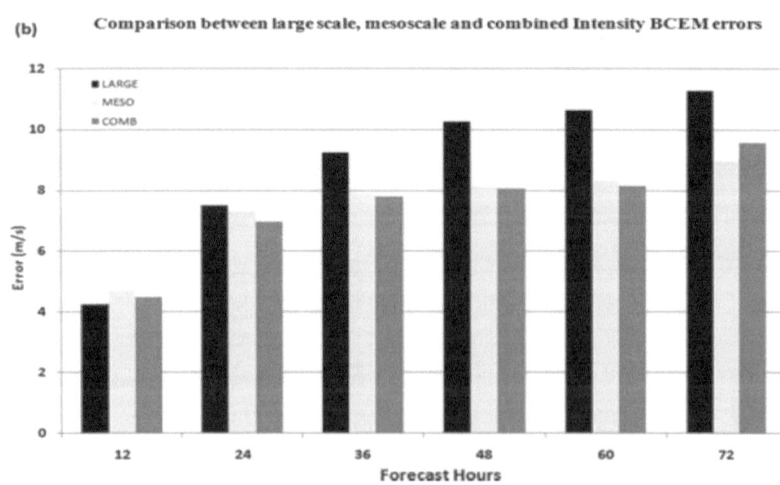

Fig. 11. Comparison of (a) track errors (km) and (b) intensity errors (m s^{-1}) between bias corrected large scale, mesoscale and using both large and mesoscale suite of models.

4.3 Comparison of Above Results with Those of a Global Suite of Models

The global suite of models consists of Global Forecasting System (GFS) of National Center for Environment Prediction (NCEP), the Navy Operational Global Prediction System (NOGAPS) and United Kingdom Met Office model (UKMet). The 72-hour forecast suggested that intensity forecast from the mesoscale suite carried lower absolute errors (except 12-hour) compared to large scale suite and overall intensity errors are reduced by as much as 2 ms^{-1} by 72 hours. This improvement in intensity error can happen because of high resolution mesoscale models. The mesoscale suite of models provide better forecasts for storm tracks compared to the large scale models for hours 48, 60 and 72 indicating that these models have large initial errors due to the spin up issue. In general it is found that skills of two suite of model are comparable at first 48h and at 72 hours mesoscale suite track is roughly 80 km less than that of the large scale models.

5. Summary

This paper reviews the application of multi model superensemble procedure to improve tropical cyclone track and intensity forecast up to day 5 (120 hours) using a number of global as well as mesoscale suite of model forecasts. In general, it is found that superensemble forecast has the highest skill both in terms of wind speed (intensity) and track in all forecast hours starting from 24 to 120 hours. Most of the results suggest that individual member model has more or less same error characteristics; however, superensemble consistently outperforms these individual model skills as well as their ensemble mean. For mesoscale suite of experiments, it is found that most of these models are having spin up issues for initial hours of forecast (i.e. 24 hours). The nature of that spin up varies from model to model; however past that spin up period, the superensemble skills improve significantly. Finally, additional research is required to improve some of the limitations (i.e. training data sets, model biases, choice of models) in order to make it more robust and reliable.

REFERENCES

Franklin, J.L., 2006: 2005 National Hurricane Center Forecast Verification Report, National Hurricane Center, Miami, FL (http://www.nhc.noaa.gov/verification/pdfs/Verification_2005.pdf).

Goerss, J.S., 2000: Tropical cyclone track forecasts using an ensemble of dynamical models. *Mon. Wea. Rev.*, **128**, 1187-1193.

Krishnamurti, T.N., M.K. Biswas, B.P. Mackey, R.G. Ellingson and P. Ruscher, 2011: Hurricane forecasts using a suite of large-scale models. *Tellus A*, **63(4)**, 727–745.

Krishnamurti. T.N., C.M. Kishtawal, Timothy E. LaRow, David R. Bachiochi, Zhan Zhang, C. Eric Williford, Sulochana Gadgil and Sajani Surendran, 1999: Improved Weather and Seasonal Climate Forecasts from Multimodel Superensemble. *Science*, **285(5433)**, 1548-1550.

Krishnamurti, T.N., C.M. Kishtawal, D.W. Shin and C. Eric Williford, 2000a: Improving Tropical Precipitation Forecasts from a Multianalysis Superensemble. *J. Climate*, **13**, 4217-4227.

Krishnamurti. T.N., C.M. Kishtawal and Z. Zhang, 2000b: Multimodel ensemble forecasts for weather and seasonal climate. *Journal of Climate*, **13**, 4196-4216.

Krishnamurti, T.N., S. Pattnaik, M.K. Biswas, M. Kramer, Ed Bensman, N. Surgi and T.S.V. Kumar, 2010: Multimodel ensemble forecasts of hurricane for a suite of mesoscale models. *Tellus*, **62A**, 633-646.

Kumar, T.S.V.V., T.N. Krishnamurti, M. Fiorino and M. Nagata, 2003: Multimodel Superensemble Forecasting of Tropical Cyclones in the Pacific. *Mon. Wea. Rev.*, **131**, 574-583.

Logutov, O.G. and A.R. Robinson, 2005: Multi-model fusion and error parameter estimation. *Q. J. Roy. Meteorol. Soc.*, **131**, 3397-3408.

Shin, D. and T. Krishnamurti, 2003a: Short- to medium-range superensemble precipitation forecasts using satellite products: 1. Deterministic forecasting. *Journal of Geophysical Research*, **108(8)**, 8383.

Shin, D. and T. Krishnamurti, 2003b: Short- to medium-range superensemble precipitation forecasts using satellite products: 2. Probabilistic forecasting. *Journal of Geophysical Research*, **108(8)**, 8384.

Rixen, M., J.W. Book, A. Carta, V. Grandi, L. Gualdesi, R. Stoner, P. Ranelli, A. Cavanna, P. Zanasca, G. Baldasserini, A. Trangeled, C. Lewis, C. Trees, R. Grasso, S. Giannechini, A. Fabiani, D. Merani, A. Berni, M. Leonard, P. Martin, C. Rowley, M. Hulbert, A. Quaid, W. Goode, R. Preller, N. Pinardi, P. Oddo, A. Guarnieri, J. Chiggiato, S. Carniel, A. Russo, M. Tudor, F. Lenartz and L. Vandenbulcke, 2009: Improved ocean prediction skill and reduced uncertainty in the coastal region from multi-model super-ensembles. *J. Mar. Syst.*, **78**, S282–S289, doi:DOI:10.1016/j.jmarsys.2009.01.014.

Rixen, M. and E. Ferreira-Coelho, 2007: Operational surface drift prediction using linear and non linear hyper-ensemble statistics on atmospheric and ocean models. *J. Mar. Syst.*, **65**, 105-121.

Vandenbulcke, L., J.-M. Beckers, F. Lenartz, A. Barth, P.-M. Poulain, M. Aidonidis, J. Meyrat, F. Ardhuin, M. Tonani, C. Fratianni, L. Torrisi, D. Pallela, J. Chiggiato, M. Tudor, J. Book, P. Martin, G. Peggion and M. Rixen, 2009: Super-ensemble techniques: Application to surface drift prediction. *Progr. Oceanogr.*, **82**, 149-167, doi:10.1016/j.pocean.2009.06.002,25 2009.

Williford, C.E., T.N. Krishnamurti, R.C. Torres, S. Cocke, Z. Christidis and T.S.V. Vijaya Kumar, 2003: Real-time multimodel Superensemble forecasts of Atlantic tropical systems of 1999. *Mon. Wea. Rev.*, **131**, 1878-1894.

Wilks, D.S., 1995: Statistical methods in the atmospheric sciences. Second edition. International Geophysics Series, Vol. 59, Academic Press. ISBN-10: 0127519653. ISBN-13: 978-0127519654.

Yun, W.T., L. Stefanova, A.K. Mitra, T.S.V. Vijay Kumar, W. Dewar and T.N. Krishnamurti, 2005: Multi-Model Synthetic Superensemble Algorithm for Seasonal Climate Prediction using DEMETER Forecasts. *Tellus*, **57**, 280-289.

20

Advanced Diagnostics for the HWRF Hurricane Modelling System

Thiago Quirino and Sundararaman G. Gopalakrishnan
NOAA/Atlantic Oceanographic and Meteorological Laboratory/
Hurricane Research Division, Miami, FL, USA
thiago.quirino@noaa.gov

1. Overview

The objective of this work is to show how to derive some useful first-order model diagnostics from the prognostic variables of numerical weather prediction models. We introduce a basic set of popular diagnostics products, along with illustrations and detailed formulae, that have been adopted into various real-time applications of both tropical cyclone forecasting and hurricane model diagnostics. The diagnostic products discussed in this chapter are categorized into two types: (1) Plots of horizontal 2D fields, and (2) Time vs. height profiles of domain-averaged forecasted quantities. The diagnostic products in each of these categories provide insight into various different factors that play important roles in the time-evolution of tropical cyclones.

Our inspiration for writing this chapter comes from having observed first-hand the utility of these diagnostic products in our own applications geared towards the development of the latest Operational HWRF model. During the active hurricane season, these products are generated in real-time for HWRF forecasts of storms residing on both the Atlantic and East-Pacific basin. The resulting graphics are then transferred to our website, the HWRF Portal, which is subsequently accessed by various researchers across the country and forecast specialists from both the National Hurricane Center (NHC) and Weather Forecasting Offices (WFO) across Florida. The HWRF Portal website is located at http://storm.aoml.noaa.gov/realtime.

Figure 1 illustrates the main interface of the HWRF Portal website. The website adopts a simple "storm-centered" product search approach. In this approach, the user first selects a storm case of interest from a list that includes both active (i.e. real-time) and retrospective cases. Next, the user selects an available configuration of the HWRF model. Finally, the user visualizes

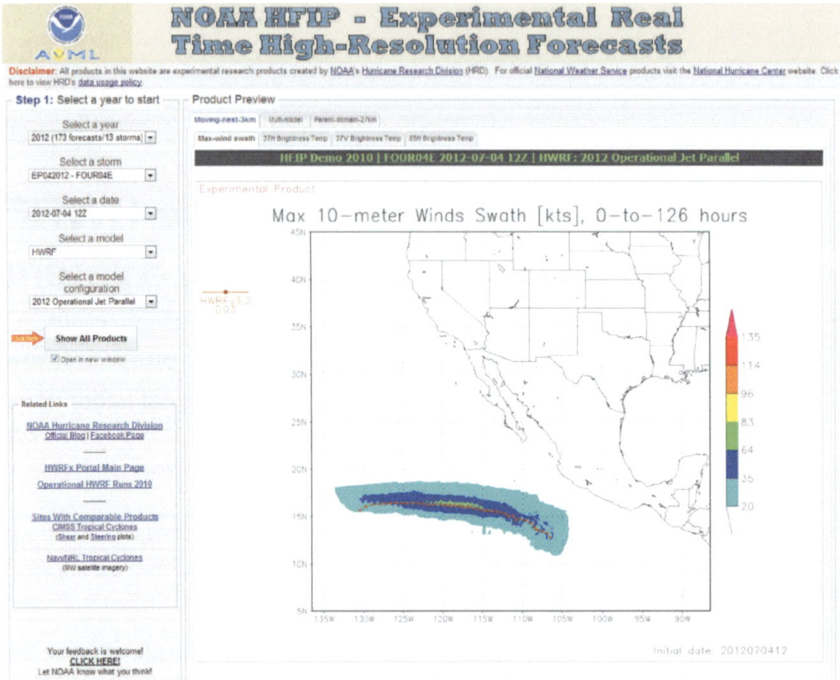

Fig. 1. The HWRF Portal website (http://storm.aoml.noaa.gov/realtime) displays various diagnostic products from real-time forecasts of the HWRF model using a *storm centered* approach to model diagnostic product search.

diagnostic products of interest for the specified forecast. In addition to hosting Operational HWRF products, the website also hosts a multitude of products derived from forecasts of our research versions of HWRF, products which are also generated in real-time as part of our participation in NOAA's Hurricane Forecast Improvement Program (HFIP) effort.

In the following sections we will discuss and derive the basic formulae of various basic diagnostic products. As we do so, we will also demonstrate the utility of each product to forecasting and model diagnostic applications through graphical illustrations. These illustrations were derived from a retrospective forecast of Hurricane EARL07L (2010) that was performed by the 2012 Operational HWRF model starting on 2010-08-26 18Z. We chose Hurricane EARL07L due to its ongoing importance as a representative case of *Rapid Intensification* (RI) over North Atlantic waters. The importance of Hurricane EARL07L is further augmented by the unprecedented number of flight-level, radar, and dropsonde observations that were collected through various NOAA and Air Force missions that took place during EARL07L's lifetime. These data sets promise to provide a vast amount of knowledge regarding the intensification processes of tropical cyclones for years to come. Furthermore, one of the reasons for choosing the HWRF forecast of EARL07L starting at 2010-08-26 18Z was due to the quality of the resulting track and intensity forecasts. Another reason was that the 5-day window covered by this specific forecast captured various crucial stages of EARL07L's development, including the initial weak stage, the RI period, and the potent Category 4 steady state that threatened extreme damage to the US east coast.

In the next section we will discuss a variety of horizontal 2D products derived from basic prognostic model fields. Then, in the subsequent section we will discuss diagnostic products based on *time vs height* profiles and their utility in the analysis of hurricane inner core processes.

2. Horizontal 2D Fields

In this section we discuss the derivation of basic model diagnostic products that correspond to 2D slices made along surfaces of either constant pressure or height across 3D model data fields. The products discussed here provide basic insight into (1) how favourable is the large-scale environment to the development of a tropical cyclone (i.e., deep layer vertical shear), (2) in which direction is the tropical cyclone being pushed toward by the large-scale environment flow (i.e., deep layer steering flow), and (3) what does the structure of a tropical cyclone, which is dynamically and thermo-dynamically driven, looks like along different vertical levels of the atmosphere (i.e., wind speed, vorticity, divergence, and inner-core temperature anomaly). To emphasize the relevance of the products discussed in this section to the analysis of the inner-core processes of tropical cyclones, most of the illustrations are focused on the inner-core region of Hurricane EARL07L. However, the products discussed in this section are useful for the analysis of both the large-scale environment as well as the inner-core of tropical cyclones.

2.1 Wind Speed

The horizontal wind speed is one of the most basic and informative diagnostic product, being generally computed for various different levels in the atmosphere. It is also one of the simplest products to derive mathematically through the application of the Pythagorean Theorem on the prognostic U (meridional wind) and V (zonal wind) variables. A plot of the wind speed (and wind vector flow) provides insight into both the dynamical and thermo-dynamical structure of a tropical cyclone by indicating both (1) where the centre of the circulation is located at different levels and (2) where the strongest winds are located relative to the centre of the circulation. Figure 2 illustrates these concepts by plotting the HWRF forecasted 10-metre wind speeds for EARL07L (shaded contour). The forecast start time is 2010-08-26 18Z.

The plot contrasts the strength and distribution of the 10-metre wind speed field at both the start and end of EARL07L's RI period (forecast hours 60 and 102, respectively). By the end of the RI period, the overlayed wind vector flow clearly indicates a well-defined circulation. For the same case,

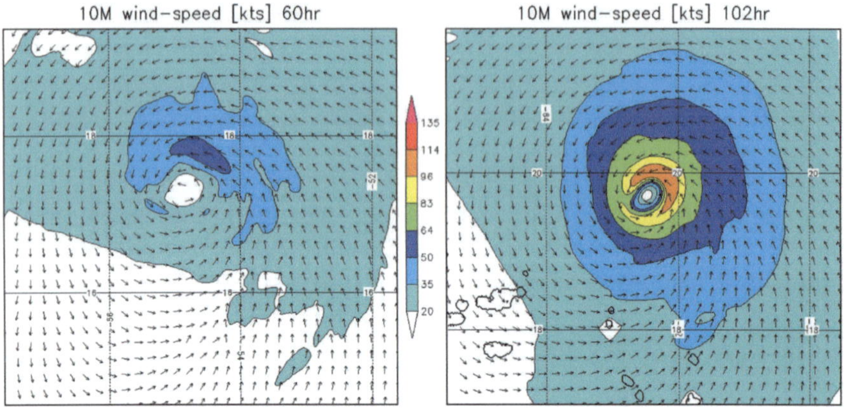

Fig. 2. HWRF forecasted 10-metre winds for EARL07L at both the start and end of the RI period. The forecast start time is 2010-08-26 18Z.

Fig. 3 compares the time evolution of the max 10-metre wind speed (in knots) for the HWRF and Decay SHIPS models versus the observed BEST.

Deriving the 10-metre wind speed for a specified forecast hour is simple, as is illustrated by the following pseudo-code. Assume that the prognostic U (meridional) and V (zonal) components of the 10-metre winds are available as 2D matrices of size M-rows by N-columns. Also, assume that U and V are

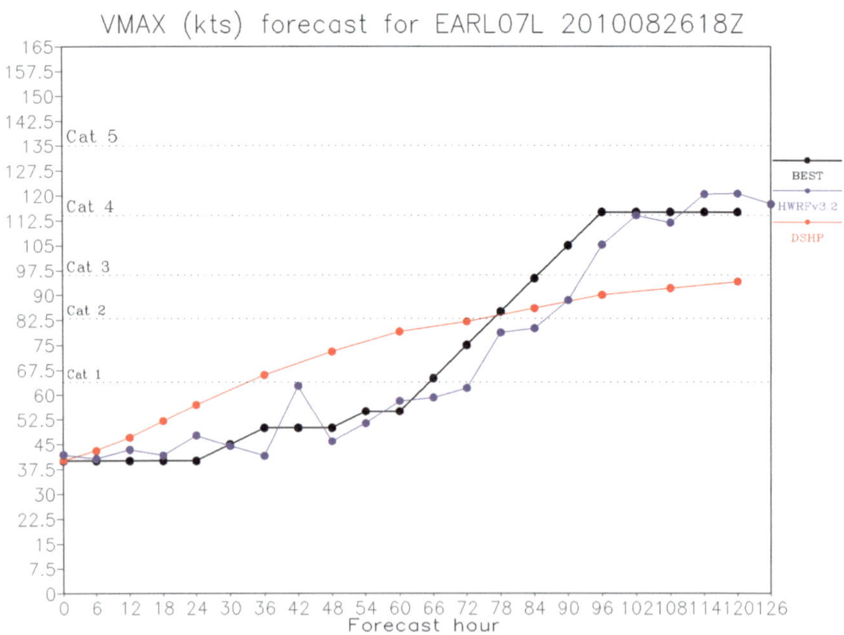

Fig. 3. Comparison of the max sustained 10-metre wind forecast for EARL07L by the HWRF and Decay SHIPS models versus the observed BEST intensity. The forecast start time is 2010-08-26 18Z.

projected onto the same rectangular grid. The wind speed S can be estimated as:

```
   Let U be a M × N matrix holding the meridional
components of the 10-metre wind.
   Let V be a M × N matrix holding the zonal components
of the 10-metre wind.
   Let S be a M × N matrix whose components are index
by the i-th row and j-th column.
   Loop, for i in 1 to M
      Loop, for j in 1 to N
            Let S(i,j) = (U(i,j)^2 + V(i,j)^2)^1/2
      End Loop
   End Loop
```

2.2 Relative Vorticity

Vorticity measures the tendency of a fluid to spin. In particular, horizontal *relative vorticity* indicates areas of locally cyclonic flow (i.e. positive in the CCW direction) within a fluid. Figure 4 illustrates the HWRF forecasted relative vorticity at the 850 mb level for EARL07L (shaded contour). The forecast start time is 2010-08-26 18Z. The plot contrasts the strength of the inner-core relative vorticity at both the start and end of EARL07L's RI period (forecast hours 60 and 102, respectively). At the end of the RI period, the inner-core region shows a well-defined centre of circulation and strong inner-core winds, which is reflected in the strong inner-core vorticity.

The following pseudo-code illustrates how horizontal relative vorticity is derived for an arbitrary forecast hour. Assume that the prognostic U

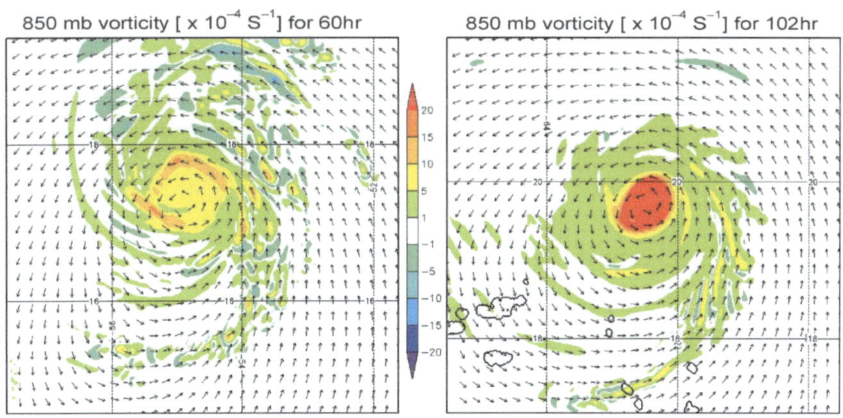

Fig. 4. HWRF forecasted 850 mb relative vorticity for EARL07L at both the start and end of the RI period. The vorticity of the inner-core region after RI is much higher, indicating a stronger and more well-defined inner-core circulation. The forecast start time is 2010-08-26 18Z.

(meridional) and *V* (zonal) components of wind are available for a specific horizontal level as 2D matrices of size *M*-rows by *N*-columns. Also, assume that *U* and *V* are projected onto an almost rectangular grid, as is often the case in models, with resolutions *dx* and *dy* along the *x* (i.e. longitude) and *y* (i.e. latitude) dimensions, respectively. The magnitude of the horizontal vorticity ζ can be estimated as follows:

```
   Let U be a M × N matrix holding the meridional
components of the wind at a horizontal level.
   Let V be a M × N matrix holding the zonal components
of the wind at a horizontal level.
   Let ζ be a M × N matrix whose components are index
by the i-th row and j-th column.
   Loop, for i in 1 to M
      IF(i = 1) Let top = i and bottom = i + 1
      ELSE IF(i = M) Let top = M - 1 and bottom = M
      ELSE Let top = i - 1 and bottom = i + 1
      Loop, for j in 1 to N
            IF( j = 1) Let left = j and right = j + 1
            ELSE IF(j = N) Let left = N - 1 and right = N
            ELSE Let left = j-1 and right = j + 1
            Let ζ(i,j) = (V(i,right) - V(i,left))/(2·dx) -
            (U(top,j) - U(bottom,j))/(2·dy)
      End Loop
   End Loop
```

2.3 Divergence

In hurricanes, divergence aloft is associated with heating, and consequently warming of the inner-core region. Thus, the strength of the upper-level divergence in a tropical cyclone is an important indicator of its intensification rate. Figure 5 illustrates the HWRF forecasted relative divergence at the 200 mb level for EARL07L (shaded contour). The forecast start time is 2010-08-26 18Z. The plot contrasts the strength and symmetry of the upper level divergence in the inner core region at both the start and end of EARL07L's RI period (forecast hours 60 and 102, respectively).

The following pseudo-code illustrates how horizontal divergence is derived for an arbitrary forecast hour. Assume that the prognostic *U* (meridional) and *V* (zonal) components of wind are available for a specific horizontal level as 2D matrices of size *M*-rows by *N*-columns. Also, assume that *U* and *V* are projected onto an almost rectangular grid, as is often the case in models, with resolutions *dx* and *dy* along the *x* (i.e. longitude) and *y* (i.e. latitude) dimensions, respectively. The magnitude of the horizontal divergence *D* can be estimated as follows:

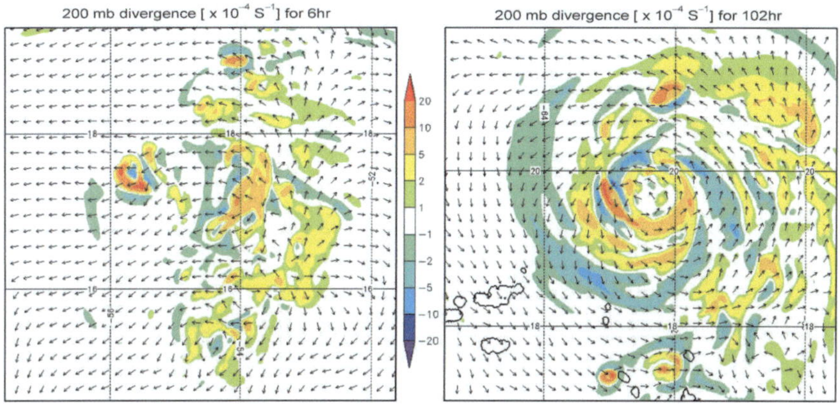

Fig. 5. HWRF forecasted 200 mb divergence for EARL07L at both the start and end of the RI period. The divergence of the inner-core region at the end of the RI period is more intense and symmetric around the storm eye when compared to the pre-RI period. The forecast start time is 2010-08-26 18Z.

```
   Let U be a M × N matrix holding the meridional
components of the wind at a horizontal level.
   Let V be a M × N matrix holding the zonal components
of the wind at a horizontal level.
   Let D be a M × N matrix whose components are index
by the i-th row and j-th column.
   Loop, for i in 1 to M
      IF(I = 1) Let top = i and bottom = i + 1
      ELSE IF(i = M) Let top = M - 1 and bottom = M
      ELSE Let top = i - 1 and bottom = i + 1
      Loop, for j in 1 to N
            IF(j = 1) Let left = j and right = j + 1
            ELSE IF(j = N) Let left = N - 1 and right = N
            ELSE Let left = j - 1 and right = j + 1
            Let D(i,j) = (U(i,right) - U(i,left)) / (2·dx) +
            (V(top,j) - V(bottom,j)) / (2·dy)
      End Loop
   End Loop
```

2.4 Inner-core Temperature Anomaly

Temperature anomaly is a measure of the relative warmth of the inner-core region of a tropical cyclone with respect to its surrounding environment. It is an important model diagnostic because it provides insight into the thermodynamical structure of the warm inner core region of tropical cyclones. Moreover, the time evolution of the warm core is associated with tropical cyclone intensification rate. Higher temperature anomalies in the inner-core region, especially aloft, translate into deeper storms with lower surface pressures, and consequently, also higher surface winds.

To compute temperature anomaly at a horizontal level of interest, it is necessary to first define a standard (or base) value of atmospheric air temperature measured at that vertical level. The base temperature represents the air temperature of the large scale environment surrounding the storm. The temperature anomaly is then computed by subtracting the chosen base air temperature value from the model forecasted air temperature at that level.

There are various ways to define the base air temperature. A common approach is to let the base air temperature vary for each forecast hour by computing it as the average temperature of the immediate environment surrounding a storm, i.e. anywhere within 300 km to 500 km from the storm centre, or within the bounds of a storm-following domain in models such as HWRF. Another approach involves using a constant base air temperature for all forecast hours that is defined as the average temperature of the immediate storm environment at the start of a forecast (i.e., forecast hour 0). This choice of base air temperature will reveal how the temperature of the inner-core evolves with respect to its initial conditions. A more general approach involves using the well-established air temperature values given in the US Standard Tropical Atmosphere Temperature table as the base air temperature. This is the approach we adopted to formulate our illustrations.

Figure 6 illustrates the HWRF forecasted temperature anomaly at both the 6 km and 10 km height levels for EARL07L (shaded contour). The forecast start time is at 2010-08-26 18Z. The plot illustrates the evolution of the warm inner-core region at both the start and end of the RI period. Particularly for this forecast, EARL07L's warmest inner-core temperature was found at the height of 10 km after the RI period. The base temperatures used for the 6 km level (249.15 Kelvin) and 10 km level (223.15 Kelvin) were taken from the US Standard Atmosphere Temperature table. Notice from the plots how the warm core evolves as a result of the RI process, with temperature changes of over 6 Kelvin. Overall, the entire storm region becomes on average warmer by at least 2 Kelvin.

The following pseudo-code shows how the temperature anomaly at a horizontal level can be easily derived for an arbitrary forecast hour. Assume that the prognostic variable T (air temperature) at a level is available as a 2D matrix of size M-rows by N-columns and that T is projected onto a rectangular grid. Also, assume that the base temperature at the corresponding level is t_b. The magnitude of the temperature anomaly T_A can be estimated as:

```
Let T be a M × N matrix holding the air temperature
at a horizontal level.
Let t_b be a scalar with the same units as T and
corresponding to the base temperature.
```

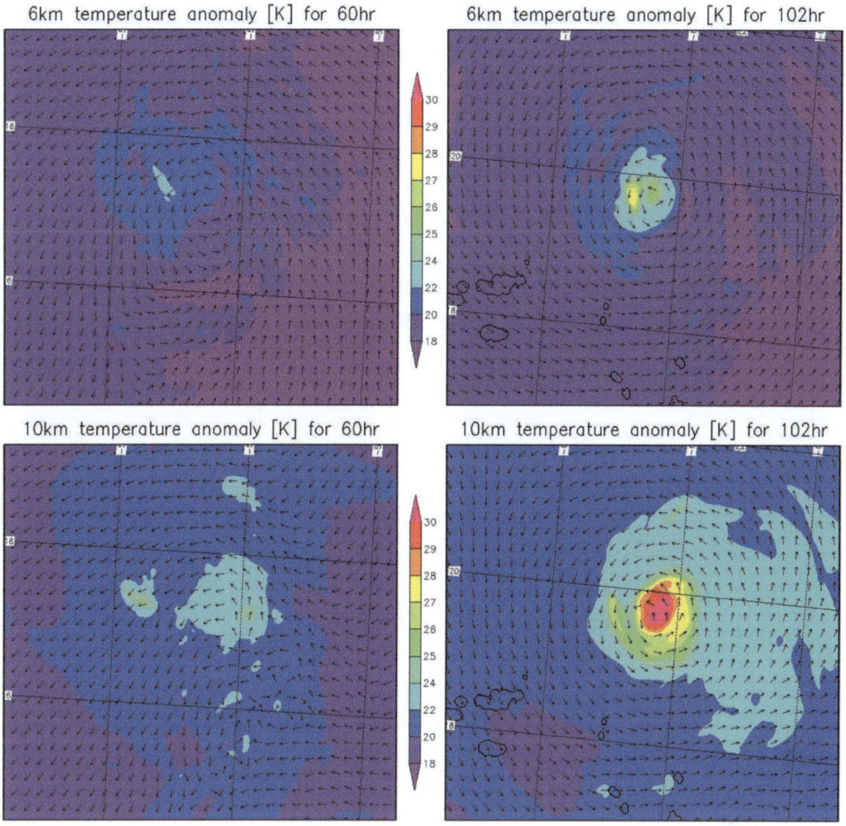

Fig. 6. The HWRF forecasted temperature anomaly at both the 6 km and 10 km height levels for EARL07L, plotted at both the start and end of the RI period. The forecast start time is 2010-08-26 18Z.

```
   Let T_A be a M × N matrix whose components are index
by the i-th row and j-th column.
   Loop, for i in 1 to M
      Loop, for j in 1 to N
            Let T_A(i,j) = T(i,j) - t_b
      End Loop
   End Loop
```

2.5 Deep Layer (850 mb to 200 mb) Vertical Shear

The vertical shear is the difference in winds (speed and direction) between two vertical levels. The shear that a tropical cyclone encounters may vary with height because of the natural stratification of winds in the atmosphere. In addition, storms in different stages of development will tend to interact with different layers of shear in the atmosphere. In general, well developed storms are more likely to interact with the *deep layer shear*, which is defined as the difference in winds between the 850 mb and 200 mb vertical levels. Shear plots are standard products used by both researchers and forecast specialists to measure how favourable an environment is for the development of tropical cyclones.

Figure 7 shows the HWRF forecasted deep layer shear (850-200 mb) for EARL07L at the onset of the RI period (forecast hour 60). The plot, which is centered at the storm eye, is projected onto a cylindrical grid. The average magnitude of the shear vector is 12 kts, a value that persists throughout the RI period. The axisymmetric component of the wind (i.e. average azimuthal wind component) was removed from both the top (200 mb) and bottom (850 mb) level winds prior to computing the shear. Removal of the axisymmetric vortex is a simple and common approach used to filter out the majority of the vortex signal (i.e. wave #0) from the wind field in order to retain mostly only the environmental flow through the vortex for the shear computations.

The following pseudo-code shows how the deep layer, storm-centered shear illustrated in Fig. 7 is derived using the winds projected onto a cylindrical grid. The pseudo-code is divided into three parts: (1) removal of the axisymmetric vortex from the tangential winds at the 850 mb and 200 mb levels, (2) computation of the Cartesian meridional and zonal winds from the radial and remaining tangential winds at the 850 mb and 200 mb levels, and (3) computation of the magnitude of the shear. Since steps (1) and (2) above need to be equally performed for the winds at both the 850 mb and 200 mb levels prior to shear computation, the pseudo-code is generalized to reflect the computations required for any arbitrary horizontal level.

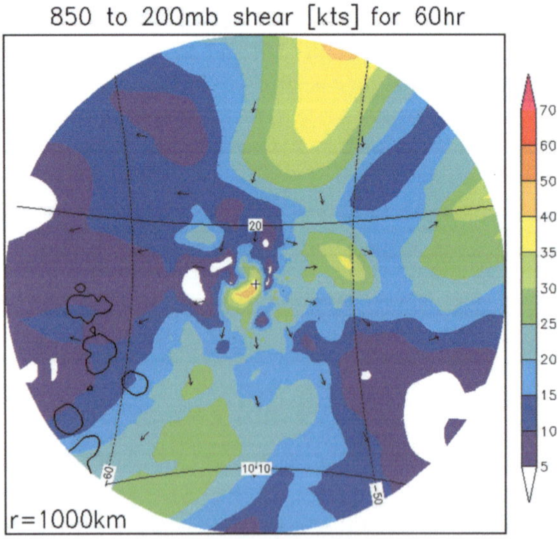

Fig. 7. HWRF forecasted deep layer shear (850-200 mb) for EARL07L at the onset of the RI period. The plot is centered at the storm eye. The forecast start time is 2010-08-26 18Z.

Assume that for a certain horizontal level, the tangential wind component V_T and radial wind component U_R are available as 2D matrices of size *M*-rows by *N*-columns, where each row corresponds to a specific radial distance and each column corresponds to an azimuth (or angle) within that radial distance; and the centre point of the cylindrical grid corresponds to row 1 and column 1. In addition, assume that the vector *A* with *N* elements holds the angle in the counter-clockwise direction relative to the east for every *j*-th azimuth in the grid. First, let's derive from V_T the average tangential wind as a vector V_{TAVE} of size *M* as follows:

```
Loop, for i in 1 to M
   Let sum = 0
   Loop, for j in 1 to N
         Let sum = sum + V_T (i,j)
   End Loop
   V_TAVE (i) = sum / M
End Loop
```

Having computed the average tangential wind, the axisymmetric vortex must then be removed from the tangential wind V_T. This results in a new matrix V_{Tnew} of size *M*-rows by *N*-columns that corresponds to an estimate of the magnitude of the tangential wind within the cylindrical grid that is due only to the large-scale environmental flow at that level:

```
Loop, for i in 1 to M
   Loop, for j in 1 to N
         Let V_Tnew (i,j) = V_T (i,j) + V_TAVE (i)
   End Loop
End Loop
```

Next, the Cartesian *U* (meridional) and *V* (zonal) components of the large-scale environmental winds can be estimated from matrices V_{Tnew}, U_R, and vector *A* as follows:

```
Loop, for i in 1 to M
   Loop, for j in 1 to N
         U(i,j) = U_R(i,j) · cos(A(j)) − V_Tnew (i,j) · sin(A(j))
         V(i,j) = U_R(i,j) · sin(A(j)) + V_Tnew (i,j) · cos(A(j))
   End Loop
End Loop
```

At this stage, the axisymmetric vortex has been removed from the tangential winds and the Cartesian wind components of every point in the *M*-by-*N* cylindrical grid have been computed. Now, to estimate the magnitude of the deep layer shear, the resulting meriodional and zonal wind components of the wind at the 200 mb level are subtracted from the corresponding wind components at the 850 mb level. Finally, the magnitudes of the resulting shear vectors are computed for each grid point. Let U_{200mb} and U_{850mb} correspond to the meridional components of the large-scale winds (after vortex removal) at the 200 mb and 850 mb levels, respectively. Similarly, let V_{200mb} and V_{850mb} correspond to the resulting zonal components of the large winds (after vortex removal) at the 200mb and 850 mb levels, respectively. The magnitude of the shear S is computed for every grid point as follows:

```
Loop, for i in 1 to M
   Loop, for j in 1 to N
         S(i,j) = ((U200mb (i,j) - U850mb (i,j))^2 + (V200mb (i,j)
         - V850mb (i,j))^2)^1/2
End Loop
End Loop
```

Researchers will at times also opt to compute the magnitude of the shear from wind fields that are projected onto a rectangular grid. However, removal of the vortex from the winds projected onto a rectangular grid is a rather non-trivial task when compared to the removal of the axisymmetric vortex from a cylindrical grid.

2.6 Deep Layer (850 mb to 250 mb) Steering Flow

Track and translation speed of tropical cyclones is known to be a result of its interaction with the large scale environment. The influence of the large scale environment on track and speed can be estimated as the magnitude of the *steering flow*, which is defined as the average magnitude of the horizontal winds along various layers of the atmosphere. The steering flow is an important model diagnostic because it provides insight about which direction the large scale environment is likely to push a storm forward. In addition, storms in different stages of development will likely interact with the large scale winds of different depths in the atmosphere. For example, weak storms are likely to interact with shallower layers of the steering flow, while stronger storms are more likely to interact with deeper layers of steering flow in the atmosphere.

Figure 8 illustrates the HWRF forecasted large-scale deep layer steering flow (averaged horizontal winds from the 850 mb to 250 mb levels) for EARL07L (shaded contour) at the onset of the RI period (forecast hour 60). The forecast start time is 2010-08-26 18Z. Notice the strong steering flow to the east of the storm, which contributes to its predominantly westward track.

Deriving the steering flow between two horizontal layers for a specified forecast hour is relatively simple. Assume that the prognostic U (meridional) and V (zonal) components of the winds are available as 3D matrices of size K-levels, M-rows, and N-columns, where levels 1 and K correspond to the lower and upper steering levels of interest, respectively. Also, assume U and V lie on the same rectangular grid. The magnitude of the steering flow S can be estimated as follows:

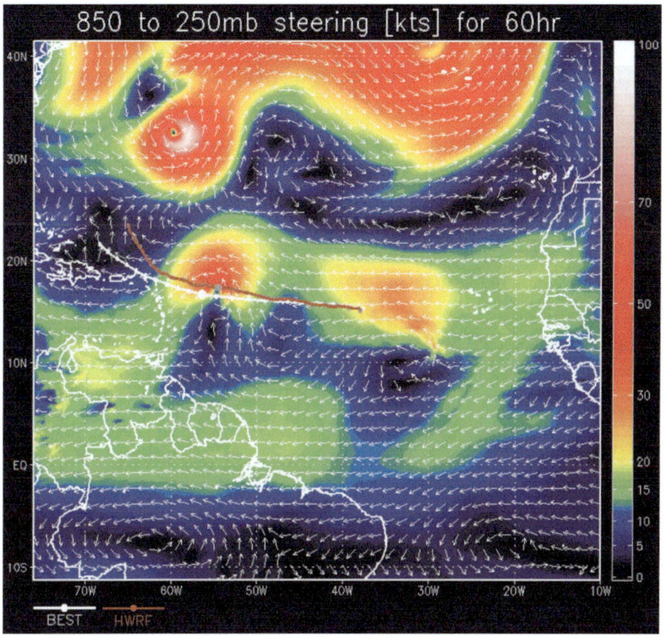

Fig. 8. HWRF forecasted large-scale deep layer steering flow (850-250 mb) for EARL07L at the onset of the RI period. The forecast start time is 2010-08-26 18Z.

Let U be a $K \times M \times N$ matrix holding the $M \times N$ meridional components of the wind at K levels of interest.

Let V be a $K \times M \times N$ matrix holding the $M \times N$ zonal components of the wind at K levels of interest.

Let S be a zero-valued $M \times N$ matrix whose components are index by the i-th row and j-th column.

```
Loop, for i in 1 to M
   Loop, for j in 1 to N
         Let sumU = 0 and sumY = 0
         Loop, for k in 1 to K
               Let sumU = sumU + U(k,i,j)
               Let sumV = sumV + V(k,i,j)
         End Loop
         Let sumU = sumU / K
         Let sumV = sumV / K
         Let S(i,j) = (sumU^2 + sumV^2)^(1/2)
   End Loop
End Loop
```

3. Time vs. Height Profiles

In this section we discuss a special type of diagnostic product called *time vs. height* profiles. As the name suggests, this type of diagnostic attempts to illustrate the *time* evolution of vertical column soundings (i.e. *height* in terms of pressure or altitude) taken from the environment of a tropical cyclone. When applied to model forecasts, column soundings can be attained by averaging 3D fields horizontally at every level within a region that is centered at the eye of a storm and which encompasses a significant portion of its circulation. Column soundings can also be attained by simply selecting vertical column from forecasted 3D fields that are located within the inner-core region of a storm.

3.1 Mean Wind Profile

Figure 9 plots the time evolution of the HWRF forecasted average winds surrounding the eye of Hurricane EARL07L. The x-axis ranges from 0 to 126 forecast hours. The y-axis corresponds to pressure levels ranging from 1000 mb to 150 mb. The wind column soundings displayed in the plot were attained by averaging the 3D wind field of HWRF's inner-most (i.e. storm-following)

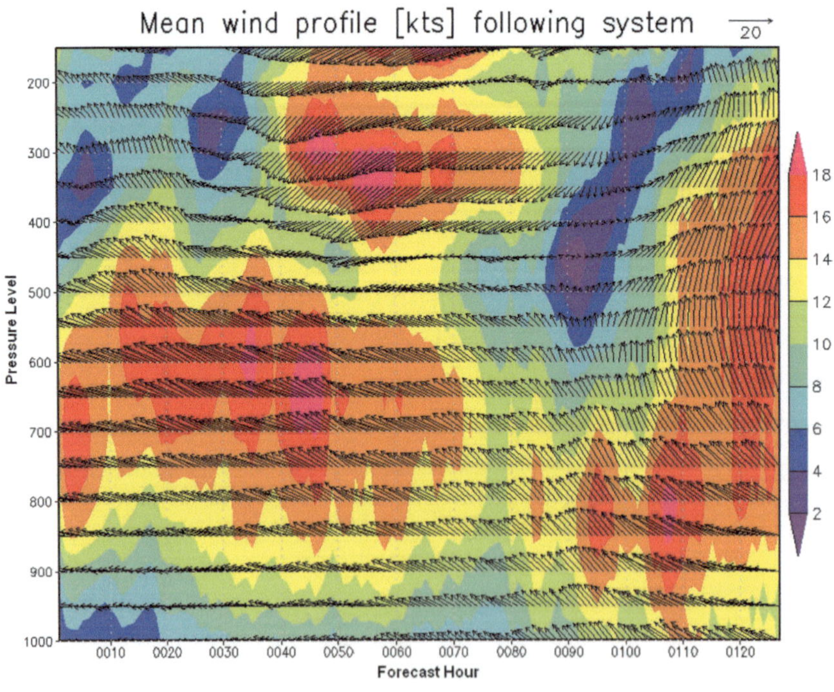

Fig. 9. Time vs. height plot of the HWRF forecasted averaged winds for EARL07L. The plot gives the averaged winds for every level between 1000 mb and 150 mb. The averaging region is approximately a 6 × 6 degrees box centered at the storm eye. The forecast start time is 2010-08-26 18Z.

domain, which has a resolution of 3 km and a size of approximately 6 × 6 degrees.

The vector field overlaying the shaded contour of winds represents the domain averaged wind vector for each pressure level and forecast hour. Note that as a result of the process of domain-averaging the wind, the axisymmetric component of the storm wind was largely removed. Thus, in essence, the overlayed wind vectors represent the average magnitude and direction (i.e. up points North and right points East) of the large scale environment wind flowing through the storm at different levels. As a result, the average large scale environmental shear between any two levels at a given forecast hour can be easily estimated from the plot as the difference between pairs of vectors residing in different levels of the same time column.

3.2 Mean Divergence Profile

Figure 10 gives an example of the utility of time vs. height plots in the analysis of modelled hurricane inner-core processes. The plot shows the HWRF forecasted averaged horizontal divergence surrounding the eye of Hurricane EARL07L. Notice from the plot that a pattern of strong upper level divergence

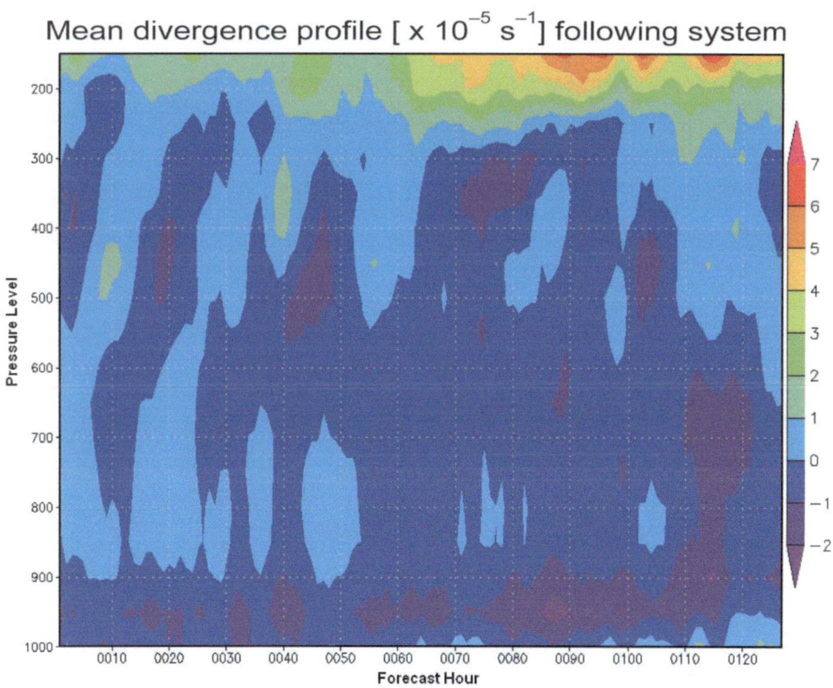

Fig. 10. Time vs. height plot of the HWRF forecasted averaged horizontal divergence for EARL07L. The plot gives the averaged horizontal divergence for every level between 1000 mb and 100 mb. The averaging region is approximately a 6 × 6 degrees box centered at the storm eye. The forecast start time is 2010-08-26 18Z.

appears at approximately forecast hour 60, coinciding with the start of the RI period that is illustrated in Fig. 3.

3.3 Temperature Anomaly Profile

Figure 11 illustrates the utility of time vs. height plots in the analysis of the thermo-dynamical structure of modelled tropical cyclones. The plot shows the time evolution of the temperature anomaly (or perturbation) at the eye of Hurricane EARL07L, measured relative to the domain averaged temperature at the start of the forecast. The plot clearly illustrates the time evolution of EARL07L's warm inner-core, delineating with precision the start of EARL07L's RI period at approximately forecast hour 60.

The following pseudo-code illustrates how the time vs. height plot of temperature anomaly given in Fig. 11 can be computed using basic prognostic variables. Assume that the prognostic air temperature variable T is projected onto a rectangular grid and available as a 4D matrix of size T-time indices, K-levels, M-rows, and N-columns. In T, let each time index from 1 to T represent a unique forecast hour and let the K levels (1 to K)

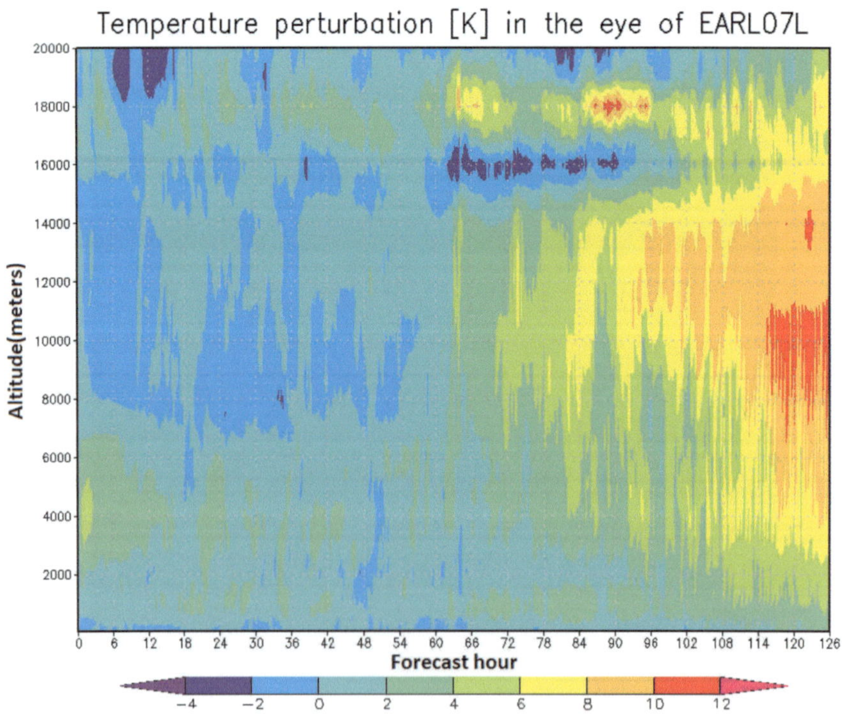

Fig. 11. Time vs. height plot of the HWRF forecasted temperature anomaly at the eye of Hurricane EARL07L. The temperature anomaly is computed relative to the domain averaged temperature at the start of the forecast. Vertical levels range from surface to 20 km. The forecast start time is 2010-08-26 18Z.

represent the available vertical pressure or height levels. First, let's compute the domain-averaged atmospheric air temperature profile at the start of the forecast. This profile will be used as the base temperature in the temperature anomaly computations. Let's denote this profile by a column vector T_0 of size K elements, where each element represents the domain averaged temperature at one vertical level:

```
Let t = 1
Loop, for k in 1 to K
   Let T_0(k) = 0
   Loop, for i in 1 to M
         Loop, for j in 1 to N
               Let T_0(k) = T_0(k) + T_(t,k,i,j)
         End Loop
End Loop
   Let T_0(k) = T_0(k) / (M·N)
End Loop
```

Now, in order to compute the temperature anomaly at the storm eye, we must first define the location of the grid column within the air temperature matrix T that corresponds to the eye of the storm. Let's assume that the column at the centre of matrix T corresponds to the eye of the storm. Now, let's define the 2D matrix T_A with size $T \times K$ as the time vs. height profile of the temperature anomaly and compute it as follows:

```
Let gridCenterX = round(N/2)
Let gridCenterY = round(M/2)
Loop, for t in 1 to T
   Loop, for k in 1 to K
         Let T_A(t,k) = T_(t, k, gridCenterY, gridCenterX) - T_0(k)
End Loop
End Loop
```

4. Conclusions

This chapter focused on demonstrating that a basic set of useful model diagnostic products can be easily derived from the most basic prognostic variables of numerical weather models. Each of the diagnostic products discussed in this chapter provides an insight into the basic understanding of the various processes playing important roles on the development of tropical cyclones. We encourage the reader to visit our HWRF Portal website, where a multitude of diagnostic products from forecasts of the operational and research versions of the HWRF model are available for both active and retrospective storms.

Acknowledgements

The authors acknowledge funding from NOAA's Hurricane Forecast Improvement Project that supported this work. The second author acknowledges the support provided by Indo-US Science and Technology Forum, NOAA's Hurricane Forecast Improvement Project and various national and international collaborators who have helped improving the operational HWRF model.

PART 5

Advances in Tropical Cyclone Research

21

Introduction to Hurricane Dynamics: Tropical Cyclone Intensification

Michael T. Montgomery
Department of Meteorology, Naval Postgraduate School
Monterey, California, USA
mtmontgo@nps.edu

1. Introduction

The broad aim of this set of two articles (chapters 21 and 22) on tropical cyclone theory is to provide graduate students and post-doctoral researchers from India and other parts of the world with state-of-the-art knowledge and understanding of tropical cyclone dynamics. It is hoped that this knowledge base will be a useful tool to help these students and post docs understand and improve the forecasts of tropical cyclone genesis and intensification in various parts of the world affected by these storms.

To me, tropical cyclones are magnificent and scientifically fascinating large-scale convective vortices that still hold many hidden secrets regarding their birth, intensification, mature evolution and decay. Observing the formation and intensification of these large-scale coherent structures fuels my scientific passion to obtain a deeper physical understanding of what many atmospheric scientists have called "the greatest storms on earth" (see examples in Fig. 1). Practical considerations, such as saving human life and property in the path of these tropical tempests are, of course, another important driving factor in the quest for knowledge about these storms. My interest in these storms spans all facets of these storms. Recent events surrounding Hurricane Sandy (2012) are a reminder that even tropical storms can wreak havoc on populated coastal communities, maritime assets and even inland populations (e.g., Lussier et al., 2015).

As coastal communities continue to grow, there is an increasing demand throughout regions of the world for more accurate tropical cyclone forecasts. Typically, the response to such a demand relies heavily on financial resources devoted to develop and apply more comprehensive numerical models using increased computer power. However, these efforts are no guarantee for success in improving tropical cyclone forecasts. It is the belief of the author that without a commensurate understanding of the dynamics and thermodynamics of these convective vortices, such efforts, even if grandiose, will not realize their full potential. It is only when the state-of-art computational methods and observations are combined with hypothesis testing and the development of conceptual models that a firm scientific foundation can be established to make measurable advances in hurricane intensity forecasts.

In coordination with aircraft and satellite observations, knowledge of the structure of mature storms progressed rapidly from the earliest hurricane reconnaissance missions just after the Second World War. Understanding of the mature stage of these tropical tempests evolved largely in parallel with the collection and analysis of these observational data (e.g., Anthes, 1982; Emanuel, 1986; Willoughby, 1995; Montgomery et al., 2006; Houze et al., 2007; Smith et al., 2008; Marks et al., 2008; Bell et al., 2008) and also see Chapters 1 and 2 by Dr. Frank D. Marks Jr. and Dr. C.M. Kishtawal respectively. Today, our understanding of the birth, intensification and mature stages of these vortices is currently undergoing dramatic changes, not yet properly conveyed in current atmospheric science textbooks. One of the objectives of this article and associated references is to address this deficiency by summarizing recent scientific progress made on the intensification, mature and genesis stages of the tropical cyclone lifecycle. This article will begin by reviewing the current state of understanding of tropical cyclone intensification and will then summarize key elements of the dynamics and thermodynamics of tropical cyclone intensification in a new intensification paradigm.

2. Conceptual Models of Tropical Cyclone Intensification

Over the past five decades substantial scientific efforts have been devoted to construct conceptual models of tropical cyclone (TC) intensification. A conceptual model (or paradigm) of TC intensification is an overarching framework of scientific ideas that may be used to direct and interpret observations, to design and develop numerical models, and to link together observations and numerical models in order to develop an improved tropical cyclone predictive system. Having a paradigm of TC intensification that correctly captures the essential dynamical and thermodynamical mechanisms occurring in nature is an essential ingredient to make forward progress in many aspects of TC research and operational activities.

The four most prominent paradigms of TC intensification are the *CISK model* (Charney and Eliassen, 1964; Ooyama, 1964; Carrier, 1971), the *cooperative intensification model* (Ooyama, 1969, 1982; Willoughby, 1995,

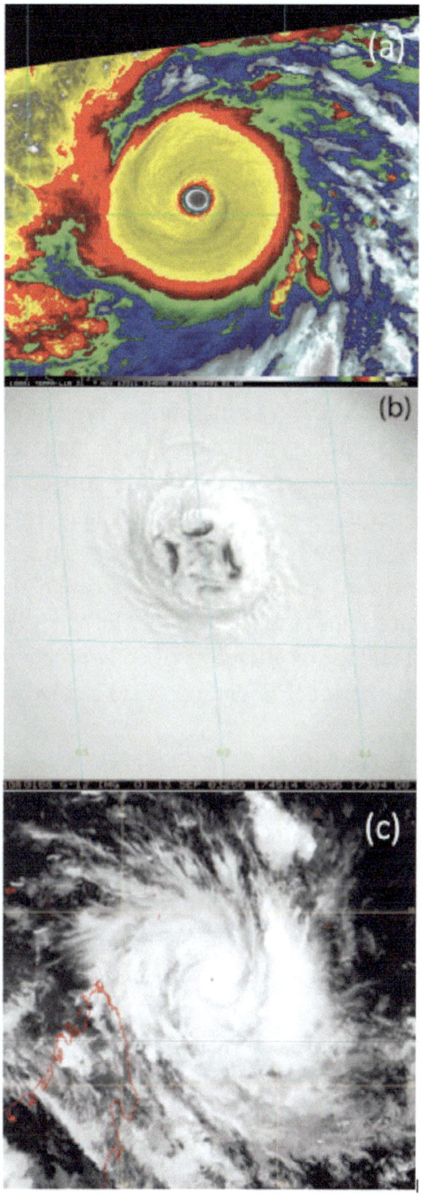

Fig. 1. Satellite imagery of (a) Typhoon Haiyan (November 07, 2013, western North Pacific), (b) Hurricane Isabel (September 13, 2003, Atlantic) and (c) Tropical Cyclone Bejisa (December 30, 2014, Northeast of Madagascar) in the southern Indian ocean. Image (a) is a colour enhanced infrared image at 13:45 UTC available at http://rammb.cira.colostate.edu. Image (b) is a visible image at 1745 UTC from GOES super-rapid-scan operations. The bottom image (c) is an infrared image at 1213 UTC from the NOAA 18 satellite. The intensity of Haiyan (strongest tropical cyclone at landfall on record), the "pinwheel" pattern of mesovortices within Isabel's eye, and the "pinhole-like" size of Bejisa´s eye are particularly striking and intriguing features.

1998), the *thermodynamic air-sea interaction* model commonly known as *Wind-Induced-Heat-Exchange* (*WISHE*) (Emanuel, 1986, 1989; Rotunno and Emanuel, 1987; Emanuel et al., 1994; Emanuel 1995, 1997; Emanuel, 2012; Holton, 2004) and the *new rotating convective model* (Nguyen et al., 2008; Smith et al., 2009; Montgomery et al., 2009; Bui et al., 2009; Fang and Zhang, 2011; Persing et al., 2013; Montgomery and Smith, 2014). The first three models were presented originally assuming axisymmetric flow (no departures from axial symmetry about rotation axis, i.e., "no azimuthal eddies"), with a simple resting tropical environment without uniform flow or vertical shear. The simple resting environment has historically served as the canonical testing ground for obtaining a basic understanding of tropical cyclone intensification. Like the first three models, the rotating convective model is illustrated most simply in a quiescent tropical environment and includes air-sea transfer of momentum and heat. However, the rotating convective model is three-dimensional and contains an azimuthally-averaged (mean-field) dynamics that is forced by the azimuthally-averaged heat and momentum sources, including eddy momentum and eddy heat fluxes and their divergence, etc. The rotating convection model will be shown to be an extended cooperative intensification model and the explicit role of eddy processes as represented by the near-cloud resolving model is diagnosed in a newly published study by Persing et al. (2013).

The four main intensification paradigms have been examined in Montgomery and Smith (2014) and the strengths and weaknesses of each of these paradigms are discussed there. *The upshot is that only the rotating convective model properly captures the simulated behaviour of intensifying tropical cyclones in three dimensions and the dependence of storm intensification on key parameters pertinent to the convective-vortex phenomenology for realistic forecast time scales.* In this lecture, only the most pertinent key points of the new paradigm and earlier models will be highlighted. To gain a broad picture of the physical elements of the new paradigm, it is first necessary to summarize the key ingredients of the first three intensification models.

2.1 The CISK, Cooperative Intensification and WISHE Paradigms

The classical, or 'conventional' model of tropical cyclone spin-up, embodied in both the CISK and cooperative intensification models, is illustrated schematically in Fig. 2. The basic tenet is that, in an azimuthally-averaged sense, deep convection in the inner-core region generates inflow in the lower troposphere. Above the frictional boundary layer, the inflowing air materially conserves its absolute angular momentum (M) and spins faster. Here, M is defined as follows:

$$M = r\langle v\rangle + fr^2/2 \quad (1)$$

where v is the tangential wind speed, f is the Coriolis parameter, r is the radius from the centre of the storm and $\langle v\rangle$ denotes the azimuthal average of tangential velocity at constant height in a storm-centered cylindrical coordinate system. The first term on the right hand side of Eq. (1) represents the *relative* angular momentum per unit mass of a moist air parcel (due to its rotation around the centre of the storm), while the second term accounts for the *planetary* angular momentum (associated with the local rotation rate of the Earth about the storm's axis of rotation at the given latitude). Strong convergence of moist air mainly in the boundary layer provides "fuel" to maintain the deep convection in the central region of the developing vortex. In the CISK-paradigm, the rate of latent heat release by deep cumulus convection is proportional to the vertically-integrated horizontal convergence of moisture through the depth of the troposphere. The bulk of this moisture convergence occurs in the boundary layer (Charney and Eliassen, 1964). In the cooperative intensification paradigm, the representation of latent heat release is more sophisticated (Montgomery and Smith (2014) provide further details on the CISK and cooperative intensification theory).

Despite its historical significance and influence, the CISK theory has attracted much criticism. An important contribution to the debate over CISK is the insightful paper by Ooyama (1982), which articulated the cooperative intensification paradigm for the spin-up process discussed in the next subsection. Among several points, Ooyama cautioned that the CISK closure and variants thereof were little more than convection in disguise, exhibiting the largest growth rates at the smallest horizontal scales and not in the scale of the storm itself. Later in the 80's and 90's, the CISK theory was critiqued

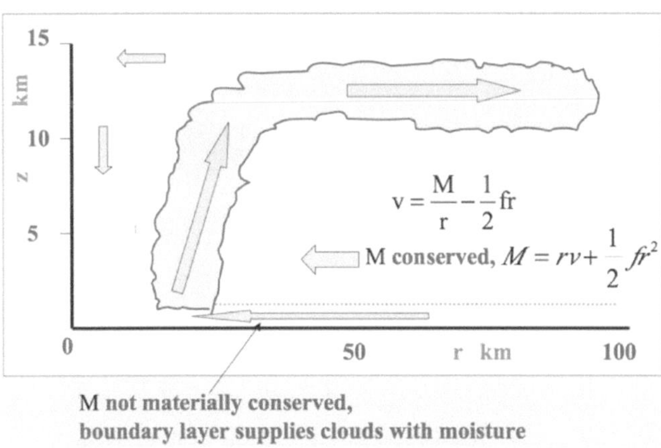

Fig. 2. Schematic diagram illustrating the CISK and the cooperative intensification models (see text for details). Adapted from Montgomery and Smith (2014).

in a number of papers by Emanuel (1986), Raymond and Emanuel (1993), Emanuel et al. (1994), Craig and Gray (1996), Ooyama (1997) and Smith (1997). In particular, Raymond and Emanuel op cit. argued persuasively that the CISK closure fundamentally violates causality because convection is not caused by the macroscale water supply.

In a landmark effort to overcome the problems with the CISK theory, a new model was developed emphasizing thermodynamic processes and focusing in particular on a postulated positive feedback loop involving the near-surface wind speed and the evaporation of water from the underlying ocean, with the evaporation rate being a function of wind speed and thermodynamic disequilibrium (Emanuel, 1989, 1995, 1997, 2003; see also Rotunno and Emanuel, 1987, p. 559; Holton, 2004). This evaporative-wind feedback mechanism is a form of air-sea interaction instability and is now commonly referred to as Wind-Induced-Heat-Exchange (WISHE) in modern textbooks. We have a number of issues with some of the popular articulations of the putative WISHE spin-up mechanism (e.g. that of Holton and Hakim, 2012) and present here our own interpretation of the mechanism summarizing that given in Montgomery et al. (2009). The feedback mechanism is illustrated schematically in Fig. 3. The pertinent variables shown in the figure are defined in the text of Montgomery and Smith (2014) and are otherwise standard. Here, thermal wind balance refers to the axisymmetric thermal wind equation in pressure coordinates that relates the radial gradient of azimuthally-averaged θ_e to the vertical shear of the azimuthally-averaged tangential velocity field.

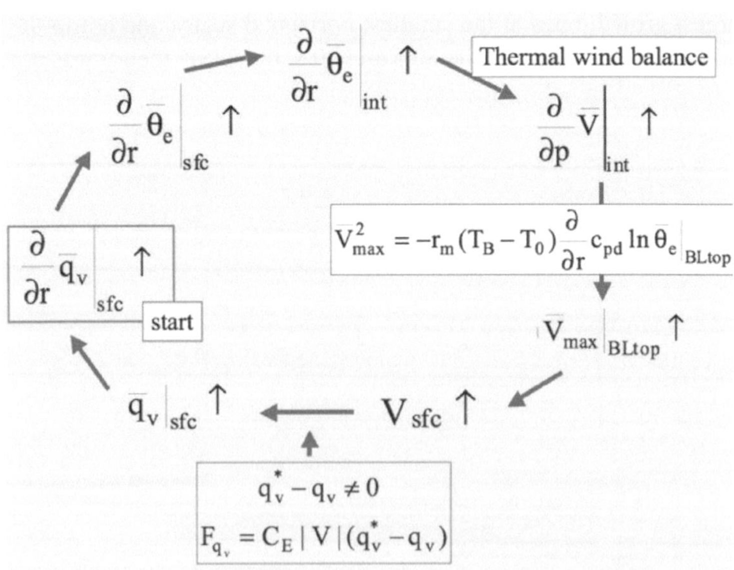

Fig. 3. Schematic diagram illustrating the WISHE feedback mechanism of tropical cyclone intensification as articulated in Montgomery et al. (2009) (see text for details).

The argument assumes that for some reason the near-surface azimuthal mean mixing ratio is increased in the core region. This increase leads to an increase in the corresponding mean radial gradient and also an increase in the mean radial gradient of θ_e throughout the boundary layer. The secondary circulation is imagined to loft this increase in θ_e into the vortex interior thereby warming the core region of the vortex. By the thermal wind relation and its corollary, obtained upon applying a Maxwell thermodynamic relation (presuming reversible thermodynamics) and then integrating this equation along a surface of constant absolute angular momentum, this increased warmth is invoked to imply an increase in the mean tangential wind at the top of the vortex boundary layer. It is understood here that θ_e values at the boundary layer top correspond to saturated values. Turbulent mixing processes in the boundary layer are assumed to communicate this wind speed increase to the surface, and thereby increase the potential for increasing the sea-to-air water vapour flux F_{q_v}. The saturation mixing ratio must increase approximately in step with the putative increase in near-surface q_v so as to maintain a thermodynamic disequilibrium. If this occurs then the surface mixing ratio will increase, thereby leading to a further increase in the mean tangential wind and so on.

2.2 Testing the WISHE Paradigm

An in-depth test of the putative role of wind-speed dependent surface heat fluxes in the intensification process was carried out by Montgomery et al. (2009) using a non-hydrostatic mesoscale numerical model without cumulus parameterization[1]. Some of the principal results of this study are summarized in Fig. 4 (adapted from their Fig. 9). When the wind-speed dependence of the surface heat fluxes is capped at 10 ms^{-1}, the maximum horizontal wind speed is a little less than in the uncapped experiments, but the characteristics of the vortex evolution are qualitatively similar, in both the pseudo-adiabatic experiments and when warm rain processes are included. In contrast, when the latent and sensible heat fluxes are suppressed altogether, the system-scale vortex does not intensify, even though there is some transient hot-tower convection and local wind-speed enhancement. Despite the non-zero CAPE present in the initial sounding, no system-scale amplification of the wind occurs. This finding rules out the possibility that vortex intensification with capped fluxes is an artifact of using an initially unstable moist atmosphere (cf. Rotunno and Emanuel, 1987). (These conclusions have been verified with further sensitivity experiments and supporting analyses using an initial sounding with nearly zero CAPE in the Appendix of Montgomery et al. (2009).) A minimum value of approximately 3 ms^{-1} was found necessary for the wind-speed cap at which the vortex can still intensify to hurricane strength in the pseudo-adiabatic configuration of the numerical model.

[1] A recent study by Montgomery et al. (2015) has revisited this issue using a state-of-the-art cloud model and affirms the conclusions of Montgomery et al. (2009).

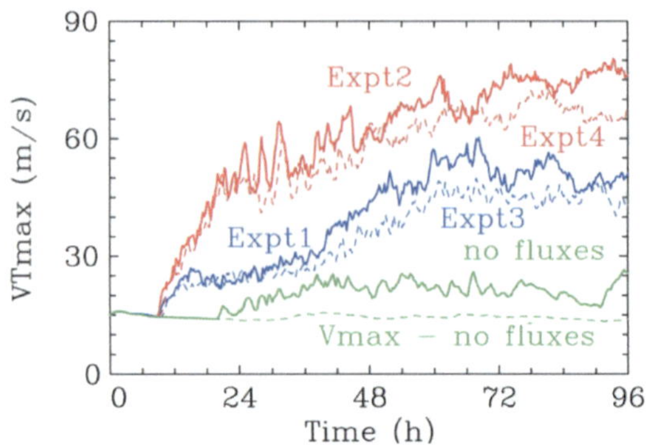

No.	*Name*	*Description*
1	Warm rain	Warm rain (control) expeirment on an f-plane.
2	Pseudo-adiabatic	Explicit latent heat release with no water loading or evaporative cooling processes.
3	Capped warm rain	Surface wind speed is capped at 10 ms^{-1} in formulae for latent and sensible heat fluxes.
4	Capped pseudo-adiabatic	As Experiment 3, but using pseudo-adiabatic option.

Fig. 4. Time-series of maximum azimuthal-mean tangential wind component (V_{max}) for experiments 1-4 (solid and dashed red/blue curves) from Montgomery et al. (2009). Shown also is the time series of the local horizontal wind speed maximum VT_{max} (solid green curve) and V_{max} (dashed green curve) of the related experiment when the surface latent- and sensible-heat fluxes are set to zero (Experiment S3; Table II of Nguyen et al. (2008)). Adapted from Montgomery et al. (2009).

When warm rain processes are included, the vortex still intensifies with a wind-speed cap of 5 ms^{-1} (not shown here, see Montgomery et al. (2009) for details). These experiments demonstrate that large latent heat fluxes in the core region are not necessary for cyclone intensification to hurricane strength and indicate that the evaporation-wind feedback process postulated in the WISHE intensification mechanism is not essential.

The take away message from these experiments is that, although the wind-speed dependent heat fluxes do augment the intensification process somewhat, the basic spin up process operates without the postulated evaporative wind-speed feedback (WISHE) process. Therefore WISHE is not essential nor the dominant intensification mechanism for these idealized configurations without uniform flow or vertical shear, configurations that have been used to underpin the WISHE feedback hypothesis.

2.3 The Rotating Convection Paradigm

The new intensification model is described in a series of papers by Nguyen (2008), Montgomery et al. (2009), Smith et al. (2009), Bui et al. (2009), and Persing et al. (2013) and highlights the presence of rotating convective structures[2] and their vorticity remnants as well as their role in organizing the vorticity structure of the core of the storm. In essence, these clouds collectively drive the spin-up of the system-scale vortex by inducing radial inflow above the boundary layer, consistent with the previous three theories. As in the earlier hypotheses, radial convergence of M above the boundary layer in conjunction with its approximate material conservation outside of regions of deep convection leads to spin-up of the bulk tangential winds and an increasing radial pressure gradient there. In contrast, this new theory stresses the fact that the spin-up of the maximum tangential winds actually occurs in the boundary layer below 1 km altitude.

In the boundary layer, M is reduced by the frictional torque, but the radial inflow is much stronger than above this layer due to both friction and an increase in the radial pressure gradient at the top of the layer resulting from elevated radial inflow atop the boundary layer. Spin-up occurs in the boundary layer when the radial inflow is large enough to converge rings of air to small radii faster than the rate at which M can be removed by friction. The amplification of the tangential wind by this mechanism naturally leads to the development of supergradient flow and an accompanying outward agradient force. The agradient force acts to arrest the radial inflow at the base of the eyewall, whereupon air parcels turn upwards and carry their elevated tangential momentum into the eyewall. This process contributes also to the spin-up of the tangential winds in the bulk vortex (Smith et al., 2009; Bui et al., 2009; Smith and Thomsen, 2010; Smith and Montgomery, 2010; Persing et al., 2013). Further elaboration of the two system-scale mechanisms that comprise the extended cooperative intensification model is presented in Montgomery and Smith (2014, in section entitled ‘An axisymmetric view of the new paradigm’). The rest of this article will present an overview of the local dynamics of the new model.

3. The Prototype Experiment for Tropical Cyclone Intensification

To fix ideas, our discussion here will be focused on *understanding* various aspects of the prototype problem for tropical cyclone intensification

[2] The term “rotating convection” is used in lieu of vortical hot towers (VHTs), which were first described in Hendricks et al. (2004) and later studied in Montgomery et al. (2006a) and Nguyen et al. (2008), among others. To avoid any potential controversy that surrounds the definition of VHTs, the term “rotating convection” will be used throughout the rest of these lectures. Usage of the term VHT has given some the impression that only deep convection (>12 km cloud depth) has strong rotation. However, a recent study by Wissmeier and Smith (2011) showed that even moderate convection (6-12 km depth) in a background rotation rate typical of the undisturbed tropical atmosphere can have a comparable impact on the stretching of low-level relative vorticity in comparison to intense deep convection. Thus, a broad definition is required for studying the aggregate impact of these convective elements on tropical cyclone intensification.

(Fig. 5), which examines the evolution of a prescribed, initially cloud free, axisymmetric, baroclinic vortex in a quiescent environment over a warm ocean on an *f*-plane. For simplicity, and for didactic reasons, the effects of an ambient flow, including those of ambient vertical shear, are not considered. Omitting the complicating effects of ambient and vertical shear flows permits a clear comparison between the new paradigm and the old paradigms (summarized above), which were historically formulated and interpreted for the case of a quiescent environment on an *f*-plane.

The numerical experiments summarized here were carried out using a modified version of the Penn State/NCAR fifth-generation Mesoscale Model (MM5 version 3.6) (Dudhia,1993; Grell et al., 1995), which is suitable for the study of idealized problems. Details are provided in Nguyen et al. (2008). (The basic results and interpretations reported herein have been verified with more state-of-the-art three-dimensional mesoscale and cloud models, such as the Weather Research and Forecasting (WRF) model (see Fang and Zhang, 2011), the hurricane version of the WRF model (called HWRF) (see Bao et al., 2012; Gopalakrishnan et al., 2011) and George Bryan's Cloud Model 1 (CM1; version 14) (see Persing et al., 2013)). The MM5 model is configured with three domains: a coarse mesh of 45 km resolution, and two, two-way-nested domains of 15 km and 5 km resolution. The domains are square, and are 5400 km, 1800 km and 600 km respectively on each side. For the experiments on the β-plane, the innermost domain moves to keep the vortex near the centre of the domain. There are 24 σ-levels in the vertical, seven of which are below

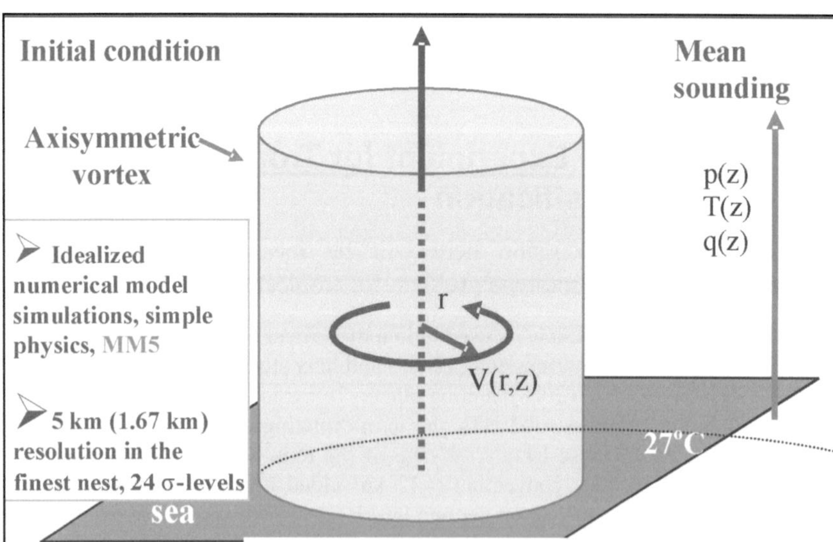

Fig. 5. The prototype experiment for understanding the dynamics and thermodynamics of tropical cyclone intensification following Nguyen et al. (2008).

850 hPa. One experiment employing a fourth domain (300 km square) has been conducted, to permit increased spatial accuracy down to 1.67 km horizontal grid spacing on the innermost domain. Another experiment employing higher vertical resolution (45 σ-levels) has been conducted also.

In order to keep the experiments as simple as possible in this presentation, the main physics options chosen are the bulk aerodynamic boundary-layer scheme (Braun and Tao, 2000) and the simplest explicit moisture scheme (the pseudo-adiabatic scheme, comprising instantaneous rainout after condensation). These schemes are applied in all domains. The sea-surface temperature is set to a constant 27 °C. The initial vortex is axisymmetric, with a maximum tangential wind speed of 15 ms^{-1} at the surface at a radius of 135 km. The strength of the swirling wind decreases sinusoidally with height, vanishing at the top model level (50 hPa). The temperature field is initialized to be in gradient wind balance with the wind field, using the method described by Smith (2006). The far-field temperature and humidity are based on Jordan's Caribbean sounding during summertime conditions (Jordan, 1958), and are shown in a skew *T*-log *p* diagram in Fig. 6.

3.1 Illustrative Solutions

The integration time for the main experiments described here is 96 h (four days). The intensification process is summarized in Fig. 6. The solid red curve in Fig. 4 shows the time series of the local horizontal wind speed in the pseudo-

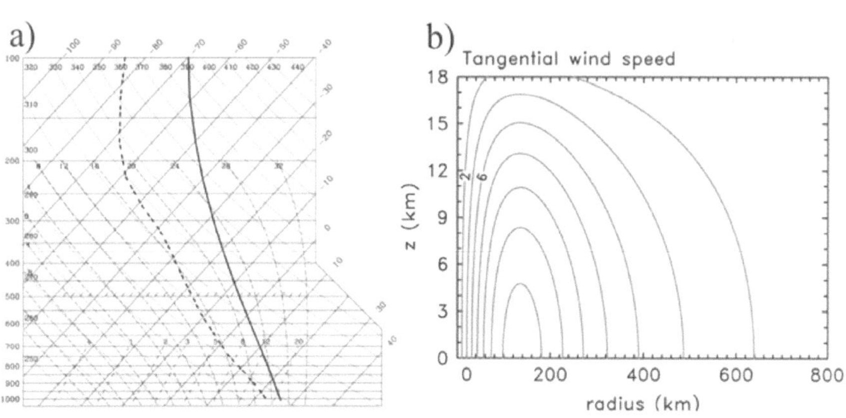

Fig. 6. Far field temperature and humidity skew *T*-log *P* diagram, showing the temperature (solid line) and dew-point temperature (dashed line) of the initial sounding at a radius of 2,250 km from the domain centre (adapted from Nguyen et al., 2008). (b) Radius-height plot of initial tangential wind velocity. Contour intervals are shown every 2 ms^{-1}. Adapted from Montgomery et al. (2009). Initial thermodynamical conditions are those in gradient wind balance with the wind field shown in part (b). The balanced fields are computed following Smith (2006), using the temperature and humidity shown in part (a) as the far field, reference sounding.

adiabatic control experiment (C0) from Nguyen et al. (2008). (This experiment is labelled as "Expt. 2" in the figure caption.) In essence, the evolution begins with a gestation period, during which the vortex slowly decays because of surface friction, but moistens in the boundary layer because of evaporation from the underlying sea surface. This period lasts approximately 9 h, during which time the minimum surface pressure rises slightly from its initial value of 1004 hPa to 1008 hPa (not shown), but the maximum tangential wind speed decreases only by a small amount (by less than 0.5 ms^{-1}). The imposition of surface friction from the initial instant leads to an azimuthal-mean inflow in the boundary layer and a mean outflow above it, the outflow accounting for the decrease in tangential wind speed through the approximate material conservation of M. The inflow is moist, and as it rises out of the boundary layer and cools, condensation progressively occurs in some grid columns interior to the corresponding radius of maximum tangential wind speed.

Latent heat release in these columns initiates deep convective updrafts with vertical velocities up to 5 ms^{-1} at 850 hPa. Figures 7 and 8 exemplify the local flow evolution, which shows the patterns of vertical velocity and the vertical component of relative vorticity at the 850 hPa and 500 hPa levels, respectively, at selected times for the control experiment of Nguyen et al. (2008). The updrafts form initially (Fig. 7, upper left panel) in an annular region inside the 135 km radius of the maximum initial tangential wind speed and their distribution shows a dominant azimuthal wavenumber-12 pattern around the circulation centre[3]. The updrafts rotate cyclonically around the vortex centre and have lifetimes on the order of an hour. As they develop, they tilt and stretch the local vorticity field and an approximate ring-like structure of intense, small-scale, vorticity dipoles emerges (lower left panel of Fig. 7). The dipoles are highly asymmetric in strength with strong cyclonic vorticity anomalies and much weaker anticyclonic vorticity anomalies. Early in the intensification phase, the strongest updrafts lie approximately in between the vorticity dipoles whereas later, they are often approximately collocated with the strong cyclonic anomalies. It is a useful reminder to point out here that these updrafts need not penetrate all the way to the tropopause to greatly enhance the local ambient rotation in the lower troposphere ($z < 3$ km) (Wissmeier and Smith, 2011).

In the simulation, during the rapid intensification phase, the number of rotating updrafts decreases from twelve at 9 h to no more than three by the end of the period of rapid intensification, but their structure remains spatially irregular and during some periods, the upward motion occupies a contiguous region around the vortex centre. The cyclonic vorticity anomalies associated with the rotating updrafts grow horizontally in scale due to merger and axisymmetrization with neighbouring cyclonic vorticity anomalies. This

[3] It turns out that the number of updrafts that form initially increases as the horizontal resolution increases, but the subsequent evolutionary picture is largely similar.

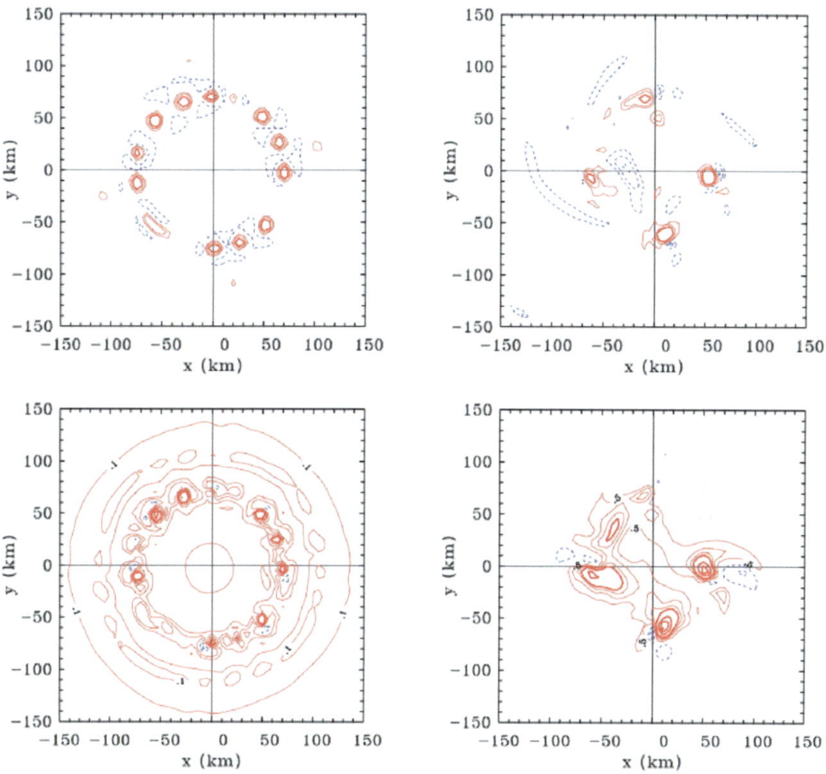

Fig. 7. Vertical velocity (upper panels) and the vertical component of relative vorticity (lower panels) at 850 hPa at 9.75 hrs (left panels) and 24 hrs (right panels) in the control experiment C0 in Nguyen et al. (2008). Contour interval for vertical velocity: tick curves 2.0 ms^{-1} and thin solid curves 0.5 ms^{-1} with highest value 1.0 ms^{-1}, thick dashed curves for negative values with interval 0.25 ms^{-1}. Contour interval for relative vorticity in left panels is 5.0×10^{-3} s^{-1} (thick curves) and 1.0×10^{-3} s^{-1} (thin curves). For the right panels is 2.0×10^{-2} s^{-1} and 5.0×10^{-3} s^{-1}, for thick and thin curves respectively. Adapted from Montgomery and Smith (2014).

upscale growth occurs in tandem with the horizontal convergence of like-sign vorticity into the circulation of the rotating updrafts. At this altitude above the frictional boundary layer, they move slowly inwards also and become segregated from the anticyclonic vorticity anomalies, which move slowly outwards relative to them. As the anticyclonic anomalies move outwards they decrease in amplitude and undergo axisymmetrization by the parent vortex. Elements of the segregation process are discussed by Nguyen et al. in their section 3.1.5.

The right panels of Fig. 7 show the situation at 24 hours. At this time there are five relatively strong updrafts (upper panel), each possessing greatly enhanced cyclonic vorticity (lower right panel), so that the initial monopole structure of the vortex is completely dwarfed by the local vorticity of the rotating updrafts.

Comparing Figs 7 and 8 shows that the rotating updrafts move inwards over time. Spiral inertia-gravity waves propagating outwards occur throughout the vortex evolution and these are especially prominent in animations of the vertical velocity fields. The pattern of strong updraughts changes continuously with time, but is mainly monopolar, dipolar, or tripolar and always asymmetric. By the end of the rapid intensification period, no more than three rotating updrafts are active around the circulation centre.

The simulated vortex, even with the simplest explicit moisture scheme (pseudo-adiabatic dynamics), exhibits many realistic features of a mature tropical cyclone (e.g., Willoughby, 1995; Kossin et al., 2002; Corbosiero et al., 2006; Marks et al., 2008) with spiral bands of convection surrounding an approximately symmetric eyewall and a central eye that is free of deep

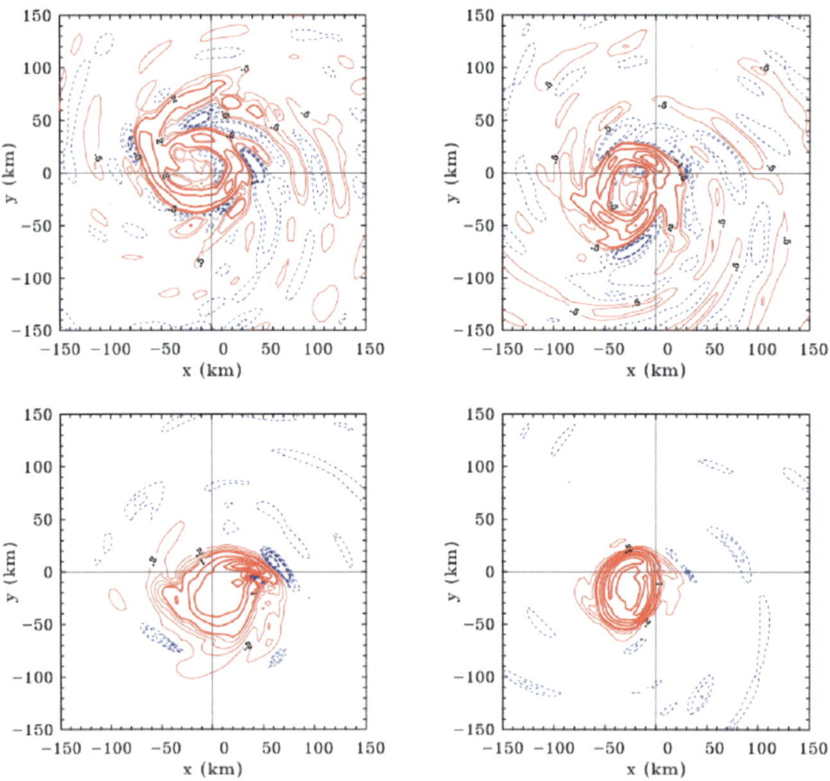

Fig. 8. Vertical velocity (upper panels) and the vertical component of relative vorticity (lower panels) at 850 hPa at 48 h (left panels) and 96 h (right panels) in the control experiment C0 in Nguyen et al. (2008). Contour intervals for vertical velocity are 4.0 ms^{-1} (thick curves, with lowest absolute value 2.0 ms^{-1}) and 0.5 ms^{-1} (thin curves, with highest absolute value 1.0 ms^{-1}). Contour intervals for relative vorticity are 1.0×10^{-2} s^{-1} (thick curves) and 1.0×10^{-3} s^{-1} (thin curves). Positive values are solid curves in red and negative values are dashed curves in blue. The zero contour is not plotted. Adapted from Montgomery and Smith (2014).

convection (Fig. 8 top row). In the mature phase, the evolution of the vortex core is characterized by an approximately axisymmetric circulation superimposed on which are:

- small, but finite-amplitude vortex Rossby waves that propagate azimuthally, radially and vertically on the mean potential vorticity gradient of the system scale vortex (Shapiro and Montgomery, 1993; Guinn and Schubert, 1993; Montgomery and Kallenbach, 1997; Möller and Montgomery, 2000; Chen and Yau, 2001; Wang, 2002a, b; Chen et al., 2003; McWilliams et al., 2003; Martinez, 2008; Martinez et al., 2011)[4];
- barotropic-baroclinic shear instabilities (Schubert et al., 1999; Nolan and Montgomery, 2000);
- eyewall mesovortices (Schubert et al., 1999; Montgomery et al., 2002; Braun et al., 2006); and
- trochoidal "wobbling" motion of the inner-core region (Nolan and Montgomery, 2000; Nolan et al., 2001).

Montgomery and Kallenbach (1997) presented rudimentary aspects of the vortex Rossby wave dynamics and related vortex amplification. Schubert et al. (1999) presented a complimentary model for barotropic shear instability and related vorticity mixing processes for vortex weakening. However, we must be mindful that the processes as presented in that work, as well as the other processes cited in the foregoing list, are generally not adiabatic and inviscid because they are often coupled to the frictional boundary layer and convection (Schecter and Montgomery, 2007; Nguyen et al., 2011). A new study devoted in part to addressing these concerns was published by Persing et al. (2013). Those authors conducted a systematic diagnosis of the simulated eddy dynamics during the intensification process for the prototype intensification problem and found that eddies can contribute positively to the intensification process. Both sheared vortex Rossby waves and rotating convective structures were found to be important elements of the spin up dynamics. However, a full thermo-mechanical explanation of the eddy dynamics is still lacking.

3.2 Intermediate Summary

From a three-dimensional fluid dynamics perspective, the intensification mechanism is via the generation of locally buoyant rotating updrafts and the corresponding convergence that the rotating updrafts induce both above and within the system-scale boundary layer. During the amplification process, Montgomery et al. (2009) found that the system-scale density temperature generally lagged behind the local density temperature within the rotating updraft cores. It is these cores that intrinsically drive the intensification

[4] The restoring mechanism for vortex Rossby waves and shear instabilities related thereto is associated with the radial and vertical gradient of dry Ertel potential vorticity of the system-scale vortex.

process until the system-scale temperature along absolute angular momentum surfaces nearly coincides with the local temperature of the rotating updraft cores. The azimuthally-averaged aspects of this spin-up process have been presented in Smith et al. (2009) and Persing et al. (2013). We refer the readers to the above mentioned references, where we present a brief comparison of the azimuthally-averaged view with the prior intensification models.

3.3 Predictability Aspects

Before closing this article we briefly discuss some of the implications of the foregoing findings on the predictability of tropical cyclone intensification. A very important practical issue is to investigate the predictability of the vortex intensity and asymmetric flow structure (e.g., Lorenz, 1969). In contrast to Lorenz (1969), the word "predictability" is used here in a qualitative sense to describe the degree of randomness of the ensuing vortex evolution. It is not unreasonable to expect that the asymmetric flow structures may be strongly influenced by deep convection, at least in the inner-core region, broadly defined as the region with radius less than twice the radius of maximum tangential wind speed. Since the pattern of deep convection is strongly influenced by the low-level moisture field, which is known to have significant variability on small space scales (e.g. Weckwerth, 2000), it is of interest to examine the variability of the inner-core asymmetries as a resultof random variations in the boundary-layer moisture distribution.

The sensitivity of the flow asymmetries to low-level moisture in an intensifying model cyclone was demonstrated in Nguyen et al. (2008) in an ensemble of ten experiments in which a small (± 0.5 g/kg) random moisture perturbation was added at every horizontal grid point in the innermost domain of the model below 900 mb. As a simple and convenient measure of the local intensity, we show here a time series of the maximum horizontal wind speed VT_{max} at 900 hPa, near the top of or just above the boundary layer. As shown in Fig. 9, the evolution in the local intensity as characterized by VT_{max} at 900 hPa is similar in all ensemble members, but there is a non-negligible spread, the maximum difference at any given time in the whole simulation being as high as 20 m/s.

The pattern of evolution of the flow asymmetries is significantly different between ensemble members also. These differences are exemplified by the relative vorticity fields of the control experiment and three randomly-chosen realizations at 24 hours from the Nguyen et al. (2008) ensemble (Figs 10 and 11). In general, inspection of the relative vorticity (Fig. 10) and vertical velocity fields (Fig. 11) of each ensemble member shows similar characteristics to those in the control experiment, the field being non-axisymmetric and still dominated by locally intense cyclonic updrafts. However, the detailed pattern of these updrafts is significantly different between the ensemble members.

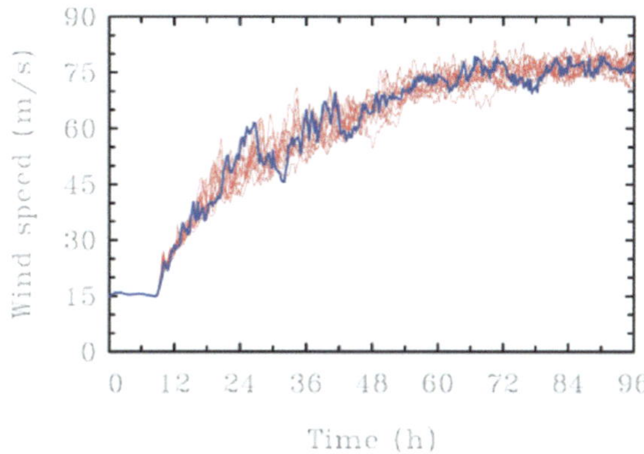

Fig. 9. Time-series of maximum total wind speed at 900 hPa in the control experiment in Nguyen et al. (2008) (blue) and in the 10-ensemble experiments therein (thin, red).

The foregoing results are supported by the calculations of Shin and Smith (2008), who performed similar ensemble calculations using the minimal three-layer hydrostatic model of Zhu and Smith (2003). Their calculations had twice the horizontal grid spacing used in Nguyen et al. (2008), and a much coarser vertical grid spacing (i.e., only three vertical levels). In fact, the calculations indicate a much larger variance between the ensembles than those in Nguyen et al., suggesting that the predictability limits may be resolution dependent also[5]. These findings suggest that the problem of estimating the "true" intensity fluctuations for real-world forecasting of these severe storms warrants further systematic investigation.

The implication of this work is as follows. Since the flow on the convective scales exhibits a degree of randomness, the convective scale asymmetries are intrinsically chaotic and unpredictable. Only the asymmetric features that survive in an ensemble average of many realizations should be regarded as robust.

---

[5] Recent work of Fang and Zhang (2011) has broadly verified the foregoing conclusions, but show a more modest variability in wind-speed intensity due to small perturbations in the initial condition. Aside from numerical model differences and spatial resolution dependencies summarized here, the difference in magnitudes of the intensity fluctuations is believed to be due in part to the retention of liquid water in the more complex microphysics option used in their WRF simulation. Nguyen et al. (2008) attributed the lower intensity found in their own 'warm-rain' simulations compared to the pseudo-adiabatic limit to a reduction in the convective instability that results from downdrafts associated with the rain process and to the reduced buoyancy in clouds on account of water loading. Therefore it seems scientifically plausible that the associated water loading in the low-to-mid troposphere that tempers the convective updrafts during intensification would temper also the maximum wind fluctuations in these more complete cloud representations.

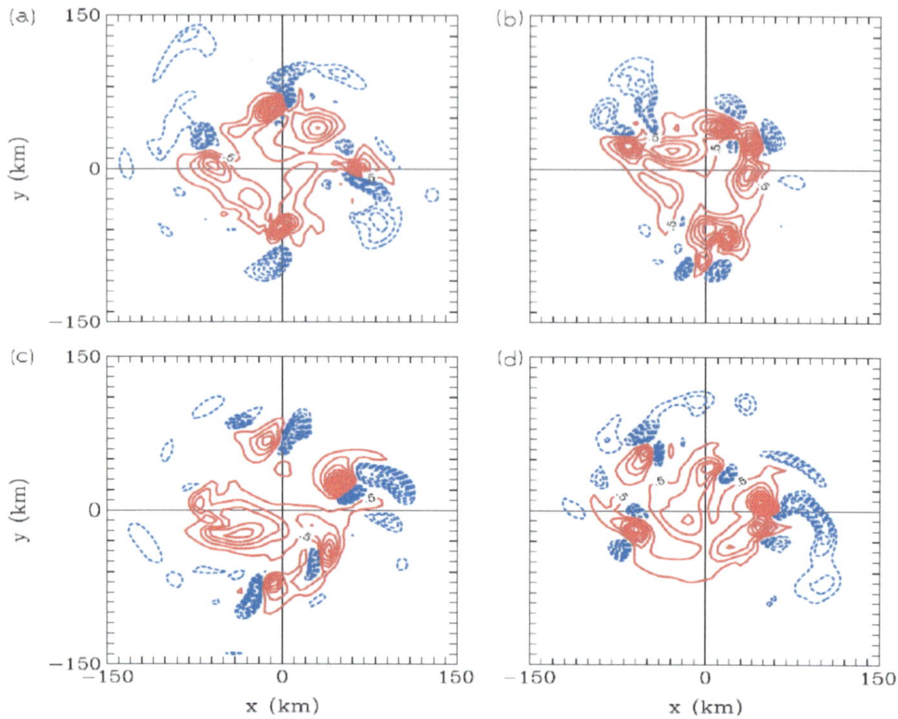

Fig. 10. Relative-vorticity field at 850 hPa of (a) the control experiment C0 in Nguyen et al. (2008), and (b, c, d) three representative realizations from the *f*-plane ensemble at 24 h by perturbing the boundary layer moisture as described in the text. Positive values are indicated by red solid contours, with a contour interval of 0.5×10^{-3} s^{-1}. Negative values are indicated by blue dashed contours, with a contour interval of 0.1×10^{-3} s^{-1}. The zero contour is not plotted.

3.4 Summary of Rotating Convection Paradigm

The foregoing simulations, both on an *f*-plane and on a β-plane (not shown here), show that the rotating updrafts dominate the intensification period at early times and that the progressive aggregation, merger and axisymmetrization of these cyclonic updrafts together with the low-level convergence they generate are prominent features of the tropical cyclone intensification process. As discussed in more detail by Montgomery and Smith (2014 and supporting refs.), the emerging flow asymmetries are not merely a manifestation of the numerical representation of the equations on a Cartesian coordinate grid mesh, but rather a reflection of a key physical process, namely moist convective instability. The results of Nguyen et al. (2008) support those of Montgomery et al. (2006) in an idealized study of the transformation of a relatively-weak, axisymmetric, middle-level, cold-cored vortex on an *f*-plane to a warm-cored vortex of tropical cyclone strength and also those of near cloud-resolving numerical simulations of tropical storm Diana (1984) (Davis and

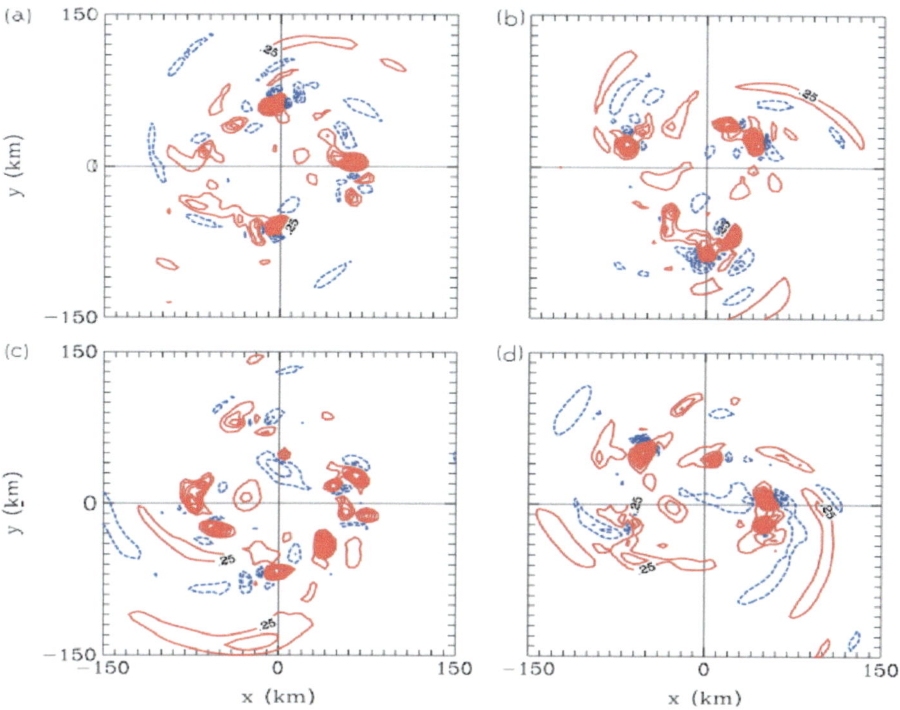

Fig. 11. Vertical-velocity field at 500 hPa of (a) the control experiment C0 in Nguyen et al. (2008), and (b, c, d) representative ensemble experiments at 24 h by perturbing boundary layer moisture as described in the text. Positive vertical velocity values are shown in red (solid contours); negative values are shown in blue (dashed contours). Contour interval is 0.25 ms^{-1} and the zero contour is not plotted.

Bosart, 2002; Hendricks et al., 2004), which drew attention to the important role of convectively-amplified vorticity in the spin-up process. In turn, the above results are supported by more recent calculations of the intensification process using a range of different cloud-representing models with a refined horizontal resolution (Fang and Zhang, 2011; Gopalakrishnan et al., 2011; Bao et al., 2012; Persing et al., 2013). All of these studies indicate that rotating convective updrafts are the basic coherent structures of the tropical cyclone intensification process.

Acknowledgements

Much of the work presented in this article has been conducted and written in collaboration with my close colleague and friend, Professor Roger K. Smith. The foregoing material is a highly abridged version of the review by Montgomery and Smith (2014) and is based on work carried out with our student and research colleagues over the past several years.

REFERENCES

Anthes, R.A., 1982: Tropical Cyclones: Their Evolution, Structure and Effects. Meteorological Monographs of the American Meteorological Society.

Bao, J.W., S.G. Gopalakrishnan, S.A. Michelson and M.T. Montgomery, 2012: Impact of physics representations in the HWRFX on simulated hurricane structure and pressure-wind relationships. *Mon. Wea. Rev.*, **140**, 3278-3299.

Bell, M.M. and M.T. Montgomery, 2008: Observed structure, evolution, and potential intensity of Category 5 Hurricane Isabel (2003) from 12 to 14 September. *Mon. Wea. Rev.*, **65**, 2025-2046.

Bui, H.H., R.K. Smith, M.T. Montgomery and J. Peng, 2009: Balanced and unbalanced aspects of tropical-cyclone intensification. *Q. J. R. Meteorol. Soc.*, **135**, 1715-1731.

Braun, S.A. M.T. Montgomery and Z. Pu, 2006: High-resolution simulation of Hurricane Bonnie (1998). Part I: The organization of eyewall vertical motion. *J. Atmos. Sci.*, **63**, 19-42.

Carrier, G.F., 1971: The intensification of hurricanes. *J. Fluid Mech.*, **49**, 145-148.

Charney, J.G. and A. Eliassen, 1964: On the growth of the hurricane depression. *J. Atmos. Sci.*, **21**, 68-75.

Chen, Y. and M.K. Yau, 2001: Spiral bands in a simulated hurricane. Part I: Vortex Rossby wave verification. *J. Atmos. Sci.*, **58**, 2128-2145.

Chen, Y., G. Brunet and M.K. Yau, 2003: Spiral Bands in a Simulated Hurricane. Part II: Wave Activity Diagnostics. *J. Atmos. Sci.*, **60**, 1240-1256.

Corbosiero, K.L., J. Molinari, A.R. Aiyyer and M.L. Black, 2006: The Structure and Evolution of Hurricane Elena (1985). Part II: Convective Asymmetries and Evidence for Vortex Rossby Waves. *Mon. Wea. Rev.*, **134**, 3073-3091.

Craig, G.C. and S.L. Gray, 1996: CISK or WISHE as a mechanism for tropical cyclone intensification. *J. Atmos. Sci.*, **53**, 3528-3540.

Davis, C.A. and L.F. Bosart, 2002: Numerical simulations of the genesis of Hurricane Diana (1984). Part II: Sensitivity of track and intensity prediction. *Mon. Wea. Rev.*, **130**, 1100-1124.

Dudhia, J. 1993: A nonhydrostatic version of the Penn State-NCAR mesoscale model: Validation tests and simulation of an Atlantic cyclone and cold front. *Mon. Wea. Rev.*, **121**, 1493-1513.

Emanuel, K.A., 1986: An air-sea interaction theory for tropical cyclones. Part I: Steady state maintenance. *J. Atmos. Sci.*, **43**, 585-604.

Emanuel, K.A., 1989: The finite amplitude nature of tropical cyclogenesis. *J. Atmos. Sci.*, **46**, 3431-3456.

Emanuel, K.A., 1994: Atmospheric convection. Oxford University Press.

Emanuel, K.A., 1995: Sensitivity of tropical cyclones to surface exchange coefficients and a revised steady-state model incorporating eye dynamics. *J. Atmos. Sci.*, **52**, 3969-3976.

Emanuel, K.A., 1997: Some aspects of hurricane inner-core dynamics and energetics. *J. Atmos. Sci.*, **54**, 1014-1026.

Emanuel, K.A., 2003: Tropical Cyclones. *Annu. Rev. Earth Planet. Sci.*, **31**, 75-104.

Emanuel, K.A., 2012: Self-Stratification of Tropical Cyclone Outflow. Part II: Implications for Storm Intensification. *J. Atmos. Sci.*, **69**, 988-996.

Fang, J. and F. Zhang, 2011: Evolution of multiscale vortices in the development of Hurricane Dolly (2008). *J. Atmos. Sci.*, **68**, 103-122.

Gopalakrishnan, S.G., F. Marks, X. Zhang, J.W. Bao, K.S. Yeh and R. Atlas, 2011: The experimental HWRF system: A study on the influence of horizontal resolution on the structure and intensity changes in tropical cyclones using an idealized framework. *Mon. Wea. Rev.*, **139**, 1762-1784.

Grell, A., J. Dudhia and D.R. Stauffer, 1995: A description of the fifth-generation Penn State/NCEP mesoscale model (MM5). NCAR Tech. Note NCAR/TN-398_STR.

Guinn, T. and W.H. Schubert, 1993: Hurricane spiral bands. *J. Atmos. Sci.*, **50**, 3380- 3403.

Hendricks, E.A., M.T. Montgomery and C.A. Davis, 2004: On the role of "vortical" hot towers in formation of tropical cyclone Diana (1984). *J. Atmos. Sci.*, **61**, 1209-1232.

Holton, J.R., 2004: An introduction to dynamic meteorology. Fourth Edition. Academic Press, London.

Holton, J.R. and G. Hakim, 2012: An introduction to dynamic meteorology. Fifth Edition. Academic Press, London.

Houze, R.A. Jr, S.S. Chen, B.F. Smull, W.C. Lee and M.M. Bell, 2007: Hurricane intensity and eyewall replacement. *Science*, **315**, 1235-1239.

Jordan, C.L. 1958: Mean soundings for the West Indies area. *J. Meteorol.*, **15**, 91-97.

Kossin, J., B.D. McNoldy and W.H. Schubert, 2002: Vortical swirls in hurricane eye clouds. *Mon. Wea. Rev.*, **130**, 3144-3149.

Lorenz, E.N., 1969: The predictability of a flow which possesses many scales of motion. *Tellus*, **21**, 289-307.

Lussier III, L.L., B. Rutherford, M.T. Montgomery, T.J. Dunkerton and M.A. Boothe, 2013: Examining the roles of the easterly wave critical layer and vorticity accretion during the tropical cyclogenesis of Hurricane Sandy. *Mon. Wea. Rev.*, **143**, 1703-1722.

Marks, F.D., P.G. Black, M.T. Montgomery and R.W. Burpee, 2008: Structure of the Eye and Eyewall of Hurricane Hugo (1989). *Mon. Wea. Rev*., **136**, 1237-1259.

Martinez, Y.H., 2008: Diagnostic study of hurricane asymmetries using empirical normal modes. Ph.D. thesis. McGill University.

Martinez, Y.H., G. Brunet, M.K. Yau and X. Wang, 2011: On the dynamics of concentric eyewall genesis: Spacetime empirical normal modes diagnosis. *J. Atmos. Sci.*, **68**, 458-476.

McWilliams, J.C., L.P. Graves and M.T. Montgomery, 2003: A formal theory for vortex Rossby waves and vortex evolution. *Geophys. Astrophys. Fluid Dynamics*, **97**, 275-309.

Möller, J.D. and M.T. Montgomery, 2000: Tropical cyclone evolution via potential vorticity anomalies in a three-dimensional balance model. *J. Atmos. Sci.*, **57**, 3366-3387.

Montgomery, M.T. and R.J. Kallenbach, 1997: A theory for vortex Rossby waves and its application to spiral bands and intensity changes in hurricanes. *Q.J.R. Meteorol. Soc.*, **123**, 435-465.

Montgomery, M.T., V.A. Vladimirov and P.V. Denissenko, 2002: An experimental study on hurricane mesovortices. *J. Fluid. Mech.*, **471**, 1-32.

Montgomery, M.T., M.E. Nicholls, T.A. Cram and A.B. Saunders, 2006: A vortical hot tower route to tropical cyclogenesis. *J. Atmos. Sci.*, **63**, 355-386.

Montgomery, M.T., S.V. Nguyen, R.K. Smith and J. Persing, 2009: Do tropical cyclones intensify by WISHE? *Q.J.R. Meteorol. Soc.*, **135**, 1697-1714.

Montgomery, M.T. and R.K. Smith, 2014: Paradigms for tropical-cyclone intensification. *Australian Meteorological and Oceanographic Journal*, Bruce Morton Memorial Volume, in press.

Montgomery, M.T., J. Persing and R.K. Smith, 2015: Putting to rest WISHE-ful misconceptions. *J. Adv. Model, Earth Syst.*, 07, DOI: 10.1002/2014MS000362.

Nguyen, S.V., R.K. Smith and M.T. Montgomery, 2008: Tropical cyclone intensification and predictability in three dimensions. *Q.J.R. Meteorol. Soc.*, **134**, 563-582.

Nguyen, C.M., M.J. Reeder, N.E. Davidson, R.K. Smith and M.T. Montgomery, 2011: Inner-core vacillation cycles during the intensification of Hurricane Katrina. *Q.J.R. Meteorol. Soc.*, **137**, 829-844.

Nolan, D.S., M.T. Montgomery and L.D. Grasso, 2001: The Wavenumber-One Instability and Trochoidal Motion of Hurricane-like Vortices. *J. Atmos. Sci.*, **58**, 3243-3270.

Nolan, D.S. and M.T. Montgomery, 2000: The algebraic growth of wavenumber one disturbances in hurricane-like vortices. *J. Atmos. Sci.*, **57**, 3514-3538.

Ooyama, K.V., 1964: A dynamical model for the study of tropical cyclone development. *Geophys. Int.*, **4**, 187-198.

Ooyama, K.V., 1969: Numerical simulation of the life cycle of tropical cyclones. *J. Atmos. Sci.*, **26**, 3-40.

Ooyama, K.V., 1982: Conceptual evolution of the theory and modeling of the tropical cyclone. *J. Meteor. Soc. Japan*, **60**, 369-380.

Ooyama, K.V., 1997: Footnotes to 'Conceptual Evolution'. Extended Abstracts, 22nd Conference on Hurricanes and Tropical Meteorology, American Meteorological Society, Boston.

Persing, J., M.T. Montgomery, J.C. McWilliams and R.K. Smith, 2013: Asymmetric and axisymmetric dynamics of tropical cyclones. *Atmos. Chem. Phys.*, **13**, 12229-12341.

Raymond, D.J. and K.A. Emanuel, 1993: The Kuo cumulus parameterization. *In:* The representation of cumulus convection in numerical models. Meteorological Monograph No. 46. American Meteorological Society, Boston, Mass. USA.

Rotunno, R. and K.A. Emanuel, 1987: An air-sea interaction theory for tropical cyclones. Part II: Evolutionary study using a nonhydrostatic axisymmetric numerical model. *J. Atmos. Sci.*, **44**, 542-561.

Schecter, D.A. and M.T. Montgomery, 2007: Waves in a cloudy vortex. *J. Atmos. Sci.*, **64**, 314-337.

Schubert, W.H., M.T. Montgomery, R.K. Taft, T.A. Guinn, S.R. Fulton, J.P. Kossin and J.P. Edwards, 1999: Polygonal eyewalls, asymmetric eye contraction, and potential vorticity mixing in hurricanes. *J. Atmos. Sci.*, **56**, 1197-1223.

Shapiro, L.J. and M.T. Montgomery, 1993: A three-dimensional balance theory for rapidly-rotating vortices. *J. Atmos. Sci.*, **50**, 3322-3335.

Shin, S. and R.K. Smith, 2008: Tropical-cyclone intensification and predictability in a minimal three-dimensional model. *Q.J.R. Meteorol. Soc.*, **134**, 1661-1671.

Smith, R.K., 1997: On the theory of CISK. *Q.J.R. Meteorol. Soc.*, **123**, 407-418.

Smith, R.K., 2006: Accurate determination of a balanced axisymmetric vortex. *Tellus*, **58A**, 98-103.

Smith, R.K., M.T. Montgomery and S. Vogl, 2008: A critique of Emanuel's hurricane model and potential intensity theory. *Q.J.R. Meteorol. Soc.*, **134**, 551-561.

Smith, R.K., M.T. Montgomery and S.V. Nguyen, 2009: Tropical cyclone spin up revisited. *Q.J.R. Meteorol. Soc.*, **135**, 1321-1335.

Smith, R.K. and G.L. Thomsen, 2010: Dependence of tropical cyclone intensification on the boundary-layer representation in a numerical model. *Q.J.R. Meteorol. Soc.*, **136,** 1671-1685.

Smith, R.K. and M.T. Montgomery, 2010: Hurricane boundary-layer theory. *Q.J.R. Meteorol. Soc.*, **136**, 1665-1670.

Wang, Y., 2002a: Vortex Rossby waves in a numerical simulated tropical cyclone. Part I: Overall structure, potential vorticity, and kinetic energy budgets. *J. Atmos. Sci.*, **59**, 1213-1238.

Wang, Y., 2002b: Vortex Rossby waves in a numerical simulated tropical cyclone. Part II: The role in tropical cyclone structure and intensity changes. *J. Atmos. Sci.*, **59**, 1239-1262.

Weckwerth, T.M., 2000: The effect of small-scale moisture variability on thunderstorm initiation. *Mon. Wea. Rev.*, **128**, 4017-4030.

Willoughby, H.E., 1995: Mature structure and evolution. *In:* Global Perspectives on Tropical Cyclones. WMO/TD No 693 (ed. R.L. Elsberry), World Meteorological Organization, Geneva.

Willoughby, H.E., 1998: Tropical cyclone eye thermodynamics. *Mon. Wea. Rev.*, **126**, 3053-3067.

Wissmeier, U. and R.K. Smith, 2011: Tropical-cyclone convection: The effects of ambient vertical vorticity. *Q.J.R. Meteorol. Soc.*, **137**, 845-857.

Zhu, H. and R.K. Smith, 2003: Effects of vertical differencing in a minimal hurricane model. *Q.J.R. Meteorol. Soc.*, **129**, 1051-1069.

22

Recent Advances in Tropical Cyclogenesis

Michael T. Montgomery
Department of Meteorology, Naval Postgraduate School
Monterey, California, USA
mtmontgo@nps.edu

1. Introduction

The genesis of tropical cyclones, hurricanes and typhoons has been regarded by some as one of the most important unsolved problems in dynamical meteorology (Emanuel, 2005) and climate (Gore, 2006). From a scientific point of view, the problem is certainly a fascinating one. This article begins with a brief synopsis of the problem of tropical cyclogenesis and then presents an overview of some recent work on the tropical cyclogenesis problem by the author and his colleagues. The research reviewed here points to a unified view of the genesis and intensification process. It provides also a basis for some new tools to aid in forecasting tropical-cyclone formation.

1.1 Terminology Defined

Before embarking on this review, it is appropriate to discuss some relevant terminology. The glossary on the Hurricane Research Division's website uses "tropical cyclone as the generic term for a nonfrontal synoptic-scale low-pressure system over tropical or sub-tropical waters with organized convection (i.e. thunderstorm activity) and a definite cyclonic surface wind circulation (Holland, 1993)." Notably, this definition does not invoke any wind threshold. The same glossary defines a tropical depression as a tropical cyclone with maximum sustained surface winds of less than 17 m s^{-1} (34 kt, 39 mph) and, in the Atlantic and Eastern Pacific Basins, a "tropical storm" as a tropical cyclone with surface winds between 17 ms^{-1} and 33 ms^{-1}.

In contrast, a universally accepted definition of tropical cyclogenesis does not exist. For example, Ritchie and Holland (1999) define genesis as: "... the series of physical processes by which a warm core, tropical-cyclone-scale vortex with maximum amplitude near the surface forms". In a recent paper, Nolan et al. (2007) use a wind speed threshold of 20 ms^{-1} to define the *time of genesis*. The formation of a tropical cyclone (TC), a phenomenon, commonly referred to as tropical cyclogenesis (hereafter, TC genesis), is a process by which some pre-existing, synoptic-scale or mesoscale weather feature in the tropics evolves so as to take on the characteristics of a tropical cyclone. The specific characteristics adopted by operational forecasting centres are as follows: a quasi-circular, closed circulation at or near the surface with the strongest circulating winds within or near the top of the atmospheric boundary layer and the presence of sustained deep moist convection near the centre of the circulation.

For the purpose of this chapter, tropical cyclogenesis will be defined as the formation of a tropical depression as described above and, like Ritchie and Holland *op. cit.*, but unlike Nolan, we impose no formal threshold on wind speed. We will refer to "intensification" as the amplification of the surface wind speed beyond the stage of tropical depression. Another issue, recognized long ago by Ooyama (1982), is the usage of the words "formation" and "genesis". For the purposes of this chapter we will use these terms interchangeably. Moreover, it will be argued later that, from the point of view of understanding the formation and intensification process, a precise definition of cyclogenesis, understood as the attainment of specific wind thresholds, is unnecessary.

1.2 Necessary Conditions for Tropical Cyclogenesis

Many disturbances travel across the tropical ocean basins during their respective tropical cyclone seasons, and yet relatively few become TCs. Since the pioneering work by Gray (1968), it has been known that a number of environmental conditions are favourable for genesis. For a cyclone to form six preconditions must be met:

1. Warm ocean waters (of at least 26.5°C) throughout a sufficient depth (unknown how deep, but at least on the order of 50 m). (Warm waters are necessary to fuel the heat engine of the tropical cyclone.)
2. An atmosphere which cools fast enough with height (is "unstable" enough) such that it encourages thunderstorm activity. (It is the thunderstorm activity that allows the heat stored in the ocean waters to be liberated for the tropical cyclone development.)
3. Relatively moist layers near the mid-troposphere (5 km). (Dry mid levels are not conducive for allowing the continuing development of widespread thunderstorm activity.)
4. A minimum distance of around 500 km from the equator. (Some of the earth's spin is generally needed to provide a background rotation that can subsequently be concentrated and amplified through the partial material conservation of absolute angular momentum. Systems can form closer to the equator, or even on the equator given a disturbance with sufficient local vertical rotation, but these are relatively rare events.)

5. A pre-existing disturbance near the surface with sufficient spin (vorticity) and inflow (convergence). (Tropical cyclones are generally not observed to form spontaneously. To develop, they require a weakly organized system with sufficient spin and low level inflow.)
6. Little change in the horizontal wind with height (low vertical wind shear, i.e. typically less than 40 km/h from surface to tropopause). (Large values of wind shear tend to disrupt the organization of the thunderstorms that are important to the inner part of a cyclone.)

Having these conditions met is necessary albeit insufficient since many disturbances that appear to have favourable conditions do not develop.

1.3 The Forecasting Problem

Over the past few decades there have been significant strides in improving tropical cyclone forecasting. Two formidable challenges that remain for TC forecasters are predicting TC intensity changes and forecasting TC genesis. The problem of forecasting TC genesis remains difficult for a number of reasons: (1) The structure and three-dimensional flow fields associated with pre-genesis systems are poorly understood, in part due to the relative lack of reconnaissance aircraft data and research campaigns compared to mature TCs; (2) TC genesis is often not well handled by operational forecast numerical models; and (3) there are competing theories for the fundamental processes involved, that culminate in the transformation of a cluster of thunderstorms into an organized tropical cyclone (Tory and Frank, 2010).

1.4 "Top down" Versus "Bottom up" Views

There are two main viewpoints that have arisen to the forefront of the TC genesis discussion in the last two decades. The first includes two distinct variations, and they will be grouped together as "top-down". Variation 1, or "Top-down merger" (Ritchie and Holland, 1997, 1993) features two mid-level vortices originating from the stratiform region of neighbouring mesoscale convective systems that interact, creating an area of enhanced cyclonic vorticity that appears to grow downwards. The downward growth of the merged circulation is a function of the local Rossby radius of deformation for asymmetric disturbances, NH/f_{local}, where N is the Brunt-Vaissalla frequency, H is an equivalent depth for internal gravity waves forced by deep convection and $f_{local} = \sqrt{(f+\zeta)(f+2v_{\tan}/r)}$ is the effective Coriolis parameter for the local environment assuming weak departures of axisymmetric flow (Shapiro and Montgomery, 1993; Hoskins et al., 1985). According to this model, if the merged circulation extends to the surface it will promote the spin-up of an organized system that can result in the initiation of a tropical cyclone.

Variation 2, or the "Top-down showerhead" theory of tropical cyclogenesis, was developed by Bister and Emanuel (1997) in their analysis of the data collected in the eastern Pacific basin during the TEXMEX experiment conducted in the summer of 1991, based in Mexico. Unlike the model of Ritchie and Holland *op. cit.*, this model requires only a single mesoscale convective vortex with maximum circulation at middle levels (near 500 hPa). This study emphasized the importance of thermodynamical processes within a so-called "mesoscale convective vortex embryo". The study proposed that the development of a cool, moist environment resulting from stratiform rain serves as the incubation region for the formation of a low-level, warm-core cyclonic vortex. The study suggested that sustained precipitation in the stratiform cloud deck together with the evaporation of rain drops below would gradually cool and saturate the layer below cloud base while transporting cyclonic vorticity downwards to the surface. The idea is that there will be an accompanying increase in near-surface winds that would increase surface moisture fluxes leading ultimately to convective destabilization. A subsequent bout of deep convection was hypothesized to induce low-level convergence and vorticity stretching, thereby increasing the low-level tangential winds and "igniting" an amplification process fuelled by the increased surface moisture fluxes (the WISHE mechanism, see Chapter 21).

Some questions about the dynamics of the "pre-ignition" process have been raised by Tory and Montgomery (2006) who noted, in particular, the inconsistency with vorticity substance impermeability between isobaric surfaces (Haynes and McIntrye, 1987). An equivalent way to understand this inconsistency is through the use of absolute angular momentum M (see Chapter 21). The downward advection of M by the downdraught is accompanied by horizontal divergence, which moves M surfaces outwards. The net effect of this process is one in which the M is materially conserved. Since the absolute circulation is proportional to M, the absolute circulation similarly will not change. Thus the hypothesized mechanism cannot increase the absolute circulation of the lower troposphere above that of the initial middle-level vortex, and cannot by itself lead to a net spin up of the low-level circulation. As we saw in Chapter 21, concerns arise also about the assumed air-sea interaction feedback known as WISHE. Notwithstanding these issues, the thermodynamical aspects of the genesis process are still important and interesting and have been investigated further in Raymond et al. (2011), Smith and Montgomery (2012), Montgomery and Smith (2012) and Wang (2012). The new thermodynamic findings in the context of the new cyclogenesis model are discussed later in Section 4 of this article.

The "top down" viewpoints have in recent years been challenged by a "Bottom-up" view of TC genesis. Although some of the basic elements of this viewpoint are not new, there are new and important elements of the theory to be discussed below. These ideas have been advanced primarily by the author and his colleagues. A new element of the theory recognizes the presence of deep cumulus convection in the form of "vortical hot towers" (VHTs) that act to concentrate and spin-up relatively large areas of near-surface vorticity (Hendricks et al., 2004; Montgomery et al., 2006). Another new element invokes the existence of a lower-tropospheric "sweet spot" within a favourable region of cyclonically recirculating flow in a layer spanning the surface to lower/middle troposphere. The latter region is referred to by the author and his colleagues as a "Pouch", the name given to a protected area of vorticity and high moisture in a lower-tropospheric critical layer that moves along with a parent easterly wave (Dunkerton et al., 2009; Montgomery et al., 2012). When an approximately aligned pouch exists in the middle and lower troposphere, and thermodynamic and kinematic conditions are favourable as outlined in the six necessary conditions above, cumulus convective activity within and near the pouch will cause the local circulation to strengthen through horizontal convergence on the system scale circulation as vortex tubes are stretched vertically locally, drawn inwards and amalgamated near the sweet spot.

The two foregoing viewpoints are not necessarily mutually exclusive, and work remains to be done to determine what pieces of the theories best fit the observations of TC genesis.

Bottom-up TC genesis was one of the foci of two recent field experiments called Tropical Cyclone Structure 2008 (TCS08; Elsberry and Harr, 2008) and PRE-Depression Investigation of Cloud systems in the Tropics (PREDICT; Montgomery et al., 2012). The PREDICT experiment was designed to study exclusively pre-depression systems and test the so-called "marsupial hypotheses" discussed further below. The overarching goal of PREDICT was to obtain a better understanding of the processes involved in TC genesis. The exclusive focus on TC genesis sets the PREDICT project apart from other TC field campaigns, such as NOAA's Intensity Forecasting Experiment (IFEX) or NASA's Genesis and Rapid Intensification Processes (GRIP) experiment, both of which have multiple goals, and provides a unique opportunity for the extensive study of pre-depression cases. Both TCS08 and PREDICT build on the TEXMEX experiment and NASA's Tropical Cloud Systems and Processes Experiment (TCSP; Halverson et al., 2007) conducted in 2005 by adding many more new genesis cases to study with consecutive sampling over many days.

2. The Crux of the Problem: The Formation of Tropical Depressions

The development of tropical depressions is inextricably linked to synoptic-scale disturbances that come in a variety of forms. The most prominent synoptic-scale disturbances in the Atlantic basin are African easterly waves. Typically, they have periods of 3-5 days and wavelengths of 2000-3000 km (e.g. Reed et al., 1977).

The parent easterly waves over Africa and the far eastern Atlantic are relatively well studied, as in the classic GATE campaign in 1974 and more recently in NASA AMMA (2006). Sometimes a vigorous, diabatically-activated wave emerging from Africa generates a tropical depression immediately, but in most instances these waves continue their westward

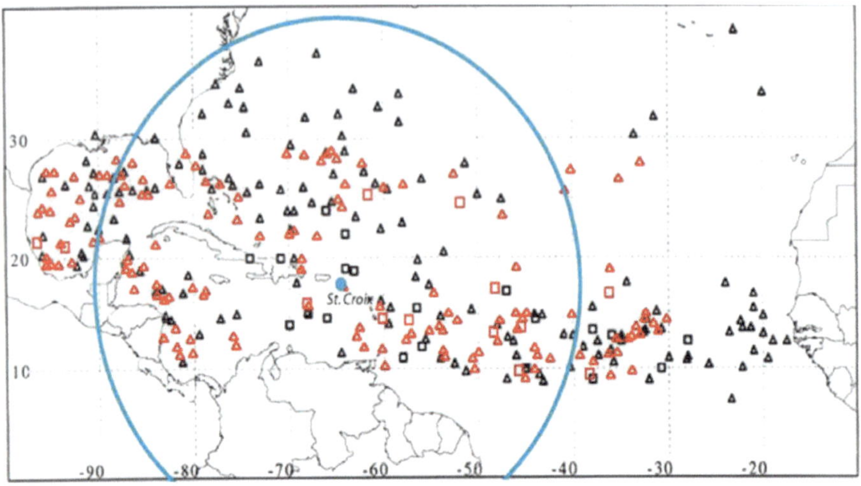

Fig. 1. First-detection locations of developing (triangles) and non-developing (squares) tropical depressions from 1975 to 2005 (1995-2005 in red), adapted from Bracken and Bosart (2000). The blue circle denotes the approximate PREDICT domain.

course harmlessly over the open ocean, or blend with new waves excited in the mid-Atlantic ITCZ. In a minority of waves, the vorticity anomalies they contain become seedlings for depression formation in the central and western Atlantic and farther west.

Figure 1 shows the detection locations of developing and non-developing tropical depressions from 1975 to 2005 based on the work of Bracken and Bosart (2000). It is apparent that there are relatively few Atlantic tropical depressions that fail to become tropical cyclones. It is well known also that most (approximately 80%) tropical waves do not become tropical depressions. This fact is supported by numerous studies (e.g., Frank, 1970; DMW09, their footnote 3). The key questions would appear to be:

- Which tropical waves (or other disturbances) will evolve into a tropical depression?
- What is different about developing waves?
- Can this difference be identified, and on what time scale?
- Why do so few disturbances develop?

2.1 The Multi-scale Nature of the Problem

The multi-scale nature of tropical cyclogenesis within tropical waves is well-known and substantial progress in further understanding the genesis sequence has been made in the past few years. The processes are illustrated schematically in Fig. 2 (after Gray, 1998). There, two length scales are illustrated, with a cluster of deep, moist convection confined to the trough of the synoptic-scale wave. Within these clusters are individual mesoscale convective systems (MCSs).

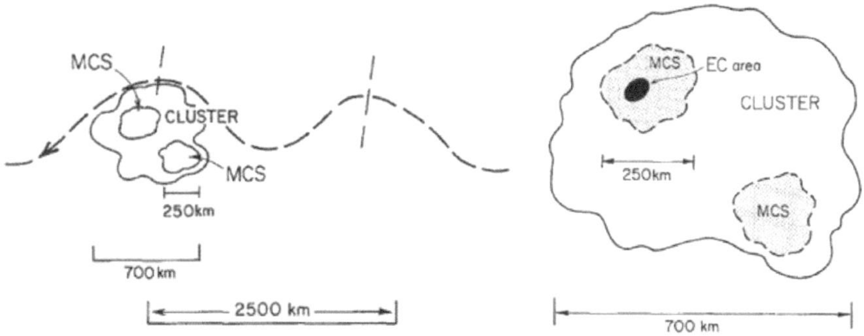

Fig. 2. (a) Schematic of synoptic-scale flow through an easterly waves (dashed) with an embedded cluster of convection in the wave trough. In (b) the cluster is shown to contain mesoscale convective systems (MCSs) and extreme convection (EC, black oval) within one of the MCSs. From Gray (1998).

As discussed by Dunkerton et al. (2009, hereafter DMW09), "...critically important processes and their multi-scale interactions thought to be involved in cyclogenesis have been a challenge to model and observe. Nature in some cases provides little advance warning of these storms and prediction of genesis beyond 48 h is generally too uncertain to be useful. Funding and technological resources are needed to remedy these deficiencies, to the extent they can be remedied, but it is unlikely that fundamental progress will be made without a quantum leap in theoretical understanding as well as the available observations on the synoptic scale need to be analyzed in a manner that is consistent with the Lagrangian[1] nature of tropical cyclogenesis. In the earliest stage of genesis, the fluid motion is mostly horizontal and quasi-conservative, punctuated by intermittent deep convection, a strongly diabatic and turbulent process. In order to fully appreciate the transport of vorticity and moist entropy by the flow, their interaction with one another, the impacts of deep convective transport and protection of the proto-vortex from hostile influences requires, among other things, an understanding of material surfaces or "Lagrangian boundaries" in the horizontal plane. This viewpoint, although used subconsciously by forecasters, is invisible to researchers working with standard meteorological products in an Eulerian or Earth-relative framework."

2.2 A Lagrangian Flow Perspective

This new Lagrangian perspective has been applied largely to tropical cyclone formation in the Atlantic and eastern Pacific basins. Some inroads have been forged also into understanding the Lagrangian flow dynamics for developing

[1] In a broad sense, Lagrangian refers to the practice of following an air parcel or a flow feature. Here we are following a tropical wave disturbance. For the present purposes we are invoking a wave-relative viewpoint following the trough axis of the wave.

and non-developing westward-propagating disturbances in the western Pacific region during the TCS08 experiment. For example, Montgomery et al. (2010b), Raymond and Lopez-Carillo (2011), Montgomery and Smith (2012) and Lussier et al. (2014) analyzed a well-observed formation case from a westward-propagating disturbance during the TCS08 experiment. Because of the strong similarities between this particular case and easterly wave formation events in the Atlantic basin, most of this chapter will focus on the formation of tropical depressions in the Atlantic basin. The formation of tropical depressions in other storm basins, originating from synoptic-scale precursors other than easterly-like waves, is a topic of active research and scientific progress using the new Lagrangian perspective will be communicated in due course.

The new cyclogenesis model (summarized below) offers a new understanding into how locally-favourable recirculation regions are generated *within synoptic-scale precursor disturbances* in the lower troposphere, and has gone a substantial way to providing answers to the foregoing questions. On the one hand, these circulation regions help protect seedling vortices from the detrimental effects of vertical and horizontal shearing deformation and from the lateral entrainment of dry air. On the other hand, they favour sustained column moistening and low-level vorticity enhancement by vortex-tube stretching in association with deep cumulus convection.

3. The Marsupial Paradigm for Tropical Easterly Waves

DMW09 proposed a new model for tropical cyclogenesis that recognizes the intrinsic multi-scale nature of the problem from synoptic, sub-synoptic, mesoscale and cloud scales. Using three independent datasets—ECMWF5 Reanalysis data, TRMM6 3B42 3-hourly precipitation and the best track data from the National Hurricane Center (NHC)— the Kelvin cat's eye within the critical layer of a tropical easterly wave, or the wave "pouch", was hypothesized to be important to tropical storm formation because:

1. Wave breaking or roll-up of cyclonic vorticity and lower-tropospheric moisture near the critical surface in the lower troposphere provides a favourable environment for the aggregation of vorticity seedlings for tropical-cyclone formation;
2. The cat's eye is a region of approximately closed circulation, where air is repeatedly moistened by deep moist convection and protected to some degree from dry air intrusion; and
3. The parent wave is maintained and possibly enhanced by diabatically-amplified mesoscale vortices within the wave.

This genesis sequence and the overarching framework for describing how such hybrid wave-vortex structures become tropical depressions is likened to the development of a marsupial infant in its mother's pouch wherein the juvenile proto-vortex is carried along by the mother parent wave until it is strengthened into a self-sustaining entity. A survey of 55 named storms in the Atlantic and eastern Pacific sectors during August–September 1998–2001 was shown to support this so-called "marsupial paradigm". Tropical cyclogenesis tended to occur near the intersection of the trough axis and the critical surface of the wave, the nominal centre of the cat's eye. The marsupial paradigm provides a useful road map for exploration of synoptic-mesoscale linkages essential to tropical cyclogenesis.

An idealized schematic of a tropical easterly wave in the "old" (traditional) and "new" (wave-pouch) flow viewpoints is sketched in Fig. 3. A deep pouch extending from the surface to the approximate 500 hPa level can protect the enhanced vortical structures generated by convection from the hostile tropical environment. Examples of such hostile environments are the dry air masses associated with the Saharan air layer or environments with strong vertical or horizontal wind shear. The sketch is drawn for the situation in which there is weak convergence into the pouch region in the lower troposphere. For such a circumstance, the pouch has an opening that allows the influx of environmental air and vorticity (Dunkerton et al., 2009; Riemer and Montgomery, 2011).

Figure 3 shows an idealized schematic of a tropical easterly wave in the "old" and "new" flow geometry viewpoints. The thick red line indicates the easterly jet maximum of the eastern- and mid-Atlantic basin. The dashed black curves represent the easterly wave's streamlines in the ground-based frame of reference, which is usually open and has an inverted-V pattern. The solid black curves delineate the wave pouch as viewed in the frame of reference moving at the same zonal speed with the easterly wave. The preferred region of persistent convection is indicated by pink shading. The thick purple and black curves represent the local critical latitude and the trough axis, respectively. The critical latitude on a particular pressure surface is defined by the locus of points satisfying $U = c\_x$, where c_x denotes the wave's zonal phase speed and U denotes the local zonal wind. The intersection of these two curves pinpoints

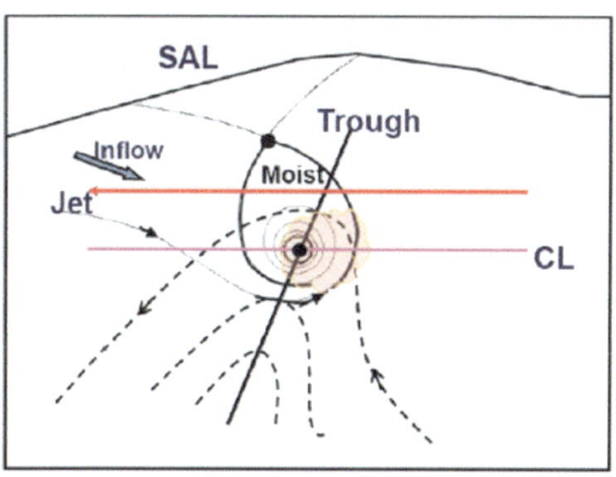

Fig. 3. Idealized schematic of a tropical easterly wave in the "old" and "new" flow geometry viewpoints. (Figure taken from Wang et al., 2010)

the pouch centre (or "sweet spot"), which Montgomery et al. (2010) and Wang et al. (2010) have shown to be the preferred location for vorticity aggregation and tropical cyclogenesis within easterly wave disturbances[1]. Relative inflow represented by a thick black arrow. (Figure taken from Wang et al., 2010.)

The marsupial model subsumes many of the prior ideas regarding tropical cyclone formation summarized in Section 1.4.

3.1 Fundamental Coherent Structures within the Pouch

Vortical hot towers (or VHTs) have been identified as fundamental coherent structures in both the tropical cyclone genesis process (Hendricks et al., 2004; Montgomery et al., 2006; Braun et al., 2010; Fang and Zhang, 2010) and the tropical-cyclone intensification process (Nguyen et al., 2008; Shin and Smith, 2008; Montgomery et al., 2009; Fang and Zhang, 2011). A widely accepted definition for these vortical convective structures does not yet exist, but the definition by DMW09 highlights the essential physical characteristics, namely, "deep moist convective clouds that rotate as an entity and/or contain updrafts that rotate in helical fashion (as in rotating Rayleigh-Bénard convection). These locally buoyant vortical plume structures amplify pre-existing cyclonic vorticity by at least an order of magnitude larger than that of the aggregate vortex." Even for background rotation rates as low as that of an undisturbed tropical atmosphere away from the equator, cloud model simulations demonstrate this tendency to amplify planetary vorticity by vortex-tube stretching on time scales of an hour (Saunders and Montgomery, 2004; Weismeir and Smith, 2011; Kilroy and Smith, 2012). These cloud model simulations indicate also that the induced horizontal circulation by a single updraught is typically no more than a few metres per second with a horizontal scale of around 10 km, and would be barely detectable by normal measurement methods in the presence of an ambient wind field. All of these results together suggest that non-shallow tropical convection away from the equator is vortical to some degree and can amplify the vertical vorticity by between one and two orders of magnitude. It is not hard to imagine, then, that the stretching of vertical vortex tubes by a developing cumulus cloud is a fundamental process. Based on this accumulating evidence, we believe that a precise quantitative definition of a VHT in terms of the degree of vorticity amplification and possibly updraught strength is not required and for this reason a precise definition will not be pursued here.

---

[1] Here, we follow tradition in tropical meteorology and adopt an equivalent-barotropic approximation of the travelling easterly wave. In this view, the mid to lower troposphere disturbance travels approximately in unison as a quasi-normal mode. Deviations of this picture can and do occur, and the Orr mechanism can become operative here. The westward propagating case analyzed as part of the TCS08 experiment is thought to be such an example whereby the middle level disturbances slowly overtakes the lower level disturbance near the time of the genesis (see Raymond and Lopez-Carillo, 2011 and Lussier et al., 2014 for details).

3.2 Observational Evidence for Vortical Convection in Pre-Storms

The discovery of VHTs in three-dimensional numerical model simulations of tropical cyclogenesis and tropical-cyclone intensification has motivated efforts to document such structures in observations. Two early studies were those of Reasor et al. (2006), who used airborne Doppler radar data to show that VHTs were present in the genesis phase of Hurricane Dolly (1996), and Sippel et al. (2006), who found evidence for VHTs during the development of Tropical Storm Alison (2001). It was not until very recently that Houze et al. (2009) presented the first detailed observational evidence of VHTs in a depression that was intensifying and which subsequently became Hurricane Ophelia (2005). The specific updraught that they documented was 10 km wide and had vertical velocities reaching 10-25 ms^{-1} in the upper portion of the updraught, the radar echo of which reached to a height of 17 km. The peak vertical velocity within this updraught exceeded 30 ms^{-1}. This updraught was contiguous with an extensive stratiform region on the order of 200 km in extent. Maximum values of vertical relative vorticity averaged over the convective region during different fly-bys of the convective region were on the order of 5-10 $\times$ 10^{-4} s^{-1} (see Houze et al., 2009, their Fig. 20).

Bell and Montgomery (2010) analyzed airborne Doppler radar observations from the recent TCS08 field campaign in the western North Pacific and found the presence of deep, buoyant and vortical convective features within a vertically-sheared, westward-moving pre-depression disturbance that later developed into Typhoon Hagupit. Raymond and Lopez-Carrillo (2011) carried out a similar analysis of data from the same field experiment, in their case for different stages during the formation and development of Typhoon Nuri and provided further observational evidence for the existence of VHT-like structures.

3.3 A Mesoscale View of Tropical Cyclogenesis within an Easterly Wave

Having discussed the fundamental coherent structures of the surface spin up process, we now move back up in scale to gain a new perspective into how the large scale and small work together on the mesoscale. Figure 4 summarizes the evolution of relative vertical vorticity and column (-integrated) relative humidity (saturation fraction (SF)) during the simulated transition of a tropical disturbance into a tropical storm. The basic formation sequence is simulated using the WRF model and is “a revisit with new eyes” of the classical Tuleya and Kurihara (1982) modelling study using the environmental characteristics of the western Atlantic tropical region during the GATE experiment (see Montgomery et al., 2010a for details). Figure 4a shows the maximum relative vorticity along each latitude near and within the wave’s cyclonic critical layer and shows cyclonic vorticity with values larger than 10 $\times$ 10^{-5} s^{-1}. Figure 4b shows the maximum column relative humidity (in %) from the surface to 500 hPa of the wave during its transition to a tropical storm.

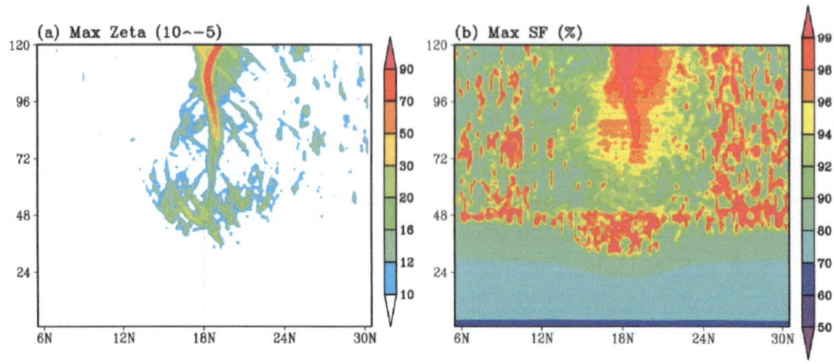

Fig. 4. Time-latitude plots of maximum vertical relative vorticity and saturation fraction for the high resolution simulation presented in Montgomery et al. (2010a): (a) depicts the time evolution of the maximum relative vorticity along each latitude. Vorticity values are depicted by the vertically shaded bar on the right. The x-axis indicates the latitude of the vorticity maximum and the y-axis is time in h (from day 0 to day 5); and (b) depicts the time evolution of the maximum saturation fraction (SF) in % from the sea surface to 500 hPa. SF values are depicted by the vertically shaded bar on the right.

Figure 4 shows that a concentrated and intense mesoscale vortex appears around 96 h (day 4) near the original critical latitude of the parent wave (18N). After this time, the vorticity intensifies primarily in a highly localized spatial region near the critical latitude of the parent wave disturbance, and the emergent vortex starts a slow northwestward drift away from the original critical latitude. The chaotic swarming of vorticity maxima about the critical latitude (prior to 96 h) followed by the highly focused vorticity concentration is a striking feature. Figure 4b shows a similar diagram of the analyzed saturation fraction. The emergent vortex, which develops into a tropical storm, forms in a region with high saturation fraction very near the intersection of the critical latitude of the parent wave and wave trough axis. Before storm formation the horizontal scale of the region of high saturation fraction is comparable to that of the region of large cyclonic vorticity. After consolidation to a single master vortex, there is a broader "funnel" of saturation fraction. This "attractor-like" property of predicted development near the sweet spot of the parent wave has been verified by the PREDICT observations described further below.

3.4 A Multiscale View of Tropical Cyclogenesis

The paper by DMW09 (p5596) offers a multiscale perspective of tropical cyclogenesis associated with tropical easterly waves: "the critical layers guarantee some measure of protection from intrusion. However, actual flow fields are transient and contain mesoscale fine structure, making the Lagrangian kinematics rather messy. A group of smaller vortices, for example, will entrain the surrounding air more readily than a single larger vortex. As for how these smaller vortices are created in the first place, there are essentially two possibilities: (i) *upscale aggregation* of mesoscale convective vortices associated with mesoscale convective systems and/or VHTs, and (ii) eddy shedding that works its way to smaller scales via a *forward enstrophy cascade*, such as might be associated with wave breaking at the critical layer. Tropical cyclogenesis evidently represents a kind of process in which the inverse energy and forward enstrophy cascades (originating respectively from cloud system and synoptic scales) collide in "spectral" space at some intermediate scale to form a diabatic vortex larger in horizontal scale than the vortices associated with individual cloud systems but substantially smaller in scale than the mother pouch created by the synoptic wave. This geophysical fluid dynamics aspect is perhaps the most fascinating and daunting of tropical cyclogenesis; one that has not yet been fully explored (owing to imitations of horizontal resolution in observations or models), but to be advanced as a framework for understanding the multi-scale nature of the problem."

Figure 5 presents an idealized schematic of this multi-scale view of the problem. Plotted on the abscissa is the natural logarithm of horizontal wavenumber with various motion systems indicated and whose horizontal wavenumber increases from left to right. Plotted on the ordinate is the natural logarithm of kinetic energy of the various motion systems. The arrows denote the direction of the two cascades. The downscale cascade from the synoptic to sub-synoptic (meso-α) scale resembles broadly the forward enstrophy cascade of quasi-two-dimensional turbulence theory in which strong jets and eddies irreversibly deform weaker eddies into filaments on progressively smaller scales (McWilliams, 2006).

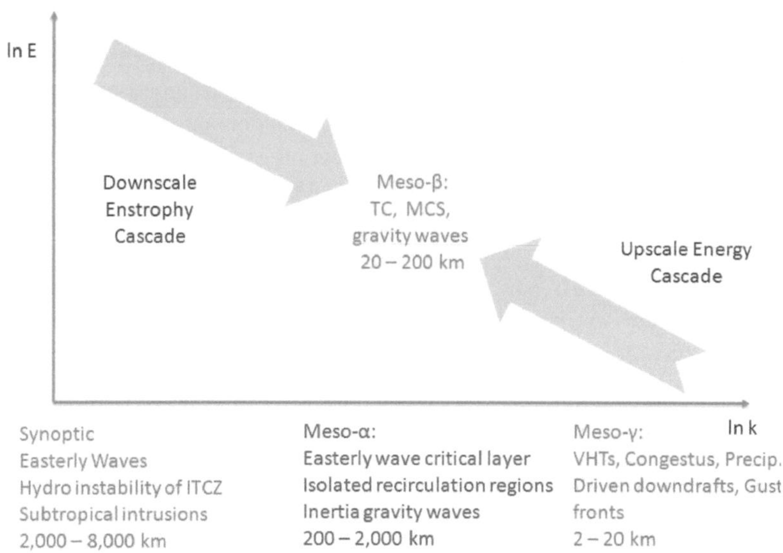

Fig. 5. A spectral view of the tropical cyclogenesis problem. The figure depicts a downscale cascade of enstrophy from the synoptic scale to the meso-α scale and an upscale cascade of energy from the cloud scale (meso-γ) to the meso-β scale. The tropical cyclone resides at the meso-β scale.

For the case of easterly waves, the forward enstrophy cascade depicted schematically in Fig. 5 is intended to include "eddy shedding events" from the mean easterly jet that create a sub-synoptic scale nonlinear critical layer, i.e., cat's eye, or pouch. The pouch circulation resides at meso-α. The upscale cascade from the cloud (meso-γ) scale to the larger (meso-β) scale of a tropical depression vortex is associated with the aggregation of the vortical convective elements in a region of high saturation fraction as described in the foregoing section. The latter cascade resembles also the inverse energy cascade of quasi-two-dimensional turbulence in which small vortical features merge to form progressively larger features, such as large eddies, Rossby waves and zonal jets (e.g., McWilliams, 2006; Vallis, 2007). An important difference between the simple two-dimensional inverse cascade model and the moist inverse cascade model sketched in Fig. 5 lies with the strong vortex-tube stretching and low-level spin up that is associated with vortical convection and the corresponding secondary circulation of individual elements that helps aggregate and concentrate the convectively-amplified vorticity. For such an inverse cascade to be maintained, energy input at small scales is necessary.

4. Highlights from the PREDICT Experiment

In the late summer of 2010, a trio of field campaigns was conducted by NASA, the National Oceanic and Atmospheric Administration (NOAA), and the National Science Foundation (NSF) to investigate tropical cyclogenesis in the Caribbean and West Atlantic and the subsequent intensification of named storms in these regions. While two of the campaigns included intensification in their portfolio of objectives, the Pre-Depression Investigation of Cloud-Systems in the Tropics (PREDICT) campaign was designed exclusively to study genesis (see Montgomery et al., 2012 for details). Priority was given to developing storms prior to their classification as tropical depressions even when mature storms were present nearby. The primary measurement platform of PREDICT was the NSF–National Center for Atmospheric Research (NCAR) Gulfstream V (GV), equipped with dropsondes and onboard sensors for meteorological variables and ice microphysics (see Table I of Montgomery et al., *op. cit.* for details). The range and speed of the GV, and the high altitude (~12–13 km) from which it could release dropsondes, were exploited to sample storm formation from Central America to the mid-Atlantic (roughly 40 west longitude) operating out of St. Croix in the U.S. Virgin Islands.

4.1 Practical Outcomes

Some of the main lessons learned, about the meso-α scale circulation during the PREDICT experiment, are as follows:

- The GV's ability to fly quickly to the target region at high altitude and safely navigate over or around deep convection proved that this sampling method, incorporated in the PREDICT proposal strategy, is highly effective for investigating tropical cyclogenesis in remote oceanic areas.

- Tropical cyclogenesis may be more predictable than previously thought. PREDICT demonstrated that genesis regions could be targeted more than two days in advance, and in some cases 4-day projections were useful. This is not to say that genesis itself was predictable on longer time scales, but our ability to anticipate the existence and approximate location of a pouch (and associated sweet spot) exceeded prior expectations of many of the PIs of the experiment. Recall that the sweet spot is the intersection of the wave trough axis and the wave critical line. This translates to an enhanced ability to anticipate the path along which genesis may occur, even though the exact timing of genesis remains uncertain due to the chaotic influence resulting from moist convection.
- A practical outcome is the realization of a trackable feature in forecast models that can be treated much the same way as a tropical cyclone centre, long before an identifiable organized storm exists! The predictability of the track of the sweet spot manifests qualitatively similar predictability as the track of a tropical cyclone. In particular, this suggests that ensembles of global and regional models should be effective in estimating a most likely path along which genesis can occur as well as providing the uncertainty in this path. Predictive skill using the pouch products exists to 72 hours and beyond, extending the 48-hour range currently employed by NHC.
- The developing cases from PREDICT show convection congealing around the sweet spot as genesis approaches (cf. Section 3.3).

Karl, a well-surveyed case during PREDICT, serves as an illustrative example of the foregoing skill using the Montgomery Group pouch products (Fig. 6). The GV sampled pre-Karl (PGI44) for five consecutive days from 10 to 14 September. During the PREDICT midday coordination sessions on

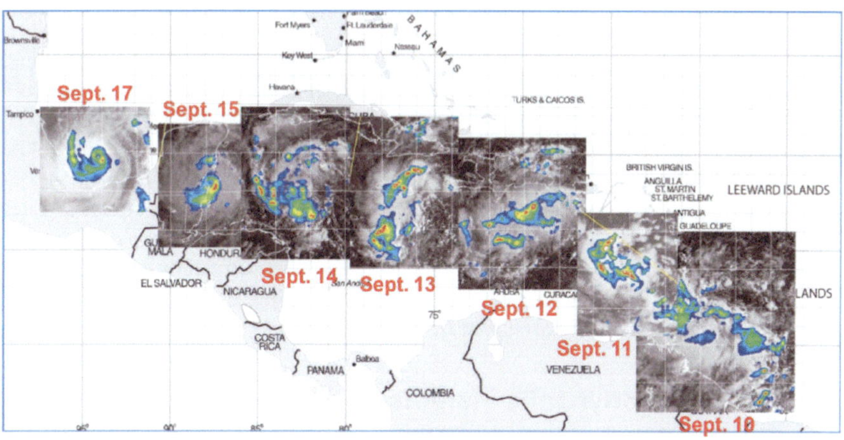

Fig. 6. 85-GHz montage (images courtesy of NRL-Monterey) for the active convection periods on each day from 10 to 17 (excluding 16 Sep) during the genesis of Karl (PG144). Note the small eye on the Yucatan coast on 15 Sep. (Figure taken from Montgomery et al., 2012).

13 September, the pouch products based on ECMWF forecasts from 0000 UTC 13 September were available for planning the following day's flight. ECMWF 36-h forecasts depicted a trough located along 82W at the flight time of 1200 UTC 14 September. In an Earth-relative frame, the circulation centre as depicted by 700-hPa streamlines was at about 17.3N and situated on the southern edge of a large region of positive values of the OW parameter (Fig. 7, left). The circulation in the co-moving frame (Fig. 7, right) is better defined than in the Earth-relative frame. The 700-hPa Earth-relative flow depicts a tropical wave with an inverted-V pattern and only weak westerly flow south of the vortex. The circulation centre in the co-moving frame of reference is located between two areas of high OW (Fig. 7, left) that appear to be wrapping around the pouch centre. A large lawnmower pattern was constructed that sampled both Earth-relative and co-moving circulations and a region beyond the central convection.

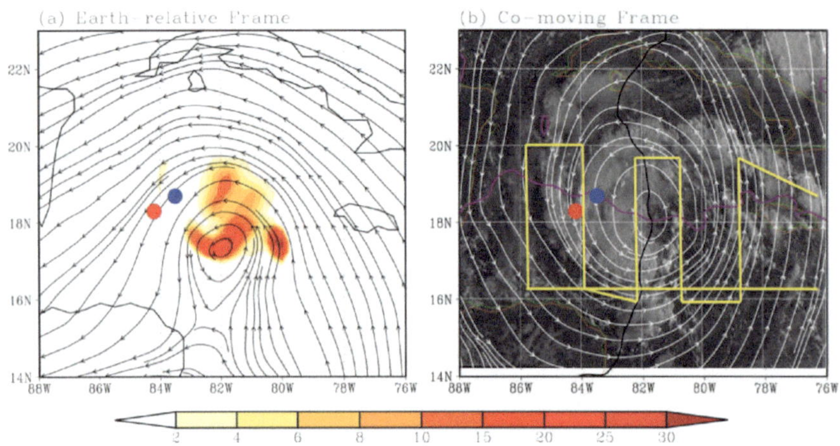

Fig. 7. ECMWF 36-h forecast of 700-hPa Earth-relative streamlines and OW (shading; units: 10^{-9} s^{-2}) centered on wave pouch PGi44/AI92 (pre-Karl) valid at 1200 UTC 14 Sep 2010. (Right) streamlines in the co-moving frame of reference (phase speed of 6.2 ms^{-1} westward), GOES visible imagery at 1225 UTC, and flight pattern of GV aircraft (yellow track). In the right figure, the black curve represents the trough axis and the purple curve the local critical latitude defined by $U = c\_x$, where c_x denotes the wave's zonal phase speed. The red dot represents the actual genesis location, and the blue dot is the ECMWF 700-hPa predicted sweet spot, defined by the intersection of the trough axis (black curve) and critical latitude (purple curve), at 2100 UTC 14 Sep 2010. (Figure taken from Montgomery et al., 2012).

4.2 New Thermodynamic Insights

The data collected during the TCS08 and PREDICT experiments provided unprecedented observations over consecutive days of developing and non-developing tropical disturbances that had been given some chance of development by professional forecasters.

4.2.1 Pouch-averaged and Regional Variations of θ_e

Perhaps the two most important thermodynamic observations in the PREDICT field campaign are that developing disturbances were found (i) to have a lower saturation deficit in mid-troposphere than non-developers, and (ii) to become increasingly moist in the days approaching genesis (Smith and Montgomery, 2012). Figure 8 shows an example from the disturbances pre-Karl and ex-Gaston. The thermodynamics of the well-observed Nuri genesis case during the TCS08 experiment are analyzed in Montgomery and Smith (2012) and the findings are consistent with that found in the PREDICT experiment discussed further below.

These observations, noted by Smith and Montgomery (2012) and Montgomery and Smith (2012), were evaluated in more detail by Wang (2012) who examined the radial dependence relative to pouch centre, and noted that the inner pouch region is moist relative to the outer pouch region, and becomes moister sooner, approaching genesis (see Fig. 9). These favourable characteristics of the inner pouch region are thought to be due to the fact that the inner-pouch region has near solid-body rotation flow and is not subject to adverse horizontal straining that is present on the outskirts of the pouch. The Okubo-Weiss parameter in Fig. 7a nicely illustrates this property for the genesis of Hurricane Karl. The minimum adverse shearing in the inner-pouch region provides a favoured region for persistent convection and convective moistening.

4.2.2 Classical and New Views on the Role of Dry Air on the Convective Scale within Pre-storm Disturbances

One of the outstanding questions that arose during the PREDICT experiment was why the ex-Gaston disturbance failed to re-develop. The dropwindsonde observations presented in Smith and Montgomery (2012) and Montgomery et al. (2012) indicate that from a thermodynamics-only perspective the most prominent difference between this non-developing system and the two developing systems (pre-Karl and pre-Matthew) was the much larger reduction of individual and pouch-averaged θ_e between the surface and a height of 3 km, typically 25 K in the non-developing system, compared with only 17 K in the developing systems (see Fig. 8).

Conventional wisdom would suggest that, for this reason, the convective downdrafts would be stronger in the non-developing system and would thereby act to suppress the development. Traditional reasoning argues that ensuing convection within a relatively dry, elevated layer of air would lead to comparatively strong downdrafts (e.g. Emanuel, 1994). From the perspective of convective dynamics, stronger downdrafts (implied by the lower relative humidity at heights between about 2 and 8 km in ex-Gaston inferred from the θ_e profiles in Figs 8(a) and 8(c) of Smith and Montgomery, *op. cit.*) would

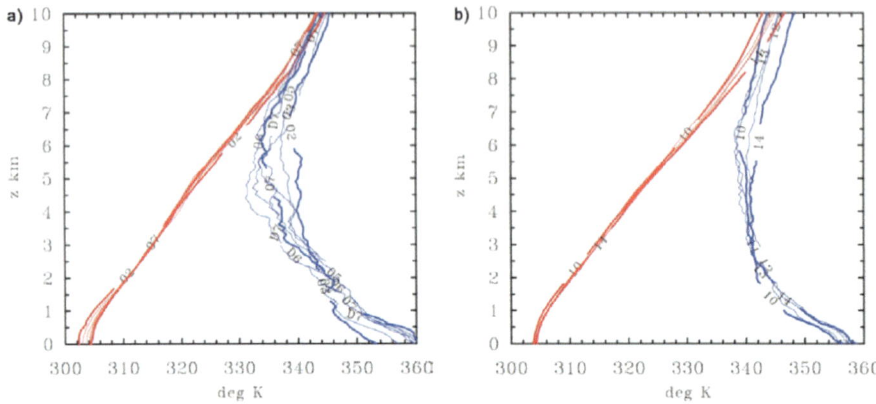

Fig. 8. Comparison of pouch-mean soundings of virtual potential temperature (θ_v) (red) and pseudo-equivalent potential temperature (θ_e) (blue) as a function of height derived from the NSF GV and NASA DC8 dropsondes in (a) the pouch of PGI38 (ex-Gaston) that failed to redevelop and (b) the pouch of PGI44 (pre-Karl) that did develop. Numbers on curves refer to the day of the flight mission in Sep 2010. The thick curves mark the first and last days of the GV flights. Curves for the NASA DC8 flights are denoted with the prefix D. (Figure taken from Montgomery et al., 2012.)

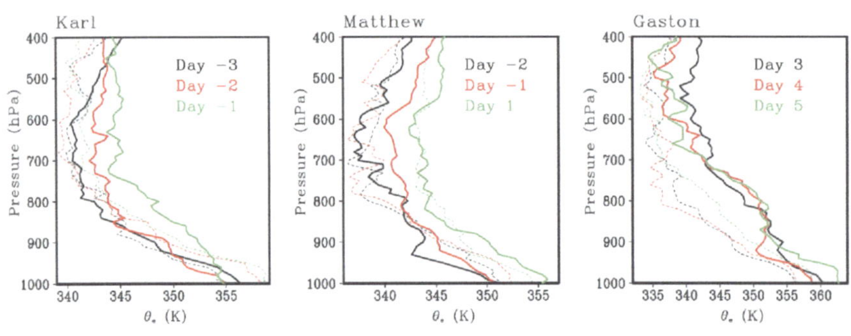

Fig. 9. Vertical profiles of θ_e for Karl, Matthew, and Gaston. Solid lines represent the averages over the inner pouch region, and dashed lines represent the averages over the outer pouch region. Different colours represent different days. (Figure taken from Wang, 2012.)

tend to import low θ_e into the boundary layer and frustrate the enhancement of boundary-layer θ_e by sea-to-air moisture fluxes. This enhancement is necessary to fuel subsequent deep convective activity. However, the thermodynamic data collected within Gaston's pouch showed a general day-to-day increase in the lower tropospheric θ_e (see Fig. 8a), associated in large part with an increase in the underlying sea-surface temperature as the disturbance moved from the western Atlantic into the Caribbean Sea. Thus it would appear that the traditional argument cannot be invoked to explain the non-re-development of Gaston.

Recent idealized high-resolution cloud model experiments conducted by Kilroy and Smith (2012) and James and Markowski (2009) have presented evidence that convective downdrafts are not strengthened by mid-level dry air, in non-supercell environments with CAPE less than about 3000 J/kg. These studies suggest that the principal effects of the dry air are to reduce the convective updraft strength and water loading, while the convective downdraft strengths are not changed appreciably. The dilution of the updraft with dry air would reduce cloud buoyancy, making the updraft less effective in amplifying vertical vorticity throughout a significant depth. These findings led Smith and Montgomery (2012) to propose an alternative hypothesis in which the drier air weakens the convective updrafts and thereby weakens the convective amplification of vertical vorticity in a layer spanning low to mid-levels necessary for tropical cyclogenesis.

4.3 A Proposed Explanation of the Non-redevelopment of ex-Gaston during PREDICT

A link between the dynamics and thermodynamics of the non-developing Gaston case was suggested by the shallowing of Gaston's pouch over the 5-day observation period. A proposed explanation of Gaston's failure to redevelop may be offered based on our work and colleagues from the PREDICT experiment by combining the new insights discussed in the foregoing subsection with recent work studying the Lagrangian flow trajectories inside and surrounding ex-Gaston's pouch.

As to the source of the dry air in ex-Gaston during the PREDICT campaign, there is evidence of a dry air mass surrounding the northern hemisphere of ex-Gaston in satellite-derived products (Montgomery et al., 2012, e.g., their Figure 5). A Lagrangian trajectory analysis indicates that the dry air documented in the dropsonde observations of Smith and Montgomery *op. cit.* is the result of a complex process of intrusion from outside of the pouch region (Rutherford and Montgomery, 2012; their Sec. 5.1.2 and their Figures 6–9, their Figure 6 is copied as Figure 10). The intrusion of dry air is associated in part with a deep-layer of vertical wind shear acting on the pouch after 2 September (Davis and Ahijevych, 2011; their Figure 11). The Lagrangian coherent structures and trajectory analyses demonstrate that the intrusion of dry air was occurring laterally within the 700 hPa and 500 hPa layer during the period from 1 to 5 September. Figure 10 illustrates the complex intrusion process suggested by the ECMWF global model analyses at the 700 hPa level. During this time the vertical structure of the pouch was significantly degraded and 5 September marks the point where the intrusion of dry air and escape of moisture through ventilation could not be overcome by convective moistening from below. By 5 September the pouch had already lost most of its vertical structure, as there was no upper level circulation remaining (Evans et al., 2012).

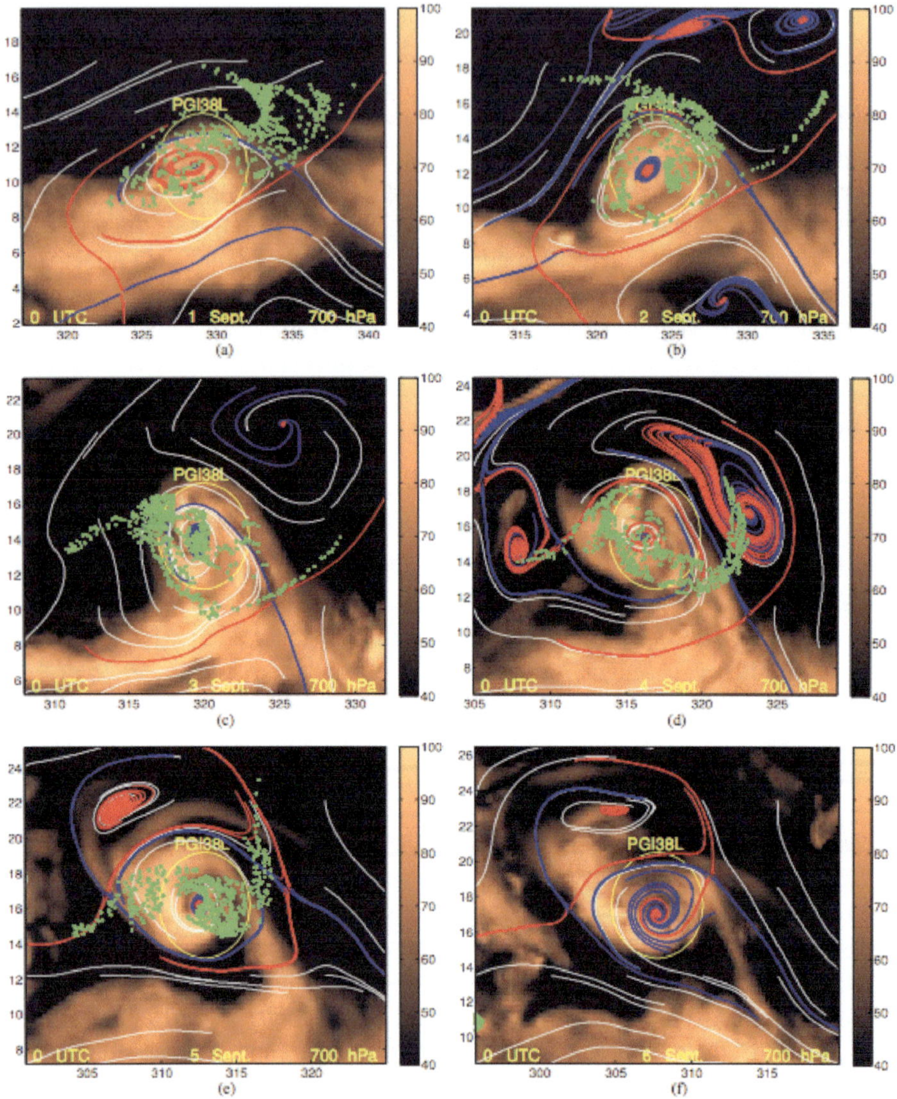

Fig. 10. The stable (red) and unstable (blue) manifolds of hyperbolic stagnation points and streamlines (white) of the co-moving frame are overlaid on the relative humidity fields (%) for Gaston from 1 September to 6 September at 700 hPa. Green dots indicate particles that will be within 3 degrees of the pouch centre in 48 h. (Figure taken from Rutherford and Montgomery, 2012.)

On the basis of the new view regarding dry air in convective dynamics discussed above, the combination of dry air intruding into a vertically sheared pouch sets the stage for weakened convective updrafts and a reduced ability to amplify vertical vorticity in the mid- and lower-troposphere on the system-scale circulation.

Recently a complimentary analysis of this case has appeared by Gjorgjievska and Raymond (2014). They suggest that the onset of the decay of ex-Gaston was caused by the observed decrease in the vertical mass flux with height at middle levels on 2 September, resulting in divergence at these levels and spindown of the mid-level vortex. These authors suggest that the spindown of the mid-level vortex was instrumental in the subsequent breakdown of the pouch and the intrusion of dry air into the core. However, why the convective mass flux profile took the form that it did is still a matter of scientific debate. These authors also suggest that the passage of the tropical storm Gaston into a region possessing a pronounced trade inversion (between 2 and 3 km height) on the periphery of the pouch was the critical factor in determining the vertical mass flux profile and the subsequent sequence of events. However, fluid dynamical considerations would suggest that wave breaking of the incipient pouch would carry the favourable properties into an environment, which may be unfavourable. Based on the foregoing findings (which are not necessarily mutually exclusive with the findings from Rutherford and Montgomery (2012)), we believe further research is needed to determine why Gaston failed to re-develop.

5. A Unified View of Tropical Cyclogenesis and Intensification

On the basis of the theoretical and observational evidence summarized in the foregoing sections, we believe that a unified view of tropical cyclogenesis and intensification is emerging. In this unified view, the separate stages proposed in previous significant studies and reviews (e.g. Frank, 1987; Emanuel, 1989; McBride, 1995; Karyampudi and Pierce, 2002; Tory and Frank, 2010) appear unnecessary. The idea that tropical cyclones in the current climate are a manifestation of a finite amplitude instability or that they are the result of some "trigger" mechanism is challenged by a new way of thinking about the basic processes of vortex spin up by vortical convection in a favourable tropical environment.

The basis for this unified view is that deep convection developing in the presence of vertical vorticity amplifies the vorticity locally by vortex tube stretching, irrespective of the strength of the updraught and the depth of convection (Wissmeier and Smith, 2010; Kilroy and Smith, 2012), and that the vortical remnants outlive the convection that produced them in the first place. The vertical remnants tend to aggregate in a quasi two-dimensional manner with a corresponding upscale energy cascade and some of these remnants will be intensified further by subsequent convective episodes. The amplification and aggregation of vorticity represents an increase in the relative circulation within a fixed circuit straddling the convective area. As the circulation progressively increases in strength, there is some increase in the surface moisture fluxes. However, it is not necessary that the moisture fluxes continue to increase with surface wind speed. The boundary layer pseudo-equivalent potential temperature, θ_e, will continue to rise as long as the air adjacent to the surface remains unsaturated relative to saturation at the sea-surface temperature and the positive entropy flux from the ocean surface overcomes downward import of low θ_e from above the boundary layer (Montgomery et al., 2009; Montgomery and Smith, 2014). The upshot is that the boundary layer θ_e will continue to increase towards the saturation value, providing air parcels acquire the needed boost in θ_e necessary for them to ascend the warmed troposphere created by prior convective events.

This unified view is consistent with the insights articulated long ago by Ooyama (1982), who wrote: "It is unrealistic to assume that the formation of an incipient vortex is triggered by a special mechanism or mechanisms, or that genesis is a discontinuous change in the normal course of atmospheric processes". "... It is far more natural to assume that genesis is a series of events, arising by chance from quantitative fluctuations of the normal disturbances, with the probability of further evolution gradually increasing as it [the process] proceeds. According to this view, the climatological and synoptic conditions do not directly determine the process of genesis, but may certainly affect the probability of its happening. With a better understanding of the mesoscale dynamics of organized convection, the range of statistical uncertainty can be narrowed down. Nevertheless, the probabilistic nature of tropical cyclogenesis is not simply due to lack of adequate data, but is rooted in the scale-dependent dynamics of the atmosphere."

The recent review of paradigms for tropical-cyclone intensification by Montgomery and Smith (2014) summarized in Chapter 21 is thought to be relevant also to understanding aspects of the emerging unified view of genesis and intensification. Although the genesis process summarized in the foregoing discussion is intrinsically non-axisymmetric, it is nonetheless insightful to adopt an axisymmetric viewpoint of this process as discussed in Smith et al. (2009) and Persing et al. (2013). A schematic for understanding the amplification of the azimuthally-averaged tangential wind field within the marsupial pouch is shown in Fig. 11. The idea is that the aggregate effects of diabatic heating associated with the vortical convective elements leads to a system-scale inflow in the lower troposphere. This diabatically-driven inflow can be represented approximately using axisymmetric balance dynamics in which the aggregate of diabatic heating, boundary layer friction and related eddy fluxes of heat and momentum force a meridional overturning circulation (Bui et al., 2009). This inflow converges azimuthal-mean absolute angular momentum, a quantity that is approximately materially conserved above the shallow frictional boundary layer, and outside regions of deep convection, so that its convergence leads to a spin up of the azimuthal-mean tangential winds. As these winds increase in strength, so does the azimuthal-mean radial inflow within the boundary layer (see e.g. Smith et al., 2009). As described above, this inflow converges moist air that has been enriched by surface fluxes from the ocean surface to "fuel" the deep convection.

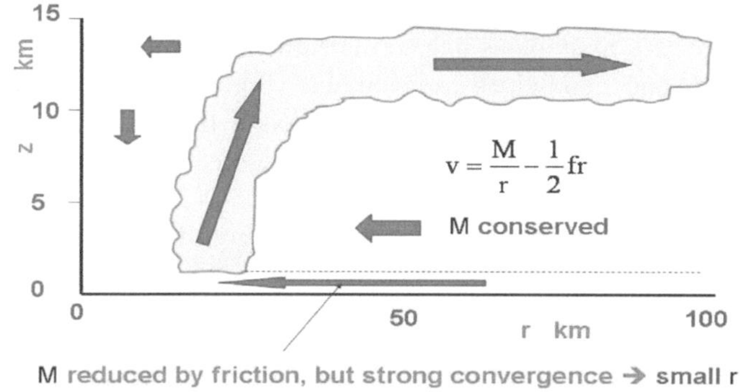

Fig. 11. Axisymmetric conceptual model of the dynamical elements of the unified view of tropical cyclogenesis and intensification as discussed in Section 5. This figure aims to convey the idea that, in an azimuthally-averaged sense, deep convection in the inner-core region induces convergence in the lower troposphere. Above the frictional boundary layer and outside regions of deep vortical convection, the inflowing air approximately materially conserves its absolute angular momentum (*M*) and spins faster. Strong convergence of moist air in the boundary layer provides moisture to "fuel" the deep convection. Although the air-parcels converging in the boundary layer lose a fraction of their *M*, they undergo much larger inward displacements and acquire a higher tangential wind speed than those converging above the boundary layer. (Figure taken from Montgomery and Smith, 2014).

The above description presumes that the boundary layer of the system-scale circulation has become well established. However, during the genesis phase when there is weak system-scale rotation, the boundary layer inflow associated with this rotation is much weaker than the inflow forced by the aggregate diabatic heating (e.g., Montgomery et al., 2006). As long as there is convergence above the boundary layer the system-scale rotation will amplify because of the convergence of absolute angular momentum. The corresponding boundary layer inflow will increase progressively. Although the air-parcels converging in the boundary layer lose a fraction of their absolute angular momentum, they undergo much larger inward displacements. A point is reached during the evolution at which the highest tangential wind speeds are found to occur in the boundary layer (Smith et al., 2009). Beyond this point in time, the boundary layer plays also a *dynamical role* in the spin up process because the amplification of the inner-core tangential winds occurs within this layer. In summary, the foregoing discussion indicates that the boundary layer exerts a progressive control on the vortex evolution as the system-scale rotation amplifies.

Acknowledgements

The author expresses his profound gratitude to his scientific colleagues and the PREDICT team for their contributions to the work summarized here. This work would not have been possible without the close collaboration of Roger

Smith, Tim Dunkerton, Zhuo Wang, Lou Lussier, Chris Davis, and Blake Rutherford. These individuals helped the author write pieces of the material surveyed here and also helped carry out analysis. Their work has been duly cited. The PREDICT experiment was supported by the U.S. National Science Foundation. The TCS08 experiment was supported by the U.S. Office of Naval Research. Finally, the author would be remiss in not thanking student M. Rajasekha for taking the initiative in assembling a first draft of my lecture materials from the 2012 summer school and helping frame some of the introductory material in the context of early pioneering work.

REFERENCES

Bell, M.M. and M.T. Montgomery, 2010: Sheared deep vortical convection in pre-depression Hagupit during TCS08. *Geophys. Res. Lett.*, **37**, L06802, doi: 10.1029/2009GL042313.

Bister, M. and K.A. Emanuel, 1997: The genesis of Hurricane Guillermo: TEXMEX analyses and a modeling study. *Mon. Wea. Rev.*, **125**, 2662-2682.

Bracken, W.E. and L.F. Bosart, 2000: The role of synoptic-scale flow during tropical cyclogenesis over the North Atlantic Ocean. *Mon. Wea. Rev.*, **128**, 353-376.

Braun, S.A., M.T. Montgomery, K. Mallen and P.D. Reasor, 2010: Simulation and interpretation of the genesis of Tropical Storm Gert (2005) as part of the NASA Tropical Cloud Systems and Processes Experiment. *J. Atmos. Sci.*, **67**, 999-1025.

Bui, H.H., R.K. Smith, M.T. Montgomery and J. Pen, 2009: Balanced and unbalanced aspects of tropical-cyclone intensification. *Q.J.R. Meteorol. Soc.*, **135,** 1715-1731.

Davis, C.A. and D.A. Ahijevych, 2012: Mesoscale Structural Evolution of Three Tropical Weather Systems Observed during PREDICT. *J. Atmos. Sci.*, **69**, 1284-1305.

Dunkerton, T.J., M.T. Montgomery and Z. Wang, 2009: Tropical cyclogenesis in a tropical wave critical layer: Easterly waves. *Atmos. Chem. Phys.*, **9**, 5587-5646.

Elsberry, R.L. and P. Harr, 2008: Tropical cyclone structure (TCS08). Field experiment scientific basis, observational platforms, and strategy. *Asia-Pacific J. Atmos. Sci.*, **44,** 1-23.

Emanuel, K.A., 2005: Divine Wind: The History and Science of Hurricanes. Oxford University Press.

Emanuel, K.A., J.D. Neelin and C.S. Bretherton, 1994: On large-scale circulations of convecting atmospheres. *Q.J.R. Meteorol. Soc.*, **120**, 1111-1143.

Emanuel, K.A., 1989: The Finite-Amplitude Nature of Tropical Cyclogenesis. *J. Atmos. Sci.*, **46**, 3431-3456.

Evans, C.A., H.J. Cordeira, C. Fritz, T.J. Galarneau Jr., S. Gjorgjievska, A. Griffin, K. Johnson, W. Komaromi, S. Monette, P. Muradyan, B. Murphy, M. Riemer, J. Sears, D. Stern, B. Tang and S. Thompson, 2012: The pre-depression investigation of cloud-sytems in the tropics (PREDICT) field campaign: Perspectives of early career scientists. *B. Am. Meteor. Soc.*, **92**, 173-187.

Fang, J. and F. Zhang, 2010: Initial development and genesis of Hurricane Dolly (2008). *J. Atmos. Sci.*, **67**, 655-672.

Fang, J. and F. Zhang, 2011: Evolution of multiscale vortices in the development of Hurricane Dolly (2008). *J. Atmos. Sci.*, **68**, 103-122.

Frank, N.L., 1970: Atlantic tropical systems of 1969. *Mon. Wea. Rev.*, **98**, 307-314.

Frank, W.M., 1987: Tropical cyclone formation: A global view of tropical cyclones. U.S. Office of Naval Research Publication.

Gjorgjievska, S. and D.J. Raymond, 2014: Interaction between dynamics and thermodynamics during tropical cyclogenesis. *Atmos. Chem. Phys.*, **14**, 3065-3082.

Gore, A., 2006: An inconvenient truth. Paramount Classics. US gross revenue to date: $49,796,507.

Gray, W.M., 1998: The formation of tropical cyclones. *Meteor. Atmos. Phys.*, **67**, 37-69.

Gray, W.M., 1968: Global view of the origin of tropical disturbances and storms. *Mon. Wea. Rev.*, **96**, 669-700.

Halverson, J., M. Black, R. Rogers, S. Braun, G. Heymsfield, D. Cecil, M. Goodman, R. Hood, A. Heymsfield, T. Krishnamurti, G. McFarquhar, M.J. Mahoney, J. Molinari, J. Turk, C. Velden, E.D. Zipser, R. Kakar and L. Zhang, 2007: NASA's Tropical Cloud Systems and Processes Experiment. *B. Am. Meteorol. Soc.*, **88**, 867-882.

Haynes, P.H. and M.E. McIntyre, 1987: On the Evolution of Vorticity and Potential Vorticity in the Presence of Diabatic Heating and Frictional or Other Forces. *J. Atmos. Sci.*, **44**, 828-841.

Hendricks, E.A., M.T. Montgomery and C.A. Davis, 2004: The role of "vortical" hot towers in the formation of Tropical Cyclone Diana (1984). *J. Atmos. Sci.*, **61**, 1209-1232.

Holland, G.H., 1993: "Ready Reckoner" – Chapter 9, Global Guide to Tropical Cyclone Forecasting. WMO/TC-No. 560, Report No. TCP-31, World Meteorological Organization: Geneva, Switzerland.

Hoskins, B.J., M.E. McIntyre and A.W. Robertson, 1985: On the use and significance of isentropic potential vorticity maps. *Quart. J. Roy. Meteorol. Soc.*, **111**, 877-946.

Houze, R.A., W.-C. Lee and M.M. Bell, 2009: Convective Contribution to the Genesis of Hurricane Ophelia (2005). *Mon. Wea. Rev.*, **137**, 2778-2800.

James, R.P. and P.M. Markowski, 2010: A numerical investigation of the effects of dry air aloft on deep convection. *Mon. Wea. Rev.*, **138**, 140-161.

Karyampudi, V.M. and H.F. Pierce, 2002: Synoptic-Scale Influence of the Saharan Air Layer on Tropical Cyclogenesis over the Eastern Atlantic. *Mon. Wea. Rev.*, **130**, 3100-3128.

Kilroy, G. and R.K. Smith, 2012: A numerical study of rotating convection during tropical cyclogénesis. **139**, 1255-1269.

Lussier, L.L. III, M.T. Montgomery and M.M. Bell, 2014: The genesis of Typhoon Nuri as observed during the Tropical Cyclone Structure 2008 (TCS08) field experiment - Part 3: Dynamics of low-level spin up during the genesis. *Atmos. Chem. Phys.*, **14**, 8795-8812.

McBride, J.L., 1995: Tropical cyclone formation. *In:* Global Perspective in Tropical Cyclones. 63-105. WMO/TD-No. 693 (Ed. R.L. Elsberry), World Meteorological Organization, Geneva.

McWilliams, J.C., 2006: Fundamentals of Geophysical Fluid Dynamics. Cambridge University Press, Cambridge.

Montgomery, M.T. and R.K. Smith, 2012: The genesis of Typhoon Nuri as observed during the Tropical Cyclone Structure 2008 (TCS08) field experiment - Part 2: Observations of the convective environment. *Atmos. Chem. Phys.*, **12**, 4001-4009.

Montgomery, M.T. and R.K. Smith, 2014: Paradigms for tropical-cyclone intensification. *Australian Meteorological and Oceanographical Journal*, Bruce Morton Memorial Volume.

Montgomery, M.T. and R.K. Smith, 2012: The genesis of Typhoon Nuri as observed during the Tropical Cyclone Structure 2008 (TCS08) field experiment. Part 2: Observations of the convective environment. *Atmos. Chem. Phys.*, **12**, 4001-4009.

Montgomery, M.T., C. Davis, T. Dunkerton, Z. Wang, C. Velden, R. Torn, S. Majumdar, F. Zhang, R.K. Smith, L. Bosart, M.M. Bell, J.S. Haase, A. Heymsfield, J. Jensen, T. Campos and M.A. Boothe, 2012: The Pre-Depression Investigation of Cloud systems in the Tropics (PREDICT) experiment: Scientific basis, new analysis tools, and some first results. *B. Am. Meteorol. Soc.*, **93**, 153-172.

Montgomery, M.T., Z. Wang and T.J. Dunkerton, 2010(a): Coarse, intermediate and high resolution numerical simulations of the transition of a tropical wave critical layer to a tropical storm. *Atmospheric Chemistry and Physics*, **10**, 10803-10827.

Montgomery, M.T., L.L. Lussier III, R.W. Moore and Z. Wang, 2010(b): The genesis of Typhoon Nuri as observed during the Tropical Cyclone Structure 2008 (TCS-08) field experiment – Part 1: The role of the easterly wave critical layer. *Atmos. Chem.Phys.*, **10**, 9879-9900.

Montgomery, M.T., V.S. Nguyen, J. Persing and R.K. Smith, 2009: Do tropical cyclones intensify by WISHE? *Q.J.R. Meteorol. Soc.*, **135**, 1697-1714.

Montgomery, M.T., M.E. Nicholls, T.A. Cram and A. Saunders, 2006: A "vertical" hot tower route to tropical cyclogenesis. *J. Atmos. Sci.*, **63**, 355-386.

Nguyen, V.S., R.K. Smith and M.T. Montgomery, 2008: Tropical-cyclone intensification and predictability in three dimensions. *Quart. J. Roy. Meteor. Soc.*, **134**, 563-582.

Nolan, D.S., 2007: What is the trigger for tropical cyclogenesis? *Aust. Meteor. Mag.*, **56**, 241-266.

Ooyama, K.V., 1982: Conceptual evolution of the theory and modeling of the tropical cyclone. *J. Meteor. Soc. Japan*, **60**, 369-380.

Persing, J., M.T. Montgomery, J.C. McWilliams and R.K. Smith, 2013: Asymmetric and axisymmetric dynamic of tropical cyclones. *Atmos. Chem. Phys.*, **13**, 12229-12341.

Raymond, D.J. and C. Lopez Carrillo, 2011: The vorticity budget of developing typhoon Nuri (2008). *Atmos. Chem. Phys.*, **11**, 147-163.

Reasor, P.D., M.T. Montgomery and L.F. Bosart, 2005: Mesoscale Observations of the Genesis of Hurricane Dolly (1996). *J. Atmos. Sci.*, **62**, 3151-3171.

Reed, R.J., D.C. Norquist and E.E. Recker, 1977: The Structure and Properties of African Wave Disturbances as Observed During Phase III of GATE. *Mon. Wea. Rev.*, **105**, 317-333.

Riemer, M. and M.T. Montgomery, 2011: Simple kinematic models for the environmental interaction of tropical cyclones in vertical wind shear. *Atmos. Chem. Phys.*, **11**, 9395-9414.

Ritchie, E.A. and G.J. Holland, 1999: Large-scale patterns associated with tropical cyclogenesis in the western Pacific. *Mon. Wea. Rev.*, **127**, 2027-2042.

Ritchie, E.A. and G.J. Holland, 1997: Scale interactions during the formation of Typhoon Irving. *Mon. Wea. Rev.*, **125**, 1377-1396.

Ritchie, E.A. and G.J. Holland, 1993: On the interaction of tropical-cyclone-scale vorticies. II: Discrete vortex patches. *Quart. J. Roy. Meteorol. Soc.*, **119**, 1363-1379.

Rutherford, B. and M.T. Montgomery, 2012: A Lagrangian analysis of a developing and non-developing disturbance observed during the PREDICT experiment. *Atmos. Chem. Phys.*, **12**, 11355-11381.

Saunders, A.B. and M.T. Montgomery, 2004: A closer look at vortical hot towers within a tropical cyclogenesis environment. Colorado State University, Atmospheric Science Bluebook.

Shapiro, L.J. and M.T. Montgomery, 1993: A three-dimensional balance theory for rapidly rotating vortices. *J. Atmos. Sci.*, **50**, 3322-3335.

Shin, and R.K. Smith, 2008: Tropical-cyclone intensification and predictability in a minimal three dimensional model. *Quart. J. Roy Met. Soc.*, **134**, 1661-1671.

Sippel, Jason A. and Fuqing Zhang, 2010: Factors Affecting the Predictability of Hurricane Humberto (2007). *J. Atmos. Sci.*, **67**, 1759-1778.

Sippel, Jason A. and Fuqing Zhang, 2008: A Probabilistic Analysis of the Dynamics and Predictability of Tropical Cyclogenesis. *J. Atmos. Sci.*, **65**, 3440-3459.

Sippel, Jason A., John W. Nielsen-Gammon and Stephen E. Allen, 2006: The Multiple-Vortex Nature of Tropical Cyclogenesis. *Mon. Wea. Rev.*, **134**, 1796-1814.

Smith, R.K. and M.T. Montgomery and S.V. Nguyen, 2009: Tropical cyclone spin up revisited. *Q.J.R. Meteorol. Soc.*, **135**, 1321-1335.

Smith, R.K. and M.T. Montgomery, 2012: Observations of the convective environment in developing and non-developing tropical disturbances. *Q.J.R. Meteorol. Soc.*, **138**, 1721-1739.

Troy, K.J. and W.M. Frank, 2010: Tropical cyclone Formation. *In:* Global Perspectives on Tropical Cyclones, from science to mitigation. J.C.L. Chang and J.D. Kepert (Editors), World Scientific.

Tory, K.J. and M.T. Montgomery, 2006: Internal influences on tropical cyclone formation. *In:* Sixth International Workshop on Tropical Cyclones, San Jose, Costa Rica. World Meteorological Organization.

Tuleya, R.E. and Y. Kurihara, 1982: A Note on the Sea Surface Temperature Sensitivity of a Numerical Model of Tropical Storm Genesis. *Mon. Wea. Rev.*, **110**, 2063-2069.

Vallis, G.K., 2007: Atmospheric and Oceanic Fluid Dynamics: Fundamentals and Large-scale Circulation. Cambridge University Press, Cambridge.

Wang, Z., 2012: Thermodynamic aspects of tropical cyclone formation. *J. Atmos. Sci.*, **69**, 2433-2451.

Wang, Z., M.T. Montgomery and T.J. Dunkerton, 2010: Genesis of Pre-hurricane Felix (2007). Part I: The Role of the Easterly Wave Critical Layer. *J. Atmos. Sci.*, **67**, 1711-1729.

Wissmeier, U. and R.K. Smith, 2011: Tropical-cyclone convection: The effects of ambient vertical vorticity. *Quart. J. Roy Met. Soc.*, **137**, 845-857.

23

The Hurricane Boundary Layer

Sundararaman G. Gopalakrishnan[1], C. Venkata Srinivas[2] and Kieran T. Bhatia[3]

[1]NOAA/Atlantic Oceanographic and Meteorological Laboratory/Hurricane Research Division, Miami, FL, USA

[2]Radiological Safety and Environmental Group, Indira Gandhi Centre for Atomic Research, Kalpakkam, India

[3]Rosenstiel School of Marine and Atmospheric Science University of Miami, Florida, USA

gopal@noaa.gov

1. Introduction

The planetary boundary layer (PBL) within the troposphere of the atmosphere regulates the transfer of heat, moisture, and momentum between the atmosphere and the surface through drag effects, turbulent diffusion, and mixing processes (Stull, 1988; Garrat, 1994). The PBL is the layer of the atmosphere closest to the Earth and is therefore directly influenced by effects of the surface (friction, heating, cooling, evapotranspiration, etc.). Therefore, almost all the weather generation processes originate in the PBL. It is variously defined as a continental PBL or marine boundary layer (MBL) according to its location and secondly, as stable, neutral, or unstable, depending upon the degree of turbulence. The Hurricane Boundary Layer (HBL)[1] is a special class of the MBL that refers to the bottom layer of the atmosphere in tropical cyclones (TCs). However, there are features unique to the HBL that has made modelling and understanding it one of the largest challenges facing the tropical cyclone forecasting community. Until recently, the risks to humans and scientific equipment associated with the extreme conditions in the HBL prevented the collection of high-quality observations in the HBL. Secondly, numerical models did not possess adequate resolution to model the HBL processes. Also there is a lack of understanding on the connection between the HBL structure and tropical cyclone intensity changes.

[1] We will postpone the precise definition on the height of the HBL to section 4. Until then we refer to the lowest 2 km as the HBL in this text.

In the mid and late 60s, two seminal contributions on the hurricane intensification process appeared in literature, both of which depicted the importance of the HBL, and continued to flourish for the next twenty five years. In a highly influential paper describing the Conditional Instability of the Second Kind (CISK) theory, Charney and Eliassen (1964) suggested an axisymmetric balance theory for the cooperative interaction between moisture convergence in the HBL, heating produced by deep convective clouds and the subsequent deepening of TCs. A similar theory, but with a different cumulus parameterization scheme was proposed independently by Ooyama in the same year. Both the theories highlighted the role of surface friction in supporting the hurricane amplification process. It should be noted that the idea was novel at that time because friction is normally associated with dissipation. As clearly stated in Charney and Eliassen (1964): "Friction performs a dual role; it acts to dissipate kinetic energy, but because of the frictional convergence in the moist surface boundary layer, it acts also to supply latent heat energy to the system." Ooyama (1969) developed a simple three layer axisymmetric model, with the upper layer representing the troposphere, lower layer within the HBL (aka the "Ekman layer" in his work) and an interfacial layer representing the air-sea layer, that captured the essence of hurricane evolution very realistically. Ooyama's model illustrated the cooperative interactions between the radial inflow or the cross isobaric flow (aka the secondary circulation when thermal lifting in the eyewall region and the subsequent adiabatic subsidence motion in the eye are included) and the primary circulation that includes the strong tangential flow. As mentioned in Montgomery and Smith (2013), two aspects of both the CISK theory and Ooyama's cooperative intensification theory that have gained wide acceptance are that (a) the main spin up of the vortex occurs via the convergence of absolute angular momentum above the boundary layer, and (b) the boundary layer plays an important role in converging moisture to sustain deep convection, but its dynamical role opposes the spin up (by the transfer of momentum between the atmosphere and the ocean). However, it should be noted that both these theories emphasized on convection, i.e., heating due to cumulus is proportional to integrated moisture convergence mainly in the lower troposphere ignoring the importance of air-sea interactions.

In what has become a highly influential paper since the 90s, Emanuel (1986) refocused the meteorological community on the importance of air-sea interaction in TCs (see Emanuel, 1991 and references therein; Emanuel et al., 1994; Emanuel, 1995). Except for the treatment of convection, wherein the undiluted pseudo-adiabatic ascent occurred along surfaces of absolute angular momentum rather than upright ascent with entrainment of middle tropospheric air, Emanuel's time-dependent model was similar to Ooyama's 1969 model (e.g. gradient wind and hydrostatic balance, wind-speed-dependent surface moisture fluxes). Nevertheless, the new paradigm for TC intensification, called the Wind Induced Surface Heat Exchange (WISHE) theory that emerged from the model, invokes a positive feedback between the near-surface wind speed and the rate of evaporation of water from the underlying ocean, which depends on the wind speed rather than moisture convergence in the HBL alone.

Despite the phenomenal insights that the theoretical models have offered on the TC intensification process, these models were not designed to advance our understanding of the HBL structure and its impact on intensity problem. Early works of Kurihara and Tuleya (1974) and Anthes and Chang (1978), in particular, noted the importance of higher resolution in the HBL. In fact, Anthes and Chang (1978) illustrated that both the intensification due to changes in the sea surface temperature as well as the structure of the HBL become more realistic in a model with at least eleven layers within HBL than with a set up with one layer. The teleconnections between the HBL and the hurricane intensification is better established now. For instance, Gopalakrishnan et al. (2011) using a version of the Hurricane Weather Forecast (HWRF) system with 11 layers within the HBL, demonstrated how the frictional convergence in the HBL helps to distribute the high entropy air from the air-sea interface. Figure 1 provides an overview on how the θ_e air from the HBL gets distributed by means of vortical thermal plumes (clouds) in a hurricane system. The study related the warm core formation to a WISHE type of a feedback in the HBL wherein the surface pressure decreases (by hydrostatic principles), resulting in increase in wind speed, increase in surface enthalpy fluxes (θ_e) and subsequently a warmer core, and consequently, a more intense system eventually reaching an equilibrium. Finally, recent works by Montgomery

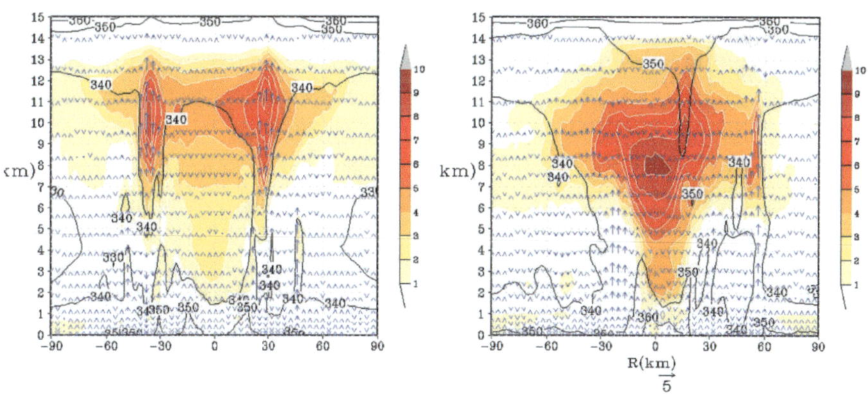

Fig. 1. Warm-core development. Depiction of the x–z cross section of the instantaneous temperature anomaly (colour shaded) across an arbitrary plane in the eyewall region at about (a) 9 and (b) 24 h for the 3-km run. The white contours are used to emphasize the nature of the vortical thermal plumes within the warmer environment discussed in Gopalakrishnan et al. (2011). As depicted by the temperature anomaly and θe (in K and contours indicated by black colour), both moisture and heat are carried to the mid- and upper troposphere by the vortical thermal plumes shown in that work. The vector indicates the direction of the vertical motion. *Courtesy:* The American Meteorological Society (Gopalakrishnan et al., 2011, *Monthly Weather Review*).

and Smith and his group members (see Montgomery and Smith, 2013 for a good review) and Gopalakrishnan et al. (2013) are starting to fuel arguments that support a more important role for the HBL in the spin up of TCs simply beyond what was discovered in either CISK or WISHE theory. These findings illustrate the significance of the acceleration of the inflow in the boundary layer and its major influence on the intensification process. In brief, if the acceleration in the HBL is large enough, instead of friction acting as a sink of angular momentum, it can act to spin up TCs. With the advancements in observations, support of such a finding is becoming more evident. Some parts of this chapter will focus on these latest developments.

While satellites have provided vital information primarily on the environment of TCs, specially-equipped NOAA aircrafts gather inner core observations that have been essential in improving our understanding of the HBL. An introduction to the history of these aircrafts is provided in Aberson et al. (2006a) and Chapter 1 by Dr. Marks provide an overview on the use of observations in TC operations and research. TCs have been observed by both reconnaissance and research aircraft since the first flight piloted by U.S. Army Air Force Lieutenant Colonel Joseph P. Duckworth on July 17, 1943. However, it was not until the mid-1970s, when the National Oceanic and Atmospheric Administration (NOAA) purchased two customized WP-3D (P-3) aircraft, that aircrafts were used to provide inner core observations. Another type of aircraft, the Gulfstream IV Special Performance (SP) Jet (aka G4) which began hurricane surveillance missions in 1997 in support of forecasters at NOAA's National Hurricane Center, flies at an altitude of 13-14 km and deploys instruments called GPS (Global Positioning System) dropwindsondes that continuously radio transmit measurements of pressure, humidity, temperature, wind direction and speed. As the dropwindsondes fall to the sea, they provide a detailed look at the synoptic scale environment of the tropical cyclone. The P3s and G4s are the backbone of TC observations in the Atlantic basin.

The observations from the P-3s have helped revolutionize understanding of TC behaviour and structure (Rogers et al., 2012). Observational studies (e.g., Marks and Houze, 1987; Houze et al., 1992; Marks et al., 1992; Gamache et al., 1993, Rogers et al., 2012) highlight that the symmetric primary circulation is characterized by a tangential wind with a maximum within the HBL, while the secondary circulation has a radial inflow whose strength is maximum within the HBL and decreases with height above it, a rising branch collocated with the eyewall and sloping radially outward with height, a layer of intense outflow above 10-km altitude. Observations have also revealed low azimuthal wave-number asymmetric structures such as shear-induced vortex tilt and unevenly distributed wind and rainfall (Marks et al., 1992; Reasor et al., 2000, 2009; Black et al., 2002), high-wavenumber asymmetries such as vortex Rossby waves and eyewall mesovortices (Aberson et al., 2006b; Montgomery et al., 2006; Marks et al., 2008), and low-level kinematic and thermodynamic structures in the inner core and rainband regions (Barnes et al., 1983; Powell, 1990; Barnes and Powell, 1995; Cione et al., 2000) (see Rogers et al., 2012 for a good summary). As early as in 60s, Smith (1968) used a simple analytical model to examine some of the features of the HBL by varying the eddy diffusivities. He concluded that inadequate knowledge of the turbulence structure prevented a realistic understanding of the layer. Thanks largely to the work of Zhang et al. (2008, 2009, 2010, 2011a, b, 2012), more recently there has been an increased focus on the HBL observations, as well. Direct measurements of enthalpy fluxes in the HBL for some TCs are now available as a result of the Coupled Boundary Layer Air-sea Transfer (CBLAST) experiments (Black et al., 2007; Zhang, 2007; Zhang et al., 2008). Flight level data sets from the P3s have been used to estimate turbulent momentum flux, turbulent kinetic energy (TKE), and vertical eddy diffusivity obtained before and during the eyewall penetrations of several TCs (Zhang et al., 2009, 2011a, b) are starting to provide wealth of information on the HBL. More recently, Zhang and Montgomery (2012) obtained observational estimation of the horizontal eddy diffusivity and mixing length in the low-level region of intense hurricanes. These data sets may be very useful for evaluating the HBL high resolution numerical models. In fact Gopalakrishnan et al. (2013), for the first time, used some of these information to evaluate the vertical eddy diffusivity in the HWRF system.

In the late 70s, Shea and Gray (1973) and Gray and Shea (1973) used a new observational technique that provided a framework for advancing our understanding of the TC structure. Observational information from 533 radial flight legs executed by NOAA aircrafts over a 13-year period (1957–69) was used to present the structural characteristics and the variability of the hurricane's inner core region. Such a technique indeed revealed that the inflow is confined to the lowest layers outside the radius of maximum wind, outflow occurs aloft, and the largest convergence occurs in the lowest levels near the radius of maximum wind, consistent with the understanding drawn from limited data from individual samples. An advantage of such a composite analysis is that it provides a general picture of several TCs. Composites of airborne Doppler radar data, collected by the NOAA Hurricane Research Division (HRD) for the past 40 years provide wealth of information on three-dimensional TC inner-core kinematic structures in a statistically robust framework. The ability to construct composites of airborne Doppler radar data has been facilitated by the development of an automated processing algorithm based on a variational scheme first described in Gamache (1997) and documented in detail in Reasor et al. (2009). Many of these studies have involved examinations of individual cases (Montgomery et al., 2006) or several TCs (Rogers et al., 2012). The structure of the HBL extensively discussed in this chapter will be based on such an analysis.

In summary, the HBL is clearly an important part of tropical cyclones and it is through the boundary layer that much of the impact on humanity and the environment occurs. Yet, even with the increased effort towards finding the "optimum" HBL scheme for high-resolution numerical models and discovering the detailed structure of the HBL through observation, the HBL is still poorly understood. This chapter aims to build on the significant advances in recent years by summarizing the available information about the HBL while highlighting the greatest challenges in increasing our understanding. The discussion of the HBL begins with a description of the structure of the HBL using multiple observational studies. In section 3, the observed dynamics in the HBL is analyzed. Section 4 shows the importance of defining an HBL height and how this definition influences tropical cyclone forecasting. In section 5, a simple slab model is utilized to gain some basic understanding of a typical HBL. Further, a brief overview on numerical modelling of the HBL is presented. In section 7, the mean and turbulent structure of the HBL as modelled using a cloud resolving modelling system (HWRF) is compared with observations. Sections 8 and 9 present some important sensitivity experiments that illustrate the importance of HBL on TC intensity changes.

2. Observed Mean Structure of the HBL

Isolating the mean structure of the HBL is important for an improved understanding in TCs and for improving predictions in numerical models, as well. The 2006 study by Montgomery et al. is an example of the importance of HBL observations for the hurricane research community. Montgomery et al. published unprecedented results on the mean vortex structure of Hurricane Isabel from observational datasets. They analyzed the kinematic and thermodynamic structure and the associated secondary circulation of Hurricane Isabel (Figs 2 and 3) by compositing GPS dropwindsonde and flight-level *in situ* data discussed in section 1. Figure 2 depicts the low-level (below 4 km) structure of Isabel from the composites. In Fig. 2a, the azimuthally averaged storm-relative tangential and radial wind velocities in the radius-height plane are depicted. The maximum tangential wind speed (~76 ms^{-1}) occurs at approximately 40 km from the storm centre and between 500-1000 m above the surface. The tangential winds increase from calm conditions in the eye of 0-5 ms^{-1} to 70 ms^{-1} at about 43 km. The composite structure indicates that there is low-level radial inflow of 20 ms^{-1} at a radius of 25 km, which suggests significant penetration of air from under the eyewall into the eye. A rapid fall in the tangential wind speed is visible inside the eye. Figure 2b shows the distribution of equivalent potential temperature (θ_e), specific absolute angular momentum ($M = rv_\lambda + 1/2\,fr^2$, where '$r$' is the radial distance from the vortex centre, 'v_λ' is the storm-relative tangential wind, f = Coriolis parameter), and secondary circulation vectors. The highest θ_e (~374 K) is located in the eye region with a gradual radial decrease in the eyewall and decreasing vertically above 2.5 km. Figure 2b also depicts an outflow at upper levels of high-entropy air, which is represented by the

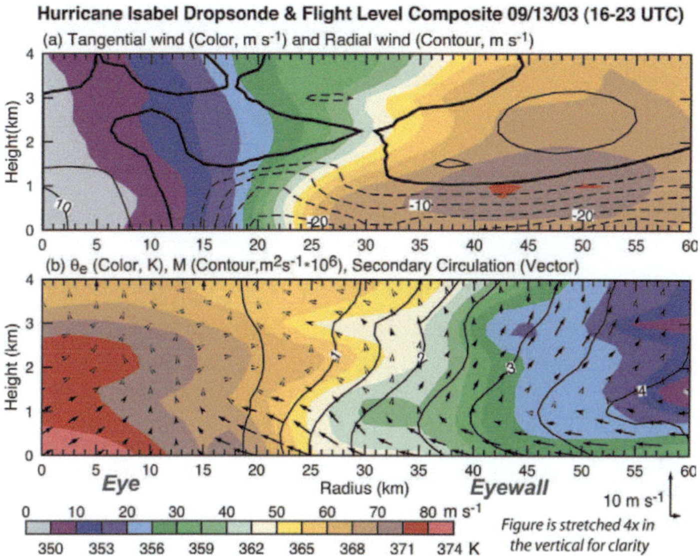

Fig. 2. Radius–height azimuthal mean structure derived from GPS dropwindsonde and flight-level data from 1600 to 2300 UTC 13 Sep. (a) Storm-relative tangential wind (colour) and radial wind (contour; ms^{-1}); (b) θe (colour; K), absolute angular momentum (contour; m^2 s^{-1} × 10^6), and transverse secondary circulation (vector). The origin (0, 0) denotes the eye centre at the ocean surface. *Courtesy:* The American Meteorological Society (Montgomery et al., 2006; BAMS).

outward bulge in θ contours and absolute angular momentum near a radius of 30 km and an altitude of 2 km.

Figure 3 (from Montgomery, 2006) shows the mean soundings constructed using the 50-m vertical resolution dropwindsonde data within four radial rings defined as the eye region ($r < 15$ km), within the eyewall region (RMW: 40 km $< r <$ 50 km), the outer core (150 km $< r <$ 250 km), and the storm environment (300 km $> r >$ 1000 km). These regions defined by Montgomery et al. represent the azimuthal, time mean low-level ($0 < z <$ 2 km) kinematic and thermodynamic structures for Hurricane Isabel. In Fig. 3a the highest tangential winds are distributed in the region between the eye and RMW; increasing sharply from the surface up to 250 m and stabilizing aloft. The maximum tangential wind speed of ~76 ms^{-1} occurs at a height between 500-1000 m above the surface. Figure 3b shows the mean radial wind for the four different regions. A common feature of the HBL is a deep inflow region near the surface which is located below 1.2 km in this figure. The mean profiles also show a weak tangential wind and radial outflow inside the eye and the presence of weak inflow in the ambient environment. Figure 3c depicts the vertical temperature distribution and Fig. 3e depicts the virtual potential temperature in the four regions. Both plots show the progressive decrease in stability from eye to the environment. The mean temperature profiles indicate a ~1°K cooling from the eyewall to the outer core and another

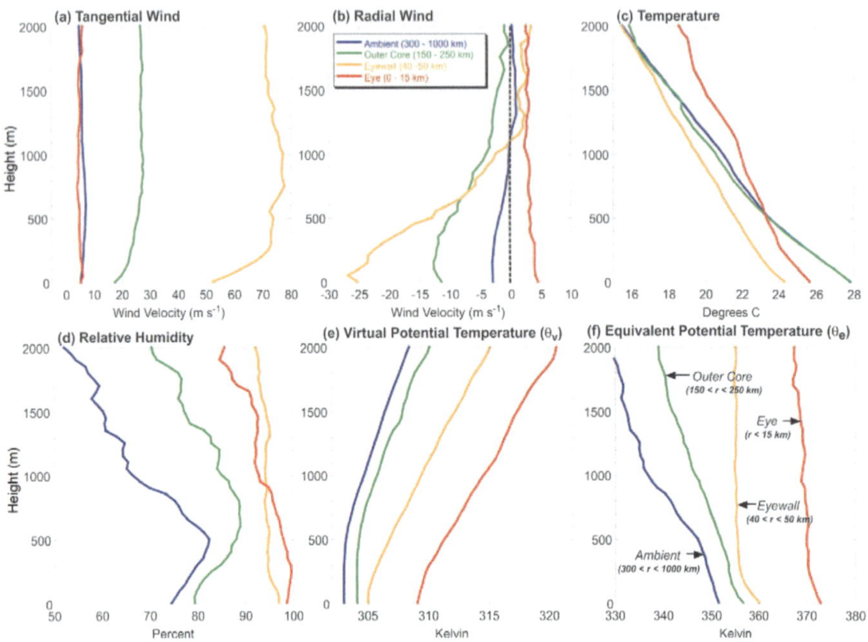

Fig. 3. Low-level (0–2 km) mean soundings from dropwindsonde observations for the eye (0–15-km radius, red), nominal eyewall (40–50-km radius, orange), outer core (150-250 km radius, green), and ambient environment (300–1000-km radius, blue) during 1600–2300 UTC 13 Sep for Hurricane Isabel. (a) Tangential wind (ms^{-1}), (b) radial wind (ms^{-1}; inflow/outflow transition indicated by dashed line), (c) temperature (°C), (d) RH (%), (e) θ_v (K), and (f) θ_e (K). *Courtesy:* The American Meteorological Society (Montgomery et al., 2006; BAMS).

3°K temperature drop off from the eye to the eyewall. In Fig. 3d, the vertical profile of relative humidity (RH) is shown. The lowest 2 km region has a RH greater than 90% (especially near and inside the RMW) which indicates the prevalence of almost saturated conditions in the HBL. In the mixed layer (~500 m deep) the ambient average equivalent potential temperature (Fig. 3f) is seen to increase from 350 to 354 K in the outer core (~200 km radius), and to an additional 3 K as the air approaches the eyewall in the HBL. This shows the extent of enthalpy gained by inflowing air. The temperature stratification in the eye and eyewall indicate statically stable (Fig. 3e) and conditionally unstable HBL (Fig. 3f).

More recently, Rogers et al. (2012) presented the observed inner core structure of mature tropical cyclones by compositing several airborne observations and Doppler radar analyses. Their study reported several common features seen in the HBL of multiple hurricanes: the presence of primary and secondary circulations, a sloping eyewall, and a reduction in tangential wind with height, a low-level inflow layer, and a region of enhanced outflow. Figure 4 shows the normalized radius-height (r^*-z) plots of different storm fields

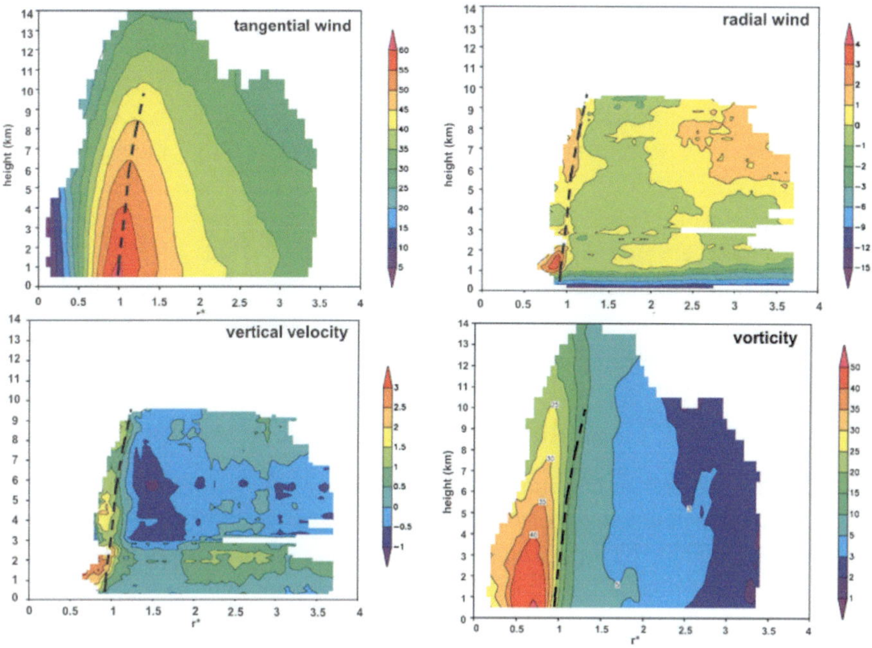

Fig. 4. Composite of (a) axisymmetric tangential wind (ms^{-1}) from swath data, (b) radial wind from profile data and (c) vertical velocity from profile data. (d) Relative vorticity (1×10^{-4} sec^{-1}) from swath data from several hurricanes after normalizing all of them with respect to the radius of maximum winds. *Courtesy:* The American Meteorological Society (Rogers et al., 2012, *Monthly Weather Review*).

obtained by compositing the P3 data sets (Rogers et al., 2012). Figure 4a depicts the axisymmetric tangential wind, which attains a maximum value of about 60 ms^{-1} at $r^* = 1$ at 1 km altitude, decreasing to 40-45 ms^{-1} by $r^* = 2$ and to 35 ms^{-1} by $r^* = 3$, all at the same altitude. The eyewall slope, an expected feature of the HBL, is also visible in Fig. 4a. The value of the peak axisymmetric tangential wind at 8 km is 78% of the value at 2-km altitude indicating substantial reduction with height. In Fig. 4b, the axisymmetric radial wind from the composites is shown. The positive values correspond to outflow and negative values correspond to inflow. Figure 4b indicates that the depth of the inflow layer varies from about 1 to 1.5 km altitude between $r^* = 1$ and 2.5. The lower tropospheric inflow reaches a peak value of 4 ms^{-1} at 0.5 km altitude within the normalized radius of maximum winds. Nevertheless, it should be emphasized that as seen in Fig. 3b and in Gopalakrishnan et al. (2013), the maximum could be as high as 25-30 ms^{-1} in individual cases. A low-level enhanced outflow is also apparent at the RMW and between an altitude of 1 to 2 km. A weak inflow is also observed above the HBL between $r^* = 1$ and 2. Several theoretical and high resolution model studies are starting to reproduce such features (e.g., Gopalakrishnan et al., 2011, 2013).

In Fig. 4c, the vertical velocity distribution is contoured in the r^*-z plane. The highest vertical velocities are distributed in the core region and diminish in the radial direction. The peak radial velocities (2-3 ms^{-1}) are located in the inner core at an altitude of 1-2 km with a second peak (about 2 ms^{-1}) at about 5 km altitude at $r^* = 1$. At $r^* = 1.5$ and a height of 3.5-9 km, the significant negative vertical motion represents a large area of strong downdrafts. Another interesting feature is the vorticity distribution in the core of the cyclones. In Fig. 4d, there are maximum axisymmetric vorticity values of 50×10^{-4} s^{-1} along the inner edge of the eyewall, specifically near $r^* = 0.75$ below the 1-km altitude. The vorticity values then decrease radially, falling to $<30 \times 10^{-4}$ s^{-1} at $r^* = 0.3$ and 1 km altitude.

3. Observed Turbulence Structure of the HBL

The surface exchange coefficient for momentum and heat/moisture, Cd and Ck are critical parameters for boundary layer parameterizations and tropical cyclone modelling. The magnitude of Cd determines how easily momentum is transferred between the atmosphere and the ocean, and it increases with sea surface roughness and wind speed. Ck determines how easily enthalpy fluxes are transferred between the ocean and the atmosphere. Malkus and Riehl (1960) presented the first evidence relating Cd to the magnitude of the tangential wind speeds observed in a hurricane. Their theory applied to steady-state hurricanes and linked the maximum swirling wind to the square root of the ratio of C_T (surface exchange coefficients for sensible heat) and Cd. Ooyama (1969) and Rosenthal (1971) published seminal studies soon after and observed similar behaviour of the maximum tangential wind on the drag coefficient. In 1986, Emanuel published an influential paper that demonstrated that the square root of the ratio of the surface exchange coefficient for enthalpy, Ck, and Cd was proportional to the maximum swirling wind. Emanuel (1995) later discovered that the development of intense hurricanes (category 3 and higher on the Saffir–Simpson scale) was impeded when the ratio Ck/Cd fell below a value of 0.75, thereby creating a threshold value of Ck/Cd below which tropical depressions will not intensify to major storms.

However, until recently, direct measurements of surface fluxes for computation of Cd and Ck in the inner-core region of intense hurricanes were non-existent. As a result, different surface layer parameterization schemes often yield a diverse range of structure and intensity predictions when applied to the same storm (see for example, Figs 9 and 16 in Braun and Tao, 2000). In 2003, Powell et al. reported an important observational finding for high wind regime (Fig. 5a). Their work suggested that Cd decreases with wind speed beyond 40 ms^{-1} rather than increasing as parameterized in numerical models at that time. Donelan et al. (2004) presented some strong evidence to confirm that at high wind speeds Cd ceases to increase with wind speed. At the same time, based on wind stress obtained from NOAA's National Centers

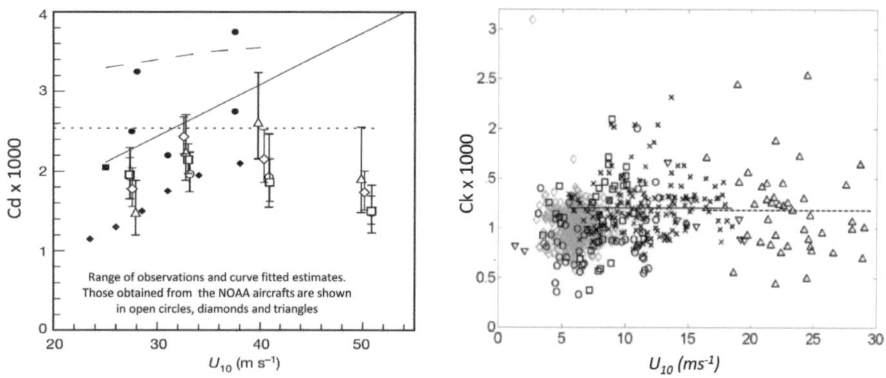

Fig. 5. Behaviour of observed surface exchange coefficients with wind speed for momentum (Cd); Adopted from Nature (Powell et al., 2003) and (b) moisture and heat (Ck) or enthalpy (Ce). *Courtesy:* The American Meteorological Society (Drennan et al., 2007. *Journal of the Atmospheric Sciences*).

for Environmental Prediction (NCEP) surface wave model WAVEWATCH-III, Moon et al. (2004a and b) also reported similar findings regarding the Cd behaviour in hurricanes. Yet, the high wind speed behaviour of Ck is still lacking strong observational support. Nevertheless, synthesis of observations from CBLAST (Black et al., 2007; Zhang, 2007; Zhang et al., 2008) provide some extended evidence on the observed behaviour of Ck in moderate wind region especially between 15-30 ms^{-1} (Fig. 5b). In addition, recent laboratory experiments (Haus et al., 2010) and theoretical studies of Montgomery et al. (2010) present some additional evidence to confirm a constant behaviour of Ck beyond wind speed of about 30 ms^{-1}. A value of about 1.2-1.3 $\times$ 10^{-3} for Ck appears to be a reasonable estimate.

Most recently, since direct turbulent flux measurements in extremely strong wind beyond 50 ms^{-1} condition are extremely difficult, Bell et al. (2012) deduced the momentum and enthalpy fluxes via absolute angular momentum and total energy budgets. An error analysis of the methodology was performed to quantify and mitigate potentially significant uncertainties resulting from unresolved budget terms and observational errors. An analysis of six missions from the 2003 CBLAST field programme in major hurricanes Fabian and Isabel was conducted using a new variation technique discussed in Bell et al. (2012). The analysis indicates a near-surface mean drag coefficient Cd of 2.4 $\times$ 10^{-3} with a 46% standard deviation and a mean enthalpy coefficient Ck of 1.0 $\times$ 10^{-3} with a 40% standard deviation for wind speeds between 52 and 72 ms^{-1}. These are the first known estimates of Ck and the ratio of enthalpy to drag coefficient Ck/Cd in major hurricanes. The results suggest that there is no significant change in the magnitude of the bulk exchange coefficients estimated at minimal hurricane wind speeds, and that the ratio Ck/Cd does not significantly increase for wind speeds greater than 50 ms^{-1}.

Above the surface layer, the mixing and diffusion in the HBL are difficult to estimate because it depends on high temporal and spatial resolution observations. Nevertheless, research aircraft (P3s) penetrated the eyewalls of category 5 Hurricane Hugo (1989) and category 4 Hurricane Allen (1980) between 1 km and the sea surface (see Marks, 1985; Marks et al., 2008). Although at fixed altitudes, estimates of turbulent momentum flux, turbulent kinetic energy (TKE), and vertical eddy diffusivity are obtained before and during the eyewall penetrations in these intense hurricanes (Zhang et al., 2011b and Fig. 6a here). Spatial scales of turbulent eddies are determined through a spectral analysis. The turbulence parameters estimated for the eyewall penetration leg were found in Zhang et al. (2011b) to be nearly an order of magnitude larger than those for the leg outside the eyewall at similar altitudes. In the low-level intense eyewall region, the horizontal length scale of the dominant turbulent eddies is found to be between 500 and 3000 m, and the corresponding vertical length scale is approximately 100 m. Zhang et al. (2011b) noted that the results can offer useful guidance in numerical weather prediction efforts aimed at improving the forecast of hurricane intensity. He also noted that, of the small sample size analyzed in this study, further analyses of the turbulent characteristics in the high-wind region of hurricanes are imperative. More recently Zhang and Montgomery (2012) also provide an estimate of horizontal eddy diffusivity and mixing length in the low-level region of intense hurricanes.

Another important parameter in the HBL is the turbulence kinetic energy (TKE). Given the three dimensional nature of turbulence in the hurricanes, it may be difficult to create such observations at high temporal and spatial resolutions. Nevertheless, some estimates are becoming available from the NOAA WP-3D tail radars. The geometry of the analysis coupled with the relatively small beam resolution at ranges 8 km from NOAA WP-3D tail radars allows for the estimation of at least sub kilometre turbulent processes without the need for instruments to resolve very high spatial and temporal scale. Although such an analysis may not be able to provide micro scale estimates of turbulence, yet may be useful in the evaluation of TKE in high resolution models. Lorsolo et al. (2010) estimated the TKE using airborne Doppler measurements from the radars for a variety of storms at different stages of their life cycle (Fig. 6b). Two-dimensional profiles of TKE available for that analysis (Fig. 6b) revealed that the strongest turbulence was generally located in convective regions, such as the eyewall, with magnitudes often exceeding 15 m^2s^{-1} and in the boundary layer with values of 5-10 m^2s^{-1} in the lowest kilometre. Further analysis showed that the strong turbulence was generally associated with strong horizontal shear of vertical and radial wind components in the eyewall and strong vertical shear of horizontal wind in the boundary layer. Mean vertical profiles of TKE decrease sharply above the hurricane boundary layer and level off at low magnitude for all

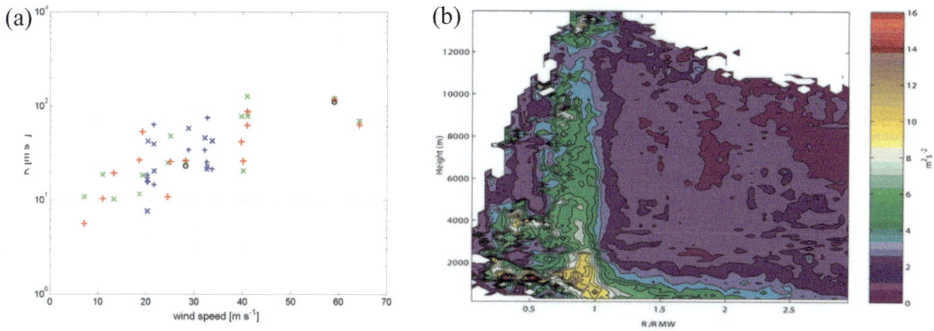

Fig. 6. (a) Estimates of Eddy diffusivity for momentum obtained from flight level data as a function of wind speed (Zhang et al., 2011a, *Monthly Weather Review*) (b) Composite estimation and mapping of Hurricane Turbulent Energy using Airborne Doppler Measurements from several hurricanes (Katrina, Rita, Isabel, Frances and Fabien). Units are in m^2s^{-1}. *Courtesy:* The American Meteorological Society (Lorsolo et al., 2010, *Monthly Weather Review*).

regions outside the radius of maximum wind. The quality of the retrieval method was evaluated and showed very good agreement with TKE values directly calculated from the three-dimensional wind components of in situ measurements obtained from flight level data.

4. A Note on the Height of the HBL

Zhang et al. (2011a) pointed out that there is no consensus in prescribing a boundary layer top in the hurricane research community. Definitions of boundary layer height have been derived from thermodynamics (e.g., Moss and Merceret, 1976; Powell, 1990; Anthes and Chang, 1978), dynamics[2] (Kepert and Wang, 2001; Bryan and Rotunno, 2009a, b; Smith et al., 2009), as well as using the critical Richardson number (e.g., Troen and Mahrt, 1986; Vogelezang and Holtslag, 1996; Bender et al., 2007; Noh et al., 2003). Zhang et al. (2011a) sought to compare the results of these different techniques by using observations of the hurricane boundary layer with 794 GPS dropsondes deployed into 13 different hurricanes. The dropsonde data are analyzed employing a composite framework, which is discussed in detail in the paper by Zhang et al. (2011). Figure 7 shows the different boundary layer height metrics, including height of the maximum tangential wind speed (h_{vmax}), the inflow layer depth (h_{infl}/δ), and the mixed layer depth (Z_i), as a function of normalized distance. Also shown in Fig. 7 is the boundary layer depth estimated using the critical bulk Richardson number method. The results shown are significant because there is considerable disparity between

[2] Depth of the inflow layer (h_{infl} or δ) is generally adopted to be the height of the hurricane boundary layer. However, significant turbulence and mixing may occur above this layer, especially in the eyewall region.

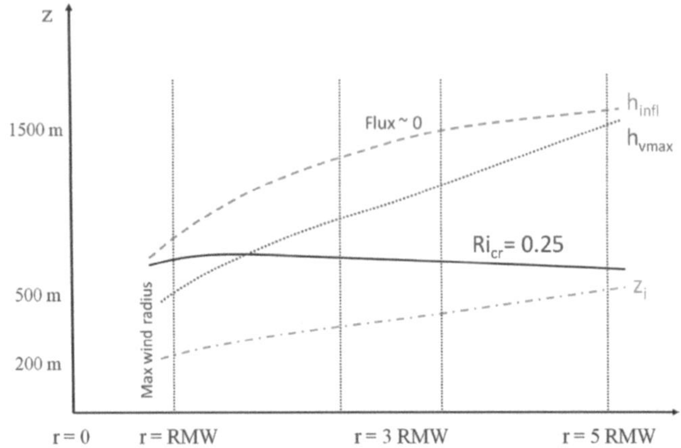

Fig. 7. Summarizes the different ways the boundary layer has been defined and provides actual values of these characteristic height scales for the hurricanes analyzed. *Courtesy:* The American Meteorological Society (Zhang et al., 2011a, *Monthly Weather Review*).

the dynamical and thermodynamic boundary layer heights. The separation of the dynamical and thermodynamic boundary layer height scales is also evident in the numerical simulations of Nolan et al. (2009a) and in the already mentioned observational study of Hurricane Isabel (Montgomery et al., 2006, see their Fig. 4).

Another notable feature of Fig. 7 is all three height scales (h_{vmax}, h_{infl}, and Z_i) appear to diminish towards the storm centre, especially in the inner-core region. It is also clear that h_{vmax} decreases more than the other two height scales. On the other hand, the three boundary layer height metrics level off in the outer-core region (3.5 RMW). Kepert (2010) hypothesized that the subsidence warming is the reason the boundary layer depth is reduced at large radius. The boundary layer depth estimated using the critical Richardson number method is fundamentally different from all the other height scales (i.e., h_{vmax}, h_{infl}, Z_i), because the depth increases toward the eyewall. Nevertheless, it should be noted that specification of the height of the boundary layer may only be of significance in steady-state (e.g., Foster, 2009), analytical (Smith and Montgomery, 2008), and slab models (such as the one described in the next section). For a typical three-dimensional TC forecasting system (such as those described in section 6), consistent with observational studies, the different height scales evolve as a consequence of both dynamical as well as thermo dynamical forcings. For instance, the height of the inflow layer depth, δ (e.g., Fig. 4b), which is important for dynamical evolution in TCs, may indeed be different if one defines the depth of the HBL in terms of mixing, Z_i (e.g. Fig. 6b).

5. Simple Slab Model of the HBL

To facilitate a better understanding of the role of drag, eddy diffusivities, and height of the HBL on the hurricane structure and intensity, it is convenient to discuss the axisymmetric inner-core dynamics in cylindrical coordinates (e.g., Zhang et al., 2001; Gopalakrishnan et al., 2011). The cylindrical coordinates (r, λ, z) are defined as follows: r is the distance from the centre of the vortex, λ is the azimuthal angle, and z is the vertical height. Using a cylindrical polar height coordinate system and recasting the horizontal equations of motion into tangential momentum, the governing equations can be written as:

$$\frac{dv_\lambda}{dt} = -\frac{1}{\rho r}\frac{\partial p}{\partial \lambda} \underbrace{- \frac{u_r v_\lambda}{r} - fu_r}_{\text{Term A}} + Dv_\lambda \tag{1}$$

$$\frac{du_r}{dt} = \underbrace{-\frac{1}{\rho}\frac{\partial p}{\partial r} + \frac{v_\lambda v_\lambda}{r} + fv_\lambda}_{\text{Term B}} + Du_r \tag{2}$$

where $$\frac{d}{dt} = \frac{\partial}{\partial t} + u_r\frac{\partial}{\partial r} + \frac{v_\lambda}{r}\frac{\partial}{\partial \lambda} + w\frac{\partial}{\partial z} \tag{3}$$

In equations 1, 2 and 3, u_r, v_λ and w are respectively the radial, tangential and vertical winds in the earth-relative transformed coordinate system, p is the pressure, and Dv_λ and Du_r are, respectively, the diffusion (or frictional force) terms in the tangential and radial directions. Equation 1 is related to the "spin up" of the vortex, and Term A is known as the generalized Coriolis term. Its significance to the "spin up" process is discussed later in section 8 with the help of the HWRF system. Equation 2 represents the net forcing related to the secondary circulation. As mentioned in the introduction, recent works by Montgomery and Smith and his group members (see Montgomery and Smith, 2014 for a good review) and Gopalakrishnan et al. (2013) are starting to fuel arguments that support a more important role for the HBL in the spin up of TCs simply beyond what was discovered in either CISK or WISHE theory based on the importance of the acceleration term in the HBL (Equations 2 and 3). We will focus on the importance of these findings using a slab model here and further in section 8.

A schema for the slab model is presented in Fig. 8. By assuming the radial pressure gradient force is in balance with the centripetal and Coriolis forces above the boundary layer and a bulk drag law at the lower boundary (i.e, Km (dU_b/dz) = Cd M U_b, where U_b denotes the horizontal component of wind in the slab boundary layer, and M represents the wind speed), for a slab boundary layer of mean thickness (i.e., inflow layer depth), δ. Equations 1 and 2 reported

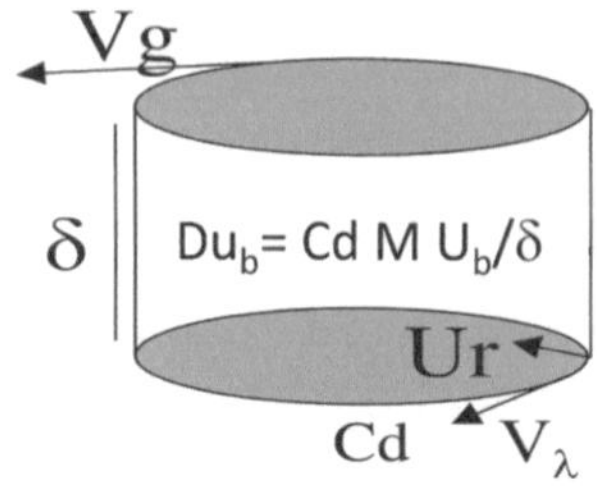

Fig. 8. A simple slab model for the HBL

in this work may be recast into the horizontal momentum equations 9 and 10 reported in Smith and Montgomery (2008). However, for the Lagrangian analysis discussed here, we replaced the radial advection term on the left side in Smith and Montgomery (2008) with total time derivative terms and solved the coupled system moving forward in time with an iterative procedure reported in Stull (2000). We also neglected the effect of updrafts in this simple model. It should be noted that in a few iterations the model will achieve a steady state. We traced the hodographs during that process. Given the tangential winds above the slab boundary layer, the drag coefficient at the lower boundary, and the mean depth of the slab layer, the tangential winds within the slab layer adjust to a new state due to the changing balance between the pressure gradient and the Coriolis, centrifugal, and friction forces within the boundary layer. In the process, a radial inflow is created. It should be noted that in a slab model, the frictional force term (in equations 2 and 3) is parameterized by Cd $M\ U_b/\delta$. This drag formulation was adopted from Smith and Montgomery (2008).

An increase in the depth of diffusion within the inflow layer, for instance, leads to a reduction in frictional force in the slab model and may be equivalent to producing a more diffusive boundary layer by increasing the eddy diffusivity. On the other hand, decreasing the depth of diffusion in the slab model is equivalent to increasing the frictional forces and, hence, the radial acceleration. For the hypothetical situation, where the eddy diffusivity is reduced to zero (not shown), although the air is in gradient wind balance above the surface layer, there would also be an infinite acceleration in the surface layer because the depth of the inflow becomes negligible. Figures 9a, b and c provide the hodographs for different tangential winds (51, 64 and 81 ms^{-1}) that are representative of those obtained for a matured storm. For each of these cases, Table 1 provides the list of sensitivity experiments performed with the slab model.

The results obtained from these simple experiments are consistent with the discussions provided in Gopalakrishnan et al. (2013) on the role of eddy diffusivity and the radial diffusion term (in equation 2). In general, for a given tangential wind, the *weaker the frictional force, the weaker the inflow*. From a Lagrangian point of view (Figs 9a, b, c), the larger the frictional force, the larger the radial acceleration as measured by the size of radial displacement. The steady state values in Table 1 also indicate that the values of the inflow from the slab model under different conditions are consistent with those from HWRF (Gopalakrishnan et al., 2013). While there is large variability

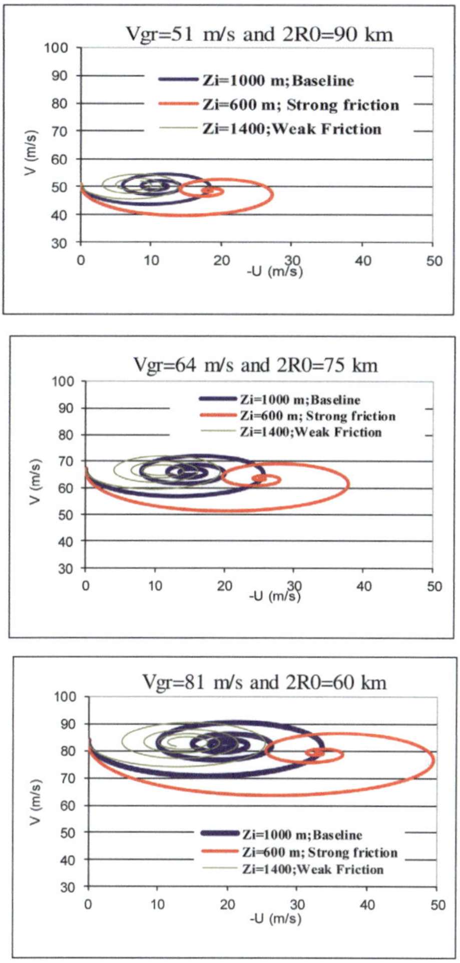

Fig. 9. Hodographs obtained for different tangential winds (51, 67 and 84 ms^{-1}) from a simple slab model that was used to provide additional insights on the imbalance of forces within the boundary layer. For each vortex, three boundary layer heights were used, namely, 1000 m (baseline), 600 m (strong frictional force) and 1400 m (weak frictional force).

in the steady state values, it is again seen that for a prescribed tangential wind, stronger frictional force (lower but non-zero vertical diffusion) leads to stronger inflow and vice-versa. Clearly, there are certain dynamical processes within HBL that cannot be modelled by neglecting the acceleration of the inflow. Apart from simply converging moisture, stronger the inflow, stronger will be the convergence of angular momentum within the HBL. In summary, frictional forces provide an additional impetus to the storm spin up process.

Table 1. List of experiments performed with the slab model

Expt No.	*Tangential wind above the slab PBL (ms^{-1})*	*Inflow layer depth (m)*	*Significance of the experiment and analogy with 3D model simulations*	*Steady state value of radial inflow (ms^{-1})*
1	51	1000	Weak vortex baseline	12
2	51	600	Weak vortex and strong frictional force	19
3	51	1400	Weak vortex and weak frictional force	9
4	64	1000	Moderate vortex baseline	16
5	64	600	Moderate vortex and strong frictional force	25
6	64	1400	Moderate vortex and weak frictional force	13
7	81	1000	Strong vortex baseline	19
8	81	600	Strong vortex and strong frictional force	33
9	81	1400	Strong vortex and weak frictional force	13

6. Boundary Layer Representation in High Resolution Hurricane Models

Numerical studies have shown that tropical cyclone intensity forecasts are very sensitive to the selection of HBL parameterization schemes (e.g., Braun and Tao, 2000; Nolan et al., 2009a, b; Smith and Thomsen, 2010). These parameterizations all aim to represent the effect of the turbulent fluxes in lower portion of the atmosphere for a tropical cyclone. Specifically, the specification of the drag coefficients (Cd and Ck) and the eddy diffusivities for momentum and heat/moisture (Km and Kh) are central to the problem. Below, we provide a brief description of NOAA's premier dynamical models available for tropical cyclone forecast guidance with special reference to the HWRF system and the boundary layer parameterizations available in the system.

The Geophysical Fluid Dynamical Laboratory (GFDL) regional hurricane prediction system originated as a research model in the 1970s (Kurihara and Tuleya, 1974; Kurihara, 1975; Tuleya and Kurihara, 1975, 1978, 1982). In the mid-1980s, the Hurricane Dynamics group at GFDL began a 10-year effort to transform their research hurricane model into an operational hurricane forecasting tool at the National Weather Service (NWS), USA. Since 1995, the GFDL hydrostatic hurricane prediction system has been used operationally by National Hurricane Center (NHC) and is regarded by NHC as one of the most skillful and reliable dynamical models for track forecasts. Since 2005, the model has demonstrated improved skill in intensity predictions, especially after the atmosphere–ocean coupled system (Ginis et al., 1999) and increases in vertical and horizontal resolution were implemented (Bender et al., 2007). A thorough summary of the modelling system and its performance is provided in Bender et al. (2007).

The Weather Research Forecasting Model (WRF) is a general purpose, multi-institutional mesoscale modelling system to address weather related problems (http://www.wrf-model.org/index.php). The Hurricane WRF (HWRF) which is a part of that framework was developed at NOAA's National Weather Service (NWS)/National Centers for Environmental Predictions (NCEP), in collaboration with NOAA's research laboratories: Atlantic Oceanographic and Atmospheric Laboratories (AOML), Earth System Research Laboratories (ESRL) and the GFDL and the University of Rhode Island to address the nation's next generation hurricane forecast problems and became an operational track and intensity guidance tool in 2007. A version of this evolving system is available at the Development Testbed Center (DTC), National Center for Atmospheric Research (NCAR), in Boulder, Colorado, and the scientific documentation (Gopalakrishnan et al., 2013) is available at http://www.dtcenter.org/HurrWRF/users/docs/scientific_documents/HWRF_final_2-2_cm.pdf.. The HWRF model is a primitive equation non-hydrostatic coupled atmosphere-ocean model with the atmospheric component formulated, at this time, with 42 levels in vertical. The model uses the Non-hydrostatic Mesoscale Model (NMM) dynamic core, including its rotated latitude-longitude projection with E-grid staggering. One of the backbone developments of the HWRF system is its flexible moving nest for hurricane application. The current operational configuration is a triply nested system. The model uses an outer domain at a horizontal grid resolution of 0.18° (about 27 km) spanning about 75° × 75°, with multiple telescopic, two-way interactive, storm following moving nest domains of about 10 × 10° with a grid spacing of 0.06° (9 km) within which is embedded a higher resolution nest covering an area of about 6 × 6° with a grid spacing of 0.02° (3 km).

In order to address the multi-scale nature of the hurricane forecasting problem, the Global Forecasting System (GFS) physics packages are used, with some modifications for higher resolution, both in the operational GFDL and the HWRF modelling systems. An extensive overview of the packages used in the HWRF system is provided in Gopalakrishnan et al. (2011) and Yeh et al. (2011) and only briefly discussed here. The Ferrier scheme (Ferrier et al., 2002) simulates latent heating to resolve the grid scale microphysical processes in the atmosphere, and the Simplified Arakawa and Schubert scheme, also known as the SAS scheme (Pan and Wu, 1995; Hong and Pan, 1998), is used to parameterize subgrid cumulus-cloud activity for the two domains at resolutions of 27 km and 9 km. The SAS convection scheme is switched off for the 3-km moving nest in the 2012 version of the HWRF system.

The HWRF and the GFDL surface layer parameterization scheme is based on the Monin-Obukhov similarity theory (Tuleya, 1994). The parameterization scheme provides estimates[3] of the surface layer exchange coefficients, Cd and Ck, for further computation of the surface layer fluxes. Figures 10a and 10b provide an estimate of Cd and Ck, respectively, obtained from one of the HWRF simulations (Gopalakrishnan et al., 2013). Clearly the wind speed dependence of Cd and constant behaviour of Ck with wind speed is well captured in the HWRF surface layer parameterization scheme.

The GFS boundary layer formulation (Hong and Pan, 1996) is used to parameterize the flux transport and subsequent mixing in the atmosphere in the HWRF and GFDL models. In this scheme, the momentum eddy diffusivity is

$$\mathrm{Km} = k\,(\mathrm{U}_*/\Phi_{\mathrm{m}})\, Z\, \{\alpha(1 - Z/Zi)^2\}, \tag{4}$$

where k is the von Karman constant (=0.4), U_* is the surface frictional velocity scale, Φ_m is the wind profile function evaluated at the top of the surface layer, Z is the height above the surface, and Zi is the depth of the turbulent PBL (also known as the mixed layer) that is determined based on the height above the ground at which the bulk Richardson number exceeds a critical value of 0.5. As discussed in Hong and Pan (1996), the eddy diffusivity for heat (and moisture), Kh, was computed from Km using the relationship of the Prandtl number. In the original formulation, which may be valid over land, $\alpha = 1.0$. Gopalakrishnan et al. (2013) used the flight-level data collected by a NOAA WP-3D research aircraft during the eyewall penetration of category 5 Hurricane Hugo (1989) at an altitude of about 450-500 m and Hurricane Allen (1980) as the basis to best match the modelled eddy diffusivities with wind speed. While reduction of the eddy diffusivity to a quarter of its original value (i.e., $\alpha = 0.25$) produced the best match with the observations, such a reduction revealed a significant decrease in the height of the inflow layer as well which, in turn, drastically impacted the size and intensity changes in the modelled TC. We will discuss these sensitivities in section 8.

In addition to the GFS physics for the surface and boundary layer, HWRF has been tested with different parameterization schemes. Bao et al. (2012) were able to use an experimental version of the HWRF system (dubbed as HWRFX) to study the influence of different boundary layer schemes on tropical cyclone development. The exact framework of the version of HWRF utilized is discussed in detail in the referenced paper. The purpose of their experiments was to demonstrate if the HWRFX model simulated hurricanes structure is susceptible to the choice of microphysics, boundary-layer, and radiation parameterization schemes. The sensitivity results are examined

[3] In advanced numerical models, the exchange coefficients are a function of stability and the roughness lengths for momentum, heat and moisture.

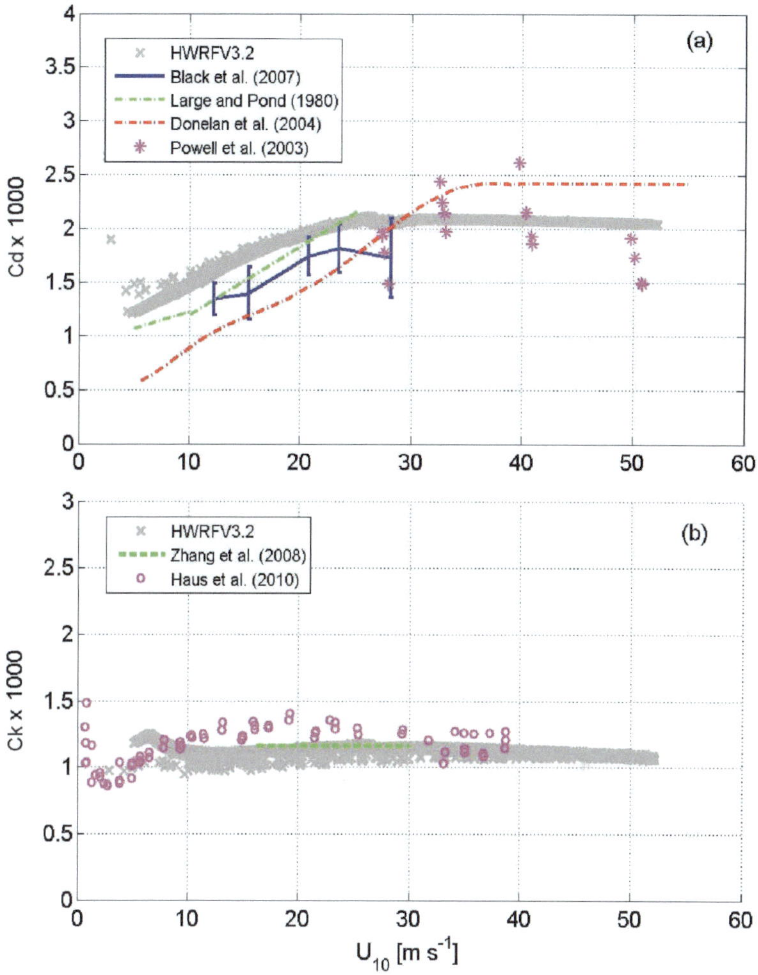

Fig. 10. The variation of (a) drag coefficient (Cd) and (b) heat and moisture exchange coefficient (Ck) with 10-m wind speeds; high resolution HWRF model with 2010 upgrades (thick gray dots) from the control experiment (Gopalakrishnan et al., 2012) compared with data from Large and Pond (1981), Powell et al. (2003), Donelan et al. (2004), Black et al. (2007), Zhang et al. (2008) and Haus et al. (2010).

using various metrics in terms of the maximum velocity and the minimum pressure, and the azimuthally averaged structural characteristics of the primary and secondary circulations. Three major results are obtained from the comparisons of the sensitivity experiments. First, different boundary layer physics parameterization schemes for vertical sub-grid turbulence mixing lead to differences not only in the intensity evolution in terms of the maximum velocity and minimum pressure, but also in the structural characteristics of the simulated tropical cyclone. Second, the surface drag coefficient is a key parameter that controls the pressure-wind relationship and the agradient force near the surface. Third, different microphysics and subgrid convection parameterization schemes, due to different realization of diabatic heating distribution, lead to significant variations in the vortex structure. All of these results indicate that the current uncertainties in the boundary level mixing, surface drag, and microphysics parameterization schemes have comparable impacts on the intensity and structure of the HWRFX model-simulated tropical cyclones. In the next few sections, the discussion will shift to the sensitivities for numerical predictions to Cd, Ck, and eddy diffusion on the structure of the HBL and their effect on the intensification processes in TCs.

Apart from operational hurricane models, cloud resolving models have more recently been used as research tools to help understand the hurricane intensity prediction problem. For example, in a series of explicit simulations of Hurricane Andrew (1992) using the fifth-generation Mesoscale Model (MM5), developed by the Pennsylvania State University (PSU) and NCAR at a resolution of about 6 km, several interesting features of the modelled hurricane (Liu et al., 1997, 1999; Yau et al., 2004; Zhang et al., 1999, 2000, 2001, 2002) compared well to the observed structure of hurricanes (Marks and Houze, 1987; Marks et al., 1992; Willoughby, 1979, 1990b, a). As verified against various observations and the post-storm derived "best track" data (positions and intensities) provided by NHC, the PSU–NCAR MM5 simulations reasonably captured many of the features of the inner-core structure of the storm.

The first attempt to assess how well boundary-layer processes are represented in the MM5 model was carried out by Braun and Tao's (2000) evaluation of the prediction of Hurricane Bob (1991). Tests were conducted to determine the sensitivity of the simulation to four planetary boundary- layer parameterizations in the model, including the bulk aerodynamic, Blackadar, medium-range forecast (MRF) model and Burk–Thompson schemes. The various schemes caused maximum winds varying by 15 ms^{-1}, but all of the calculations overestimated the observed intensity of the storm. At the time of their study, the boundary layer was thought of mainly as providing the moisture source to fuel the eyewall convection but Braun and Tao discovered that the boundary played a more central role in hurricane structure predictions. Nolan et al. (2009a, b) evaluated the PBL parameterization scheme in TCs by comparing in situ observations and high resolution simulations of hurricane Isabel (2003). The two HBL schemes used in their study, i.e., Yonsei University (YSU) and the Mellor–Yamada–Janjic (MYJ) parameterization scheme reproduced the HBL structure in the outer core region remarkably well. More recently, Smith and Thomsen (2010) also analyzed the MM5 model boundary layer schemes but specifically to investigate the dependence of vortex spin-up on the choice of the boundary-layer parameterization scheme. They compared the five available schemes for the MM5 model (the Mellor-Yamada-based scheme was added to the MM5 suite) but were not able to advocate a particular scheme. Still, deficiencies were highlighted in the different boundary layer formulations and certain schemes led to unrealistic inflow depths, inflow speeds, RMWs, and tangential wind speeds.

7. Sensitivity Experiments with Drag Coefficients

In 2010, Montgomery et al. investigated the sensitivity of tropical cyclone intensity to the surface drag coefficient. As part of their study, Montgomery et al. used the MM5 model, whose specifications are described in detail in their paper, to analyze the sensitivity of tropical cyclone models to a varying surface drag coefficient while keeping the coefficient of enthalpy transfer between sea and air unchanged and near observed values from the CBLAST experiment. They found that the intensification rate of the vortex and the intensity increase with increasing Cd up to approximately 2×10^{-3}. When Cd is increased further, no significant difference in the intensification rate or intensity is found until a threshold of approximately 1.3×10^{-2}, beyond which the intensity decreases. Another major conclusion was that the simulated intensity does not depend markedly on the critical value of the ratio Ck/Cd below a value of approximately 0.7, as was suggested by Emanuel (1995). In fact, Montgomery et al. (2010) observed weak vortices intensifying into major hurricane intensity for Ck/Cd values as small as 0.1. Bryan (2012) highlighted that the different conclusions between Montgomery et al.'s study and other similar modelling studies (e.g., Emanuel, 1995; Bryan and Rotunno, 2009a, b) are attributable to the relatively short integration time of Montgomery et al.'s simulations. Their simulated tropical cyclones are probably not quasi-steady after four days of integration, and the values of maximum winds likely do not represent the maximum possible values.

It may be easy to understand the role of Ck on the intensification process because an increase in enthalpy fluxes, which is the chief source of energy, should lead to stronger TCs and a decrease should weaken the hurricanes. However, the effects of Cd may be more complex to comprehend because the frictional drag should weaken the tangential winds in TCs (Bryan, 2011), while, at the same time, results in section 5 show an increase in Cd leads to a stronger inflow. Stronger inflow could lead to large moisture flux convergence and subsequently a more intense storm. Nevertheless, as shown in Smith et al. (2013), depending on the PBL schemes, the tangential winds may either strengthen or weaken. Sensitivity experiments were carried out with the HWRF system to find out the influence of a range of drag coefficients that may potentially influence the TC intensification process. Figure 11 provides the azimuthally averaged, radius height cross section of the 10-m wind (top row) and radial inflow at the first model level (bottom row). Clearly, the strength of the inflow and the size of the storm depend on Cd. A weaker Cd produces a weaker inflow. Nevertheless, the decrease in Cd to half the functional value

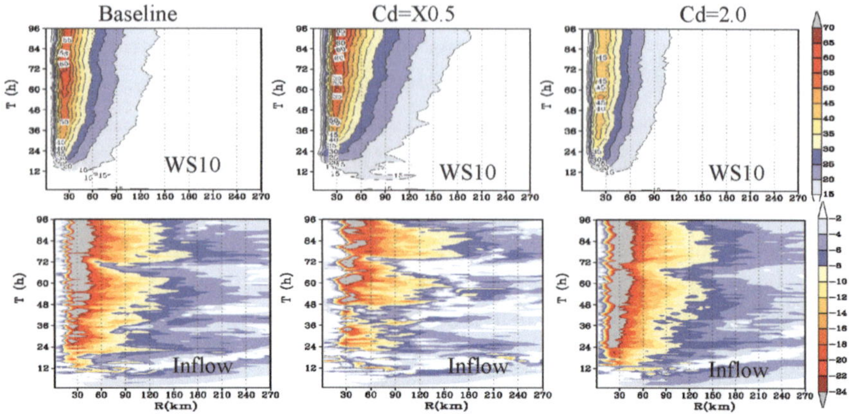

Fig. 11. Top row: radius-time cross section of azimuthally averaged wind speed at 10 m level. Bottom row; radius-time cross section of the azimuthally averaged radial winds at an altitude of about 30 m obtained from the HWRF system from (a): the baseline; (b): when drag is decreased to half its original value and (c): when drag is increased to twice the baseline.

provided in Fig. 10a has marginal effects on the tangential wind. However, a stronger Cd produced a smaller storm and a stronger inflow, but also a weaker storm.

8. Influence of Eddy Diffusivity on Structure and Intensification of TCs

In section 5, a simple slab model that parameterized diffusion in terms of frictional drag and height of the boundary layer was discussed to gain some fundamental understanding on the influence of diffusion on the inflow. It should be noted that in those experiments, Cd was held close to observations. Recently, through a series of numerical experiments using the HWRF system in an idealized framework, Gopalakrishnan et al. (2013) demonstrated and documented some very large impacts of vertical diffusion on the structure and subsequent intensification of TCs. Such sensitivities have been noticed in numerical models, especially when different PBL parameterization schemes are used. However, this study, for the first time, addressed and explained the fundamental source of variability in TC intensity and structure prediction that may arise due to vertical diffusion. The MRF/GFS PBL scheme was used to understand the influence of vertical diffusion on the dynamics of TCs. Flight-level data collected by a NOAA WP-3D research aircraft during the eyewall penetration of category 5 Hurricane Hugo (1989) at an altitude of about 450-500 m (Marks et al. 2008) and Hurricane Allen (1980) (Marks, 1985) were used as the basis to best match the modelled vertical eddy diffusivities with wind speed. Cd and Ck kept consistent with observations (Fig. 12). The following are the salient results from this study:

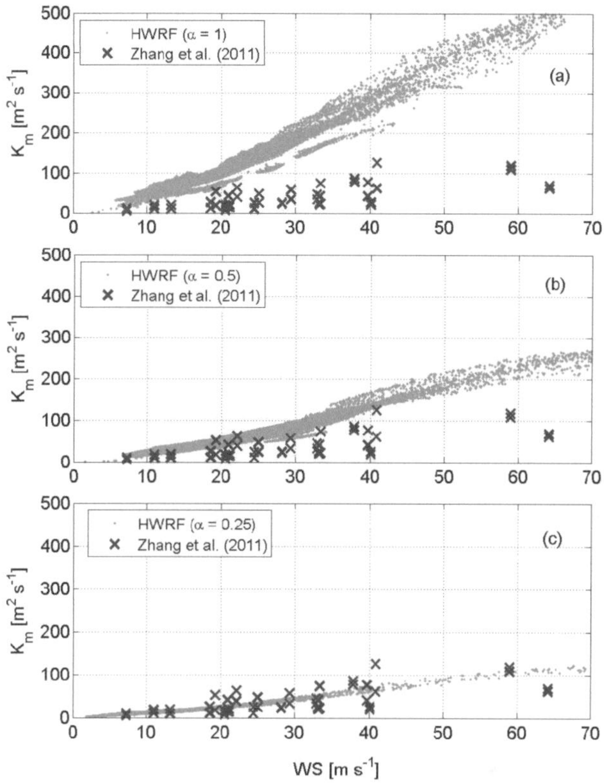

Fig. 12. The variation of the eddy diffusivity coefficient, Km, with 10-m wind speeds; high resolution HWRF model outputs from the (a) control experiment (α = 1), (b) Km reduced to half (α = 0.50), and (c) Km reduced to a quarter (α = 0.25) compared with data from Zhang et al. (2011a) obtained from flight-level data between 450-500 m. Modeled outputs were compared between the same levels. Thanks to the American Meteorological Society (Gopalakrishnan et al., 2012, Journal of the Atmospheric Sciences).

1. As mentioned in section 6, the HWRF model uses the GFS boundary layer parameterization scheme with alpha value set to 1 (equation 4). However this parameter can be adjusted. We made some experiments and found that such an adjustment may be consistent with the log law in the surface layer. Reduction of eddy diffusivities, Km (and Kh) in equation 4, to 25% of its original value (α = 0.25) produces reasonable diffusion coefficients consistent with observations for all wind speeds ranging from about 10 ms^{-1} to about 60 ms^{-1} in the MRF scheme (Fig. 12).
2. Reduction of Km (and Kh) has a significant influence on the structure, size and evolution of the vortex. With Km set to 25% of its original value, the depth of the inflow layer is reduced from ~2 km to about 1 km, again, more consistent with the observations. Figure 13, for instance,

shows the tangentially-averaged, 6-hourly time averaged, radius-height cross section of the secondary circulation centered around the 93rd hour of the simulations and represents the mean structure of a mature storm for the control run (A100), a of 0.50 (A050), and a of 0.25 (A025), respectively from Gopalakrishnan et al. (2013). As indicated by the vectors and contours in Fig. 13, the circulation in the A100 simulation is characterized by a deep layer of radial inflow a couple of kilometres deep[4] in the lower troposphere and a layer of intense outflow at about 14 km height (not shown).

The rising branch of the secondary circulation, depicted by green contours and located in the eyewall region, slopes radially outward. A *very weak* outflow is observed above the shallow inflow layer due to mass conservation. The model is able to reproduce the same features as reported in Gopalakrishnan et al. (2011). However, a comparison with the available observational data from Zhang et al. (2011b) indicates that the simulated inflow layer depth is at least twice the thickness of that observed. Nevertheless, more reasonable modelled estimates are produced after decreasing the α parameter. A comparison between A100 with A050 and A025 (Figs 14a, b, c) indicates a systematic reduction of the inflow depth with a decrease in Km and an increase in the strength of the radial winds. The inflow layer in A100 is about 2.5 km, while it is about 1.5 km in A050 and reduces to the observed value of about 1 km in A025. As the depth of the inflow layer decreases, the strength of the inflow increases. While the maximum radial wind in A100 is on an average about 6-9 ms^{-1}, it is 9-15 ms^{-1} in A050 and 27-33 ms^{-1} in A025. The radius of the maximum winds also shrinks in size. While the radius at the first model level is 26 km in A100, it is 22 km in A050 and further reduces to 15 km in A025. A comparison of the modelled structure of the HBL (Fig. 13c) with observations (Fig. 4b) indicates that the high resolution model such as the HWRF is able to produce consistent HBL.

3. Larger diffusion negates the gradients between the surface layer and the HBL, leading to weaker frictional forces and weaker inflow (Fig. 14a). On the contrary, a reduction in diffusion leads to a reduction in the dissipation of angular momentum in turn, leads to enhanced HBL inflow (Fig. 14c), enhanced spin-up in HBL, enhanced convergence of moisture in HBL, enhanced convection in the eyewall and, subsequently, enhanced feedback to the HBL inflow and a stronger hurricane.

   In a hypothetical situation where the diffusivity is reduced to zero, radial acceleration in the boundary layer will be negligible and the rings of air will be in gradient wind balance (i.e., Term B in equation 2)

---

[4] The depth of the inflow in this study was taken to be the height where the radial wind velocity is reduced to about 3 ms^{-1}. Significant inflow above the boundary layer is not uncommon (e.g., Willoughby 1979).

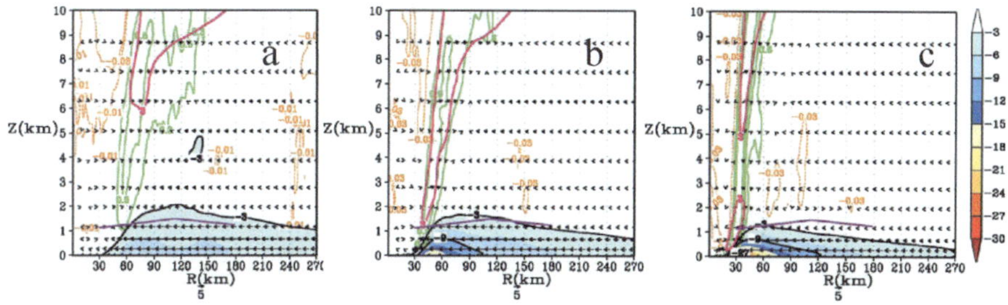

Fig. 13. Tangentially averaged, 6-hourly time averaged, radius-height cross-section of the secondary circulation at 93 hours for (a) control (α = 1), (b) Km reduced to half (α = 0.50), and (c) Km reduced to a quarter (α = 0.25). The black solid contour indicates the inflow, and the magenta contours show the outflow of the radial wind component (in ms^{-1}). Inflow is also shaded in colours. Only the lower 10 km has been expanded for convenience. The vertical velocity (in ms^{-1}) is shown by contour lines with updrafts indicated by green and weak subsidence indicated by yellow. Because the distribution of vertical velocity is skewed, please note that the scales are unequally spaced. For convenience, the vector field obtained that compounded the tangentially averaged vertical and the radial velocity components is also provided here. Presented in purple colour are the estimates of the inflow layer depth from Zhang et al. (2011b). *Courtesy:* The American Meteorological Society (Gopalakrishnan et al., 2013, *Journal of the Atmospheric Sciences*).

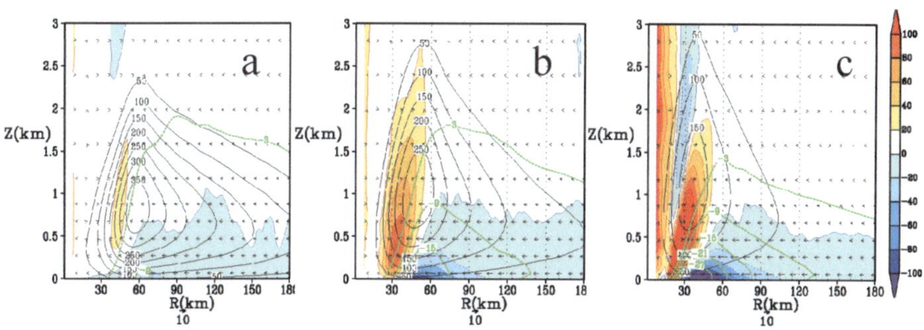

Fig. 14. Tangentially averaged, 6-hourly time averaged, radius-height cross-section of Eddy diffusivity (Km) at 93 hours for (a) control (α = 1), (b) Km reduced to half (α = 0.50), and (c) Km reduced to a quarter (α = 0.25). Superposed in colour are the forcing terms related to gradient wind imbalance within the boundary layer (i.e., Term B in equation 3). Units of the forcing term are in ms^{-1} h^{-1}. The inflow is shown in green background contours. For convenience, the vector field obtained that compounded the tangentially averaged vertical and the radial velocity components is also provided here. For the sake of clarity only the first 3 km above the ground is shown here. *Courtesy:* The American Meteorological Society (Gopalakrishnan et al., 2013, *Journal of the Atmospheric Sciences*).

conserving absolute angular momentum even above the shallow surface layer. There will be no inflow above this layer. However, Km controls and modulates the radial frictional force (diffusion term in equation

2), and subsequently the radial acceleration and strength of the inflow especially in the eye wall region above the surface layer by creating or annulling gradients in the mean flow. Figure 14 shows tangentially averaged, 6-hourly time averaged, radius-height cross-section of eddy diffusivity (Km) at 93 hours for (a) control ($\alpha = 1$), (b) Km reduced to half ($\alpha = 0.50$), and (c) Km reduced to a quarter ($\alpha = 0.25$). The maximum turbulence and eddy diffusion is simulated in the eyewall region. Lorsolo et al. (2010), for the first time, reported some estimates of turbulent kinetic energy in hurricanes using airborne Doppler radar. They found that the strongest turbulence was generally located in convective regions, such as the eyewall and the mean vertical profiles of Turbulence Kinetic Energy decrease sharply above the mixed layer and level off at low magnitude for all regions outside the radius of maximum wind (Fig. 5). Further, superposed in colour (Fig. 14) are the forcing terms related to gradient wind imbalance within the boundary layer (i.e., Term B in equation 2). Units of the forcing term are in ms^{-1} h^{-1}. We find that within the first 500 m above the ground level and between radius of about 30 km to 120 km, where the effect of inflow is the maximum, the eddy diffusivity (Km) has a maximum of 300 m^2s^{-1} for $\alpha = 1$ case; 150-200 m^2s^{-1} for $\alpha = 0.5$ case and 50-100 m^2s^{-1} for $\alpha = 0.25$ case. It may be noted that the observed Km values from flight level data at around 450-500 m matches well for $\alpha = 0.25$ case (Fig. 13). Also, clearly, in the eye wall region, the run with $\alpha = 0.25$ shows the maximum imbalance. The winds are strongly sub-gradient (pressure gradient term dominates the centripetal force in equation 3) and, subsequently, leads to a stronger acceleration near the eye wall region. One could notice (Fig. 14c) an overshoot of near-surface inflow that leads to super gradient winds (indicated in red). This overshoot is again consistent with observations shown in Fig. 5b. Whereas, large vertical diffusion ($\alpha = 1$) annals any possible gradients that may be created in the control run (Fig. 14a). *In other words, weaker the frictional force in the mixed layer, weaker is the radial inflow*. Thus, the influence of the frictional forces on the inflow is well supported by results obtained from a simple slab model for the hurricane boundary layer discussed in section 5.

4. Gopalakrishnan et al. (2013) illustrated that inflow was observed to be stronger as inflow depth decreases. Stronger inflow not only increased the spin of the storm, enhancing the generalized Coriolis term in the equations of motion for tangential velocity (i.e., Term A in equation 1), but also resulted in enhanced θ_e in the boundary layer, a stronger and warmer core and, subsequently, a stronger storm (not shown here). More importantly, rapid acceleration of the inflow not only produced a stronger

outflow at the top of the inflow layer, more consistent with observations (Fig. 5b), but also a smaller and more intense inner core (Fig. 15). For instance, while the radius of maximum wind speed at the first model level was 42 km in the simulation using the original MRF scheme, it decreased to about 31 km when the value of eddy diffusivity was decreased to a quarter of the original value (Fig. 15). Similarly, Figs 15c, d, and e indicate that A025 produces the strongest hurricane. Also, the maximum tangential wind at any location in the storm during its lifetime for A100 was 51 ms^{-1}, whereas it was about 84 ms^{-1} for A025.

5. The structure of the PBL and the axisymmetric "spin up" of the idealized vortex discussed here appear to be more consistent with some of the observational (Zhang et al., 2011a, b) and theoretical modelling works (Montgomery and Smith, 2012) that use a different parameterization scheme for vertical diffusion (e.g., Nolan et al., 2009a, b) when Km (and Kh) in the MRF scheme is reduced to 25% of its original value.

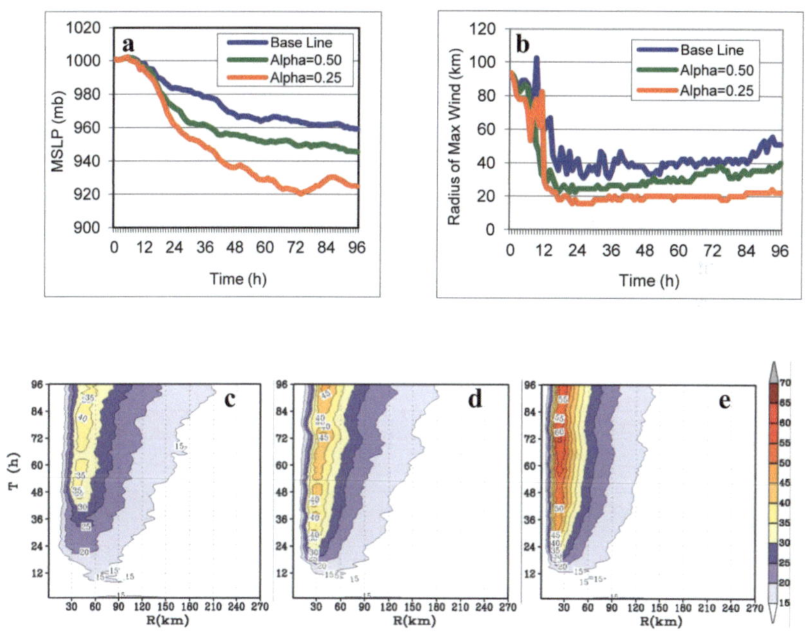

Fig. 15. Time history of the intensification process in an idealized storm for the three simulations provided in Table 1: (a) minimum mean sea level pressure in hPa, (b) radius of maximum wind at 10 m; Hovemoller diagram of the axisymmetric mean wind at a height of 10 m for (c) baseline simulation ($\alpha = 1$), (d) Km reduced to half ($\alpha = 0.50$), and (e) Km reduced to a quarter ($\alpha = 0.25$).

Courtsy: The American Meteorological Society (Gopalakrishnan et al., 2013, *Journal of the Atmospheric Sciences*).

9. Further Advances in Structure Predictions

In an effort to compare the composite structure of the observed storm discussed in section 2, beyond the structure of secondary circulations discussed in section 8, we did some comparisons of the primary circulation (tangential wind), and the vorticity and patterns from the model with those composites (Fig. 16). It appears that at the time of writing this text book, the mean structures of the TCs are indeed well reproduced by high resolution models. However, it should be emphasized that only a simplified view mostly related to axisymmetric, mean structure of the HBL and a bird eye view of turbulence as represented in high resolution models was presented in this chapter. The role of the drag and enthalpy coefficient (Cd and Ck) along with diffusion and how that is treated in numerical models was discussed.

Modelling the air sea interface in the HBL is perhaps the most challenging problem in the future. While it is well known that an increase in the SSTs should fuel a stronger hurricane because of the increase in both latent and sensible heat fluxes and ocean coupled models are essential to improve intensity predictions and coupled models are starting to prove a value for improving TC predictions (Bender, 2007; Yablonsky and Ginis, 2013), it appears that both waves as well as sea spray might have a very huge impact on the intensification process. Several pioneering works using the GFDL model by Isaac Ginis and

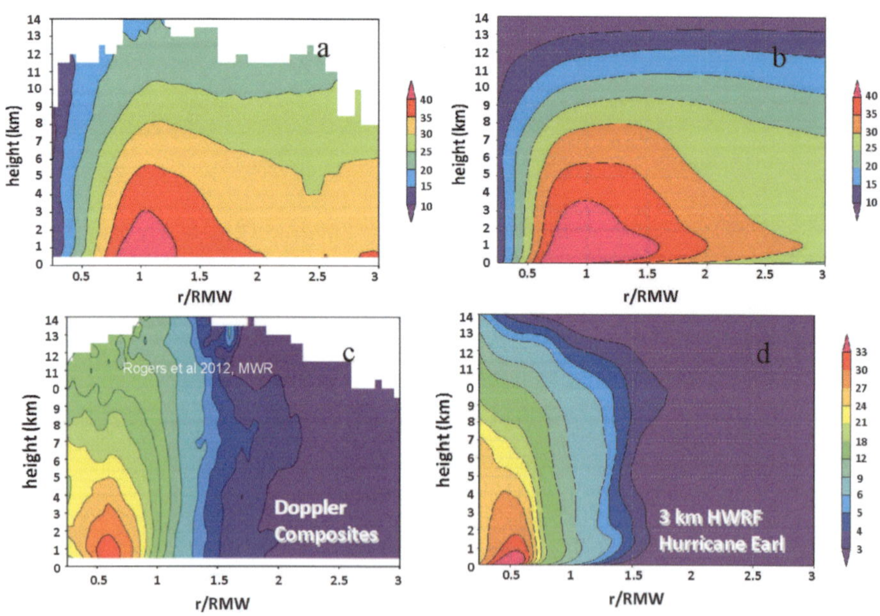

Fig. 16. Comparisons of Doppler composites with 3 km HWRF model simulations. Azimuthally averaged normalized radius height cross section of (a) and (b) : Tangential wind speed and (c) and (d) vorticity.

his group members (Moon et al., 2007, 2008; Bender et al., 2007; Fan et al., 2009a, b) are starting to prove the importance of ocean waves under hurricane conditions. Air-sea (momentum, heat, moisture) fluxes and turbulent mixing above/below the air-sea interface in tropical cyclones are greatly modified by surface waves. Figure 17 presents the complex interactions between waves and hurricanes. Size and asymmetries in the structure of TCs might have a very huge influence on ocean waves which in turn affects the roughness (i.e., Cd) and subsequently the momentum transfer between the atmosphere and the ocean surface.

While the rough, yet, non-breaking waves may act as a sink of angular momentum in the HBL, the breaking waves may counter those effects by the creation of sea spray that may have some significant influence on the enthalpy fluxes (both sensible as well as latent) and subsequently on the intensification process in TCs. The life cycle of an individual spray drop is complex (Andreas and Emanuel, 2001). Nevertheless, Fairall et al. (1994) developed a parameterization scheme for numerical models that accounts for the following spray drop processes: (a) the spray drop gives off sensible heat in cooling from the ocean temperature (Ts) to the air temperature (Ta); (b) the spray drop evaporates, which results in a spray-air latent heat flux; (c) as the drop evaporates, the drop temperature cools to what is essentially a wet-bulb temperature (Tw), and the spray drop temperature becomes lower than the temperature of the surrounding environment, which results in a negative sensible heat flux from air to spray because Tw < Ta; (d) while the drop is evaporating, the ambient moisture field nears saturation, which decreases the saturation mixing ratio gradient between the drop and the air and causes a decrease in the spray-air latent heat flux term; and (e) the drop is constantly releasing or absorbing sensible heat, which in turn modifies the near-surface ambient temperature and then affects the spray-air sensible heat flux. The

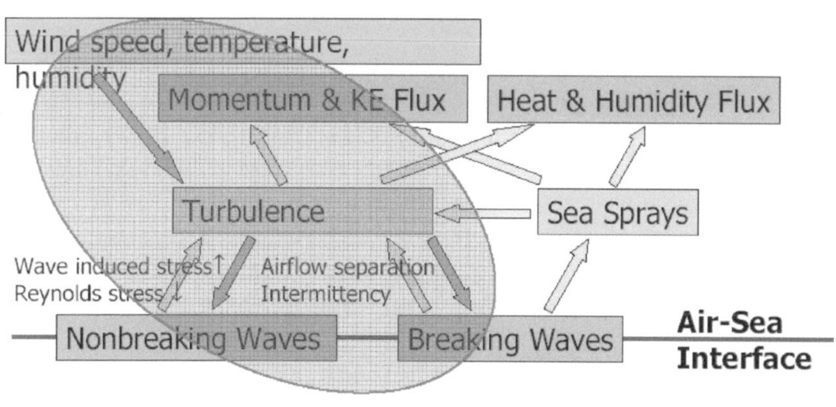

Fig. 17. Complex nature of air-sea interactions in a hurricane system (Isaac Ginis, personal communication).

changes to the fluxes in the HBL due to the above processes are usually parameterized in terms of Ck.

Bao et al. (2000) used the MM5 modelling system, the University of Colorado version of the Princeton Ocean Model, and the ocean surface gravity wave model developed by the Wave Model Development and Implementation Group to simulate air-sea interaction under high wind conditions. The results from the experiments with and without sea spray show that the inclusion of sea spray evaporation can significantly increase hurricane intensity in a coupled air-sea model when the part of the spray that evaporates is only a small fraction of the total spray mass. In this case the heat required for spray evaporation comes from the ocean. When the fraction of sea spray that evaporates increases, so that the evaporation extracts heat from the atmosphere and cools the lower atmospheric boundary layer, the impact of sea spray evaporation on increasing hurricane intensity diminishes. There are several works following Bao et al. (2000) that have all presented the importance of waves on the TC intensification problem. However, since such interactions occur at an observationally inaccessible region of the TCs at this time, those results pose challenges for verification. Nevertheless, unmanned aircraft system (UAS) may be useful in the future for further model verifications at the air-sea interface.

Another important area of research and an unsolved problem is the role of horizontal diffusion in the region of high gradients such as the eyewall of hurricanes. Bryan and Rotunno (2009a, b) used an axisymmetric, non-hydrostatic numerical model to evaluate the maximum possible intensity of tropical cyclones. At a resolution of 1 km or less, the authors found significant sensitivity to the specification of turbulence intensity in their model. In particular, the authors concluded that the lateral diffusion in the model limits maximum intensity because it reduces the radial gradient of angular momentum, and turbulence also reduces radial gradients of θe (scalar). Unlike the vertical diffusion process which is parameterized in terms of physical effects, the application of lateral diffusion in atmospheric models in general has always been a subject of debate since it is not clear to what extent one can model horizontal diffusion. Although it is not feasible to run direct numerical simulations even to understand the turbulence structure of TC at this time, we believe that the use of observations to improve parameterization scheme in NWP applications especially for TC is a realizable goal. Recent studies from Zhang and Montgomery (2012) also provide an estimate of horizontal eddy diffusivity and mixing length in low-level region of intense hurricane. We are in the process of evaluating both the vertical and horizontal diffusivity in the HWRF system for real cases, as well.

Organized turbulence, and consequently, roll convection in the boundary layer occurs over heterogeneous land surface when the scales of such heterogeneities in the order of tens of kilometres. Formation of rolls in the day time convective boundary layer over land have been well documented in several studies (e.g., Gopalakrishnan et al., 2000; Gopalakrishnan and Avissar, 2000). More recently, various researchers (see Foster, 2005) have identified coherent linear features (rolls/streaks) in the HBL, as well. These features lead to relatively small-scale velocity perturbations in the general wind field, and yet their structural characteristics, evolution, intensity and overall effect on the surface level wind field are still largely unknown. Because of the large vertical extent and strong intensity, these features can have a significant influence on the vertical momentum, sensible heat and moisture fluxes in hurricanes and therefore may have a potential for influencing the hurricane intensity. However, the roll vortices are not explicitly represented in existing hurricane models because of their small spatial scales. In order to parameterize the effect of roll vortices in hurricane models we must know the favourable conditions for their formation and their impact on the boundary layer structure.

Acknowledgements

The authors acknowledge funding from NOAA's Hurricane Forecast Improvement Project that supported this work. The author acknowledges the support provided by Indo-US Science and Technology Forum, NOAA's Hurricane Forecast Improvement Project and various national and international collaborators who have helped improving the operational HWRF model. Thanks are due to Ms. Lisa Bucci for offering editorial support and to Drs. J-W.Bao, Jun Zhang, Hua Chen, Frank Marks and Mr. Joshua Alland for providing the internal review for this manuscript.

REFERENCES

Aberson, S.D., M.L. Black, R.A. Black, J.J. Cione, C.W. Landsea, F.D. Marks and R.W. Burpee, 2006a: Thirty years of tropical cyclone research with the NOAA P-3 aircraft. *Bull. Amer. Meteor. Soc.*, **87**, 1039-1055.

Aberson, S.D., M.L. Black, M.T. Montgomery and M. Bell, 2006b: Hurricane Isabel (2003): New insights into the physics of intense storms. Part II: Extreme localized wind. *Bull. Amer. Meteor. Soc.*, **87**, 1349-1354.

Andreas, E.L. and K.A. Emanuel, 2001: Effects of sea spray on tropical cyclone intensity. *J. Atmos. Sci.*, **58(24)**, 3741-3751.

Anthes, R.A. and S.W. Chang, 1978: Response of the hurricane boundary layer to changes of sea-surface temperature in a numerical model. *J. Atmos. Sci.*, **35**, 1240-1255.

Barnes, G.M. and M.D. Powell, 1995: Evolution of the inflow boundary layer of Hurricane Gilbert (1988). *Mon. Wea. Rev.*, **123**, 2348-2368.

Bao, J.-W., J.M. Wilczak, J.-K. Choi and L.H. Kantha, 2000: Numerical simulations of air-sea interaction under high-wind conditions using a coupled model: A study of hurricane development. *Mon. Wea. Rev.*, **128**, 2190-2210.

Bender, M.A., I. Ginis, R.E. Tuleya, B. Thomas and T. Marchok, 2007: The operational GFDL coupled hurricane-ocean prediction system and a summary of its performance. *Mon. Wea. Rev.*, **135**, 3965-3989.

Black, M.L., J.F. Gamache, F.D. Marks, C.E. Samsury and H.E. Willoughby, 2002: Eastern Pacific Hurricanes Jimena of 1991 and Olivia of 1994: The effect of vertical shear on structure and intensity. *Mon. Wea. Rev.*, **130**, 2291-2312.

Black, P.G. et al., 2007: Air-sea exchange in hurricanes: Synthesis of observations from the Coupled Boundary Layer Air-Sea Transfer Experiment. *Bull. Amer. Meteor. Soc.*, **88**, 357-374.

Braun, S.A. and W.-K. Tao, 2000: Sensitivity of high-resolution simulations of Hurricane Bob (1991) to planetary boundary layer parameterizations. *Mon. Wea. Rev.*, **128**, 3941-3961.

Bryan, G.H. and R. Rotunno, 2009a: The maximum intensity of tropical cyclones in axisymmetric numerical model simulations. *Mon. Wea. Rev.*, **137**, 1770-1789.

Bryan, G.H. and R. Rotunno, 2009b: Evaluation of an analytical model for the maximum intensity of tropical cyclones. *J. Atmos. Sci.*, **66**, 3042-3060.

Bryan, George H., 2011: Comments on Sensitivity of tropical-cyclone models to the surface drag-coefficient http://www.mmm.ucar.edu/people/bryan/Papers/bryan_2011_qjrms_submitted.pdf

Cione, J.J., P.G. Black and S.H. Houston, 2000: Surface observations in the hurricane environment. *Mon. Wea. Rev.*, **128**, 1550-1561.

Donelan, M.A., B.K. Haus, N. Reul, W.J. Plant, M. Stiassnie, H.C. Graber, O.B. Brown and E.S. Saltzman, 2004: On the limiting aerodynamic roughness of the ocean in very strong winds. *Geophys. Res. Lett.*, **31**, L18306, doi:10.1029/2004GL019460.

Drennan, W.M., J.A. Zhang, J.R. French, C. Mccormick and P.G. Black, 2007: Turbulent Fluxes in the Hurricane Boundary Layer. Part II: Latent Heat Flux. *J. Atmos. Sci.*, **64**,1103-1115.

Emanuel, K.A., 1986: An air-sea interaction theory for tropical cyclones. Part I: Steady-state maintenance. *J. Atmos. Sci.*, **43**, 585-604.

Emanuel, K.A., 1995: Sensitivity of tropical cyclones to surface exchange coefficients and a revised steady-state model incorporating eye dynamics. *J. Atmos. Sci.*, **52**, 3969-3976.

Emanuel, K.A., 2003: A century of scientific progress. *In:* Hurricane! Coping with Disaster. R. Simpson, R. Anthes and M. Garstang (Eds). Special Publication, Vol. 55, Amer. Geophys. Union, 177-216.

Fan, Y., I. Ginis, T. Hara, C.W. Wright and E. Walsh, 2009b: Numerical simulations and observations of surface wave fields under an extreme tropical cyclone. *J. Phys. Oceanogr.*, **39**, 2097-2116.

Fan, Y., I. Ginis and T. Hara, 2009a: The effect of wind-wave-current interaction on air-sea momentum fluxes and ocean response in tropical cyclones. *J. Phys. Oceanogr.*, **39**, 1019-1034.

Foster, R.C., 2005: Why rolls are prevalent in the hurricane boundary layer. *J. Atmos. Sci.*, **62**, 2647-2661.

Gamache, J.F., 1997: Evaluation of a fully three-dimensional variational Doppler analysis technique. Preprints, 28th Conf. on Radar Meteorology, Austin, TX, Amer. Meteor. Soc., 422-423.

Gamache, J.F., R.A. Houze Jr. and F.D. Marks Jr., 1993: Dual-aircraft investigation of the inner core of Hurricane Norbert. Part III: Water budget. *J. Atmos. Sci.*, **50**, 3221-3243.

Ginis, I., W. Shen and M.A. Bender, 1999: Performance evaluation of the GFDL coupled hurricane ocean prediction system in the Atlantic basin. Preprints, 23rd Conf. on Hurricanes and Tropical Meteorology, Dallas, TX, Amer. Meteor. Soc., 607-610.

Gopalakrishnan, G., S. Baidya Roy and R. Avissar, 2000: An evaluation of the scale at which topographical features affect the convective boundary layer using large eddy simulations. *J. Atmos. Sci. (USA)*, **57(2)**, 334-351.

Gopalakrishnan, S.G. and R. Avissar, 2000: An LES study of the impacts of land surface heterogeneity on dispersion in the convective boundary layer. *J. Atmos. Sci. (USA)*, **57(2)**, 352-371.

Gopalakrishnan, S.G., F.D. Marks, X. Zhang, J.-W. Bao, K.-S. Yeh and R. Atlas, 2011: The experimental HWRF system: A study on the influence of horizontal resolution on the structure and intensity changes in tropical cyclones using an idealized framework. *Mon. Wea. Rev.*, **139**, 1762-1784.

Gopalakrishnan, S.G., S. Goldenberg, T. Quirino, F. Marks, X. Zhang, K.-S. Yeh, R. Atlas and V. Tallapragada, 2012: Towards improving high-resolution numerical hurricane forecasting: Influence of model horizontal grid resolution, initialization, and physics. *Wea. Forecasting*, **27**, 647-666.

Kepert, J.D. and Y. Wang, 2001: The dynamics of boundary layer jets within the tropical cyclone core. Part II: Nonlinear enhancement. *J. Atmos. Sci.*, **58**, 2485-2501.

Kepert, J.D., 2010: Comparing slab and height-resolving models of the tropical cyclone boundary layer. Part I: Comparing the simulations. *Quart. J. Roy. Meteor. Soc.*, **136**, 1689-1699.

Kurihara, Y. and R.E. Tuleya, 1974: Structure of a tropical cyclone developed in a three-dimensional numerical simulation. *J. Atmos. Sci.*, **31**, 893-919.

Lorsolo, S., J. Zhang, F.D. Marks and J.F. Gamache, 2010: Estimation and Mapping of Hurricane Turbulent Energy Using Airborne Doppler Measurements. *Mon. Wea. Rev.*, **138**, 3656-3670.

Malkus, J.S. and H. Riehl, 1960: On the dynamics and energy transformations in steady-state hurricanes. *Tellus*, **12**, 1-20.

Marks, F.D., Jr., 1985: Evolution of the structure of precipitation in Hurricane Allen (1980). *Mon. Wea. Rev.*, **113**, 909-930.

Marks, F.D. and R.A. Houze Jr., 1987: Inner core structure of Hurricane Alicia from airborne Doppler radar observations. J. Atmos.Sci., 44, 1296–1317.

Marks, F.D. and L.K. Shay, 1998: Landfalling tropical cyclones: Forecast problems and associated research opportunities. *Bull. Amer. Meteor. Soc.*, **79**, 305-323.

Marks, F.D., R.A. Houze Jr. and J.F. Gamache, 1992: Dual-aircraft investigation of the inner core of Hurricane Norbert. Part I: Kinematic structure. *J. Atmos. Sci.*, **49**, 919-942.

Marks, F.D., P.G. Black, M.T. Montgomery and R.W. Burpee, 2008: Structure of the eye and eyewall of Hurricane Hugo (1989). *Mon. Wea. Rev.*, **136**, 1237-1259.

Montgomery, M.T., M.M. Bell, S.D. Aberson and M.L. Black, 2006: New insights into the Physics of Intense storms. Part I: Mean Vortex Structure and Maximum Intensity Estimates. *Bull. Amer. Meteor. Soc.*, **87**, 1335-1347.

Montgomery, M.T., R.K. Smith and S. Nguyen, 2010: Sensitivity of tropical cyclone models to the surface drag coefficient. *Quart. J. Roy. Meteor. Soc*, **136**, 1945-1953.

Montgomery, M.T. and Roger K. Smith, 2013: Paradigms for tropical cyclone intensification, 2013. Tropical Cyclone Research Report TCRR, **1**, 1-30. Meteorological Institute, Ludwig Maximilians University of Munich.

Moon, I.-J., T. Hara, I. Ginis, S.E. Belcher and H. Tolman, 2004a: Effect of surface waves on air–sea momentum exchange. Part I: Effect of mature and growing seas. *J. Atmos. Sci.*, **61**, 2321-2333.

Moon, I.-J., I. Ginis and T. Hara, 2004b: Effect of surface waves on air–sea momentum exchange. Part II: Behavior of drag coefficient under tropical cyclones. *J. Atmos. Sci.*, **61**, 2334-2348.

Moon, I.-J., I. Ginis and T. Hara, 2008: Impact of the reduced drag coefficient on ocean wave modeling under hurricane conditions. *Mon. Wea. Rev.*, **136**, 1217-1223.

Moon, I.-J., I. Ginis, T. Hara and B. Thomas, 2007: A physics-based parameterization of air-sea momentum flux at high wind speeds and its impact on hurricane intensity predictions. *Mon. Wea. Rev.*, **135**, 2869-2878.

Moss, M.S. and F.J. Merceret, 1976: A note on several low-layer features of Hurricane Eloise (1975). *Mon. Wea. Rev.*, **104**, 967-971.

Noh, Y., W.G. Cheon, S.Y. Hong and S. Raasch, 2003: Improvement of the K-profile model for the planetary boundary layer based on large eddy simulation data. *Bound.-Layer Meteor.*, **107**, 401-427.

Nolan, D.S., J.A. Zhang and D.P. Stern, 2009a: Evaluation of planetary boundary layer parameterizations in tropical cyclones by comparison of in-situ data and high-resolution simulations of Hurricane Isabel (2003). Part I: Initialization, maximum winds, and outer core boundary layer structure. *Mon. Wea. Rev.*, **137**, 3651-3674.

Nolan, D.S., J.A. Zhang and D.P. Stern, 2009b: Evaluation of planetary boundary layer parameterizations in tropical cyclones by comparison of in-situ data and high-resolution simulations of Hurricane Isabel (2003). Part II: Inner core boundary layer and eyewall structure. *Mon. Wea. Rev.*, **137**, 3675-3698.

Ooyama, K., 1969: Numerical simulation of the life cycle of tropical cyclones. *J. Atmos. Sci.*, **26**, 3-40.

Persing, J. and M.T. Montgomery, 2003: Hurricane superintensity. *J. Atmos. Sci.*, **60**, 2349-2371.

Powell, M.D., 1990: Boundary layer structure and dynamics in outer hurricane rainbands. *Mon. Wea. Rev.*, **118**, 918-938.

Powell, M.D., P.J. Vickery and T. Reinhold, 2003: Reduced drag coefficient for high wind speeds in tropical cyclones. *Nature*, **422**, 279-283.

Reasor, P.D., M.T. Montgomery, F.D. Marks Jr. and J.F. Gamache, 2000: Low-wavenumber structure and evolution of the hurricane inner core observed by airborne dual-Doppler radar. *Mon. Wea. Rev.*, **128**, 1653-1680.

Rogers, R., S. Lorsolo, P. Reasor, J. Ganache and F.D. Marks, 2012: Multiscale analysis of mature tropical cyclone structure from airborne Doppler composites. *Monthly Weather Review*, **140(1)**, 77-99.

Shea, D.J. and W.M. Gray, 1973: The hurricane's inner core region. I. Symmetric and asymmetric structure. *J. Atmos. Sci.*, **30**, 1544-1564.

Smith, R.K., 1968: The surface boundary layer of a hurricane. *Tellus*, **20**, 473-483.

Smith, R.K. and M.T. Montgomery, 2008: Balanced depth-averaged boundary layers used in hurricane models. *Quart. J. Roy Met. Soc.*, **134**, 1385-1395.

Smith, R.K. and G.L. Thomsen, 2010: Dependence of tropical-cyclone intensification on the boundary-layer representation in a numerical model. *Quart. J. Roy. Meteror. Soc.*, **136**, 1671-1685.

Smith, R.K., M.T. Montgomery and N.V. Sang, 2009: Tropical cyclone spin-up revisited. *Quart. J. Roy. Meteor. Soc.*, **135**, 1321-1335.

Smith, R.K., Michael T. Montgomery and Gerald L. Thomsen, 2013: Sensitivity of tropical-cyclone models to the surface drag coefficient in different boundary-layer schemes. *Quarterly Journal of the Royal Meteorological Societyn/a*-n/a.

Stull, R.B., 2000: Meteorology for Scientists and Engineers. Second Edition, Brooks/ Cole, Thomson Learning, Pacific Grove.

Stull, R.B., 1988: An introduction to boundary layer meteorology. Kluwer, Dordrecht, Holland.

Troen, I. and L. Mahrt, 1986: A simple model of the atmospheric boundary layer: Sensitivity to surface evaporation. *Bound.-Layer Meteor.*, **37**, 129-148.

Tuleya, R.E. and Y. Kurihara, 1978: A numerical simulation of the landfall of tropical cyclones. *J. Atmos. Sci.*, **35**, 242-257.

Tuleya, R.E., 1994: Tropical storm development and decay. Sensitivity to surface boundary conditions. *Mon. Wea. Rev.*, **122**, 291-304.

Vogelezang, D.H.P. and A.A.M. Holtslag, 1996: Evaluation and model impacts of alternative boundary-layer height formulations. *Bound.-Layer Meteor.*, **81**, 245-269.

Yablonsky, R.M. and I. Ginis, 2013: Impact of a warm ocean eddy's circulation on hurricane-induced sea surface cooling with implications for hurricane intensity. *Mon. Wea. Rev.*, **141**, 997-1021.

Zhang, D.-L., Y. Liu and M.K. Yau, 1999: Surface winds at landfall of Hurricane Andrew (1992) – A reply.

Zhang, D.-L., Y. Liu and M.K. Yau, 2000: A multiscale numerical study of Hurricane Andrew (1992). Part III: Dynamically-induced vertical motion. *Monthly Weather Review*, **128**, 3772-3788.

Zhang, D.-L., Y. Liu and M.K. Yau, 2001: A multiscale numerical study of Hurricane Andrew (1992). Part IV: Unbalanced flows. *Monthly Weather Review*, **129**, 92-107.

Zhang, D.-L., Y. Liu and M.K. Yau, 2002: A multiscale numerical study of Hurricane Andrew (1992). Part V: Inner-core thermodynamics. *Monthly Weather Review* , **130**, 2745-2763.

Zhang, J.A., W.M. Drennan, P.G. Black and J.R. French, 2009: Turbulence structure of the hurricane boundary layer between the outer rainbands. *J. Atmos. Sci.*, **66**, 2455-2467.

Zhang, J.A., 2010: Estimation of dissipative heating using low-level *in situ* aircraft observations in the hurricane boundary layer. *J. Atmos. Sci.*, **67**, 1853-1862.

Zhang, J., 2007: An airborne investigation of the atmospheric boundary layer structure in the hurricane force wind regime. Doctoral Dissertation, University of Miami.

Zhang, J.A., P.G. Black, J.R. French and W.M. Drennan, 2008: First direct measurements of enthalpy flux in the hurricane boundary layer: The CBLAST results. *Geophys. Res. Lett.*, **35**, L14813, doi:10.1029/2008GL034374.

Zhang, J.A., R.F. Rogers, D.S. Nolan and F.D. Marks, 2011a: On the characteristic height scales of the hurricane boundary layer. *Mon. Wea. Rev.*, **139**, 2523-2535.
Zhang, J.A., F.D. Marks, M.T. Montgomery and S. Lorsolo, 2011b: An estimation of turbulent characteristics in the low-level region of intense Hurricanes Allen (1980) and Hugo (1989). *Mon. Wea. Rev.*, **139**, 1447-1462.
Zhang, J.A. and M.T. Montgomery, 2012: Observational estimates of the horizontal eddy diffusivity and mixing length in the low-level region of intense hurricanes. *J. Atmos. Sci.*, **69**, 1306-1316.

24

Balanced Dynamics in Tropical Cyclones

Chanh Q. Kieu[1] and Da-Lin Zhang[2]
[1]Environmental Modeling Center, NOAA/NWS/NCEP
College Park, Maryland
[2]Department of Atmospheric and Oceanic Science, University of Maryland, College Park, Maryland, USA
dalin@atmos.umd.edu

1. Introduction

The Earth's atmosphere is generally a stochastic dynamical system in which physical processes take place within a wide range of temporal and spatial scales. Dynamical systems associated with small-scale perturbations often have a characteristically low predictability due to our lacked understanding of various physical processes and their feedback mechanisms. In contrast, the atmospheric motion at the meso- to synoptic-scales tends to possess a slow manifold with more predictable behaviours. It is this slow manifold component, often referred to as the *balanced flow* to be distinguished from the fast smaller-scale motion, that is more of our interest because it plays a critical role in evolution of common weather systems such as midlatitude baroclinic disturbances, mesoscale convective systems (MCSs), or tropical cyclones (TCs). Any weather system that differs too far from a balanced state will undergo a brief period of rapid adjustment, i.e., the so-called adjustment processes. From the balanced perspective, the evolution of mesoscale and larger-scale systems can be viewed as a series of continuous balanced adjustments, often referred to as the quasi-balanced dynamics. Herein, the quasi-balanced dynamic is simply that the mean flows are in a near-balanced state, but the (weak) superimposed perturbation flows are not, which are often related to the development of secondary circulation beyond the framework of balanced appropriations.

The roles of balanced flows could hardly be underestimated, because this type of flows is important in various practical situations. For example, it is the balanced flows associated with an upper-level trough that are responsible for potential surface cyclogenesis. Another example is how a TC can influence the track and intensity of another TC nearby. The concept of balanced flows is also useful for understanding unbalanced adjustment, and for developing tools for wave filtering, numerical weather prediction, or 4D-variational data assimilation. Note that the term "balanced flows" could involve different underlying dynamics, depending on the type of approximations that are of most concern. For example, balanced flows could mean geostrophic balance if they occur at the large scales or they could mean gradient wind balance if associated with MCSs.

Given a mesoscale weather system, such as an upper-level trough or an MCS, two fundamental questions that one usually needs to answer are: (1) How will the mesoscale system induce changes to other larger- or smaller-scale flows? and (2) What is the distribution of quasi-balanced motion associated with a given balanced state? The first question concerns primarily with the balanced response associated with MCSs and it can be tackled by using several balanced approximations. As will be seen in the next sections, the procedure to obtain the balanced flows can be formulated mathematically as an elliptic problem. The second question is more relevant to the vertical (transverse) circulation, which is more reserved for a very specific axisymmetric system such as TCs or extratropical cyclones. These types of axisymmetric configurations are important as we could treat the transverse circulations as being superimposed on a given primary circulation. Answering these questions could provide a good knowledge of balanced flows rather than instantaneous response because it is the balanced flows that persist long enough to contribute to the larger-scale flows during adjustment processes.

In this chapter, we will present some theoretical formulation and associated techniques to obtain the balanced dynamics from the perspective of potential vorticity (PV). Section 2 describes the general formulation of the balanced dynamics in the context of PV inversion, which is one of a few problems in dynamical meteorology that are well-defined in terms of an elliptic equation. Section 3 shows some theoretical basis for mesoscale PV inversion by assuming an axisymmetric configuration in the isentropic coordinate. Section 4 presents the quasi-geostrophic (QG) simplifications of the PV dynamics. Section 5 describes in detail the algorithms and applications of the PV inversion for the nonlinear balance system from given PV anomalies and boundary conditions. The quasi-balanced dynamics and its various approximations will be presented in the next chapter.

2. Mesoscale Formulation

The balanced dynamics can be formulated from several different perspectives such as the PV perspective or directly from the first principle equations under some balanced constraints (see Hack and Schubert, 1985; Hoskins et al., 1985, hereinafter H85; Davis and Emanuel, 1991, hereinafter DE91; Shapiro and Montgomery, 1993; Shapiro and Willoughby, 2001; Montgomery and Franklin, 2004; Wang and Zhang, 2003, hereinafter WZ03). Among different approaches, the PV approach for studying MCSs appears to dominate due partly to its succinct physical content, and partly to its invertibility principle. For the sake of clarity and self-contained reference, the PV approach will be adopted herein to study the balanced dynamics.

To take advantage of the pressure coordinate in which air density in the pressure gradient force is absorbed in the definition of the geopotential height while retaining the meaning of the vertical dimension as a natural height, we will work in the pseudo-height coordinate defined as follows:

$$z = \left(1 - \frac{\pi}{c_p}\right) z_a ,$$

where $\pi \equiv c_p(p/p_0)^{R_d/c_p}$ is the Exner function, $z_a = (c_p/R_d)H_s$ with $H_s = p_0/\rho_0 \cdot g = R_d\,\theta_0/g$ is the scale height, and θ_0 is the reference potential temperature. This pseudo-height coordinates are mostly useful for designing numerical computations as will be seen in section 4. In the pseudo-height coordinate system, the primitive equations have the following forms (see DE91):

$$\frac{\partial u}{\partial t} + u\frac{\partial u}{\partial x} + v\frac{\partial u}{\partial y} + \omega\frac{\partial u}{\partial z} - fv = -\frac{\partial \phi}{\partial x} + F_x, \tag{1}$$

$$\frac{\partial v}{\partial t} + u\frac{\partial v}{\partial x} + v\frac{\partial v}{\partial y} + \omega\frac{\partial v}{\partial z} + fu = -\frac{\partial \phi}{\partial y} + F_y, \tag{2}$$

$$\frac{\partial \phi}{\partial z} = -\frac{g}{\theta_a}\theta_p, \tag{3}$$

$$\frac{\partial \theta_p}{\partial t} + u\frac{\partial \theta_p}{\partial x} + v\frac{\partial \theta_p}{\partial y} + \omega\frac{\partial \theta_p}{\partial z} = J, \tag{4}$$

$$\frac{\partial u}{\partial x} + \frac{\partial v}{\partial y} + \frac{\partial \omega}{\partial z} + \frac{c_p}{R_d(z_a - z)}\omega = 0, \tag{5}$$

where u, v and ω are three components of wind vectors in the pseudo-height coordinates, θ_p is the potential temperature perturbation with respect to a reference θ_0, ϕ is geopotential, f is the Coriolis parameter, (F_x, F_y) are the frictional forcing, and J is the diabatic heating source. Take $\partial(1)/\partial x + \partial(2)/\partial y$ to obtain the divergent equation as follows

$$\frac{dD}{dt} + \left[\left(\frac{\partial u}{\partial x}\right)^2 + 2\frac{\partial v \partial u}{\partial x \partial y} + \left(\frac{\partial v}{\partial y}\right)^2\right] + \left[\frac{\partial \omega}{\partial x}\frac{\partial u}{\partial z} + \frac{\partial \omega}{\partial y}\frac{\partial v}{\partial z}\right]$$

$$= -\nabla^2\Phi + f\zeta + v\frac{\partial f}{\partial x} - u\frac{\partial f}{\partial y} + \nabla_h \cdot F_r \tag{6}$$

where $D = \frac{\partial u}{\partial x} + \frac{\partial v}{\partial y}$ is divergence, $\zeta = \frac{\partial v}{\partial x} - \frac{\partial u}{\partial y}$ is the relative vorticity in the vertical direction, $F_r \equiv (F_x, F_y)$ and ∇_h is the horizontal divergent operator. Scale analyses of various terms on Eq. (6) show that the temporal change of the divergence and tilting terms, i.e., the first and third terms on the left hand side (LHS) of Eq. (6), are of secondary importance to the divergent equation for most practical situations (see Holton, 2004). As a result, Eq. (6) can be reduced to

$$\left(\frac{\partial u}{\partial x}\right)^2 + 2\frac{\partial v \partial u}{\partial x \partial y} + \left(\frac{\partial v}{\partial y}\right)^2 = -\nabla^2\phi + f\zeta + v\frac{\partial f}{\partial x} - u\frac{\partial f}{\partial y} + \nabla_h \cdot F_r \tag{7}$$

Using the Helmholtz theorem, the horizontal wind field $\mathbf{V}_h$ can always be decomposed into a rotational and a divergent component (Majda and Bertozzi, 2002):

$$\mathbf{V}_h = (u, v) = \mathbf{V}_\psi + \mathbf{V}_\chi = \mathbf{k} \times \nabla\psi + \nabla\chi \tag{8}$$

Substituting expression (8) into Eq. (7) and restricting the resulting equation to the order of $O(\mathbf{R}_\psi)$, where $\mathbf{R}_\psi \equiv \mathbf{V}_\psi / Lf_0$ is defined as the Rossby number for rotational wind, we obtain the nonlinear balanced (NLB) equation:

$$\nabla^2 \Phi = \nabla_h (f \nabla_h \Psi) + 2J\left(\frac{\partial \psi}{\partial x}, \frac{\partial \psi}{\partial y}\right) + \nabla_h . F_r \tag{9}$$

where $J(\partial\psi/\partial x, \partial\psi/\partial y)$ is the Jacobian operator. In deriving the NLB Eq. (9), terms of order $O(\mathbf{V}_\chi / Lf_0)$ have been neglected. It should be emphasized that this approximation is applied only to the divergent equation (7), and this neglect does not mean that $\mathbf{V}_\chi$ is completely ignored. In the subsequent formulation of the quasi-balanced dynamics, the divergent wind component $\mathbf{V}_\chi$ will be recovered in an equation for the transverse circulation (see DE91; WZ03). As shown in the next chapter, it is the divergent component $\mathbf{V}_\chi$ that drives the evolution of the quasi-balanced systems.

To close the system for the two unknowns ψ and ϕ, we manipulate the hydrostatic Eq. (3) in the form:

$$\theta_p = \frac{\partial \phi}{\partial \pi} \tag{10}$$

Given a three-dimensional (3D) distribution of the potential temperature perturbation θ_p, one could in principle obtain the balanced geopotential height ϕ directly by integrating (10). Using equation (9), the balanced streamfunction ψ will then follow. In reality, information from potential temperature alone does not contain dynamical properties of any meteorological system such as upper-level troughs, MCSs, or TCs. Thus, we cannot use θ_p to characterize the physical content of a system. In this regard, we are looking for a quantity that could represent both the dynamical and thermodynamic characteristics of balanced systems. Of several tracer variables, PV is a unique dynamical variable that could address the above issues. The most general definition of PV is

$$P_V = \frac{1}{\rho}\eta \cdot \nabla\theta_p \tag{11}$$

where ρ is the density of air, and $\boldsymbol{\eta} = 2\boldsymbol{\Omega} + \nabla \times \mathbf{V}$ is the absolute vorticity vector. It is evident that PV contains the 3D information of potential temperature, the absolute vorticity and the stratification of the atmosphere. An important property of the above PV definition is that *in the absence of diabatic heating and friction, PV is conserved along the flow trajectory*. In particular, the time-dependence of the volume-integrated PV can be casted in a flux form such that any change of the total PV within the volume is determined by the flux of PV at the boundaries (Haynes and McIntyre, 1987). Furthermore, *the product of PV and air density*, the so-called PV substance as defined by Haynes and McIntyre (1990), can be shown to be impermeable across isentropic surfaces. That is, the PV substance within two bounded isentropic surfaces can only be changed by its flux at the lateral boundaries. In the presence of diabatic heating, however, Kieu and Zhang (2012) have demonstrated that PV substance is very likely to break through isentropic surfaces at the points where the atmosphere stratification could evolve into a layer of neutral stability (e.g., in the presence of strong diabatic heating related to deep convection or radiative cooling). Thus, PV substance is not generally conserved within two bounded isentropic surfaces in the presence of intense diabatic heating.

Given the definition of PV, we can transform it into a more suitable form for studying the balanced dynamics by using the hydrostatic equation to replace θ_p in Eq. (11) by ϕ as follows:

$$P_V = \frac{1}{r(z)}\frac{\theta_0}{g}\left[(f + \nabla_h^2\psi)\frac{\partial^2\phi}{\partial z^2} - \frac{\partial^2\psi}{\partial x\partial z}\frac{\partial^2\phi}{\partial x\partial z} - \frac{\partial^2\psi}{\partial y\partial z}\frac{\partial^2\phi}{\partial y\partial z}\right], \tag{12}$$

where $r(z) = \rho_0(P/P_0)^{R/Cp}$ is pseudo-density. Note that vertical velocity has been neglected in Eq. (12). Note also that the vanishing contribution from the χ-component wind in Eq. (12) is due solely to the curl operator ($\nabla \times \nabla\chi = 0$), not because of the nondivergent approximation as in deriving Eq. (9). This point is emphasized as one will see in deriving of the quasi-balanced dynamics in the next chapter that exclusion of such nondivergent component could simplify the quasi-balanced dynamics dramatically, rendering the quasi-balanced dynamics much more tractable.

Equations (9) and (12) together form a close system for ψ and ϕ. That is, given the distribution of PV and appropriate boundary conditions, Eqs (9) and

(12) could be solved simultaneously to find ψ and ϕ. As will be seen later, the main advantage of the PV viewpoint is that it can be used to classify quantities of different meteorological origins, such as upper-level troughs, mesoscale convective vortices (MCVs), or environmental PV anomalies. Furnished with an efficient algorithm to perform PV inversion, one could in principle examine the impacts of any PV anomaly. This lies at the heart of the PV balanced dynamics, for which we turn into in the next section.

3. IPV Inversion and the PV-induced Flows

Because of its high nonlinearity, the balanced equations system consisting of Eqs (9) and (12) may not possess a unique solution. Only with certain approximations or under symmetrical configurations, the balanced equations system could be well-posed. To familiarize readers with algorithms to invert the balanced flows from a given PV field, we proceed with a simplified PV inversion problem by assuming an axisymmetric configuration in the isentropic (θ) coordinates. An important advantage of adopting the θ-coordinates is that the vertical motion vanishes in the absence of diabatic heating, i.e., $\dot{\theta}= 0$), so the PV in the θ-coordinates or the isentropic PV (IPV), P_θ, does not contain horizontal components. That is, IPV under the hydrostatic approximation is defined as (see H85)

$$P_\theta = -g\frac{(\zeta_\theta + f)}{\frac{\partial p}{\partial \theta}} = (\zeta_\theta + f)/\sigma \tag{13}$$

where $\sigma = -(1/g)\,\partial p/\partial\theta$ is the static stability parameter, and ζ_θ is the vertical component of relative vorticity in the axisymmetric-isentropic (r, θ)-coordinates given by

$$\zeta_\theta = \frac{1}{r}\frac{\partial rV}{\partial r}$$

where V is the tangential wind. It is apparent from (13) that PV is large (small) in the stratosphere (troposphere) due to the presence of strong (weak) static stability. In the troposphere, the magnitude of PV for a typical mid-latitude, synoptic-scale disturbance ($\zeta << f$) can be estimated by

$$P_\theta = -g\,(\zeta + f)\,\partial\theta/\partial p = -(10\ \mathrm{ms^{-2}})\,(10^{-4}\ \mathrm{s^{-1}})\,(-10\ \mathrm{K}/10000\ \mathrm{Pa})$$
$$= 10^{-6}\ \mathrm{m^2s^{-1}K\ kg^{-1}} = 1\ \mathrm{PVU},$$

assuming that $\theta = 350$ K (or $T = -52$°C) and $\theta = 290$ K (or $T = 20$°C) are located near 200 and 1000 hPa, respectively. Thus, PV values of greater than 2 PVU are usually defined as the height of the tropopause (Fig. 1). However, in the presence of intense diabatic heating, TCs could have PV values as large as over 50 PVUs in the eyewall (see Yau et al., 2004).

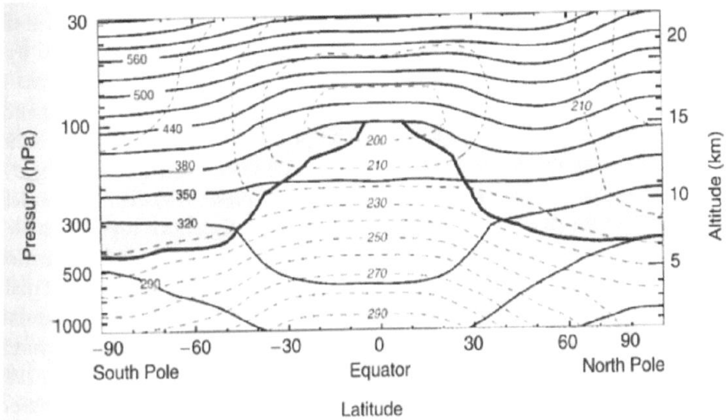

Fig. 1. Latitude-height cross section of the monthly and zonally mean temperature (dashed) and potential temperature (thin solid) with 2 PVU contour (heavy solid) for Jan 1993 (from Holton et al., 1995).

An important property of IPV that it is conserved following the motion in an inviscid, adiabatic flow. Recall that, in the (r, θ) coordinates, the NLB equation becomes the gradient-wind balance equation,

$$V^2/r + f v - \partial M/\partial r = 0 \tag{14}$$

where M is the Montgomery streamfunction, $M = gh + c_p T$, and obtained from the coordinate transformation $[(\alpha\nabla_z p = \alpha\nabla_\theta p - \alpha\partial p/\partial z\ \nabla_\theta h = \nabla_\theta (c_p T + gh)]$. Likewise, the hydrostatic balance in the θ-coordinate system is expressed as

$$\partial M/\partial\theta = c_p \pi \tag{15}$$

Differentiating (14) with respect to θ and (15) with respect to r gives

$$\partial^2 M/\partial\theta\partial r = (f + 2V/r)\ \partial V/\partial\theta,$$

$$\partial^2 M/\partial r\ \partial\theta = c_p\ (\partial\pi/\partial p)\ (\partial p/\partial r).$$

By eliminating M from the above two equations, we obtain the thermal wind relation

$$\partial V/\partial\theta = c_p\ (\partial\pi/\partial p)\ (\partial p/\partial r)\ /(f + 2V/r). \tag{16}$$

Equation (16) relates the vertical wind shear in θ-coordinates ($\partial V/\partial\theta$) to the radial pressure gradient on an isentropic surface ($\partial p/\partial r$); the latter is equivalent to the radial temperature gradient on an isobaric surface. To cast the IPV definition (13) in a more suitable form for subsequent inversion, we re-write Eq. (13) as

$$\sigma P_\theta = \frac{1}{r}\frac{\partial rV}{\partial r} + f \tag{17}$$

The derivative of Eq. (17) with respect to r is

$$\sigma\frac{\partial P_\theta}{\partial r}+\frac{(\zeta_\theta+f)}{\sigma}\frac{\partial\sigma}{\partial r}=\frac{\partial}{\partial r}\left(\frac{1}{r}\frac{\partial rV}{\partial r}\right)+\frac{\partial f}{\partial r} \tag{18}$$

By taking derivative of σ with respect to r and making use of Eq. (16), we obtain

$$\frac{\partial\sigma}{\partial r}=-\frac{1}{g}\frac{\partial^2 p}{\partial\theta\partial r}=-\frac{1}{g}\frac{\partial}{\partial\theta}\left(\frac{f+\frac{2V}{r}}{c_p\frac{\partial\pi}{\partial p}}\frac{\partial V}{\partial\theta}\right)$$

Use of the above expression for $\partial\sigma/\partial r$ and Eq. (18), we arrive at a second-order PDE for the tangential wind $V(r, \theta)$ that is forced by the radial gradient of IPV

$$\frac{\partial}{\partial r}\left(\frac{1}{r}\frac{\partial rV}{\partial r}\right)+\left[\frac{\zeta_\theta+f}{g\sigma}\right]\frac{\partial}{\partial\theta}\left(\frac{f+\frac{2V}{r}}{c_p\frac{\partial p}{\partial p}}\frac{\partial V}{\partial\theta}\right)+\frac{\partial f}{\partial r}=\sigma\frac{\partial P_\theta}{\partial r} \tag{19}$$

Note that Eq. (19) is the second-order PDE with the unknown variable V. Apparently, Eq. (19) is elliptic if

$$-4\,(\zeta_\theta+f)(f+2V/r)/(g\sigma c_p\partial\pi/\partial p)=-\,4\,P_\theta(f+2V/r)/\,(g\sigma c_p\partial\pi/\partial p)<0$$

or

$$P_\theta\,(f+2V/r)>0. \tag{20}$$

This implies that Eq. (19) is elliptic if the atmosphere is statically stable ($\sigma > 0$), inertially stable ($\zeta_\theta + f > 0; f + 2V/r > 0$) and symmetrically stable ($P_\theta >$ 0). According to the invertibility principle, the axisymmetric tangential wind field could be obtained from (19), given the proper boundary conditions for $V(r, \theta)$ and the IPV forcing on the RHS of (19). Given the solution $V(r, \theta)$, the temperature and pressure fields can be calculated by using the gradient wind balance and the hydrostatic equation. Note however that, Eq. (19) is highly nonlinear when ζ_θ, σ and $f + 2V/r$ are functions of r or θ. In order to use the superposition principle, we have to find a way to linearize Eq. (19) and then invert the IPV field.

To this end, suppose that we can decompose the total IPV field into two parts (see H85):

$$P_\theta\,(x, y, \theta) = P_{\text{Ref}}\,(\theta) + P'(x, y, \theta),$$

in which the reference state is defined as

$$P_{\text{Ref}}\,(\theta) = -\,g f_0\partial\theta_0/\partial p,$$

$$\zeta_\theta = f_0,\ \sigma = \sigma_{\text{Ref}}\,(\theta),\ \text{and}\ \pi = \pi_{\text{Ref}}\,(\theta),$$

Equation (19) can then be written as

$$\frac{\partial}{\partial r}\left(\frac{1}{r}\frac{\partial rV}{\partial r}\right)+\frac{f_0^2}{g\sigma_{\text{Ref}}}\frac{\partial}{\partial\theta}\left(\frac{1}{c_p\dfrac{\partial\pi_{\text{Ref}}}{\partial p}}\frac{\partial V}{\partial\theta}\right)=\sigma_{\text{Ref}}\frac{\partial P_\theta}{\partial r} \tag{21}$$

Now Eq. (21) is a linear PDE, so the total response $[V(r, \theta)]$ to n IPV anomalies $[\sigma_{\text{Ref}}\partial P_{\theta 1}/\partial r,\ \sigma_{\text{Ref}}\partial P_{\theta 2}/\partial r,\ \ldots\ \sigma_{\text{Ref}}\partial\ P_{\theta n}/\partial r]$ is simply the sum of n individual responses $(V_1 + V_2 + \ldots + V_n)$. Once V is inverted, the axisymmetric pressure and temperature fields can be obtained from the gradient balance and hydrostatic equations (see H85 for more details).

Given an isolated IPV anomaly with a scale of L near the tropopause, our next question is: To what extent will it be felt by the flow above and below? This is in contrast to the Rossby radius of deformation problem that applies to the horizontal scale of atmospheric disturbances. To analyze in more details the scaling influence of an IPV distribution, let assume that the vertical scale of the influence is $\Delta\theta$, Eq. (21) reveals the following scaling relationship:

$$\left(\frac{1}{r}\frac{\partial rV}{\partial r}\right)\sim\frac{U}{L^2}$$

$$\frac{f_0^2}{g\sigma_{ref}}\frac{\partial}{\partial\theta}\left(\frac{1}{c_p\partial\pi/\partial p}\frac{\partial V}{\partial\theta}\right)\sim\frac{f_0^2}{g\sigma_{ref}c_p\partial\pi_{ref}/\partial p\left[\dfrac{U}{(\Delta\theta)^2}\right]}$$

which gives:

$$\Delta\theta\sim f_0 LP_\theta/(g\sigma_{\text{Ref}}c_p\partial\pi/\partial p)^{1/2} \tag{22}$$

It is apparent from Eq. (22) that the vertical influence of an IPV anomaly, the so-called penetration depth, is proportional to the horizontal length scale L of the IPV anomaly and inversely proportional to the square root of the static stability. In the z-coordinates, the penetration depth could be written as $\Delta z = f_0 L/N$. The above scale analysis provides an important theoretical basis for PV inversion, as will be seen later. To see the magnitude of the penetration depth, plugging in some typical values:

$$g\sigma_{\text{Ref}} = -\,\partial p/\partial\theta = -\,(-\,10000\ \text{Pa}/10\ \text{K}) = 1000\ \text{Pa m}^{-1}\text{s}^2\ \text{K}^{-1}$$

$$c_p\partial\pi_{\text{Ref}}/\partial p = (R/p)\ \pi_{\text{Ref}} = RT/(p\theta) = 1/(\rho\theta) = 1/(1\ \text{kg m}^{-3}\ 300\ \text{K})$$

$$L = 1000\ \text{km}$$

we find the penetration depth:

$$\Delta\theta\sim(10^{-4}\ \text{s}^{-1})\,(10^6\ \text{m})\ [(1\ \text{kg m}^{-3}\ 300\ \text{K})/(1000\ \text{kg ms}^{-2}\ \text{m}^{-1}\text{s}^2\ \text{K}^{-1})]^{1/2}$$

$$\sim(100\ \text{ms}^{-1})(0.55\ \text{ms}^{-1}\ \text{K}^{-1})\sim 55\ \text{K}.$$

This is roughly as deep as the average depth of the troposphere. Thus, a synoptic-scale IPV anomaly near the tropopause can induce a balanced wind-mass perturbation all the way down to the surface.

The dependence of the wind perturbation intensity can also be estimated from the RHS of Eq. (21):

$$U/L^2 \sim f_0^{\,2}/(g\sigma_{\text{Ref}}c_p\partial\pi_{\text{Ref}}/\partial p)\,[U/(\Delta\theta)^2] \sim \sigma_{\text{Ref}}P_{\text{Ref}}/L \tag{23}$$

or

$$U \sim \sigma_{\text{Ref}}P_{\text{Ref}}\,L. \tag{24}$$

This implies that small-scale IPV anomalies tend to induce weak wind perturbations, and the vertical influence would occur in a shallow layer, based on Eq. (24); the opposite is also true for large-scale IPV anomalies with the same strength. Figure 2 shows how surface cyclogenesis may occur as an upper-level PV anomaly approaches to a surface baroclinic zone, based on the above discussions.

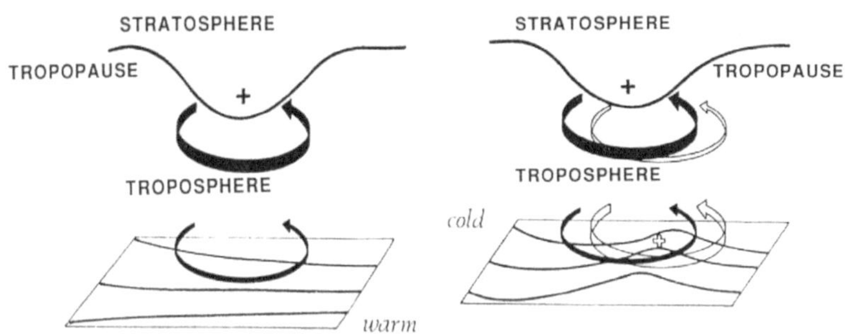

Fig. 2. Surface cyclogenesis associated with the approaching of an upper-level PV anomaly (from Hoskin et al., 1985)

4. Quasi-geostrophic PV Inversion

In sections 2 and 3, we have shown the balanced dynamics from the PV framework under the NLB and gradient wind balance, respectively. As mentioned earlier, the balanced dynamics should be, however, understood in a broader sense of any slow manifold of the atmospheric flows rather than restricted specifically to a given constraint. The simplest balance system in this regard is the geostrophic balance, which approximates the large-scale flow well in the midlatitudes in the presence of little curvature effects.

Since the geostrophic balance is better manifested in the isobaric coordinates, we will present the geostrophic balanced dynamics in this coordinate system. One may have noted that we have frequently changed to different vertical coordinates for different approximations. For the nonlinear formulation, we have used the pseudo-height coordinate, which is convenient for designing efficient numerical algorithms. In the absence of diabatic forcing, we have worked with the isentropic coordinates, which has an advantage of eliminating the vertical velocity to maintain the constant θ. On the other hand, the isobaric coordinates have an incompressible form for the continuity equation, which helps simplify the derivations substantially.

It should be mentioned that the definition of PV according to Eq. (11) is not coordinate-invariant but has to be modified as one goes from one vertical coordinate to another (see H85). This is because the definition of relative vorticity as $\overline{V}$ is not a vector but a pseudo tensor of weight 1. Therefore, as one performs a coordinate transformation, it may not preserve its physical meaning in general because of the appearance of the Jacobian determinant[1]. For most meteorological applications in which only vertical component of the coordinate (i.e., the z-component) is changed, the vertical relative vorticity often preserves its meaning but the horizontal components do not. This can be seen clearly by considering, for example, the x-component of the relative vorticity in several different coordinates. The naïve definition of the horizontal component of the relative vorticity would give:

$$\zeta_x = \frac{\partial w}{\partial y} - \frac{\partial v}{\partial z} \quad \text{in } z\text{-coordinates}$$

$$\zeta_x = \frac{\partial \omega}{\partial y} - \frac{\partial v}{\partial p} \quad \text{in } p\text{-coordinates}$$

$$\zeta_x = \frac{\partial \dot{\theta}}{\partial y} - \frac{\partial v}{\partial \theta} \quad \text{in } \theta\text{-coordinates}$$

Apparently, the dimensions of this x-component of the relative vorticity in different coordinates do not match, revealing the different forms of the relative vorticity in different coordinate systems. A more careful definition of the relative vorticity would have to distinguish between covariant and contravariant vector components to give the relative vorticity a proper physical meaning. Because of such complication involved in the PV concept when changing coordinate system, one should expect that the PV dynamics and explanation have to be modified for different coordinates. Of course, this does not mean that dynamical processes are changing from one coordinate system to another. It simply means that the PV concept has different approximations in different coordinate systems.

[1] One can see this directly by doing a simple coordinate transformation from z to p-coordinates as follows. Define the coordinate transformation as: $x_p = x_z$; $y_p = y_z$; $p = p\,(x_z, y_z, z)$; $t_p = t_z$, and then the Jacobian determinant will be $J = \left|\partial p/\partial z\right|$. Under this coordinate transformation, the relative vorticity defined as $\zeta_i \equiv \varepsilon_{ijk}\,\partial u_j/\partial x_k$, where ε_{ijk} is Levi-Civia symbol, will transform to $\zeta_i^p = \left|J\right| \partial x_j^p/\partial x_i \zeta_i^z$. Apparently, $|J| \neq 1$, and thus the horizontal component of the relative vorticity in z-coordinates is no longer preserved as it is transformed to p-coordinates. Therefore, the correct definition of PV in curvilinear coordinate needs to take into account properly the pseudo tensor properties of the coordinate transformation.

With the above in mind, we may follow Holton (2004) and define the quasi-geostrophic PV (QGPV) under the QG approximations as follows:

$$P_{QG} = \frac{\nabla^2 \phi}{f_0} + f + \frac{\partial}{\partial p}\left(\frac{f_0}{\sigma}\frac{\partial \phi}{\partial p}\right) \tag{25}$$

Although the dimension of P_{QG} is not the same as that defined in (11), it is customary to call P_{QG} the QGPV as it contains both the dynamical information through the first two terms on the RHS of Eq. (25), and thermodynamic information through the last RHS term in Eq. (25). Thus, the succinct physical content of the QGPV is essentially the same as that in (11), despite their different dimensions.

In the absence of diabatic heating, it is possible to show that QGPV is conserved along the flow trajectory, i.e.,

$$\frac{dP_{QG}}{dt} \equiv \left(\frac{\partial}{\partial t} + V_g \frac{\partial}{\partial t}\right) P_{QG} = 0, \tag{26}$$

in which the material derivatives are taken with respect to the geostrophic winds. An attractive feature of the QGPV is that Eq. (25) is exactly a linear second-order PDE, which allows for a complete inversion of the Poison equation, given a 3D distribution of P_{QG} and appropriate boundary conditions. If the LHS of Eq. (25) is known, a 3D geopotential height field can be inverted from P_{QG} and the temperature and wind distributions are subsequently obtained via the hydrostatic and geostrophic balance equations. This is the essence of the invertibility principle in the QG framework. In this case, we may state that the height field ϕ (x, y, p) is induced by QGPV, much like a charge distribution inducing an electric field. Together, Eqs (25) and (26) can be used to forecast the evolution of the balanced geopotential field completely. Specifically, Eq. (26) provides the evolution of the QGPV, while Eq. (25) determines the 3D flow field within the entire domain by inverting the Poison equation. Despite the advantage of the QGPV inversion is its simplicity and absence of ambiguity, its main weakness is its inaccuracy as the Rossby number of atmospheric flows becomes sufficiently large, e.g., close to O(1), which is typical at the mesoscale.

Note that unlike the PV definition given by Eq. (12), there is no nonlinear coupling between geopotential and streamfunction. This is because the geostrophic wind is determined entirely by the geopotential distribution that there is no need to introduce the streamfunction ψ and the potential function χ to derive the flow field. The gradient wind balance is essentially replaced by the geostrophic approximation such that the system of Eqs (9) and (12) can now be reduced to a single elliptic equation.

As an illustration of determining the QG balanced evolution from Eqs (25) and (26), consider a simple example given in Holton (2004) as follows. Assume that the lower boundary is set as p_0 = 1000 hPa, and that ω vanishes at the lower boundary surface so that the boundary condition for adiabatic motion can be determined by the thermodynamic equation on the 1000 hPa surface as follows:

$$\left(\frac{\partial}{\partial t}+V_g\frac{\partial}{\partial t}\right)\theta=-\left(\frac{\partial}{\partial t}+V_g\frac{\partial}{\partial t}\right)\frac{\partial\phi}{\partial p}=0 \tag{27}$$

In this situation, the evolution of the balanced flows can be derived from the system of Eqs (25)-(27) by

- Inverting Eq. (25) to obtain the balanced geopotential height within the 3D domain;
- Advancing the PV equation (26) and the boundary condition according to Eq. (27) to obtain the new distribution of QGPV and boundary for the new time step; and
- Repeating step 1 to obtain the new balanced flows and step 2 to advance the balanced flow in time.

The above procedures of obtaining the balanced flows show that the linear characteristics and the conservation of the QGPV along the geostrophic flows could help simplify drastically the balanced dynamics.

5. Piecewise PV Inversion for TCs

In the preceding sections, we have analyzed several different simplifications of the NLB and PV equations. These analyses are useful for understanding their pertinent balanced flows. However, for TC applications, the balanced dynamics are more complicated and involve a wide range of nonlinear interactions that may not be simply studied with the equations shown in the previous sections, particularly if we have to proceed with the nonlinear system of Eqs (9) and (12). In the next, we show how Eqs (9) and (12) can be solved using the piecewise inversion technique proposed by DE91.

Note that although our goal is to obtain solutions for the nonlinear PV system, in many situations we are often more interested in examining the balanced responses of a PV anomaly (PVA) than the full 3D PV per se (see DE91; Wu and Emanuel, 1995a, b; Morgan, 1999; Huo et al., 2001; Shapiro and Franklin, 1999; WZ03; KZ10). This is because it is the balanced flows associated with individual PVAs, such as those associated with the eyewall or rainbands, that are closely related to the variations of the TC intensity. Thus, we will focus more attention on inverting PVAs to obtain their induced balanced flows rather than inverting the whole 3D PV field itself.

Because it is important to develop an efficient way to decompose the total PV into a mean state and an anomaly field, in the next we present the method developed by DE91 for the piecewise PV inversion, in which the total PV field is partitioned into a mean state and a perturbation field as follows:

$$P_V=\overline{P_V}+P_V',\psi=\overline{\psi}+\psi',\text{and }\phi=\overline{\phi}+\phi'.$$

A major advantage of DE91's method is that there is no requirement for the perturbation fields to be the second order approximation with respect to the mean state (i.e., the conditions $\overline{P_V} \gg P_V$, $\overline{\psi} \gg \psi'$, $\overline{\phi} \gg \phi'$. are not needed). Provided that the mean fields $\overline{P_V}$, $\overline{\Psi}$ and $\overline{\phi}$ satisfy the balanced equations (9), the partition of the total fields into mean and perturbations part is always valid. The procedures of solving the NLB system of Eqs (12) and (14) can be divided into two steps as follows:

Step 1. Define a Balanced Basic State

Because the mean flows have to satisfy a particular balanced constraint in order for the PVA inversion to be applicable, it is necessary to develop a systematic method to obtain these mean flows. There are several methods to define a meaningful balanced state. For example DE91, and Wu and Emanuel (1995a, b) used a temporally averaged PV field over a period of five days to obtain the mean PV $\overline{P_V}$. Equations (9) and (12) will then be assumed for the mean balanced $\overline{\Psi}$ and $\overline{\phi}$, and an iteration of these equations can be conducted until they converge to obtain $\overline{\Psi}$ and $\overline{\phi}$ over the whole domain. This approach is straightforward but in reality hard to apply directly because inverting the total PV for MCSs is too sensitive to the boundary condition and first guess field. Using a different approach, WZ03, Shapiro (1996), Shapiro and Franklin (1999) define the balanced fields by taking the azimuthal average of $\overline{\Psi}$. The balance equation (9) is then inverted to obtain $\overline{\phi}$, and finally $\overline{P_V}$ is calculated from Eq. (12). In this way, the fields $\overline{P_V}$ and $\overline{\phi}$ are in balance with respect to the azimuthally-averaged $\overline{\Psi}$. However, as noted by WZ03, this approach, when applied to TCs, may lead to some negative static stability near the top of the PBL where the tangential wind is maximized and the subsequent PVA inversion may not converge. In addition, the azimuthal average results in a mean field possessing nearly all the major characteristics of TC that PVA inversion provides little additional interesting information. This can be seen when the mean tangential wind is at the same order of magnitude as the total field. As a result, this approach would result in a small residual perturbation after subtracting the mean PV field from the total PV field, causing the PVA to become too small without any significant characteristics.

An alternative way to obtain the balanced mean state is to define first $\overline{\phi}$. Instead of taking the azimuthal average of streamfunction $\overline{\Psi}$, one can smooth out the total field ϕ as many times as one wishes. Because the inherent feature of the geopotential within the TC inner core region is roughly an U-shaped structures with the minimum at the storm centre, we can control the intensity of the mean vortex easily by modifying the U-shaped pattern as much as we want; the shallower the U-shaped pattern, the weaker the mean vortex would be. By doing so, the PVA, defined as the difference between ϕ and $\overline{\phi}$, can be isolated as much as one wishes, rendering the PVA inversion more meaningful.

Given such a mean balanced $\overline{\Psi}$ distribution, we obtain next $\overline{\Psi}$ from $\overline{\phi}$ by using the NLB equation (9). This inversion, however, needs a special treatment due to the special form of the Eq. (9). An efficient algorithm is to start from Eq. (9),

$$\nabla^2\phi = \nabla(f\nabla\psi) + 2J\left(\frac{\partial\psi}{\partial x}, \frac{\partial\psi}{\partial y}\right) + \nabla F_f. \tag{28}$$

and expand the RHS of Eq. (28) as follows:

$$\nabla^2\phi = f\nabla^2\psi + \frac{\partial f}{\partial x}\frac{\partial\psi}{\partial x} + \frac{\partial f}{\partial y}\frac{\partial\psi}{\partial y} + 2\left(\frac{\partial^2\psi}{\partial x^2}\frac{\partial^2\psi}{\partial y^2} - \left(\frac{\partial^2\psi}{\partial x\partial y}\right)^2\right) + \nabla F_f. \tag{29}$$

Rewrite the fourth term on the RHS of (29) as:

$$\frac{\partial^2\psi}{\partial x^2}\frac{\partial^2\psi}{\partial y^2} = \frac{1}{4}\left[\left(\frac{\partial^2\psi}{\partial x^2} + \frac{\partial^2\psi}{\partial y^2}\right)^2 - \left(\frac{\partial^2\psi}{\partial x^2} - \frac{\partial^2\psi}{\partial y^2}\right)^2\right], \tag{30}$$

and plug Eq. (30) into Eq. (29) to obtain

$$(\nabla^2\psi)^2 + 2f\nabla^2\psi + 2\left(\frac{\partial f}{\partial x}\frac{\partial\psi}{\partial x} + \frac{\partial f}{\partial y}\frac{\partial\psi}{\partial y}\right) - \left(\frac{\partial^2\psi}{\partial x^2} - \frac{\partial^2\psi}{\partial y^2}\right)^2 - \left(4\frac{\partial^2\psi}{\partial x\partial y}\right)^2 + 2\nabla F_f - 2\nabla^2\phi = 0. \tag{31}$$

This is a quadratic equation for the Laplacian of ψ. Thus, we can solve for $\nabla^2\psi$ to yield:

$$\nabla^2\psi = -f + \left\{f^2 + \left(\frac{\partial^2\psi}{\partial x^2} - \frac{\partial^2\psi}{\partial y^2}\right)^2 + \left(2\frac{\partial^2\psi}{\partial x\partial y}\right)^2 + 2\nabla^2\phi - 2\nabla F_f - 2\left(\frac{\partial f}{\partial x}\frac{\partial\psi}{\partial x} + \frac{\partial f}{\partial y}\frac{\partial\psi}{\partial y}\right)\right\}^{\frac{1}{2}} \tag{32}$$

The plus sign is chosen, because it results in a familiar geostrophic approximation: $\nabla^2\psi = \nabla^2\phi/f$ for the weak field approximation. Equation (32) will then be solved iteratively,

$$\nabla^2\psi^{n+1} = -f + \left\{f^2 + \left(\frac{\partial^2\psi^n}{\partial x^2} - \frac{\partial^2\psi^n}{\partial y^2}\right)^2 + \left(2\frac{\partial^2\psi^n}{\partial x\partial y}\right)^2 + 2\nabla^2\phi - 2\nabla F_f - 2\left(\frac{\partial f}{\partial x}\frac{\partial\psi^n}{\partial x} + \frac{\partial f}{\partial y}\frac{\partial\psi^n}{\partial y}\right)\right\}^{\frac{1}{2}} \tag{33}$$

The iteration will converge when $|\psi^{n+1} - \psi^n|$ reaches some small threshold. The convergent condition for iterating Eq. (33) is

$$f^2 + 2\nabla^2 \phi > 0 \tag{34}$$

Our experiments with the simulation of Hurricane Bonnie (1998) using the MM5 model during the storm mature stage show that less than 10% of total model grid points violate the condition (34). A simple treatment for the violation of (34) is to assume a continuous solution over these points due to its global smoothness.

After obtaining $\overline{\Psi}$ from $\overline{\phi}$, the mean PV field $\overline{P_V}$ is computed simply by using Eq. (12). By construction, all the fields are thus in balance without inverting Eqs (9) and (12), which is difficult to control. Note that although it is not required to invert Eqs (9) and (12), boundary conditions are still needed to calculate the mean PV field, $\overline{P_V}$. This is because calculating $\overline{P_V}$ from (12) is restricted to the interior points only. The boundary conditions are required to complete the definition of $\overline{P_V}$. An advantage of this method is that we can attain the mean field $\overline{\phi}$ as flat as possible, depending on each specific situation. This allows us to investigate PVAs more effectively since it is feasible to separate a large PV perturbation in the same dynamically balanced way as in the previous studies (see Kieu and Zhang, 2011).

Step 2. Perform Piecewise PV Inversion

With the total (ψ, ϕ) and mean field $(\overline{\Psi}, \overline{\phi})$ available, the total PVA P'_V can be computed easily by taking $P'_V = P_V - \overline{P_V}$. Our final task is to invert P'_V to find the associated perturbation streamfunction ψ' and geopotential ϕ'. It turns out that the total PVA P'_V is still too general to obtain any meaningful balanced perturbation in many cases. For example, it would be of more interest to examine the balanced response of a PVA associated with an outer eyewall or a nearby upper-level PVA to the TC development than the entire PVA within the domain. Thus, it is of interest to partition the total PVA into smaller pieces P'_i where $i \in [1, N]$. Our final goal is to find the balanced perturbation fields ψ'_k and ϕ'_k associated with a particular piece of PVA P'_k.

Because of the nonlinearity of the equation system of (9) and (12), the perturbation ψ'_k and ϕ'_k are not only determined by the PV piece P'_k but also influenced by other PVA pieces. This means that given a piece of PVA P'_k, it is not enough to find ψ'_k and ϕ'_k associated with P'_k by applying the invertibility principle to P'_k alone. Only in case of one piece (i.e., $N = 1$) it is possible to invert the piece P'_k to find ψ'_k and ϕ'_k. For more general cases with $N > 1$, we are confronted with the problem of how to obtain balanced perturbation fields ψ'_k and ϕ'_k associated with the piece of PVA P'_k such that the sum of all perturbations ψ'_i and ϕ'_i will be equal to those inverted from the total PVA P'_V. Namely, the following relations must hold:

$$\psi' = \sum_{i=1}^{N} \psi'_i \quad \text{and} \quad \phi' = \sum_{i=1}^{N} \phi'_i \quad \text{if} \quad P'_V = \sum_{i=1}^{N} P'_i$$

where ψ' and ϕ' are obtained by inverting the total PV anomaly P'_V. It turns out that there are infinite ways to partition PVA into small pieces that could satisfy the above superimposition principle (see DE91 for more discussion). Here, we will follow DE91's method to invert a PVA, the so-called piecewise PV inversion technique. Given a total PVA P'_V, we partition it into N pieces $P'_V = \sum_{i=1}^{N} P'_i$. DE91's approach assumes that the i-th perturbation fields ψ_i' and ϕ_i' associated with piece P'_i are given by the following system:

$$\nabla^2\phi'_i = \nabla(f\nabla\psi'_i) + 2\left(\frac{\partial^2\psi^*}{\partial x^2}\frac{\partial^2\psi'_i}{\partial y^2} - 2\frac{\partial^2\psi^*}{\partial x\partial y}\frac{\partial^2\psi'_i}{\partial x\partial y} + \frac{\partial^2\psi^*}{\partial y^2}\frac{\partial^2\psi'_i}{\partial x^2}\right) \tag{35}$$

$$P'_i = (f+\nabla^2\psi^*)\frac{\partial^2\phi'_i}{\partial z^2} + \nabla^2\psi'_i\frac{\partial^2\phi^*}{\partial z^2} - \left(\frac{\partial^2\psi^*}{\partial x\partial z}\frac{\partial^2\phi'_i}{\partial x\partial z} - \frac{\partial^2\phi^*}{\partial x\partial z}\frac{\partial^2\psi'_i}{\partial x\partial z}\right) - \left(\frac{\partial^2\psi^*}{\partial y\partial z}\frac{\partial^2\phi'_i}{\partial y\partial z} - \frac{\partial^2\phi^*}{\partial y\partial z}\frac{\partial^2\psi'_i}{\partial y\partial z}\right) \tag{36}$$

where $(\phi,\psi)^* = (\bar{\phi},\bar{\psi}) + \frac{1}{2}\sum_{i=1}^{N}(\phi'_i,\psi'_i)$. A quick check of the sum of Eqs (35) and (36) over $i = 1, N$ shows that the superimposition principle is ensured. Recall that $\overline{P_V}$, $\bar{\Psi}$ and $\bar{\phi}$ in Eqs (35) and (36) have to satisfy both Eqs (9) and (12) such that (35) and (36) can guarantee the superimposition principle. The price we have to pay for the requirement that the sum of each inverted PV piece P'_i be equal to the inverted total PV anomaly P'_V is that all ψ'_j and ϕ'_j with $j \neq i$ must be known in advance in order to find ψ'_i and ϕ'_i. Because these perturbation fields ψ'_j and ϕ'_j are unknown, a system of 2N nonlinear equations in the form of Eqs (35) and (36) in principle need to be solved simultaneously.

In their early work of PVA inversion, WZ03 have presented a method for one-piece PV inversion by combining (35) and (36) to obtain a new set of equations in which a new parameter ε is introduced. By allowing this free parameter to vary from 0 to 1, it is possible to control the convergence of the system of Eqs (35)-(36) effectively. KZ10 further extended this technique to a piecewise inversion for cases with an arbitrary number of PVA pieces ($N > 1$). Following KZ10, we multiply Eq. (35) by a parameter ε ($0 < \varepsilon < 1$), and add it into Eq. (36). Using the explicit expression for ψ^* and ϕ^*, after some manipulations, we get:

$$(\varepsilon f + \frac{\partial^2\bar{\phi}}{\partial z^2})\nabla^2\psi'_i = P'_i - (f+\nabla^2\bar{\psi})\frac{\partial^2\phi'_i}{\partial z^2} + \frac{\partial^2\bar{\psi}}{\partial x\partial z}\frac{\partial^2\phi'_i}{\partial x\partial z} + \frac{\partial^2\bar{\phi}}{\partial x\partial z}\frac{\partial^2\psi'_i}{\partial x\partial z} + \frac{\partial^2\bar{\psi}}{\partial y\partial z}\frac{\partial^2\phi'_i}{\partial y\partial z} + \frac{\partial^2\bar{\phi}}{\partial y\partial z}\frac{\partial^2\psi'_i}{\partial y\partial z} - \frac{1}{2}(\sum_{n=1}\nabla^2\psi'_n\frac{\partial^2\phi'_i}{\partial z^2} + \nabla^2\psi'_i\sum_{n=1}\frac{\partial^2\phi'_n}{\partial z^2})$$

$$+\frac{1}{2}(\sum_{n=1}\frac{\partial^2\psi'_n}{\partial x\partial z}\frac{\partial^2\phi'_i}{\partial x\partial z}+\sum_{n=1}\frac{\partial^2\phi'_n}{\partial x\partial z}\frac{\partial^2\psi'_i}{\partial x\partial z})+\frac{1}{2}(\sum_{n=1}\frac{\partial^2\psi'_n}{\partial y\partial z}\frac{\partial^2\phi'_i}{\partial y\partial z}+\sum_{n=1}\frac{\partial^2\phi'_n}{\partial y\partial z}\frac{\partial^2\psi'_i}{\partial y\partial z})$$
$$+\varepsilon[\nabla^2\phi'_i-\beta\frac{\partial\psi'_i}{\partial y}-2(\frac{\partial^2\overline{\psi}}{\partial x^2}\frac{\partial^2\psi'_i}{\partial y^2}-2\frac{\partial^2\overline{\psi}}{\partial x\partial y}\frac{\partial^2\psi'_i}{\partial x\partial y}+\frac{\partial^2\overline{\psi}}{\partial y^2}\frac{\partial^2\psi'_i}{\partial x^2})$$
$$-(\sum_{n=1}\frac{\partial^2\psi'_n}{\partial x^2}\frac{\partial^2\psi'_i}{\partial y^2}-2\sum_{n=1}\frac{\partial^2\psi'_n}{\partial x\partial y}\frac{\partial^2\psi'_i}{\partial x\partial y}+\sum_{n=1}\frac{\partial^2\psi'_n}{\partial y^2}\frac{\partial^2\psi'_i}{\partial x^2})+\frac{\partial\square_x}{\partial x}+\frac{\partial\square_y}{\partial y}\square \quad (37)$$

A second equation for ψ'_i and ϕ'_i can be obtained by subtracting Eq. (35) from (36) and rearranging:

$$\nabla^2\phi'_i+(f+\nabla^2\overline{\psi})\frac{\partial^2\phi'_i}{\partial z^2}=P'_i-\nabla^2\psi'_i\frac{\partial^2\overline{\phi}}{\partial z^2}+f\nabla^2\psi'_i+\beta\frac{\partial\psi'_i}{\partial y}+\frac{\partial F_x}{\partial x}+\frac{\partial F_y}{\partial y}$$
$$-\frac{1}{2}(\sum_{n=1}\nabla^2\psi'_n\frac{\partial^2\phi'_i}{\partial z^2}+\nabla^2\psi'_i\sum_{n=1}\frac{\partial^2\phi'_n}{\partial z^2})+\frac{\partial^2\overline{\psi}}{\partial x\partial z}\frac{\partial^2\phi'_i}{\partial x\partial z}+\frac{\partial^2\overline{\phi}}{\partial x\partial z}\frac{\partial^2\psi'_i}{\partial x\partial z}+\frac{\partial^2\overline{\psi}}{\partial y\partial z}\frac{\partial^2\phi'_i}{\partial y\partial z}$$
$$+\frac{\partial^2\overline{\phi}}{\partial y\partial z}\frac{\partial^2\psi'_i}{\partial y\partial z}+\frac{1}{2}(\sum_{n=1}\frac{\partial^2\psi'_n}{\partial x\partial z}\frac{\partial^2\phi'_i}{\partial x\partial z}+\sum_{n=1}\frac{\partial^2\phi'_n}{\partial x\partial z}\frac{\partial^2\psi'_i}{\partial x\partial z})+\frac{1}{2}(\sum_{n=1}\frac{\partial^2\psi'_n}{\partial y\partial z}\frac{\partial^2\phi'_i}{\partial y\partial z}$$
$$+\sum_{n=1}\frac{\partial^2\phi'_n}{\partial y\partial z}\frac{\partial^2\psi'_i}{\partial y\partial z})+2(\frac{\partial^2\overline{\psi}}{\partial x^2}\frac{\partial^2\psi'_i}{\partial y^2}-2\frac{\partial^2\overline{\psi}}{\partial x\partial y}\frac{\partial^2\psi'_i}{\partial x\partial y}+\frac{\partial^2\overline{\psi}}{\partial y^2}\frac{\partial^2\psi'_i}{\partial x^2})$$
$$+2(\frac{1}{2}\sum_{n=1}\frac{\partial^2\psi'_n}{\partial x^2}\frac{\partial^2\psi'_i}{\partial y^2}-2\frac{1}{2}\sum_{n=1}\frac{\partial^2\psi'_n}{\partial x\partial y}\frac{\partial^2\psi'_i}{\partial x\partial y}+\frac{1}{2}\sum_{n=1}\frac{\partial^2\psi'_n}{\partial y^2}\frac{\partial^2\psi'_i}{\partial x^2}) \quad (38)$$

Note that although the system of Eqs (37) and (38) are applied to the case of N PVA pieces, the elliptic condition for Eq. (37) is still the same as that for the case of one PVA piece (using the same elliptic criteria of equation (37) as in WZ03) and it is given by:

$$\left(\varepsilon f+2\varepsilon\frac{\partial^2\overline{\psi}}{\partial x^2}+\frac{\partial^2\overline{\phi}}{\partial z^2}\right)\left(\varepsilon f+2\varepsilon\frac{\partial^2\overline{\psi}}{\partial y^2}+\frac{\partial^2\overline{\phi}}{\partial z^2}\right)-4\varepsilon^2\left(\frac{\partial^2\overline{\psi}}{\partial x\partial y}\right)^2>0 \quad (39)$$

An advantage of this ε-multiplication is that the parameter of ε can be varied appropriately so that iteration of Eqs (37) and (38) will converge. As seen from (39), this ε parameter is expected to be small in order for the elliptic condition (39) to be satisfied. This is because the first product term on the LHS of (39) is proportional to ε and thus decreases slower than the last term, which is second order as ε approaches zero. For $N = 1$, Eqs (37) and (38) will be exactly identical to Eqs (37) and (38) in WZ03.

It should be mentioned that, in deriving Eqs (35) and (36) and the subsequent Eqs (37) and (38), the smallness of ψ' or ϕ' with respect to $\overline{\psi}$ and $\overline{\phi}$ is not pre-assumed. The only requirement in deriving these equations is the condition that $\overline{\psi}$ and $\overline{\phi}$ are the solutions of Eqs (9) and (12), and the partition Eqs (37) and (38) follow automatically without appealing to their linearization. The validity of Eqs (37)-(38) is always assured if the mean fields $\overline{\psi}$ and $\overline{\phi}$ are chosen to satisfy Eqs (9) and (12) *a priori*. In this sense, the smallness of the perturbations as compared to the mean is not required for the piecewise PV inversion. In practice, we notice, however, that the condition of $\psi' << \overline{\psi}$ and $\phi' << \overline{\phi}$ is still imposed to ensure the sufficient convergence of Eqs (37)-(38) during their iterations.

Another important aspect of the piecewise inversion is to define the boundary conditions properly. Since the piecewise inversion is essentially to solve elliptic equations by iteration, such boundary issues have to be handled properly. The most common lateral boundary conditions used for piecewise PV inversion are Dirichlet boundaries for ψ' and ϕ' along the lateral boundaries, where ϕ' matches the simulated output (or from observations) and ψ' will be given by:

$$\frac{\partial \psi'}{\partial s} = -V_h \cdot n + \frac{\oint V_h \cdot n \cdot dl}{\oint dl} \tag{40}$$

where l is a path along the lateral boundaries, **s** is a vector tangent to that path and **n** is a normal vector of that path. The bottom ($z = z_B$) and top ($z = z_T$) boundaries for ψ' and ϕ' will be given by Neumann conditions:

$$\left.\frac{\partial \phi'}{\partial z}\right|_{z=z_B} = g\frac{\theta'_B}{\theta_0}\left.\frac{\partial \phi'}{\partial z}\right|_{z=z_B} = g\frac{\theta'_T}{\theta_0} \tag{41}$$

(see DE91 or WZ03 for more detail). If the domain is sufficiently large, one can assume homogenous boundary conditions for ψ' and ϕ'.

As an illustration for application of the piecewise PVA inversion, consider an idealized balanced vortex embedded in the atmosphere at rest. To ensure this idealized vortex as realistic as possible, outputs of the 3D fields from a simulation of Hurricane Andrew (1992) using the MM5 model (WZ03) during the mature stage are used to create the vortex. The modelled small-scale noises are eliminated by temporally averaging between 56 and 57h into integration. The mean balance is constructed by taking the azimuthal average of tangential winds from which the mean streamfunction $\overline{\Psi}$ is obtained, using the Poisson equation for the relative vorticity (i.e., $\nabla^2 \overline{\Psi} = \zeta$). Next, the balanced mean geopotential $\overline{\phi}$ is solved using Eq. (9), and the mean PV $\overline{P_V}$ is computed using Eq. (12). As discussed in Step 1, such a sequence of calculations of the mean fields ensures the validation of the balanced requirement for the piecewise formulation (37)-(38). Given the mean balanced vortex, two idealized rings of PVAs that mimic the outer and inner eyewall during the double eyewalls stage are superimposed on the vortex. This is to investigate the mutual balanced interaction of these eyewalls during the eyewall replacement cycle that is often observed when TCs possess a double eyewalls configuration (Chen and Yau 2001; KZ10). Our focus will be mainly on the inversion of these PVA pieces, as they are of our major interest rather than the balanced vortex itself.

Figure 3 shows the perturbation tangential flows and perturbation potential temperature associated with the inner PVA (hereinafter referred to as p_i). As expected, p_i induces significant cyclonic flows around the vortex centre with the maximum induced tangential wind perturbation of ~14 m s^{-1}. In agreement with the balanced cyclonic flows, p_i also induces a low-pressure system that peaks at the bottom of its PVA volume with a warm core above a cold dome. Such a vertical distribution of the PVAs-induced warmth and coldness enhances the local static stability in their influencing layers, thus limiting the further development of the storm.

Although the observed warm core in the eye and the minimum central pressure are mostly attributed to the balanced constraint by the mean vortex, Fig. 3 shows that the balanced potential temperature perturbation induced by p_i appears to contribute also to the enhancement of the warm core. Such balanced contribution associated the prescribed eyewall PVA is hydrostatically related to the low pressure anomalies induced by p_i at the centre, indicating that TC intensity fluctuations could be closely connected to any processes leading to an aggregation of PVAs within the eyewall.

Figure 4 shows further the horizontal distribution of pressure perturbations and tangential flows associated with the outer PVA (hereinafter p_o) at three selected levels. One can see that the outer eyewall induces a balanced low-pressure volume within a radius of about 180 km, with a surface low of 10 hPa in the hurricane centre and a surface low-pressure ring in the outer eyewall. Corresponding to this low-pressure volume is a deep layer of cyclonic perturbation and flows outside the outer eyewall, which is similar to that shown in Fig. 3. Of importance is that the p_o-induced flows inside the outer eyewall are anticyclonic with respect to the inner eyewall. As a result, the tangential wind is reduced in the region between the inner and the outer eyewall, leading to formation of a secondary RMW. As the outer eyewall wraps around the inner eyewall and contracts radially inward, the inner RMW would diminish as observed during the eyewall replacement cycle. This could help explain why two RMWs often co-exist during the eyewall replacement period (see Figs 14 and 18 in Zhu et al., 2004). Apparently, the stronger the outer eyewall is, the more intense is the induced surface pressure perturbation, and the stronger (weaker) is the tangential wind outside the outer (inner) eyewall.

Similar to the p_o-induced pressure perturbation, its induced potential temperature perturbation θ' shows a widespread area (r ~180 km) of warming aloft and cooling below, with a peak warm anomaly ring above at the inner edge of the outer eyewall and a weaker warm anomaly centered in the eye, which are hydrostatically consistent with the pressure perturbation field

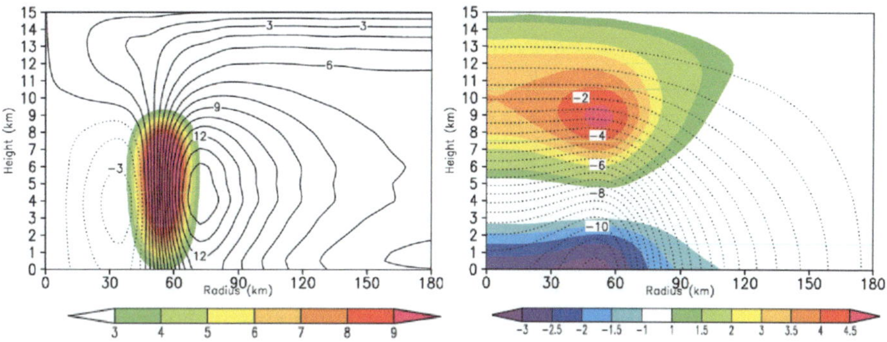

Fig. 3. Radius-height cross sections associated with the inner eyewall PVA of (left panel) the perturbation tangential winds (contoured, m s^{-1}) and the corresponding PV piece (shaded, PVU); and (right panel) the perturbation pressure (contoured, hPa) and the potential temperature perturbations (shaded, K). From Kieu and Zhang (2011).

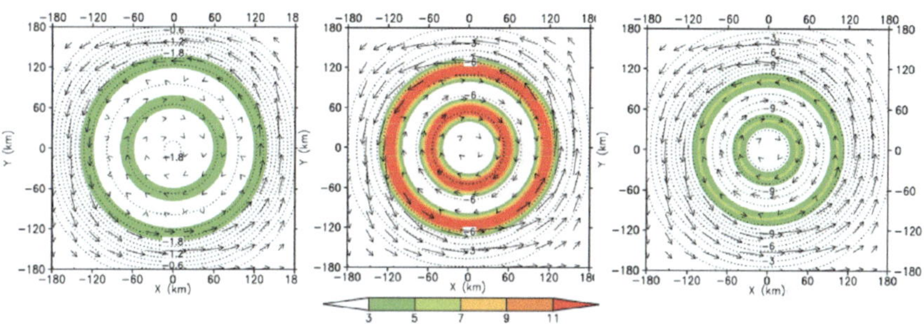

Fig. 4. Horizontal distribution of the pressure perturbation (contoured) associated with the outer PVA piece p_o (shaded at intervals of 3 PVUs) at (a) $z = 10$ km, (b) $z = 3$ km, and (c) $z = 0.5$ km. The perturbation pressures are dashed at intervals of 0.2 hPa in (a), and 1 hPa in (b) and (c). Superimposed are the p_o-induced perturbation tangential flows (ms^{-1}).

(Fig. 4). As the lower pressure and thermal anomaly rings are superimposed to those associated with the pre-existing inner eyewall, the radial pressure and thermal gradients may be reduced or even reversed between the two eyewalls, thereby reducing the vertical shear of tangential winds in the inner eyewall. More details of the balanced interaction between these PVAs can be found in KZ10.

It should be noted that although the magnitudes of the balanced perturbation flows induced by the inner and the outer eyewalls may appear relatively weak compared to the mean flows, they are dynamically persistent and will last as long as the double eyewall configuration is maintained. Because of the balanced impacts, the mean cyclonic flows will experience continuous deceleration in the vicinity of the moat area during the eyewall replacement cycle. As one could imagine, such inimical impacts could also be held for any closed rainband, or more or less spiral rainbands. This example of the piecewise inversion of the eyewall PVA demonstrates the usefulness of the PV inversion technique for understanding the balanced responses as well as interactions between different PVAs within TCs.

6. Concluding Remarks

In this chapter, we have studied the mesoscale balanced dynamics from the PV perspective, starting from the nonlinear formulations and then simplifying them under the gradient and geostrophic approximations. In the absence of diabatic heating, the balanced dynamics based on the PV invertibility principle allow one to obtain balanced response of mass and wind fields to a given PV perturbation. This balanced response was demonstrated to be very useful in examining the vertical influence of a PV anomaly embedded within a circular background flow. To help readers understand further the PV inversion concept, the geostrophic balance was used to demonstrate that the PV equation could be reduced to a very simple formulation, the so-called QGPV that allows for complete inversion of the PV equation. Because of the conservation of QGPV along the flow trajectory, the combination of the PV inversion and the PV equation could give a complete picture of the associated balanced response that is otherwise not readily to extract from full physics models.

For mesoscale circulations associated with TCs, the nonlinear balance renders the PV inversion more complicated. However, through a carefully defined basic state, mesoscale PV inversion techniques, with proper linearization, have been developed and successfully applied to various TC cases and idealized analyses. Results show that the piecewise PV inversion could be an important tool in examining the interaction of different PVAs with the mean vortices or the interaction among themselves in TCs. As mentioned before, these balanced flows are obtained after assuming the absence of diabatic heating and friction. However, latent heat release and boundary-layer processes are potential source and sink for many MCSs, including TCs, and they will affect the balanced mean flows through secondary circulations. This will be the subject of the quasi-balanced dynamics in the next chapter.

Problems

P1. Derive the hydrostatical relationship (10). *Hint: using the Eq. (3) and definition of the pseudo-height coordinate.*

P2. Derive the PV expression (13) in the isentropic coordinates. *Hint: first write the PV definition in the pressure coordinates using the hydrostatic equation, then transform it to the isentropic coordinates.*

P3. Show that the QGPV defined in (25) can be considered as a simplification of Ertel's PV in the isobaric coordinate defined by H85 as: $Q = -g\,(f\mathbf{k}$

$+ \nabla_p.\mathbf{V}).\nabla_p\theta$, where ∇_p is the 3D gradient operator in (xyp) coordinate. *Hint: use the thermal wind relationship in the p-coordinate and the chain rule for derivatives.*

REFERENCES

Chen, Y.-S. and M.K. Yau, 2001: Spiral bands in a simulated Hurricane. Part I: Vortex Rossby Wave verification. *J. Atmos. Sci.*, **58**, 2128-2145.

Davis, C.A., 1992: Piecewise potential vorticity inversion. *J. Atmos. Sci.*, **49**, 1397-1411.

Davis, C.A. and K. Emanuel, 1991: Potential vorticity diagnostics of cyclogenesis. *Mon. Wea. Rev.*, **119**, 424-439.

Frank, W. M. and E.A. Ritchie, 2001: Effects of vertical wind shear on the intensity and structure of numerically simulated hurricanes. *Mon. Wea. Rev.*, **129**, 2249-2269.

Gall, R.L., J. Tuttle and P. Hildebrand, 1997: Small-scale spiral bands observed in Hurricanes Andrew, Hugo and Erin. *Mon. Wea. Rev.*, **126**, 1749-1766.

Huo, Z.-H., D.-L. Zhang and J.R. Gyakum, 1998: An application of potential vorticity inversion to improving the numerical prediction of the March 1993 Superstorm. *Mon. Wea. Rev.*, **126**, 424-436.

Huo, Z.-H., D.-L. Zhang and J.R. Gyakum, 1999a: The interaction of potential vorticity anomalies in extratropical cyclogenesis. Part I: Static piecewise inversion. *Mon. Wea. Rev.*, **127**, 2546-2561.

Huo, Z.-H., D.-L. Zhang and J.R. Gyakum, 1999b: The interaction of potential vorticity anomalies in extratropical cyclogenesis. Part II: Sensitivity to initial perturbations. *Mon. Wea. Rev.*, **127**, 2563-2575.

Hoskins, B.J., M.E. McIntyre and A.W. Robertson, 1985: On the use and significance of isentropic potential vorticity maps. *Quart. J. Roy. Meteor. Soc.*, **111**, 877-946.

Jones, S.C., 1995: The evolution of Vortices in Vertical Shear. I: Initially barotropic Vortices. *Q.J.R. Meterorol. Soc.*, **121**, 821-851.

Kieu, C.Q. and D.-L. Zhang., 2010: A piecewise potential vorticity inversion algorithm and its application to hurricane inner-core anomalies. *J. Atmos. Sci.*, **67**, 1745-1758.

Kieu, C.Q. and D.-L. Zhang, 2012: Is the isentropic surface always impermeable to the potential vorticity substance? *Adv. Atmos. Sci.*, **29**, doi: 10.1007/s00376-011-0227-0.

Krishnamurti, T.N., 1968: A diagnostic balance model for studies of weather systems of low and high latitudes, Rossby number less than 1. *Mon. Wea. Rev.*, **96**, 197-207.

Liu, Y., D.-L. Zhang and M.K. Yau, 1999: A multiscale numerical study of Hurricane Andrew (1992). Part II: Kinematics and inner-core structures. *Mon. Wea. Rev.*, **127**, 2597-2616.

Möller, J.D. and L.J. Shapiro, 2002: Balanced contributions to the intensification of Hurricane Opal as diagnosed from a GFDL model forecast. *Mon. Wea. Rev.*, **130**, 1866-1881.

Möller, J.D. and S.C. Jones, 1998: Potential vorticity inversion for tropical cyclones using the Asymmetric Balance Theory. *J. Atmos. Sci.*, **55**, 259.

Montgomery, M.T. and J.L. Franklin, 1998: An assessment of the balance approximation in hurricanes. *J. Atmos. Sci.*, **55**, 2193-2200.

Morgan, M.C., 1999: Using Piecewise Potential Vorticity Inversion to Diagnose Frontogenesis. Part I: A partitioning of the **Q** vector applied to diagnosing surface frontogenesis and vertical motion. *Mon. Wea. Rev.*, **127**, 2796-2821.

Shapiro, L.J., 1996: The motion of Hurricane Gloria: A potential vorticity diagnosis. *Mon. Wea. Rev.*, **124**, 2497-2508.

Shapiro, L.J. and M.T. Montgomery, 1993: A three-dimensional balance theory for rapidly rotating vortices. *J. Atmos. Sci.*, **50**, 3322-3335.

Shapiro, L.J. and J.L. Franklin, 1995: Potential vorticity in Hurricane Gloria. *Mon. Wea. Rev.*, **123**, 1465-1475.

Shapiro, L.J. and J.L. Franklin, 1999: Potential vorticity asymmetries and tropical cyclone motion. *Mon. Wea. Rev.*, **127**, 124-131.

Wang, X.-B. and D.-L. Zhang, 2003: Potential vorticity diagnosis of a simulated hurricane. Part I: Formulation and quasi-balanced flow. *J. Atmos. Sci.*, **60**, 1593-1607.

Wang, Y. and G.J. Holland, 1996: Tropical cyclone motion and evolution in vertical shear. *J. Atmos. Sci.*, **53**, 3313-3332.

Wu, C.-C. and K.A. Emanuel, 1995a: Potential vorticity diagnostics of hurricane movement. Part I: A case study of Hurricane Bob (1991). *Mon. Wea. Rev.*, **123**, 69-92.

Wu, C.-C. and K.A. Emanuel, 1995b: Potential vorticity diagnostics of hurricane movement. Part II: Tropical storm Ana (1991) and Hurricane Andrew (1992). *Mon. Wea. Rev.*, **123**, 93-109.

Yau, M.K., Y.-B. Liu, D.-L. Zhang and Y.-S. Chen, 2004: A multiscale numerical study of Hurricane Andrew (1992). Part VI: Small-scale inner-core structures and wind streaks. *Mon. Wea. Rev.*, **132**, 1410-1433.

Zhang, D.-L., Y. Liu and M.K. Yau, 2000: A multiscale numerical study of Hurricane Andrew (1992). Part III: Dynamically induced vertical motion. *Mon. Wea. Rev.*, **128**, 3772-3788.

Zhang, D.-L., Y. Liu and M.K. Yau, 2001: A multiscale numerical study of Hurricane Andrew (1992). Part IV: Unbalanced flows. *Mon. Wea. Rev.*, **129**, 92-107.

Zhang, D.-L., Y. Liu and M.K. Yau, 2002: A multiscale numerical study of Hurricane Andrew (1992). Part V: Inner-core thermodynamics. *Mon. Wea. Rev.*, **130**, 2745-2763.

Zhang, D.-L. and C.K. Kieu, 2005: Shear-forced vertical circulations in tropical cyclones. *Geophys. Res. Lett.*, **32**, L13822, doi:10.1029/2005GL023146.

Zhu, T., D.-L. Zhang and F. Weng, 2004: Numerical simulation of Hurricane Bonnie (1998). Part I: Eyewall evolution and intensity changes. *Mon. Wea. Rev.*, **132**, 225-241.

25

Quasi-balanced Dynamics in Tropical Cyclones

Chanh Q. Kieu[1] and Da-Lin Zhang[2]
[1]Environmental Modeling Center, NOAA/NWS/NCEP
College Park, Maryland
[2]Department of Atmospheric and Oceanic Science
University of Maryland, College Park, Maryland, USA
dalin@atmos.umd.edu

1. Introduction

In chapter 24, we have examined balanced dynamics from the potential vorticity (PV) perspective with different approximations. Strictly speaking, the balanced dynamics in mesoscale convective system (MCSs) or tropical cyclones (TCs) concerns mostly with horizontal flows, commonly known as the primary circulations. As a result, the balanced flows contain no vertical circulation (i.e., vertical motion and across-isobaric flows) and no time-dependent component. For TCs with high degree of axisymmetry, such a vertical circulation, consisting of convergent inflow in the planetary boundary layer (PBL), slantwise upward motion in the eyewall, and divergent outflow in the upper troposphere (referred to as secondary or transverse circulation, SC), plays an important role during the development of TCs. This is mainly because it is the SC that is responsible for the continuous adjustment of balanced flows in time.

Generally, the evolution of a dynamically balanced system under continuous adjustments by SCs is referred to as "quasi-balanced" dynamics. In practical applications, any pure balanced theory that does not take into account SCs is considered incomplete, because after all it is the SCs that account for severe weather events, such as the formation of hail, wind gusts, tornados, and heavy rainfall. However, including the SCs that contain the divergent wind component and vertical motion into a balanced framework presents some new complexity, as inclusion of the divergent wind component would break the conventional balance definition. Thus, certain approximations and justifications are required to develop consistent dynamics under the quasi-balanced constraints. In this regard, it would be more appropriate to consider the three-dimensional (3D) flow consisting of a balanced nondivergent component, and a divergent component, in which the nondivergent component associated with the balanced flow is presumably projected onto 'slow manifold' while the divergent component acts as perturbations superimposing on top of the balanced component. In this chapter we will examine different formulations and applications of the quasi-balanced dynamics, starting from the simplest quasi-geostrophic (QG) dynamics to a more general formulation for mesoscale applications that render studies of the quasi-balanced dynamics more practical.

2. Quasi-geostrophic Dynamics

In a nutshell, the QG dynamics assumes that the basic-state flow is in geostrophic balance and its temporal changes result from small ageostrophic processes. As an introduction to the quasi-balanced framework, in this section, we will first follow the presentation by Holton (2004) to derive the quasi-balanced dynamics for the geostrophic basic state. More complete treatment of the quasi-balanced dynamics for meso-scale systems will be examined in the next sections.

Under the geostrophic approximation, the horizontal wind components, $\vec{V}_g = (u_g, v_g)$ in the isobaric coordinates can be defined as follows:

$$f_0 u_g = -\frac{\partial \phi}{\partial y}, \text{ and } f_0 v_g = \frac{\partial \phi}{\partial x}. \tag{1}$$

where ϕ is geopotential and f_0 is a reference Coriolis parameter. Decompose the horizontal wind into the geostrophic and ageostrophic components as

$$\vec{V}_h = \vec{V}_g + \vec{V}_a,$$

along with an assumption that $\vec{V}_g \gg \vec{V}_a$, the total derivative can be approximated as:

$$\frac{D}{Dt} \equiv \frac{\partial}{\partial t} + (\vec{V}_h \cdot \nabla) + \omega \frac{\partial}{\partial p} \approx \frac{\partial}{\partial t} + (\vec{V}_g \cdot \nabla) \equiv \frac{D_g}{D_t}.$$

Given definition of the geostrophic wind as in (1), we have the following identity,

$$f\vec{k} \times \vec{V} + \nabla\phi = (f_0 + \beta y)\vec{k} \times (\vec{V}_g + \vec{V}_a) - f_0 \vec{k} \times \vec{V}_g = f_0 \vec{k} \times \vec{V}_a + \beta y \vec{k} \times \vec{V}_g, \tag{2}$$

and the horizontal momentum equations thus become:

$$\frac{D_g u_g}{Dt} = f_0 v_a + \beta y f_0 v_g \tag{3}$$

$$\frac{D_g v_g}{Dt} = f_0 u_a - \beta y f_0 u_g \tag{4}$$

Combining Eqs. (3) and (4) gives the QG vorticity equation as follows:

$$\frac{D_g \xi_g}{Dt} = f_0 \left(\frac{\partial u_a}{\partial x} + \frac{\partial v_a}{\partial y} \right) - \beta v_g . \tag{5}$$

Because the geostrophic wind is nondivergent, the continuity equation in the isobaric coordinates reduces to:

$$\frac{\partial u_a}{\partial x} + \frac{\partial v_a}{\partial y} + \frac{\partial \omega}{\partial p} = 0. \tag{6}$$

With the use of (6) in Eq. (5), the QG vorticity equation can be rewritten as:

$$\frac{\partial \xi_g}{\partial t} = -\vec{V}_g \cdot \nabla(\xi_g + f) + f_0 \frac{\partial \omega}{\partial p}. \tag{7}$$

Note that because of the geostrophic wind relationship, $\zeta_g = \nabla^2 \phi / f_0$, the QG vorticity equation can be further manipulated in the form:

$$\frac{1}{f_0} \nabla^2 \chi = -\vec{V}_g \cdot \nabla(\frac{1}{f_0} \nabla^2 \chi + f) + f_0 \frac{\partial \omega}{\partial p}. \tag{8}$$

where $\chi \equiv \partial \phi / \partial t$. To eliminate χ, we use the hydrostatic equation:

$$-\frac{RT}{p} = \frac{\partial \phi}{\partial p} \tag{9}$$

and then substitute Eq. (9) into the thermodynamics equation to obtain a second relationship between ϕ and χ as follows:

$$\left(\frac{\partial}{\partial t} + \vec{V}_g \cdot \nabla \right) \left(-\frac{\partial \phi}{\partial p} \right) - \sigma \omega = \frac{kJ}{p} \tag{10}$$

where $\sigma \equiv -(RT_0/p)(d\ln\theta_0/dp)$ is the static stability, $k = R/c_p$ and θ_0 is the reference potential temperature. The typical magnitude of σ is $\sim 2.5 \times 10^{-6}$ m^2 Pa^{-2} s^{-2} that we will hereinafter assume to be a constant for the sake of simplicity. Take ∇(10) and combine with Eq. (8) to eliminate $\nabla^2 \chi$, we arrive finally at the QG ω-equation as follows:

$$\left(\nabla^2 + \frac{f_0^2}{\sigma} \frac{\partial^2}{\partial p^2} \right) \omega = \frac{f_0}{\sigma} \frac{\partial}{\partial p} \left[V_g \cdot \nabla \left(\frac{1}{f_0} \nabla^2 \phi + f \right) \right] + \frac{1}{\sigma} \nabla^2 \left[V_g \cdot \nabla \left(-\frac{\partial \phi}{\partial p} \right) \right] - \frac{k}{\sigma p} \nabla^2 J \tag{11}$$

A comparison of the QG ω-equation (11) to the fully nonlinear version in which the advection of the divergent flow is retained [see Eq. (23) below] shows that use of the geostrophic approximation for the basic state together with neglect of the tilting terms have reduced the nonlinearity of the ω-equation dramatically. Furthermore, ignoring the advection by the ageostrophic component in the momentum equations also helps simplify the forcing terms such that the physical content of the ω-equation becomes more apparent. To see the physical meaning of the forcings on the RHS of Eq. (11), one expands the first two forcing terms and neglect the higher order terms such that Eq. (11) is reduced to (see Holton, 2004)

$$\left(\nabla^2 + \frac{f_0^2}{\sigma}\frac{\partial^2}{\partial p^2}\right)\omega \sim \frac{f_0}{\sigma}\left[\frac{\partial V_g}{\partial p}\cdot\nabla\left(\frac{1}{f_0}\nabla^2\phi + f\right)\right] - \frac{k}{\sigma p}\nabla^2 J \qquad (12)$$

The first term on the RHS of Eq. (12) represents the vorticity advection by thermal wind (i.e., differential geostrophic wind) and the second term represents impacts of diabatic heating. The physics of the first term is most obvious if one considers a situation in which an idealized baroclinic system is developing as illustrated in Fig. 1. Because of the westward tilt of the system with height for this type of baroclinic system, the 500-hPa geopotential field tends to lead the isotherm pattern. Since the thermal wind (i.e., $\partial V_g/\partial p$) is parallel to the isotherms, the first term on the RHS of (12) is evaluated by following the absolute vorticity along the isotherms. A careful analysis of the vorticity advection at 500 hPa in Fig. 1 shows that the positive relative vorticity advection tends to attain the maximum value above the surface low, while the relative vorticity advection is negative and strongest above the surface high. Because of this, the vertical motion at 500 hPa, which is forced by the positive (negative) advection of vorticity by the thermal wind will be downward (upward) to the eastern side of the 500-hPa ridge (trough) for a short-wave system that ensures $\zeta > f$. The second term on the RHS of Eq. (12) represents the effects of diabatic heating, which causes thermally

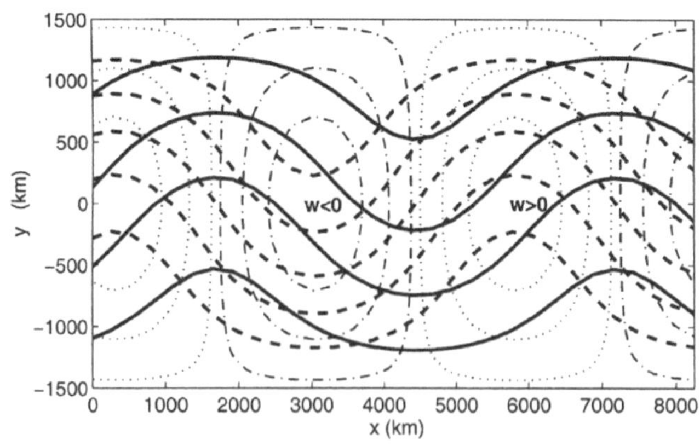

Fig. 1. Plane view of the 500-hPa height contours (solid lines), isotherms (dashed lines), and vertical motion field ($w > 0$ dash-dot lines, $w < 0$ dotted lines) for a developing synoptic-scale system (*Courtesy:* Holton, 2004).

direct circulations. That is, the positive heating (i.e. warming) will generate ascending motion ($J > 0$ implies $\omega < 0$) and vice versa. See Holton (2004) for more detailed examples.

An attractive property of Eq. (11) is that it involves only spatial derivatives. As a result, the only information required to diagnose vertical motion is instantaneous observation of the geopotential and diabatic heating. In this sense, Eq. (11) is essentially a diagnostic equation for ω. Unlike the continuity equation approach, the ω-equation provides a method of estimating vertical motion that does not depend on observations of the ageostrophic wind which is difficult to obtain directly. However, a disadvantage of Eq. (11) is that it does rely on higher order derivatives than in the other methods of estimating vertical motion. In many situations, accurately estimating such terms from noisy observational data can be difficult.

3. Sawyer-Eliassen Equation

While the ω-equation (12) contains some insightful physical implications, it should be noted that the application of this QG balance is confined only within a large-scale framework. The geostrophic approximation is generally not applicable for MCSs, especially for weather systems that possess some degree of axisymmetry for which the impact of the centrifugal force cannot be ignored. In this section, we will examine a appropriate balance approximation for TCs in which the underlying dynamics is the gradient wind balance instead of the geostrophic balance.

Under the gradient wind balance, the ω-equation has to be re-derived, because the balanced dynamics now involves a new contribution from the centrifugal forcing. With the gradient wind balance approximation, the formulation of the ω-equation can be established in terms of a well-posed elliptic equation, the so-called Sawyer-Eliassen (SE) equation, for a transverse streamfunction ψ that is defined in the radius-height cross section. There are numerous references on the formulation of the SE equation (see Eliassen, 1955; Krishnamurti, 1968; Schubert and Hack, 1983; DE91; Shapiro and Montgomery, 1993; Wang and Zhang, 2003, hereafter WZ03). For the sake of clarity, we follow the presentation by Schubert and Hack (1982) to derive the SE equation as their presentation is most pedagogically straightforward.

We start from the system of equations in the cylindrical coordinates as follows:

$$\left\{\frac{\partial u}{\partial t}+u\frac{\partial u}{\partial r}+w\frac{\partial u}{\partial z}\right\}-v\left(f_0+\frac{v}{r}\right)+\frac{\partial \phi}{\partial r}=F_u, \tag{13}$$

$$\frac{\partial v}{\partial t}+u\eta+w\frac{\partial v}{\partial z}=F_v, \tag{14}$$

$$\left\{\frac{\partial w}{\partial t}+u\frac{\partial w}{\partial r}+w\frac{\partial w}{\partial z}\right\}-g\frac{\theta}{\theta_0}+\frac{\partial \phi}{\partial z}=F_w, \tag{15}$$

$$\frac{1}{\theta_0}\frac{\partial ru}{\partial r}+\frac{1}{\rho_0}\frac{\partial \rho_0 w}{\partial z}=0, \tag{16}$$

$$\frac{1}{\theta_0}\left[\frac{\partial\theta}{\partial t}+u\frac{\partial\theta}{\partial r}\right]+w\frac{N^2}{g}=\frac{J}{c_p T_0}+F_\theta, \tag{17}$$

$$\frac{\partial q}{\partial t}+u\frac{\partial q}{\partial r}+w\frac{\partial q}{\partial z}=F_q, \tag{18}$$

where J is diabatic heating, $F_{u,v,w}$ are frictional forcings, $F_{\theta,q}$ represent the sources/sinks of moisture and temperature, $\eta = f_0 + \frac{1}{r}\frac{\partial rv}{\partial r}$, and all other variables bear similar meteorological meaning as in Chapter 24. Note that the reference density ρ_0 is a function of z because of the atmospheric stratification, and it is related to the reference temperature T_0 via the hydrostatic equation. Here, the azimuthal axisymmetry has been assumed in the system of Eqs (13)-(18). We assume further that the basic state is under the gradient wind balance and hydrostatical balance, i.e.,

$$f_0 v+\frac{v^2}{r}=\frac{\partial\phi}{\partial r}, \text{ Gradient wind balance – radial equation} \tag{13$'$}$$

$$\frac{\partial\phi}{\partial z}=g\frac{\theta}{\theta_0}, \text{ Hydrostatic balance} \tag{15$'$}$$

Note that Eqs (16) and (13′) encapsulate the fundamental difference between the QG and the SE formulation. While the continuity equation (16) implies the existence of the purely transverse winds that can be represented by a streamfunction, the gradient wind approximation (13′) is critical in simplifying the quasi-balanced dynamics of the system, because it discards entirely the non-balanced or agradient components of the tangential wind (recall that in the QG approximation, the non-balanced components are represented by the ageostrophic wind, or equivalently the nondivergent wind). We first combine Eqs (13′) and (15′) to obtain the thermal wind relationship as follows

$$\left(f_0+\frac{2v}{r}\right)\frac{\partial v}{\partial z}=\xi S=\frac{g}{\theta_0}\frac{\partial\theta}{\partial r} \tag{19}$$

From the continuity equation (16), there must exist a local streamfunction ψ such that:

$$u=-\frac{1}{r\rho_0}\frac{\partial\psi}{\partial z},\ w=\frac{1}{r\rho_0}\frac{\partial\psi}{\partial r}, \tag{20}$$

Take $\frac{\partial(19)}{\partial t}$ and eliminate the local derivative of v and θ using Eqs (14) and (17), we arrive at the celebrated SE equation.

$$\frac{\partial}{\partial r}\left(\frac{A}{r\rho_0}\frac{\partial\psi}{\partial r}+\frac{B}{r\rho_0}\frac{\partial\psi}{\partial z}\right)+\frac{\partial}{\partial z}\left(\frac{C}{r\rho_0}\frac{\partial\psi}{\partial z}+\frac{B}{r\rho_0}\frac{\partial\psi}{\partial r}\right)$$

$$=-\frac{\partial(\xi F_v)}{\partial z}+\frac{g}{c_p T_0}\frac{\partial J}{\partial r} \tag{21}$$

where

$$A = N^2,$$

$$B=-\xi S=-\left(f_0+\frac{2v}{r}\right)\frac{\partial v}{\partial z}=-\frac{g}{\theta_0}\frac{\partial\theta}{\partial r},$$

$$C=\xi\eta=\left(f_0+\frac{2v}{r}\right)\left(f_0+\frac{1}{r}\frac{\partial rv}{\partial r}\right)\equiv I^2,$$

If one defines the absolute angular momentum $M = rv + \frac{1}{2}fr^2$, the coefficient C can be re-written also as:

$$C = I^2 = \frac{1}{r^2}\frac{\partial M^2}{\partial r}$$

Physically, the coefficient A represents static stability, B baroclinity, and I^2 inertial stability. The typical distribution of the coefficients A, B, and C for TCs are shown in Fig. 2. A common characteristic of these coefficients is that they are highly compacted within the inner core region. While the squared inertial stability (I^2) is peaked near the radius of maximum wind from surface to ~ 300 hPa, the baroclinity (B) is the largest at the upper levels where the warm core is typically peaked. Likewise, the static stability (A) shows a dual peak: one at the upper level, and the other in the lower half of the troposphere. The lower peak is an inherent characteristic related to TC warm core structure that is not observed under the normal conditions. Note that for the SE equation Eq. (21) to be elliptic equation, the following condition has to be met:

$$AC - B^2 > 0$$

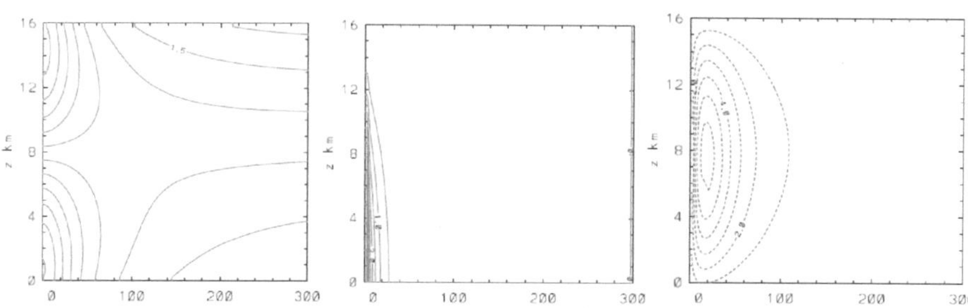

Fig. 2. Height-radius cross sections of (a) the buoyancy frequency squared or static stability (i.e., coefficient A, unit: 10^{-4} s^{-2}), (b) the inertial frequency squared (coefficient C, unit: 10^{-5} s^{-2}), and (c) the radial gradient of the buoyancy (coefficient B, unit: 10^{-6} s^{-2}) (Smith, 2007).

In general, it is hard, if not at all, to isolate the roles of each coefficient as the Poisson characteristics of the SE equation renders any inverted solution smooth and global over the entire domain. The essence of these coefficients can be illustrated by imposing a bubble-like source of heating to the SE equation at some distance from the vortex centre [i.e., prescribed by a compact-supported function $J(r, z)$ as suggested by Shapiro and Willoughby, 1983; Shubert and Hack, 1983]. Figure 3 shows an example of the response to such external forcing approach after inverting the SE equation. One can see that the SC induced by a heat source is primarily vertical with two cells of circulation that are counter-rotating and extend outside the source over the whole domain. The flow emerges from the source, spreads outward through a large volume surrounding it, and converges back into it from below. One can see from this example how the inertial stability coefficient C can influence the compensating subsidence surrounding heat-induced updrafts.

A more detailed sensitivity analysis of the SE equation shows that the coefficient C determines the radial scale of the circulation dipoles seen in Fig. 3 in a way similar to the coefficient A (i.e., static stability) controlling the vertical scale. For a fixed coefficient A, the transverse streamfunction ψ tends to be elongated vertically when C is large, and horizontally when C is small (Fig. 3). Physically, the vertical gradients of the absolute angular momentum associated with the vertical shear of the primary circulation tilt the updraft through a heat source, because the path of least resistance for the rising air lies along the surfaces of constant absolute angular momentum. Likewise, horizontal temperature gradients associated with the vertical shear deflect the flow through momentum sources from the horizontal direction, because the path of least resistance in this case lies along isentropic surfaces.

With the SE equation and some further inversion techniques to be discussed in the next section, the quasi-balanced evolution of a hurricane-like vortex can be determined completely by the system of Eqs (14), (17) and (21). Specifically, given a basic vortex state in the gradient balance, a diabatic heating source, and appropriate boundary conditions for the streamfunction,

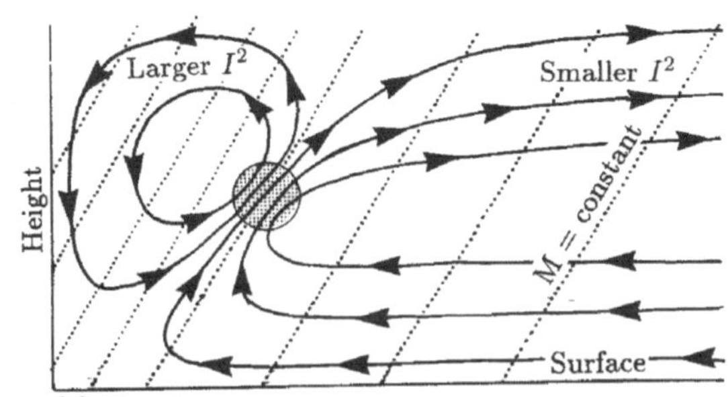

Fig. 3. Secondary circulation induced in a balanced warm-core vortex by a heat source showing the distortion induced by variation in inertial stability I^2, thermodynamic stability N^2, and baroclinity B (Smith, 2007).

one can in principle solve Eq. (21) for ψ by directly inverting it as long as the elliptic condition is met. The secondary circulation can then be retrieved by simply using relationship (20), which is then substituted into Eqs (14) and (17) to find v and θ after advancing in time. The coefficients A, B and C are then re-computed and the above steps are repeated for the next time integration. This provides a complete description of the quasi-balanced evolution of any balanced vortex.

Note that under the barotropic approximations, coefficients A, B and C can be considered as constants, allowing one to obtain analytical solutions for ψ with some specific heating function. As demonstrated by Schubert et al. (2011), the resulting solutions could provide some insight into the inner-core dynamics of TCs during the mature stage. For example, a diabatic heating source applied at the outer edge of the RMW could induce a regime of strong ascending motion in the heated region and a compensating subsidence in the eye and outside the inner core region for a barotropic vortex (Fig. 4). A similar heating function, however, would result in a vertical shift of the peak vertical motion to a higher altitude for a baroclinic vortex (cf. Figs 4a and 4b), which is expected due to the decrease of the static stability in the lower atmosphere as compared to that in the barotropic vortex. More detailed analysis of the quasi-balanced response to different heating profiles can be found in Schubert et al. (2009).

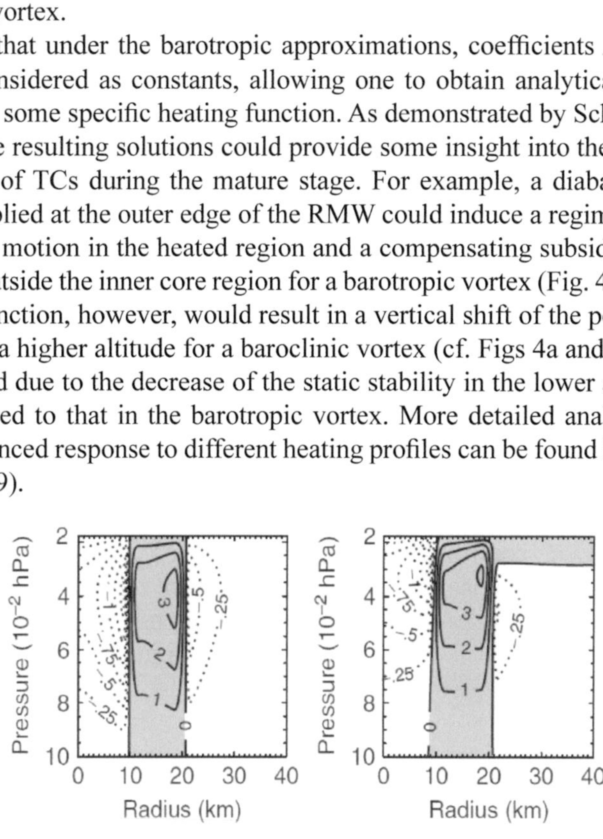

Fig. 4. Radius-height cross section of the vertical motion (contours) in response to a top-hat profile of diabatic heating in barotropic vortex (left panel); and baroclinic vortex (right panel) (Schubert et al., 2009).

4. Mesoscale Quasi-balanced Formulations

Various approaches have been adopted to derive a complete quasi-balanced system, which dated back to Sawyer-Eliassen (1955) and subsequently Charney and Eliassen (1964), Yanai (1965), Ooyama (1972), Krishnamurti (1968), Willoughby (1985), Hack and Schubert (1985), DE91 and WZ03. In this section, we follow the derivations presented by WZ03 to obtain the full quasi-balanced model for TCs. To derive the quasi-balanced equations in a complete form and be consistent with the derivations of the PV balanced dynamics, we start again from the vorticity equation in the pseudo-height coordinates. A simple combination of Eqs (1) and (2) in chapter 24 results in the vorticity equation in the pseudo-height coordinate as follows:

$$\frac{\partial \zeta}{\partial t} = -(V_{\psi} + V_{\chi}) \cdot \nabla_h \eta - \omega \frac{\partial \eta}{\partial z} - \eta \nabla_h V_h + \frac{\partial \omega}{\partial y}\frac{\partial u}{\partial z} - \frac{\partial \omega}{\partial x}\frac{\partial v}{\partial z} + \frac{\partial F_x}{\partial x} - \frac{\partial F_y}{\partial y} \tag{22}$$

where a decomposition of the horizontal wind component into the divergent and nondivergent parts has been used. Next, take $g/\theta_0 . \nabla_h^2$ (Eq. (4) in chapter 24) – ∂^2(Eq. (9) in chapter 24) / $\partial t \partial z - f\partial(8)/\partial z$ to obtain the following quasi-balanced ω-equation:

$$\begin{aligned}
&\nabla^2\left(\frac{\partial^2 \phi}{\partial z^2}\omega\right) + f\eta\frac{\partial}{\partial z}\left\{(z_a - z)^{-\mu}\frac{\partial}{\partial z}\left[(z_a - z)^{\mu}\right]\omega\right\} - f\frac{\partial}{\partial z}\left(\frac{\partial \omega}{\partial x}\frac{\partial^2 \psi}{\partial x \partial z} + \frac{\partial \omega}{\partial y}\frac{\partial^2 \psi}{\partial y \partial z}\right) \\
&- f\frac{\partial}{\partial z}\left(\frac{\partial \omega}{\partial x}\frac{\partial^2 \chi}{\partial y \partial z} - \frac{\partial \omega}{\partial y}\frac{\partial^2 \chi}{\partial x \partial z}\right) - \left(f\frac{\partial \eta}{\partial z}\frac{\mu}{z_a - z} + f\frac{\partial^2 \eta}{\partial z^2}\right)\omega \\
&= f\frac{\partial}{\partial z}[\mathrm{V}_h \bullet \nabla \eta] - \nabla^2[\mathrm{V}_h \bullet \nabla \frac{\partial \phi}{\partial z}] - \beta\frac{\partial^3 \psi}{\partial t \partial y \partial z} - 2\frac{\partial^2}{\partial t \partial z}\left(\frac{\partial^2 \psi}{\partial x^2}\frac{\partial^2 \psi}{\partial y^2} - \frac{\partial^2 \psi}{\partial x \partial y}\frac{\partial^2 \psi}{\partial x \partial y}\right) \\
&+ \frac{g}{\theta_0}\nabla^2 J - f\frac{\partial}{\partial z}\left(\frac{\partial F_y}{\partial x} - \frac{\partial F_x}{\partial y}\right) - \frac{\partial^2}{\partial t \partial z}\left(\frac{\partial F_x}{\partial x} - \frac{\partial F_y}{\partial y}\right)
\end{aligned} \tag{23}$$

where $\mu = c_d/R_d$ and (ψ, ϕ) again are the streamfunction and geopotential height. The dimensional form of Eq. (23) is very similar to that derived by Krishnamurti (1968) and DE91 except for the use of the pseudo-height coordinates and inclusion of the velocity potential χ and the frictional forcing. Mathematically, the quasi-balanced vertical motion is determined by the forcing terms on the RHS of Eq. (23), which are the differential vorticity advection, the Laplacian of thermal advection by nondivergent and divergent winds, the differential deformation or Jacobian term, the β effect, latent heating, and the effects of friction, respectively. The first four forcing terms are more dynamically related and diagnosed from the nonlinear balanced equation (see chapter 24 on balanced dynamics in tropical cyclones for details). The last three terms are the major sources that determine the magnitude of divergent winds and secondary circulations in TCs. Note that the quasi-balanced vertical motion obtained from Eq. (23) is always associated with the pre-existing balanced flows. Therefore, its magnitude is expected to be smaller than the instantaneous vertical motion derived directly from the continuity equation, which is spontaneous and fluctuates rapidly with time, and so it is virtually impossible to predict precisely.

Because of the existence of three local tendency terms, Eq. (23) is not a fully diagnostic equation; that is, we must know three tendency terms $\partial\psi/\partial t$, $\partial F_x/\partial t$ and $\partial F_y/\partial t$ in advance to find ψ. The simplest way to obtain the term $\partial\psi/\partial t$ is to iterate directly the vorticity Eq. (22) in the form:

$$\nabla^2 \frac{\partial \psi}{\partial t} = -V_h \cdot \nabla \eta - \omega \nabla^2 \frac{\partial \psi}{\partial z} - \eta \nabla^2 \chi + k \cdot \frac{\partial V_h}{\partial z} \times \nabla \omega + \frac{\partial F_y}{\partial x} - \frac{\partial F_x}{\partial y} \tag{24}$$

The two other frictional terms $\partial F_x/\partial t$ and $\partial F_y/\partial t$ can be obtained directly from the model output and are considered as external forcing for the quasi-balanced system. Because of the additional variable χ included in Eq. (24), we need to use the continuity equation to close Eqs (23)-(24). For computational purpose, we rearrange the continuity equation (10) in chapter 24 of in the form:

$$\nabla^2 \chi = -(z_a - z)^{-\mu} \frac{\partial}{\partial z}[(z_a - z)^{\mu} \omega] \tag{25}$$

Thus, if the diabatic term J and frictional contributions (F_x, F_y) are given (e.g., from a model simulation), Eqs (23)-(25) form a closed set of equations in ϕ, χ and $\partial\psi/\partial t$. The system of Eqs (23)-(25) is called the quasi-balanced ω-system, which allows one to obtain the quasi-balanced vertical motion and horizontal divergent wind with respect to a given balanced primary circulation. Together with the balanced equations (9) and (12) in chapter 24, the entire equations system is called the PV-omega dynamical system. As mentioned in WZ03, the inversion of the PV-omega equation is practically doable and it is applicable to various analyses of the balanced dynamics of TCs. Together with meaningful partitioning of PV anomalies (PVAs), one can study various aspects of the PV balanced and quasi-balanced influences during the TC development. For example, a recent study by Kieu and Zhang (2010) on the piecewise inversion of PVAs associated with different hurricane rainbands and eyewalls shows that formation of an outer eyewall tends to be inimical to development of the inner eyewall and results inevitably in collapse of the inner eyewall with time during an eyewall replacement process. Some other applications of the piecewise PV inversion can be found in DE91, Shapiro (1996), Huo et al. (1998, 1999a, b), Möller and Shapiro (2002), Wu and Emanuel (2001), WZ03, or Egger (2008).

5. Quasi-balanced Inversion

Given the fact that the ω-equation (23) is the second-order PDE, it would be desirable to solve this ω-equation directly to obtain vertical motion by iteration techniques. In reality, the inversion of such a nonlinear ω-equation is mostly restricted to TCs, because the TC strong rotational flows could ensure the ellipticity over most of the domain.

Because of the dominant rotation of the TC horizontal flows, it turns out that the elliptic condition for the quasi-balanced equations (23)-(24) is well satisfied within a TC except in the inner core region near the top of the PBL where strong super gradient flows occur. A common treatment for this non-elliptic problem is to assume smooth continuity of the inverted solution over the entire domain such that solutions over the non-elliptic discrete points can be approximated as continuous extension of the smooth solutions. Assuming that the underlying balanced dynamics and appropriate boundary condition for χ are given, the non-elliptic points will be set to a small positive value such that the system (23)-(25) becomes elliptic over the entire domain. Under this assumption, procedures of solving Eqs (23)-(25) can be encapsulated as follows:

- Step 1: obtain the frictional tendency terms $\partial F_x/\partial t$ and $\partial F_y/\partial t$, and the diabatic heating term J directly from model outputs;
- Step 2: prescribe a first guess for the set of the unknowns $\frac{\partial \psi^{(0)}}{\partial t}$, $\omega^{(0)}$ and $\chi^{(0)}$ over the entire domain;
- Step 3: iterate Eq. (23) to obtain $\omega^{(1)}$, (24) for $\frac{\partial \psi^{(1)}}{\partial t}$, and (25) for $\chi^{(1)}$; and
- Step 4: plug the new set of the unknowns ($\omega^{(1)}$, $\frac{\partial \psi^{(1)}}{\partial t}$, $\chi^{(1)}$) onto the RHS of Eqs (23)-(25) and repeat steps 2 and 3 until the difference between two consecutive iterations becomes sufficiently small, i.e., $\left| \omega^{(i+1)} - \omega^{(i)} \right| < \varepsilon$, where ε is a prescribed accuracy.

The above numerical approach can be sped up greatly by using some effective algorithm for inverting the Poisson equation. Our numerous experiments with the quasi-balanced inversion show that the converged solution can be achieved after ~80 iterations. Of course, there is no guarantee that the above iterating algorithm will converge for all cases. However, for TCs with sufficient axisymmetry, there does seem to exist a limit that the system tends to converge into. One can verify if the converged limit is the right solution by testing different first guesses to see if the final solutions with different initial guesses are close to each other. In case all converged limits with different first guesses are very similar, it is a good indicator that the correct solutions have been obtained.

As an illustration of the application of the quasi-balanced inversion, we examine the vertical structures of the total quasi-balanced secondary circulation and the flow divergence that are inverted from Eqs (23)-(25) forced by a given diabatic heating source and frictional forcing (hereafter referred to as the forced secondary circulation or FSC). The balanced vortex is obtained from a simulation of Hurricane Andrew (1992), using the MM5 model similar to those described in the chapter 24 (see a detailed description of the model experiments and configuration in WZ03). One can see from Fig. 5 that the FSCs forced by all of the forcing terms on the RHS of Eq. (23) exhibit an asymmetric development of slantwise updrafts along with the divergence in the western and eastern parts of the eyewall, respectively. Even though the updrafts increase with height up to 9 km, the divergent and convergent flows are somewhat separated by the updrafts core for the layers above 2 km. Consistent with that of diabatic heating in the eyewall, the convergence-divergence has a slantwise pattern along the eyewalls where the air converges into the core of updrafts from the layers below in the outer region and diverges aloft in the inner region.

It is of interest to see that the quasi-balanced inversion could capture the generation of weak subsidence with convergence aloft and divergence below in the eye where latent heating is absent (Fig. 5a). Consistent with what is found in Liu et al. (1999), the upper-level convergence develops as a result of divergent outflows from the eyewall into the eye. In particular, the absence of a deep layer of narrow downdrafts right at the inner edge of the eyewall suggests that these downdrafts may be of unbalanced nature that could not be taken into account by the quasi-balanced inversion.

To see the role of each individual forcing in the quasi-balanced inversion, Figs 5b-d show the FSCs associated with isolated contributions of latent heating, frictional and dry dynamical processes, respectively. As expected from the thermodynamics equation, the distribution of updrafts coincides well with that of latent heat release in the eyewall and outer rainbands (Fig. 5b), which accounts for more than 60% of the radial inflow in the PBL, and almost all the outflow in the upper levels. This clearly demonstrates that the latent heating is the dominant factor in determining[1] most of the vertical motion in

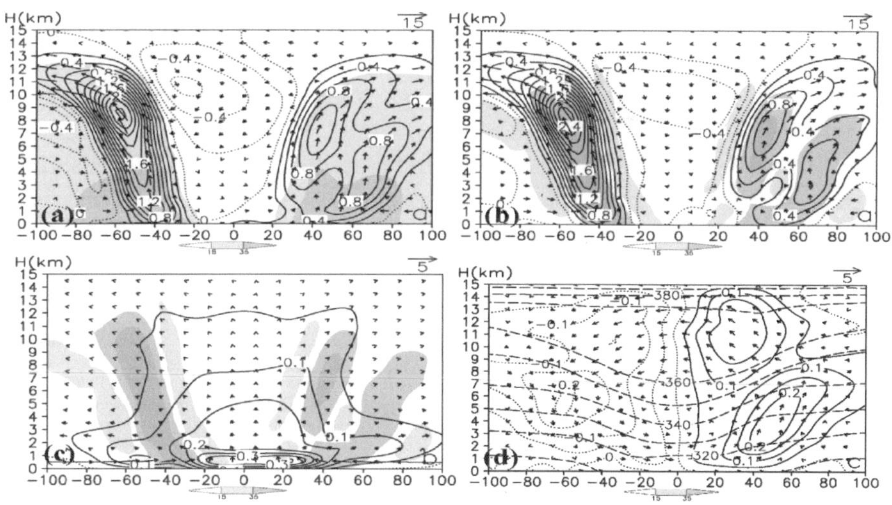

Fig. 5. West-east vertical cross sections through the storm centre of quasi-balanced (system-relative) in-plane flow vectors and vertical motion (contours) that are forced by (a) the total forcing; (b) latent heating forcing; (c) frictional forcing; and (d) dry-dynamics forcing. Shadings in (a) denote simulated reflectivity and in (b)-(d) denote the latent heating or cooling rates in K h^{-1} (see the scales). Thick-dashed lines in (d) are isentropic surfaces at intervals of 10 K. Note that the vertical components of the velocity vectors have been amplified by a factor of 5. From Zhang and Kieu (2005).

[1] Here, the term "determining" is understood from the mathematical viewpoint of the quasi-balanced inversion, and it has no cause-and-effect implication.

the eyewall. Furthermore, the vertical motions forced by the latent heating in the eyewall could explain most of the compensating subsidence from the tropopause to the surface in the eye, with the magnitudes as large as 0.4 ms^{-1} below the tropopause (Fig. 5b).

Unlike the response to the diabatic heating forcing, the FSC induced by the frictional forcing exhibits a different picture with most of the radial inflows confined within the PBL (Fig. 5c). Specifically, the friction-induced inflows increase inward until the RMW, and begin to ascend in the eyewall. On average, the radial inflow could account for nearly half of the PBL convergence near the surface, whereas the ascending motion reaches a peak value near the top of the PBL in the eye. Although this ascending motion decreases upward, it can reach the upper outflow layer to offset some portion of the eye subsidence induced by latent heat release in the eyewall. This PBL-FSC is somewhat similar to the circulation associated with Ekman pumping that causes the spin-down of a cyclonic vortex (see Holton, 2004). This friction-induced secondary circulation appears to be responsible for transporting partly the PBL high equivalent potential temperature air into the eye centre where the core of high equivalent potential temperature air is often present. It should be noted, however, that this friction-induced FSC differs from that produced by Eliassen's balanced model (see Willoughby, 1979) in which the PBL pumping is actually peaked in the eyewall. This difference is likely related to the specification of surface friction as the bottom boundary condition as well as background balanced vortex.

In contrast to the response to the frictional forcing or diabatic heating, the FSC by the dry-dynamical processes shows a deep, anticlockwise vertical circulation across the eye with a clockwise transverse circulation across the vortex centre (Fig. 5d). This result explains why the latent heating induced vertical motion in the west (east) semicircle is about 10% stronger (20% weaker) than that in the total vertical motion (cf. Figs 5a, b). Such a wavenumber-1 vertical motion couplet also appears in the horizontal (x, y) planes, although the updraft (downdraft) core "shifts" from the east-northeast (southwest) at $H = 2$ km to southeast (northwest) at $H = 12$ km in the eyewall. As explained in Zhang and Kieu (2006), such dipole of FSCs in the horizontal distribution is a consequence of the vertical wind shear that rotates with height.

In particular, the response of the FSC to the dry-dynamical forcing seems to increase linearly with the imposed vertical wind shear with ascending motion on the upshear side and descending motion on the downshear side. Associated with the vertical motion couplet is a pair of well-defined divergence and convergence centres as determined by the mass continuity. In the study of Zhang and Kieu (2006), they showed that the wavenumber-1 vertical motion couplet results from the interaction of the TC vortex with the vertical shear of horizontal winds, as documented by many observational, theoretical and modelling studies (e.g., Raymond, 1992; Marks et al., 1992; Franklin et al.,

1993; Frank and Ritchie, 1999, 2001; Molinari et al., 2002; Black et al., 2002; Zhu et al., 2004). The well-organized horizontal divergent flows with opposite directions at the lower and upper levels represent an important feedback of the shear-FSC to the larger-scale shear flows as presented in Zhang and Kieu (2006). This helps explain why TCs sometimes are very resilient when embedded in strong shear environment.

The above example of the different response of the secondary circulation by different forcing terms illustrates the powerful feature of the quasi-balanced PV ω-equation inversion in TC studies. Some other more detailed application of the PV ω-equation inversion can be also found in Kieu and Zhang (2012) and numerous references therein.

6. Summary

In this chapter, we have examined the quasi-balanced dynamics under different underlying balanced approximations. The quasi-balanced system was first considered from the simple quasi-geostrophic perspective in which the large-scale horizontal circulation was approximated by the geostrophic balance. It was shown that by neglecting the ageostrophic advection of horizontal momentum, the secondary circulation can be diagnosed from a well-posed elliptic equation, i.e., the ω-equation, that involves only spatial derivatives. Because the ω-equation does not involve time derivative, the only information required to diagnose the vertical motion is instantaneous observation of geopotential heights and diabatic heating without any external frictional tendency forcings or tendency of a transverse streamfunction. This simplifies the ω-equation drastically, but on the other hand, poses some challenges in reliably estimating the higher-order derivatives terms from observations.

Under the gradient wind and hydrostatic approximation for mesoscale processes, the quasi-balanced dynamics has been examined further in the form of another elliptic equation for the secondary circulation, the the so-called Sawyer-Eliassen equation. A particular interesting characteristic of the SE formulation is the explicit dependence of the elliptic coefficients on the storm dynamical and thermodynamic structures. Such a dependence allows for examining different responses of the secondary circulation to different stages of the storm intensity and structures. Unlike the inversion of the QG ω-equation that requires inversion over a 3D domain, the SE equation is formulated in the radius-height cross section, rendering it more apparent for examining the *axisymmetrical* behaviours of TC flows. The SE equation is therefore most useful for TC studies in which the quasi-balanced dynamics associated with the transverse circulation can be extracted for different types of external forcings such diabatic heating at various locations or frictional convergence.

To deliver a more complete picture of the quasi-balanced dynamics for practical applications in which the divergent components *cannot* be neglected, we have provided in sections 4 and 5 a comprehensive formulation of the quasi-balanced dynamics tailored particularly for TC dynamics from the primitive equations in the pseudo-height coordinate. By separating the total wind into a divergent and a non-divergent component, the governing equation for the divergent component and its associated vertical motion can be formulated in terms of a PV ω-equation. Given a diabatic heating source and other frictional tendency forcing terms, it is possible to invert this PV ω-equation to obtain a 3D distribution of secondary circulations over a limited domain. This type of 3D inversion is valuable for investigation of quasi-balanced responses related to PV anomalies that are highly asymmetric such as those originated from upper level troughs or TC rainbands. Combined with the PV balanced inversion presented in Chapter 24, the quasi-balanced inversion shown in this chapter provides a complete description of the 3D structures as well as evolution of the TC balanced dynamics.

Problems

P1. Show that the QG omega-equation (21) and the ω equation (25) are equivalent under the geostrophic approximations.

P2. Derive the continuity equation (4) from the original continuity equation in the pseudo-height coordinate given in Chapter 24.

REFERENCES

Charney, J.G. and A.A. Eliassen, 1964: On the growth of the hurricane depression. *J. Atmos. Sci.*, **21**, 68-75.

Davis, C.A., 1992: Piecewise potential vorticity inversion. *J. Atmos. Sci.*, **49**, 1397-1411.

Davis, C.A. and K. Emanuel, 1991: Potential vorticity diagnostics of cyclogenesis. *Mon. Wea. Rev.*, **119**, 424-439.

Frank, W.M. and E.A. Ritchie, 2001: Effects of vertical wind shear on the intensity and structure of numerically simulated hurricanes. *Mon. Wea. Rev.*, **129**, 2249-2269.

Huo, Z.-H., D.-L. Zhang and J.R. Gyakum, 1998: An application of potential vorticity inversion to improving the numerical prediction of the March 1993 superstorm. *Mon. Wea. Rev.*, **126**, 424-436.

Huo, Z.-H., D.-L. Zhang and J.R. Gyakum, 1999a: The interaction of potential vorticity anomalies in extratropical cyclogenesis. Part I: Static piecewise inversion. *Mon. Wea. Rev.*, **127**, 2546-2561.

Huo, Z.-H., D.-L. Zhang and J.R. Gyakum, 1999b: The interaction of potential vorticity anomalies in extratropical cyclogenesis. Part II: Sensitivity to initial perturbations. *Mon. Wea. Rev.*, **127**, 2563-2575.

Hoskins, B.J., M.E. McIntyre and A.W. Robertson, 1985: On the use and significance of isentropic potential vorticity maps. *Quart. J. Roy. Meteor. Soc.*, **111**, 877-946.

Kieu, C.Q. and D.-L. Zhang, 2010: A piecewise potential vorticity inversion algorithm and its application to hurricane inner-core anomalies. *J. Atmos. Sci.*, **67**, 1745-1758.

Krishnamurti, T.N., 1968: A diagnostic balance model for studies of weather systems of low and high latitudes, Rossby number less than 1. *Mon. Wea. Rev.*, **96**, 197-207.

Liu, Y., D.-L. Zhang and M.K. Yau, 1999: A multiscale numerical study of Hurricane Andrew (1992). Part II: Kinematics and inner-core structures. *Mon. Wea. Rev.*, **127**, 2597-2616.

Molinari, J., S. Skubis, D. Vollaro, F. Alsheimer and H.E. Willoughby, 1998: Potential vorticity analysis of tropical cyclone intensification. *J. Atmos. Sci.*, **55**, 2632-2644.

Möller, J.D. and L.J. Shapiro, 2002: Balanced contributions to the intensification of Hurricane Opal as diagnosed from a GFDL model forecast. *Mon. Wea. Rev.*, **130**, 1866-1881.

Ooyama, K., 1969: Numerical simulation of the life cycle of tropical cyclones. *J. Atmos. Sci.*, **26**, 3-40.

Shapiro, L.J., 1996: The motion of Hurricane Gloria: A potential vorticity diagnosis. *Mon. Wea. Rev.*, **124**, 2497-2508.

Shapiro, L.J. and J.L. Franklin, 1999: Potential vorticity asymmetries and tropical cyclone motion. *Mon. Wea. Rev.*, **127**, 124-131.

Shapiro, L.J. and H.E. Willoughby, 1982: The response of balanced hurricanes to local sources of heat and momentum. *J. Atmos. Sci.*, **39**, 378-394.

Schubert, W.H. and J.J. Hack, 1983: Transformed Eliassen balanced vortex model. *J. Atmos. Sci.*, **40**, 1571-1583.

Schubert, W.H., C.M. Rozoff, J.L. Vigh, B.D. McNoldy and J.P. Kossin, 2007: On the distribution of subsidence in the hurricane eye. *Quart. J. Roy. Meteor. Soc.*, **133**, 595-605.

Smith, R.K., 2007: Balanced dynamical theories. Lecture Note. Available online at: http://www. meteo.physik.uni-muenchen.de/~roger/Tropical_Cyclones/080508_TCs_Chapter_3.pdf.

Wang, X.-B. and D.-L. Zhang, 2003: Potential vorticity diagnosis of a simulated hurricane. Part I: Formulation and quasi-balanced flow. *J. Atmos. Sci.*, **60**, 1593-1607.

Wu, C.-C. and K.A. Emanuel, 1995a: Potential vorticity diagnostics of hurricane movement. Part I: A case study of Hurricane Bob (1991). *Mon. Wea. Rev.*, **123**, 69-92.

Yau, M.K., Y.-B. Liu, D.-L. Zhang and Y.-S. Chen, 2004: A multiscale numerical study of Hurricane Andrew (1992). Part VI: Small-scale inner-core structures and wind streaks. *Mon. Wea. Rev.*, **132**, 1410-1433.

Zhang, D.-L., Y. Liu and M.K. Yau, 2000: A multiscale numerical study of Hurricane Andrew (1992). Part III: Dynamically induced vertical motion. *Mon. Wea. Rev.*, **128**, 3772-3788.

Zhang, D.-L., Y. Liu and M.K. Yau, 2001: A multiscale numerical study of Hurricane Andrew (1992). Part IV: Unbalanced flows. *Mon. Wea. Rev.*, **129**, 92-107.

Zhang, D.-L., Y. Liu and M.K. Yau, 2002: A multiscale numerical study of Hurricane Andrew (1992). Part V: Inner-core thermodynamics. *Mon. Wea. Rev.*, **130**, 2745-2763.

Zhang, D.-L. and C.K. Kieu, 2005: Shear-forced vertical circulations in tropical cyclones. *Geophys. Res. Lett.*, **32**, L13822.

Yanai, M., 1964: Formation of tropical cyclones. *Rev. Geophys.*, **2**, 367-414.

Zhu, T., D.-L. Zhang and F. Weng, 2004: Numerical simulation of Hurricane Bonnie (1998). Part I: Eyewall evolution and intensity changes. *Mon. Wea. Rev.*, **132**, 225-241.

26

Using Budgets to Study Tropical Cyclones

Wallace Hogsett[1] and Da-Lin Zhang[2]

[1]Hydrometeorological Prediction Center, NOAA/NWS/NCEP
College Park, Maryland

[2]Department of Atmospheric and Oceanic Science, University of Maryland, College Park, Maryland

dalin@atmos.umd.edu

1. Introduction

In this chapter, the general formulation and methodology of budget studies for tropical cyclones (TCs) are reviewed. Budget calculations are very useful for understanding the dynamical and thermodynamical processes taking place in TCs, because they could provide not only the relative magnitudes of the processes in budget equations but also insight into the most fundamental processes leading to the intensity and intensity changes as well as structural changes of TCs. Budget studies have evolved dramatically in the past 50 years, in response not only to increases in observational data quality and availability, but also to rapidly increasing computing power for numerical model simulations. Accordingly, the scales of interest to researchers have progressed downward with time, from early studies on the general synoptic-scale to recent mesoscale processes. More recent budget studies have begun to focus on even smaller scale features such as individual convective elements, which remain virtually unobservable within the TC inner core. Many of the early budget studies were limited by assumptions required because of data limitations, but today's numerical models allow more rigorous and detailed investigation of dynamic processes taking place in TCs.

Early budget studies focused on the observational diagnosis of heat and moisture budgets, in an attempt to understand the relationship between tropical convective cloud clusters and the large-scale environment. Yanai et al. (1973) used rawinsonde data to calculate the spatially and temporally averaged apparent heating source (Q_1), moisture sink (Q_2), and radiative heating (Q_R) profiles, to ascertain the impact of cloud clusters on the environment. Because diabatic heating rates were not directly observable, the net diabatic heating was diagnosed as a residual of the large-scale budgets. The Yanai study concluded that subsidence warming between convective cloud clusters and evaporative cooling are critical to the heat balance of the large-scale tropical environment. Investigation of the internal structures of cloud clusters, however, was not possible.

Johnson (1984) expanded the work of Yanai et al. (1973) by partitioning the apparent heat source (Q_1) into mesoscale and cumulus scale heating to show distinct heating profiles between the two regimes. Other studies have used the distribution of hydrometeors (Tao et al., 1990) and the vertical motion profiles of MCSs to diagnose heating rates (Anthes, 1982; Mapes and Houze, 1995). However, all of these studies acknowledge that the errors so obtained may be quite large and suggest that better data quality and higher resolution may yield additional information about heat and moisture budgets. In general, data limitations in space and time limit observational budget studies. Observational TC budget studies have been conducted (e.g. Hawkins and Rubsam, 1968), for quantities including momentum, moisture and various energies. Although still limited by data, these early budget studies began to clarify which areas of a TC were critical to the TC evolution.

As numerical models advanced and began to resolve the eye, eyewall and spiral rainbands in TCs, though marginally, it became possible to perform budget studies for a variety of quantities using more comprehensive model data. Tuleya and Kurihara (1975) calculated various energy budgets from their three-dimensional TC model by partitioning the budget terms into both mean and eddy components. In addition, Kurihara and Bender (1982) conducted area-averaged heat budgets in the TC eye. In these early modelling studies many processes were parameterized, and the TCs studied were highly idealized. However, these early modelling studies were the first to begin an investigation of the TC inner core.

TC budget studies have continued to evolve dramatically during the past several decades as high-resolution simulations enable more accurate calculations of budget terms that are difficult to observe. Now, with wide use of cloud-permitting models, there is no need to diagnose the residual terms (i.e., the apparent heat and moisture sources/sinks). Moreover, high-resolution numerical models allow investigation into very specific regions of a TC that are well resolved by the models, including the eye, eyewall, spiral rainbands, and maritime boundary layer (MBL).

In general, through evaluating the contribution of different physical processes taking place in TCs, budget studies allow researchers to verify some competing hypotheses or develop new hypotheses on important processes that may have been impossible to investigate in prior work. For example, diabatic heating is still not directly observable by any instrument, but it can be calculated directly from model output.

In this chapter, we present the basic methodologies and a few budget formulations. A review of some recent work, primarily conducted by Zhang and collaborators on explicit, cloud-permitting simulations of Hurricanes Andrew (1992) and Bonnie (1998) is given. TC budget studies by other investigators are also reviewed. Section 2 shows general procedures for budget studies, including spatial and temporal averaging. Section 3 presents the budget equations and some relevant findings on momentum budgets. Thermodynamic and energy budget studies are reviewed in sections 4 and 5, respectively. Concluding remarks are given in the final section.

2. Budget Analysis Procedures

Spatial and temporal averaging is particularly important for budget studies to ensure that transient features do not dominate the results and the figures presented are representative of the TC circulations under study. Because the TC circulations are approximately circular, it is more convenient to work within a cylindrical coordinate framework, in which the origin is the ocean surface at the location of minimum surface pressure. Of course, in a mature TC the origin is typically the centre of the eye. The analysis procedures presented below follow Marks et al. (1992) and Liu et al. (1999). Any scalar variable, $X(r, \theta, z)$ may be decomposed:

$$X = X_s + X_a, \tag{1}$$

where the symmetric component, X_s, is given by:

$$X_s(r, \theta, z) = \frac{1}{2\pi}\int_0^{2\pi} X(r, \theta, z)\, d\theta, \tag{2}$$

and the asymmetric component, X_a, is simply:

$$X_a(r, \theta, z) = X - X_s. \tag{3}$$

The symmetric component is most easily visualized using a radius-height diagram, which will be presented frequently in subsequent sections. If the analysis is performed correctly, the azimuthal mean of the asymmetric component would nearly vanish. This scalar analysis is often used to show the mean structure of TC variables such as individual wind components, temperature and moisture.

For vector fields such as the horizontal wind velocity, $\mathbf{V}$, it is often desirable to calculate a volume-integrated horizontal velocity ($\mathbf{V}_m$) to approximate the translational speed of the TC:

$$\mathbf{V}_m = \frac{\int_{900hPa}^{150hPa}\int_0^{2\pi}\int_0^{150km} \mathbf{V} r\, dr\, d\theta\, dp}{\int_{900hPa}^{150hPa}\int_0^{2\pi}\int_0^{150km} r\, dr\, d\theta\, dp} \tag{4}$$

The mean horizontal velocity may then be subtracted from the total velocity to yield the TC-relative velocity field:

$$\mathbf{V}' = \mathbf{V} - \mathbf{V}_m. \tag{5}$$

The TC-relative horizontal velocity vector, $\mathbf{V}'$, may then be decomposed into tangential (V) and radial (U) components:

$$V' = V'_s + V'_a, \text{ and} \tag{6a}$$

$$U' = U'_s + U'_a \tag{6b}$$

where the axisymmetric components (V'_s, U'_s) and the asymmetric components (V'_a, U'_a) may each be calculated in a manner similar to the simple scalar example shown above. In Fig. 1a, an example radius-height diagram of the azimuthal and temporal mean horizontal wind speed, superimposed with in-plan flow vectors, is shown. Note that the typical TC characteristics, including an outward-sloping radius of maximum wind (RMW) and a wind maximum near the top of the MBL, are captured. In order to see better the secondary circulation of a TC, vertical motion that is used to construct the vertical cross-sectional flow vectors is typically multiplied by an arbitrary factor, e.g., typically 10. Details of the model configuration used in Fig. 1 are provided in Liu et al. (1997, 1999).

It is important to note that the bounds chosen for the radial and vertical integrals in Eq. (4) should be considered as case-dependent and variable. If one is investigating a relatively weak TC, in which the circulation does not extend vertically to the tropopause, integrating to 150 hPa may yield unreasonable results for the mean velocity calculation. Similarly, the radial extent of TCs varies greatly, and sometimes 150 km may be insufficient to capture the entire TC circulation.

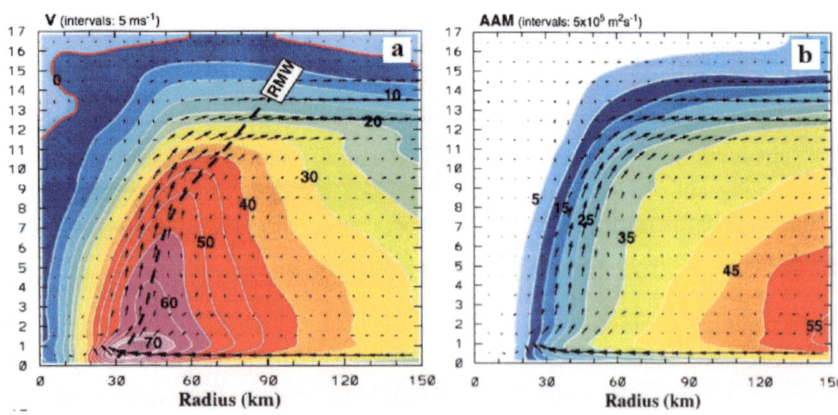

Fig. 1. Radius–height cross section of the (a) temporally (1-hr) and azimuthally averaged tangential winds (V) at intervals of 5 m s^{-1}, and (b) absolute angular momentum (AAM, solid) at intervals of 5×10^5 m^2 s^{-1} , superposed with the in-plane azimuthally averaged flow vectors (U, W), taken from the 56–57-h integration (Hurricane Andrew, 2000–2100 UTC 23 Aug 1992). The radius of maximum wind (RMW) is denoted by the heavy dashed line. Adapted from Zhang et al. (2001).

An important consideration in some cases is that the centre of the TC circulation may not occur at the same horizontal location at each vertical level. For example, the vertical axis of the circulation may be tilted or discontinuous; in such a case, the assumption of an upright TC may be invalid. In essence, one must use caution and carefully consider each case independently. The analysis procedures presented above yield the best results for mature TCs with approximately upright circulations.

Because of the too limited space, in this review we focus primarily on the symmetric structures, i.e., those that emerge after azimuthal averaging. The symmetric structures and associated budget calculations could provide insight into the most fundamental processes in determining the intensity and intensity changes of TCs. However, higher-wave number asymmetric structures that remain after subtracting the azimuthal mean from the total field likely play important roles in the time evolution of important TC processes. To understand the processes such as shear-induced circulations, localized deep convective processes, one must consider the contributions of asymmetric structures. Much of this research is ongoing.

Of course, the symmetric and asymmetric processes are related. For example, asymmetric deep convective processes may serve to intensify the primary TC circulation, thus projecting onto the symmetric circulation. Although not reviewed herein, the relationship between the symmetric and asymmetric processes must be considered to gain a full understanding of TCs. However, the decomposition of fields into symmetric and asymmetric components eases the study of the complex TC phenomenon.

3. Horizontal Momentum Budgets

Let us start from the horizontal momentum and absolute angular momentum (M) budgets, including some relevant work on the topic. The formulation below follows Zhang et al. (2001), and the results shown are from the cloud-permitting MM5 simulation of Hurricane Andrew (1992) with the finest grid resolution of 6 km. As in the previous section, the formulation is in the cylindrical coordinate system, with the origin at the location of minimum surface pressure.

The total change in radial (U) and azimuthal (V) momentum may be written as

$$\frac{dU}{dt} = -\frac{1}{\rho}\frac{\partial p}{\partial r} + \frac{V^2}{r} + fV + 2\Omega\cos\phi\, W\cos\lambda + U_D \tag{7}$$

$$\frac{dV}{dt} = -\frac{1}{\rho}\frac{\partial p}{\partial \lambda} - \frac{UV^2}{r} - fU + 2\Omega\cos\phi\, W\sin\lambda + V_D \tag{8}$$

where $$\frac{d}{dt} = \frac{\delta}{\delta t} + \frac{U'\partial}{\partial r} + \frac{V'}{r}\frac{\partial}{\partial \lambda} + W\frac{\partial}{\partial z},$$

and where U, V and W are the earth-relative radial, azimuthal and vertical winds. Storm-relative wind components are denoted as primed quantities. ϕ is latitude, and Ω is the angular velocity of Earth. From left-to-right, the forces responsible for horizontal accelerations on the RHS of Eqs (7) and (8) are the pressure gradient force, the centrifugal force, the Coriolis force, and frictional effects (including both boundary layer and numerical diffusion).

Because of its near-conservative property, it is desirable to develop a budget equation for M. So, we may rewrite the azimuthal momentum equation as a prognostic equation for M:

$$\frac{dM}{dt} = -\frac{1}{\rho}\frac{\partial p}{\partial \lambda} + 2r\,\Omega \cos\phi\, W \sin\lambda + v_m \beta \frac{r^2}{2} + V_D, \tag{9}$$

where

$$\mathrm{M} = r\left(V + \frac{fr}{2}\right), \tag{10}$$

and wherein Eq. (9), β is the latitudinal variation of the Coriolis parameter, and v_m denotes the projections of U and V onto the longitudinal axes. Terms on the RHS of Eq. (9) demonstrate that the total change in M is associated with pressure, Coriolis, beta, and frictional torques, respectively.

The azimuthal and temporal mean M is given in Fig. 1b, showing that M decreases radially inward and exhibits a large vertical gradient in the MBL due to surface friction. M increases in the upper-level outflow layer due to vertical and radial advection. Analysis of the M budget by Zhang et al. (2001) indicates that the pressure torque dominates the RHS of Eq. (9) above the MBL, where M is approximately conserved in a quasi-axisymmetric TC. Tangential winds in the TC core spin up as a result of both radially inward M transport in the MBL and upward transport of M by intense updrafts. The inward M transport in the MBL is the primary source of TC spin-up. Zhang et al. found that the system-averaged amplification depends on the excess of upward M transport over that of radial transport in the eyewall. During the spin-up stage, the upward transport tends to exceed the spindown associated with the upper-level outflow. Later, using an azimuthal momentum budget, Yau et al. (2004) diagnosed the formation of localized wind streaks in the eyewall with the simulated Andrew data but at the finest grid size of 2 km. Results reveal that small-scale Lagrangian acceleration/deceleration of the azimuthal flow near the surface is closely related to the radial M advection per unit radius from the outer region, while above the MBL it is related to the vertical advection of the azimuthal wind, in addition to the radial M advection per unit radius.

Zhang et al. (2001) also investigated the relative degree of gradient balance in the inner core of the TC using the radial momentum budget (Eq. 7). Results indicate that radial inflow in the MBL transports more M into the eyewall than the frictional dissipation. As the inflow approaches the RMW, the inflow decelerates because the centrifugal force exceeds the radial pressure gradient force (cf. Figs 2a and 2b). Because the inflow must then rise,

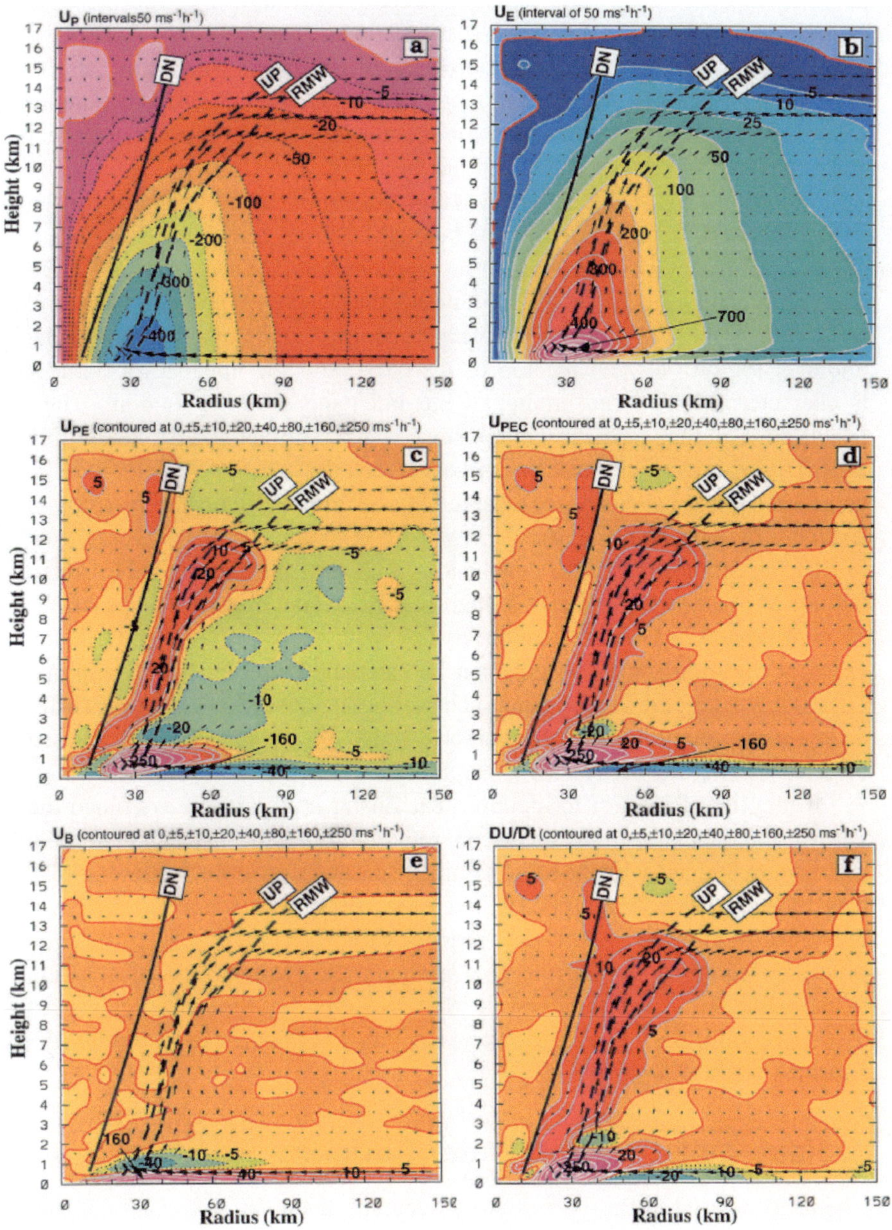

Fig. 2. As in Fig. 1, but for the radial momentum budget (unit: m s^{-1} h^{-1}): (a) the radial pressure gradient force (PGF$_R$, U_P), (b) the centrifugal force (U_E), (c) the cyclostrophic force imbalance ($U_{PE} = U_P + U_E$), (d) the gradient force imbalance ($U_{PEC} = U_{PE} + U_C$), (e) the diffusion and boundary layer effects (U_B), and (f) the Lagrangian tendency (dU/dt). Adapted from Zhang et al. (2001).

updrafts transport M upward, which increases the local centrifugal force and causes supergradient acceleration radially outward. This outflow results in the development of a supergradient outflow jet at the top of the MBL. Despite these effects, the local changes in tangential winds are small due to intense advection in the eyewall. Various components of the radial momentum budget and associated force imbalances are shown in Fig. 2.

Yang et al. (2011) compared the M and radial momentum budgets between an over-ocean TC and a landfalling simulated TC. They found that the gradient wind imbalance associated with the supergradient outflow jet above the MBL was significantly enhanced during landfall due to the enhanced mixing and friction over land. They also found that in contrast to the over-ocean TC, the midlevel region between the top of the MBL and the upper-level outflow is characterized by large gradient wind imbalances during landfall. Their results suggest that gradient wind balance is generally invalid for landfalling TCs. This is a stark contrast to over-ocean steady-state TCs, which may be considered quasi-balanced (see Chapter 24 on *Balanced Dynamics in Tropical Cyclones* by Kieu and Zhang).

Of importance is that the conclusions obtained above are possible due to the use of dynamically consistent high-resolution model data. The processes so investigated are virtually unobservable with today's means, but all the terms in the momentum budgets can be calculated explicitly with minimal uncertainty using the model data. Such model diagnostic budget studies offer an important compliment to observations, which may be subsequently focused on areas of interest, such as the supergradient outflow jet.

It may be of interest to mention that Zhang et al. (2000) examined the vertical force balance in the inner-core region through the analysis of vertical momentum budgets with the same simulated data of Hurricane Andrew (1992). The divergence equation is used to invert the three-dimensional buoyancy and dynamically induced perturbation pressures in order to gain insight into the processes leading to the subsidence warming in the eye and the vertical lifting in the outer portion of the eyewall. Results show that the centrifugal force associated with the eyewall rotation plays an important role in forcing the subsidence warming in the eye.

4. Heat, Moisture and Water Budgets

The heat released in the TC primarily occurs in deep convective clouds that comprise a relatively small portion of the TC circulation and are thus difficult to comprehensively observe. The microphysical and thermodynamic processes ongoing within convection are undoubtedly critical to TC evolution. It is well known that despite intense latent heat release (LHR) in the TC eyewall, little temperature increase is observed in the eyewall. On the other hand, the eye of the TC, in which virtually no deep convection occurs, is the warmest region of the warm-core TC.

As discussed in the Introduction, the early observational budget studies relied on the residual term to diagnose diabatic heating and thus could not address these complexities of the TC inner core. However, contemporary model simulations that include explicit microphysics open the door to more thorough thermodynamic budget studies, which can be valuable tools in understanding the development and evolution of such TC characteristics.

In this section, we focus on thermodynamic budgets calculated by Zhang et al. (2002) using the numerical simulation of Hurricane Andrew (1992). We also provide a review of the water budgets of mature TCs by Braun et al. (2006) and Yang et al. (2011). It is well known that much latent heat is acquired from the warm ocean and transported inward toward the TC core, and that great amounts of heat are released in convective clouds in the eyewall and rainbands. With explicit simulations, it is possible to directly calculate the tendencies of water vapour (q_v), potential temperature (θ), and equivalent potential temperature (θ_e), which are given below, respectively,

$$\frac{dq_v}{dt} = -(q_{CON} - q_{CEV} - q_{REV} + q_{DEP} - q_{SUB}) + q_{BLD}, \tag{11}$$

$$\begin{aligned}\frac{d\theta}{dt} = {} & \frac{L_v}{C_p}\pi(q_{CON} - q_{CEV} - q_{REV}) + \frac{L_f}{C_p}\pi(q_{FRZ} - q_{MLT}) \\ & + \frac{L_s}{C_p}\pi(q_{DEP} - q_{SUB}) + \theta_{RAD} + \theta_{BLD},\end{aligned} \tag{12}$$

$$\frac{d\theta_e}{dt} = \left[\frac{L_f}{C_p}\pi(q_{FRZ} - q_{MLT} + q_{DEP} - q_{SUB})\right] \times \exp\left(\frac{L_v q}{C_p T_L}\pi\right) + \theta_{eRAD} + \theta_{eBLD}, \tag{13}$$

where the terms on the RHS of Eqs. (11)-(13) represent the tendencies associated with condensation (CON), cloud evaporation (CEV), rain evaporation (REV), deposition (DEP), sublimation (SUB), freezing (FRZ), melting (MLT), MBL and diffusive effects (BLD), and radiative effects (RAD). L_c, L_f and L_s are the latent heat of condensation, fusion and sublimation, respectively, and

$$\pi = \left(\frac{p_0}{p}\right)^{R/C_p}, \text{ and}$$

$$\frac{d}{dt} = \frac{\partial}{\partial t} + U\frac{\partial}{\partial r} + \frac{V}{r}\frac{\partial}{\partial \lambda} + W\frac{\partial}{\partial z},$$

where the horizontal velocity components are TC-relative. All terms in Eqs (11)-(13) are obtained directly from the model output. Details on each physical process are available in Liu et al. (1997) and Tao and Simpson (1973). Alternative formulations of the budget equations are discussed below.

The budget terms in the heat and moisture equations may be calculated individually, but Zhang et al. found it more convenient to group the terms by physical process. For example, all terms associated with moist convection were grouped as "cloud" processes to simplify the analysis. Such a simplification is often convenient when conducting budget studies of complex processes.

The dominant heat budget terms in the TC core are LHR in the eyewall, cooling associated with sublimation and evaporation at the eye-eyewall interface, and the heating via fluxes from the underlying ocean. The LHR in the eyewall, although dominated by warm rain, is shifted slightly vertically upward due to depositional and freezing heating in the mid-tropospheric eyewall. The θ advection also plays an important role, particularly in the eyewall where adiabatic cooling almost entirely counteracts the intense LHR heat source. This near-balance explains why little heating is observed in the eyewall despite the large heat source. Vertical advection actually leads to a net cooling in the eyewall, while the eye achieves a net warming due to downward advection and adiabatic warming. The resulting thermal gradient between the eye and eyewall is gradually reduced by thermal advection across the eye-eyewall interface.

The moisture budget terms are intimately related to the heat budget terms, with opposite signs. The only major difference is the source of latent heat at the warm ocean surface. It is found that θ_e is conserved above the MBL due to the slantwise orientation of the eyewall. The gradient at the eye-eyewall interface is constantly increased as the TC intensifies due to downward advection of θ_e in the eye and upward transport of θ_e by updrafts. In some sense, the TC inner core is thermodynamically maintained by a combination of latent heat flux from the ocean surface, the downward advection of high-θ_e air in the upper troposphere, and the outward advection in the outflow regions.

Although the heat and moisture budget results are not shown thoroughly herein, Fig. 3 demonstrates several different ways to present budget results graphically. In addition to the height-radius diagrams presented in the sections above, there are many alternatives. For example, Zhang et al. (2002) take a height-azimuth cross section in the eyewall region (Fig. 3a) to show the vertical structure of net diabatic heating within the eyewall. The fields are not only temporally averaged to show characteristic structures, but also the height-azimuth plane slants outward with height to account for the slanting eyewall.

Horizontal cross sections near the height of peak vertical motion are instructive to show the contrast between intense heating in the eyewall and latent cooling at the inner edge of the eyewall (Fig. 3b). The vertical level is chosen as 8 km, i.e., approximately the level of peak vertical motion. Additionally, one may construct area-averaged vertical profiles of budget terms to show the level of peak heating (Fig. 3c). By plotting a variety of carefully constructed figures, one may obtain a clearer understanding of the processes ongoing in the TC inner core.

The ultimate energy source for the TC is water vapour. Some of the water vapour is acquired via evaporation from the warm ocean near the inner core. However, most of the vapour is transported toward the inner core from outer

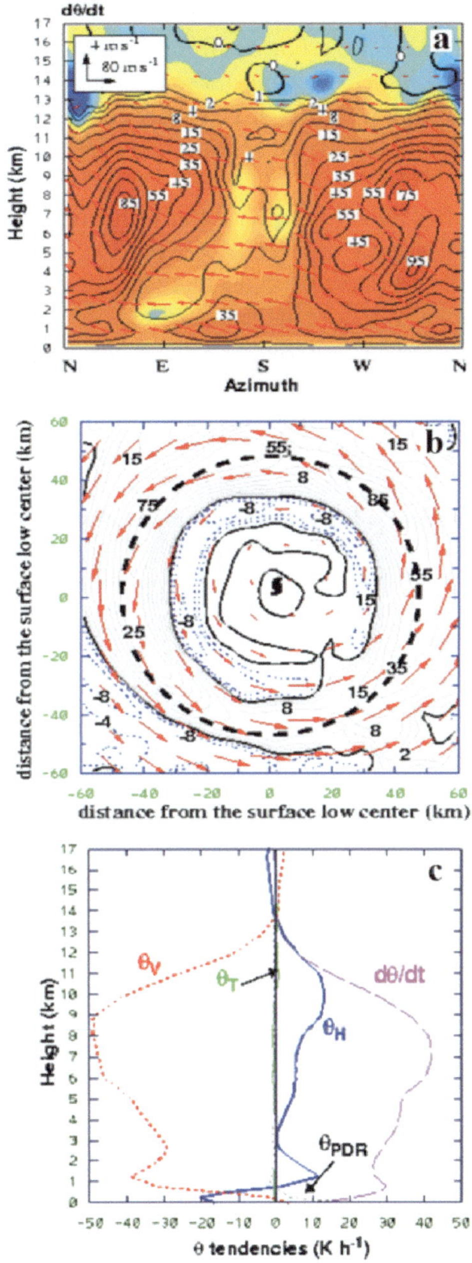

Fig. 3. Net diabatic heating rate ($d\theta/dt$, K hr^{-1}) plotted as a (a) height-azimuth cross section in the outward-slanting eyewall, (b) horizontal cross section near the location of peak vertical motion, and (c) area-averaged vertical profile in the inner core. In (a) and (b), the in-plane flow vectors are superimposed. In (c) the profiles of local tendency (θ_T), horizontal (θ_H) and vertical (θ_V) advection, and MBL diffusion and radiative effects (θ_{PDR}) are also shown. Adapted from Zhang et al. (2002).

regions of the TC circulation. Braun (2006) found that 96% of the water vapour consumed by the TC originates from outside the inner core.

Of course, in addition to heat and moisture, water is also a critical variable involved in the development of TCs. The water budget formulation, including budget equations for cloud (liquid and ice) and precipitation (rain, snow and graupel), is presented by Braun (2006) and Yang et al. (2011). Braun (2006) presents the budget equations in flux form, compared to the advective form presented above. The only difference between the two formulations is that advective processes are captured in the total derivative in the advective form and as the total flux divergence in the flux form.

Condensation in deep updrafts consumes water vapour and creates cloud droplets, the concentration of which is maximized in the melting layer (Fig. 4c). The cloud droplets are advected upward by deep eyewall updrafts and coalesced into raindrops above the melting layer. Raindrops are formed and precipitated primarily within the inner core and in outer stratiform bands. Braun (2006) found that convective processes dominate in the inner core,

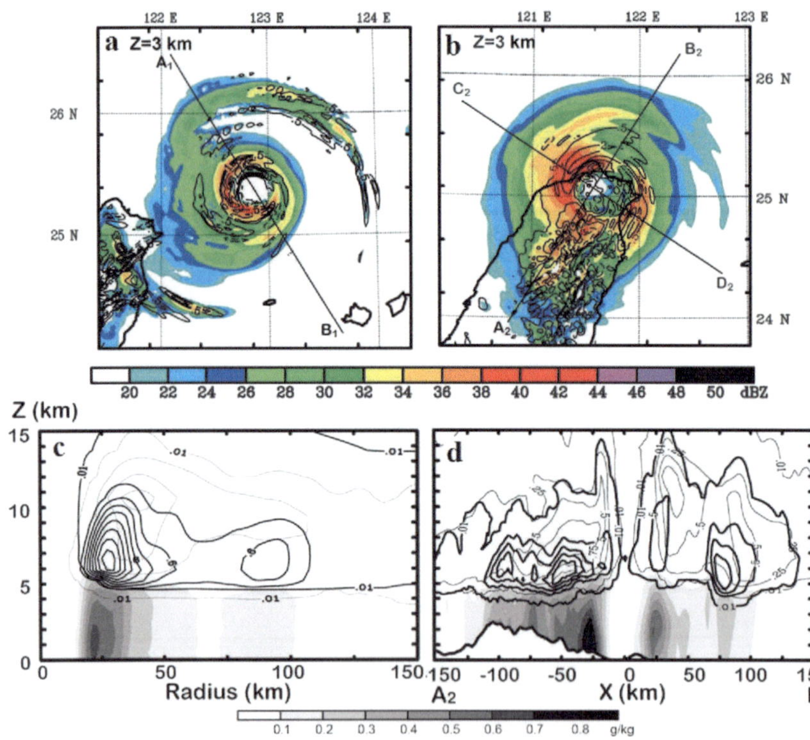

Fig. 4. Adapted from Yang et al. (2011), in (a) and (b) simulated radar reflectivities at $Z = 3$ km (shaded) and vertical velocities (contoured) are shown before and during landfall. In (c), the azimuthally averaged rain (shaded), graupel (thin contours), and snow (thin contours) are shown at the same time as (a). In (d), a southwest-northeast vertical cross section, as in (c) but during the landfall stage.

while stratiform processes dominate in the outer regions. Most of the water generated within the inner core is consumed and precipitated, and only a small amount of snow is transported by upper-level outflow to the outer regions of the TC circulation.

Yang et al. (2011) investigated the differences in the condensate species between the over-ocean stage (Figs 4a, c) and the landfalling stage (Figs 4b, d). They found that the increased frictional convergence over land leads to more than 20% increased storm-total condensate compared to the over-ocean stage. The enhanced secondary circulation is associated with enhanced vapour transport in the low levels and an increase in snow and ice transport outward in the upper levels. In agreement with Braun (2006), Yang et al. found that precipitation in the inner core falls out as quickly as it is produced.

5. Energy Budgets

The TC is a highly energetic phenomenon that acquires much of its power from latent heat extracted from the warm underlying ocean surface. Energy budget studies can provide insight into the processes by which great amounts of kinetic energy (KE) are concentrated in the TC inner core. These processes can be understood by analyzing energy conversions, such as those associated with LHR.

Several observational studies have conducted energy budget studies of TCs using available observations, e.g., Frank (1977) and McBride (1981), but similar to the budget studies of heat and moisture discussed in the Introduction, data limitations prevented the investigation of inner-core processes. Early modelling studies of TC energetics with parameterized convection, such as that by Tuleya and Kurihara (1975), leverage the dynamically consistent dataset from those in which the budget terms are explicitly calculated.

In this section, we review the recent work by Hogsett and Zhang (2009), who presented a framework for performing energy budget studies using high-resolution numerical model output of Hurricane Bonnie (1998) that was strongly sheared by the synoptic flow. The intensity evolution of the simulated Bonnie is shown in Fig. 5a. Features of interest associated with this TC include an eyewall replacement cycle (ERC) and notable fluctuations in maximum wind speed. A complete description of the storm and model configuration is provided in Zhu et al. (2004).

Although it is outside the scope of this review, of interest is that the large and high-frequency intensity fluctuations in Fig. 5a are associated with physical processes within the inner core. Hogsett and Zhang (2009) found that they are closely related to the growth and decay of low-wavenumber deep convective features that propagate around the eyewall region. These asymmetric features are generated by vertical shear and likely play some roles in the general intensification of the TC. These resulting localized intensity fluctuations project onto the symmetric circulation, effectively intensifying the

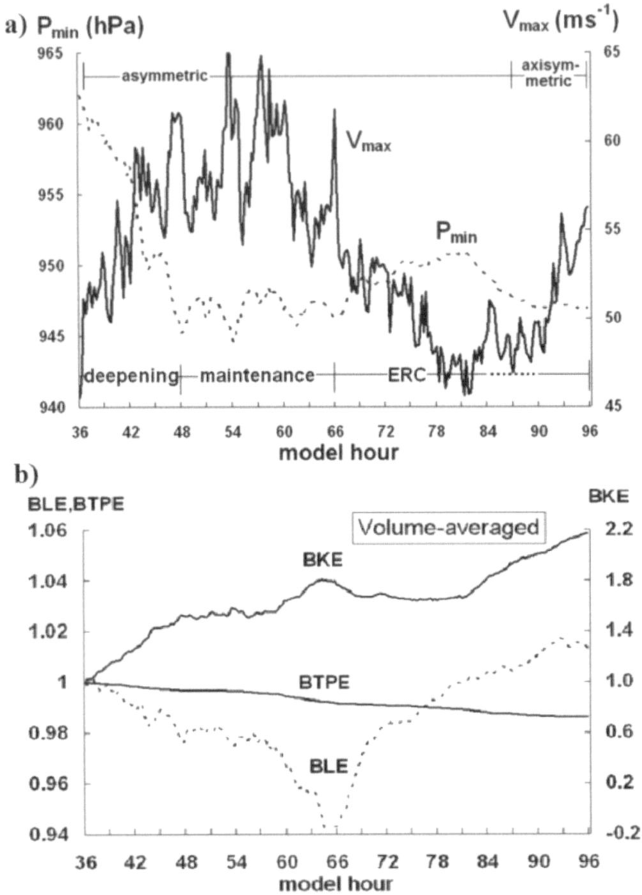

Fig. 5. (a) Time series (at 15-min intervals) of the sea-level minimum central pressure (dashed; hPa) and the maximum surface winds (solid; m s^{-1}) from the 36–96-h simulation. The three characteristic stages of deepening, maintenance and eyewall replacement (ERC) are delineated (bottom line), as well as the asymmetric and axisymmetric stages (top line). The dotted segment of the ERC stage (84–90 h) denotes the period in which the most rapid transition to axisymmetry occurs. (b) Time series (at 15-min intervals) of the volume-averaged (a) BKE (right ordinate), BLE and BTPE (left ordinate) following the storm. Each time series has been normalized such that each 0.01 represents a 1% change from the initial value of the energy, that is, 1.28 $\times 10^2$ (BKE), 7.89 $\times 10^3$ (BLE), and 9.92 $\times 10^4$ J m^{-3} (BTPE). Adapted from Hogsett and Zhang (2009).

mean flow. To capture such features, it is important to acquire high temporal frequency model output when conducting budget studies. Otherwise, such processes may be inadequately captured.

Total energy is conserved in the atmosphere, and atmospheric processes simply represent conversions between the various forms of energy. Here we present the three forms of energy:

$$\mathrm{LE} = \rho L_v q_v, \tag{14}$$

$$\mathrm{TPE} = \rho(c_v T + gz) \text{ and} \tag{15}$$

$$\mathrm{KE} = \rho \frac{\mathbf{V}_h \cdot \mathbf{V}_h}{2}, \tag{16}$$

where LE, TPE and KE refer to latent, total potential and kinetic energies, respectively. Note that the vertical velocity is disregarded in calculating KE because it is typically an order of magnitude smaller than horizontal velocity in the TC core and its exclusion simplifies the energy conversion analysis below.

Using the methodology described in section 2 above, the scalar energy quantities may be volume-averaged or storm-scale integrated by treating the simulated 3D TC as a control volume in a quasi-Lagrangian framework, in order to examine the time rate of changes in energetics at the storm scale. Figure 5b shows the time evolution of the volume-integrated energies, denoted as "bulk" energies, e.g., bulk KE or BKE. During the initial deepening stage, the BKE increases and BLE (i.e., bulk LE) decreases, corresponding to an intensifying TC that is consuming environmental LE via LHR. However, as an ERC commences, the BKE temporarily decreases and BLE rapidly increases.

The budget equations for total change in the various energies can be derived by taking the total derivative of Eqs (14)-(16), yielding with some manipulation:

$$\frac{d}{dt} LE = -\rho L_v (Q_{con} - Q_{ev}) + LHF + LE \frac{d \ln \rho}{dt}, \tag{17}$$

$$\frac{d}{dt} TPE = -\nabla \cdot (\rho V) + \mathbf{V}_h \cdot \nabla p' + \rho L_v (Q_{con} - Q_{ev}) + SHF + TPE \frac{d \ln \rho}{dt}, \tag{18}$$

$$\frac{d}{dt} KE = -\mathbf{V}_h \cdot \nabla p' - \mathbf{V}_h \cdot \mathrm{F} + KE \frac{d \ln \rho}{dt}, \tag{19}$$

where the terms on the RHS represent sources and sinks contributing to the total energy change. Of interest is that several terms are shared between the tendency equations with opposite signs. These terms are referred to as conversion terms and represent physical processes by which energy is transferred between different forms.

The equations for the total change in each energy are useful for analyzing the structures and tendency locally, but for acquiring an understanding of the energy balance of the TC system as a whole, it is convenient to construct volume-averaged energy budgets. To do so, we use the formulation by Dutton (1973) in natural coordinates,

$$\frac{d}{dt}\left[\int_{V(\tau)} \alpha dV\right] = \int_{V(\tau)} \frac{\partial \alpha}{\partial t} dV + \int_S \alpha U \cdot ndS$$
$$= \int_{V(\tau)} E_S dV + \int_S \alpha(U - V) \cdot ndS \tag{20}$$

where U represents the horizontal translational velocity of the TC, α represents any scalar energy, E_S is the energy tendency term, and the last term on the RHS represents the energy flux divergence. When (20) is combined with Eqs (17)-(19), some manipulation yields the volume-integrated forms of the energy equations. As an example, we focus here on only the volume-integrated KE and LE equations:

$$\frac{d}{dt} BKE = -\int V_h \cdot \nabla \mathrm{p}' dV - \int V_h \cdot \mathrm{F} dV - \int \nabla \cdot \mathrm{KE(V - U)\,dV}, \tag{21}$$

$$\frac{d}{dt} BLE = -\int \rho L_v \,(Q_{con} - Q_{ev})\, dV - \int LHF\; dV - \int \nabla \cdot \mathrm{LE(V - U)\,dV} \tag{22}$$

where the processes on the RHS of Eq. (21) contributing to the total change in BKE are cross-isobaric horizontal flow (KGEN), frictional dissipation (KDIS), and KE flux at the boundaries of the control volume (KFLX), respectively. The processes contributing to the BLE changes are associated with heating due to phase changes (LHR), latent heat flux from the ocean surface (LHF), and LE flux occurring at the lateral boundaries of the control volume (LFLX), respectively.

The above bulk form of budgets may also be applied to other quantities, such as the storm-scale mass, cloud hydrometeors, vorticity, and potential vorticity (PV). For example, Kieu and Zhang (2009) used a volume-integrated PV budget to examine the genesis of Tropical Storm Eugene (2005) due to the merging of vortices, and the PV flux in the Intertropical Convergence Zone (ITCZ). They found the doubling of the bulk PV during the genesis as a result of the internal dynamical forcing and the continuous PV supply from the ITCZ. Without the latter, Eugene would be shorter-lived after the merger under the influence of intense vertical shear and colder sea-surface temperature. Then, they showed that the flux form of the vorticity budget equation by Haynes and McIntyre (1987) could provide better insight into the vorticity generation processes than that is possible using the classical vorticity equation. Through the bulk vorticity budget calculations, Kieu and Zhang (2009) showed that unlike the previously hypothesized downward growth associated with merging vortices, the most rapid vorticity growth rate is found in the bottom layers of the merger due to the frictional convergence, thereby confirming the previous finding that tropical cyclogenesis tends to occur from the bottom upward (Zhang and Bao, 1996; Hendricks et al., 2004; Montgomery et al., 2006).

Figure 6 shows the evolution of the BKE and BLE budgets. The tendencies (KTEN and LTEN) are explicitly calculated from the time series in Fig. 5.

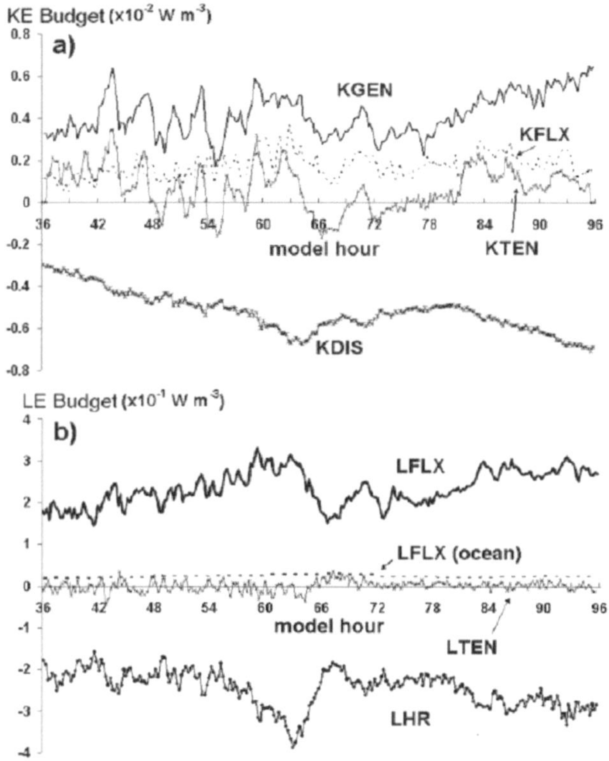

Fig. 6. Time series (at 15-min intervals) of the volume-averaged (a) BKE and (b) BLE tendency (both in W m^{-3}) and their budgets, following the storm. Adapted from Hogsett and Zhang (2009).

Comparison of the RHS residual and the explicit tendencies (not shown) reveals that the budget terms are sufficient to capture the changes in the respective bulk energies. In other words, the RHS and LHS of the equations are calculated independently and shown to be similar. Demonstrating the independence between the RHS and LHS is an important aspect of budget studies. Only after seeing little differences between the LHS and RHS of a budget equation, one can be assured about the quality of a budget analysis.

The BKE balance (Fig. 6a) is largely between the source of KE that occurs due to cross-isobaric flow in the MBL and the sink of KE occurring due to frictional processes in the MBL. Of note, however, is the net inward transport (i.e. positive tendency) of KE occurring at the lateral boundaries. Without this inward transport, KTEN would be negative, suggesting TC intensification may not occur. In addition, it is relevant to note that significant fluctuations occur only in the KGEN time series, which suggests that the high-frequency fluctuations in the TC intensity (Fig. 5a) are associated with the KGEN term. Conversely, the KDIS term shows very little fluctuation, likely because dissipation is associated with the mean TC vortex, while KGEN may be locally enhanced by transient convective processes.

Similar to the BKE budget, the BLE balance (Fig. 6b) occurs largely between two terms: the source of BLE via fluxes at the lateral boundaries of the control volume, and the sink of BLE occurring via diabatic heating. A very interesting characteristic of the BLE budget is that the LHF occurring at the ocean surface, while always positive, is an order of magnitude smaller than LFLX. This indicates that the dominant source of BLE originates from outside the control volume. As mentioned in the prior section, Braun (2006) reached a similar conclusion regarding water vapour transport in his water budget study. However, without the small LHF, the BLE tendency would likely be negative. This suggests that LHF serves as a very important source of moisture to counteract the rapid consumption of BLE in deep convective processes.

Although bulk energy budgets provide insight into the TC system as a whole, it is also possible to investigate the spatial structure of budget terms. Of particular value of such budget analysis is the inner-core convective processes that presumably impact TC intensity but are often unobservable. Hogsett and Zhang (2009) provide details on this analysis, as well as the use of Fourier decomposition to separate mean and eddy contributions of each energy.

6. Concluding Remarks

In this chapter, we have reviewed a limited number of recent budget studies in TC research, primarily using cloud-permitting models. Many budget studies have been performed in the past several decades. The earliest studies utilized sparse observational data and thus required many approximations and were subject to various sources of uncertainty. In the current era of high-resolution, and now cloud-permitting TC simulations, many of the constraints that limited prior studies have been lifted. Budget terms are now calculated directly from the model output data, eliminating the need to depend on residual terms to close the budget.

The use of numerical simulations to conduct budget studies has facilitated a better understanding of different processes taking place in TCs. These may include the quasi-balanced dynamics, vertical force balance in the eyewall and eye, absolute angular momentum in the MBL and layers above, the supergradient outflow jet, the inner-core energetics and thermodynamics, the magnitude and distribution of diabatic heating in the eyewall, the different distributions of hydrometeors during different development stages, and different magnitudes of the processes between the over-ocean and the landfalling stages of TCs. Many of the above processes and structures could not be obtained by any remote-sensing and in situ instruments of today. It should be mentioned, though, that the results presented herein are generally from budget studies of intense mature TCs. Thus, some caution should be taken when studying weaker or atypical TCs. However, volume-integrated, or "bulk" budgets may be used to understand the dynamics of less organized TCs, including those during their genesis stage.

While there are many advantages to conduct budget studies using high-resolution, dynamically consistent model output from the simulated TCs, it is important to mention limitations of the model budget studies. That is, the results so obtained are sensitive to model errors. In particular, the cloud microphysical schemes in cloud-permitting simulations may not adequately represent the full spectrum of microphysical species and the time evolution of conversions among species. As an example, these models tend to produce the radar reflectivity field that is higher than the observed (Liu et al., 1997; Zhu et al., 2004; Braun, 2006). Accordingly, the LHR associated with phase changes, the vertical motion fields, the rainfall, the associated energy conversions etc., are subject to uncertainty. The existence of such error sources emphasizes the importance of verifying the model simulation with all available observations (e.g., intensity and intensity changes, size, and structure) prior to performing detailed budget analyses. Other areas of large uncertainty include the MBL, air-sea interaction including dissipative heating, cloud-radiation interaction, in-cloud turbulence mixing, and the processes above, in and below the upper-level outflow layer (Zhang and Altshuler, 1999; Zhang, 2010; Zhang et al., 2011; Gopalakrishnan et al., 2012).

Despite the above-mentioned limitations, the use of budget studies has led to a more comprehensive understanding of TC dynamics and energetics. Continued effort in these research areas is clearly needed to clarify some of the remaining uncertainties of TC dynamics and energetics. In addition, eyewall replacement cycles, tropical cyclogenesis, rapid intensity changes, interaction between asymmetric and symmetric components of TC vortices and the other phenomena, are features or issues that have not yet been solved unequivocally. Budget studies have the potential to address some of these issues so that TC intensity and intensity changes could be better understood and may be more accurately predicted in the future.

REFERENCES

Anthes, R.A., 1982: Tropical Cyclones—Their Evolution, Structure, and Effects. Meteor. Monogr., No. 41. Amer. Meteor. Soc.

Braun, S.A., 2006: High-resolution simulation of Hurricane Bonnie (1998). Part II: Water budget. *J. Atmos. Sci.*, **63**, 43-64.

Dutton, J.A., 1976: The Ceaseless Wind. McGraw-Hill.

Frank, W.M., 1977: The structure and energetics of the tropical cyclone II. Dynamics and energetics. *Mon. Wea. Rev.*, **105**, 1136-1150.

Gopalakrishnan, S.G., S. Goldenberg et al., 2012: Toward Improving High-Resolution Numerical Hurricane Forecasting: Influence of Model Horizontal Grid Resolution, Initialization, and Physics. *Wea. Forecasting*, **27**, 647-666.

Hawkins, H.F. and D.T. Rubsam, 1968: Hurricane Hilda, 1964. *Mon. Wea. Rev.*, **96**, 617-636.

Hendricks, E.A., M.T. Montgomery and C.A. Davis, 2004: The role of "vortical" hot towers in the formation of tropical cyclone Diana (1984). *J. Atmos. Sci.*, **61**, 1209-1232.

Haynes, P. and M. McIntyre, 1987: On the evolution of vorticity and potential vorticity in the presence of diabatic heating and frictional or other forces. *J. Atmos. Sci.*, **44**, 828-841.

Hogsett, W. and D.-L. Zhang, 2009: Numerical simulation of Hurricane Bonnie (1998). Part III: Energetics. *J. Atmos. Sci.*, **66**, 2678-2696.

Johnson, R.H., 1984: Partitioning tropical heat and moisture budgets into cumulus and mesoscale components: Implications for cumulus parameterization. *Mon. Wea. Rev.*, **112**, 1590-1601.

Kieu, C.Q. and D.-L. Zhang, 2009: Genesis of Tropical Storm Eugene (2005) from merging vortices associated with ITCZ breakdowns. Part II: Roles of vortex merger and ambient potential vorticity. *J. Atmos. Sci.*, **66**, 1980-1996.

Liu, Y., D.-L. Zhang and M.K. Yau, 1997: A multiscale numerical study of Hurricane Andrew (1992). Part I: Explicit simulation and verification. *Mon. Wea. Rev.*, **125**, 3073-3093.

Liu, Y., D.-L. Zhang and M.K. Yau, 1999: A Multiscale Numerical Study of Hurricane Andrew (1992). Part II: Kinematics and Inner-Core Structures. *Mon. Wea. Rev.*, **127**, 2597-2616.

Marks, F.D. and R.A. Houze Jr., 1987: Inner-core structure of Hurricane Alicia from airborne Doppler-radar observations. *J. Atmos. Sci.*, **44**, 1296-1317.

Marks, F.D., R.A. Houze Jr. and J.F. Gamache, 1992: Dual-aircraft investigation of the inner core of Hurricane Norbert. Part I: Kinematic structure. *J. Atmos. Sci.*, **49**, 919-942.

Mapes, B.E. and R.A. Houze Jr., 1995: Diabatic divergence profiles in western Pacific mesoscale convective systems. *J. Atmos. Sci.*, **52**, 1807-1828.

McBride, J.L., 1981: Observational analysis of tropical cyclone formation. Part III: Budget analysis. *J. Atmos. Sci.*, **38**, 1152-1166.

Montgomery, M.T., M.E. Nicholls, T.A. Cram and A.B. Saunders, 2006: A vortical hot tower route to tropical cyclogenesis. *J. Atmos. Sci.*, **63**, 355-386.

Tao, W-K., S. Lang, M. McCumber, R. Adler and R. Penc, 1990: An algorithm to estimate the heating budget from vertical hydrometeor profiles. *J. Appl. Meteor.*, **29**, 1232-1244.

Tuleya, R.E. and Y. Kurihara, 1975: The energy and angular momentum budgets of a three-dimensional tropical cyclone model. *J. Atmos. Sci.*, **32**, 287-301.

Yanai, M., S. Esbensen and J.-H. Chu, 1973: Determination of Bulk Properties of Tropical Cloud Clusters from Large-Scale Heat and Moisture Budgets. *J. Atmos. Sci.*, **30**, 611-627.

Yang, M.-J., T.-C. Chen Wang, Y. Zhang and C.-Y. Weng, 2011: Momentum budget evolution of Typhoon Nari (2001) during the landfall process. *Terr., Atmos., and Oceanic Sci.*, **22**, 595-612.

Yang, M.-J., S.A. Braun and D.-S. Chen, 2011: Water budget of Typhoon Nari (2001). *Mon. Wea. Rev.*, **139**, 3809-3828.

Yau, M.K., Y. Liu, D.-L. Zhang and Y. Chen, 2004: A multiscale numerical study of Hurricane Andrew (1992). Part VI: Small-scale inner-core structures and wind streaks. *Mon. Wea. Rev.*, **132**, 1410-1433.

Zhang, D.-L. and N. Bao, 1996: Oceanic cyclogenesis as induced by a mesoscale convective system moving offshore. Part II: Genesis and thermodynamic transformation. *Mon. Wea. Rev.*, **124**, 2206-2226.

Zhang, D.-L. and E. Altshuler, 1999: The effects of dissipative heating on hurricane intensity. *Mon. Wea. Rev.*, **127**, 3032-3038.

Zhang, D.-L., Y. Liu and M.K. Yau, 2000: A multiscale numerical study of Hurricane Andrew (1992). Part III: Dynamically-induced vertical motion. *Mon. Wea. Rev.*, **128**, 3772-3788.

Zhang, D.-L., Y. Liu and M.K. Yau, 2001: A multiscale numerical study of Hurricane Andrew (1992). Part IV: Unbalanced flows. *Mon. Wea. Rev.,***129**, 92–107.

Zhang, D.-L., Y. Liu and M.K. Yau, 2002: A multiscale numerical study of Hurricane Andrew (1992). Part V: Inner-core thermodynamics. *Mon. Wea. Rev.*, **130**, 2745-2763.

Zhang, J.A., 2010: Estimation of dissipative heating using low-level in-situ aircraft observations in the hurricane boundary layer. *J. Atmos. Sci.*, **67**, 1853-1862.

Zhang, J.A., F.D. Marks et al., 2011: Estimation of turbulence characteristics in the low-level region of intense Hurricanes allen (1980) and Hugo (1989). *Monthly Weather Review*, **139**, 1447-1462.

Zhu, T., D.-L. Zhang and F. Weng, 2004: Numerical simulation of Hurricane Bonnie (1998). Part I: Eyewall evolution and intensity changes. *Mon. Wea. Rev.*, **132**, 225-241.

27

The Storm Surge Prediction over Bay of Bengal and Arabian Sea: A Review

Sujata Pattanayak[1], U.C. Mohanty[1] and S.K. Dube[2]
[1]SEOCS, Indian Institute of Technology Bhubaneswar, Odisha, India
[2]Amity University Rajasthan, Jaipur, India
sujata05@gmail.com

1. Introduction

Storm surge is an abnormal rise of sea level generated by a cyclone, over and above the predicted astronomical tide. Storm surges associated with severe tropical cyclones are by far the most damaging rather than rain and strong winds. Death and destruction arise directly from the intense winds that are characteristic of tropical cyclones blowing over a large surface of water. If bounded by a shallow basin, these winds cause the sea water to pile up on the coast and lead to sudden inundation and flooding of coastal regions. In addition, flooding of the river deltas occurs from the combined effects of tides and surges from the sea, which penetrate into the rivers, because at the same time, excess water in the rivers due to heavy rains from the cyclone is trying to flow through the rivers into the sea. Almost all the loss of lives and most of the damage from a tropical cyclone are attributable to the storm surge generated by the cyclones. Thus, the real-time monitoring and warning of storm surges is of great concern. It is necessary that the problem of the storm surge be seriously addressed by the countries of the various regions through collective efforts and in an integrated manner.

Most of the countries around the North Indian Ocean are threatened by storm surges associated with approaching severe tropical cyclones. The destruction due to the storm surge flooding is a serious concern along the coastal regions of India, Bangladesh, Myanmar, Pakistan, Sri Lanka and Oman. Storm surges cause heavy loss of lives and property damage to the coastal structures and losses of agriculture which lead to annual economic losses in these countries.

1.1 Storm Surges in the Bay of Bengal and Arabian Sea

Storm surges are extremely serious along the east coast of India, Bangladesh,

Myanmar and Sri Lanka, though Sri Lanka is less affected by the storm surges. The destruction due to storm surge flooding is a major cause of loss of life and property for a land falling TC. The main factors contributing to disastrous storm surges along the east coast of India, Bangladesh and Myanmar are: (1) shallow bathymetry, (2) convergence of the bay, (3) high astronomical tides, (4) low-lying delta region entrenched with a large number of river systems and (5) concentrated coastal population etc. About 90% of the damage due to land falling tropical cyclones is mainly attributed to the coastal inundation with saline sea water (Dube et al., 2009). Intrusion of saline water increases soil salinity and reducing its fertility in coastal areas. The surge height depends on the strength of wind, bathymetry and offshore coastal topographic nature of the landfall point. The storm surges become most devastating if landfall time coincides with the time of high tide. Additionally, the low-lying delta regions such as the Gangetic-delta and Irrawaddy-delta endure severe storm surges in comparison to many other shallow regions along the Bay of Bengal.

Although the frequency of storms and storm surges is less over the Arabian Sea than the Bay of Bengal, major destructive surges can occur occasionally that caused a huge damage to the area of Saurashtra, Gulf of Kutch, Gulf of Cambay, Karachi, Karwar etc. A detailed review of the problem of storm surges in the Bay of Bengal and Arabian Sea is given by Murty (1984), Murty et al. (1986), Das (1994 a, b), Dube et al. (1997) etc.

1.2 Destruction Potential

Though the tropical cyclones occur in many parts of the world, the Bay of Bengal is one of the ideal tropical cyclogenesis area. Statistics show that only about 7% of the global tropical cyclones form over the Bay of Bengal and the Arabian Sea, which is very less in comparison with other basins like northwest Pacific, southwest Pacific, northwest Atlantic, etc. Though the frequency is relatively less over Indian seas, out of 10 recorded deadliest cases with very heavy loss of life (ranging from about 40,000 to well over 300,000) over the world, eight cases were formed in the Bay of Bengal and Arabian Sea in the past 300 years. And the world's highest recorded storm surge of 45 feet occurred over the Bay of Bengal in 1876 due to Bakherganj cyclone near Meghna Estuary, Bangladesh (WMO, 2009). These disasters are mainly due to long, funnel-shaped low-lying coastal region (shallow bathymetry) and dense population. Further, according to Mohanty (1994), the region between 5°N–20°N and 87°E–91°E over the Bay of Bengal is more suitable region for the formation of cyclonic disturbances. Figure 1 demonstrates the classification of the Indian coast as of the vulnerability due to storm surges. It is basically of three types and distinguish as per the water level crossed in the coastline. Mohapatra et al. (2012) classify the marine districts of Indian coast for cyclone hazards and found that the cyclone hazard proneness factor is very high for the districts of Nellore, East Godavari and Krishna in Andhra Pradesh; Yanam in Puducherry; Balasore, Bhadrak, Kendrapada and Jagatsinghpur in

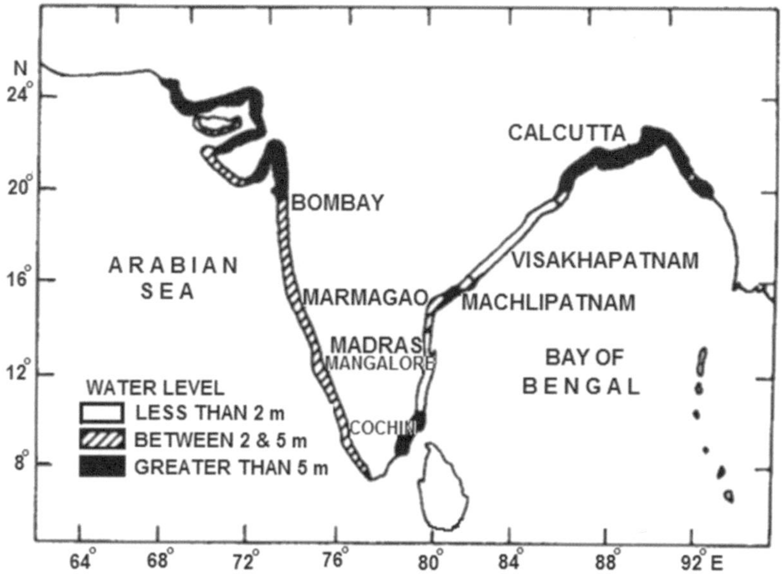

Fig. 1. Classification of the Indian coast into three types of vulnerability to storm surges.

Orissa; and South and North 24 Parganas, Medinipur and Kolkata in West Bengal.

1.3 Factors Influence Storm Surge

There are several factors that contribute to the amount of surge a given storm produces at a given location.

(i) Central pressure: Lower pressure will produce a high surge; however the central pressure is the minimal contributor compared to the other factors.

(ii) Storm intensity: Stronger winds will produce a higher surge. Storm surge is caused primarily by the strong winds in a hurricane or tropical storm.

(iii) Size: A larger storm will produce higher surge. There are two reasons for this. First, the winds in a larger storm are pushing on a larger area of the ocean. Second, the strong winds in a larger storm will tend to affect an area longer than a smaller storm.

(iv) Storm forward speed: On the open coast, a faster storm will produce a higher surge. However, a higher surge is produced in bays, sounds, and other enclosed bodies of water with a slower storm.

(v) Angle of approach to coast: The angle at which a storm approaches a coastline can affect how much surge is generated. A storm that moves onshore perpendicular to the coast is more likely to produce a higher storm surge than a storm that moves parallel to the coast or moves inland at an oblique angle.

(vi) Shape of the coastline: Storm surge will be higher when a hurricane makes landfall on a concave coastline (curved inward, such as Apalachee Bay in Florida) as opposed to a convex coastline (curved outward, such as the Outer Banks of North Carolina).

(vii) Width and slope of the ocean bottom: Higher storm surge occurs with wide, gently sloping continental shelves, while lower storm surge occurs with narrow, steeply sloping shelves. Areas along the Gulf Coast, especially Louisiana and Mississippi, are particularly vulnerable to storm surge because the ocean floor gradually deepens offshore. Conversely, areas such as the east coast of Florida have a steeper shelf, and storm surge is not as high.

(viii) Local features: Storm surge is highly dependent on local features and barriers that will affect the flow of water. A good example is the coast of North Carolina, which has the complexities of such features as barrier islands, inlets, sounds, bays and rivers.

1.4 Observing and Measuring Storm Surge

There are different methods available for observing and measuring storm surges. However, each of these methods have certain limitations. These involve:

(i) The instruments often fail during an event.

(ii) Data is often tied to different reference levels, and the conversions are complicated.

(iii) Each method has different or even unknown error characteristics.

(iv) Data may be measuring different things (still-water vs. wave run-up).

Instead of these complexities, observing and measuring storm surge is foremost important to the operational forecasters. These methods include:

1.4.1 Tide Station

Tide stations measure the variation in water level along the coast. Since tidal cycles are predictable, storm surge can be calculated by subtracting what the water level would have been in the absence of the storm from the measured water level. NOAA's National Ocean Service (NOS) maintains a network of approximately 175 tide stations throughout the United States, which serve as the foundation for NOAA's tide prediction products. They are generally located in areas that are sheltered from waves. This enables a measurement of the 'still-water' height, or the height of the water when it is not disturbed by waves. In India, as part of the Indian Tsunami Early Warning System, a real time network of tide gauges has been established by Survey of India (SOI) and National Institute of Ocean Technology (NIOT). The network is designed to measure and monitor the progress of tsunamis from the two tsunami-genic source regions of Indian Ocean in real time. The network comprises 50 (36 by SOI and 14 by NIOT) state-of-the-art tide gauges transmitting real time data through satellite communication to Indian National Centre for Ocean

Information Services (INCOIS) at Hyderabad, SOI at Dehradun and NIOT at Chennai simultaneously for processing and interpretation. In addition to Indian tide gauge network, the real time tide data from the Global Sea Level Observing System (GLOSS), in Indian Ocean is being received at INCOIS via internet.

Prospective

- Available in real time
- Generally located in areas sheltered from waves and are able to measure 'still-water'
- Traditionally the most reliable way of measuring surge

Constraints

- Limited number of stations along the coast, so there is often no real-time validation of storm surge in the most vulnerable areas
- Often fail at the height of an event due to loss of electrical power or damage

1.4.2 High Water Marks

High water marks are lines found on trees and structures marking the highest elevation of the water surface for a flood event, created by foam, seeds, or other debris. Survey crews are deployed after a storm to locate and record reliable high water marks, usually through GPS methods.

Prospective

- Traditionally the best method for capturing the highest surge from an event

Constraints

- Are not available in real time
- Are perishable, so surveys need to be conducted as soon as possible after a storm
- Can be subjective
- Generally include the effects of wave run-up and setup, and only a small percentage represent 'still-water'

1.4.3 Pressure Sensors

Pressure sensors from the United States Geological Survey are temporary barometric pressure sensors that provide information about storm surge duration, times of arrival and retreat, and maximum depths. The sensors are installed around posts and other structures before the arrival of a hurricane.

Prospective

- Are able to provide timing information that cannot be obtained from high water marks
- Can be deployed ahead of a storm at the locations of the highest expected surge

Constraints

- Information from the sensors is not available in real time
- Can include the effects of waves, which will provide an overestimation of the storm surge height
- May be difficult to recover the instruments after a storm

2. Data Used in Surge Prediction Models

The basic input information to a surge prediction models includes oceanographic parameters, meteorological parameters including storm characteristics, hydrological input, basin characteristics, and coastal geometry, wind stress, sea bed friction and information regarding astronomical tides. These input parameters strongly influence the surge development. The Bay of Bengal is very shallow with sharp variations in seabed contours. This shallow bathymetry significantly influences the surge prediction. A brief description on these input parameters are as follows.

2.1 Oceanographic Parameters

The oceanographic parameters include information on (i) bathymetry, (ii) astronomical tides and (iii) inshore currents in closed region. The northern Bay of Bengal is very shallow and is characterized by sharp changes in sea bed contours. The shallow bathymetry will considerably change the surge height in this region. Hence accurate bathymetry data is a necessary input in improving accurate prediction of surges. Astronomical tides and the inshore currents also influence the surge prediction through non-linear interaction with shallow water.

2.2 Meteorological Parameters

The meteorological input parameters are one of the major input for the surge prediction models. This includes: (i) pressure drop, (ii) vector motion of the storm, (iii) point of landfall, (iv) storm duration, (v) maximum sustainable wind and (vi) the radius of maximum sustained wind. The essential input information for a surge computation is the wind stress and calculated from pressure drop, maximum sustained wind and radius of maximum wind. The ocean model needs the forcing from wind stress parameter and is proportional to the square of the speed of the wind. The error in wind estimation leads to the wrong surge prediction. With the development and improvement in high resolution mesoscale models and the advancement in the tropical cyclone track and intensity prediction, the wind stress parameter for the surge prediction may be provided from the dynamical models. However, the experiments on this aspect leads to the error of $\pm 10\%$ in the surge prediction.

2.3 Basin Characteristics and Coastal Geometry

The high surge area predominantly depends on the coastal geometry of the basin. A study on the effect of the coastal geometry on location of peak surge

(Dube et al., 1982) suggests that the curving coast not only shift the peak surge position, but also affect its height. Also the surge prediction depends on the representation of coastline. The conventional method of coastline representation by orthogonal straight line segments tend to over-reflect the wind driven water (Johns et al., 1981).

2.4 Hydrological Input

The main hydrological information are:

(i) River discharge in the sea and
(ii) Rainfall distribution.

The river-ocean coupled mathematical models suggest that the discharge of fresh water carried by the rivers may modify the surge situation, especially in the northern Bay of Bengal, since the largest river system Ganga-Brahmaputra-Meghna, joins the sea (Dube et al., 1986). Another dynamic effect of these inlets and estuaries is the potentially deep inland penetration of surge originating in the sea. The impact of heavy precipitation, associated with tropical cyclone, on surge height has not been considered in the models for the Indian seas. However, it is expected to influence the surge amplitude. Hence, for the development of any forecasting storm surge model for the Indian sea, these factors need to be considered.

2.5 Surface and Bottom Stress

The surface stress and sea-bed friction is usually parameterized by conventional quadratic law. The bottom stress is computed from the depth-integrated current using conventional quadratic law with a constant coefficient of 0.0026 (Johns et al., 1983a, b). Wind is the main generating mechanism for the surge development, the height of the wave depends on the strength of the wind.

3. Different Methods of Storm Surge Prediction

There are different methods of predicting storm surges and a brief description of these are as follows.

3.1 Empirical Methods

Empirical formulae have been devised for many coastal regions of the world by correlating the sea surface elevation with sea level pressure, the strength and direction of the prevailing wind, etc. However, the empirical methods necessitates a large series of observational data for better correlation. The empirical methods are also developed for the storm surge prediction in the Bay of Bengal (Rao and Majumdar, 1966; Chaudhury and Ali, 1974; Qayyum, 1983; Das et al., 1978). Rao and Majumdar (1966) used simple empirical relation to prepare a nomogram to estimate the amplitude of surge from wind data. Rao (1968) used this nomogram to classify coast of India into three types of vulnerability due to storm surges.

3.2 Numerical Storm Surge Prediction

The numerical storm surge prediction technique consists of solving the hydrodynamic equations governing the motion in the sea at a discrete set of points in space and at discrete instants of time. In India, this numerical storm surge prediction was pioneered by Das (1972). In this method, a two-dimensional linear model and a telescoping grid is used and the surge generated by an idealized cyclone crossed the Bangladesh coast. Das et al. (1974) extended this study for application to the coast of West Bengal and north Odisha. Here the nomograms for the storm surge as a function of the storm intensity and speed of movement of cyclones. Ghosh (1977) has evolved an objective method to predict the storm surges on the entire east coast of India north of 10 °N. He developed the nomograms based on the model outputs of Jelesnianski (1972). The nomograms were prepared separately for the northern part and for the remaining part of the coast. The nomograms for the storm surges as a function of the storm intensity and the speed of movement of cyclone and is presented in Fig. 2. John and Ali (1980) and John (1981) have developed a non-linear model for the simulation of surges along the Bangladesh coast in which the dynamic effects of the Ganga-Brahmaputra-Meghna river system as well as those of islands near the Meghna estuary have been incorporated.

In all these storm surge prediction models for the Bay of Bengal the coastal boundary is approximated by conventional orthogonal straight line segments. Modelling experiments have shown that the surge is sensitive to the way of construction of the coastal boundary. An effective means of achieving greater accuracy is a precise representation of the shorelines which are curved and/or broken by bays or estuaries. Henry and Murty (1982) and Murty and Henry (1983) developed a series of numerical models for tides and surges in the Bay of Bengal. Besides using regular grids, they also used an irregular triangular grid which has advantage of representing coastal geometry with greater accuracy. Johns et al. (1981) used the curvilinear boundary representation depending on the application of conformal mapping for the storm surge prediction over Bay of Bengal. Das et al. (1983) and Dube et al. (1985) developed numerical storm surge prediction model for the Bay of Bengal, which is designed to treat with

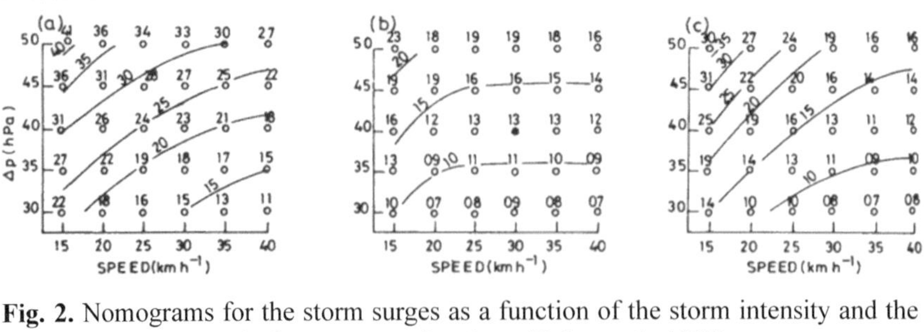

Fig. 2. Nomograms for the storm surges as a function of the storm intensity and the speed of movement of cyclone (Dube et al., 1997).

curved coasts. The treatment of the coastal boundaries involve a procedure leading to a realistic curvilinear representation of both the western and eastern sides of the Bay of Bengal. This coastal representation has another advantage of taking finer resolution automatically in the shallow regions of the northern Bay. The model is fully non-linear and has successfully been used to simulate the surges generated by several severe cyclonic storms.

3.3 Operational Models

The operational numerical storm surge models have been developed and are being routinely used for several coastal regions of the world, such as, North Sea (Flather, 1976; Flather and Proctor, 1983), the Gulf of Mexico and Atlantic coast (Jelesnianski and Chen, 1979), Hong Kong, China etc. A review of these models is given in Jiping et al. (1990), Jelesnianski (1989), Murthy (1984) and Sundermann and Lenz (1983). Most of these models require large computing power. Therefore, for routine forecasting of storm surges, especially providing multiple forecast scenarios, the models can't be used in the absence of access to sufficient computing facility. To overcome this difficulty, most of the operational agencies use the nomogram methods of Jelesnianski (1972) for prediction of storm surges associated with tropical cyclones. The nomograms have been developed from modelling studies of two large number of bathymetries and approach angles. However, certain limitations are involved in the preparation of the nomograms. The correction graph is necessary to provide a valid forecast. The correction graph is drawn between the surges that were observed in the case studies as well as those that were reported at other times and the computed surges (from the nomograms), produced by cyclones having the same pressure drops, radii of curvature and approximate angles of incidence.

The second WMO International Workshop on Tropical Cyclones (WMO, 1990) recommended the use of personal computers to implement the storm surge forecasting system. In India, the dynamic storm surge model was developed by IIT Delhi (Dube et al., 1985). Thus, the efforts have been made towards the adaptation of the IIT Delhi surge model on the personal computers for its use as an operational system (Dube et al., 1994). The first version of this storm surge prediction system has been implemented at IIT Madras.

In the above prediction system the surge is generated by a cyclone, tracking across the analysis area. In view of the strong associated winds and consequently high values of wind stress, the forcing due to barometric changes have been neglected. Thus, the surface wind field associated with tropical cyclone is the primary requirement for modelling storm surges. The wind field at the sea surface is derived by using a dynamic storm model developed by Jelesnianski and Taylor (1973). This storm model uses radius of maximum wind and pressure drop as the basic input parameters. The model basically consists of a trajectory model and a wind speed profile approximation scheme.

The trajectory model represents the balance between pressure gradient, centrifugal, Coriolis and surface frictional forces for a stationary storm.

4. Models Used in Storm Surge Prediction

In this section, two types of numerical models based on finite-difference (IITD storm surge) model and finite-element (ADCIRC) are presented. The geometry of the model domain is extracted from digital world map of National Geophysical Data Center (NGDC), and corresponding bathymetry is obtained from the General Bathymetric Chart of the Oceans (GEBCO) 1-min global bathymetric grid. The storm surge model requires the wind stress forcing as the basic input to the model which is described in the wind forcing module in section 5.

4.1 IITD Storm Surge Model

It is vertically-integrated numerical storm surge prediction model developed at IIT Delhi (Dube et al., 1985) and used with coarse and fine spatial resolutions near the coast to compute the surge height. In formulation of this model, a system of rectangular Cartesian coordinates is used. This model has been adopted for the surge prediction which is used as menu-driven stand-alone system. The details of the model and the numerical solution procedure are described in Das et al. (1983) and Dube et al. (1985). The flowchart of the IIT Delhi storm surge prediction system is presented in Fig. 3.

The depth-averaged equations of continuity and momentum for the dynamical processes are given in the flux form by Dube et al. (1985) and Chittibabu et al. (2000). The model is fully nonlinear system with staggered grid and is integrated with semi-explicit finite difference scheme. The staggered grid consists of three distinct types of computational points on which the sea surface elevations and the zonal and meridional components of depth averaged currents are computed by satisfying the CFL criterion. The model is forced by wind stress and quadratic bottom friction. It is to be noted that the non-linear advection terms have significant effect in the shallow coastal waters of the head Bay of Bengal.

The treatment of the coastal boundaries in the model involves a procedure leading to a realistic curvilinear representation of both the western and the eastern sides of the Bay of Bengal. This coastal representation has another added advantage of taking automatically into account the finer resolutions in the shallow regions of the northern Bay. A desirable feature of storm surge simulation scheme is the ability to incorporate increased resolution adjacent to the coastline. This has been achieved by using a variable grid which leads to a substantial refinement of resolution near the coastline and a coarser resolution in the deeper waters. This type of grid assists in the incorporation of a more detailed bathymetric specification in the important coastal region.

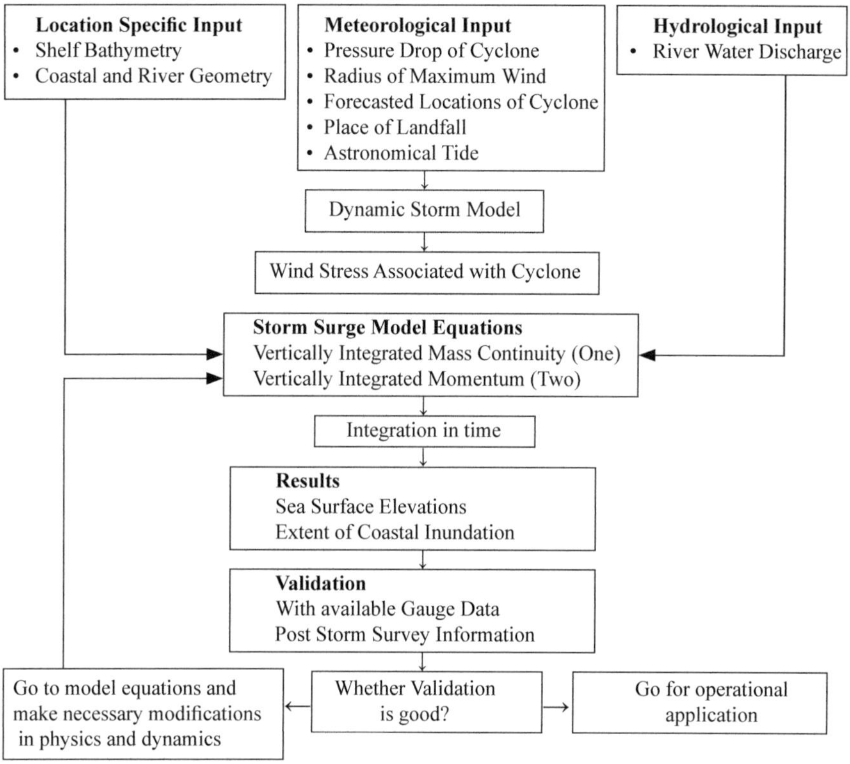

Fig. 3. IIT Delhi storm surge prediction system.

In formulation of this model, a system of rectangular Cartesian coordinates is used. The basic governing equations are as follows.

$$\frac{\partial \zeta}{\partial t}+\frac{\partial \tilde{u}}{\partial x}+\frac{\partial \tilde{v}}{\partial y}=0 \tag{1}$$

$$\frac{\partial \tilde{u}}{\partial t}+\frac{\partial}{\partial x}(u\tilde{u})+\frac{\partial}{\partial y}(v\tilde{u})-f\tilde{v}=-g(\zeta+h)\frac{\partial \zeta}{\partial x}+\frac{F_S}{\rho}-\frac{C_f\,\tilde{u}}{(\zeta+h)}\left(u^2+v^2\right)^{\frac{1}{2}} \tag{2}$$

$$\frac{\partial \tilde{v}}{\partial t}+\frac{\partial}{\partial x}(u\tilde{v})+\frac{\partial}{\partial y}(v\tilde{v})+f\tilde{u}=-g(\zeta+h)\frac{\partial \zeta}{\partial x}+\frac{G_S}{\rho}-\frac{C_f\,\tilde{v}}{(\zeta+h)}\left(u^2+v^2\right)^{\frac{1}{2}} \tag{3}$$

where $(\tilde{u},\tilde{v})=(\zeta+h)(u,v)$, u, v—averaged components of velocity (ms^{-1}) in x, y directions, respectively, ζ—sea surface elevation (m) above the mean water level, h—depth of the water (m), t—time (sec), ρ—density of the sea water, f—Coriolis parameter (= 2ωsinφ), g—acceleration due to gravity, F_s, G_s—x- and y-components of the surface wind stress, C_f—bottom friction coefficient (= 2.6×10^{-3}).

The surface stresses are parameterized using a conventional quadratic law (Johns et al., 1985a, b)

$$\left(F_S, G_S\right) = C_d \rho_a \left(u_a^2 + v_a^2\right)^{\frac{1}{2}} \left(u_a, v_a\right) \tag{4}$$

where $C_d = 2.8 \times 10^{-3}$ is the surface drag coefficient, ρ_a is the density of the air and u_a, v_a are the x- and y-components of the wind at 10 metre.

At the open-sea boundaries, a radiation boundary condition (Heaps, 1973) is applied. A conditionally stable semi-explicit finite-difference scheme with staggered grid is used for the numerical solution of the model equations.

$$v + \left(\frac{g}{h}\right)^{\frac{1}{2}} \zeta = 0 \text{ at } y = 0\text{, southern open sea boundary} \tag{5}$$

$$v - \left(\frac{g}{h}\right)^{\frac{1}{2}} \zeta = 0 \text{ at the northern open sea boundary} \tag{6}$$

$$u - \left(\frac{g}{h}\right)^{\frac{1}{2}} \zeta = 0 \text{ at the eastern open sea boundary} \tag{7}$$

$$u + \left(\frac{g}{h}\right)^{\frac{1}{2}} \zeta = 0 \text{ at the western open sea boundary} \tag{8}$$

Application of a radiation condition, as mentioned above, at the open sea boundary of a model allows the propagation of energy (disturbances) only outwards from the interior in the form of a simple progressive wave. It also helps to eliminate the transient response more quickly, as a result of the frictional dissipation in the system. Concerning its effectiveness, Flather (1976) notes that the application of such a radiation condition in the numerical model may remove the unrealistically large currents and grid-scale oscillations in the vicinity of the open boundary. It is assumed that the motion in the sea is generated from the initial state of the rest, so that $\zeta = u = v = 0$, everywhere for $t = 0$.

4.2 ADCIRC Model

In order to properly describe the physics of storm surges, a numerical model must resolve coastal features that affect storm surge generation and propagation. Therefore, a model domain must incorporate complex coastal geometries (bathymetry and topography), large gradients in bathymetry along the continental shelf, that permit reasonable boundary conditions (i.e., tidal elevations). A finite-element based model is ideal for such type of computations, as its flexibility allows for large spatial scales to be represented in the domain while permitting higher degree of grid refinement near the landward boundary. Blain et al. (1994) showed in a grid convergence study that near-coastal resolution is the most critical factor for accuracy of storm surge computations. In this article, we use a framework of advanced circulation (ADCIRC) model that allows storm surge computations to be run in a distributed grid environment. The framework integrates domain-specific tools and standard grid and portal tools to provide an integrated environment for forecasting and information dissemination. The framework was developed as a component of the Southeastern Universities Research Application's (SURA) Southeastern Coastal Ocean Observing and Prediction (SCOOP) program.

The ADCIRC is a parallel two-dimensional, depth-averaged, finite-element hydrodynamic circulation model (Westerink et al., 1992, 1993) for simulations of circulation and cyclone surge propagation in coastal areas. ADCIRC is ideal for coasts with large inter-tidal zones or occasionally flooded areas, because it can handle wet and dry grid points (Winner and Naomi, 2005). The ADCIRC model has both 2D and 3D versions, where both versions are barotropic in nature. Grenier et al. (1995) made a comparison between the performance of the 2D and 3D versions and concluded that as far as reproducing the major tidal constituents, both these versions produced identical results. The ADCIRC model had been extensively used to study water surface elevation and circulation features for shelves, marginal seas, and estuaries by Luettich et al. (1992). Model simulations for same geographical location were attempted with coarse and fine grid resolution near the coast to compare these results with simulations from ADCIRC model. The main emphasis of this study is to understand the impact of near-coastal grid resolution and coastal geometry on the overall computation of surge profiles.

The detailed description of the finite-element-based hydrodynamic model ADCIRC is given in Luettich et al. (1992) and Grenier et al. (1995). The governing model equations comprise the depth integrated equations for mass and momentum conservation, subject to incompressibility, Boussinesq, and hydrostatic pressure approximations. A set of conservation equations in primitive form expressed in the Cartesian coordinate system is incorporated in the model (Flather, 1988; Kolar et al., 1994). The basic governing equations for continuity and momentum used in the model are given below (Westerink et al., 1994).

$$\frac{\partial \zeta}{\partial t} + \frac{\partial UH}{\partial x} + \frac{\partial VH}{\partial y} = 0 \tag{9}$$

$$\frac{\partial U}{\partial t} + U\frac{\partial U}{\partial x} + V\frac{\partial U}{\partial y} - fV = -\frac{\partial}{\partial x}\left[\frac{p_s}{\rho_0} + g(\zeta - \alpha\eta)\right] + \frac{1}{H}M_x + \frac{\tau s_x}{\rho_0 H} - \tau_* U \tag{10}$$

$$\frac{\partial V}{\partial t} + U\frac{\partial V}{\partial x} + V\frac{\partial V}{\partial y} + fU = -\frac{\partial}{\partial y}\left[\frac{p_s}{\rho_0} + g(\zeta - \alpha\eta)\right] + \frac{1}{H}M_y + \frac{\tau s_y}{\rho_0 H} - \tau_* V \tag{11}$$

where t = time; x, y are horizontal coordinates, aligned in the east and north directions, respectively; ζ = free-surface elevation, relative to the geoid; U, V = depth-averaged horizontal velocity in the x- and y-directions; H = total water column depth, $h + \zeta$; h = bathymetric depth relative to the geoid; $f = 2\Omega \sin \phi$ = Coriolis parameter; Ω = angular speed of the earth; ϕ = degrees latitude; Ps = atmospheric pressure at the free surface; g = acceleration due to gravity; η = Newtonian equilibrium tide potential; α = earth elasticity factor; ρ_0 = reference density of water; τs_x, τs_y applied free-surface stress in the x- and y-directions; $\tau_* = C_f \dfrac{\sqrt{U^2+V^2}}{H}$ = bottom stress; C_f = bottom friction coefficient; and

$$M_x = Eh_2\left[\frac{\partial^2 UH}{\partial x^2} + \frac{\partial^2 UH}{\partial x^2}\right]$$ = depth-integrated momentum dispersion in the x-direction; (12)

$$M_y = Eh_2\left[\frac{\partial^2 VH}{\partial x^2} + \frac{\partial^2 VH}{\partial y^2}\right]$$ = depth-integrated momentum dispersion in the y-direction; (13)

and Eh_2 = horizontal eddy viscosity.

The governing equations are discretized in space by linear finite-elements and in time by a finite-difference scheme (Luettich et al., 1992). The finite-element solution to the shallow water equations gives rise to spurious modes and numerical instabilities. Hence, it becomes necessary to reformulate the equations into a form that provides a stable solution in its finite element representation. Therefore, ADCIRC-2D employs the generalized wave continuity equation (GWCE), together with the momentum conservation equations, to eliminate this problem (Westerink et al., 1994). The result is a noise-free solution that is used to solve for the deviation from the geoid (ζ) and the velocities in the x- and y-directions.The finite-element mesh was constructed using the software package Surface Modeling System (SMS). The program automatically generates a grid with a low resolution in the deeper central region and with a high resolution near the coast. The model parameters for the storm surge simulations are set as follows: (i) spherical coordinate system for the study area; (ii) model execution from a cold start for a 3-day period with 0.5-day as ramp time; (iii) meteorological forcing (wind stress and pressure) are provided as input to model; and (iv) minimum depth of 0.5 m sets for wet and dry elements with horizontal eddy viscosity coefficient of 5 $m^2 s^{-1}$.

5. Wind Forcing Module

Two main forces, which generate storm surges on the sea surface, are surface wind stress and pressure gradient force. These act parallel and normal to the sea surface respectively and their relative importance depends on the water depth. In view of the strong associated winds, the forcing due to barometric changes is neglected. Thus, the surface wind field associated with a TC is the primary requirement for modelling of storm surges. For this purpose, the wind field at the sea surface is derived by using a dynamic storm model (Jelesnianski and Taylor, 1973). To obtain a dynamic wind profile, initially a stationary symmetric model wind profile is taken, and then correction is applied to approximate the asymmetry due to storm motion. The vector equation which rules the horizontal motion of wind flow near the sea surface in the area of storm is

$$\frac{d\mathbf{V}_g}{dt} = -\frac{1}{\rho_a}\,\text{grad}\,p + f\,\mathbf{V}_g \times \mathbf{k} + \mathbf{F} \tag{14}$$

where $\mathbf{k}$ is the vertical unit vector, $\mathbf{V}_g$ is the wind velocity, ρ_a is the density of air, f is the Coriolis parameter, p is the pressure of atmosphere and $\mathbf{F}$ is the horizontal frictional force per unit mass.

Assume that the pressure and the wind fields move forward without change at the velocity of storm. For a stationary symmetric storm, from equation (14) we get, at a distance r from the centre of the cyclone (Myers and Malkin, 1961; Ueno, 1981)

$$\left.\begin{aligned} \frac{dp}{dr} &= \rho_a \mathbf{V}_g^2 \left(\frac{\mathbf{K}_s}{\sin\phi} - \frac{1}{\mathbf{V}_g}\frac{d\mathbf{V}_g}{dr} \right) \\ \frac{1}{\rho_a}\frac{dp}{dr}\cos\phi &= f\mathbf{V}_g + \frac{\mathbf{V}_g^2}{r}\cos\phi - \mathbf{V}_g^2\frac{d\phi}{dr}\sin\phi + \mathbf{K}_n\mathbf{V}_g^2 \end{aligned}\right\} \tag{15}$$

where $\varphi(r)$ is the inflow angle of wind i.e. the angle towards the storm centre. $\mathbf{K}_s$ and $\mathbf{K}_n$ are empirically determined coefficients representing stress coefficients in the directions opposite and to the right of the wind respectively. These stresses are given by the coefficient times the square of the wind speed. By eliminating pressure from equation (15), we can get

$$\frac{du}{dr} = \frac{\mathbf{K}_s u}{\sin\varphi} - f\,r - \mathbf{K}_n \mathbf{V}_g\, r \tag{16}$$

where $u = \mathbf{V}_g\, r \cos\varphi$

Equation (16) can be solved numerically to get the distribution of $u(r)$ or $\varphi(r)$. Then integrating equation (15), we can get the distribution of $p(r)$. The wind profile formulation as provided by Jelesnianski (1972) is

$$\mathbf{V}_g(r) = \mathbf{V}_R \frac{2\mathbf{R}_m r}{(\mathbf{R}_m^2 + r^2)} \tag{17}$$

where $\mathbf{V}_R$ is the value of the maximum wind speed and $\mathbf{R}_m$ is the distance from the centre where the maximum wind speed appears. The value of $\mathbf{R}_m$ is usually fixed from the synoptic map, while the value of $\mathbf{V}_R$ is determined by solving the equations (15) and (16) numerically.

6. Validation Experiments

In order to validate the models, several simulation experiments have been performed by using the data of severe cyclonic storms hitting the coastal regions of the Bay of Bengal and the Arabian Sea. Also the simulated seasurface elevations are compared with available observations from local tide gauges wherever possible, with post storm survey estimates of Meteorological Departments of India, Bangladesh, Myanmar, Oman, Pakistan and Sri Lanka. Unisys Weather Information Services estimate the intensity of tropical cyclones and the associated range of surge heights based on the post-analysis of all available satellite images, surface data, upper air data, and radar data. This information has also been used for validation.

The numerical experiments carried out using location-specific models to simulate the surges generated by Orissa Super cyclone (1999, India), Sidr (2007, Bangladesh), Nargis (2008, Myanmar), Gonu (2007, Oman), Karachi cyclone (1999, Pakistan) and Rameswaram cyclone (1964, Srilanka) that struck the coasts along the North Indian Ocean (Bay of Bengal and Arabian Sea) are presented. As a very special case, the very severe cyclonic storm "Phalin (2013)" the strongest cyclone that hit the eastern coast of India, Odisha state, since the supercyclone of 1999 is also discussed. The model computed surges are in good agreement with the available observational estimates. The storm surge model requires the wind stress forcing as the basic input to the model. For this purpose, the wind stress is computed by using a dynamic storm model of Jelesnianski and Taylor (1973). In order to obtain a dynamic wind profile in the storm model, initially a stationary symmetric model wind profile is taken, and then correction is applied to approximate the asymmetry due to the storm motion. The storm model represents a balance among pressure gradient, centrifugal, Coriolis, and surface frictional forces for a stationary storm.

6.1 Orissa Super Cyclone (October 1999)

A depression formed in the Bay of Bengal near 12°N and 98.5°E at 0600 UTC of October 25, 1999. It became a cyclonic storm in the early hours of October 26 and located at 13.5 °N and 96.5 °E. The cyclone moved further in a northwesterly direction and lay centered at 16 °N and 92 °E. It became a severe cyclonic storm with a core of hurricane winds on October 27 at 0300 UTC. The cyclone further intensified into a very severe cyclonic storm and lay centered at 17.5°N and 89.5 °E on October 28. Cyclone crossed the Orissa coast of India close to Paradip between 0430 and 0630 UTC of 29 October. For the track land falling near Paradip, the pressure drop is 98 hPa and radius of maximum winds is 40 km.

For this experiment, the tentative best track of the cyclone is taken from the IMD the cyclone crossed few kilometres south of Paradip. The model is integrated with 98 hPa and 40 km radius of maximum winds. Figure 4 shows the model-computed peak surge envelope along the Orissa coast. It can be seen that a maximum surge of about 7.8 m occurred close to the landfall point (Dube et al., 2000). The coastal regions between Konark and Chandbali are affected by a surge of more than 5 m. Post-storm survey reports also show that the surge is more than 7 m, near Paradip.

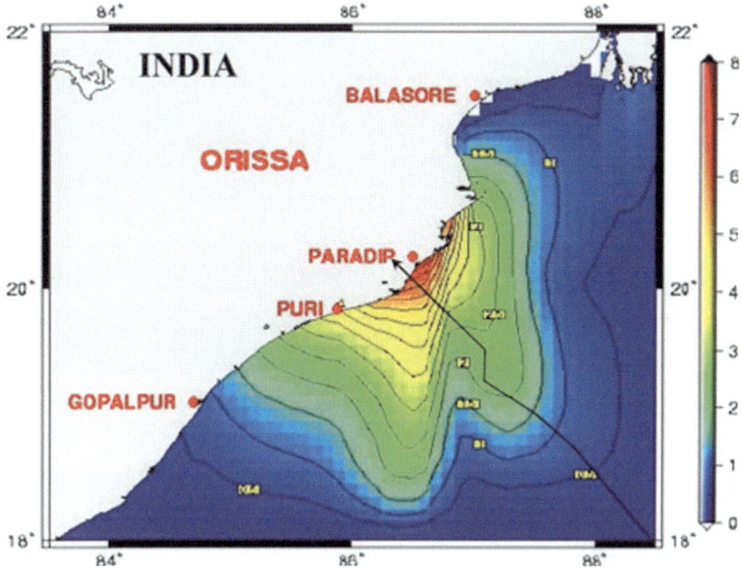

Fig. 4. Peak surge contours (m) associated with Orissa Super cyclone, 1999 (Dube et al., 2000).

6.2 Phailin (October 2013)

A well-marked low pressure was seen over the Andaman Sea on 7 October 2013 from a remnant of unnamed cyclonic circulation over the South China Sea (RSMC, 2013). It intensified into a depression (8.7–13.9 ms^{-1} or 17–27 kt) and then a deep depression (14–17 ms^{-1} or 28–33 kt) on 8 October and further intensified into cyclonic storm Phailin (17–24 ms^{-1} or 34–47 kt) on 9 October. Moving in the northwestward direction, it intensified into a very severe cyclonic storm (VSCS) and crossed the Odisha, India, coast around 1700 UTC 12 October 2013 with winds in excess of 59 ms^{-1} (115 kt), taking a similar course as the 1999 Orissa supercyclone that made landfall with 69 ms^{-1} (134 kt) peak winds. The damage due to the 1999 Orissa supercyclone was approximately $4.5 billion U.S. dollars and resulted in approximately 10,000 fatalities (Kalsi, 2006). In the case of Phailin, the reported death toll was 22, and property loss was considerably minimized. Thus, despite the similarity in the two storms in terms of track, intensity, and lead time, the story of devastation and casualties was fortunately not repeated (Mohanty et al., 2015). Figure 5 represents the IIT-BBS predicted surge height with a lead time of 84 hrs of landfall of the system.

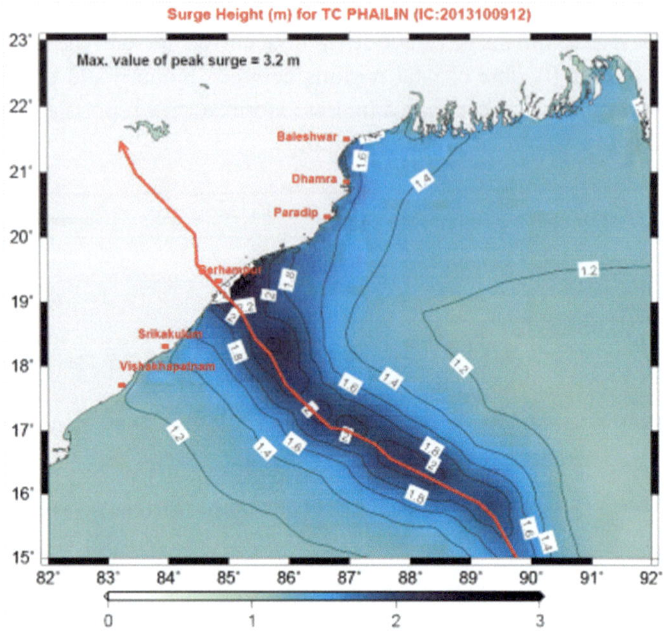

Fig. 5. Peak surge contours (m) associated with Phailin, 2013 (Mohanty et al., 2015).

6.3 Sidr (November 2007)

The depression formed over southeast Bay of Bengal and adjoining Andaman Sea intensified into a deep depression and moved northwestward. It lay centered at 1800 UTC of November 11, 2007 near 10.5 °N and 91.5 °E, about 200 km southwest of Port Blair. The system moved slightly westward and intensified into a cyclonic storm "Sidr" and lay centered at 0300 UTC of November 12 over southeast Bay of Bengal near 10.5 °N and 91.0 °E. It moved northwestward and further intensified into a severe cyclonic storm and lay centered at 1200 UTC of November 12 over southeast Bay of Bengal near 11.5 °N, 90.0 °E, about 300 km west of Port Blair. The severe cyclonic storm "Sidr" over southeast Bay of Bengal remained practically stationary and intensified into a very severe cyclonic storm, lay centered at 1800 UTC of November 12 over southeast Bay of Bengal near 11.5 °N and 90 °E.

The very severe cyclonic storm "Sidr" over southeast Bay of Bengal moved northwestward in the early hours of November 13 near 12 °N and 89.5 °E and further moved slightly northward and lay centered at 0900 UTC of November 13 near 13 °N 89.5 °E. The very severe cyclonic storm "Sidr" over east-central and adjoining west-central Bay of Bengal moved further northward and lay centered at 2100 UTC of November 14 over east-central and adjoining west-central Bay of Bengal near 17 °N and 89 °E, about 650 km south of Kolkata. The "sidr" over north Bay of Bengal moved northeastward and crossed west Bangladesh coast near 89.8 °E around 1700 UTC of the

November 15 and lay centered at 1800 UTC over Bangladesh near 22.5 °N and 90.5 °E, about 100 km south of Dhaka. The system further moved in a north-northeasterly direction and weakened gradually.

The Unisys (2007) estimated the lowest central pressure of the cyclone to be 944 hPa and the maximum sustained winds of 60–69 ms^{-1}. Numerical experiments are carried with a pressure drop of 68 hPa and with a radius of maximum wind of 25 km. The landfall surge contours computed by the model for the November 2007 Sidr cyclone are shown in Fig. 6. The maximum surge of about 5.8 m is simulated to be near Mongla Port, and a surge of about 1.5 m is computed at Kuakata. The maximum computed surge height of 5.8 m is in agreement with the highest value in the range of surge estimates provided by Unisys.

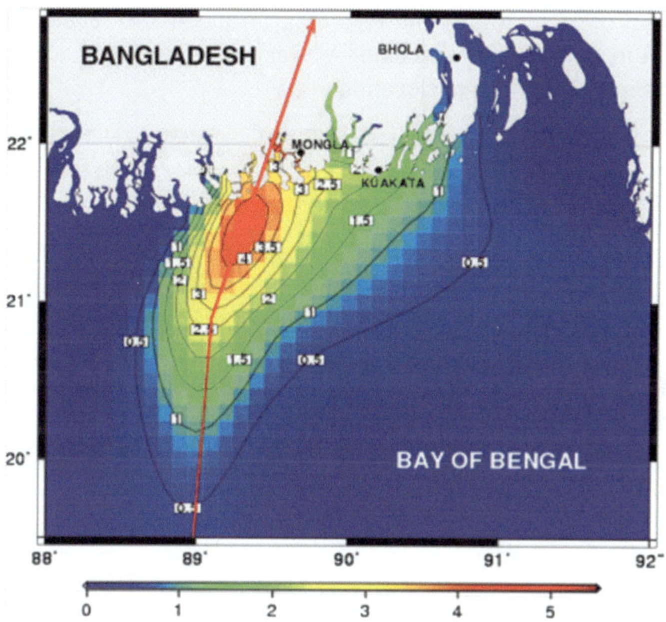

Fig. 6. Peak surge contours (m) associated with Sidr, 2007 (Dube et al., 2009).

6.4 Nargis (April 2008)

In the last week of April, an area of low pressure was detected over the Bay of Bengal about 1,150 km east–southeast of Chennai, India. At 0300 UTC on April 27, Indian Meteorological Department (IMD) classified the system as a depression, and nine hours later, the system intensified into a deep depression. At 0000 UTC on April 28, the system became a cyclonic Storm Nargis when it was located about 550 km east of Chennai. On April 28, the motion of Nargis became nearly stationary when located between ridges to its northwest and southeast. The system gained further strength to become a severe cyclonic storm by April 28 at 0600 UTC. On May 1, after turning nearly due eastward, the system continued to gain strength and attained a maximum wind speed of 59 ms^{-1} on May 2 at 0600 UTC, as it approached the coast of Burma. Around 1200 UTC on May 2, cyclone Nargis made landfall in the Ayeyarwady Division of Myanmar. Early on May 3, it quickly weakened after turning to the northeast toward the rugged terrain near the Myanmar–Thailand border.

The model is integrated with a pressure drop of 65 hPa and radius of maximum winds of 25 km. The model-computed surge contours along the coast of Myanmar are shown in Fig. 7. It may be seen that a maximum surge of 4.5 m occurred close to the landfall point. The Deltaic region of Ayeyarwady is affected by surges between 2.5 and 4 m. The Myanmar coast from Pyapon to Yangon is flooded with a surge of more than 2 m. The computed surge values at Pegu and Moulmein are 2.5 m and 1.5 m, respectively. During this cyclone, the surge of the order of 4 m was reported by Unisys (2008). The Department of Meteorology and Hydrology, Yangon also reported the surge of about 4 m at the Deltaic region of Ayeyarwady. This is in good agreement with our simulated sea level elevations.

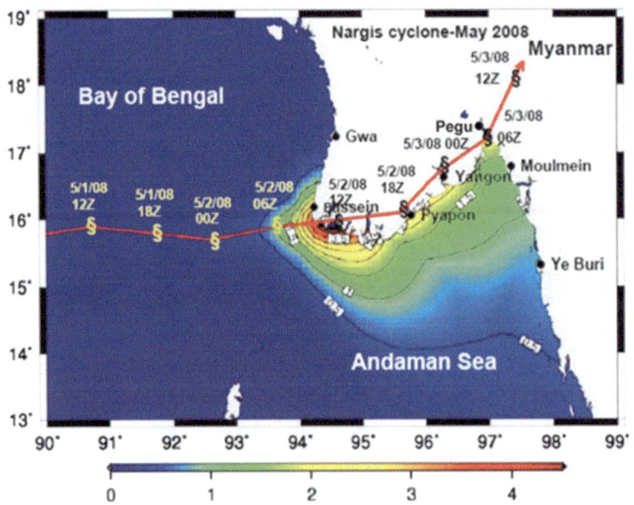

Fig. 7. Peak surge contours (m) associated with Nargis, 2008 (Dube et al., 2009).

6.5 Gonu (June 2007)

A depression was formed over the eastern Arabian Sea in the late hours of June 1, 2007. Moving westward, it steadily intensified into a deep depression on the early hours of June 2. The system intensified into a cyclonic storm "Gonu" near 15.1 °N and 67.7 °E on the same day when it was located 760 km southwest of Mumbai, India. Moving northwestward, it rapidly intensified into a severe cyclonic storm on June 3 and lay centered at 16.8 °N and 67.4 °E at 0600 UTC. The system further intensified into a very severe cyclonic storm on late hours of June 3. It became a very severe cyclonic storm with a core of hurricane winds near 19.2 °N and 64.9 °E at 0600 UTC on June 4. Gonu strengthened further and became a super cyclonic storm at 1800 UTC on June 4 with sustained winds reaching up to 240 km h^{-1} (150 mph) and pressure of 920 hPa. The super cyclonic storm Gonu weakened into very severe cyclonic storm and lay centered at 2100 UTC of June 4 over northwest Arabian sea and adjoining central and northeast Arabian Sea near 20.5 °N and 63.5 °E, about 600 km southwest of Dwarka (India). The cyclone gradually weakened as it continued tracking northwestward, and over a period of 24 hours, the intensity decreased by 95 km h^{-1}. On June 6, the cyclone turned to the north-northwest to severe cyclonic storm status and later to cyclonic status early on June 7, and Gonu crossed the Makran coast in Iran.

The track of the cyclone (Fig. 8) and the relevant data are taken from the Unisys Weather Information Services (2007). Numerical experiments are carried out with a pressure drop of 98 hPa and radius of maximum wind as 40 km. The surge contours computed by the model are shown in Fig. 8. A maximum surge of 2.6 m is simulated near the eastern tip of Oman due to the close proximity of the storm track. A surge of about 2.6 m and 2.5 m are simulated at Sur and Nazwa Fins, respectively. The peak surge value at Muscat is around 2 m. Also, it is seen that a coastal stretch of about 100 km to the northeastern Oman is flooded with surges of about 1 m. The computed maximum surge value of 2.6 m is comparable with the estimated range of surge heights provided by Unisys Weather Information Services (2007).

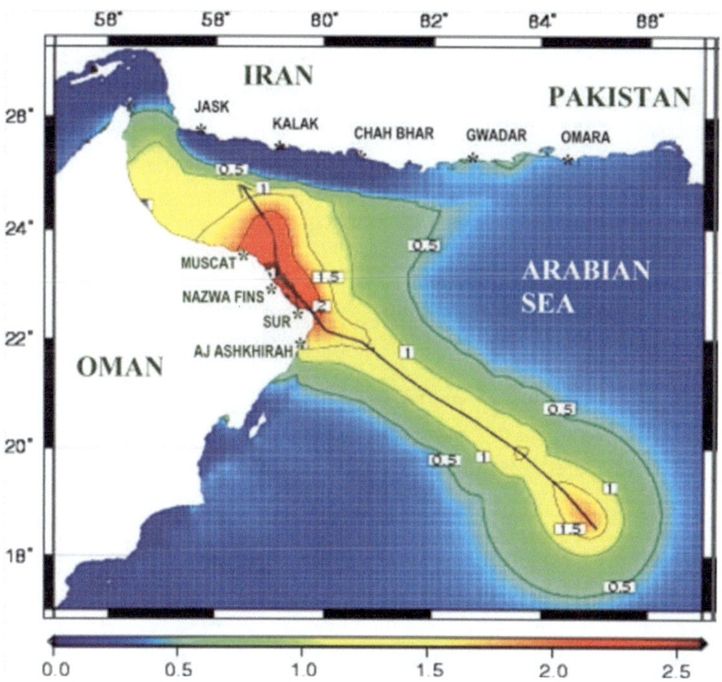

Fig. 8. Peak surge contours (m) associated with Gonu, 2007 (Dube et al., 2009).

6.6 Karachi Cyclone (May 1999)

A depression developed over the eastern Arabian Sea on May 15, 1999 at 0000 UTC near 12°N, 72.5°E and continued until 1800 UTC of the same day. It intensified into a deep depression by May 16 at 0000 UTC and further intensified into a cyclonic storm by 1200 UTC. The system gained further strength to become a severe cyclonic storm by May 17 at 0600 UTC. The system continued to gain strength and attained a maximum wind speed of 57 ms^{-1} on May 19 at 0000 UTC, which it maintained through to the early hours of May 20. The cyclone made landfall at 0600 UTC on May 20 near Wari creek at about 23.6°N and 68.2°E. In this experiment, the cyclone track (Fig. 9) and the relevant data are taken from the Unisys Weather Information Services (1999). Using these data, numerical experiments are carried out with a pressure drop (the difference between the central and ambient pressures) of 40 hPa and radius of maximum winds of 30 km. The surge contours computed by the model at the time of landfall are shown in Fig. 9. It may be seen that a maximum surge of 3.8 m is predicted near the Bhitiaro creek, and the surges at Wari creek and southern Kori creek are 3.4 m and 2 m, respectively (Jain et al., 2006b). The surge of about 0.5 m can be seen near the southern part of Karachi. Unisys Weather Information Services estimate the intensity of tropical cyclones and the associated range of surge heights based on the post-analysis of all available satellite images, surface data, upper air data, and radar data. It is found that the maximum computed surge height of 3.8 m is comparable with the highest value in the range of surge estimates provided by Unisys (1999).

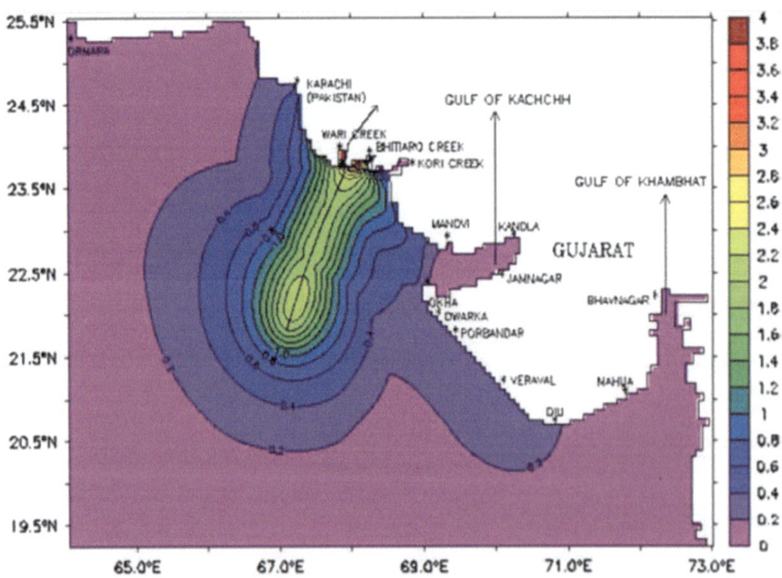

Fig. 9. Peak surge contours (m) associated with Pakistan cyclone, 1999 (Jain et al., 2006).

6.7 Rameswaram Cyclone (December 1964)

Rameswaram cyclone was one of the severest storms, which affected Sri Lanka and the southern Indian peninsula. The track of the storm is shown in Fig. 10. It may be seen from Fig. 10 that the storm crossed the Sri Lankan coast about 50 km north of Trincomalee on December 22 at about 0600 UTC. Then, it further moved northwestward and struck the Indian coast about 30 km south of Tondi on December 23 at about 0600 UTC. The reported maximum wind speed associated with the storm during the period was 60 ms^{-1}. The model is integrated ahead in time up to 72 h with a pressure drop of 60 hPa and a radius of maximum wind of 30 km. The peak surge contours are shown in Fig. 10. A maximum surge of about 5.4 m is computed near Tondi, which is shown in the figure (Chittibabu et al., 2002). Heavy inland flooding in the region, with an estimated peak elevation of about 5 m, was reported by Rao and Mazumdar (1966). The predicted and reported maximum sea surface elevations are in good agreement. It can also be seen that the coastal region to the north of Mannar Island along the west coast of Sri Lanka is flooded with a surge of more than 4 m.

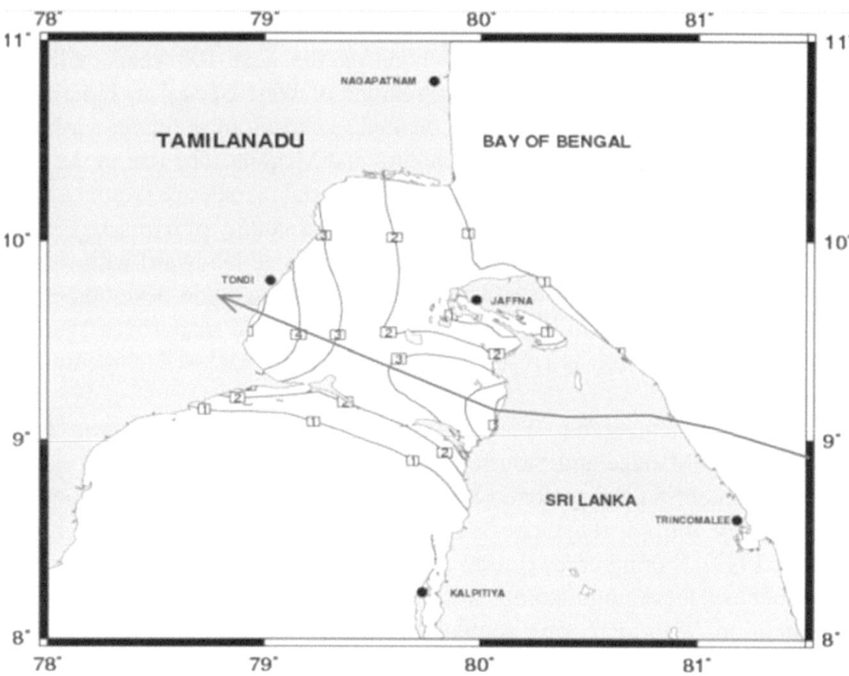

Fig. 10. Peak surge contours (m) associated with Rameswaram cyclone, 1964 (Chittibabu et al., 2002).

7. Prediction of Inland Inundation

While the accurate prediction of surge is very important, it is also necessary to have an idea of the extent of inland inundation of surge from the sea, so that the coastal stretch upto which significant flooding is expected can be estimated for the purpose of evacuation. Keeping this in view a model has been developed towards the prediction of inland inundation along the east coast of India and Bangladesh (Johns et al., 1982; Dube et al., 1986).

Frequently, the lateral boundaries in numerical storm surge prediction models are taken to be vertical side-walls through which there is no flux of water. In actuality, however, the water will usually move continuously inland and the use of idealized vertical side-walls may lead to misrepresentation of the surge development. Thus, instead of using the conventional solid wall boundary at the coast, the model uses a continuously deforming lateral fluid boundary and includes the onshore topography in order to route the storm surges over the land. The IITD storm surge model has been tested with idealized coastal topography along the east coast of India and Bangladesh coast.

8. Sea Level Rise and Storm Surge

The issue of the sea level rise as a consequence of climatic warming and the greenhouse effect is of extremely grave concern for Bangladesh. The projected worldwide rise of sea level during the next 100 years will be particularly hard-felt in the low-lying regions of West Bengal in India and Bangladesh. These regions have been created as a result of sediment washing down the major rivers Ganga, Brahmaputra and Meghna. The rise in the sea level would inundate lowlying areas, destroy coastal marshes & swamps, and erode shoreline, cause coastal flooding and increase salinity of rivers, bays and aquifers. Mangroves and other vegetation will migrate landward with rising sea-level, but such migration will be inhibited by man-made developments. Thus the protective buffering effects of coral reefs and mangroves against cyclones will be reduced as a result of sea level rise and increased temperatures (Broadus, 1993).

Rising air temperature resulting from the greenhouse warming will increase SST (Manabe and Stouffer, 1980). An increase in the surface water area of warmer SST for cyclone generation is expected to increase frequency and intensity of tropical cyclones and also the duration of the cyclone season (Raper, 1993). Storm surges caused by rising water resulting from landward movement of these huge storms would also increase in frequency, severity and duration. With increasing population densities in the future, through the combined effects of natural population increase and loss of coastal zone due to sea level rise, the exposure to severe storm surges will only increase.

Broadus et al. (1986) and Broadus (1993) have made a detailed study on the potential consequences of sea-level rise in the deltaic regions of the Nile River in Egypt and the complex Ganga-Brahmaputra Meghna river system in Bangladesh. In the case of Bangladesh the authors have presented different transgression scenarios as a result of the projected rise in the sea level. They have also discussed the problem of exposure of population to storm surges as a result of sea-level rise. Economic considerations vis-a-vis sea level rise are also detailed in the study. Dube and Rao (1991) used a continuously deforming shore-line model to determine the effect of sea-level rise on storm surges affecting the east coast of India. Their model uses an idealized on shore topography to compute the inland inundation associated with sea-level rise and storm surges. Flather and Khandker (1993) used numerical model to determine the effect of an increase in mean sea level on tidal storm surge and combined tide and storm surge levels for Bangladesh, while Brammer (1993) made an impact assessment of sea-level rise and geographical complexities for the Ganga-Brahmaputra-Meghna delta of Bangladesh. In a recent paper Sinha et al. (1996) studied the impact of sea 1evel rise on the tidal circulation in the Hooghly estuary.

9. Recent Developments

The recent developments on storm surge prediction over North Indian Ocean are briefly summarized as follows.

9.1 Observational Network

Real time observations on meteorological and oceanographic parameters over Indian seas are improved to enhance the operational capability of surge prediction. This includes:

(i) Installation of data buoys in the coastal and deep waters of the Bay of Bengal and the Arabian Sea.
(ii) Installation of high precision sea-level gauges at selected locations along the east and west coasts of India, Bangladesh, Myanmar and Sri Lanka.
(iii) The meteorological sea level data obtained in this manner will be helpful in the prediction of natural hazards, such as storm surges, tsunamis and set up due to wind waves.
(iv) Satellite altimeter will complement the data obtained from tide gauges but cannot replace it satisfactorily because of the lack of needed precision for assessment of sea-level rise.

9.2 Mesoscale Model

The advancement on the prediction of storm surges over North Indian Ocean is achieved with the development of coupled atmosphere-ocean mesoscale system (Pattanayak, 2012). In Atlantic, this coupled system is used for the surge and tide prediction. In India, the one-way coupled system is developed by Pattanayak et al. (2015) and is used in real time for the storm surge prediction over Bay of Bengal. This one-way coupled system is designed with the atmospheric model, Weather Research Forecasting (WRF) system and the IITD storm surge model. As stated above, the basic meteorological input to the IITD storm surge model is the wind stress. Thus, the WRF model simulated surface wind is converted to wind-stress employing the relationship proposed by Garrett (1977). In order to provide the WRF simulated wind at each grid point of IITD storm surge model, the WRF model is configured with same horizontal resolution and domain coverage as of the surge model. With this one-way coupled system, it is possible to predict the storm surges associated with tropical cyclones with a lead time of 72 hours with reasonable accuracy. However, the tidal effect is yet to be included in this approach.

10. Risk Assessment and Disaster Preparedness

Accurate assessment is an essential step for measures to reduce risk. Following steps are suggested for making risk assessment:

(i) Mapping of cyclone hazards giving information on matters such as location, frequency, duration and severity including wind speeds, inland inundation, etc.
(ii) Assessment about the vulnerability of people and property to mapped cyclone/surge hazard.
(iii) Although the vulnerability analysis is a 'site-specific' problem even though, as tropical cyclones may threat a number of neighbouring countries, a regional approach to the problem may be very useful.

Risk reduction due to cyclone hazard also requires active preparedness plan which include the following:

(i) Educating the community in the areas of risk about the hazard.
(ii) Practical training about how to reduce damage in the event of cyclone hazard.
(iii) Promoting ingenious ways for coping with the consequences.

11. Future Perspectives

(a) Storm surge predictions are affected by the error in tropical cyclone predictions in terms of both their tracks and intensities. Taking this into account, ensemble and probabilistic methods and outputs should be considered for use in operational storm surge forecast.
(b) As mesoscale NWP models with high resolution are having ability to solve tropical cyclone wind fields, use of the results from these models in tropical storm surge modelling should be investigated.
(c) Total water level is the combined effect of storm surge, wave set-up, and high tide; therefore, accurate prediction of wind waves and tidal height together with their nonlinear interaction with the storm surge in the model is essential. Johns et al. (1985a, b), and more recently, Sinha et al. (1996, 2008) have shown that the non-linear interaction of surge and tide may significantly modify the evolution of surges. However, prediction methods for wave setup are not well established yet. Therefore, further studies of wave setup prediction method are needed.

(d) While accurate prediction of surge is important, it is also necessary to have an estimate of the coastal stretch likely to be inundated for effective evacuations. In order to achieve this, development of real-time ocean–river coupled models are required for the regions where they do not exist. For the Bay of Bengal region, this aspect is very important as one of the world's largest river systems (Ganga–Brahmputra–Meghana) joins the head Bay of Bengal. Dube et al. (1986, 2005) have studied in detail about the penetration of surge through rivers and its interaction with river runoff. It would also be necessary to develop very high-resolution coastal inundation models to estimate flooding of the low-lying coastal regions in the event of storm surges. Currently, efforts are being made to use FE (finite element) models for the purpose.
(e) Another area that requires attention is the impact of climate change and a possible sea level rise and changes in the frequency and intensity of storms. These factors may change flooding risks from storm surges, especially in the low-lying regions of the North Indian Ocean such as Bangladesh and Maldives.
(f) Detailed time histories and data dossiers of individual storm surges should be prepared by the concerned countries, which will enable calibration of storm surge models and improvement of prediction techniques. These data are also helpful to assess the potential and susceptibility of the coastlines. Estimates of storm surge potential from historic records are also valuable for efficient administration of cyclone mitigation plans to determine the safety and economics of coastal construction and installations.
(g) It would also be appropriate to store all the pertinent data when a given storm affects an area, i.e., inundation maps, high-water marks, etc. It is also important to mention that now GIS (geographical information system) work is a common tool for most researchers, the design and creation of a GIS that contains precipitation, stream flow, and cyclone track data would be very valuable.

Conclusions

A brief review on the storm surge prediction over Bay of Bengal and Arabian Sea is described. The main factors that contribute to the amount of storm surge with its destruction potential are illustrated. It is to be noted that with wide enhancement of tide gauges and high precision seafloor pressure sensors, the storm surge measurement is improved.

Discussing different approaches in storm surge prediction, the dynamic storm surge prediction technique overruled the statistical and empirical approach. The numerical storm surge prediction technique consists of solving the hydrodynamic equations governing the motion in the sea at a discrete set of points in space and at discrete instants of time.

The recent developments in non-linear dynamic models in storm surge prediction are described. Two types of numerical models based on finite-difference (IITD storm surge) model and finite-element (ADCIRC) are presented. The geometry of the model domain is extracted from digital world map of National Geophysical Data Center (NGDC), and corresponding bathymetry is obtained from the General Bathymetric Chart of the Oceans (GEBCO) 1-min global bathymetric grid. The storm surge model requires the wind stress forcing as the basic input to the model. A dynamic storm model developed by Jelesnianski and Taylor (1973) is used for computation of surface winds associated with cyclonic storms. Only meteorological inputs required for the model are positions of the cyclone, pressure drop and radii of maximum winds at any fixed interval of time. The models have been validated with the available data of storm surges generated by severe cyclonic storms. The models are able to simulate surge heights, which are in broad agreement with the estimated values provided by Unisys and Meteorological Services of different countries. The results emphasize the suitability of a fine resolution location-specific model for an operational prediction of surges in the countries of the North Indian Ocean. The TCP of WMO supported technology transfer to run and make operational the IITD storm surge models for Bangladesh, Myanmar, Sri Lanka, Pakistan and Oman.

In this review, the cyclonic storm is the sole driving force for the dynamical processes in the sea. However, the tides and wave setup have not been included in the present study. Therefore, the non-linear interaction of surge and the tide has not been studied. Such an interaction may be significant if the occurrence of the surge coincides with that of the high tide (Johns et al., 1985; Sinha et al., 1996, 2008).

In general, if there is a break in the coast, such as a river, it provides an additional path for the water to escape into the river, instead of getting piled up. The numerical models used in this study do not take into account the effect of rivers that communicate with the Arabian Sea and Bay of Bengal. However, the discharge of the fresh water carried by the river may modify the surge height along the coasts. The heavy precipitation associated with cyclone as well as seaward flow of river is expected to modify, to some extent, the evolution of the surge.

The recent advancement in observational network and developments of one-way and fully coupled Atmosphere-Ocean models enhance the storm surge prediction over North Indian Ocean with a lead time of 72 hours. This leads to the successful risk management, disaster mitigation strategy and planning for the policy makers, researchers, scientists and moreover to the local people to save their livelihood and the society.

REFERENCES

Blain, C.A., J.J. Westerink and R.A. Luettich, 1994: Domain and grid sensitivity studies for hurricane storm surge predictions. *In:* Peters, A. et al. (ed.). Computational methods in water resources X. Heidelberg.

Brammer, H., 1993: Geographical complexities of detailed impact assessment for Ganges-Brahmaputra-Meghna delta of Bangladesh. *In:* Effects of changes in Stratospheric Ozone and Global Climate. Vol. 4: Sea Level Rise (Editor, G. Titus). UNEP/EPA.

Broadus, J.M., 1993: Possible impacts of, and adjustments to, sea level rise: The case of Bangladesh and Egypt. *In:* Effects of changes in Stratospheric Ozone and Global Climate. Vol. 4: Sea Level Rise (Editor, J.G. Titus). UNEP/EPA.

Broadus, J.M., J.D. Milliman, S.F. Edwards, D.G. Aubrey and F. Gable, 1986: Rising sea level and damming of rivers: Possible effects in Egypt and Bangladesh. *In:* Effects of changes in Stratospheric Ozone and Global Climate. Vol. 4: Sea Level Rise (Editor, J.G. Titus). UNEP/EPA.

Chaudhary, A.M. and A. Ali, 1974: Prediction of maximum surge heights associated with cyclones affecting Bangladesh—A preliminary investigation. *Nucl. Sci. Appl.*, **7(B)**, 118-120.

Chittibabu, P., S.K. Dube, A.D. Rao, P.C. Sinha and T.S. Murty, 2000: Numerical simulation of extreme sea levels using location specific high resolution model for Gujarat coast of India. *Mar Geod*, **23**, 133-142.

Chittibabu, P., S.K. Dube, A.D. Rao, P.C. Sinha and T.S. Murty, 2002: Numerical simulation of extreme sea levels for the Tamilnadu (India) and Sri Lanka coasts. *Mar Geod*, **25**, 235-244. Doi: 10.1080/01490410290051554.

Das, P.K., 1972: A prediction model for storm surges in the Bay of Bengal. *Nature*, **239**, 211-213.

Das, P.K., 1994a: Prediction of storm surges in the Bay of Bengal. *Proc. Indian Natn. Sci. Acad.*, **60**, 513-533.

Das, P.K., 1994b: On the Prediction of Storm Surges. *Sadhna*, **19**, 583-595.

Das, P.K., S.K. Dube, U.C. Mohanty, P.C. Sinha and A.D. Rao, 1983: Numerical simulation of the surge generated by the June 1982 Orissa cyclone. *Mausam*, **34**, 359-366.

Das, P.K., M. Miyazaki and C.P. Jelesnianski, 1978: Present techniques of storm surge prediction. WMO Report No. 500, Geneva.

Das, P.K., M.C. Sinha and V. Balasubramanyam, 1974: Storm Surges in the Bay of Bengal. *Quart. J. Roy. Met. Soc.*, **100**, 437-449.

Dube, S.K., Indu Jain, A.D. Rao and T.S. Murty, 2009: Storm surge modelling for the Bay of Bengal and Arabian Sea. *Natural Hazards*, **51**, 3-27.

Dube, S.K., P. Chittibabu, A.D. Rao, P.C. Sinha and T.S. Murty, 2000: Sea levels and coastal inundation due to tropical cyclones in Indian coastal regions of Andhra and Orissa. *Mar Geod*, **23**, 65-73. doi:10.1080/01490410050030643.

Dube, S.K., A.D. Rao, P.C. Sinha, T.S. Murty and N. Bahulayan, 1997: Storm surge in the Bay of Bengal and Arabian Sea: The problem and its prediction. *Mausam* (New Delhi), **48**, 283-304.

Dube, S.K., P.C. Sinha, A.D. Rao, I. Jain and N. Agnihotri, 2005: Effect of Mahanadi river on the development of storm surge along the Orissa coast of India: A numerical study. *Pure Appl Geophys*, **162**, 1673-1688. Doi: 10.1007/s00024-005-2688-5.

Dube, S.K. and A.D. Rao, 1991: Implications of sea level rise on coastal flooding due to storm surges in the Bay of Bengal. *Proc. Ind. Nat. Sci. Acad. Part A, Physical Sciences*, **57**, 567-572.

Dube, S.K., A.D. Rao, P.C. Sinha and P. Chittibabu, 1994: A real time storm surge prediction system: An Application to east coast of India. *Proc. Indian Natn. Sci. Acad.*, **60**, 157-170.

Dube, S.K., P.C. Sinha and A.D. Rao, 1982: The effect of coastal geometry on the location of peak surge. *Mausam*, **33**, 445-450.

Dube, S.K., P.C. Sinha and G.D. Roy, 1985: The numerical simulation of storm surges along the Bangladesh coast. *Dyn. Atmos. Oceans.*, **9**, 121-133.

Dube, S.K., P.C. Sinha and G.D. Roy, 1986: The effect of continuously deforming coastline on the numerical simulation of storm surges in Bangladesh. *Math. Comput. Simul.*, **28**, 41-56.

Flather, R.A., 1988: A numerical model investigation of tides and diurnal-period continental shelf waves along Vancouver Island. *J Phys Ocean*, **18**, 115-139.

Flather, R.A. and H. Khandker, 1993: The storm surge problem and possible effects of sea level changes on coastal flooding in the Bay of Bengal. *In:* Effects of changes in Stratospheric Ozone and Global climate. Vol. 4: Sea Level Rise (Editor, J.G. Titus). UNEP/EPA.

Flather, R.A. and R. Proctor, 1983: Prediction of North-sea storm surges using numerical models: Recent developments in UK. *In:* North sea dynamics (Ed. Sinderman/lenz). Springer Verlag, Berlin Heidelberg.

Flather, R.A., 1976: Practical aspects of the use of numerical models for surge prediction. Rep. No. 30, Inst. Ocean. Sci., Bidston, U.K.

Garrett, J.R., 1977: Review of the drag coefficients over oceans and continents. *Mon Weather Rev*, **105**, 915-929.

Ghosh, S.K., 1977: Prediction of storm surges on the coast of India. *Ind. J. Meteor. Geophys.*, **2S**, 157-168.

Grenier, R.R. Jr, R.A. Luettich Jr and J.J. Westerink, 1995: A comparison of the nonlinear frictional characteristics of two-dimensional and three dimensional models of a shallow tidal embayment. *J Geophys Res*, **100(C7)**, 13719-13736.

Heaps, N.S., 1973: Three-dimensional numerical model for the Irish Sea. *Geophys J R Astron Soc*, **35**, 99-120.

Henry, R.F. and T.S. Murty, 1982: Tides in the Bay of Bengal. Proc. Int. Conf. Comput. Methods Exp. Meas. Washington, DC. (Editors, G.A. Keramidas and C.A. Brebbia), June 30-July 2, 1982, Springer Verlag, New York.

Jain, I., P. Chittibabu, N. Agnihotri, S.K. Dube, P.C. Sinha and A.D. Rao, 2006: Numerical storm surge prediction model for Gujarat coast of India and adjoining Pakistan coast. *Nat Hazards*, **42**, 67-73. doi:10.1007/s11069-006-9060-7.

Jelesnianski, C.P., 1972: SPLASH I: Landfall storms. NOAA Technical Memorandum. NWS, TDL-46, Washington, DC.

Jelesnianski, C.P. and A.D. Taylor, 1973: A preliminary view of storm surges, before and after storm modifications. NOAA Technical Memorandum. ERL, WMPO-3, National Oceanic and Atmospheric Administration, U.S. Department of Commerce.

Jelesnianski, C.P. and J. Chen, 1979: SLOSH (Sea Lake and Overland Surges from Hurricanes). Report of TDL, NWS, Silver Spring, MD.

Jelesnianski, C.P., 1989: Storm surge and sea-state. Topic 5.2, Second International Workshop on Tropical Cyclones, Manila, December 1989. WMOffD-No. 319. WMO, Geneva.

Jiping, C., T.S. Murthy, B. Chenglan, M.I. El-Sabh and L. Fengshu, 1990: Storm Surges: Observations and Modelling. China Ocean Press.

Johns, B., A.D. Rao, S.K. Dube and P.C. Sinha, 1985a: Numerical modelling of the tide surge interaction in the Bay of Bengal. *Philos Trans R Soc Lond A*, **313**, 507-535. doi:10.1098/rsta.1985.0002.

Johns, B. and A. Ali, 1980: The numerical modelling of storm surges in the Bay of Bengal. *Quart. J. Roy Met. Soc.*, **106**, 1-8.

Johns, B., 1981: Numerical simulation of storm surges in the Bay of Bengal. *In:* Monsoon Dynamics (Ed. M.J. Lighthill and R.P. Pearce). Cambridge University Press.

Johns, B., S.K. Dube, U.C. Mohanty and P.C. Sinha, 1981: Numerical simulation of the surge generated by the 1977 Andhra Cyclone. *Quart. J. Roy. Met. Soc.*, **107**, 915-934.

Johns, B., S.K. Dube, P.C. Sinha, U.C. Mohanty and A.D. Rao, 1982: The simulation of continuously deforming lateral boundary in problems involving the shallow water equations. *Comp. Fluids*, **10**, 105-116.

Johns, B., P.C. Sinha, S.K. Dube, U.C. Mohanty and A.D. Rao, 1983a: Simulation of storm surges using a three-dimensional numerical model: An application to the 1977 Andhra Cyclone. *Quart. J. Roy. Met. Soc.*, **109**, 211-224.

Johns, B., P.C. Sinha, S.K. Dube, U.C. Mohanty and A.D. Rao, 1983b: On the effect of bathymetry in numerical storm surge simulation experiments. *Computers and Fluids*, **11**, 161-174.

Johns, B., P.C. Sinha, S.K. Dube, U.C. Mohanty and A.D. Rao, 1985b: On the effect of bathymetry in numerical storm surge simulation experiments. *Mathematical Modelling*, **6(1)**, 79.

Kolar, R.L., W.G. Gray, J.J. Westerink and R.A. Luettich, 1994: Shallow water modeling in spherical coordinates: Equation formulation, numerical implementation, and application. *J Hydraul Res*, **32**, 3-24.

Luettich, R.A. Jr, J.J. Westerink and N.W. Scheffner, 1992: ADCIRC: An advanced three dimensional circulation model for shelves coasts and estuaries. Report 1: Theory and methodology of ADCIRC-2DDI and ADCIRC-3DL. Dredging Research Program Technical Report DRP-92-6. U.S. Army Engineers Waterways Experiment Station, Vicksburg.

Manabe, S. and R.V. Stouffer, 1980: Sensitivity of a global climate model to an increase of CO_2 concentration in the atmosphere. *J. Geophys. Res.*, **85**, 5529-5554.

Mohanty, U.C., 1994: Tropical cyclones in the Bay of Bengal and deterministic methods for prediction of their trajectories. *Sadhana*, **19(4)**, 567-582.

Mohanty, U.C., Krishna K. Osuri, Vijay Tallapragada, Frank D. Marks, Sujata Pattanayak, M. Mohapatra, L.S. Rathore, S.G. Gopalakrishnan and Dev Niyogi, 2015: A Great Escape from the Bay of Bengal 'Super Sapphire-Phailin' Tropical Cyclone – A case of improved weather forecast and societal response for disaster mitigation. *Earth Interaction*, DOI: 10.1175/EI-D-14-0032.1.

Mohapatra, M., G.S. Mandal, B.K. Bandyopadhyay, Ajit Tyagi and U.C. Mohanty, 2012: Classification of cyclone hazard prone districts of India. *Natural Hazards*, **63**, 1601-1620.

Murty, T.S. and R.F. Henry, 1983: Tides in the Bay of Bengal. *J. Geophys. Res.*, **88**, 6069-6076.

Murty, T.S., 1984: Storm Surges: Meteorological Ocean Tide. Department of Fisheries and Oceans, Ottawa, Canada.

Murty, T.S., R.A. Flather and R.F. Henry, 1986: The storm surge problem in the Bay of Bengal. *Prog. Oceanog.*, **16**, 195-233.

Myers, V.A. and W. Malkin, 1961: Some properties of hurricane wind fields as deduced from trajectories. *Nat Hurric Res Proj Rep*, **49**, 5-9.

Pattanayak, S., 2012: Simulation of track and intensity of tropical cyclones over North Indian Ocean and associated storm surges. PhD Thesis.

Pattanayak, S., U.C. Mohanty and A.D. Rao, 2015: Simulation of storm surges in the Bay of Bengal using 1-way coupling between WRF and IITD storm surge model. *Marine Geodesy* (in review)

Qayyum, M.F., 1983: Prediction of storm surges for Bangladesh coast by empirical method: Results and discussions. WMO ESCAP l0th Session Panel on Tropical Cyclones.

Rao, N.S.B. and S. Mazumdar, 1966: A technique for forecasting storm waves. *Indian J Meteorol Geophys*, **17**, 333-346.

Rao, N.S.B., 1968: On some aspects of local and tropical storms in India. Ph.D. Thesis, Univ. of Jadavpur.

Raper, S.C.B., 1993: Observational data on the relationships between climatic change and the frequency and magnitude of severe tropical storms. *In:* Effects of changes in Stratospheric Ozone and Global climate. Vol. 4: Sea level Rise (Editor, J.G. Titus) UNEP/EPA.

Sinha, P.C., I. Jain, N. Bhardwaj, A.D. Rao and S.K. Dube, 2008: Numerical modeling of tide-surge interaction along Orissa coast of India. *Nat Hazards*, **45**, 413-427. Doi: 10.1007/s11069-007-9176-4.

Sinha, P.C., Y.R. Rao, S.K. Dube and A.D. Rao, 1996: Numerical investigation of tide-surge interaction in Hooghly Estuary, India. *Mar Geod*, **19**, 235-255. Doi: 10.1080/01490419609388082.

Sundermann, J. and W. Lenz, 1983: North Sea Dynamics. Springer-Verlag, Berlin. Heidelberg.

Ueno, T., 1981: Numerical computations of the storm surges in Tosa Bay. *J Oceanogr Soc Jpn*, **37**, 61-73.

Unisys Hurricane Database, 2008: Unisys Corporation, http://weather.unisys.com/hurricane/n_indian/index.html

Unisys Hurricane Database, 2007: Unisys Corporation, http://weather.unisys.com/hurricane/indian_oc/index.html

Unisys Weather Information Services, 1999: http://weather.unisys.com/hurricane/indian_oc/(1999)/

Westerink, J.J., C.A. Blain, R.A. Luettich and N.W. Scheffner, 1994: ADCIRC: An advanced three dimensional circulation model for shelves coasts and estuaries. Report 2: User's manual for ADCIRC-2DDI. Dredging Research Program Technical Report DRP-92-6, U.S. Army Engineers Waterways Experiment Station, Vicksburg, MS, USA.

Westerink, J.J., R.A. Luettich Jr and N.W. Scheffner, 1993: ADCIRC: An advanced three dimensional circulation model for shelves coasts and estuaries. Report 3: Development of a tidal constituent data base for the Western North Atlantic and

Gulf of Mexico. Dredging Research Program Technical Report DRP-92-6. U.S. Army Engineers Waterways Experiment Station, Vicksburg.

Westerink, J.J., R.A. Luettich, A.M. Baptista, N.W. Scheffner and P. Farrar, 1992: Tide and storm surge predictions using finite element model. *ASCE J Hydraul Eng*, **118(10)**, 1373-1390.

Winner, H. and A. Naomi, 2005: Advanced circulation model—A hurricane protection design tool. *In:* Solutions to coastal disasters 2005. ASCE, Reston,

World Meteorological Organization, 1990: WMO Second International Workshop on Tropical Cyclones (IWTC-II). WMO/TD No. 83, WMO, Geneva.

WMO, 2009: Tropical cyclone operational plan for North Indian Ocean, Tropical cyclone programme (TCP) Report No. TCP-21, 2009 Edition. WMO Geneva, WMO/TD No. 84.

Index

About the Editors

Prof. Uma Charan Mohanty is a Visiting Professor, School of Earth, Ocean and Climate Sciences, Indian Institute of Technology Bhubaneswar. Prior to this he has about 34 years of experience in teaching and research at Centre for Atmospheric Science, Indian Institute of Technology Delhi. Prof Mohanty received his B.Sc and M.Sc degree in Physics from Utkal University, Bhubaneswar and Doctoral degree in Tropical Meteorology from Odessa Hydro-Meteorological Institute, USSR in 1978. His research interests are tropical cyclones, monsoon dynamics, numerical weather predictions across the scales for severe weather events in tropics, extended range prediction of tropical rainfall and regional climate modeling. Prof. Mohanty has guided 35 PhD students and has authored more than 250 publications in peer reviewed national and international journals and more than 60 publications in proceedings of Conferences/Symposia, book chapters and reviewed technical reports. Prof. Mohanty has received several awards and honours for his significant research contribution in atmospheric sciences such as the most prestigious "Shanti Swarup Bhatnagar Prize" (1993), National Award in Atmospheric Science and Technology (2013) and Sir Gilbert Walker Gold Medal (2009) for his significant contribution to atmospheric sciences in India. He has become fellow of all the four national academies of sciences and engineering in India, namely, Indian National Science Academy (INSA), Indian Academy of Sciences (IASc), National Academy of Sciences, India (NASI) and Indian National Academy of Engineering (INAE).

Dr. Sundararaman G. Gopalakrishnan is a research meteorologist at the Hurricane Research Division of Atlantic Oceanographic Meteorological Laboratory, NOAA and the principal architect of the Hurricane Weather Research and Forecasting (HWRF) model. His research involves simulating a variety of complex, non-linear, scale interacting systems starting from dry thermals (Large Eddy Simulations) to hurricanes; examining the mesoscale structures and evolution as well as the mechanism(s) whereby they develop; testing theories, hypotheses and various near-surface model physical representations; and finally interpreting, to the extent possible, the modeled and the observed behaviour of these systems. He has over 50 publications in peer-reviewed international journals, several of which have been cited in standard text books for mesoscale meteorology. In the past he has served as an Associate Editor for the *Monthly Weather Review* and *Weather and Forecasting*. Gopal is the head of the modeling group at the Division where he supervises and mentors advanced scientists and students at graduate as well as post-doc levels. His recent awards include "OAR Scientific Employee of Year", 2013; "South Florida Federal Executive Board - Scientific Employee of the Year", 2014 and "Department of Commerce Gold Medal", 2015. He is the co-lead on the NOAA's next generation High Impact Weather Prediction Project (HIWPP) titled "Hurricane Nesting", Co-Developmental Manager for Hurricane Forecast Improvement Project (HFIP) and is leading the Research to Operational model development activities at AOML. He also serves in the Science Advisory Board at the Developmental Test Bed Center, UCAR, Boulder.

Contributors

Dr. Alexandre Fierro is a research scientist with the National Severe Storms Laboratory (NSSL/NOAA) and the Cooperative Institute for Mesoscale Meteorological Studies (CIMMS/OU). He has a wide scope of research interests with main focuses directed towards numerical modeling of tropical cyclones, tropical convection, and cloud electrification. In the recent years, Dr. Fierro also has placed some emphasis on studies related to climate variability and, most importantly, has developed and evaluated various techniques aimed at assimilating total lightning data in convection allowing numerical weather prediction models.

Dr. Ashish Routray is currently working as Scientist at National Centre for Medium Range Weather Forecasting (NCMRWF), Ministry of Earth Sciences, Govt. of India since 2009. He has adequately contributed towards understanding and prediction of extreme weather events using mesoscale model and data impact studies mainly Indian Doppler Weather Radar (DWR) and radiance observations using advance data assimilation system over Indian monsoon region. He has published more than 25 research papers and book chapters in international and national peer reviewed journals/volume books. His research focuses on mesoscale modeling and data assimilation, monsoonal studies, study of extreme weather events and regional climate modeling.

Dr. Chandra M. Kishtawal is the group director of Atmospheric and Oceanic Sciences in Space Applications Centre of Indian Space Research Organisation (ISRO). His research involves application of satellite observations in numerical weather prediction with emphasis on the prediction of tropical cyclones and Indian monsoon. He also contributed to the retrieval of atmospheric motion vector winds from Kalpana and INSAT-3D satellites. At present he is the project director of INSAT-3D and INSAT-3D-R Utilization Project. Dr. Kishtawal has to his credit, more than 100 publications in various national and international journals. He was a visiting scientist to the Florida State University in 1999-2000, Purdue University in 2009-10, and to National Center for Atmospheric Research, Boulder in 2014.

Prof. Da-Lin Zhang works with Department of Atmospheric and Oceanic Science, University of Maryland, Maryland. His research involves simulating various tropical cyclones and other severe convective systems; examining the dynamics and meso-β-scale structures and evolution as well as the mechanism(s) whereby they develop; testing theories, hypotheses and various model physical representations; and finally interpreting, to the extent possible, the observed behaviours in these weather systems. He has authored more than 160 peer-rewiewed publications and served as the Editor of *Weather and Forecasting*. He received the C.L. Meisinger Award of the American Meteorological Society (AMS) in 1991 “for pioneering work in the understanding and numerical modeling of mesoscale convective weather systems”, and became an AMS Fellow in 2006.

Prof. Dev Niyogi is a Professor at Purdue University, USA with joint appointments in Department of Agronomy - Crops, Soils, and Environmental Sciences, and Department of Earth, Atmospheric, and Planetary Sciences. He is also the Indiana State Climatologist. His research interests include multiscale analysis of land-atmosphere interactions on heavy rain processes. Other topical areas include analysis of severe weather and hydro climatic extremes, urban climate and smart city framework, development of climate resiliency tools, and the Indian monsoons and land falling tropical cyclones. He has developed over more than 30 successful research projects, funded through a variety of competitive federal research grants—NSF, NASA, Joint Center for Satellite Data Assimilation, DOE, and USDA/NIFA. He has chaired and been member of many scientific and professional committees related to land surface processes and land-atmosphere interactions. He has been recognized through Purdue ‘Seeds for Success’ award, the University Faculty Scholar recognition, the NSF’s coveted CAREER award, the USDA NIFA Partnership award. He has published more than 150 research papers in peer reviewed international journals, been a guest editor for topical issues on landuse/land cover change feedbacks based journals/volume books. His research has been cited by over 5500 times, with a h-index and i-index over 40 and 100 respectively, and popular media such as CNN and NASA news release.

Dr. Frank D. Marks is the Director and Meteorologist with the Hurricane Research Division at the Atlantic Oceanographic and Meteorology Laboratories for NOAA. He is also an Adjunct Professor at the University of Miami Rosenstiel School for Marine and Atmospheric Sciences. Marks' research interests include analyzing meteorological remote sensing data sets particularly in tropical cyclones, and he has published more than 100 refereed articles and book chapters on these topics. Earlier Marks received a B.S. in Meteorology from Belknap College and an M.S. and Sc.D. in Meteorology from the Massachusetts Institute of Technology.

Dr. Fuqing Zhang is a Professor of Meteorology and Director, Center for Advanced Data Assimilation and Predictability Techniques at Penn State University. His research contributes to the fundamental understanding of dynamics and predictability of weather and climate at various scales, to the design of advanced data assimilation techniques and to the innovative use of radar and satellite observations that lead to significant advancement in the analysis and prediction of severe weather and tropical cyclones.

Dr. Jian-Wen Bao received his M.S. in 1990 and PhD. in 1993 from the Department of Meteorology at The Pennsylvania State University. He is currently a research meteorologist in the Physical Sciences Division of NOAA's Earth System Research Laboratory. Since joining NOAA in 1994, he has been working on the development and improvement of physics parameterization schemes in numerical weather prediction models, particularly the impact of air-sea interaction on tropical storm development. He led a team effort to evaluate the performance numerical weather prediction models in air quality simulations and prediction. Recently, he has been responsible for developing and testing the physics component of the NOAA's Earth System Research Laboratory-global model. He is also leading an effort to develop and improve the grid-resolved and sub-grid cloud parameterization schemes in numerical weather prediction models.

Dr. Krishna K. Osuri is a gold medalist in M.Sc (Meteorology) and he has his M.Tech and PhD degrees from Andhra University in 2006 and 2012 respectively and is now serving as Assistant Professor in the Department of Earth and Atmospheric Sciences, National Institute of Technology (NIT), Rourkela, India. His research interests include understanding the processes and improving the prediction of extreme weather events such as cyclones, severe thunderstorms and heavy rain events. His main research interests are: (i) Numerical Weather Prediction, (ii) Data Assimilation techniques, (iii) Tropical Cyclones and (iv) Land Surface Modeling. He was a visiting scientist to Purdue University, Hurricane Research Division (HRD) and Jet Propulsion Laboratory (JPL), NASA. He has published about 21 publications in peer reviewed national and international journals.

Dr. M. Mohapatra is the head of Weather Forecasting Services of the India Meteorological Department. He is also the head of WMO recognized Regional Specialized Meteorological Centre-Tropical Cyclone, New Delhi. He has made significant contribution in improvement of early warning services of IMD, especially the cyclone forecasting and warning services over the North Indian Ocean and south Asian region. The author of more than 55 research papers published in peer reviewed journals, he has received a number of recognitions including 25th Biennial Mausam Award and Young Scientist Award of the Ministry of Earth Sciences, Government of India for his research contributions in the field of atmospheric sciences.

Dr. Michael Montgomery is a professor of tropical meteorology in the Department of Meteorology at the Naval Postgraduate School. His meteorological research and teaching interests span dynamic meteorology, synoptic meteorology, geophysical fluid dynamics, geophysical turbulence, theoretical, experimental, and computational fluid dynamics, and applied mathematics. Meteorological research expertise in tropical cyclones (hurricanes/typhoons), mid-latitude cyclones, polar lows, fronts, mesoscale convective vortices, tornadoes, stratospheric circumpolar vortex, and other atmospheric vortices. Dr. Montgomery has authored about 140 publications and is serving as the Editor of *Dynamics of Atmospheres and Oceans.*

Dr. Noel E. Davidson is Principal Research Scientist with Research and Development Division, Bureau of Meteorology, Australia. His main research interests are: 1. Objective analysis and prediction for the tropics, 2. Vortex initialization and prediction of tropical cyclones, 3. Extratropical influences on the Australian monsoon, 4. Environmental influences on tropical cyclone behaviour, and 5. Understanding and prediction of extreme rainfall and the rapid intensification of tropical cyclones. He has written more than 60 peer-reviewed articles in international journals on these topics.

Dr. Richard M. Yablonsky, a Senior Scientist and Manager at catastrophe modeling company AIR Worldwide in Boston, MA, obtained a PhD. in Oceanography from the University of Rhode Island (2009) and three degrees from North Carolina State University: M.S. in Atmospheric Science (2004), B.S. in Meteorology (2002), and B.A. in Chemistry (2002). He is currently involved with risk modeling of storm surge, tsunami, and tropical cyclone wind and flood. Prior to joining AIR Worldwide in 2014, Yablonsky researched hurricane-ocean interaction and contributed to the development of the ocean component of the HWRF model as a Marine Research Scientist at the University of Rhode Island's Graduate School of Oceanography in Narragansett, RI and a Visiting Scientist at the Developmental Testbed Center in Boulder, CO.

Dr. Roy Bhowmik is the Dy. Director General for Numerical Weather Prediction at the India Meteorological Department. He has published 95 research publications in national and international journals on Atmospheric Sciences, 15 Science Reports/Monographs and produced five PhD. awardees. His scope of research interest has been Numerical Weather Prediction of Monsoon, Tropical Cyclone and meso-scale convective features. Some of his important contributions are: (a) Development and implementation of a Multi-Model Ensemble (MME) based district level weather forecast system for Agro-advisory Service of India, (b) Development of a statistical-dynamical method for cyclone forecasting, (c) Processing of Doppler Weather Radar data for now-casting and mesoscale modeling and (d) Implementation of Polar WRF for Maitri Region in Antarctica. He is involved with the National Monsoon Mission Program as the project Director of IMD. His contribution to WMO Technical Commission for Basic System for the period 2009-12 was recognized by WMO.

Prof. Shishir K. Dube, former Director of Indian Institute of Technology Kharagpur and former Professor at the Centre for Atmospheric Sciences, Indian Institute of Technology (IIT) Delhi, is currently the Vice Chancellor of the Amity University Rajasthan, Jaipur, India. His research interests include Numerical Storm Surge Modeling. He is internationally recognized for his pioneering contributions in the field of storm surge prediction in the Bay of Bengal and the Arabian Sea. He is credited with the development of real time operational surge prediction systems which under the auspices of World Meteorological Organization (WMO) have been transferred to the National Weather Services of Bangladesh, India, Maldives, Myanmar, Oman, Pakistan, Sri Lanka and Thailand for their operational use.

Dr. Sujata Pattanayak completed her PhD. degree in Atmospheric Sciences in 2012 from IIT Delhi. At present, she is working as Fast Track Young Scientist of Department of Science and Technology, Govt. of India at IIT Bhubaneswar. The improvement of track and intensity prediction of tropical cyclones with data assimilation system and intense vortex initialization are the main thrust areas of her research. She also works on real time prediction of storm surges associated with land falling tropical cyclones with 1-way coupled mesoscale modeling system.

Dr. Thiago Quirino is a research computer scientist in the Numerical Modeling Group of the US National Oceanic and Atmospheric Administration's (NOAA) Hurricane Research Division (HRD) where he works in the development of multiple operational hurricane models. His research interests include numerical modeling of tropical cyclones, high performance computing, applied evolutionary search and optimization, machine learning, and data mining. He has a PhD. in electrical and computer engineering from the University of Miami.

Dr. T.N. Krishnamurti is Professor Emeritus, PhD. from University of Chicago, Department of Earth, Ocean and Atmospheric Science, Florida State University, USA. His research interests are high resolution hurricane forecast (tracks, landfall, and intensity), monsoon forecasts on short, medium range, and monthly time scale and studies of intra seasonal and inter annual variability of the tropical atmosphere. His phenomenological interests include hurricanes, monsoons, jet streams, and the meteorology of arid zones. In 1985, the American Meteorological Society awarded him the Carl-Gustaf Rossby Research Medal. He won the International Meteorological Organization Prize of the World Meteorological Organization in 1996. In 2012 the Indian Meteorological Society gave him their Sir Gilbert Walker Gold Medal. A symposium in his honor was held in 2012 as part of the American Meteorological Society annual meeting in New Orleans.

Dr. Vijay Tallapragada holds a PhD. in Tropical Meteorology, M.Tech. Degree in Atmospheric Science, and M.S. Degree in Meteorology from Andhra University, India. He has been a Meteorologist and Hurricane Team Leader with the Environmental Modeling Center of the National Oceanic and Atmospheric Administration's National Centers for Environmental Prediction since

2006, and Chief of the Global Climate and Weather Modeling Branch since July 2015. Dr. Tallapragada led the implementation of the Hurricane Weather Research and Forecast modeling system that has evolved into a unique regional model providing operational global tropical cyclone forecast guidance. Dr. Tallapragada is serving as the Development Manager for NOAA's Hurricane Forecast Improvement Project and is also in-charge for development and operational implementation of NOAA's Next Generation Global Prediction System. He received National Weather Service Isaac Cline Award in 2013, and the United States Department of Commerce Gold Medal in 2014. He has more than 80 refereed publications and supervised five PhD. students.

Resource experts and participants of the Advanced Indo-US Workshop and Colloquium.

MIX
Papier aus verantwortungsvollen Quellen
Paper from responsible sources
FSC® C105338

If you have any concerns about our products,
you can contact us on
ProductSafety@springernature.com

In case Publisher is established outside the EU,
the EU authorized representative is:
Springer Nature Customer Service Center GmbH
Europaplatz 3, 69115 Heidelberg, Germany

Printed by Libri Plureos GmbH
in Hamburg, Germany